REPRESENTATIVE
ELEMENTS

NOBLE
GASES

| | | | | | 8A |
| | | | | | 18 |

	3A	4A	5A	6A	7A	2
	13	14	15	16	17	**He**
						4.003

TION ELEMENTS

					5	6	7	8	9	10
					B	**C**	**N**	**O**	**F**	**Ne**
					10.81	12.01	14.01	16.00	19.00	20.18

	8B			1B	2B	13	14	15	16	17	18
						Al	**Si**	**P**	**S**	**Cl**	**Ar**
	8	9	10	11	12	26.98	28.09	30.97	32.06	35.45	39.95

	26	27	28	29	30	31	32	33	34	35	36
	Fe	**Co**	**Ni**	**Cu**	**Zn**	**Ga**	**Ge**	**As**	**Se**	**Br**	**Kr**
94	55.85	58.93	58.69	63.55	65.38	69.72	72.59	74.92	78.96	79.90	83.80

	44	45	46	47	48	49	50	51	52	53	54
	Ru	**Rh**	**Pd**	**Ag**	**Cd**	**In**	**Sn**	**Sb**	**Te**	**I**	**Xe**
)	101.1	102.9	106.4	107.9	112.4	114.8	118.7	121.8	127.6	126.9	131.3

	76	77	78	79	80	81	82	83	84	85	86
	Os	**Ir**	**Pt**	**Au**	**Hg**	**Tl**	**Pb**	**Bi**	**Po**	**At**	**Rn**
.2	190.2	192.2	195.1	197.0	200.6	204.4	207.2	209.0	(209)	(210)	(222)

7	108	109
s	**Uno**	**Une**
2)	(265)	(266)

Metals ←⎯⎯⎯→ Nonmetals

f

INNER TRANSITION ELEMENTS

	61	62	63	64	65	66	67	68	69	70	71
	Pm	**Sm**	**Eu**	**Gd**	**Tb**	**Dy**	**Ho**	**Er**	**Tm**	**Yb**	**Lu**
.2	(145)	150.4	152.0	157.3	158.9	162.5	164.9	167.3	168.9	173.0	175.0

	93	94	95	96	97	98	99	100	101	102	103
	Np	**Pu**	**Am**	**Cm**	**Bk**	**Cf**	**Es**	**Fm**	**Md**	**No**	**Lr**
.0	(237)	(244)	(243)	(247)	(247)	(251)	(252)	(257)	(258)	(259)	(260)

An Introduction to Physical Science

An Introduction to
Physical Science

Sixth Edition

James T. Shipman
Ohio University

Jerry D. Wilson
Lander College

D. C. Heath and Company
Lexington, Massachusetts   Toronto

Acquisitions Editors: Mary Le Quesne and Kent Porter
Developmental Editors: Mary Le Quesne and Laura Goeselt
Production Editor: Bryan Woodhouse
Designer: Henry Rachlin
Production Coordinator: Lisa Arcese
Photo Researcher: Wendy Johnson

Cover: Satellite photo taken during the *Challenger 6* mission (October 5–13, 1984), showing a southwestern view of the Greater Himalayas, bordering on the Karakoram Range. India is to the left, Pakistan to the right, and China in the foreground. The valley of the Indus River is in the right background, the fabled valley of Kashmir is near the right edge of the photo, and the great peaks of the Karakoram Range are near the lower edge. (NASA/Johnson Space Center)

To Mary Le Quesne, whose editorial insight and encouragement were our ongoing source of inspiration

Preface

In today's world, a knowledge of physical science and an understanding of modern technology grow daily more important. For this reason, our purpose in revising *An Introduction to Physical Science* for this Sixth Edition was twofold—to stimulate student interest in the sciences and to present the information and skills needed to cope in a technological society.

This textbook, written for the first-year college nonscience major, presents basic concepts in the five major areas of physical science: physics, chemistry, astronomy, geology, and meteorology. We make these concepts easily accessible to students by developing them in a logical rather than a chronological fashion and by discussing them in the context of everyday experience. Chapter 1 begins with the fundamental concepts of measurement. From these fundamentals, we move on progressively to the concepts of motion, force, energy, wave motion, heat, electricity, magnetism, and modern physics. These concepts are then used to develop the principles of chemistry, astronomy, meteorology, and geology. The text is readily adaptable to either a one- or two-semester course, as its past success has demonstrated.

We have treated each discipline both descriptively and quantitatively, with the relative emphasis on each of these two approaches left to the discretion of the instructor. To those who wish to emphasize the descriptive approach in teaching physical science, we recommend using only the Questions at the end of each chapter and omitting the Exercises. In this edition, we have placed most derivations of equations in the Appendix to provide the quantitative emphasis for those who desire it without distracting those who do not.

Changes in the Sixth Edition

This edition of *An Introduction to Physical Science* incorporates the suggestions of helpful instructors and represents a significant modernization of the text. We have reduced the coverage of physics from eleven chapters to nine, reflecting the wishes of instructors using the text, who found previous editions more comprehensive and quantitative than their students required. The book now covers atomic physics and quantum mechanics in one chapter, and we have moved relativity from the text to the Appendix.

Also per instructors' requests, the chapters on geology now begin with an examination of minerals and rocks. We have expanded the coverage of geological time, which now appears in its own chapter. We have consolidated the discussion of air, land, and water pollution and placed it in the meteorology section, reinforcing for students the close relationship between the actions of people and their far-reaching effects on the environment. Environmental issues retain their past emphasis, with continued coverage of climatic change and global warming, acid rain, and the effects of tornadoes.

We have emphasized historic and special-interest topics in this edition by placing them in Highlight boxes. This feature focuses on topics of particular interest to the student that are generally not included in classroom lectures.

An important change in this edition is the full-color presentation throughout the text. Physical science lends itself well to color description, and this edition takes full advantage of four-color technology. We have converted all of the drawings to color, and approximately 90 percent of the photographs now appear in color, as well. Moreover, the conversion to color has given us the opportunity to provide completely new illustration, and forms the basis of a modern and accessible text. Color incorporated into the text design provides students with numerous pedagogical tools; laws are highlighted, and important concepts and definitions receive special treatment to aid the student in both learning and reviewing key topics.

Supplements

We are pleased that the supplements have received special attention in this revision, representing a great improvement in support for both students and instructors. These supplements include:

- The *Study Guide*, by James T. Shipman of Ohio University, Jerry D. Wilson of Lander College, and

Clyde D. Baker, also of Ohio University. The Sixth Edition *Study Guide* is a significant revision of previous editions. Each chapter now includes study goals, discussion, review questions, solved problems, multiple-choice questions, and a quiz. The Math Review remains at the end of the *Study Guide*, as in the previous editions.

- The *Instructor's Guide*, by James T. Shipman and Jerry D. Wilson, has been thoroughly updated and revised. This edition includes a new Teaching Aids section for each of the five sciences and up-to-date audiovisual resources. Each chapter includes a brief discussion, suggested demonstrations, answers to text questions and solutions for exercises, and answers to the *Study Guide* quizzes.

- The *Laboratory Guide*, by James T. Shipman, contains two new experiments, for a total of 47. Each experiment includes an introduction, learning objectives, a list of required apparatus, a detailed procedure for collecting data (requiring students to generate tables and graphs and to perform calculations), and questions about the experiment.

- The *Instructor's Resource Manual* for the *Laboratory Guide*, also by James T. Shipman, now includes an integrated equipment list to assist instructors in planning experiments. Additional data and calculations are provided for most experiments, as well as answers to questions, a discussion of each experiment, and additional questions.

- The *Instructor's Test Bank*, available to adopters, offers a printed version of more than 1900 questions. They are available in completion, multiple-choice, and short exercise formats.

- The *Test Bank* questions are also available in a new computerized testing program, *HeathTest Plus*. Instructors can produce chapter tests, midterms, and final exams easily and with excellent graphics capability. Instructors may edit existing questions or add new ones as desired or preview questions on screen and add them to the text with a single key stroke. *HeathTest Plus* is available for IBM and Macintosh computers.

- The Transparencies—68 one-, two- and four-color—illustrate important concepts from the text. They are available to adopters of the Sixth Edition.

Acknowledgments

We wish to thank our colleagues and students for the many contributions made to this Sixth Edition of *An Introduction to Physical Science*. We would also like to thank the following reviewers for their suggestions and comments:

Steven R. Addison, University of Central Arkansas; Richard M. Bowers, Weatherford College; Susan G. Burlingham, Pratt Community College; Robert W. Childers, Tulsa Junior College; Donald Dalrymple, North Central Missouri College; Paul D. Lee, Louisiana State University; L. Whit Marks, Central State University; Eugene D. Miller, Luzerne County Community College; J. Ronald Mowry, Harrisburg Area Community College; Ronald H. Orcutt, East Tennessee State University; Charles Ostrander, Hudson Valley Community College; Surendra N. Pandey, Albany State College; Harry D. Powell, East Tennessee State University; Pushpa Ramakrishna, Chandler Gilbert Community College; Thomas G. Read, Martin Methodist College; John E. Rives, University of Georgia; David Robertson, Alcorn State University; Aaron W. Todd, Middle Tennessee State University; Alan Ziv, Arizona State University.

We are grateful to those individuals and organizations who contributed photographs, illustrations and other information used in this text. We are also indebted to the D. C. Heath staff for their dedicated and conscientious efforts in producing this Sixth Edition. In particular, we thank Mary Le Quesne, acquisitions editor; Laura Goeselt, developmental editor; Bryan Woodhouse, production editor; Henry Rachlin, book designer; Wendy Johnson, photo researcher; and Patricia Wakeley, supplements developmental editor. Finally, we acknowledge the contributions of Genny Shipman, Ruth Hodges, Clyde Baker, and John Smith to various aspects of the writing and production of this book and its supplements.

We welcome comments from students and instructors of physical science, and invite you to forward your impressions and suggestions.

J. T. S.
J. D. W.

About the Authors

After serving in the U.S. Navy during World War II, **James T. Shipman** attended Ohio University. He received his B.S. and M.S. degrees from Ohio and an honorary Ph.D. from Chubu University in Japan. For 25 years, he taught physics and physical science at Ohio, where he was also Department Chair for five years (1968–73). He has also taught physics and/or physical science at Clark College, Fairmont State College, and Salem College.

Professor Shipman has received and served as director of several National Science Foundation education grants. His research activities have included studies of cosmic rays, ionospheric radio-propagation, and high-frequency radio tracking. He is a member of the American Association of Physics Teachers, and he has served as president of the Appalachian Section. He has written a number of other books, including *Fundamental Concepts of Physical Science, Concepts of Modern Physics and Chemistry,* and the *Physical Science Laboratory Manual.* A translation of the physics section of *An Introduction to Physical Science* is currently used in Japan.

Doctor Shipman is now Professor Emeritus of Physics at Ohio University, where he remains active in writing and continues an active role. He is a member of the Board of Visitors of the College of Arts and Sciences and serves on the National Campaign Council. He is now semiretired and living in West Virginia, his native state.

Jerry D. Wilson, a native of Ohio, is Professor of Physics and Chair of the Division of Science and Mathematics at Lander College in Greenwood, South Carolina. He received his B.S. from Ohio University, a M.S. from Union College, and in 1970 a Ph.D. from Ohio University. As a doctoral graduate student, he taught physical science and held the faculty rank of Instructor. He wrote his portion of the first edition of *An Introduction to Physical Science* at that time. The text was originally published locally in three sections before being published nationally *in toto* in 1971 by D. C. Heath and Company.

Being primarily interested in science teaching, Dr. Wilson has continued to write and currently has three other texts and a laboratory manual in publication in various editions—*Physics: A Practical and Conceptual Approach, Technical College Physics,* and *College Physics.* His popular laboratory manual, *Physics Laboratory Experiments,* published by D. C. Heath, is now in its third edition. He has also written articles for professional journals and presented a number of papers at various scientific-organization meetings. He is the contributing editor for the Favorite Demonstration column in *The Journal of College Science Teaching,* and in his spare time (of which there is little), he writes a weekly column, The Science Corner, for local newspapers.

Brief Table of Contents

Contents

Introduction

This book will introduce you to the various disciplines of physical science, the basic laws that govern each of them, and some of the history of their development. This exploration will enrich your perspective on how scientific knowledge has grown throughout the course of human history, how science influences the world we live in today, and how it could be employed in the future.

The English word *science* is derived from the Latin *scientia*, meaning knowledge. Physical science is the organized knowledge of our physical environment and the methods used to obtain it. Physical science is classified into five major divisions: *physics*, the science of matter and energy; *chemistry*, the science of matter and its changes; *astronomy*, the science of the universe beyond our planet; *meteorology*, the science of climate and weather; and *geology*, the science of Earth and its history. Physical science studies the nonliving matter in the universe, whereas biological science studies the living matter.

Because we live in a highly technological age, we tend myopically to view science as an exclusive product of the twentieth century. An appreciation of the physical sciences, however, dates back to the beginning of the human race. Although the earliest humans did not have sophisticated tools with which to view the universe, they did have a curiosity about the world around them and a compelling need to survive in a harsh environment. The movement of the stars, the passing of the seasons, and the need to make tools, create fire, and predict the weather using the clues of the wind and the clouds grew out of such a curiosity or need and was addressed through observation of Earth and sky. Every aspect of physical science you study in this book has its roots in mankind's first observations.

Indeed, observation forms the basis of all scientific knowledge, even in the modern world. Scientific knowledge is cumulative, and if the earliest humans had not asked questions and made observations, our own knowledge of the physical sciences could not exist. Over the entire history of the human race, people have gathered information about physical science in very much the same way the first humans did. But although observation

has remained the first step in unveiling scientific discovery throughout history, the wealth of scientific knowledge has developed and advanced, and each new discovery yields the possibility for many more.

The activities of scientists are complex, and although they may not follow a given set of rules, scientists do approach problems in a systematic way. There is no single well-defined method or procedure for uncovering the secrets of nature; however, scientists find that some loose guidelines are helpful for isolating a phenomenon, determining the best way to observe it, and drawing conclusions from their observations. The scientific method is the main way that scientists define a problem and seek its solution. In general terms, the scientific method includes the following:

1. **Observation of phenomena and recording of facts.** *Phenomena* are defined as whatever happens in the environment, and *facts* are accurate descriptions of what is observed.

2. **Formulation of a theory from the generalization of the phenomena.** A *theory* is a description of a certain behavior of nature that extends beyond what has been observed—usually stated in general terms.

3. **Prediction of new data and new phenomena based on the theory.** A theory comprises a general scheme of thought that explains the nature or behavior of the phenomena and correlates the known facts in such a manner that new thoughts and relationships initiate the prediction of new phenomena. Einstein's theory of relativity and the kinetic theory of gases are examples.

4. **Experimentation to confirm the new data or phenomena predicted by the theory.**

5. **Confirmation, modification, or disposal of the theory.** Further predictions are made. Steps 4 and 5 are then repeated.

The scientific method is no magic formula, but rather a formulation of the thought processes that carry one from questions (observations of phenomena) to solutions (explanations of phenomena). Researchers in all disciplines of physical science employ its steps to increase scientific

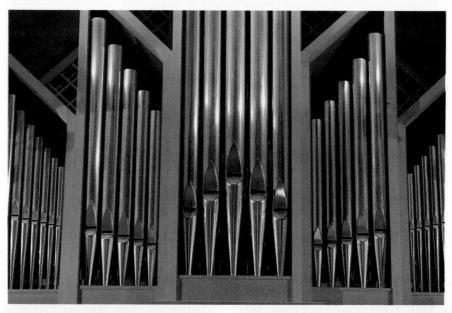

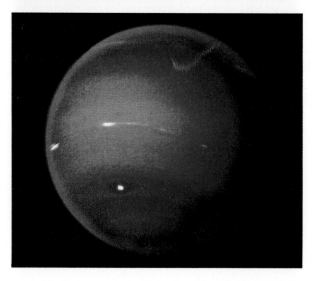

knowledge, not only on the edge of modern technology but also in very mundane ways. We hear daily about advances in technology, some of which will be discussed in this book. But it is more important that you use the concepts in this book to increase your understanding of how the scientific method affects your day-to-day life. The digital watch on your wrist, the compact disc player in your room, the optical scanner in your supermarket, and the calculator as small as a credit card all resulted from scientists employing the scientific method to discover new technology and apply it usefully.

We hope you will not only become knowledgeable about how the scientific method applies to your life but also about why scientific discovery appears to take such a haphazard course. An understanding of how scientists learn should give you a better grasp of how technological change occurs and why our knowledge seems to be increasing rapidly in one area yet slowly in another. As you read about the many scientific discoveries discussed in the text, you should realize that major discoveries often come to light by accident, and years of intense research may yield few results. While we may discuss the scientific method as discrete and organized steps in the pursuit of knowledge, the reality of scientific discovery rarely follows a predictable course.

The world in which we live changes constantly, and people are constantly seeking to discover and understand these changes. Scientific research results in advances in technology and affects all aspects of our daily life. To cope with a rapidly evolving society, each individual needs to know and understand the physical concepts that make technological advancement possible. Complex issues such as industrialization, pollution control and cleanup, space exploration, and the search for new energy sources will therefore demand a basic understanding of basic physical concepts and their modern applications.

An Introduction to Physical Science

CHAPTER 1

Measurement

> When you can measure what you are speaking about and express it in numbers,
> you know something about it; but when you cannot measure it, when you cannot
> express it in numbers, your knowledge is of a meager and unsatisfactory kind.
>
> —Lord Kelvin

A FIRST STEP in understanding our environment is to find out about the physical world through measurements. Over the centuries human beings have developed increasingly sophisticated methods of measurements, and scientists make use of the most advanced of these. The student should gain both a working knowledge of these methods and an awareness of their limitations.

We are continually making measurements in our daily life. Each day we plan our work, play, and rest schedules as a function of time. With watches and clocks we measure the time for events to take place. Every 10 years we take the census and determine (measure) the population. We count our money, our food, the minutes, hours, days, and years of our life.

Some of us keep accurate measurements of food and drugs taken into the body because of illness. Many lives depend on accurate measurements being made by the medical doctor, laboratory technician, and pharmacist in the diagnosis and treatment of disease.

Meteorologists measure the many elements (temperature, pressure, humidity, precipitation, wind) that make up the weather. This information is relayed to millions by the communications media, which must at all times measure all phases of their operation to stay within standards designed to protect the rights of others.

Human beings, in their efforts to understand their total environment, must measure the very small and the very large. Scientists probe farther inward to examine smaller and smaller particles, and explore outward to discover a larger and larger universe.

At one time it was thought that all things could be measured with exact certainty. However, as we measured smaller and smaller objects, it became evident that the very act of measuring distorted the measurement. This uncertainty in making measurements of the very small will be discussed in detail in Chapter 9.

◄ Scientists measure population density using satellite photos like this one of the Baltimore, Maryland, area. Regions with concentrations of people and buildings generate heat and appear blue.

The ability of the scientist to know and predict is a function of accurate measurements. From accurate measurements taken on the moon's surface, the geologist obtains new knowledge for understanding continental drift on Earth.

Our conquest of the moon and planets is largely due to our ability to make accurate measurements on the hardware that makes up the space vehicle and to program computers that make accurate and continuous measurements of position, velocity, and the numerous other factors involved in space travel.

These examples show the relevance of measurement and underscore our need to know the concepts of measurement. Understanding measurement is the first step in the understanding of our physical environment.

1.1 The Senses

Our environment stimulates our senses, either directly or indirectly. The five senses (sight, hearing, smell, touch, taste) make possible our knowledge of the environment. Therefore, they are a good starting point in the study and understanding of the physical world.

Most information about our environment comes through sight. This information is not always a true representation of the facts because the eyes, and therefore the mind, can be fooled. There are many well-known optical illusions, such as those in Fig. 1.1. Many people are quite convinced that what they see in such drawings actually exists as they perceive it.

Hearing ranks second to sight in supplying the brain with information about the external world. The senses of touch, taste, and smell, although very important for good health and happiness, rank well below sight and hearing in providing environmental information.

All the senses can be deceived and thus provide false information about our environment. Anyone who has gone to the beach to swim during the early morning hours when the air is cold knows how warm the water feels to the body. Later in the afternoon when the air becomes hot, the water, which has remained at practically the same temperature, feels cold. Thus if we were asked

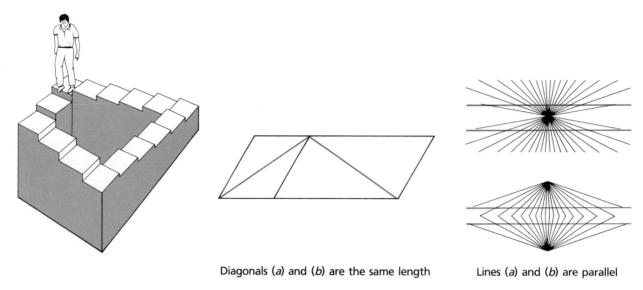

Diagonals (*a*) and (*b*) are the same length   Lines (*a*) and (*b*) are parallel

Figure 1.1 Some optical illusions.
We can be deceived by what we see.

to judge the temperature of the water, our answers would vary according to the temperature of the air.

Not only can the senses be deceived, but they also have their limitations. The unaided eye is unable to distinguish the stars of our galaxy from the planets of our solar system. In fact, the word *planet* meant "wandering star" in Greek.

The other senses have similar limitations. For instance, if you were asked to distinguish between two identical sounds produced $\frac{1}{100}$ second apart, you would certainly fail.

The handicaps of the senses can be conquered by close scrutiny of phenomena with measuring instruments. For example, in the diagram of optical illusions (Fig. 1.1), diagonals (*a*) and (*b*) can be measured with a ruler, and their length accurately determined.

We extend our ability to measure our environment with many specialized instruments, or tools. But these, too, have their limitations. The wristwatch is a precision instrument, but it cannot be used to measure time intervals of less than $\frac{1}{10}$ second.

Even the most precise instruments have limitations. More accurate information about our physical environment can be obtained by comparing the measurement of one instrument against another. Later in our study of the microscopic world we shall learn of other limitations concerning the measurement of physical quantities.

1.2 Concepts and Fundamental Quantities

The development of a logical system for understanding the physical sciences begins with the formulation of fundamental concepts. A **concept** is a meaningful idea that can be used to describe phenomena. The "phenomena" are defined as whatever happens in the environment. Concepts must be conceived and developed so that they explain, with clarity, the observed phenomena.

The meaning of a concept is established by a working definition. The definition may be stated in words, symbolic notation, or by means of a mathematical formula. In whatever way the definitions are stated, we learn them, explain them, and interpret their meaning with respect to our environment.

Physical characteristics concerning the phenomena can be expressed in terms of fundamental quantities, and our comprehension of the physical world is based on these fundamental quantities. Presently, scientists have conceived four quantities they specify as fundamental. They are **length, mass, time,** and **electric charge.** These quantities are fundamental in the sense that they form the foundation for other quantities needed to describe and understand the physical sciences.

How would you describe your environment? As you begin to observe, you would ask questions like

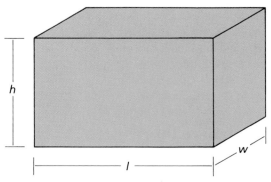

Figure 1.2
The dimensions of a box are commonly given in terms of its length (1), width (w), and height (h), but all are measurements of length.

"Where is the bookstore?" "When does it open?" "How much do you want to buy?" and so forth. These questions of "Where?" "When?" and "How much?" refer to the basic concepts of space, time, and matter.

The description of space might refer to a location or to the size of an object. To measure locations and sizes, we use the fundamental quantity of **length,** which is defined as the measurement of space in any direction.

Space has three dimensions, each of which can be measured by a length (see Fig. 1.2). The three dimensions can easily be seen by considering a rectangular object. It has a length, width, and height, but each of these dimensions is a length. A sphere, such as the Earth, has a radius, a diameter, and a circumference. Again, all of these dimensions are easily described by a length measurement.

Once we know where something is, we are frequently interested in what is happening to it. "Is the car moving?" "When will the next plane leave?" "What day will you be going home?"—all of these questions can be answered using the fundamental quantity of time.

Each of us has an idea of what time is, but we would probably find it difficult to define or to explain it. Some terms that are often used in referring to time are duration, period, or interval. We will define **time** as the continuous, forward flowing of events.

Without events or happenings of some sort, there would be no perceived time (Fig. 1.3). The mind has no innate awareness of time, merely the awareness of events taking place in time. That is, we do not perceive time as such, only the events that take place in time.

Einstein, in his theory of relativity, has shown that space and time are inextricably linked together into what

Figure 1.3
Alberto Salazar crossing the finish line at the 1980 NYC marathon. The time for this event was 2 hr, 9 min, and 40 s.

is called space-time. In his theory, time joins the three dimensions of space as a fourth dimension. For the most part, however, we can use our intuitive ideas, which tend to regard space and time as separate fundamental quantities.

When we ask questions concerning the amount of matter, we need a third fundamental quantity known as mass. To define mass precisely, we need to understand the concepts of force and acceleration. These concepts are discussed in Chapters 2 and 3. For now, let us simply say that **mass** refers to the amount of matter an object contains.

Since we live on Earth, many of us tend to measure matter in terms of weight. However, in the metric system of measurement (see Section 1.3), weight is not a fundamental quantity. An astronaut who weighs 210 pounds on Earth will weigh $\frac{1}{6}$ of that amount, or 35

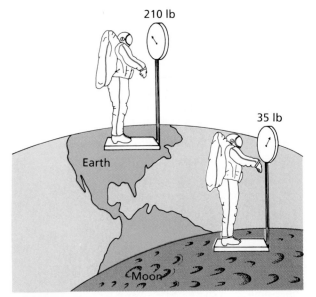

Figure 1.4
The weight of an astronaut on the moon is $\frac{1}{6}$ what it is on Earth. The mass, however, is the same.

pounds, on the moon (Fig. 1.4). However, the astronaut's mass will be the same on the Earth and the moon.

Weight is related to the force of gravity, which changes depending on where we are in the universe. On the other hand, mass is a fundamental quantity, which remains the same throughout the universe. In Chapter 3 we will discuss in detail the relationship between mass and weight.

A fourth fundamental quantity is electric charge. There are two kinds of **electric charge,** called positive and negative. Two positive charges or two negative charges repel each other, while opposite charges attract (Fig. 1.5).

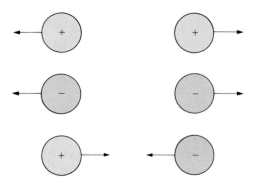

Figure 1.5
Like electrical charges repel. Unlike electrical charges attract.

Electric charge is an important property of matter since all atoms are composed of electrically charged particles and most matter is composed of atoms. Electric current is simply the flow of electrical charge. The concepts of electric charge and electric forces will be discussed in Chapter 8.

1.3 Standard Units

In order to measure the fundamental quantities and their various combinations, we need to compare them with reference, or standard, measurements. Each reference is called a standard unit. A **standard unit** is a fixed and reproducible value for the purpose of taking accurate measurements. There are several systems of measurement used around the world. Each system uses different standard units.

The United States is one of the few nations that uses the British engineering system of measurement. The **British system** uses the familiar unit of foot for length. It uses the pound as the unit of weight and is thus called a gravitational system of measurement.

The unit of mass in the British system is the slug, which weighs approximately 32 pounds on the Earth's surface. All systems of measurement use the second as the standard unit of time, and the coulomb is usually used as the standard unit of charge.

The **metric system** of measurement is much simpler than the British system because converting from one unit to another can be accomplished in the metric system by using factors of 10. For example, in the metric system 1 kilometer is 1000 meters, whereas in the British system 1 mile is 5280 feet. Memorizing the various conversions, such as 12 inches in one foot, 3 feet in one yard, 5280 feet in one mile, etc., makes the British system unwieldy compared with the simplicity of the metric system.

There are actually two metric systems in common use. One is called the mks system; the other is the cgs system. The letters **mks** stand for *m*eter, *k*ilogram, and *s*econd, while **cgs** stands for *c*entimeter, *g*ram, and *s*econd. Table 1.1 lists the standard units for the mks, cgs, and British systems of measurement.

The standard unit of length in the meter-kilogram-second (mks) system is the **meter** (from the Greek *metron,* "to measure"), which was originally intended to be one ten-millionth of the distance from the Earth's equator to the geographic north pole (Fig. 1.6). The unit was first adopted by the French in the 1790s, and it is now used in scientific measurements of length throughout the world.

Table 1.1 Standard Units for the Metric and British Systems of Measurement

	Absolute Systems		Gravitational System
Fundamental Concept	*Metric (mks)*	*Metric (cgs)*	*British*
Length	Meter (m)	Centimeter (cm)	Foot (ft)
Mass	Kilogram (kg)	Gram (g)	Slug
Time	Second (s)	Second (s)	Second (s)
Electric charge	Coulomb (C)	Coulomb (C)	Coulomb (C)

From 1889 to 1960 the standard meter was a platinum-iridium bar kept in the vaults at the International Bureau of Weights and Measures near Paris, France. However, the stability of this bar was questioned, and so new standards were fixed in 1960 and, most recently, in 1983. The current definition of the meter links it to the speed of light in a vacuum. One meter is defined to be the distance light travels in 1/299,792,458 of a second. That is, the speed of light in a vacuum is defined to be 299,792,458 meters per second.

From this basic standard other units of length are defined. For example, the millimeter is defined as $\frac{1}{1000}$ meter, the centimeter as $\frac{1}{100}$ meter, and the kilometer as 1000 meters.

After a concept has been defined and explained, an example needs to be given to indicate that the concept is understood and can be used in respect to the environment. Here are some examples of length: The length of a football field is 100 yards, or 91.35 meters. The long dimension of this textbook is 26 centimeters. The distance from New York to Washington, D.C., is 236 miles. Note that all of these examples include a number and a unit.

The standard unit of time in both the metric system and the British system is the second. For many years the **second** was defined as a fractional part (1/86,400) of the average solar day. The average day was used because the length of the day (as measured when the Sun is directly overhead) varies slightly, since the Earth's path around the Sun is not a perfect circle.

Today's scientists use an atomic definition of the second, based upon the vibrations of a cesium 133 atom as it radiates a certain wavelength of light. The **second** is defined as the duration of 9,192,631,770 cycles of the radiation associated with a specified transition of the cesium 133 atom.

Here are some examples of time: The science class met for 50 minutes. The period of a simple pendulum is 2.8 seconds. The student drove from New York to Washington, D.C., in 4 hours.

Although the meter and second are now accurately defined in terms of the speed of light and atoms, the definition of the kilogram is not so precise. It is associated with the meter.

Originally, one **gram** was defined as the mass of one cubic centimeter of pure water at its maximum density. The **kilogram** was taken to be 1000 grams, or the mass of 1000 cm³ of water. Since one liter is 1000 cm³, a kilogram was taken to be the mass of one liter of water at its maximum density.

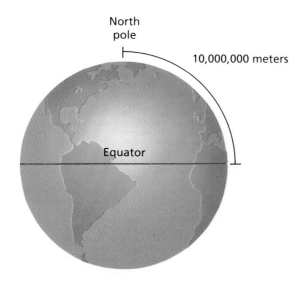

North pole

10,000,000 meters

Equator

Figure 1.6
The meter was originally defined so that the distance from the North Pole to the equator would be 10,000,000 meters.

In everyday terms, a kilogram is the mass of one liter of water. Because most drinkable liquids (soft drinks, milk, juices, etc.) have about the same density as water, a liter of any soft drink will have a mass of approximately one kilogram. The liter is illustrated in Fig. 1.7. Also, since a liter is approximately a quart (1 L = 1.056 qt), a quart of milk has a mass of approximately one kilogram.

A definition of the kilogram in terms of the properties of water is not exact enough for precision measurements. Currently, the kilogram is defined to be the mass of a cylinder of platinum-iridium kept at the International Bureau of Weights and Measures near Paris. The U.S. prototype is kept at the Institute of Standards and Technology in Washington, D.C. (Fig. 1.8).

When giving examples of mass, be careful to use mass units and not weight units. Here are some examples of mass: One liter of water has a mass of one kilogram. The mass of this textbook is 1.4 kilograms. Sarah has a mass of 50 kilograms.

Most countries now use the metric system. In fact, the United States is the only major country not to use the metric system officially. However, the use of metric units such as grams, kilograms, liters, meters, and kilometers is becoming more common in the United States. Figure 1.9 shows a highway sign given in metric units. Table 1.2 shows the vast range of length, mass, and time, expressed in powers-of-10 notation. Powers of 10 is explained in Section 1.7.

HIGHLIGHT

The Metric System

Historians generally agree that the metric system originated with Gabriel Moulton, a French mathematician, when in 1670 he proposed a comprehensive decimal system based upon a physical quantity of nature and not the human anatomy. Moulton proposed that the fundamental base unit of length be equal to one minute of arc of a great circle of the Earth, and a pendulum constructed with this length be used to define the unit of time.

Over one hundred twenty years later, in the 1790s during the French Revolution (1789–1799), the French Academy of Science recommended the adoption of a decimal system with a unit of length equal to one ten-millionth the distance on the surface of the Earth from the equator to the north pole. The unit of length was named the "metre" from the Greek *metron*, meaning "to measure." A cubing of this length was proposed to be a unit of volume, and a unit of mass was proposed to be equal to the amount of pure water needed to fill the cube. Thus an easy usable system based on multiples or submultiples of 10 and defined on a single base unit

related to a physical quantity of nature was established.

The metric system as originally conceived had problems, especially with standard units. To solve these problems, the French government in 1870 legalized a conference to work out standards for a unified measurement system. Five years later on May 20, 1875, the Treaty of the Meter was signed in Paris by 17 nations, including the United States.

The Treaty established a General Conference on Weights and Measures as the supreme authority for all actions. The treaty also established an International Committee of Weights and Measures with the responsibility for the supervision of the International Bureau of Weights and Measures—a permanent laboratory and world center of scientific metrology (science of measurement).

The United States officially adopted the metric system in 1893, but there were no mandatory requirements, and the British units have continued to be used. The United States Congress enacted Public Law 94–168, the Metric

Conversion Act of 1975, that stated that "the policy of the United States shall be to coordinate and plan the increasing use of the metric system in the United States and to establish a United States Metric Board to coordinate the voluntary conversion to the metric system." However, no mandatory requirements were made, and the United States continues to use the British units of measurement.

In 1960 a modified metric system consisting of six basic standard units (meter, kilogram, second, ampere, kelvin, and candela) was established by the 11th General Conference of Weights and Measures. Today this system is known as the International System of Units, or SI for short (SI is an abbreviation for Le Système International d'Unitès).

The metric mass unit (kilogram) adopted in 1960 did not meet the needs of chemistry, so the 14th General Conference meeting in 1971 established the seventh base unit, the mole, for the amount of substance. See Appendix I for the seven base units and their definitions.

Figure 1.7
A one-liter bottle of soft drink will have a mass of approximately one kilogram.

Figure 1.8
Prototype kilogram number 20 is the United States standard unit of mass. The prototype is a platinum-iridium cylinder, 39 mm in diameter and 39 mm high.

Figure 1.9
Highway sign showing British and metric units for the maximum speed limit.

Table 1.2 Some Values of Length, Mass, and Time (Values Are Approximate and Rounded to Nearest Power of 10)

Length (meters)		Mass (kilograms)		Time (seconds)	
Radius of known universe	10^{25}	Known universe	10^{51}	Time for light waves to reach Earth from most distant quasar	10^{17}
Diameter of Milky Way	10^{21}	Milky Way galaxy	10^{41}		
One light year	10^{16}	Sun	10^{30}	Half-life of uranium 235	10^{16}
Distance from Earth to Sun	10^{11}	Earth	10^{23}	Half-life of carbon 14	10^{11}
Distance light travels in one second	10^{8}	Man	10^{2}	One day	10^{5}
		One liter of water	10^{0}	One class session	10^{3}
Diameter of Earth	10^{7}	One dime	10^{-3}	One minute	10^{2}
Length of football field	10^{2}	Postage stamp	10^{-5}	Time between heartbeats	10^{0}
Width of hand	10^{-2}	Red blood cell	10^{-12}	Time a discharged bullet travels the barrel of a rifle	10^{-3}
Thickness of paper	10^{-4}	Iron atom	10^{-25}		
Diameter of hydrogen atom	10^{-9}	Proton	10^{-27}	Time for a beam of light to travel the length of a football field	10^{-6}
Diameter of proton	10^{-15}	Electron	10^{-30}	Half-life of polonium 212	10^{-7}
				Time for the electron to revolve once around nucleus of the hydrogen atom	10^{-15}
				Time for a proton to revolve once in the nucleus	10^{-22}

To make comprehension and the exchange of ideas among people of different nations as simple as possible, the International System of Units (SI) was established. The **SI** is a modernized version of the metric system and contains seven base units: the meter (m), the kilogram (kg), the second (s); the ampere (A), to measure rate of flow of electric charge; the kelvin (K), to measure temperature; the mole (mol), to measure the amount of substance; and the candela (cd), to measure luminous intensity. A definition of each of these seven units is given in Appendix I. All other SI units are derived from the base units and two supplementary units (radian and steradian: a measure of plane and solid angle, respectively) by means of the established scientific laws relating the respective physical quantities.

1.4 Derived Quantities and Conversion Factors

Fundamental quantities are basic quantities from which other quantities are derived. That is, **derived quantities** are formed by combining one or more of the fundamental quantities. For example, area, volume, speed, and density are derived quantities, and they are defined as follows:

$$\text{area} = \text{length}^2$$

$$\text{volume} = \text{length}^3$$

$$\text{speed} = \frac{\text{length}}{\text{time}}$$

$$\text{density} = \frac{\text{mass}}{\text{volume}} = \frac{\text{mass}}{\text{length}^3}$$

The fundamental quantities are combined to derive most of the terms used by the scientist. The combinations will become evident as you continue your study of physical science. Some examples follow.

Derived Quantity	Standard Unit
Area (length²)	m², cm², ft², etc.
Volume (length³)	m³, cm³, ft³, etc.
Density (mass per volume)	kg/m³, g/cm³

Density refers to how compact or crowded a substance is. In more formal language, **density** is the amount of mass located in a definite volume, or simply the mass

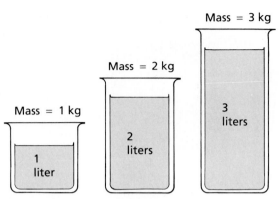

Figure 1.10 Equal densities.
Whether you have one, two, or three liters of pure water, (with masses of one, two, or three kilograms), the density is the same. In this example, the density of water in each of the three containers is 1 kg/liter.

per unit volume. Thus, something with a mass of 20 kg that occupies a volume of 5 m³ has a density of 20 kg/5 m³ = 4 kg/m³. In this example the mass is measured in kilograms and the volume in cubic meters.

If mass is uniformly distributed throughout the volume, then the density of the matter will remain constant. Figure 1.10 shows that if you have a uniform substance, such as water, the density remains the same no matter how much of the substance you have.

The density of water is 1 g/cm³, 1 kg/L, or 1000 kg/m³. If density is expressed in units of grams per cubic centimeter, we can compare any density with that of water. For example, a rock might have a density of 3.3 g/cm³, pure iron has a density of 7.9 g/cm³, and the earth as a whole has an average density of 5.5 g/cm³.

Densities of liquids such as blood or alcohol can be measured by means of a hydrometer. A **hydrometer** consists of a weighted glass bulb that floats in the liquid. The higher the glass bulb floats, the greater the density of the liquid is.

When a medical technologist checks a sample of urine, one test he or she runs is for density (Fig. 1.11). Urine has a density in a healthy person of from 1.015 to 1.030 g/cm³. That is, it consists mostly of water and dissolved salts. When the density is greater or less than this normal range, the urine may have an excess or deficiency of dissolved salts, perhaps caused by an illness.

A hydrometer is used to test for antifreeze in a car radiator. The hydrometer is calibrated directly in degrees rather than actual density. The closer the density is to 1.00 g/cm³, the closer the antifreeze and water solution is to being pure water. When the density corresponds

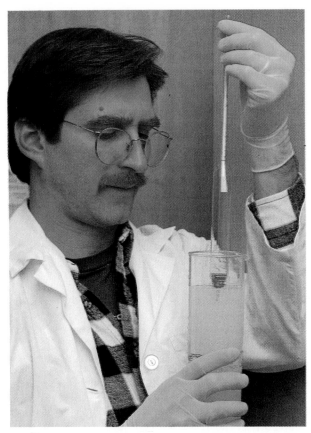

Figure 1.11
A hydrometer is used to measure the density of a liquid.

to 1.00 g/cm³, the hydrometer will read a temperature of 0°C or 32°F, the freezing point of water. The further the mixture is from being pure water, the lower the temperature reading will be.

The units on both sides of an equation must always be similar. For example, we cannot equate two quantities with fundamentally different units. We cannot equate something with units of velocity (e.g., meters per second) to something with units of area (e.g., square meters).

When a combination of units gets very complicated, we frequently give it a name of its own. Consider the following examples, which will be discussed in later chapters:

$$\text{joule} = \text{kg} \times \text{m}^2/\text{s}^2$$

$$\text{newton} = \text{kg} \times \text{m}/\text{s}^2$$

$$\text{watt} = \text{kg} \times \text{m}^2/\text{s}^3$$

There are many more, but the point is that it is easier to talk about watts than kg × m²/s³. When a particular

combination of units is important, it is usually given its own name.

Frequently, we want to convert from one system of units to another in order to make comparisons. For instance, we frequently want to make comparisons between the metric and the British systems. Many of these conversion factors are listed on the inside back cover. For instance,

$$1 \text{ in} = 2.54 \text{ cm}$$

This conversion factor is frequently used. For example, if you want to know your height in centimeters and you are 5 ft 6 in tall, or 66 in, your height in centimeters is given by

$$66 \text{ in} = 66 \text{ in} \times 2.54 \frac{\text{cm}}{\text{in}} = 167.6 \text{ cm}$$

In the metric system your height is 167.6 cm. If you ask someone what his or her height is and the reply is a number such as 165 or 180 or 190, the height is given in centimeters. To convert these numbers to inches, you have to divide by 2.54.

You must perform a similar exercise to convert mass to weight or vice versa on the Earth's surface. Strictly speaking, mass and weight refer to two different quantities. They are not equal, but a given mass does have an equivalent weight on Earth. The appropriate conversion factor on the Earth's surface is

$$1 \text{ kg mass} = 2.2 \text{ lb weight}$$

To find your mass in kilograms, simply divide your weight by 2.2 (actually 2.2 lb/kg). For instance, if you weigh 132 lb, then your mass in kilograms is

$$132 \text{ lb} \times \frac{1 \text{ kg}}{2.2 \text{ lb}} = 60 \text{ kg}$$

If a person's mass is known in kilograms, his or her equivalent weight on Earth can be found by multiplying by 2.2 (actually 2.2 lb/kg). For example, a mass of 70 kg is an equivalent weight of

$$70 \text{ kg} \times 2.2 \frac{\text{lb}}{\text{kg}} = 154 \text{ lb}$$

Often we are concerned with converting from the metric unit of speed, which is meters per second, to the more familiar miles per hour. The necessary conversion factor is

$$1 \text{ m/s} = 2.24 \text{ mi/h}$$

Figure 1.12
The speedometer mounted in most automobiles today is calibrated in mph and km/h.

To convert from meters per second to miles per hour, we simply multiply by 2.24 (actually 2.24 mi/h ÷ m/s). As an example, 20 m/s can be converted as follows:

$$20 \text{ m/s} \times 2.24 \frac{\text{mi/h}}{\text{m/s}} = 44.8 \text{ mi/h}$$

In Fig. 1.12 a speedometer is calibrated in both mi/h and km/h. In Table 1.3 we summarize how to convert the quantities we have discussed. A more extensive list is given on the inside back cover.

Table 1.3 Conversion of Some Common Units

To convert from inches to centimeters, multiply by 2.54.

To convert from centimeters to inches, divide by 2.54.

To convert from kilograms to pounds, multiply by 2.2.

To convert from pounds to kilograms, divide by 2.2.

To convert from meters per second to miles per hour, multiply by 2.24.

To convert from miles per hour to meters per second, divide by 2.24.

1.5 Measurement of Circles

The units of time probably originated with the Babylonians, who reckoned the year as 360 days. They were aware that, in any circle, a chord that is equal to the radius subtends an arc equal to 60 degrees (usually designated 60°), as shown in Fig. 1.13.

This interesting property may have been the basis of their sexagesimal number system and perhaps accounts for the division of the **degree** into 60 minutes, and minutes into 60 seconds (see Fig. 1.14). Because the Babylonians used the apparent motion of the Sun to tell the passing of daylight hours, their sexagesimal system may also be the basis for our present method for reckoning time, that is, 60 seconds equal 1 minute and 60 minutes equal 1 hour.

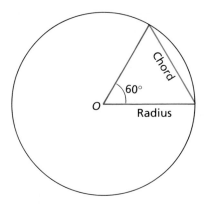

Figure 1.13
A chord that is equal to the radius subtends an angle of 60°.

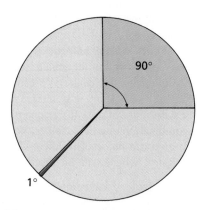

Figure 1.14
A circle is divided into 360°. A right angle has 90°. Each degree is divided into 60 minutes of angle (designated 60′), and each minute of angle is divided into 60 seconds of angle (designated 60″). Thus there are 3600 seconds of angle in 1 degree of angle.

The early Greeks were extremely interested in geometry and the properties of circles. They were particularly interested in the ratio of the circumference of a circle to its diameter. This ratio they designated as π (pi). We can write $\boldsymbol{\pi}$ as

$$\pi = \frac{\text{circumference of a circle}}{\text{diameter of a circle}} \qquad (1.1)$$

This ratio is always the same for every circle. The value of π is given by

$$\pi = 3.14159\ldots \qquad (1.2)$$

and is usually rounded to 3.14 or $\frac{22}{7}$. From the definition of π, we find that the circumference of a circle is just $\pi \times$ diameter. But since the diameter is twice the radius, we get

$$\text{circumference} = 2 \times \pi \times \text{radius} \qquad (1.3)$$

EXAMPLE 1

The Earth goes around the Sun in approximately a circular orbit with a radius of 93 million miles. What is the circumference of the Earth's orbit?

Solution   The circumference is given by

$$\text{circumference} = 2 \times 3.14 \times 93{,}000{,}000 \text{ mi}$$
$$= 584{,}040{,}000 \text{ mi}$$

We see, then, that each year the Earth travels over 500 million miles in its journey around the Sun.

The quantity π is also important when we find the area of a circle. The area of a circle is given by

$$\text{area} = \pi \times (\text{radius})^2$$

or

$$A = \pi r^2$$

1.6 Experimental Error

Making measurements of physical quantities requires the use of some device that will give a number denoting the ratio of the observed quantity to one of the known standard units. A **measurement** is a comparison of the unknown physical quantity with the standard unit.

The process of taking any measurement by any known means always involves some uncertainty. This uncertainty is usually called **experimental error.** In taking measurements, we try to keep the experimental error to a minimum.

Experimental errors are classified as either systematic or random. Systematic errors are always in the same direction; that is, the magnitude, or size, of the number

obtained is always too small or too large, as with a watch that always runs too fast or a speedometer that always shows too few miles per hour.

The observer may also be the cause of systematic errors, because of bad vision or other difficulties with his or her senses. An example of systematic experimental error is shown in Fig. 1.15. Not all such errors are so apparent, however.

Random errors are in either direction and result from accidental variations of the observed physical quantity or of the measuring instruments. Random errors are caused by small variations in any direct or indirect physical quantity associated with the measurement. They can result from such causes as temperature or pressure variations.

When random errors are small, the measurement is said to have high precision. **Precision** refers to the degree of reproducibility of a measurement, that is, to the maximum possible error of the measurement, and may be expressed as a plus or minus correction. For example, if the length of the sample is expressed as 44.4 cm $\pm$ 0.1 cm (read "plus or minus $\frac{1}{10}$ of 1 centimeter"), we know the sample is somewhere between 44.3 and 44.5 cm long.

A measurement having high precision does not necessarily have high accuracy. For example, one could determine the boiling point of a liquid with a precision of $\pm 0.01°C$ ("plus or minus $\frac{1}{100}$ of 1 degree Celsius"),

Figure 1.15
If the observer does not realize that the zero point of the measuring instrument is incorrect, he or she will be making a systematic error.

but any impurity in the liquid would prevent the accurate determination of the true boiling point of the liquid.

With experimental error always appearing in our measurements, how close an approximation can we obtain in making a measurement of any physical quantity; that is, how close can we come to the true value? The term **accuracy** refers to how close the measurement comes to the true value. How is the true value determined? The true value is the value currently accepted by the best scientists in the field. This value is subject to change as measurements are made with better methods and better instruments.

It is obvious from the preceding discussion that the terms *precision* and *accuracy* should never be used interchangeably. They have different meanings and should be used only where they apply.

For example, suppose a manufacturer advertises a thermometer that will measure to the nearest hundredth of a degree Celsius. Thus, using this thermometer, one can measure with a precision of $\pm 0.01°C$. Now suppose an experimenter wants to determine the boiling point of pure water to the nearest 0.01°C with this thermometer. If the experimenter has pure water, he or she can perform the experiment and obtain an accurate answer. If, however, the water is not pure, an "accurate" measurement of the boiling point of pure water cannot be made, even though a high "precision" thermometer is used.

Since error is impossible to eliminate, methods have been devised to calculate the amount of error that exists in a given measurement. Calculation of the amount of error is done by either one of two methods, depending on the circumstances.

If there is an accepted or true value of the physical quantity, then a calculation known as percentage error is made. **Percentage error** is defined as the ratio of the absolute difference* between the experimental and the accepted values to the accepted value, expressed as a percentage. This term can be written as

percentage error

$$= \frac{|\text{exp. value} - \text{accepted value}|}{\text{accepted value}} \times 100$$

If there is no accepted value, then a percentage difference is obtained. **Percentage difference** is defined

as the ratio of the absolute difference between the experimental values to an average of the experimental values, expressed as a percentage. This term can be written as

percentage difference

$$= \frac{|\text{1st exp. value} - \text{2nd exp. value}|}{\text{average of the two values}} \times 100$$

A more useful value of measurement in experimental work is often the average value. For example, if we wish to find the length of a table, a good procedure is to take three separate measurements, find the sum of the three measurements, and then divide by 3 to obtain the average. When three or more measurements are made, the percentage difference is computed, using the average value and the measured values farthest from the average.

A better understanding of errors will be obtained when measurements are taken on real objects and the accuracy of the measurements determined. Generally, this kind of activity occurs in the laboratory.

1.7 Powers-of-10 Notation

In physical science many numbers are very big or very small. In order to express very big or very small numbers, we frequently use the **powers-of-10** notation. When the number 10 is squared or cubed, we get

$$10^2 = 10 \times 10 = 100$$

$$10^3 = 10 \times 10 \times 10 = 1000$$

You can see that the number of zeroes is just equal to the power of 10. As an example, 10^{23} is a 1 followed by 23 zeroes.

Negative powers of 10 can also be used. For example,

$$10^{-2} = \frac{1}{10^2} = \frac{1}{100} = 0.01$$

We see that if a number has a negative exponent, we shift the decimal place to the left once for each power of 10. Thus one micrometer, which is 10^{-6} m, equals 0.000001 m.

We can also multiply numbers by powers of 10. Table 1.4 shows a wealth of examples of various large and small numbers expressed in powers of 10 notation.

There are many standard prefixes that are used to represent powers of 10. These prefixes are listed in Table 1.5. The prefixes are frequently used in expressing var-

---

* "Absolute difference" is the result obtained when the smaller value is subtracted from the larger. It is designated by two vertical lines enclosing the subtraction.

Table 1.4 Numbers Expressed in Powers of 10 Notation

Number	Powers of 10 Notation
0.025	2.5×10^{-2}
0.0000408	4.08×10^{-5}
0.0000001	1×10^{-7}
0.0000000000000000016	1.6×10^{-19}
247	2.47×10^2
186,000	1.86×10^5
4,705,000	4.705×10^6
9,000,000,000	9×10^9
30,000,000,000	3×10^{10}
602,300,000,000,000,000,000,000	6.023×10^{23}

Table 1.5 Prefixes Representing Powers of 10

Multiple	Name	Abbreviation
10^{18}	exa	E
10^{15}	peta	P
10^{12}	tera	T
10^9	giga	G
10^6	mega	M
10^3	kilo	k
10^2	hecto	h
10	deka	da
10^{-1}	deci	d
10^{-2}	centi	c
10^{-3}	milli	m
10^{-6}	micro	μ
10^{-9}	nano	n
10^{-12}	pico	p
10^{-15}	femto	f
10^{-18}	atto	a

ious units. The most important are **mega, kilo, milli,** and **micro.** Some examples are

$$1\ mega\text{ton} = 10^6 \text{ tons}$$

$$1\ kilo\text{meter} = 10^3 \text{ meters}$$

$$1\ milli\text{gram} = 10^{-3} \text{ grams}$$

$$1\ micro\text{second} = 10^{-6} \text{ seconds}$$

We can represent a number in powers of 10 notation many different ways—all correct. For example, the distance from the Earth to the Sun is 93 million miles. This data can be represented as 93,000,000 miles, or 93×10^6 miles, or 9.3×10^7 miles, or 0.93×10^8 miles, etc. Scientists generally pick the power of 10 so that it is multiplied by a number between 1 and 10. Thus 9.3×10^7 miles would probably be chosen. However, any of the given representations of 93 million miles is correct.

We can see from the above example that changing the power of 10 causes a change in the number it multiplies. When the power of 10 is increased by one, the decimal must be moved one space to the left; when the power of 10 is decreased by one, the decimal must be moved one space to the right. Examples are

$$16 \times 10^{-6} = 1.6 \times 10^{-5} = 0.16 \times 10^{-4}$$

$$24 \times 10^3 = 2.4 \times 10^4 = 0.24 \times 10^5$$

$$13 \text{ mg} = 13 \times 10^{-3} \text{ g} = 1.3 \times 10^{-2} \text{ g}$$

Example 2 illustrates the use of powers of 10 notation and conversion factors in the solution of a problem. See Appendix IV, Example 3, for additional information on using conversion factors.

EXAMPLE 2

Using the powers of 10 notation and conversion factors, calculate the number of seconds a college student, age 20 years, has lived.

Solution

20 years = 20 years

$$20 \text{ years} = 2.0 \times 10\ \cancel{\text{years}} \times 3.65 \times 10^2\ \frac{\text{days}}{\cancel{\text{year}}}$$

(cancel years)

$$20 \text{ years} = 7.3 \times 10^3\ \cancel{\text{days}} \times 2.4 \times 10\ \frac{\text{hours}}{\cancel{\text{day}}}$$

(cancel days)

$$20 \text{ years} = 1.752 \times 10^5\ \cancel{\text{h}} \times 6.0 \times 10\ \frac{\text{min}}{\cancel{\text{h}}}$$

(cancel hours)

$$20 \text{ years} = 1.0512 \times 10^7\ \cancel{\text{min}} \times 6.0 \times 10\ \frac{\text{s}}{\cancel{\text{min}}}$$

(cancel minutes)

$$20 \text{ years} = 6.3072 \times 10^8 \text{ s}$$

Knowing how to express a measured quantity by using powers of 10 notation is not sufficient for solving problems in the physical sciences. One must know how to use the notation in the simple operations of addition, subtraction, multiplication, and division. See Appendix III for rules and examples.

Learning Objectives

After reading and studying this chapter, you should be able to do the following without referring to the text:

1. State four fundamental quantities of nature.

2. State the units in which these fundamental quantities are measured in three different systems of measurements.

3. Explain the concept of density and tell how it is different from mass.

4. State how the gram, meter, and second are defined.

5. Be able to convert from centimeters to inches, from pounds to kilograms, and from meters per second to miles per hour, or vice versa, using Table 1.2.

6. State the origin of π.

7. State the formulas for the circumference and area of a circle if the radius is known.

8. Distinguish between percentage error and percentage difference.

9. Express any number in powers of 10 notation.

10. State the meaning of the prefixes *mega, kilo, milli,* and *micro.*

11. Name the seven base units of the International System of Units (SI).

12. Define and explain the important words and terms listed below.

Important Words and Terms

fundamental quantities	mks system	kilogram	percentage error
concept	cgs system	SI	percentage difference
length	meter	derived quantities	powers of 10
time	experimental error	density	mega
mass	precision	hydrometer	kilo
electric charge	accuracy	degree	milli
standard unit	second	π	micro
British system	gram	measurement	

Questions

The Senses

1. Which of our five senses provides us with the most knowledge of our physical world?

2. State some limitations of our senses to obtain accurate information concerning our environment.

3. Do all measurements ultimately depend on our senses?

Concepts and Fundamental Quantities

4. What are the four fundamental quantities we seek to measure?

5. Define, explain, and give an example of (a) length, (b) mass, and (c) time.

6. Would time exist if there were no motion?

7. Are there four dimensions or three?

8. (a) Does a 60-kg astronaut have the same mass on the moon as on the Earth? (b) Does he have the same weight?

Standard Units

9. What is the origin of the (a) meter, (b) kilogram, and (c) second?

10. Which of the fundamental quantities is still based on an artifact?

11. Why have new standards of length and time recently been adopted?

12. Which is more basic, the standard unit of length or of time? Explain your answer.

13. State some advantages of the metric system of measurement over the British system.

14. When was the International System of Units (SI) adopted?

Derived Quantities and Conversion Factors

15. Distinguish between fundamental quantities and derived quantities.

16. Define *density* in terms of fundamental quantities.

17. State the standard units for measuring density in the mks system of units.

18. (a) Which is more dense, a kilogram of iron or a kilogram of feathers? (b) Which has more mass?

19. How is the antifreeze in a car radiator tested?

Measurement of Circles

20. How does the area of a circle vary with the radius?

21. How many seconds of angle are there in one degree of angle?

22. What is the origin of π?

Experimental Error

23. Define *measurement* and state some limitations in obtaining accurate measurements.

24. Distinguish between accuracy and precision.

25. Why is a very precise stopwatch sometimes inaccurate in timing a 100-m dash?

26. Give an example of a situation when you would use percentage difference instead of percentage error.

Powers-of-10 Notation

27. Why would you write Avogadro's number (6.02×10^{23}) in powers of 10 notation?

28. 1×10^7 is _____ times larger than 1×10^4.

Exercises

Concepts and Fundamental Quantities

1. An astronaut has a mass of 60 kg.
   (a) What is this mass on the Earth and on the moon?
   (b) What is this weight on the Earth and on the moon, in pounds?

Standard Units

2. What is the mass of 3 L of water?

3. What is the distance, in meters, light travels in 1/299,792,458 s?

4. How many seconds are in one day?

Derived Quantities and Conversion Factors

5. Compute the density in grams per cubic centimeter of a rock that has a mass of 1 kg and a volume of 280 cm³.
   Answer: 3.57 g/cm³

6. What is the volume of a piece of iron (density 7.86 g/cm³) that has a mass of 2.30 kg? *Answer:* 292.6 cm³

7. Compute the height in feet and inches of a person who is 200 cm tall. *Answer:* 6 ft 7 in.

8. Compute the height in centimeters of a woman who is 5 ft 6 in tall.

9. Compute the mass in kilograms of a man who weighs 150 lb. *Answer:* 68.2 kg

10. Compute the weight in pounds of a 2-kg package.

11. Compute the speed in meters per second of an auto traveling at 40 mi/h. *Answer:* 17.9 m/s

12. Compute the speed in miles per hour of an auto traveling at 30 m/s.

Measurement of Circles

13. The moon is approximately 240,000 mi from the Earth. Assume the moon travels in a circle around the Earth (this assumption is not quite correct).

(a) What is the circumference of its path around the Earth?
(b) What is the area of this circle, in square miles?
(c) Express your answer to (b) in powers of 10 notation.
   Answer: (c) 1.81×10^{11} mi²

14. Compute the area in square inches of a pizza with a diameter of (a) 7 in, (b) 10 in, and (c) 14 in. (*Note:* Be sure to realize the diameter is given, not the radius.) (b) The 14-in pizza is how many times as big as the 7-in pizza? *Answer:* (b) 4 times as big

Experimental Error

15. When Eratosthenes measured the circumference of the Earth over two thousand years ago, he got an answer of 24,000 mi. What was his percentage error if the correct answer is 24,900 mi? *Answer:* 3.6%

16. A student measures π to be 3.16. What is the percentage error of this measurement?

Powers-of-10 Notation

17. Fill in the blank with the correct power of 10.
   (a) 0.15 megaton = 1.5 $\times$ _____ tons
   (b) 10.5 kilovolts = 1.05 $\times$ _____ volts
   (c) 72 milligrams = 7.2 $\times$ _____ grams
   (d) 0.65 microwatt = 6.5 $\times$ _____ watts
   Answer: (c) 7.2×10^{-2} g

18. State the following quantities in powers of 10 notation.
   (a) 24 megawatts
   (b) 16 kilotons
   (c) 31 micrograms
   (d) 0.1 millimeter *Answer:* (c) 31×10^{-6} g

19. Using the powers of 10 notation and conversion factors, calculate the number of seconds in the average lifetime (70 years) of a human.

20. Using the powers of 10 notation and conversion factors, calculate the number of seconds you have lived.

Motion

> **Man is the shuttle, to whose winding quest**
> **And passage through these looms**
> **God order'd motion, but ordain'd no rest.**
>
> —Henry Vaughn

MOTION IS EVERYWHERE. We walk to class. We drive to the store. Birds fly. The wind blows the trees. The rivers flow. Even the continents drift. In the larger environment the Earth rotates on its axis. The moon revolves around the Earth. The Earth revolves around the Sun. The Sun moves in the galaxy. The galaxies move with respect to one another.

This chapter focuses on the description of motion in our environment, with definitions and discussion of terms such as *speed, velocity,* and *acceleration.* We will study these concepts without considering the forces involved, reserving that discussion for Chapter 3.

Two basic kinds of motion are straight-line motion and circular motion. We experience examples of these each day. For instance, we know that driving around a curve gives a different sensation from driving in a straight line. Understanding acceleration is the key to understanding these basic kinds of motion.

2.1 Straight-Line Motion

The term **position** refers to the location of an object. To designate the position of an object, we must give or imply a reference point or position. For example, the entrance to campus is 1.6 km (1 mi) from the intersection with the traffic light. The book is on the table. Atlanta is in Georgia. The Cartesian coordinates of the point on the graph are $(x, y) = (2.0$ cm, 3.0 cm$)$.

If an object changes its position, we say that motion has occurred. When an object is undergoing a continuous change in position, we say the object is moving or is in **motion.**

Because position is relative, motion must also be relative. For instance, the statement "The student is walking 2 meters per second" indicates that the student is changing position at the rate of 2 meters for each second

of time, relative to the sidewalk, ground, or floor. Similarly, if we say, "The car is traveling at the rate of 40 miles per hour," we are using the road as the frame of reference for the car's motion.

An example of straight-line motion is an automobile traveling north on a level highway. The motion of the automobile may or may not be at a constant rate. In either case the motion is described by using the fundamental units of length and time.

That length and time describe motion is evident in racing. For example, as shown in Fig. 2.1, runners try to run a certain length in the shortest possible time. Combining length and time to give the *time rate of change of position* is the basis of describing motion in terms of speed and velocity, as discussed in the following section.

2.2 Speed and Velocity

The terms *speed* and *velocity* are often used interchangeably. In physical science, however, these terms have distinct meanings. The basic difference is that one is a scalar quantity and one is a vector quantity. Let's generally distinguish between scalars and vectors now, because other terms will fall into these categories during the course of our study. The distinction is simple. A **scalar** quantity is one that has magnitude or size only (plus the unit of measurement). For example, you may be traveling in a car at 55 mi/h (about 90 km/h). This figure is your speed, which is a scalar quantity—magnitude (and units) only.

A **vector** quantity, on the other hand, is one that has magnitude *and* direction. For example, suppose you are traveling 55 mi/h *north.* This quantity describes your velocity, which is a vector quantity—magnitude (and units) *plus* direction. Because they give direction, also, vector quantities give more information than scalar quantities.

Now let's look more closely at speed and velocity as used in the description of motion. The **average speed** of an object is the total distance traveled divided by the

---

◄ **People in motion at an intersection.**

Figure 2.1 Motion
We describe motion in terms of distance and time. Here, the runners try to run a certain distance in the shortest possible time.

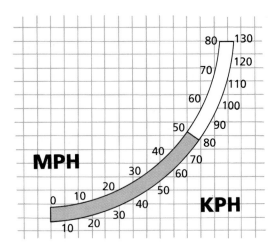

Figure 2.2 Instantaneous speed.
The speed indicated on an automobile speedometer is an example of instantaneous speed—the speed the car is going at a particular instant. Notice that the speedometer is calibrated in both mi/h and km/h, which is quite common on speedometers.

time spent in traveling the total distance. In formula form we have

$$\text{speed} = \frac{\text{distance traveled}}{\text{time to travel distance}}$$

or $$v = \frac{d}{t}$$ (2.1)

Note that d and t are length and time *intervals*. They are sometimes written Δd and Δt to explicitly indicate that they are intervals. The Δ (delta) means "change in" or "difference in"; for example, $\Delta t = t - t_o$, where t_o and t are the original and final times (on the clock), respectively. (If $t_o = 0$, then $\Delta t = t$.) The speed is an average, taken over an appreciable time interval. Speed is somewhat analogous to the average class grade on an exam in a class with a certain number of students.

The speed at any instant of time may be different from the average speed. The **instantaneous speed** of an object is its speed at that instant of time (Δt being extremely small). A common example of nearly instantaneous speed is the speed registered on an automobile speedometer (Fig. 2.2). This value is the speed at which the automobile is traveling right then, or instaneously.

Average velocity is the displacement divided by the total travel time, where **displacement** is the straight-line distance between the initial and final positions, with direction toward the final position—a vector quantity. For straight-line motion in one direction, speed and

velocity are very similar. Their magnitudes are the same (because the lengths of the distance and displacement are the same). The distinction between them in this case is that a direction must be specified for the velocity.

As you might guess, there is also **instantaneous velocity**, which is the velocity at any instant of time. For example, a car's instantaneous speedometer reading plus the direction it is traveling at that instant give its instantaneous velocity. Of course, the speed and/or the direction of the car may and usually does change. This motion is then accelerated motion, which will be discussed in the following sections.

If the velocity is *constant* or *uniform*, then we don't have to worry about changes. Suppose an airplane is flying at a constant speed of 320 km/h (200 mi/h) directly eastward. Then the airplane has a constant velocity and flies in a straight line. (Why?) Also for this special case, you should be able to convince yourself that the instantaneous velocity and average velocity are the same. (By analogy, think about everyone in your class getting the same test score. How do the class average and individual scores compare?)

Let's look at some examples of speed and velocity.

EXAMPLE 1 _____

Describe the motion of the car in Fig. 2.3.

Solution We see in the figure that the car travels

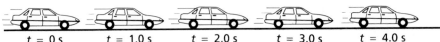

20 m   20 m   20 m   20 m

$t = 0$ s $t = 1.0$ s $t = 2.0$ s $t = 3.0$ s $t = 4.0$ s

Figure 2.3 Constant velocity. The car travels equal distances in equal periods of time in straight-line motion. With a constant speed and a constant direction, the velocity of the car is constant.

80 m in 4.0 s, so it has a speed of

$$v = \frac{d}{t} = \frac{80 \text{ m}}{4.0 \text{ s}} = 20 \text{ m/s}.$$

Notice that the car has a constant or uniform speed and travels 20 m each second. If the motion is in one direction or in a straight line, then the velocity of the car is also constant.

EXAMPLE 2

The speed of light in space is on the order of 186,000 mi/s, or 3.0×10^8 m/s. How long does it take the Sun's rays to reach the Earth?

Solution As you may know, the Earth is about 93 million miles from the Sun. If we rearrange Eq. 2.1, we get $t = d/v$; and with the data we get

$$t = \frac{d}{v} = \frac{93,000,000 \text{ mi}}{186,000 \text{ mi/s}}$$
$$= 500 \text{ s}$$

From this example we realize that although light travels very fast, it still takes 500 s, or about 8.3 min, to arrive at the Earth after leaving the Sun (see Fig. 2.4). Here again we are working with a constant speed and velocity.

QUESTION

If an object has a constant speed, does it also have a constant velocity?

Answer Not always. If an object travels with a constant speed in a straight line, then it also has a constant velocity—that is, constant speed *and* constant direction. However, an object may move with a constant speed in a curved path. In this case the velocity is not constant because the direction of the motion is continually changing. (The following example illustrates this principle.)

EXAMPLE 3

What is the average speed of the Earth as it orbits the Sun?

Solution The Earth orbits the Sun in a nearly circular orbit in a time, or period, of one year. The distance it travels is just the circumference of its circular orbit (which we assume to be the case to a good approximation). Recall from Chapter 1 that the circumference of a circle

is $2\pi r$, where r is the radius, or 93 million miles in this case, which is the mean distance of the Earth from the Sun (see Fig. 2.4). The time it takes to travel this distance is one year, or about 365.25 days.

Putting this data into Eq. 2.1 and multiplying the time by 24 h/day to convert to hours, we have

$$v = \frac{d}{t} = \frac{2\pi r}{t}$$
$$= \frac{2\pi \times 93,000,000 \text{ mi}}{365.25 \times 24 \text{ h}}$$
$$= 66,700 \text{ mi/h}$$

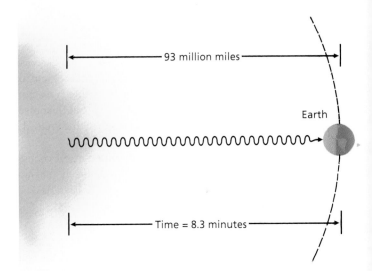

93 million miles

Earth

Time = 8.3 minutes

Figure 2.4 Traveling at the speed of light. Although light travels about 186,000 miles per second, it still takes over 8 minutes for light from the Sun to reach us. See Example 2.

How about that! The solution to Example 3 shows that the Earth, *and all of us*, are traveling through space at a speed of 66,700 mi/h (or 18.5 mi/s). Even though this value is an exceedingly high and relatively constant speed, the velocity is continually changing because the direction of the motion is continually changing. We don't generally sense or notice this great speed because of the small relative motions (apparent motions) of the stars.

Think about how you know you are in motion when riding in a perfectly smooth-riding and quiet car. You see trees and other fixed objects "moving" in relative motion.

Also, we do not sense any change in velocity because the change is too small. We can generally sense changes in motion if they are appreciable. Think about being in the smooth-riding car again and being blindfolded. You would be able to tell if the car suddenly went faster, slowed down, or went around a sharp curve, all of which are changes in velocity. A change in velocity is called an acceleration and is the topic of the following sections.

2.3 Acceleration

When you drive down a straight interstate highway and suddenly increase your speed, say from 20 m/s (45 mi/h) to 29 m/s (65 mi/h), you feel as though you are being forced back against the seat. When driving fast on a circular cloverleaf, you feel forced to the outside of the circle. These experiences result from changes in velocity.

There are three ways that you can change the velocity of an object. You can (1) increase or (2) decrease its magnitude when traveling in a straight line, and/or (3) you can change the direction of the velocity vector. When any of these changes occur, we say that the object is accelerating. The faster the change in the velocity occurs, the greater the acceleration is.

Acceleration is defined as the time rate of change of velocity. If we take the symbol Δ to mean "change in," the formula for acceleration can be written as

$$\text{acceleration} = \frac{\text{change in velocity}}{\text{time for change to occur}} = \frac{\Delta v}{t}$$

Of course, the change in the velocity is just the final velocity v_f minus the original velocity v_o (note that these are *instantaneous* velocities). Thus, in symbols, we can define the acceleration as

$$a = \frac{\Delta v}{t} = \frac{v_f - v_o}{t} \tag{2.2}$$

The units of acceleration in the SI are (m/s)/s, or m/s². These units may be confusing at first. Keep in mind that an acceleration is a measure of a change in velocity during a given time period.

Consider an acceleration of 9.8 m/s². This value means that the velocity changes by 9.8 m/s each second. Thus for straight-line motion, as the number of seconds increases, the velocity goes from 0 to 9.8 m/s dur-

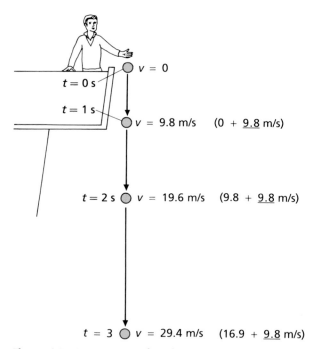

Figure 2.5 Constant acceleration.
For a constant downward acceleration of 9.8 m/s², the velocity increases by 9.8 m/s each second. The increasing lengths of the arrows indicate an increasing velocity.

ing the 1st second, to 19.6 m/s (9.8 m/s + 9.8 m/s) during the 2nd second, to 29.4 m/s (19.6 m/s + 9.8 m/s) during the 3rd second, and so forth, adding 9.8 m/s each second. This sequence is illustrated in Fig. 2.5 for a falling object that falls with an acceleration of 9.8 m/s².

We can also rewrite Eq. 2.2 to give a formula for the final velocity of an object if its original velocity and acceleration are known.

$$v_f - v_o = at$$

or

$$v_f = v_o + at \tag{2.3}$$

This formula is useful for working problems in which the quantities a, v_o, and t are all known and we wish to find v_f. If the original velocity $v_o = 0$, then

$$v_f = at \tag{2.4}$$

Since velocity is a vector quantity, acceleration is also a vector quantity. For an object in straight-line motion, the acceleration (vector) may be in the same direction as the velocity (vector) *or* the acceleration may be in the opposite direction of the velocity (Fig. 2.6). In the first instance the acceleration causes the object to

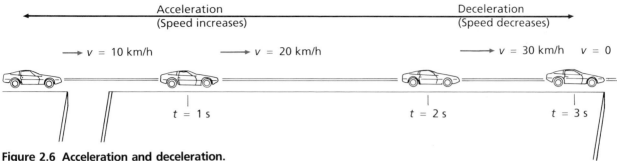

Figure 2.6 Acceleration and deceleration.
When the acceleration is in the same direction as the velocity of an object in straight-line motion, its speed increases. If the acceleration is in the opposite direction of the velocity, there is a deceleration and the speed decreases.

speed up and the velocity increases. If the velocity and acceleration are in opposite directions, then the acceleration slows down the object, which is sometimes called a *deceleration*.

Let's consider an example. What is the common name for the gas pedal of a car? Right, the *accelerator*. When you push down on the accelerator, you speed up (increase the magnitude of the velocity). But when you let up on the accelerator, you slow down, or "decelerate." Putting on the brakes will give an even greater deceleration. (Maybe we should call the brake pedal the "decelerator.")

QUESTION

Would it be appropriate to call the steering wheel of a car an "accelerator"?

Answer Yes, in a sense. As we noted above, acceleration is a change in velocity, and this change can result from a change in the vector magnitude and/or *direction*. Pushing on the gas pedal or the brake pedal can change the magnitude of a car's velocity, and turning the steering wheel can change its direction. So you could correctly call the steering wheel of a car a (direction) "accelerator."

In general, we will consider primarily constant or uniform accelerations. There is one very special constant acceleration associated with the acceleration of falling objects. The **acceleration of gravity** at the Earth's surface is directed downward and is denoted by the letter g. Its magnitude in the SI system is

$$g = 9.80 \text{ m/s}^2$$

This value corresponds to 980 cm/s² or about 32 ft/s².

The acceleration of gravity varies slightly depending on such factors as how far you are from the equator and how high up you are. However, the variations are very small, and for our purposes we will take g to be the same everywhere on the Earth's surface.

The great Italian physicist Galileo Galilei (1564–1642) was one of the first scientists to assert that all objects fall downward with the same acceleration. Of course, this assertion assumes that frictional effects are negligible. We can state Galileo's principle as follows:

> If frictional effects can be disregarded, every freely falling object near the Earth's surface accelerates downward at the same rate, regardless of the mass of the object.

One can illustrate the assertion experimentally by dropping a small mass, such as a coin, and a larger mass, such as a ball, at the same time from the same height. They will hit the floor, as best as can be judged, at the same time (negligible air friction). Legend has it that Galileo himself performed such experiments. (See the chapter Highlight.)

The effect of air resistance or friction can be demonstrated by dropping a piece of tissue paper and a coin. The air friction will prevent the tissue paper from falling as fast as the coin. If the tissue paper is wadded up into a very small ball to minimize air friction, it will fall at the same acceleration as the coin.

On the moon there is no atmosphere, so there is no air friction. One of the astronauts dropped a feather and a hammer at the same time (see Fig. 2.7). They both hit the surface of the moon at the same time because neither the feather nor the hammer was slowed by air friction. This experiment shows that Galileo's assertion applies on the moon as well as on the Earth. Of course, on the moon all objects fall at a slower rate than do objects on the Earth's surface, because the acceleration of gravity is less on the moon than on the Earth.

Figure 2.7 No air resistance.
Astronaut Scott demonstrated that a feather and a hammer fall at the same rate on the moon. There is no air on the moon and hence no air friction to slow the feather down, so it accelerates at the same rate as the hammer.

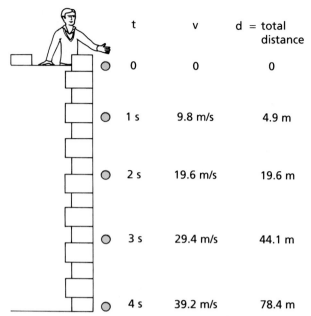

	t	v	d = total distance
	0	0	0
	1 s	9.8 m/s	4.9 m
	2 s	19.6 m/s	19.6 m
	3 s	29.4 m/s	44.1 m
	4 s	39.2 m/s	78.4 m

Figure 2.8 Time, velocity, and distance.
The velocity and distance traveled by a freely falling object at the end of the first few seconds. The velocity increases by 9.8 m/s each second. The distance traveled is computed by using the formula $d = \frac{1}{2}gt^2$ when the object is dropped from rest.

The velocity of a freely falling object increases 9.8 m/s each second, or increases uniformly with time ($v_f = at$, Eq. 2.4). But how about the distance covered each second? Distance covered is not uniform because the object speeds up. The distance a dropped object travels downward with time can be computed from the formula* $d = \frac{1}{2}gt^2$.

Distance and velocity for a falling object are illustrated in Fig. 2.8. From the figure we see that at the end of the first second the object has fallen 4.9 m. At the end of the second second, the total distance fallen is 19.6 m. At the end of the third second, the object has fallen 44.1 m, which is about the height of a 10- or 11-story building, and it is falling at a speed of 29.4 m/s, which is about 65 mi/h—pretty fast!

If we throw an object straight upward, it slows down, or the velocity decreases 9.8 m/s each second. In this case the velocity and acceleration are in opposite directions and there is a deceleration (see Fig. 2.9). The object slows down, or its velocity decreases, until it stops instantaneously at its maximum height. Then it

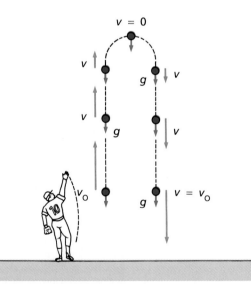

Figure 2.9 Up and down.
An object projected straight upward slows down, because the acceleration is in the direction opposite to that of the velocity, and the object stops for an instant at its maximum height ($v = 0$). It then accelerates downward and returns to the starting point with a velocity equal and opposite to the initial velocity. (The downward path is displaced in the figure for clarity.)

---

* In general, the distance an object travels when it starts at an original velocity v_o and is accelerated at an acceleration a is

$$d = v_o t + \frac{1}{2}at^2$$

starts to fall downward as though it were dropped from this height. The travel time upward is the same as the travel time downward to the original starting position. Also, the object returns with the same speed as it had initially. For example, if an object is thrown upward with an initial speed of 29.4 m/s, it will return with a velocity of 29.4 m/s downward. You should be able to conclude that it would travel 3 s upward, to a maximum height of 44.1 m and return to its starting point in another 3 s. (*Hint:* See Fig. 2.8.)

HIGHLIGHT

Galileo and the Leaning Tower of Pisa

There is a popular story that Galileo dropped stones or cannonballs of different masses from the top of the Tower of Pisa to experimentally determine whether objects fall at the same rate (Fig. 2.10). Whether this legend is true is questionable.

Galileo did indeed question Aristotle's view that objects fell because of their "earthiness"; and the heavier or more earthy an object, the faster it would fall in seeking its "natural" place at the center of the Earth. His ideas are evident in the following excerpts from his writings.*

How ridiculous is this opinion of Aristotle is clearer than light. Who ever would believe, for example, that . . . if two stones were flung at the same moment from a high tower, one stone twice the size of the other, . . . that when the smaller was halfway down the larger had already reached the ground?

And:

Aristotle says that "an iron ball of one hundred pounds falling a height of one hundred cubits reaches the ground before a one-pound ball has fallen a single cubit." I say that they arrive at the same time.

---
* From L. Cooper, *Aristotle, Galileo, and the Tower of Pisa* (Ithaca, N.Y.: Cornell University Press, 1935).

Figure 2.10 Free fall.
All freely falling objects near the Earth's surface have an acceleration of g = 9.8 m/s². Galileo is alleged to have shown this by dropping cannon balls of different masses from the Leaning Tower of Pisa. Over short distances air resistance can be neglected, so the balls would have struck the ground at the same time.

Although Galileo mentions a *high tower*, the Tower of Pisa is not mentioned in his writings, and there is no

Figure 2.11 Galileo Galilei (1564–1642).
The motion of objects was one of his many scientific pursuits.

independent record of such an experiment. The first account of a Tower of Pisa experiment is found in a biography written by one of his pupils over ten years after Galileo's death. Fact or fiction? No one really knows. What we do know is that all freely falling objects near the Earth's surface fall with the same acceleration.

2.4 Acceleration in Uniform Circular Motion

An object in uniform circular motion, has a constant speed. For example, a car going around a circular track at a uniform rate of 90 km/h (55 mi/h) has a constant speed. However, the velocity of the object is *not* constant, because the velocity is continually changing direction. Since there is a change in velocity, there is an acceleration.

This acceleration cannot be in the direction of the instantaneous motion or velocity; otherwise, the object would speed up and the motion would not be uniform. So where is the acceleration? Because it causes a change in direction that keeps the object in a circular path, the acceleration is actually perpendicular, or at a right angle, to the velocity vector.

Consider a car traveling in uniform circular motion, as illustrated in Fig. 2.12. At any point the instantaneous velocity is tangential to the curve (at an angle of 90° to a radial line at that point). After a short time the velocity vector has changed (direction). The change in velocity Δv is given by a vector triangle, as illustrated in the figure.

This change is an average over a time interval Δt, but notice how the Δv vector generally points inward toward the center of the circle. For instantaneous measurement this generalization is true, so for an object in uniform circular motion the acceleration is toward the center of the circle. This acceleration is called **centripetal acceleration** (*centripetal* means "center seeking").

Even though it is traveling at a constant speed, an object in uniform circular motion must have an inward acceleration. For a car this acceleration is supplied by friction on the tires. Should a car hit an icy spot on a curved road, it slides outward, because the centripetal acceleration is not great enough to keep it in a circular path. Similarly, if you swing a ball on a string in a horizontal circle around your head, you must pull inward on the string to supply the necessary centripetal acceleration in order to change the direction of the ball to keep it in a circular path. If you let go of the string or it breaks, the ball will have an initial tangential velocity and fly off in that direction (and downward because of the acceleration due to gravity).

In general, whenever an object moves in a circle of radius r with a constant speed v, the magnitude of the centripetal acceleration a_c is given by the formula*

$$a_c = \frac{v^2}{r} \qquad (2.5)$$

Note that centripetal acceleration increases as the square of the speed—double the speed and the acceleration increases by a factor of 4. Also, the smaller the radius, the greater is the centripetal acceleration needed to keep an object in circular motion for a given speed.

EXAMPLE 4

Determine the magnitude of the acceleration of a car going 12 m/s (about 27 mi/h) on a cloverleaf with a radius of 50 m (Fig. 2.13).

Solution The centripetal acceleration is given by Eq. 2.5:

$$a_c = \frac{v^2}{r} = \frac{(12 \text{ m/s})^2}{50 \text{ m}}$$
$$= 2.9 \text{ m/s}^2$$

This value of 2.9 m/s² is about 30% of the acceleration due to gravity, $g = 9.8$ m/s², and is a fairly large acceleration. We experience the effects of such an acceleration when riding in a car going in a circular path. But recall from Example 3 that the Earth travels at a relatively constant speed of 66,700 mi/h (about 30,000 or 3×10^4 m/s) in a nearly circular orbit about the Sun.

---

* See Appendix IV for a derivation.

Figure 2.12 Centripetal acceleration.
The car traveling with a constant speed on a circular track is accelerating, because its velocity is changing (direction). The acceleration is toward the center of the circle and is called centripetal (center-seeking) acceleration.

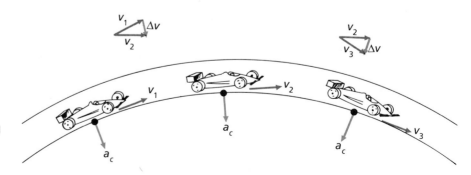

This speed is much faster than the 27 mi/h speed of the car in Example 4. To see why we don't sense the change in velocity associated with this motion, let's compute the centripetal acceleration of the Earth (and us) in its orbit about the Sun. The radius of the orbit, or the distance of the Earth from the Sun, is 93 million miles, or 1.5×10^8 km = 1.5×10^{11} m. Then, by Eq. 2.5,

$$a_c = \frac{v^2}{r}$$
$$= \frac{(3.0 \times 10^4 \text{ m/s})^2}{1.5 \times 10^{11} \text{ m}}$$
$$= 0.0060 \text{ m/s}^2$$

which is a very small acceleration—less than $\frac{1}{10}$ of a percent of g.

Figure 2.13 Frictional centripetal acceleration.
The inward acceleration necessary for a vehicle to go around a curve is, for the most part, supplied by friction on the tires. On a banked curve, some of the acceleration is supplied by gravity.

2.5 Projectile Motion

An object thrown horizontally will fall down at the same rate as an object that is dropped. The velocity in the horizontal direction does not affect the velocity and acceleration in the vertical direction.

As an example, consider a rifle that is fired horizontally (see Fig. 2.14). If a bullet is dropped simultaneously, the bullet fired from the rifle will hit the ground at the same time as the dropped bullet. Of course, this experiment assumes that frictional effects are negligible and the ground is level. The bullets will hit simultaneously because the horizontal velocity of the fired bullet has no effect on its vertical motion. The vertical motion of the fired bullet is identical to that of the dropped bullet.

Occasionally, a sports announcer claims that a hard-throwing quarterback can throw the football so many yards "on a line," meaning a straight line. This statement, of course, must be false. All objects thrown horizontally begin falling as soon as they are thrown.

Figure 2.15 illustrates the trajectories of thrown footballs, baseballs, or other objects. Paths A and B represent fast and slow horizontal throws. Path C represents the trajectory of a typical football pass. The ball is thrown slightly upward, not horizontally. Path A more closely approximates a line than does path B. Because of gravity, however, straight-line projectile motion cannot be achieved on the Earth's surface.

If a ball or other object is thrown at various angles, the path that it takes will depend on the angle at which it is thrown. If the ball is tossed gently, air friction will be negligible, and the path will resemble one of those shown in Fig. 2.16.

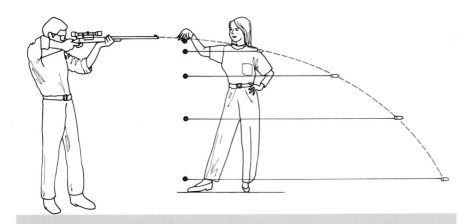

Figure 2.14 Same vertical motions.
If a rifle is fired horizontally and a bullet is dropped simultaneously, both bullets will hit the ground at the same time, because the vertical motions are identical. (Diagram not to scale.)

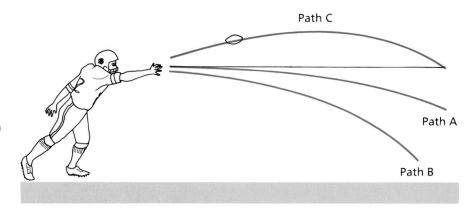

Figure 2.15 Trajectories.
The trajectories of thrown footballs are shown. Path A represents a harder throw than that of Path B. Path C represents the trajectory of a typical football pass. The ball is thrown slightly upward rather than directly horizontal.

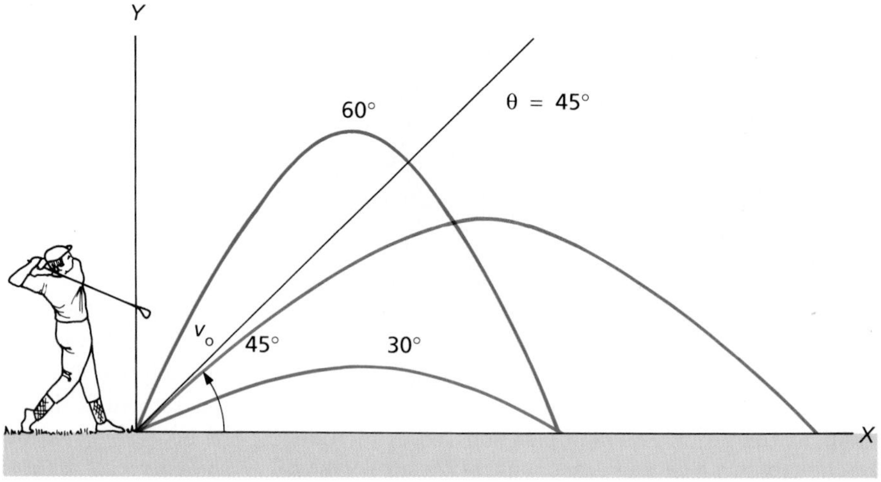

Figure 2.16 Maximum range.
The maximum range for a projectile with a given initial speed is at a projection angle of 45°. This assumes there is no air friction.

Figure 2.16 shows that the range, or horizontal distance the object travels, is maximum when the object is projected at an angle of 45° relative to level ground. Notice in the figure that for a given initial speed, projections at complementary angles—for example, 30° and 60°—have the same range.

With little or no air friction, projectiles have symmetrical paths. However, when a ball or object is thrown or hit hard, air friction comes into effect. In this case the projectile path resembles one of those shown in Fig. 2.17 and is no longer symmetrical. Air friction reduces the velocity of the projectile, particularly in the hori-

zontal direction. As a result, the maximum range now occurs at an angle less than 45°.

Athletes such as football quarterbacks and baseball players are aware of the best angle at which to throw in order to get the maximum distance. A good golfing drive also depends on the angle at which the ball is hit. Of course, in most of these instances there are other considerations, such as spin. This result is a consideration in track and field events such as discus and javelin throwing. Figure 2.18 shows an athlete hurling a javelin at an angle of less than 45° in order to get the maximum distance.

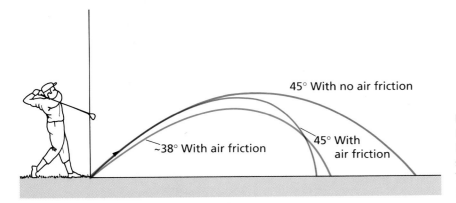

Figure 2.17 Effects of air friction.
Long football passes or hard-hit baseballs follow trajectories similar to those shown here. The air friction reduces the range.

Figure 2.18 Going for the distance.

Learning Objectives

After reading and studying this chapter, you should be able to do the following without referring to the text:

1. Distinguish between scalar and vector quantities, and give an example of each.

2. Distinguish between speed and velocity.

3. Distinguish between average and instantaneous velocity, and explain why they have the same value for an object in uniform straight-line motion.

4. Describe how acceleration is related to velocity, and explain the meaning of a negative acceleration.

5. Give the units of speed, velocity, linear acceleration, and centripetal acceleration in the SI and British systems.

6. State Galileo's observation about freely falling objects.

7. Explain how an object in uniform circular motion is accelerated yet travels with a constant speed.

8. Explain why we experience the acceleration effects when traveling in a car going around a curve at moderate speed but do not experience the effect of the Earth revolving about the Sun (at about 66,000 mi/h).

9. Explain why two bullets, the first dropped and the second simultaneously fired horizontally, will hit the ground at the same time (if friction could be neglected).

10. Define and explain the important words and terms listed in the next section.

Important Words and Terms

motion	vector	instantaneous velocity
position	velocity	acceleration
speed	displacement	centripetal acceleration
scalar	instantaneous speed	acceleration due to gravity

Questions

Straight-Line Motion

1. What is needed to designate the position of an object? Analyze each of the examples given at the beginning of Section 2.1.

2. How is position described on a Cartesian graph?

3. Define the term *motion*.

4. What is meant by the time rate of change of position?

Speed and Velocity

5. Explain the difference between scalar and vector quantities.

6. What is the difference between distance and displacement? How are these quantities associated with speed and velocity?

7. Is the speedometer reading of a car a measurement of instantaneous speed, average speed, instantaneous velocity, or average velocity? Explain.

8. A jogger jogs two blocks directly north.
   (a) How do the jogger's average speed and the magnitude of the average velocity compare?
   (b) If the return jog is over the same path, how do the average speed and velocity magnitude compare for the total trip?

9. Can the average speed and instantaneous speed of an object be the same, that is, have the same value? How about average velocity and instantaneous velocity?

10. If we are moving at high speed through space as the Earth revolves about the Sun, why don't we generally sense the motion?

Acceleration

11. What changes when there is an acceleration?

12. Can an object have an instantaneous velocity of 9.8 m/s in one direction and simultaneously have an acceleration of 9.8 m/s² in the same or opposite direction? Explain.

13. When Galileo dropped two objects from a high place, they hit the ground at almost exactly the same time. Why was there a slight difference in when they hit?

14. A ball is dropped from the top of a building with a height of 100 m (32 floors).
   (a) What is the ball's velocity after 1 s?
   (b) After 2 s?

15. Can an object have an instantaneous velocity of zero and still be accelerating? Explain. (*Hint:* Think about an object projected vertically upward at its maximum height.)

Acceleration in Uniform Circular Motion

16. Can a car be moving at a constant 30 mi/h and still be accelerating? Explain.

17. What does the term *centripetal* mean?

18. What would happen to an object in uniform circular motion if the centripetal acceleration were reduced or went to zero?

19. Are we accelerating due to the Earth's spinning on its axis? Can you sense the motion of the Earth's spin? Explain.

20. What is the direction of the acceleration vector due to the Earth spinning on its axis?

Projectile Motion

21. For projectile motion, what quantities are constant? (Neglect air friction.)

22. On what does the range of a projectile depend?

23. Can a baseball pitcher throw a fastball on a straight, horizontal line? Why or why not?

24. Does a football quarterback throw a long pass at a greater or smaller angle than a short pass? Explain.

25. Taking into account air resistance or friction, how would you throw a ball to get the maximum range?

Exercises

Speed and Velocity

1. At a track and field meet a runner runs the 100-m dash in 15 s. What is the runner's average speed?

   Answer: 6.7 m/s

2. The moon is about 240,000 mi from the Earth, and an astronaut on the moon communicates by using radio waves that travel at the speed of light, which is 186,000

mi/s. How long does it take the astronaut's signal to reach the Earth? *Answer:* 1.3 s

3. When Jupiter is 6.0×10^{11} m away from Earth, how long does it take a radio signal traveling at the speed of light (3.0×10^8 m/s) to reach us from a satellite passing near Jupiter?

4. A light-year is the distance light travels in one year. How many miles are there in a light-year if light travels at 186,000 mi/s?

5. A car is driven 2.0 h at 60 mi/h.
   (a) How far did the car travel in this time?
   The car is then driven 2.0 h at 40 mi/h.
   (b) How far did the car travel in 2.0 h?
   (c) Find the average speed for the entire trip.
   Answer: (c) 50 mi/h

6. (a) A car is driven 120 mi in 2.0 h. What is the average speed?
   (b) The return trip takes 3.0 h. What is the average return speed?
   (c) The total distance traveled for the entire trip is 240 mi, and the total time is 5.0 h. What is the average speed for the entire trip? (*Note:* This problem is different from Problem 5, because more time is spent traveling at the slower speed.) *Answer:* (c) 48 mi/h

7. An airplane flying directly eastward at a constant rate travels 300 km in 2.0 h.
   (a) What is the average velocity of the plane?
   (b) What is its instantaneous velocity?

8. The Earth's circumference is approximately 25,000 mi. A person on the equator spins with the Earth, traveling in a circle with a circumference of 25,000 mi in one day.
   (a) Compare the person's speed in mi/h and mi/s.
   (b) Convert this answer to m/s. *Answer:* (b) 466 m/s

Acceleration

9. A sprinter starting from rest on a straight and level track is able to achieve a speed of 12 m/s in a time of 4.0 s. What is the sprinter's average acceleration?

10. A car travels on a straight road.
    (a) Starting from rest, it is going 30 mi/h (44 ft/s) at the end of 5.0 s. What is the car's average acceleration?
    (b) In 4 more seconds, the car is going 60 mi/h (88 ft/s). What is the car's average acceleration for this time period?
    (c) The car then slows down to 45 mi/h (66 ft/s) in 3.0 s. What is the average acceleration for this time period?
    (d) What is the overall average acceleration for the total time? *Answer:* (d) 5.5 ft/s²

11. A ball is dropped.
    (a) What is its initial speed?
    (b) What is its initial acceleration?

12. A stone is dropped from rest and 1 s elapses. What is the (a) velocity and (b) acceleration at the end of 1 s?

13. A ball is dropped from the top of a tall building. Neglect air friction.
    (a) What is its speed at the end of 2.0 s?
    (b) How far has it fallen in this time?
    Answer: (a) 19.6 m/s (64 ft/s)
    (b) 19.6 m (64 ft)

14. A drag racer accelerates from rest at a constant rate of 4.0 m/s².
    (a) How fast will the racer be going in 3.5 s?
    (b) How far will it have traveled in this time?
    Answer: (b) 24.5 m

15. A stone is thrown downward with an initial speed of 2.0 m/s.
    (a) What is its speed at the end of 3.0 s?
    (b) How far below its starting point will it be at this time? (*Hint:* See the formula in the chapter footnote.)
    Answer: (b) 50.1 m

Acceleration in Uniform Circular Motion

16. A person drives a car around a circular cloverleaf at 10 m/s (22.4 mi/h). The cloverleaf's radius is 70 m (230 ft).
    (a) What is the acceleration in m/s²?
    (b) What is the direction of the acceleration?
    (c) Compare this answer with the acceleration due to gravity. Would you be able to sense the acceleration calculated in part (a)?

17. A race car goes around a circular track with a diameter of 1.0 km at a constant speed of 90 km/h (25 m/s). What are the magnitude and direction of the car's acceleration?

18. Refer to Problem 8, part (b). As the Earth spins, the speed of a person at the equator is 466 m/s.
    (a) Calculate the acceleration of the person in m/s². Use 6.4×10^6 m (approximately 4000 mi) as the radius of the Earth.
    (b) What is the direction of the acceleration?
    (c) Compare this answer with 9.8 m/s², the acceleration due to gravity. *Answer:* (a) 0.034 m/s²

Projectile Motion

19. If you drop an object from a height of 1.225 m (4 ft), it will hit the ground in $\frac{1}{2}$ s. If you throw a baseball horizontally at 30 m/s (67.2 mi/h) from the same height, how long will it take to hit the ground?

20. An airplane flies horizontally west at a speed of 200 m/s (448 mi/h). A package of supplies is dropped from the plane to some stranded campers.
    (a) What is the package's initial velocity? Give both speed and direction.
    (b) What are the magnitude and direction of the package's acceleration?

CHAPTER 3

Force and Motion

> **The whole burden of philosophy seems to consist in this—from the phenomena of motions to investigate the forces of nature, and from these forces to explain the other phenomena.**
>
> —Sir Isaac Newton

WHAT IS FORCE? What is it that sets a moving object in motion? What stops it? These are basic questions to be dealt with in this chapter on force and motion. We will center our efforts on Newton's three laws of motion, his law of universal gravitation, and the laws of conservation of linear and angular momentum.

Galileo (1564–1642) was one of the first scientists to make a formal statement concerning objects at rest and in motion. But it remained for Sir Isaac Newton, who was born the year Galileo died, to actually formulate the laws of motion mathematically and to explain the phenomena of moving objects on the surface of the Earth plus the motion of the planets and other celestial bodies.

Newton was only 25 years old when he formulated his discoveries in mathematics and mechanics. His book *Mathematical Principles of Natural Philosophy* (commonly referred to as the *Principia*), published in 1687 when he was 45, is considered by many to be the most important publication in the history of physics. Certainly it established Newton as one of the greatest scientists of all time.

3.1 Newton's First Law of Motion

Before the time of Galileo and Newton, scientists had asked themselves, "What is the natural state of motion?" According to Aristotle's theory of motion, which had prevailed some fifteen centuries after his death, an object required a force to keep it in motion. That is, the natural state of an object was one of rest, with the exception of celestial bodies, which were naturally in motion. It was easy to observe that moving objects tended to slow down and come to rest, so a natural state of being at rest must have seemed logical to Aristotle.

◄ Force, motion, and momentum in action during a football game.

Galileo studied motion using a ball rolling onto a level surface from an inclined plane. The smoother he made the surface, the farther the ball would roll (Fig. 3.1). He reasoned that if he could make a very long surface perfectly smooth, there would be nothing to stop the ball; and it would continue in motion indefinitely or until something stopped it. Thus, contrary to Aristotle, Galileo concluded that objects could *naturally* remain in motion rather than come to rest.

Newton also recognized this phenomenon and incorporated Galileo's result in his **first law of motion.** This law can be stated in several different ways. One of these is as follows:

> **An object will remain at rest or in uniform motion in a straight line unless acted upon by an external, unbalanced force.**

Uniform motion in a straight line means that the velocity is constant. Thus another way to state Newton's first law is to say that the natural state of motion is at a constant velocity. If the constant velocity is equal to zero, we say that a body is at rest.

Because of the ever-present forces of friction and gravity on Earth, it is difficult to observe an object in a *natural* state of constant velocity. However, in free space where there is no friction and negligible gravitational attraction, an object initially in motion maintains

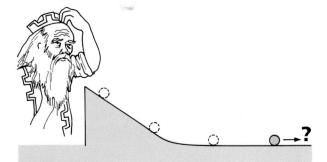

Figure 3.1 Motion without resistance.
If the level surface could be made perfectly smooth, how far would the ball travel?

a constant velocity. For example, after being projected on its way, an interplanetary satellite approximates this condition quite well between planets. On going out of the solar system, as some *Pioneer* and *Voyager* spacecrafts have done, they keep on going with a constant velocity until an external, unbalanced force alters this velocity.

Let's focus on what is meant by an external, unbalanced force. Newton's first law indicates that if an object is to speed up, slow down, and/or change direction (e.g., as in circular motion), then such a force is required. But first, what is a force? We all have an intuitive feeling for the concept of a force. A force is simply a push or pull. In terms of the first law, we might say that a **force** is a quantity that is capable of producing motion or a change in motion.

This statement does not necessarily mean that a force produces a change in motion, only that the *capability* is there. Here is where the term *unbalanced* comes in. In the tug-of-war shown in Fig. 3.2a, forces are being applied, but there is no motion. The forces in this case are balanced, that is, equal and opposite, and in effect cancel each other out. Motion occurs when the applied forces are unbalanced, that is, when they are not equal and opposite, as in Fig. 3.2b. Since forces have direction

(a)

(b)

Figure 3.2 Balanced and unbalanced forces.
(a) When two forces are equal and opposite, they are balanced and there is no net force and no motion if the system is initially at rest. (b) When F_2 is greater than F_1, the forces are unbalanced and motion occurs.

as well as magnitude, they are *vector* quantities. Forces may act on an object in different directions; but for a change in motion or an acceleration to occur, there must be an unbalanced or *net* force for a net effect or change.

An *external* force is an applied force—that is, one applied on or to the object or system. There are also internal forces. For example, suppose the object is an automobile and you are a passenger traveling inside. You can push (apply a force) on the floor or the dashboard, but there is no effect on the car's velocity because this push is an internal force.

Motion and Inertia

Galileo also introduced another concept. It appeared that objects had a behavior or a property of maintaining a state of motion. That is, there was a resistance to changes in motion. Similarly, if an object was at rest, it seemed to "want" to remain so. Galileo called this property inertia, and we say that **inertia** is the property of matter that resists changes in motion.

Newton went one step further and related the concept of inertia to something that could be measured—mass. That is, **mass** is a measure of inertia. The greater the mass of an object, the greater its inertia is, and vice versa.

As an example of the relationship of mass and inertia, suppose you horizontally push two different people on swings initially at rest, one a very large man and the other a small child (Fig. 3.3a). You'd quickly find that it

was more difficult to get the adult moving. That is, there would be a noticeable difference in the resistance to motion between the man and the child. Also, once you got them swinging and then tried to stop the motions, you'd notice a difference in the resistance to a change in motion again. Being more massive, the man has greater inertia.

Another example or demonstration of inertia is shown in Fig. 3.3b. The stack of coins has inertia that resists being moved when at rest. If the paper is quickly jerked, the inertia of the coins will prevent them from toppling. The small external force is not sufficient to overcome the inertia of the coins. You may have pulled a magazine from the bottom of a stack with a similar action.

Because of the relationship of motion and inertia, Newton's first law is sometimes called the *law of inertia*. Newton's first law may be used to describe some of the observed effects concerning circular motion and centripetal acceleration. Consider traveling down a straight and level street in a car at a constant velocity and making a turn at a corner (Fig. 3.4). The centripetal acceleration allowing the car to go around the circular path is provided by the friction on the tires of the car. Riding in the car, you tend to go in a straight line, in conformity with Newton's first law. As the car makes the turn, you feel as though you are being thrown outward (an effect we all have probably experienced). However, you are just maintaining your state of motion, until the car turns in front of you and the door pushes against you. This

(a)

(b)

Figure 3.3 Mass and inertia.
(a) An external, applied force is needed to put an object in motion, and the motion depends on the mass or inertia of the object. (b) Because of inertia, the paper can be removed from beneath the stack of quarters without toppling it by giving the paper a quick jerk.

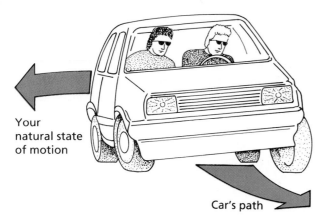

Your natural state of motion

Car's path

Figure 3.4 Newton's first law in action.
When a car initially traveling in a straight line goes around a corner, the people in it tend to continue in the original direction, according to Newton's first law. This gives them the sensation of being thrown outward. The door, seatbelt, or steering wheel provides the necessary centripetal acceleration needed to go around the curve with the car.

force supplies the inward centripetal acceleration you need to go around the curve with the car. The acceleration may be supplied by a seat belt if you are buckled up.

Speaking of seat belts, suppose you were in the front seat of a car traveling at a high speed down a straight road and the driver suddenly puts on the brakes for an emergency stop. What would happen, according to Newton's first law, if you were not wearing a seat belt? The friction on the seat of your pants would not be enough to appreciably change your motion, so you'd keep on moving while the car was coming to a stop. The next external, unbalanced force you'd experience would not be pleasant. Newton's first law makes a good case for buckling up.

3.2 Newton's Second Law of Motion

In our initial study of motion (Chapter 2), acceleration was defined as the time rate of the change of velocity ($\Delta v/\Delta t$). Nothing was said about what *causes* acceleration, only that a change in velocity was required. So what causes an acceleration? Newton's first law answers the question. If an external, unbalanced force is required to produce a change in velocity, then we see that an unbalanced force causes an acceleration.

Newton was aware of this result, but he went further and also related acceleration to inertia or mass. Because inertia is the resistance to a *change* in motion, a reasonable assumption is that the greater the inertia or mass of an object, the smaller the change in motion or velocity is when a force is applied. Such insight was typical of Newton in his many contributions to science. (See the chapter Highlight.) Hence we have this summary:

1. The acceleration produced by an unbalanced force acting on an object (or mass) is directly proportional to the magnitude of the force ($a \propto F$) and in the direction of the force.

2. The acceleration of an object being acted on by an unbalanced force is inversely proportional to the mass of the object ($a \propto 1/m$).

Combining these effects of force and mass on acceleration, we have a statement of **Newton's second law of motion,** which may be expressed simply and conveniently as

$$\text{acceleration} \propto \frac{\text{unbalanced force}}{\text{mass}}$$

or
$$a \propto \frac{F}{m}$$

where $\propto$ is a proportionality sign.

These relationships are illustrated in Fig. 3.5. Notice that if you double the force acting on a mass, the acceleration doubles (direct proportion, $a \propto F$). If, however, the same force is applied to twice as much mass, the acceleration is one-half as much (inverse proportion, $a \propto 1/m$).

When the appropriate units are used, we can write Newton's second law in equation form as

$$a = \frac{F}{m}$$

which is commonly written as

$$F = ma \qquad\qquad (3.1)$$

(force = mass × acceleration)

This equation is a mathematical statement of Newton's second law. Notice from Fig. 3.5 that the mass m is the *total* mass of the system or all of the mass that is accelerated. A system may be two or more separate masses, as will be shown in a later example. Also, F is the net or unbalanced force, which may be the vector sum of

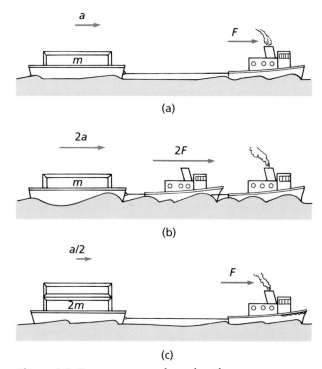

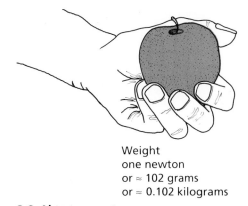

Weight
one newton
or ≈ 102 grams
or ≈ 0.102 kilograms

Figure 3.6 About a newton.
An average-sized apple weighs about 1 newton (or 3.6 oz or 0.225 lb).

Figure 3.5 Force, mass, and acceleration.
(a) An unbalanced force acting on a mass produces an acceleration. (b) If the force is doubled, the acceleration is doubled. (c) If the mass is doubled, the acceleration is decreased by one-half.

two or more forces. Unless otherwise stated, a general reference to force will mean an *unbalanced* force.

In the metric (SI) system the units of mass and acceleration are kg and m/s^2, respectively, so force has units of $F = ma = $ kg·m/s^2 But this derived unit is given the special name of **newton** (abbreviated N), for an obvious reason, and force is expressed in units of newtons.* One newton (1 N) is the force necessary to accelerate 1 kg at a rate of 1 m/s^2.

In the British system the unit of force is the pound (lb). This unit too has derived units of mass times acceleration (ft/s^2). The unit of mass is the slug, which is rarely used, and we will not employ it in this text. Recall that the British system is a gravitational or force system, and things are weighed in pounds (force). As will be seen shortly, **weight** is a force and is expressed in newtons in the metric system. Equivalent weight units in

---

* Another metric unit of force is the dyne. The units of mass and length are the gram and centimeter in this case, and 1 dyne = 1 g × 1 cm/s^2. The dyne is a small unit, only 1/100,000 as large as a newton.

the two systems are shown in Fig. 3.6. If you are familiar with the story that Newton gained insight by observing (or being struck by) a falling apple while meditating on the concept of gravity, you can easily remember that an average apple weighs about one newton.

Here's an example using $F = ma$ that illustrates that F is the net or unbalanced force and m is the total mass.

EXAMPLE 1 _____

Forces are applied to two blocks connected by a string resting on a frictionless surface, as illustrated in Fig. 3.7. If the mass of each block is 1.0 kg and the mass of the string is negligible, what is the acceleration of the system?

Solution The unbalanced or net force is the difference of the forces. Effectively, F_1 cancels part of F_2, and there is a net force $F_{net} = F_2 - F_1$ in the direction of F_2. The total mass of the system, or the total mass being accelerated, is $m = m_1 + m_2$. (*cont.*, p. 39)

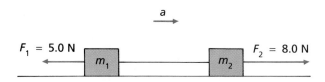

Figure 3.7 Net force and total mass.
See Example 1.

HIGHLIGHT

Isaac Newton

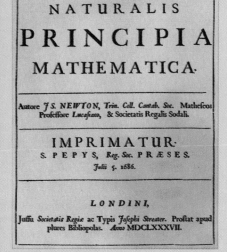

(a)   (b)

Figure 3.8 The man and the book.
(a) Sir Isaac Newton (1642–1727). (b) The title page of the *Principia*. Can you read the Roman numerals at the bottom of the page giving the year of publication?

saac Newton (Fig. 3.8) and Albert Einstein are usually considered to be the two greatest scientists in history. Newton's laws of motion were one of many contributions he made to a variety of subjects in physics. Newton was born on Christmas day in 1642 in the village of Woolsthorpe in Lincolnshire, England. He showed no particular genius in his early schooling, but fortunately a teacher encouraged him to pursue his edcuation, and in 1661 he entered Trinity College at Cambridge.

Four years later he received his degree and planned to continue studying for a masters degree. But an epidemic of the bubonic plague broke out and the university was closed. Newton returned to Woolsthorpe, and in the next two years he laid the groundwork for many of his contributions in physics, mathematics, and astronomy. In Newton's own words, "I was in the prime of my age for invention, and minded mathematics and philosophy [science] more than any time since."

Over the next 20 years Newton was very productive, and at the age of 45 he published his famous treatise, *Philosophiae Naturalis Principia Mathematica* [*Mathematical Principles of Natural Philosophy*]*, or *Principia* for short. In this book he set forth his laws of motion, along with the theory of gravitation. The publication of the *Principia* was fi-

* Physics was once called natural philosophy.

nanced by a friend, Edmund Halley, who used Newton's theories to predict the return of the comet that bears his name.

Newton was reportedly a shy man, but he often got into disputes about his theories and achievements. A famous one is his dispute with Gottfried Leibniz about who first developed calculus. Newton was elected to Parliament and later appointed Master of the Mint, where he supervised the task of recoining the English currency. He was knighted by Queen Anne in 1705.

Sir Isaac Newton died in 1727 at the age of 85 and was buried with honor in Westminster Abbey. Some insight about this austere bachelor and giant of

science may be gleaned from the following excerpts from his writings:

I seem to have been only a boy playing on the seashore and diverting myself in now and then finding a smoother pebble or a prettier shell than ordinary, whilst the great ocean of truth lay all undiscovered before me.

If I have been able to see farther than some, it is because I have stood on the shoulders of giants.

About Newton, the poet Alexander Pope wrote:

Nature and Nature's laws lay hid in night God said, Let Newton be! and all was light.

Hence, using the force values given in the figure, we get

$$a = \frac{F_{net}}{m} = \frac{F_2 - F_1}{m_1 + m_2}$$

$$= \frac{8.0\ N - 5.0\ N}{1.0\ kg + 1.0\ kg} = 1.5\ m/s^2$$

in the direction of the net force.

Question: What would happen if the surface were not frictionless and the mass of the string were not negligible? Right. You'd have another force opposing the motion in the direction of F_1 and there would be another mass, m_3, contribution to the total mass.

In our examination of Newton's second law of motion, we find that the first law is expressed in the second. Consider the equation form of the second law, $F = ma$. If the force on an object is zero, then its acceleration is zero. That is, there is no change in velocity, and the object must be moving with a constant velocity or be at rest. This is essentially what the first law states, in sort of a reverse fashion.

One may gain an insight into $F = ma$ by considering gas mileage. Cars get better gas mileage on the highway than in the city. Why? Newton's second law provides an explanation.

Since $F = ma$, every time a car is accelerated, it must burn fuel to supply the unbalanced force causing the acceleration. When a car is driven in city traffic, it must be accelerated back up to speed after it slows down or stops. All this acceleration burns a lot of gasoline.

It does not take as much fuel to keep a car going at a constant speed down a level highway. The less often you accelerate your car, the better the mileage you will get.

Mass and Weight

This is a good place to make a firm distinction between mass and weight. These quantities were generally defined previously: *Mass* refers to the amount of matter an object contains (it is also a measure of inertia), and *weight* is related to the force of gravity (i.e., the gravitational force acting on an object). However, the quantities are related, and Newton's second law clearly shows this relationship.

On the surface of the Earth, where the acceleration due to gravity is relatively constant ($g = 9.8$ m/s^2), the weight force w on an object with a mass m is given by the formula

$$w = mg \tag{3.2}$$

weight = mass × acceleration (due to gravity)

$$(F = ma)$$

Notice that this equation is a special case of $F = ma$, where different symbols have been used to distinguish this special case.

Now you can easily see why mass is the fundamental quantity. Generally, a physical object has the same amount of matter, and so it has the same or a constant mass. But the weight of an object may differ, depending on the value of g. For example, on the surface of the moon the acceleration due to gravity (g_m) is one-sixth what it is on the surface of the Earth ($g_m = g/6 = (9.8$ m/s$^2)/6 = 1.6$ m/s^2). So an object will have the same mass on the moon as on the Earth, but its weight will be different.

EXAMPLE 2

What is the weight of a 1-kg mass on (a) the Earth and (b) the moon?

Solution (a) Using Eq. 3.2, we get

$$w = mg = (1.0\ kg)(9.8\ m/s^2) = 9.8\ N$$

(b) On the moon where $g_m = g/6 = 1.6$ m/s^2, we have

$$w = mg_m = (1.0\ kg)(1.6\ m/s^2) = 1.6\ N$$

Notice that the mass is the same in both cases, but the weights are different (different g's).

When we have an unknown mass or object, we "weigh" it on a scale. A scale may be calibrated in mass units (kilograms or grams) or in weight units (N or lb; see Fig. 3.9). Although "weighing" implies weight, we

Figure 3.9 Mass and weight.
A mass of 1 kg is suspended on the scale, which has a weight of 9.8 N. This is equivalent to a weight of 2.2 lb.

generally say that the "weight" of an object is 1 kg or 2.2 lb (9.8 N).

EXAMPLE 3

What is the upward tension force in a string supporting a 2.0-kg mass, as illustrated in Fig. 3.10a?

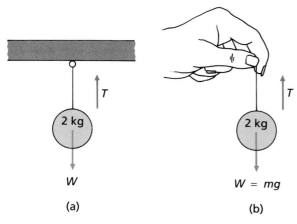

Figure 3.10
Balanced forces.

Solution   Since the mass isn't going anywhere, its acceleration is zero, which means that the net force acting on it is zero ($F = ma$). The downward weight force must then be balanced by an upward force supplied by the string.

You can get a feel for this situation if you think about holding a 2-kg (4.4-lb) mass stationary. You have to supply an upward force to balance the downward weight force (Fig. 3.10b). Then the weight force is

$$w = mg$$
$$= 2.0 \text{ kg} \times 9.8 \text{ m/s}^2$$
$$= 19.6 \text{ N}$$

and the tension force in the string has the same magnitude in the opposite direction.

Finally, an object in free-fall has an unbalanced force acting on it of $F = w = mg$ (downward). So why do objects in free-fall all descend at the same rate, as stated in the last chapter? Even Aristotle thought a heavy object would fall faster than a lighter one. Newton's laws explain.

Suppose there were two objects in free-fall, one with twice the mass of the other, as illustrated in Fig. 3.11.

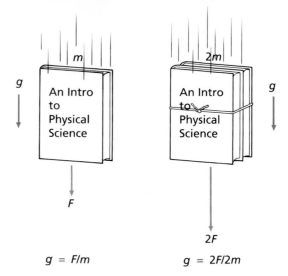

Figure 3.11 Acceleration due to gravity.
The acceleration due to gravity is independent of mass; thus it is the same for all falling objects. An object with twice the mass of another has twice the gravitational force acting upon it, but it also has twice the inertia and falls at the same rate.

According to Newton's second law, the more massive object would have twice the weight, or gravitational force of attraction. However, by Newton's first law, the more massive object has twice the inertia, so it needs twice the force to fall (accelerate) at the same rate.

3.3 Newton's Law of Gravitation

The law governing the gravitational force of attraction between two particles was formulated by Newton from his studies on planetary motion. Known as the **law of universal gravitation**, it may be stated as follows:

> **Every particle in the universe attracts every other particle with a force that is directly proportional to the product of their masses and inversely proportional to the square of the distance between them.**

Suppose the masses of two particles are designated as m_1 and m_2, and the distance between them is r (Fig. 3.12a). Then we may write this statement in symbol form as

$$F \propto \frac{m_1 m_2}{r^2}$$

where $\propto$ is a proportionality sign.

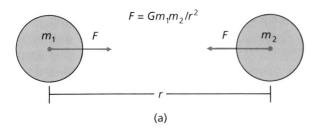

$$F = Gm_1m_2/r^2$$

(a)

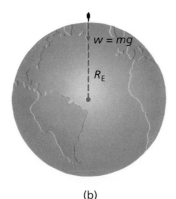

(b)

Figure 3.12 Law of gravitation.
(a) Two particles gravitationally attract each other, and the magnitude of the forces is given by Newton's law of gravitation. (b) For a homogeneous or uniform sphere, the force acts as though all the mass of the sphere were concentrated at its center. To a good approximation, our weight forces function as though all of the mass of the Earth were concentrated at its center, and the distance between the "masses" is the radius of the Earth.

An object is made up of a lot of point particles, so the gravitational force on an object is the vector sum of all the particle forces. This computation can be quite complicated, but one simple and convenient case is a homogeneous* sphere. In this case the net force acts as though all the mass were concentrated at the center of the sphere, and the separation distance for the mutual interaction with another object is measured from there. If we assume that the Earth, other planets, and the Sun are spheres with uniform mass distributions, we can apply the law of gravitation to such bodies (Fig. 3.12b).

Newton's law of universal gravitation may be written in equation form by putting in an appropriate constant (of proportionality):

$$F = \frac{Gm_1m_2}{r^2} \tag{3.3}$$

---

* *Homogeneous* means that the mass particles are distributed evenly throughout the object.

Table 3.1 What is G?

$G = 6.67 \times 10^{-11}\,\text{N} \times \text{m}^2/\text{kg}^2$
G is a very small quantity.
G is a universal constant. It is thought to be the same throughout the universe.
G is not g (g is the acceleration due to gravity).
G is not a force.
G is not an acceleration.

---

where G is a constant called the universal gravitational constant—universal because we believe it is the same throughout the universe, as is the law of gravitation.

Newton used his law of gravitation and his second law of motion to show that gravity supplied the necessary centripetal acceleration and force on the moon for it to move in its nearly circular orbit about the Earth. However, he did the derivation in ratio form because he did not know or experimentally measure the value of G. This value was measured some seventy years after Newton's death by an English scientist, Henry Cavendish, who used a very delicate balance to measure the force between two masses. The value of G is very small; it is given in Table 3.1, along with what G is and is not. (The units of G are such that when it is used in Eq. 3.3, the gravitational force is in newtons. Try it and see.)

When we drop something, the force of gravity, or acceleration due to gravity, is evident. Yet if there is a gravitational force of attraction between every two objects, why don't you feel this attraction between yourself and this textbook? (No pun intended.) Indeed, there is a force of attraction between you and this text, but it is so small that you don't notice it. Gravitational forces can be very small, as the following example illustrates.

EXAMPLE 4

Two objects with masses of 1.0 kg and 2.0 kg are 1.0 m apart. What is the gravitational force between them?

Solution   The magnitude of the force is given by Eq. 3.3:

$$F = \frac{Gm_1m_2}{r^2}$$

$$= \frac{(6.67 \times 10^{-11}\,\text{N}\cdot\text{m}^2/\text{kg}^2)(1.0\,\text{kg})(2.0\,\text{kg})}{(1.0\,\text{m})^2}$$

$$= 1.3 \times 10^{-10}\,\text{N}$$

(cont., p. 42)

This number is very, very small. A grain of sand would weigh more. For an appreciable gravitational force to exist between two masses, at least one of the masses must be relatively large (see Fig. 3.13).

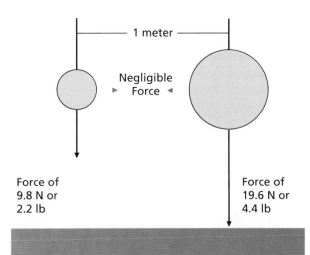

1 meter

Negligible
► Force ◄

Force of
9.8 N or
2.2 lb

Force of
19.6 N or
4.4 lb

Figure 3.13 The amount of mass makes a difference. A 1 kg mass and a 2 kg mass a distance of 1 m apart have a negligible mutual gravitational attraction ($\approx 10^{-10}$ N or 10^{-11} lb). However, because the Earth's mass is quite large, the masses are attracted to the Earth with forces of 9.8 N (or 2.2 lb) and 19.6 N (or 4.4 lb). These forces due to the Earth's gravitation are the weights of the masses.

Newton's law of gravitation gives us an insight about why the acceleration due to gravity at the Earth's surface is constant and why this acceleration does not depend on the mass of an object. (Recall that all objects in free-fall have the same acceleration, or fall at the same rate.) For a mass on the Earth's surface the distance between the masses is essentially the radius of the Earth. The force of gravity on an object of mass m is then

$$F = \frac{GmM_E}{R_E^2}$$

where M_E and R_E are the mass and radius of the Earth, respectively.

This force is just the object's weight, so we can write

$$w = mg = \frac{GmM_E}{R_E^2}$$

Canceling the m's, we have an expression for g:

$$g = \frac{GM_E}{R_E^2} \qquad (3.4)$$

Notice that since the mass of the object m cancels out of the equation, it does not appear in the expression for g. Thus g is independent of the mass of the object, or is the same for all objects. Hence all objects in free-fall have the same acceleration (resulting from the force of gravity).

Also, from Eq. 3.4 we can see why g is relatively constant near the Earth's surface. The term G is a constant, and in general, so are the mass and radius of the Earth (M_E and R_E). Hence to a good approximation, g is a constant near the Earth's surface or over the normal heights of falling objects. The force of gravity on an object (Eq. 3.3) and g do decrease with increasing distance or altitude (increasing r). However, at an altitude of 1.6 km (1 mi) the acceleration due to gravity, and hence an object's weight ($w = mg$), is only 0.05% less than g at the Earth's surface. At an altitude of 100 mi (160 km) g is still 95% of its surface value. The force between two masses is said to be an infinite-range force. That is, the only way to get it to approach zero is to separate the masses by a distance approaching infinity ($r \to \infty$). No matter what the altitude is or how far an object is from Earth, there is still a force resulting from gravity. In general, we may write Eq. 3.4 as

$$g = \frac{GM}{r^2} \qquad (3.5)$$

where M is the mass of any spherical uniform object, r is the distance from its center.

This equation can be used to find the acceleration due to gravity on the moon and other planets. For example, if the mass and radius of the moon are used, then we get the acceleration due to gravity on the surface of the moon, g_m. As noted earlier, this acceleration turns out to be one-sixth of that on Earth ($g_m = g/6$), basically because the mass of the moon is much smaller than that of Earth. Hence things weigh only one-sixth as much on the moon as on the Earth. If you want to lose weight fast, go to the moon. (See Fig. 3.14.) Similarly, the acceleration due to gravity on another planet depends on the mass of the planet and its volume or size. In general, objects would weigh more on bigger planets, such as Jupiter, than on smaller planets, such as Mercury.

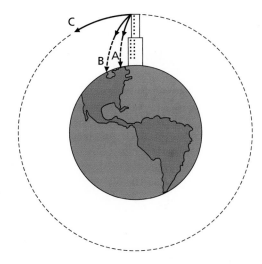

Figure 3.15 Earth orbit.
An imaginary situation of projecting objects from a very tall building. If there were no atmosphere and you "threw" an object hard enough it would "fall around" the Earth as illustrated for path C. The moon "falls around" the Earth in a similar manner.

Figure 3.14 Strong man.
An astronaut on the moon carries equipment that may weigh about 136 kg (300 lb) on Earth. On the moon, where the acceleration due to gravity is $\frac{1}{6}$ that on Earth, the mass of the equipment is the same, but it has a weight of only about 22.7 kg (50 lb). The astronaut also experiences a weight reduction. For example, a 82-kg (180 lb) astronaut on Earth would be a 13.6-kg (30-lb) astronaut on the moon.

Another interesting thing to note about Eq. 3.4 is that it provides a way to compute the mass of the Earth. Knowing the values of g, G, and R_E, we can find M_E easily. All of the quantities were known once G was determined in 1797. Scientists were then able to calculate the Earth's mass, which turns out to be about 6×10^{24} kg.

While on the subject of gravity, let's take a look at orbiting satellites, which are quite common today. When a satellite is put into orbit, it must be given a sufficient tangential velocity, which depends on the altitude of its orbit. To help understand this idea, imagine projecting objects from a *very* tall building (Fig. 3.15). If not thrown hard enough, the objects would fall to the Earth. However, with sufficient velocity an object would go into a circular orbit. Essentially, it would "fall" around the Earth. (Newton himself imagined a similar situation of firing cannonballs from the top of a high mountain.)

Gravity supplies the **centripetal force** or acceleration that keeps satellites in orbit ($F_c = ma_c = mv^2/r$). This law applies also to the Earth's only natural satellite, the moon. With astronauts in space we now hear the terms *zero g* and *weightlessness* (Fig. 3.16). These terms are not really true descriptions. Gravity certainly acts on an astronaut in an orbiting spacecraft. Otherwise, the astronaut (and spacecraft) would not remain in orbit. Since gravity is acting, the astronaut by definition has weight.

The reason an astronaut floats in the spacecraft and feels "weightless" is that the spacecraft and the astronaut are both "falling" toward the Earth. Imagine yourself in a freely falling elevator standing on a scale. The scale would read zero because it is falling as fast as you are. However, you are not weightless and g is not zero (as you would find out on reaching the bottom of the elevator shaft).

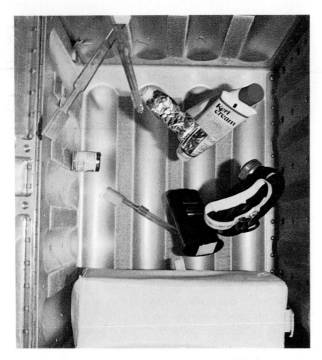

Figure 3.16 Floating around.
Toothbrushes and toothpaste float about in a cabinet in Skylab because of a ''zero g'' (gravity) or ''weightlessness.'' Actually, the weights or gravitational attractions of the Earth on the objects keep them in orbit with Skylab. (Note the Velcro strips on the sides of the cabinet that keep some objects from floating.)

Figure 3.17 Newton's third law in action.
The rockets and exhaust gases exert a force on each other and are accelerated in opposite directions.

3.4 Newton's Third Law of Motion

Newton's third law of motion is sometimes called the law of action and reaction. **Newton's third law** states:

> **For every action there is an equal and opposite reaction.**

In this statement the words *action* and *reaction* refer to forces. A more precise statement would be the following:

> **Whenever one object exerts a force upon a second object, the second object exerts an equal and opposite force upon the first object.**

Rocket motion (Fig. 3.17) is an example of Newton's third law. The exhaust gas is accelerated from the base of the rocket, and the rocket accelerates in the opposite direction. A common misconception is that the exhaust gas pushes against the launch pad to accelerate the rocket. This is nonsense. If this misconception were true, there would be no space travel, since there is nothing to push against in space. The correct explanation is one of action (gas going out the back) and reaction (rocket propelled forward). The gas (or a gas particle) exerts a force on the rocket, and the rocket exerts a force on the gas.

Another illustration was given by Newton: "If you press on a stone with your finger, the finger is also pressed upon by the stone."

Newton's third law can be expressed mathematically as

$$F_1 = -F_2 \qquad (3.6)$$

where F_1 = force exerted on object 1 by object 2,
       F_2 = force exerted on object 2 by object 1.

The minus sign indicates that F_2 is in the opposite direction from F_1. Using Newton's second law, we find

$$m_1 a_1 = -m_2 a_2 \qquad (3.7)$$

From Eq. 3.6, we see that if m_2 is much larger than m_1, then a_1 is much larger than a_2.

As an example, consider dropping a book to the floor. As the book falls, it has a force acting on it (Earth's gravity) that causes it to accelerate. What is the equal and opposite force? The equal and opposite force is the force of the book's gravitational pull on the Earth. Technically, the Earth accelerates upward to meet the book! However, since the Earth's mass is so huge compared with the book's, the Earth's acceleration is minuscule and cannot be measured.

There are many other examples of action and reaction occurring in everyday life. In fact, from Newton's third law it is clear that every time there is a force, there must also be a reaction force.

Understanding the difference between Newton's third law and the concept of balanced forces is important. If a book weighing 15 N lies on a table, the book exerts a force of 15 N down on the table (Fig. 3.18). The table, in turn, exerts an upward force of 15 N on the book. This action-reaction satisfies Newton's third law.

Now focus on the 15-N book alone. What are the two forces acting on the book? First, there is the 15-N force of the Earth's gravity acting down. Second, there is the 15-N upward force that the table exerts on the book. From Newton's second law, since there is no unbalanced force, the acceleration of the book is zero.

If the book is dropped, it falls to the floor. In terms of Newton's second law, there is only one force acting on the book. This force is the Earth's 15-N force of gravity. This single unbalanced force causes the book to accelerate downward.

Newton's third law relates two equal and opposite forces that act on two separate objects. Newton's second law concerns how forces acting on a single object can cause an acceleration. If these forces acting on a single object are equal and opposite, then there will be no acceleration.

In some instances the reaction force of Newton's third-law force pair may not be immediately recognized, but it's there. As illustrated in Fig. 3.19, a convenient technique is to consider forces in analogous situations. We just don't usually think of a ceiling exerting a force on a suspended ball or a wall exerting a force on a car.

3.5 **Momentum**

Linear Momentum

Stopping a speeding bullet is difficult because it has a high velocity. Stopping a slowly moving oil tanker is also difficult because it has a large mass. In general, the combination of mass and velocity is called the linear momentum. That is,

$$\textbf{linear momentum} = \text{mass} \times \text{velocity}$$

or in symbols,

$$p = mv$$

Both a speeding bullet and a slowly moving oil tanker have a large linear momentum. We can compare them

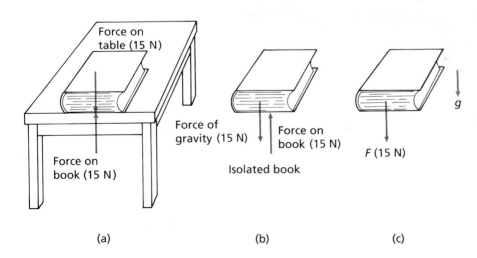

(a)   (b)   (c)

Figure 3.18 Action and reaction.
(a) Equal and opposite forces on *different* objects. The book exerts a force on the table and the table exerts a force on the book. (b) Two equal and opposite forces act on the book, because it is not moving. These are *not* the force pair of Newton's third law. Why? (c) In free fall, only the force of gravity acts on the book.

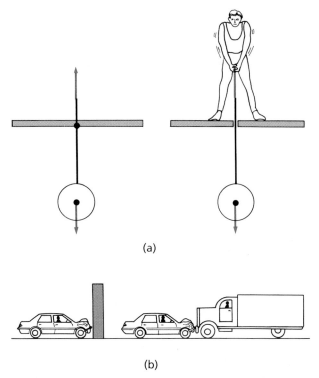

(a)

(b)

Man jumps forward

Boat moves backward

Figure 3.20 Conservation of linear momentum.
Initially at rest, the total momentum of the system is
zero. When the man jumps toward the shore (an internal
force), the boat moves in the opposite direction to conserve the linear momentum.

Figure 3.19 Finding the forces.
(a) The ceiling exerts an upward force on the mass via the
string, as can be shown by letting someone support the
mass. (b) The wall exerts a force on the car. A truck could
be substituted for the wall with the same effect.

by actually calculating the linear momentum of each.

Both mass and velocity are involved in momentum.
A small car and a large truck both traveling at 50 km/h
may have the same velocity, but the truck has more momentum because it has a much larger mass. If we have
a system of masses, we find the linear momentum of
the system by adding up the linear momentum vectors
of all the individual masses.

The linear momentum of a system is important because if there are no external unbalanced forces, it is
conserved; it does not change with time. This property
makes it extremely important in analyzing the motion
of various systems. The **law of conservation of linear
momentum** can be stated as follows:

> **The total linear momentum of an isolated sys-
> tem remains the same if there is no external,
> unbalanced force acting on the system.**

Suppose you are standing in a boat near the shore;
you and the boat are the system (Fig. 3.20). Let the boat

be stationary, so the total linear momentum of the system is zero. (No motion, no linear momentum.) If you
jumped toward the shore, you would notice something
immediately—namely, the boat would move in the opposite direction. The boat moves because the force you
exerted in jumping is an *internal* force, so the total linear
momentum of the system is conserved and must remain
zero. You have momentum in one direction, and to cancel
this, the boat must have an equal and opposite momentum. Note that momentum is a vector quantity, as
is velocity, and momentum vectors can add to zero (as
can force vectors).

We looked at jet propulsion or rockets in terms of
Newton's third law. This phenomenon can also be explained in terms of linear momentum. The burning of
the rocket fuel gives energy by which *internal* work is
done, and hence internal forces act. As a result, the
exhaust gas goes out the back of the rocket with momentum in that direction, and the rocket goes in the
opposite direction to conserve linear momentum. Here
the exhaust gas molecules have small masses and large
velocities, while the rocket has a large mass and a relatively small velocity. Also, there are many, many gas
molecules. You can and probably have demonstrated
this rocket effect by blowing up a balloon and letting
it go. Without a guidance system the balloon zigzags
wildly, but the air comes out the back and the balloon
is "jet" propelled.

Angular Momentum

Another important quantity that Newton found to be conserved is angular momentum. Angular momentum arises when objects go in paths around a fixed point. Consider a planet going around the Sun in an elliptical orbit, as illustrated in Fig. 3.21. The **angular momentum** is given by the product mvr, where m is the mass, v is the velocity, and r is the distance of the object from the center of motion, which in this case is the Sun.

The linear momentum of a system can be changed by the introduction of an external unbalanced force. Similarly, the angular momentum of a system can be changed by an external unbalanced torque. A **torque** is a twisting effect caused by one or more forces. For example, in Fig. 3.22 a torque on a steering wheel is caused by two equal and opposite forces acting on different parts of the steering wheel. These forces cause a twist, or torque, and cause the steering wheel to turn. In general, a torque tends to produce a rotational motion.

There is also a conservation law for angular momentum. The **law of conservation of angular momentum** states:

> **The angular momentum of an object remains constant if there is no external, unbalanced torque acting on it.**

That is,
$$mv_1r_1 = mv_2r_2 \qquad (3.8)$$

In our example of a planet the angular momentum, mvr, remains the same because the gravitational attraction is internal to the system. As the planet gets closer to the Sun, r decreases, so the speed v increases. For this reason, a planet moves faster when it is closer to the Sun.

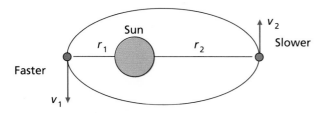

Figure 3.21 Angular momentum.
The angular momentum of a planet going around the Sun is given by mv_1r_1 and mv_2r_2 at different points in the orbit. Because angular momentum is conserved, $mv_1r_1 = mv_2r_2$. As the planet comes closer to the Sun, the radial distance r decreases, so v must increase. Similarly, the speed decreases when r increases. Thus, the planet moves fastest when closest to the Sun and slowest when farthest from the Sun.

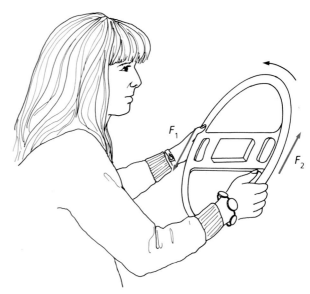

Figure 3.22 Torque.
A torque is a twisting action that produces rotational motion. This is analogous to a force producing linear motion.

(Although approximated as circles, planet orbits are slightly elliptical, so they do move at different speeds in different parts of their orbits.) The same is true for comets in elliptical orbits about the Sun.

EXAMPLE 5

A comet at its farthest point from the Sun is 900 million miles away and traveling at 6000 mi/h. What is its speed at its closest point to the Sun, which is 30 million miles away?

Solution We know v_2, r_2, and r_1, so we can calculate v_1 as follows:

$$mv_1r_1 = mv_2r_2$$

or
$$v_1r_1 = v_2r_2$$

or
$$v_1 = \frac{v_2r_2}{r_1}$$
$$= \frac{(6.0 \times 10^3 \text{ mi/h})(900 \times 10^6 \text{ mi})}{30 \times 10^6 \text{ mi}}$$
$$= 180{,}000 \text{ mi/h}$$

Thus we see that the comet moves much faster when it is close to the Sun than it does when it is far away from the Sun.

Another example of the conservation of angular momentum is demonstrated in Fig. 3.23. Ice skaters use the principle to spin faster on the ice. The skater extends both arms and perhaps one leg and obtains a slow

Figure 3.23 Conservation of angular momentum.
Going into a spin with arms and leg outstretched is analogous to having the equivalent mass at a large *r* value.

Bringing the arms over the head and leg inward causes the speed to increase and a fast spin results (but not a change in skaters).

rotation; then the arms and leg are drawn inward and rapid angular velocity is gained because of the decrease in the radial distance of the mass.

You can demonstrate the conservation of angular momentum by performing the following experiment: Take a small object like a piece of chalk and tie a length of string about a foot long securely to the chalk. Hold the other end of the string by your forefinger and thumb and whirl the chalk around in a vertical circle. Extend your forefinger so that the string will wind itself around this finger, and observe the increase in speed of the chalk as the distance between the chalk and finger becomes smaller.

The law of conservation of angular momentum would cause a helicopter with a single rotor to rotate. For conservation of angular momentum, the body of the helicopter would have to rotate in the direction opposite that of the rotor. To prevent rotation, large helicopters have two oppositely rotating rotors (Fig. 3.24). More commonly, however, a smaller helicopter will have a small "anti-torque" rotor on the tail, which counteracts the rotation of the body of the helicopter.

Figure 3.24 Conserve that angular momentum.
Large helicopters have two overhead rotors that rotate in opposite directions so as to balance the angular momentum.

Figure 3.24 (cont.)
(b) Small helicopters with one overhead rotor have an "antitorque" tail rotor to balance the angular momentum and prevent the rotation of the helicopter body.

Learning Objectives

After reading and studying this chapter, you should be able to do the following without referring to the text:

1. State Newton's first law of motion and explain what the natural state of motion is.

2. Define the terms *force*, *inertia*, and *mass*, and give the SI and British units of each.

3. State Newton's second law of motion in words and with an equation.

4. Define the term *weight*, and give its units in the SI and British system.

5. Explain why any two objects will drop at the same rate no matter what they weigh, if air friction can be neglected.

6. Explain the differences between G and g.

7. State and give an example of Newton's third law of motion.

8. Define *linear momentum* and *angular momentum*, and state and give an example of the law of conservation of each.

9. Define and explain the important words and terms listed in the next section.

Important Words and Terms

Newton's first law of motion	weight	linear momentum
force	g	law of conservation of linear momentum
inertia	Newton's law of universal gravitation	angular momentum
mass	G	torque
Newton's second law of motion	centripetal force	law of conservation of angular momentum
newton	Newton's third law of motion	

Questions

Newton's First Law of Motion

1. What is the natural state of motion?

2. What is meant when we say that a person has a lot of inertia?

3. An old party trick is to pull a tablecloth from beneath dishes and glasses on a table. Explain how this trick can be done without upsetting or pulling the dishes and glasses with the cloth.

4. Why do you feel thrown to the outside when a car in which you are riding goes around a curve, even though there is no force acting on you in that direction?

5. Explain the principle of automobile seat belts in terms of Newton's first law.

Newton's Second Law of Motion

6. Explain the relationship between (a) force and acceleration and (b) mass and acceleration.

7. (a) Can an object be at rest if forces are being applied to it? Explain.
   (b) If no forces are acting on an object, can the object be in motion? Explain.

8. Explain each of the terms in $F = ma$ in detail.

9. (a) What is a newton?
   (b) What is the weight of a 3-N hammer?

10. A 10-lb rock and a 1-lb rock are dropped simultaneously from the same height.
    (a) Some say that because the 10-lb rock has 10 times as much force acting on it, it should reach the ground first. Do you agree?
    (b) Describe the situation if the rocks were dropped by an astronaut on the moon.

Newton's Law of Gravitation

11. Discuss the quantities "big G" and "little g," and tell how they are different.

12. In an application of $F = Gm_1m_2/r^2$ to a mass on the Earth's surface, what is r?

13. The gravitational force is said to have an infinite range. What does this term mean?

14. How can the acceleration due to gravity be taken to be constant near the Earth's surface, when g varies with altitude?

15. (a) Are astronauts in a spacecraft orbiting the Earth "weightless"?
    (b) Is "zero g" possible?

Newton's Third Law of Motion

16. Use Newton's third law to explain how a rocket blasts off.

17. If there is an equal and opposite reaction for every force, how can an object be accelerated when the vector sum of these forces is zero?

18. If an object is accelerating, does Newton's third law apply? Explain.

19. Explain the kick of a rifle or shotgun in terms of Newton's third law. Do the masses of the gun and the bullet or shot make a difference?

Momentum

20. What is meant when someone says that an athletic team has a lot of momentum or loses its momentum?

21. Explain how the law of conservation of linear momentum follows directly from Newton's first law of motion.

22. A key on a string is whirled around your forefinger. Explain what happens as the string wraps around your finger and why.

23. After diving from a high board, divers "tuck" in and spin before straightening out to cleave the water. How does tucking in accomplish spinning?

24. Would you take a ride in a helicopter with only one rotor? If you did, what would happen after lift-off?

Exercises

Newton's Second Law of Motion

1. Determine the force necessary to give an object with a mass of 4 kg an acceleration of 6 m/s^2. *Answer:* 24 N

2. A force of 2800 N is exerted on a rifle bullet with a mass of 0.014 kg. What will be the bullet's acceleration?

3. What is the unbalanced force on a car moving with a constant velocity of 20 m/s (45 mi/h)?

4. What is the weight of a 6-kg package of nails?
   Answer: 58.8 N

5. What is the force, in newtons, acting on a 6-kg package of nails that falls off a roof and is on its way to the ground?

6. (a) Calculate the weight, in newtons, of a person who weighs 140 lb.
   (b) What is your weight in newtons?

7. Compute the weight, in newtons, of a 500-g package of cereal.

8. A vertical fishing line supports a 5-kg fish when it is held up by the proud angler for a picture. Compute the tension in the line. *Answer:* 49 N

9. A 2.0-kg block of material sits fixed on top of a 3.0-kg block on a level surface.
   (a) If the surface is frictionless and a horizontal force of 10 N is applied to the bottom block, what is the acceleration of the system?
   (b) If there were a constant frictional force of 4.0 N between the bottom block and the surface, what would be the acceleration in this case? *Answer:* (b) 1.2 m/s^2

Newton's Law of Gravitation

10. (a) What is the force of gravity between two 1000-kg cars when separated by a distance of 50 m on an interstate highway?
    (b) How does this force compare with the weight of a car?

11. If the separation distance between two masses (a) tripled and (b) decreased by one-half, how would the force of gravity between the masses be affected?

12. What is the acceleration due to gravity at an altitude of 1000 km (620 mi) above the Earth's surface? (*Hint:* Use $R_E = 6.37 \times 10^6$ m and form a ratio.)
    Answer: $g = 0.75 \, g_E = 7.35$ m/s^2

13. Compute the acceleration due to gravity at the Earth's surface in mks units. (Use $M_E = 5.96 \times 10^{24}$ kg and $R_E = 6.37 \times 10^6$ m.)

14. Show that the acceleration due to gravity on the surface of the moon is about one-sixth of that on the Earth's surface. (*Hint:* Use $M_m = 7.3 \times 10^{22}$ kg, $R_m = 1.74 \times 10^6$ m, and the Earth data in Exercise 12 in a ratio.)

15. (a) Determine the weight on the moon of a person whose weight on Earth is 180 lb.
    (b) What would be your weight on the moon?
    Answer: (a) 30 lb

16. An astronaut on the moon places a package on a scale and finds its weight to be 18 N.
    (a) What would be its weight on Earth?
    (b) What is the mass of the package on the moon?
    Answer: (a) 108 N

Momentum

17. Calculate the momentum of the following:
    (a) A truck with a mass of 15,000 kg that is traveling at 20 m/s (45 mi/h) eastward
    (b) A small car with a mass of 900 kg traveling at 30 m/s (67 mi/h) north   *Answer:* (a) 300,000 kg·m/s, east

18. What is the magnitude of the angular momentum of a 1000-kg car going at a constant speed of 90 km/h (25 m/s) around a circular track with a radius of 100 m?

19. A comet goes around the Sun in an elliptical orbit. At its farthest point, 600 million miles from the Sun, it is traveling at a speed of 15,000 mi/h. How fast is it traveling at its closest approach to the Sun at a distance of 100 million miles?   *Answer:* 90 million mi/h

20. An asteroid going around the Sun in an elliptical orbit at its farthest point is a distance of 8×10^{11} m from the Sun and is traveling with a speed of 3×10^4 m/s. How fast is it traveling when at its closest approach of 2×10^{11} m from the Sun?

CHAPTER 4

Work and Energy

> I like work; it fascinates me.
> I can sit and look at it for hours.
> —Jerome K. Jerome

THE COMMON MEANING of the word *work* refers to the accomplishment of some task or job. When work is done, energy has been expended. Hence work and energy are related.

A student performs a certain amount of work during the day and becomes tired. He or she must obtain rest and food in order to continue the work. We know that rest alone is not sufficient to keep a person going; thus the food must serve as fuel to supply the necessary energy.

The technical meaning of the word *work* is quite different from the common meaning. A student standing at rest and holding several books is doing no work, although he or she will feel tired after a time. Technically speaking, work is accomplished only when a force acts through a distance.

Energy is sometimes called stored work, and our mastery of energy has produced today's modern civilization. From the control of fire to the control of nuclear energy, we have advanced our standard of living through our ability to use and control the flow of energy. Energy takes many forms, such as mechanical, heat, chemical, electrical, radiant, nuclear, and gravitational energy. These forms, as we shall learn, are classified in the more general categories of kinetic and potential energy.

Our main source of energy is the Sun, which radiates into space each day an enormous amount of radiant energy, only a small portion of which is received by the Earth. Radiant energy supports all life on our planet and provides us with the means for making nature work for the Earth's inhabitants.

4.1 Work

As a physical concept, work is the result of the action of a force. We define the **work** done by a force F as the product of the magnitude of the force and the parallel distance through which the force acts. This concept may be written as

$$\text{work} = \text{force} \times \text{parallel distance}$$

or
$$W = Fd \tag{4.1}$$

Hence work requires not only a force but also motion or the movement of an object through a distance.

Figures 4.1, 4.2, and 4.3 illustrate the concept of work. In Fig. 4.1 a force is being applied to the wall, but no work is being done since the wall is not moving through any distance. Figure 4.2 shows an object being moved through a distance d by a force F, and the work done is the product of the force and distance. The illustration shows the force F and the distance d parallel to one another. When F and d are not parallel to one another, the component of the force F that is parallel to the distance d is used to calculate work done.

For example, when a person is mowing the lawn with a push-type lawn mower (Fig. 4.3), only the force component parallel to the lawn (the horizontal force) is

$W = Fd = 0$

F

Figure 4.1 No work done.
A force is applied to the wall, but no work is done, because there is no movement ($d = 0$).

◄ The potential energy of dammed water can be used to do work. At Hoover Dam, shown here, the water turns turbines to generate electricity.

$$W = Fd$$

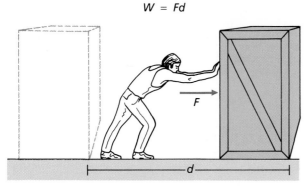

Figure 4.2 Work being done.
An applied force *F* acts through a distance *d*. The work equals the force times the distance the object moves in the direction of the applied force.

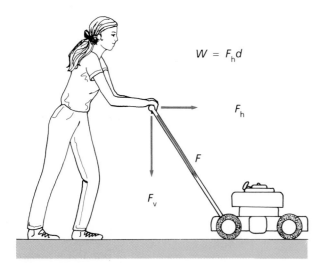

$$W = F_h d$$

F_h

F

F_v

Figure 4.3 Work and no work.
Only the horizontal component F_h of the applied force *F* does work, because it is in the direction of the motion. The vertical component F_v does no work, because $d = 0$ in that direction.

used to accomplish work. The vertical component of the force is doing no useful work, since this force is tending to push the lawn mower into the ground. When the applied force is perpendicular to the distance through which the object is moved, no work is done.

One of the most important properties of work is that it is a scalar quantity. Both the force and the parallel distance have directions associated with them, but work

Table 4.1 Work Units (Energy Units)

System	Force × Distance Units $W = F \times d$	Special or Common Name
SI	newton × meter (N·m)	joule (J)
British	pound × foot	foot × pound (ft·lb)

does not. Work is expressed only as a magnitude with the proper units. It has no direction associated with it.

Since work is the product of a force and a distance, the units of work are the units of force times length. In the SI system the units of work are the newton-meter (N·m, force × length). This unit is given the special name of **joule** (abbreviated J and pronounced "jool"). One joule is the amount of work done by a force of 1 N acting through a distance of 1 m.* Similarly, the units of force times length in the British system are pound-foot. For some reason, however, the units are commonly listed in reverse, and we express work in **foot-pound** (ft·lb) units. A force of 1 lb acting through a distance of 1 ft does 1 foot-pound of work. The units of work are summarized in Table 4.1.

Work can be done on objects in many different ways, such as work against inertia, against gravity, and against friction.

(a) *Work against inertia.* We have learned that any object at rest remains at rest unless acted upon by some outside force; or if the object is in motion, it will remain in motion at constant velocity in a straight line unless acted upon by some outside force (Newton's first law).

Newton referred to the properties of an object to stay at rest or in constant straight-line motion as *inertia.* When a force is applied to change the velocity or direction of an object, work is done against inertia (see Fig. 4.4). A mass is shown at rest on a frictionless air table. If the velocity of the mass is to increase from 0 to 1 m/s, a force must be applied through the distance *d*, thus performing work. The work done has been against inertia.

(b) *Work against gravity.* If we lift an object at a (slow) constant velocity, there is no net force on it,

---

* For the smaller metric force unit of dyne, the amount of work done by a force of 1 dyne acting through a distance of 1 cm is a dyne·cm, or an erg. One erg is 10^{-7} J.

$$W = Fd$$

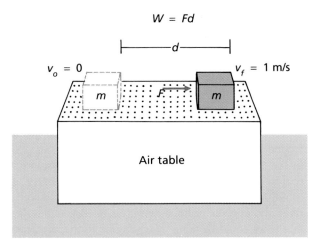

Figure 4.4 Work done against inertia.
A force *F* is applied to move the mass from rest to a velocity of 1 m/s. The air table provides an almost frictionless surface for the mass.

Figure 4.5 Work done against gravity.
The upward lift force *F* is equal to the weight force *mg*. In lifting the weights a height *h*, the work done is *mgh*.

because it is not accelerating. The weight of the object *mg* presses down, and we push up with an equal and opposite force. The distance parallel to the applied upward force is the height *h* that we lift the object (see Fig. 4.5).

Thus the work done against gravity is

$$W = Fd$$
$$W = mgh \qquad (4.2)$$

If we lift a 10-kg mass a distance of 2 m, the work done is

$$W = mgh$$
$$= 10 \text{ kg} \times 9.8 \text{ m/s}^2 \times 2 \text{ m}$$
$$= 196 \text{ J}$$

The concept of work against gravity helps to explain why it is much more tiring to walk up stairs than it is to walk on a level surface. When you walk on level ground, your center of gravity remains approximately constant, and you do very little lifting of your body (Fig. 4.6). Of course, there is some work being done by your muscles as you walk, but it is not enough to tire you very much. In fact, most healthy people can easily walk several miles on a level surface.

When you walk up stairs, you have to lift your whole body up. You do an additional amount of work equal to *mgh*, where *mg* is your weight and *h* is the height you go up.

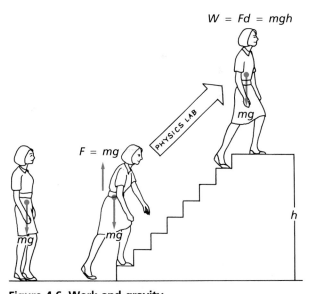

Figure 4.6 Work and gravity.
Walking up stairs requires more work than walking on a level surface. When you go up a distance *h*, the work you do is increased by an amount *mgh* over what you would do if you walked on a level surface.

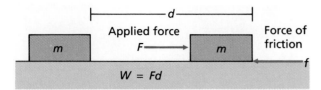

Figure 4.7 Work done against friction.
The mass is being moved at a constant velocity by the force F against the force of friction.

(c) *Work against friction.* Friction always opposes motion. Hence to move something, we have to apply a force. In doing so, work is done against friction. As illustrated in Fig. 4.7, if an object is moved at a constant velocity, the applied force is equal and opposite to the frictional force. The work done by the applied force against friction is $W = Fd$.

4.2 Power

When a family moves into a second-floor apartment, a lot of work must be done to carry their belongings up the stairs. In fact, each time the steps are climbed, the movers must carry not only the furniture, books, and so on, but also their own weight up the stairs.

If the movers do all the work in 3 h, they have not worked as rapidly as if the work had been done in 1 h. The same amount of work is done in each case, but the rate at which work is done is different.

The rate of doing work is called **power.** Power is calculated by dividing the work done by the time required to do it. Power can be written as

$$\text{power} = \frac{\text{work}}{\text{time}}$$

or
$$P = \frac{W}{t} \qquad (4.3)$$

Because work is the product of force and distance ($W = Fd$), power may be written in terms of these quantities:

$$P = \frac{W}{t} = \frac{Fd}{t} \qquad (4.4)$$

In SI units, in which work is measured in joules and time in seconds, power has the units of joule per second (J/s). This unit is given the special name of **watt** (W), and 1 W = 1 J/s.

In the British system the unit of power is foot-pound per second (ft·lb/s). However, a larger unit, the **horsepower** (hp) is commonly used to rate the power of motors and engines, and

$$1 \text{ horsepower (hp)} = 550 \text{ ft·lb/s} = 746 \text{ W}$$

The horsepower unit was originated by James Watt, a Scottish engineer who developed an improved steam engine and after whom the SI unit of power is named. In the 1700s horses were used in coal mines to bring coal to the surface and to power water pumps. In trying to sell his steam engine to replace horses, Watt cleverly rated the engines in horsepower so as to compare the rate work could be done by an engine to the rate work could be done by an average horse.

The greater the power of an engine or motor, the faster it can do work—that is, it can do more work in a given time. For instance, a 2-hp motor can do twice as much work as a 1-hp motor in the same time. Or the 2-hp motor can do the same amount of work as a 1-hp motor in half the time.

The following example shows how power is calculated in a typical situation.

EXAMPLE 1

A force of 50 N is required to push a student's stalled motorcycle 10 m along a flat road in 20 s. Calculate the power, in watts.

Solution   Using Eq. 4.3, we get

$$P = \frac{W}{t}$$

$$= \frac{50 \text{ N} \times 10 \text{ m}}{20 \text{ s}}$$

$$= 25 \text{ W}$$

One should be careful not to confuse the meaning of the letter W. In the formula $P = W/t$, W stands for "work." In the statement "$P = 25$ W," the W stands for "watts." Also, recall that w stands for "weight."

In the next section we will see that work produces a change in energy. Thus power can be thought of as the energy produced or consumed divided by the time taken, and we can write

$$\text{power} = \frac{\text{energy produced or consumed}}{\text{time taken}}$$

or
$$P = \frac{E}{t}$$

From this equation we can see that

$$E = P \times t$$

This formula is useful in computing the amount of energy consumed in a home. In particular, since energy is power times time, a watt × hour (Wh) is a unit of energy. For a larger unit of energy, a kilowatt (kW) may be used, and 1000 watts × hour is a **kilowatt-hour** (kWh). Remember that a kWh is an energy unit, not a power unit.

Electric power companies charge for the electricity consumed in units of kilowatt-hours. Hence we really pay the power company for the amount of energy consumed (which we use to do work). See Fig. 4.8. The following example illustrates how the energy consumed may be calculated.

EXAMPLE 2

A 1-hp motor runs for 10 h. How much energy is consumed?

Solution One horsepower is 746 W, or about 0.75 kW. Then

$$E = P \times t$$
$$= 0.75 \text{ kW} \times 10 \text{ h} = 7.5 \text{ kWh}$$

In this case electrical energy is consumed when the motor is running or doing work. We often complain about our electric bills. In the United States the cost of electricity ranges from about 8¢ to 15¢ per kWh. So to run the motor for 10 h at a rate of 10¢/kWh costs 75¢. That's pretty cheap for 10 h of work output. (Electrical energy is discussed further in Chapter 8.)

Figure 4.8 Energy consumption.
Electrical energy is consumed as the motor does work. Although we commonly refer to an electrical power company, we pay for energy in units of kilowatt-hours (kWh).

4.3 Kinetic Energy and Potential Energy

When work is done against inertia, an object's speed is changed. When work is done against gravity, an object's height is changed. When work is done against friction, heat is produced, and there is usually a change in the temperature. In all these examples something is changed when work is done.

The concept of energy helps us to unify all the possible changes that occur when work is done. When work is done, there is a change in energy. The amount of work done is equal to the change in energy.

Energy is one of the most fundamental concepts in science. It is a quantity that is possessed by objects: Work is *done*; objects *have* energy. When work is done on a system, the amount of energy possessed by the system changes. **Energy** is the ability to do work or simply is stored work.

Basically, work is the process by which energy is transferred from one object to another. An object with energy can do work on another object and give it energy. The amount of energy expended is equal to the work done, so you should not be surprised to learn that work and energy have the same units. In the SI system energy (and work) is measured in joules.

Work and energy are scalar quantities. That is, they have no direction associated with them. Thus various amounts of energy can be added and subtracted as numbers. Forces can become complicated because of their vector or directional nature. In contrast, energy, because it is a scalar quantity, is easy to use in computational problems.

Energy occurs in many forms. Two of the most fundamental forms of energy are kinetic energy and potential energy. Let's take a look at them.

Kinetic energy is energy a body possesses because of its motion or, simply, is the energy of motion. As we learned, work requires motion, so when work causes a change in motion, there is a corresponding change in kinetic energy.

The kinetic energy of an object can be written as

$$\text{kinetic energy} = \tfrac{1}{2} \times \text{mass} \times (\text{velocity})^2$$

or
$$E_k = \tfrac{1}{2}mv^2 \tag{4.5}$$

This equation gives the amount of kinetic energy an object has when traveling with a velocity v, or the change in the kinetic energy when the object is accelerated from rest ($E_k = 0$) to a velocity v. It would be equal to the amount of work done on the object if all the work became kinetic energy. Also, the kinetic energy is equal to the work that would have to be done to bring the object to rest. This result can be seen from the equations in Chapter 2, where we had $d = \frac{1}{2}at^2$ and $v = at$. The work done against inertia is $W = Fd$, and since $F = ma$, we get

$$W = Fd = mad$$
$$= ma(\tfrac{1}{2}at^2) = \tfrac{1}{2}m(at)^2$$
$$= \tfrac{1}{2}mv^2$$

As an example of something doing work to create kinetic energy, consider a pitcher throwing a baseball (Fig. 4.9). The amount of work needed to accelerate a baseball from rest to a speed v is just equal to the baseball's kinetic energy, $\frac{1}{2}mv^2$.

When work is done on a moving body, the work changes the kinetic energy of the body. We get the formula

$$\text{work} = \text{change in kinetic energy}$$
$$= E_{k_2} - E_{k_1}$$

This is an important formula. It shows the relationship between work and kinetic energy.

Because the kinetic energy is proportional to the velocity squared, doubling the velocity will cause a four-fold increase in kinetic energy. This relationship can be seen if we form the ratio $E_{k_2}/E_{k_1} = v_2^2/v_1^2 = (v_2/v_1)^2$, and with $v_2 = 2v_1$ or $v_2/v_1 = 2$, we have $E_{k_2}/E_{k_1} = (2)^2 = 4$.

EXAMPLE 3

By what factor is the kinetic energy increased when the speed of a car is increased from 30 km/h to 50 km/h?

Solution To solve this problem, we note that the kinetic energy is proportional to the square of the speed. Therefore, the kinetic energy increases by a factor of (50 km/h)²/(30 km/h)², or 2.8.

To stop a moving automobile takes an amount of work equal to the kinetic energy of the vehicle. In other words, the work needed to stop an object is equal to the work originally exerted to get it moving. The work necessary to stop an automobile is generally supplied by friction.

When cars travel on highways, the braking distance can be defined as the distance a car travels once its brakes are applied. The work done to stop a moving car is equal to the braking force times the braking distance. If the braking force is assumed to be constant, then the braking distance is proportional to the initial

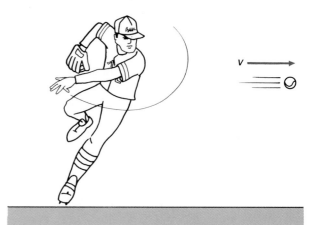

Figure 4.9 Work and energy.
The work necessary to increase the velocity of a mass is equal to the increase in the kinetic energy. (We assume that no energy is lost.)

kinetic energy of the car. Thus from Example 3 we see that the braking distance for a car going 50 km/h ($\approx$ 30 mi/h) is 2.8 times that for a car going 30 km/h ($\approx$ 20 mi/h).

This braking distance concept explains why school zones have relatively low speed limits, commonly 20 mi/h. For instance, if the braking distance of a car traveling at 30 km/h (20 mi/h) is 8 m, then at 50 km/h (30 mi/h) the braking distance is 2.8 × 8 m = 22 m (Fig. 4.10). The driver's reaction time is also a consideration here. As this simple calculation shows, if a driver exceeds the speed limit in a school zone, he or she may not be able to stop for a child darting into the street.

Potential energy is the energy a body has because of its position or location or, simply, the energy of position. Work is done in changing the position of an object, and hence there is a change in energy.

For example, if a book (mass = 1 kg) at rest on the floor is lifted a height of 1 m to the top of a table, work is done. Work is done against gravity; and as shown previously (Eq. 4.2), this work is equal to

$$W = mgh$$

where mg is the weight of the object ($w = mg$),
h is the height through which the book is lifted.

The amount of work done in lifting the book is then

$$W = mgh$$
$$= (1 \text{ kg})(9.8 \text{ m/s}^2)(1 \text{ m}) = 9.8 \text{ J}$$

With work being done, the energy of the book changed

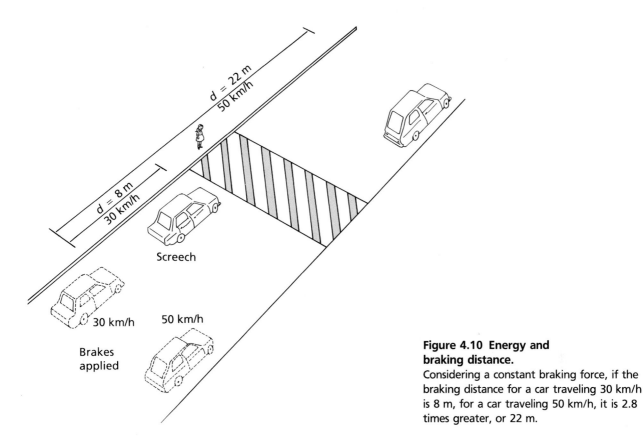

Figure 4.10 Energy and braking distance.
Considering a constant braking force, if the braking distance for a car traveling 30 km/h is 8 m, for a car traveling 50 km/h, it is 2.8 times greater, or 22 m.

(increased), and the book on the table has energy and the ability to do work because of its height or position. This energy is called **gravitational potential energy.** If the book were allowed to fall back to the floor, it could do work, such as crunching something.

The gravitational potential energy E_p is equal to the work done, so we may write

gravitational potential energy = weight × height

or
$$E_p = mgh \qquad (4.6)$$

When work is done by or against gravity, the potential energy changes, and

$$
\begin{aligned}
\text{work} &= \text{change in potential energy} \\
&= E_{p_2} - E_{p_1} \\
&= mgh_2 - mgh_1 \\
&= mg(h_2 - h_1) \\
&= mg\,\Delta h
\end{aligned}
$$

As with the kinetic energy of an object with a particular velocity, an object has a value of potential energy at a particular height or position. Work is done when there is a *change* in position, so keep in mind that the h in Eq. 4.6 is really a height *difference* (Δh).

Also, the work done is independent of path and depends only on the initial and final positions or difference in height (see Fig. 4.11). The force necessary to lift an object from the floor to tabletop height is the force needed to overcome gravity, which acts downward. Therefore, the applied force used to lift the object must be upward, no matter what path is taken to arrive at tabletop height. (Notice in Fig. 4.11 that h is really a height difference.)

The value of the gravitational potential energy at a particular position depends on the reference point, that is, the reference or zero point from which the height is measured. Near the surface of the Earth where the acceleration due to gravity is relatively constant, the designation of a zero position or height is arbitrary. Any point will do. Using an arbitrary zero point for gravitational potential energy is like using a point other than the zero mark on a meterstick to measure length (Fig. 4.12). This practice gives rise to negative positions and displacements, such as + and − positions on the axis of a Cartesian graph (see Fig. 4.12).

Height displacements may be positive or negative relative to the zero reference point, but notice that the height *difference* or change in potential energy between

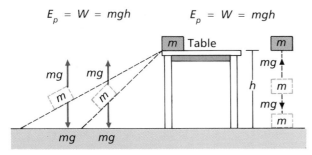

$$E_p = W = mgh \qquad E_p = W = mgh$$

Figure 4.11 Independent of path.
The work done in placing a mass on the table is independent of the path taken and depends only on the initial and final positions, or difference in height.

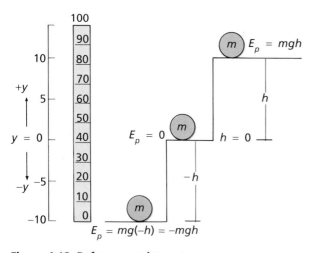

Figure 4.12 Reference point.
The reference point for measuring heights is arbitrary. For example, the zero reference position may be that on a Cartesian axis or at one end of a meter stick (left). *Right:* For positions below a chosen zero reference point (Cartesian *y* axis), the potential energy is negative because of the negative distance. Referenced to the zero end of the meter stick, the potential energy values would be positive. The important point is that the energy differences are the same for any reference.

two positions is the same in both cases (and in any case). A negative ($-$) *h* gives a negative potential energy. A negative potential energy is somewhat like being in a potential "well" or like being in a hole or a well shaft, since we usually designate $h = 0$ at the Earth's surface.

There are other types of potential energy besides gravitational potential energy. For example, when a spring is compressed or stretched, work is done (against the spring force); and the spring has potential energy or a change in potential as a result of the change in length (position). Also, work is done when a bowstring is drawn back. The bow and the bowstring bend and acquire potential energy. This potential energy is then capable of doing work on the arrow, which produces motion or kinetic energy. Note how work is a process of transferring energy.

4.4 Conservation of Energy

Although the meaning is always the same, the law of conservation of energy (or, simply, the conservation of energy) can be stated in many different ways. Examples are: *The total energy of an isolated system remains constant; energy can be neither created nor destroyed; in changing from one form to another, energy is always conserved.*

The **conservation of energy,** as mentioned above, states: The total energy of an isolated system remains constant. Thus, although energy may be changed from one form to another, energy is not lost from the system. A *system* is something enclosed within boundaries, which may be real or imaginary, and *isolated* means that nothing from the outside affects the system (and vice versa).

For example, the students in a classroom might be considered a system. The students may move around in the room, but if no one left or entered, then the number of students would be conserved ("law of conservation of students"). We sometimes say that the total energy of the universe is conserved. That is, the universe is the largest isolated system we can think of, and all the energy in the universe is in it somewhere in some form.

In equation form we state the law of conservation of energy as

$$(\text{total energy})_{\text{time 1}} = (\text{total energy})_{\text{time 2}} \qquad (4.7)$$

That is, the total energy does not change with time. If the only forms of energy involved are kinetic and potential energy, we can write

$$(E_k + E_p)_{t_1} = (E_k + E_p)_{t_2}$$

or $\qquad (\tfrac{1}{2}mv^2 + mgh)_{t_1} = (\tfrac{1}{2}mv^2 + mgh)_{t_2}$

In writing this equation, we are assuming that no energy is lost in the form of heat energy due to frictional effects. These heat energy effects are frequently important and are discussed in Chapter 5. However, in first introducing the conservation of energy, we can conveniently consider frictional effects to be negligible.

EXAMPLE 4

A girl on a sled starts from rest down a snowy slope from a vertical height of 25 m above level ground (Fig. 4.13). What will be the potential and kinetic energies of the girl and the sled at the heights indicated in the figure if she and the sled have a total mass of 50 kg? (Assume no losses due to friction.)

Solution In this case the girl and the sled are the system. According to the law of conservation of energy, the total energy (E_T) will be the same everywhere along the slope. When girl and sled are at rest at the top of the hill, the total energy of the system is all potential energy ($E_T = E_p$), since $v = 0$ and $E_k = 0$.

At any point on the slope the potential energy will be $E_p = mgh$, where $mg = 50$ kg $\times$ 9.8 m/s^2, or 490 N, and h is the height. We can find the potential energy at heights 25 m, 15 m, 10 m, and 0 m.

$$E_p = mgh$$

At $h = 25$ m: $E_p = 490$ N $\times$ 25 m $= 12{,}250$ J

At $h = 15$ m: $E_p = 490$ N $\times$ 15 m $= 7350$ J

At $h = 10$ m: $E_p = 490$ N $\times$ 10 m $= 4900$ J

At $h = 0$ m: $E_p = 490$ N $\times$ 0 m $= 0$ J

Because the total energy is conserved or constant, the kinetic energy (E_k) at any point can be found from the equation $E_T = E_k + E_p$. Therefore, $E_k = E_T - E_p = 12{,}250$ J $- E_p$, because we know that $E_T = E_p$ at the top of the hill. (Check and see if the formula gives $E_k = 0$ at the top of the hill.) A summary of the results is given in Table 4.2.

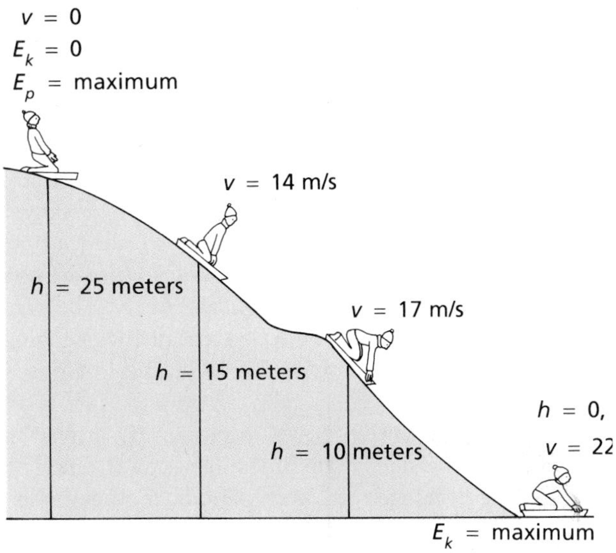

From Table 4.2 or Fig. 4.13 for Example 4, notice that as the sled travels down the hill (decreasing h), the potential energy becomes less and the kinetic energy becomes greater (increasing v). That is, potential energy is converted into kinetic energy. (Note that the shape of the slope makes no difference in these calculations;

Figure 4.13 The conservation of energy.
The total energy of the girl and the sled at the top of the hill is all potential energy, since they are at rest. As the sled and the rider move down the hill, the potential energy decreases and the kinetic energy increases (height decreases and speed increases). Assuming no loss to friction, at the bottom of the hill the total energy is all kinetic energy with the velocity being a maximum, and the potential energy is zero. See Example 4.

only the height at a given point is important. Recall that gravitational energy is independent of path.)

At the bottom of the slope, where h is zero, all the energy of the system is in the form of kinetic energy and the velocity of the sled is at a maximum. The magnitude of this velocity may be found by using the conservation of energy. We may write

$$(E_T)_{top} = (E_T)_{bottom}$$

or $\quad (\tfrac{1}{2}mv^2 + mgh)_{top} = (\tfrac{1}{2}mv^2 + mgh)_{bottom}$

and $\quad (0 + mgh)_{top} = (\tfrac{1}{2}mv^2 + 0)_{bottom}$

because $v = 0$ at the top and $h = 0$ at the bottom. Then canceling the m's in the equation to get

$$\tfrac{1}{2}v^2 = gh$$

and solving for v gives

$$v = \sqrt{2gh} \qquad (4.8)$$

Remember that h is the *change* in height. For the total slope this change is 25 m. Equation 4.8 will give the velocity at any point on the slope if h is considered

Table 4.2 Energy Summary for Example 4

Height (m)	E_T (J)	E_p (J)	E_k (J)	(v, m/s)
25	12,250	12,250	0	(0)
15	12,250	7,350	4,900	(14)
10	12,250	4,900	7,350	(17)
0	12,250	0	12,250	(22)

to be the change or decrease in height *from the top down.* For example, at a height of 15 m we use the decrease in height of 10 m in the equation:

$$v = \sqrt{2gh} = \sqrt{2(9.8 \text{ m/s}^2)(10 \text{ m})} = 14 \text{ m/s}$$

The simple pendulum (Fig. 4.14) also illustrates many of the same features of the conservation of energy. A **simple pendulum** consists of a mass (called a bob) attached to a light string and supported so that the bob can swing back and forth, or oscillate, freely. The pendulum is called "simple" because most of the swinging mass is in the bob. The string, for all practical purposes, can be neglected.

The pendulum can be set in motion by displacing the bob to one side (angle θ) and releasing it. The bob will swing back and forth with periodic motion, undergoing energy transformations.

When the bob is at position a or c, the velocity will be momentarily zero, and the kinetic energy will be zero. The potential energy, however, will be maximum, because the mass is at its maximum height above the lowest part of the swing, or reference level.

When the bob swings down from position a, the potential energy will decrease and the kinetic energy will increase correspondingly as the mass gains velocity. When the bob is at point b, it will have zero potential energy, because h will have become zero. The kinetic energy at this point will be maximum, and the mass will be traveling at its maximum velocity.

As the bob swings upward toward point c, the velocity will start to decrease, and the potential energy will begin to increase. As the bob swings higher, the velocity will continue to decrease until it stops mo-

mentarily at point c, where the kinetic energy will have decreased to zero and the potential energy will have reached the maximum. Thus as the pendulum swings back and forth, the energy is converted from potential to kinetic and back to potential.

Neglecting functional effects, the velocity of the bob at point b can be calculated by applying the law of conservation of energy as follows:

At points a or c, $E_T = E_p$ and $E_k = 0$.

At points b or d, $E_T = E_k$ and $E_p = 0$.

Therefore, E_k at the bottom of the swing equals E_p at the top of the swing:

$$(E_k)_{\text{bottom}} = (E_p)_{\text{top}}$$

And the derivation proceeds as for Eq. 4.8. Thus the velocity of the pendulum at the bottom of its swing is

$$v = \sqrt{2gh}.$$

EXAMPLE 5

A child swings on a rope that is 4 m long. She is 1.5 m above the ground at the highest point and 0.5 m above the ground at the lowest point. What are the magnitudes of her velocities at the highest and lowest points?

Solution   At the highest point the energy is all potential, and the velocity is zero because the kinetic energy is zero. At the lowest point the potential energy, $E_p = mgh$, has been converted into kinetic energy, and the velocity is a maximum and is given by Eq. 4.8. In this equation we must remember that h stands for the *change* in height, which in this problem is 1.5 m − 0.5 m, or 1.0 m.

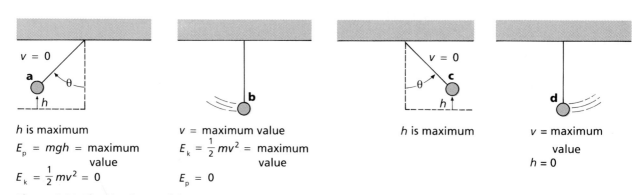

h is maximum

$E_p = mgh = $ maximum value

$E_k = \frac{1}{2}mv^2 = 0$

$v = $ maximum value

$E_k = \frac{1}{2}mv^2 = $ maximum value

$E_p = 0$

h is maximum

$v = $ maximum value

$h = 0$

Figure 4.14 The simple pendulum.
The pendulum swinging back and forth involves the transformations between kinetic and potential energies.

The magnitude of the child's maximum velocity is then

$$v = \sqrt{2gh}$$
$$= \sqrt{2 \times (9.8 \text{ m/s}^2) \times 1.0 \text{ m}}$$
$$= 4.4 \text{ m/s}$$

This speed corresponds to about 10 mi/h.

4.5 Forms and Sources of Energy

If we consider the conservation of energy to its fullest, we have to look at the examples of Section 4.4 in more detail. For instance, we know that both the sled and the pendulum will finally stop, and we must ask the question, Where did the energy go?

Once again, friction is involved. In most practical situations the kinetic or potential energy of large objects eventually ends up as heat energy. Heat energy will be studied at some length in Chapter 5, but for now let us say that heat energy is kinetic and potential energy on a microscopic level. As things get hot, the energies of the atoms and molecules that make up different substances increase.

There are many forms of energy. Besides the macroscopic kinetic and potential energy of objects, there are the following forms of energy: heat, radiant, sound, chemical, nuclear, and others. All of these are related to microscopic kinetic and potential energies that act together to give macroscopic effects. In addition, we will see in Chapter 10 that mass itself is a form of energy.

The main unifying concept when we consider energy is the conservation of energy. We cannot create or destroy the energy of a single system. However, we can transform it from one form to another.

On a global scale the source of practically all of our energy is the Sun. Wood, coal, petroleum, and natural gas are indirectly a result of the Sun's radiant energy. This energy source should be with us for another several billions of years. The Earth's solar energy balance is discussed in more detail in Chapter 20.

As we go about our daily lives, each of us is constantly using energy and giving off energy in the form of body heat. The source of this energy is food (Fig. 4.15). An average adult radiates heat energy at about the same rate as a 100-W light bulb. This heat radiation explains why a crowded room soon gets hot. In winter our clothing helps to keep this heat energy from escaping. In summer the evaporation of perspiration helps us to give off the heat energy that our body produces.

The commercial sources of our energy on a national scale are mainly coal, oil (petroleum), and natural gas.

Figure 4.15 Refueling.
The source of human energy is food. This "fuel" is used in the work of performing body functions, given off as heat, or stored for later use.

Figure 4.16 shows the percentage consumption for each of these sources over about three decades. Nuclear and hydroelectric energy are the only other significant commercial sources of energy. The federal government has projected that the total energy supply will increase slowly.

Coal is our most abundant domestic fuel supply, with an estimated 300- to 400-year reserve at the current consumption rate. Coal has been mined in the Midwest (Pennsylvania, Ohio, West Virginia, etc.) for many years. In the 1970s additional large supplies were found in western states (Montana, North Dakota, Wyoming, etc.). Coal was once the dominant energy source for transportation when steam locomotives chugged across the country. But steam engines were replaced by diesel locomotives (diesel oil is a product of petroleum). Coal is now used primarily for the generation of electricity (Fig. 14.17).

Unfortunately, coal mining and coal use bring with them a variety of environmental problems. Coal generally contains sulfur compounds. In mining, these compounds lead to acid mine drainage, which pollutes lakes and streams. A problem of greater extent comes from the gaseous effluents from coal-fired power plants and other coal burning. Although these effluents are difficult to track, scientists now generally agree that the sulfur dioxide from the combustion of coal is a major source of acid rain. These environmental problems will be considered more fully in Chapter 24.

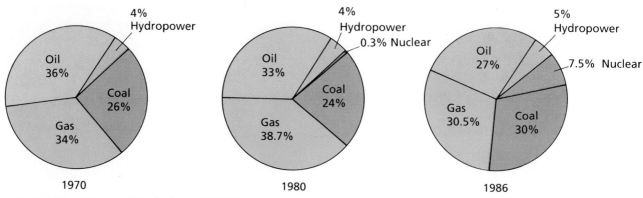

Figure 4.16 Comparative fuel consumption.
The graphs show the relative percentages of fuel consumption in the United States. Notice how the relative use of nuclear energy and coal has increased.

Figure 4.17 Coal-fired electrical generation plant.
The mound of coal in the foreground will be burned to convert chemical energy into electrical energy.

Petroleum (oil) is our most widely used source of energy. It is mainly used in transportation (Fig. 4.18), but it is also widely used in residential, commercial, and industrial sectors of the economy. Petroleum is used not only as a fuel but also as a starting material for the petrochemical industry. A variety of materials, including synthetic plastics, are made from petroleum.

A large percentage of our oil supply (40–50 percent) is imported each year. Our dependence on foreign oil is a cause of concern because of volatile international relations, the balance of payments, the national debt, and the environment (see Chapter 24).

Natural gas is a clean-burning fuel used in home heating as well as in business and industry. The burning of natural gas results in a minimum of pollution problems, but domestic reserves are relatively small compared with our reserves of coal and oil. Recently, large deposits of natural gas have been discovered deep underground, thanks to advances in drilling techniques. It is hoped that more reserves will be found, so that natural gas will supply an increasingly larger proportion of our total energy supply.

Hydropower (water power) also presents a minimum of pollution problems. However, most of the fea-

Figure 4.18 Petroleum consumption.
The main use of petroleum is in transportation, in the
form of gasoline, diesel oil, lubricating greases, and so on.

sible locations for the generation of hydroelectric energy
have been utilized. We don't have any more rivers to
dam without disrupting their normal flow and causing
environmental problems.

Nuclear energy is used in nuclear power plants to
produce electricity. It is a controversial energy source.
Its proponents point out the lack of pollution, the reasonable cost, the ability to "breed" more energy, and
the safety record of nuclear power plants (with the exceptions of Three Mile Island and Chernobyl). Its opponents point out the possibility of a nuclear accident
and gross environmental contamination and the problem
of waste disposal, including old, decommissioned plants.
Some of these aspects will be discussed in Chapter 10.

Even so, nuclear energy has become increasingly
important over the years. Some 15–20 percent of the
electricity in the United States is generated by nuclear
energy. In Europe, France generates more than 70 percent of its electricity by using nuclear energy, and several
other countries are not far behind.

All of these forms of energy go into satisfying a
growing demand. Perhaps you're wondering where all
of this energy goes and who consumes the most. Figure

4.19 gives a general breakdown of use by sector. As
the figure shows, there are four major energy-consuming
sectors, and the consumption of each is in the 20 percent
range.

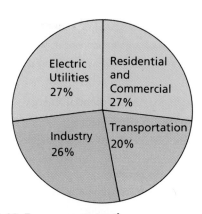

Figure 4.19 Energy consumption.
These graphs show relative consumption of energy by
various sectors of the economy.

Learning Objectives

After reading and studying this chapter, you should be able to do the following without referring to the text:

1. Define the terms *work*, *energy*, and *power*, and give their units in the SI system.

2. Distinguish between kinetic energy and potential energy.

3. Explain whether energy is a scalar or vector quantity, and explain why potential energy can be positive or negative.

4. State the principle of conservation of energy, and explain why it is important.

5. List at least five different forms of energy.

6. List the main sources and the major uses of energy in the United States.

7. Define and explain the important words and terms listed in the next section.

Important Words and Terms

work	horsepower	potential energy
joule	kilowatt-hour	gravitational potential energy
foot-pound	energy	conservation of energy
power	kinetic energy	simple pendulum
watt		

Questions

Work

1. What is required for work to be done?

2. What are the units of work?

3. A weight lifter holds 300 lb over his head. Is he doing work on the weights? Did he do work on the weights? Explain.

4. What is meant by doing work *against* (a) inertia, (b) gravity, and (c) friction?

Power

5. Persons A and B do the same job, but B takes a longer time. Who expends the greater amount of power?

6. What are the units of power?

7. What does a greater power rating mean in terms of (a) the amount of work that can be done in a given time and (b) how fast a given amount of work can be done?

8. What is the unit for which we are charged by the electric company? Are we paying for energy or power?

Kinetic Energy and Potential Energy

9. Define energy in general terms.

10. Distinguish between kinetic energy and potential energy.

11. Explain how the braking distance of an automobile is related to its kinetic energy.

12. Can you have energy without motion? Explain.

13. An object has a certain amount of potential energy at one location and a greater amount at another location. What is the significance of this difference?

14. An object is said to have negative potential energy. Not liking to work with negative numbers, can you change this value without moving the object? Explain.

15. Why do pole vaulters run so fast before vaulting? How about high jumpers?

Conservation of Energy

16. What is meant by a system?

17. When is the total energy conserved, and what does this mean?

18. A mass suspended on a spring is pulled down and released. Assuming the total energy to be conserved, describe the oscillatory motion of the mass in terms of its kinetic and potential energies.

19. When is the conservation of energy invalid?

Forms and Sources of Energy

20. On the average, how much energy do you radiate in an hour?

21. List five different forms of energy other than kinetic energy and potential energy.

22. What are the major sources of energy in the United States?

23. Discuss the problems, limitations, and advantages of coal, oil, natural gas, nuclear power, and hydropower as energy sources.

24. What are the four major energy-consuming sectors in our country, and how do they compare in relative consumption?

Exercises

Work

1. A worker pushes with a force of 50 N against an upright sheet of plywood on a truck to keep it from falling over while the truck moves 2.0 m. How much work is done by the worker?

2. A 5-"kilo" bag of sugar is on a counter.
   (a) How much work is required to put the bag on a shelf a distance of 0.75 m above the counter?
   (b) What would the answer be if this work were done in a space station on the moon?   *Answer:* (b) 6.1 J

3. A 6-kg box sits on a tabletop. How much work is required to move the box 1.5 m along the table if the force of friction is 24 N?

4. A man pushes a lawn mower on a level lawn with a force of 80 N. If 60 percent of this force is directed horizontally, how much work is done by the man in pushing the mower a distance of 10 m?   *Answer:* 480 J

Power

5. If the man in Exercise 4 pushes the mower 10 m in 45 s, how much power was expended?

6. A student weighing 600 N climbs a stairway (vertical height of 4.0 m) in 20 s.
   (a) How much work was done?
   (b) What was the power output of the student?

7. What is the horsepower necessary to carry 20 lb of books to a height of 22 ft in 40 s?   *Answer:* 0.02 hp

8. How many kilowatt-hours of energy are consumed by a 1200-W hair dryer that is run for 5.0 min?
   Answer: 0.1 kWh

Kinetic Energy and Potential Energy

9. (a) What is the kinetic energy of a 1000-kg automobile traveling at 90 km/h (56 mi/h)?
   (b) How much work would have to be done to bring the automobile to a stop?   *Answer:* (a) 3.1×10^5 J

10. By what factor is the kinetic energy increased when the speed of a car is increased from 48 km/h (30 mi/h) to 64 km/h (40 mi/h)? What is the factor of decrease in going from 40 mi/h back to 30 mi/h?

11. What is the potential energy of a child's swing with a mass of 3 kg that is stationary and 0.5 m above the ground?

12. What is the potential energy of a 3-kg mass at the bottom of a well 10 m deep? Explain the value and sign of your answer.

13. How much work is required to lift a 3-kg mass from the bottom of 10-m-deep well?   *Answer:* 294 J

Conservation of Energy

14. Answer the following questions for a comet in an elliptical orbit about the Sun.
   (a) Where is the comet's kinetic energy greatest?
   (b) Where is its potential energy greatest?
   (c) Where is its total energy greatest?
   (*Hint:* Refer to the section on angular momentum in Chapter 3.)

15. A student is swinging on a rope that is 5 m long. At the lowest point he is 0.5 m above the ground. At the highest point he is 1.5 m above the ground. What is the student's speed (a) at the highest point and (b) at the lowest point?
   Answer: (b) 4.4 m/s

16. A sled and rider with a combined weight of 700 N are at the top of hill that is 5 m high.
   (a) What is their total energy at the top of the hill? (Use the bottom of the hill as a reference.)
   (b) Assuming there is no friction, what is the total energy halfway down the hill?   *Answer:* (a) 3500 J

17. A 1.0-kg rock is dropped from a height of 6.0 m
   (a) At what height is the kinetic energy twice its potential energy?
   (b) What is the velocity of the rock at this height?
   Answer: (b) 8.9 m/s, downward

18. How many watts are given off in a crowded room with three 100-W light bulbs and 30 people?

CHAPTER **5**

Temperature and Heat

> **Then cold and hot and moist and dry,**
> **In order to their stations leap.**
> —John Dryden

THE HEATING EFFECTS produced by fire, the sensation received when a piece of ice is held, and the warmth produced when hands are rubbed together are well known. Explaining what takes place in each of these cases, however, is not easy. Both heat and temperature are commonly used when referring to hotness or coldness. However, heat and temperature are not the same thing. They have different and distinct meanings, as we shall see.

Count Rumford (1753–1814), an American (born Benjamin Thompson), was one of the first to recognize the relation between mechanical work and heat. In 1789 he published the results of an experiment in which he used a blunt boring tool to drill a cannon barrel immersed in water. The temperature of the cannon and water rose to the boiling point of water in two and one-half hours. This result convinced Rumford that large quantities of heat were produced by friction. He concluded that heat was a form of energy and appeared to be due to the motion of the drill.

Later James Prescott Joule (1818–1889), an English scientist, determined the quantitative relationship between mechanical energy and heat. He also established the important concept of conservation of energy with respect to heat and originated many of the basic ideas for the kinetic theory of gases.

The concepts of heat and temperature play an important part in our daily lives. We like our coffee hot and our ice cream cold. The temperature of our living and working quarters must be carefully adjusted to our bodies' heat demands. The daily temperature reading is perhaps the most important part of a weather report. How cold or how hot it will be affects the clothes we wear and the plans we make.

We will learn in detail in Chapter 20 how the Sun provides heat to our Earth. The heat balance between various parts of the Earth and its atmosphere gives rise to wind, rain, and other weather phenomena. The thermal pollution of rivers that can be caused by hot water from power plants is a cause for concern because of its effect on the ecology of these rivers.

On a cosmic scale, the temperature of various stars gives clues to their ages and the origin of the universe. What is temperature? What is heat? What causes heat? How is heat transferred? The answers to these questions will be examined in this chapter. Knowing these answers, we can explain many phenomena occurring around us.

5.1 Temperature

Temperature tells us whether something is hot or cold. In fact, we can say that temperature is a *relative* measure of hotness or coldness. For example, if the water in one cup has a higher temperature than the water in another cup, then we know that the water in the first cup is hotter. But it would be colder than a cup of water with an even higher temperature. Hot and cold are *relative* terms.

On the molecular level we find that temperature depends on the kinetic energy of the molecules of a substance. The molecules of all substances are in constant motion. This observation is true even for solids in which the molecules are held together by intermolecular forces, which are sometimes likened to springs. The molecules oscillate back and forth about their equilibrium positions.

In general, the greater the temperature of a substance, the greater the motion of its molecules. On this basis we say that **temperature** is a measure of the average kinetic energy of the molecules of a substance.

Temperature is a measure of hotness and coldness but not necessarily a measure of heat. A simple experiment will illustrate: Place ice cubes and water in a pan and heat the pan. As heat is applied, the ice melts. However, when we check the temperature of the ice and water mixture with a thermometer, we find that the temperature remained the same. Even with continued heating (and stirring) there is no change in temperature *until all the ice is melted*. Obviously, heat was added, but

---

◀ A crowded beach scene reflects temperature and heat.

it didn't change the temperature. We see, then, that temperature does not necessarily measure heat.

Humans have temperature perception in the sense of touch. However, this perception is unreliable and may vary a great deal for different people. Our sense of touch doesn't allow us to measure temperature accurately or quantitatively. The quantitative measurement of temperature is accomplished through the use of a thermometer. A **thermometer** is an instrument that utilizes the physical properties of materials for the purpose of accurately determining temperature. The temperature-dependent material property most commonly used to measure temperature is **thermal expansion.** Nearly all substances expand with increasing temperature (and contract with decreasing temperature).

The change in the linear dimensions or volume of a substance is quite small, but the effects of thermal expansion can be made evident by using special arrangements. For example, a bimetallic strip is made of pieces of different metals bonded together (Fig. 5.1). When it is heated, one metal expands more than the other, and the strip bends toward the metal with the smaller thermal expansion. As illustrated in Fig. 5.1, the strip can be calibrated with a scale to measure temperature. Bimetallic strips in the form of a coil or helix are used in dial-type thermometers.

The most common type of thermometer is the liquid-in-glass thermometer, with which you are probably familiar. It consists of a glass bulb on a glass stem with a capillary bore and sealed upper end. A liquid in the bulb (usually mercury or alcohol colored with a red dye to make it visible) expands on heating, and a column of liquid is forced up the capillary tube. The glass also expands, but the liquid expands much more.

Thermometers are calibrated so that numerical values can be assigned to different temperatures. The calibration requires two reference points and a choice of unit. By analogy, think of constructing a stick to measure length. You need two marks or reference points, and then you must divide the interval between the marks

Figure 5.1 Bimetallic strip.
Because of different degrees of thermal expansion, a bimetallic strip of two different metals bends when heated. The degree of deflection is proportional to the temperature, and a calibrated scale could be added for temperature readings.

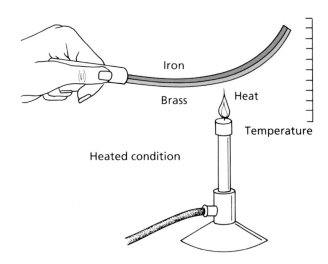

Iron

Brass

Initial temperature condition

Iron

Brass   Heat

Temperature

Heated condition

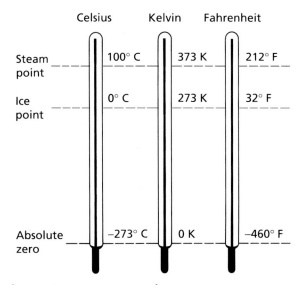

Figure 5.2 Temperature scales.
The common temperature scales are the Fahrenheit and Celsius scales. These scales have 180- and 100-degree intervals, respectively, between their ice and steam points. A third scale, the Kelvin scale, is primarily used in scientific work and takes zero at the lower limit of temperature—absolute zero. The unit or interval on the absolute Kelvin scale is the kelvin (K). Absolute zero on the Celsius and Fahrenheit scales are −273°C and −460°F, respectively.

into sections or units. Thus you might use 100 cm between the reference marks, indicating the length of a meter. Two common reference points for a temperature scale are the ice and steam points of water. The **ice point** is the temperature of a mixture of ice and water at one atmosphere of pressure. The **steam point** is the temperature at which pure water boils at one atmosphere of pressure. Common names for the ice and steam points are freezing and boiling points, respectively.

Two common temperature scales are the Fahrenheit and Celsius scales (Fig. 5.2). The **Fahrenheit scale** has an ice point of 32° (read "32 degrees") and a steam point of 212°. The interval between the ice and steam points is evenly divided into 180 units. Each unit is called a degree. Thus a **degree Fahrenheit**, abbreviated °F, is $\frac{1}{180}$ of the temperature change between the ice point and the steam point.*

The **Celsius scale** is based on an ice point of 0° and a steam point of 100°. There are 100 equal units or divisions between these points. A **degree Celsius,** abbreviated °C, is thus $\frac{1}{100}$ of the temperature change between the ice point and steam point. So a degree Celsius is almost twice as large as a degree Fahrenheit (100 degrees Celsius and 180 degrees Fahrenheit for the same temperature interval). At one time, the Celsius scale was commonly called the "centigrade" scale. Centigrade (*cent*, "one hundred" (as in *cent*ury) and *grade*, German for "degree") referred to the 100 degrees between the ice and steam points. The scale is now usually referred to correctly as the Celsius scale.*

There is no known upper limit of temperature; however, there is a lower limit. The lower limit occurs at about −273°C or −460°F, and it is called absolute zero. Another temperature scale, called the **Kelvin scale,** has its zero temperature at this absolute limit (see Fig. 5.2).[†] This scale is sometimes called the absolute temperature scale. The unit of the Kelvin scale is the **kelvin,** abbreviated K (not °K), and it has the same magnitude as a degree Celsius. Notice that since the Kelvin scale has absolute zero as its lowest reading, there can be no negative Kelvin or absolute temperatures.

Because the kelvin and degree Celsius are equal intervals, we can easily convert from the Celsius scale to the Kelvin scale. We simply add 273 to the Celsius temperature. In equation form we have

$$T_K = T_C + 273 \qquad (5.1)$$

For example, a temperature of 0°C equals 273 K, and a Celsius temperature of 27°C is equal to 300 K ($T_K = T_C + 273 = 27 + 273 = 300$ K).

Converting from Fahrenheit to Celsius and vice versa is also quite easy. The formulas for these conversions are

$$T_F = \tfrac{9}{5}T_C + 32 \quad \text{(Celsius } T_C \text{ to Fahrenheit } T_F)$$

and

$$T_C = \tfrac{5}{9}(T_F - 32) \quad \text{(Fahrenheit } T_F \text{ to Celsius } T_C) \qquad (5.2)$$

As examples, try converting 100°C and 32°F to their equivalent temperatures on the other scales. (You already

---

* The Fahrenheit scale is named after Daniel Fahrenheit (1686–1736), a German scientist, who developed one of the first mercury thermometers. In case you're wondering why he chose 32° and 212° for the ice and steam points on his scale, he didn't. Fahrenheit used two other reference points, and the ice and steam points of water came out to be 32° and 212°F on his scale.

---

* After its originator Anders Celsius (1701–1744), a Swedish astronomer. Oddly enough, Celsius originally chose the ice point as 100° and the steam point as 0°, but these references were later changed.

[†] Named after Lord Kelvin (William Thomson, 1824–1907), a British physicist who developed it.

know the answers.) Remember that on these scales you can have negative temperatures (as opposed to no negative values on the Kelvin scale). For example, suppose it is a very chilly 5°F outside. This value is equivalent to a temperature of −15°C, which may seem even chillier. (Use the second equation to verify this result.) Just to show the equations are equivalent, try changing −15°C to the temperature in Fahrenheit.

5.2 Heat

Heat is a form of energy. So far we have studied two kinds of energy—kinetic energy and potential energy. In a way, heat is a third kind of energy. However, if we were to examine the microscopic sources of heat energy, we would discover that the kinetic and potential energies of molecules are the ultimate sources of heat. Thus heat can be viewed as either another form of energy or as the manifestation of molecular kinetic and potential energies on a macroscopic scale. In any event, heat is energy.

We commonly say that an object contains heat, or that we add or remove heat. In the strictest sense, a body contains internal energy and heat is energy *transferred* from or to it *as the result of a temperature difference.*

Energy transfer involves work. For example, when we do work, there is frequently a substantial amount of heating caused by friction. The conservation of energy tells us that the work done plus the heat energy produced must equal the original amount of energy available. That is,

total energy = work done + heat produced
(usually by friction)

or
$$E_T = W + H$$

A sled and the child on it, having a total mass of 20 kg, slide down a hill that is 4 m high, starting from rest. Their final velocity is 8 m/s. How much heat was produced by the friction between the sled runners and the snow?

Solution The total energy initially is potential energy, *mgh*. The work done against inertia is $\frac{1}{2}mv^2$, where *v* is the final velocity. The amount of heat produced is then

$$H = E_T - W = mgh - \tfrac{1}{2}mv^2$$
$$= (20 \text{ kg} \times 9.8 \text{ m/s}^2 \times 4 \text{ m})$$
$$\quad - [\tfrac{1}{2} \times 20 \text{ kg} \times (8 \text{ m/s})^2]$$
$$= 144 \text{ J}$$

Notice that as energy, heat has the unit of joule (J). A traditional unit for measuring heat energy is the calorie. A **calorie** (cal) is defined as the amount of heat necessary to raise one gram of pure liquid water by one degree Celsius (from 14.5° to 15.5°C) at normal atmospheric pressure. In terms of the SI energy unit,

$$1 \text{ cal} = 4.186 \text{ J}$$

Heat measurements have been commonly made in calories rather than joules. We are now in a transition period going from calories to joules, so both will be used in this book.

The calorie that we have defined is not the same as the one used when discussing diets and nutrition. A kilocalorie is 1000 calories as we have defined it. A **kilocalorie** is the amount of heat necessary to raise the temperature of one kilogram of water one degree Celsius. A diet calorie (Cal) is equal to one kilocalorie and is commonly written with a capital C to avoid confusion. We sometimes refer to a "big" Calorie and a "little" calorie.

1 food Calorie = 1000 calories

1 food Calorie = 4186 joules

A food Calorie indicates the amount of intrinsic energy a food contains.

The unit of heat in the British system of units is the British thermal unit, or Btu. One **Btu** is the amount of heat required to raise one pound of water one degree Fahrenheit at normal atmospheric pressure.

Some relations between Btu, Calories, joules, and kWh are given in Table 5.1.

As we have seen in the measurement of temperature, one effect of heating a material is that of expansion. Almost all matter—solids, liquids and gases—expands when heated. As a general rule, a substance expands when heated and contracts when cooled. The most important exception to this rule is water. If water is frozen, it expands. That is, ice at 0°C occupies a larger volume than the same mass of water at 0°C.

Table 5.1 Relationships Among Some Common Energy Units

1 cal = 4.186 J
1 kcal = 4186 J = 3.97 Btu = 0.00116 kWh
1 Btu = 1055 J = 0.25 kcal = 0.00029 kWh
1 kWh = 3.6 × 10⁶ J = 860 kcal = 3413 Btu

Figure 5.3 Thermal expansion.
Expansion joints are built into bridges and elevated roadways to allow for the expansion of the steel girders caused by varying temperatures.

The change in length or volume of a substance due to a change in temperature is a major factor in the design and construction of items ranging from steel bridges and automobiles to watches and dental cements. The cracks in a highway are designed so that in summer the concrete will not buckle as it expands due to the heat. Expansion joints are designed into bridges for the same reason (Fig. 5.3). Heat-expansion characteristics are used to control such things as the flow of water in car radiators and the flow of heat in homes through the operation of thermostats.

5.3 Specific Heat and Latent Heat

Heat and temperature, although different, are intimately related. In general, when heat is added to a substance, the temperature increases. For example, suppose you added equal amounts of heat to equal masses of iron and aluminum. How do you think their temperatures

would change? You might be surprised to find that if the temperature of the iron increased by 100°C, the corresponding temperature change in the aluminum would be only 48°C. You would have to add more than twice the amount of heat to the aluminum to get the same temperature change as for an equal mass of iron.

This result reflects the fact that the internal forces of the materials are different (different intermolecular "springs," so to speak). In aluminum more energy goes into internal potential energy rather than into kinetic energy, which is manifested as temperature.

We express this difference in terms of specific heat.

> **The specific heat of a substance is the amount of heat, necessary to raise the temperature of one kilogram of a substance one degree Celsius.**

By definition, one kilocalorie is the amount of heat that raises the temperature of one kilogram of water one degree Celsius, so water's specific heat is 1.0 kcal/kg·°C. Other substances require different amounts of heat to raise the temperature of one kilogram of the substance by one degree.

Notice that the units of specific heat are kcal/kg·°C (amount of heat per unit mass per degree change in temperature). The SI units are J/kg·°C. The specific heats of a few common substances are given in Table 5.2. The greater the specific heat of a substance, the greater is the amount of heat required to raise the temperature of

Table 5.2 Specific Heat

Substance	Specific Heat (20°C)	
	kcal/kg·°C	J/kg·°C
Aluminum	0.22	920
Copper	0.092	385
Glass	0.16	670
Human body (average)	0.83	3470
Ice	0.50	2100
Iron	0.105	440
Silver	0.056	230
Soil (average)	0.25	1050
Steam (at constant volume)	0.50	2100
Water	1.000	4186
Wood	0.4	1700

a unit mass. Put another way, the greater the specific heat, the greater is its capacity for heat (given equal masses and temperature change). In fact, the full technical name for specific heat is specific heat capacity.

The amount of heat necessary to change the temperature of a given amount of a substance depends on three factors: the mass (m), the specific heat (designated by c), and the temperature change (ΔT). In equation form we write

$$\begin{array}{c} \text{amount of} \\ \text{heat to change} \\ \text{temperature} \end{array} = \text{mass} \times \begin{array}{c} \text{specific} \\ \text{heat} \end{array} \times \begin{array}{c} \text{temperature} \\ \text{change} \end{array}$$

or
$$H = m\,c\,\Delta T \qquad\qquad (5.3)$$

This equation applies to a substance that does not undergo a phase change when heat is added (for example, changing from a solid to a liquid). As mentioned earlier and as will be considered in detail in the next section, the temperature of a substance that is changing phase does not change when heat is added (or removed).

EXAMPLE 2

How much heat does it take to heat a bathtub full of water (80 kg) from 12°C (about 54°F) to 42°C (about 108°F)?

Solution   Using Eq. 5.3, we get

$$\begin{aligned} H &= mc\,\Delta T \\ &= 80 \text{ kg} \times (1.0 \text{ kcal/kg·°C}) \times (42 - 12)°C \\ &= 2400 \text{ kcal} \end{aligned}$$

Each kilocalorie corresponds to 0.00116 kWh, so this amount of heat is

$$H = 2400 \text{ kcal} \times 0.00116 \text{ kWh/kcal} = 2.8 \text{ kWh}$$

At 10¢ per kWh the cost is 2.8 × 10¢ = 28¢. For four people each taking a bath a day for one month, it would cost 4 × 30 × $0.28 = $33.60 to heat the water.

Substances found in our environment are usually classified as a solid, a liquid, or a gas. These forms are called phases of matter. When heat is added (or removed) from a substance, it may undergo a change of phase. For example, when water is heated sufficiently, it changes to steam (or when enough heat is removed, it changes to ice).

As we know, water changes to steam at a temperature of 100°C (or 212°F) under normal atmospheric pressure. If we keep adding heat to a quantity of water at 100°C, it continues to boil with the conversion of liquid to gas, but the temperature remains constant. Here is a case of adding heat to a substance without a resulting temperature change. Where then does the energy go?

On a molecular level, when a substance goes from a liquid to a gas, work must be done to break the molecular bonds and to separate the molecules. The molecules of a gas are relatively farther apart than the molecules in a liquid. Hence during a phase change the heat energy must go into the work of separating the molecules and not into increasing the molecular kinetic energy, which would increase the temperature. (Phase changes are discussed in more detail in Section 5.6.) The heat associated with a phase change is called **latent heat** (*latent* means "hidden").

Referring to Fig. 5.4, let's go through the process of heating a substance and changing phases from solid to liquid to gas. In the lower left-hand corner the substance is represented in the solid phase. As heat is applied, the temperature rises. When point A is reached, adding more heat does not change the temperature. Instead, the heat energy goes into changing the solid into a liquid. The amount of heat necessary to change one kilogram of a solid into a liquid at the same temperature is called the **latent heat of fusion** of the substance. In Fig. 5.4 this heat is simply the amount of heat necessary to go from A to B.

When point B has been reached, the substance is all liquid. The temperature of the substance at which this change from solid to liquid takes place is known as the **melting point.** After point B, further heating again causes a rise in temperature. The temperature continues to rise as heat is added until point C is reached.

From B to C the substance is in the liquid phase. When point C is reached, adding more heat does not change the temperature. The added heat now goes into changing the liquid into a gas. The amount of heat necessary to change one kilogram of a liquid into a gas is called the **latent heat of vaporization** of the substance. In Fig. 5.4 this heat is simply the amount of heat necessary to go from C to D.

When point D has been reached, the substance is all in the gas phase. The temperature of the substance at which this change from liquid to gas phase occurs is known as the **boiling point.** After point D is reached, further heating again causes a rise in the temperature.

In some instances a substance can change directly from the solid phase to the gaseous phase. This change is called **sublimation.** Examples are dry ice (solid CO_2), mothballs, and solid air fresheners.

From Fig. 5.4 we get a better understanding of the

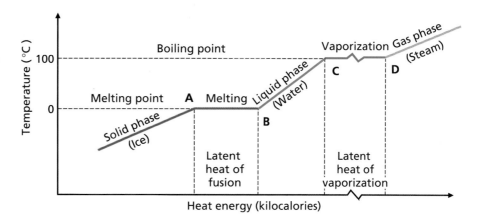

Figure 5.4 Graph of temperature versus heat energy for a typical substance.
If the typical substance were water, the solid, liquid, and gas phases would be ice, water, and steam as indicated. Notice that during phase changes heat is added but the temperature does not change. See text for detailed description.

difference between temperature and heat. We see that the temperature of a body rises as heat is added only if the substance is not undergoing a change in phase.

When heat is added to ice at 0°C, the ice melts without changing its temperature. The more ice there is, the more heat is needed to melt it. In general, the heat needed to change a solid into a liquid is just the mass of the substance times its heat of fusion. So we can write

$$\text{heat needed to melt a substance} = \text{mass} \times \text{latent heat of fusion}$$

or
$$H = mH_f \qquad (5.4)$$

Similarly, if we want to change a liquid into a gas at the boiling point, the amount of heat needed can be written as

$$\text{heat needed to boil a substance} = \text{mass} \times \text{latent heat of vaporization}$$

or
$$H = mH_v \qquad (5.5)$$

For water, the latent heats are

$$H_f = 80 \text{ kcal/kg} = 335{,}000 \text{ J/kg}$$
$$H_v = 540 \text{ kcal/kg} = 2.26 \times 10^6 \text{ J/kg}$$

These values are very large numbers compared with the specific heat. Notice that it takes 80 times more heat energy to melt 1 kg of ice than to raise the temperature of 1 kg of water by 1°C. Similarly, boiling 1 kg of water takes 540 times as much energy as raising its temperature by 1°C. Also notice that it takes almost 7 times as much

energy to change a kilogram of water to steam than it takes to change a kilogram of ice to water.

QUESTION

Why does it take so much more energy to change water to steam than to change ice to water?

Answer More work must be done in the liquid-gas phase change. In the conversion of liquid water to steam the intermolecular bonds holding the molecules together as a liquid are broken, and the molecules are separated by relatively large distances.

EXAMPLE 3

Calculate the amount of heat necessary to change 0.200 kg of ice at 0°C into water at 2°C.

Solution The total heat necessary is found in two steps. The ice melts at 0°C, and then it warms up from 0°C to 2°C.

$$H = H_{\text{melt ice}} + H_{\text{change } T}$$
$$= mH_f + mc\, \Delta T$$
$$= 0.200 \text{ kg} \times 80 \text{ kcal/kg}$$
$$\qquad + 0.200 \text{ kg} \times 1 \text{ kcal/kg·°C} \times (2 - 0)\text{°C}$$
$$= 16.4 \text{ kcal}$$

Changing phase is a two-way street. That is, heat may be removed from a substance to change its phase. For example, 540 kcal of heat would have to be removed from a kilogram of steam at 100°C to change it to water at 100°C. Just as 540 kcal is added to convert a kilogram of water to steam, 540 kcal is given up in the reverse process. For this reason burns from steam are generally more severe than those from boiling water. When the steam condenses, it gives up 540 kcal/kg of latent or "hidden" heat.

This process is simply conservation of energy, which applies to all forms of energy, including heat energy. The conservation of energy is a useful concept in dealing with heat transfer. For instance, in Example 3 the energy to melt the ice must have come from somewhere. Suppose that the piece of ice had been put into warm water and that this water and the water from the melted ice had a final temperature of 2°C. Then by the conservation of energy,

heat lost by warm water = heat to melt ice
+ heat to warm water

Suppose the warm water had a mass of 0.82 kg and an initial temperature of 22°C. Then at a final temperature of 2°C, the heat lost by the water is

$$H = mc \, \Delta T = mc(T_f - T_i)$$
$$= (0.82 \text{ kg})(1.0 \text{ kcal/kg·°C})(2°C - 22°C)$$
$$= -16.4 \text{ kcal}$$

The minus sign indicates that heat was lost (initial temperature greater than final temperature). Notice that this result is the same amount of heat gained by the ice and the water from the melted ice in Example 3.

Another example of the use of the conservation of energy is in the determination of the specific heat of an unknown substance. If a hot metal pellet is dropped into cool water in a glass, we can say, assuming that no heat is lost to the surrounding air,

heat lost by metal = heat gained by water
+ heat gained by glass

If the masses and the initial and final temperatures are all known, the specific heat of the metal can be found.

EXAMPLE 4

A 0.045-kg aluminum pellet heated to 98.5°C is dropped into an 0.080-kg glass beaker ($c = 0.20$ kcal/kg·°C) containing 0.200 kg of water at 20°C. The final maximum equilibrium temperature is 23.5°C. Determine the specific heat of aluminum.

Solution

$$H_{\text{aluminum}} = H_{\text{water}} + H_{\text{glass}}$$

or $\quad (mc \, \Delta T)_{\text{aluminum}} = (mc \, \Delta T)_{\text{water}} + (mc \, \Delta T)_{\text{glass}}$

$$0.045 \text{ kg} \times c_{\text{Al}} \times (98.5°C - 23.5°C)$$
$$= 0.200 \text{ kg} \times 1 \text{ kcal/kg·°C} \times (23.5°C - 20.0°C)$$
$$+ 0.080 \text{ kg} \times 0.20 \text{ kcal/kg·°C} \times (23.5°C - 20.0°C)$$

Since everything is in the proper units, we can now solve for c, the specific heat of aluminum, and its units will be kcal/kg·°C. Solving yields

$$c = 0.22 \text{ kcal/kg·°C}$$

5.4 Thermodynamics

Thermodynamics means the dynamics of heat—that is, the production, flow, and conversion of heat into work. We use heat energy, either directly or indirectly, to do most of the work that is done in everyday life. The operation of heat engines, such as internal-combustion engines, and of refrigerators is based on the laws of thermodynamics. These laws are important because they give relations between heat energy, work, and the directions that thermodynamics may occur.

Because one aspect of thermodynamics is concerned with heat-energy transfer, accounting for the energy involved in a thermodynamic process is important. This accounting is done by the principle of the conservation of energy, which, as you may recall, is that energy can neither be created nor destroyed. The first law of thermodynamics is simply the principle of conservation of energy applied to thermodynamic processes. For example, consider heating a balloon. As energy is added to the system (the balloon and the air inside), the temperature increases and the balloon expands. The temperature of the air inside the balloon increases because some of the heat goes into the internal energy of the air. Some of the energy also goes into doing the work of expanding the balloon. Keeping account of the energy, for the **first law of thermodynamics,** we may write, in general,

heat added to the system	=	increase in internal energy of the system	+	work done by the system

or $\qquad H = \Delta E_i + W \qquad$ (5.6)

According to this mathematical statement of the first law of thermodynamics:

> **Heat added to a closed system goes into the internal energy of the system and/or doing work.**

Internal energy will be considered in more detail in a later section. To help in understanding this concept now, think of adding heat to a gas in a rigid container. In this case the energy all goes into the internal energy of the system and increasing the temperature of the gas. No work is done ($W = 0$), and by the first law, $H = \Delta E_i$.

QUESTION

Suppose a balloon is put into a refrigerator freezer and it shrinks. How does the first law of thermodynamics apply here?

Answer The energy-conserving first law applies whether heat is added to or removed from a system. Here the heat removed is represented by a negative quantity ($-H$). Because the temperature of the air in the balloon decreases. so does its internal energy ($-\Delta E_i$). Also, if the work done by the balloon is positive when it expands, then when it shrinks, the work is negative ($-W$). So the balloon system loses internal energy and work is done on it (by the pressure of the atmosphere).

Another good example of the first law is the heat engine. A **heat engine** is a device that converts heat energy into work. There are many different types of heat engines: gasoline engines on lawn mowers and in cars, diesel engines in trucks, and steam engines in old-time locomotives. They all operate on the same principle. Heat input—for example, from the combustion of fuel—and some of the energy, but not all, goes into doing useful work.

In thermodynamics we are not concerned with the components of an engine but, rather, its general operation. We may represent a heat engine schematically as illustrated in Fig. 5.5. A heat engine operates between, or there is energy transfer from, a high-temperature reservoir and a low-temperature reservoir. These reservoirs are systems from which heat may be readily absorbed and to which heat may be readily expelled. In the process the engine uses some of the input energy to do work. Notice that the widths of the heat and work paths in the figure are in keeping with the conservation of energy:

$$\text{heat in} = \text{work} + \text{heat out,}$$

or

$$\text{work} = \text{heat in} - \text{heat out}$$

Normally, a heat engine operates in a cycle for continuous energy output.

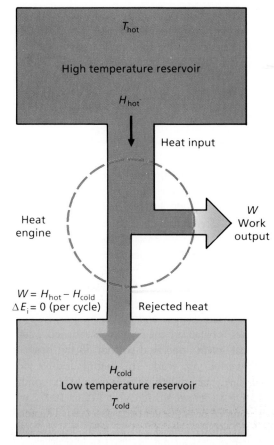

Figure 5.5 Schematic diagram of a heat engine. A heat engine takes heat, H_{hot}, from a high-temperature reservoir, converts some of it to useful work, W, and rejects the remainder, H_{cold}, to a low-temperature reservoir.

In a cyclical heat engine the system comes back to its original state. Thus the temperature and internal energy of the system are unchanged. Since $\Delta E_i = 0$, the first law of thermodynamics (Eq. 5.6) then becomes

$$H = \Delta E_i + W$$
$$= W$$

Because the net heat added to the system is the heat input from the high-temperature reservoir (H_{hot}) minus the heat rejected to the low-temperature reservoir (H_{cold}), we can write this equation as

$$H_{hot} - H_{cold} = W$$

or

$$H_{hot} = H_{cold} + W \qquad (5.7)$$

The conversion of heat energy into work is expressed in terms of thermal efficiency. Similar to mechanical efficiency, it is a ratio of the work output and the energy input. That is,

$$\textbf{thermal efficiency} = \frac{\text{work output}}{\text{heat input}} \times 100\%$$

or

$$\text{eff}_{\text{th}} = \frac{W}{H_{\text{hot}}} \times 100\%$$

For example, if a heat engine absorbs 1000 J each cycle and rejects 400 J or does 600 J of work, then it has an efficiency of 600/1000 = 0.60, or 60%. This efficiency is quite high. The efficiency of an automobile is on the order of 15%. That is, 85% of the energy from fuel combustion is wasted or goes into doing nonessential work not associated with moving the car, for example, running a tape player.

Thus the first law of thermodynamics is concerned with the conservation of energy. As long as the energy check sheet is balanced, the first law is satisfied. Suppose that a heat engine operated so that *all* the heat input were converted into work. This engine doesn't violate the first law, but something is wrong. This condition would give a thermal efficiency of 1.0, or 100%, which doesn't happen or has never been observed. The **second law of thermodynamics** tells us what can and cannot happen thermodynamically and can be stated in several ways. One of these statements applies to this situation:

> **No heat engine operating in a cycle can convert heat energy completely into work.**

Another way of saying this is that no heat engine operating in a cycle can have 100 percent efficiency.

A heat engine must lose some heat, but what is the best or maximum efficiency that can be obtained? We know it isn't 100 percent. It can be shown that the maximum or ideal efficiency is determined only by the high and low temperatures of the reservoirs and nothing else. This ideal efficiency is given by

$$\text{ideal eff} = \frac{T_{\text{hot}} - T_{\text{cold}}}{T_{\text{hot}}} \times 100\%$$

$$= 1 - \frac{T_{\text{cold}}}{T_{\text{hot}}} \times 100\%$$

where T_{hot} and T_{cold} are the *absolute* temperatures of the high-temperature reservoir and the low-temperature reservoir, respectively.

Don't forget that the temperatures must be expressed as absolute temperatures on the Kelvin scale.

The actual efficiency of a heat engine will always be less than its ideal efficiency. The ideal efficiency sets an upper limit, but this limit can never be achieved.

EXAMPLE 5

What is the ideal efficiency of a coal-fired steam power plant operating between temperatures of 300°C and 100°C?

Solution First, use Eq. 5.1 ($T_K = T_C + 273$) to convert the temperatures to their Kelvin equivalents:

$$T_{\text{hot}} = 300 + 273 = 573 \text{ K}$$

and

$$T_{\text{cold}} = 100 + 273 = 373 \text{ K}$$

Then

$$\text{ideal eff} = 1 - \frac{T_{\text{cold}}}{T_{\text{hot}}} \times 100\%$$

$$= 1 - \frac{373 \text{ K}}{573 \text{ K}} \times 100\% = 35\%$$

When other losses are taken into account, an actual efficiency of 32% or so is typical for a power plant.

Notice from Eq. 5.9 that the second law forbids a cold reservoir of $T_{\text{cold}} = 0$ K. If absolute zero could be used, then we could have an ideal efficiency of 100 percent, which would violate the second law. Actually, a temperature of absolute zero cannot be attained. That is, thermodynamically,

> **it is impossible to obtain a temperature of absolute zero.**

This result is sometimes called the **third law of thermodynamics.**

Scientists have tried to reach absolute zero and have come within one-millionth of a degree. However, the third law has still never been violated experimentally. It becomes more difficult to lower the temperature of a material (pump heat from it) the closer the temperature gets to absolute zero. The difficulty increases with each step, to the point where an infinite amount of work would be required to reach the very bottom of the temperature scale.

Here is another statement of the second law:

> **It is impossible for heat to flow spontaneously from a colder body to a hotter body.**

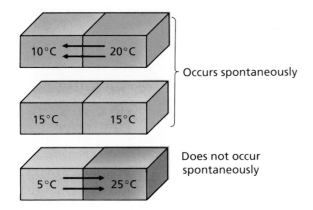

Occurs spontaneously

Does not occur spontaneously

Figure 5.6 Heat flow.
When in thermal contact, heat flows spontaneously from a hotter object to a colder object until they are at the same temperature—i.e., in thermal equilibrium. Heat never flows spontaneously from a colder object to a hotter object. That is, a cold object never gets colder when placed in thermal contact with a warmer object.

This phenomenon is well known. If a cold object and a hot object are placed in contact, the hot object cools down and the cold object warms up (Fig. 5.6). The reverse never happens, even though there's nothing in the first law that indicates that it could not happen. If heat did flow spontaneously from a colder object to a hotter one, it would be like a ball rolling up a hill to a higher potential on its own accord. In all of our experience, the second law seems to hold.

Of course, we can get a ball to roll up a hill by applying a force and doing work on it. Similarly, we can get heat to flow up the "temperature hill," so to speak, by doing work, which is the principle of a **heat pump.** The schematic diagram of a heat pump is shown in Fig. 5.7. Work input is required to "pump" heat energy from a low-temperature reservoir to a high-temperature reservoir. Essentially, it is the reverse of a heat engine.

Refrigerators and air conditioners are examples of heat pumps. Heat is transferred from the inside volume of a refrigerator to the outside through the compressor doing work on a gas (and the expenditure of electrical energy). The heat transferred to the room (high-temperature reservoir) is much greater than the electrical energy used to perform the work. Similarly, an air conditioner transfers heat from the inside of a home or car to the outside.

The heat pumps used for home heating and cooling have a descriptive name. In the summer they operate as air conditioners. In the winter heat is extracted from the outside air or from a water source and pumped inside the home for heating. Heat pumps are used extensively in the South, where the climate is milder. In places with very cold winter months an auxiliary heating unit (usually an electric heater) must be used to supply extra heat when needed. A heat pump is generally more expensive than a regular furnace, but there are long-term advantages. There are no fuel costs associated with a heat pump (other than the cost of the electricity to supply the work input), because it takes heat from the air or from water in a reservoir or well.

The second law can also be expressed in terms of entropy. **Entropy** is a measure of the disorder of a system. When heat is added to an object, its entropy increases because the added energy increases the disordered motion of the molecules. As a natural process takes place, the disorder increases. For example, when a solid melts, the molecules are freer to move in a random motion in the liquid phase than in the solid phase. Likewise, when evaporation takes place, there is greater disorder and an increase in entropy.

In terms of entropy, the second law of thermodynamics can be stated as:

> **The entropy of an isolated system never decreases.**

Processes that are left to themselves tend to become more and more disordered, never the reverse. A student's dormitory room naturally becomes disordered . . . never the reverse. Of course, the room can be cleaned and items put in order, and the entropy *of the room system* decreases. But to put things back in order, someone must expend energy, with a greater entropy increase than the room's entropy decrease. We sometimes say that *the total entropy of the universe increases in every natural process.*

Here's another long-term implication of the second law of thermodynamics: Heat naturally flows from a region of higher temperature to one of lower temperature. In terms of order, heat energy is more "orderly" when it is concentrated. When transferred naturally to a region of lower temperature, it is "spread out" or more "disorderly," and the entropy increases. Hence the uni-

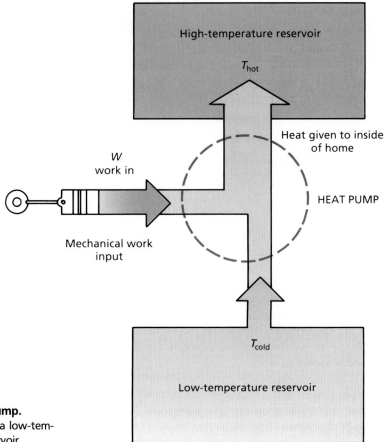

Figure 5.7 Schematic diagram of a heat pump.
Work input is required to "pump" heat from a low-temperature reservoir to a high-temperature reservoir.

verse—the stars and galaxies—should eventually cool down to a final common temperature when the entropy of the universe has reached maximum. This possible fate of the universe, billions of years from now, is sometimes referred to as its "heat death."

5.5 Heat Transfer

Because heat is transferred energy, how the transfer is done is an important consideration. Heat transfer is accomplished by three methods: conduction, convection, and radiation.

Conduction is the transfer of heat by molecular activity. The kinetic energy of molecules is transferred from one molecule to another through collision. How well a substance conducts heat depends on the molecular bonding. Solids are generally the best thermal conduc-

tors, with metals being very good conductors. In addition to molecular collisions, there are a large number of "free" electrons (not permanently bound) in a metal that can move around. These electrons contribute significantly to heat transfer, or thermal conductivity. The **thermal conductivity** of a substance is a measure of its ability to conduct heat. As listed in Table 5.3, metals have relatively high thermal conductivities.

Liquids and gases are in general relatively poor thermal conductors. Liquids are better than gases because their molecules are closer together and collide more often. Gases are poor conductors because their molecules are relatively far apart and conductive collisions do not occur as often. Substances that are poor thermal conductors are sometimes referred to as **thermal insulators.**

We make cooking pots and pans out of metals, so heat is readily conducted to the foods inside. Pot holders, on the other hand, are made out of cloth, a poor thermal

Table 5.3 Thermal Conductivities of Some Common Substances

Substance	W/°C·m*
Silver	425
Copper	390
Iron	80
Brick	3.5
Floor tile	0.7
Water	0.6
Glass	0.4
Wood	0.2
Cotton	0.08
Styrofoam	0.033
Air	0.026
Vacuum	0

* Note that W/°C = (J/s)/°C is the rate of heat flow per temperature difference $(\Delta H/\Delta t)/\Delta T$, where W represents watt. The length unit (m) arises from dimensional considerations of the conductor (area and thickness).

conductor, for obvious reasons (Fig. 5.8). Many solids, such as cloth, wood, and plastic foam (Styrofoam), are porous and have large numbers of air (gas) spaces that add to their poor conductivity. For example, thermal underwear and Styrofoam coolers depend on this property, as does fiberglass insulation used in the walls and attics of our homes.

Figure 5.8 Thermal insulator.
Potholders are made of cloth, a poor thermal conductor, thus preventing heat from being conducted to the hand, causing a burn.

The transfer of heat by **convection** requires the movement of a substance, or mass, from one position to another. The movement of air or water is an example of heat transfer by convection.

Most homes are heated by convection (movement of hot air). The air is heated at the furnace, then circulated throughout the house by way of metal ducts. When the air has "lost its heat," it passes through a cold-air return on its way back to the furnace to be reheated and recirculated. See Fig. 5.9.

The warm-air vents in a room are usually in the floor, as are the cold-air return ducts, but on opposite sides of a room. The warm air entering the room rises (being "lighter," or less dense, and therefore buoyant). As a result, cooler air is forced toward the floor, and convection cycles are set up in the room that promote even heating. Some of the cooler air returns to the furnace for reheating and recirculation. The transfer of heat within a region or volume by means of convection currents as set up in a room is similar to the way heat is distributed in the Earth's atmosphere (Chapter 20).

The transfer of heat by convection and conduction requires a material medium for the process to take place. Heat from the Sun reaches the Earth via electromagnetic waves. The process of transferring heat energy through space by means of electromagnetic waves is known as **radiation.** Electromagnetic waves carry energy and can travel through a vacuum. The heat we get from the Sun is transmitted through the vacuum of space by radiation.

Another example of heat transfer by radiation is heat from an open fire or a fireplace. We can readily feel the warmth of the fire on exposed skin. Yet the air is a poor conductor; moreover, the air warmed by the fire is rising (up the chimney in a fireplace). Therefore, the only mechanism for appreciable heat transfer is radiation.

In general, we find that dark objects are good absorbers of radiation, while light-colored objects are poor absorbers and good reflectors. For this reason, we commonly wear light-colored clothing in the summer so as to be cooler. In the winter we generally wear dark-colored clothes to take advantage of the absorption of the reduced solar radiation.

An application using all three methods of heat transfer is the Thermos bottle, which is used to keep liquids either hot or cold (Fig. 5.10). Actually, a knowledge of the methods of heat transfer is used to *prevent* the transfer of heat is this case. The sealed, double-walled glass bottle is partially evacuated. Glass is a relatively poor conductor, and any heat conducted through a wall (from the outside in or the inside out) will find the partial vacuum an even greater thermal insulator. Also, heat

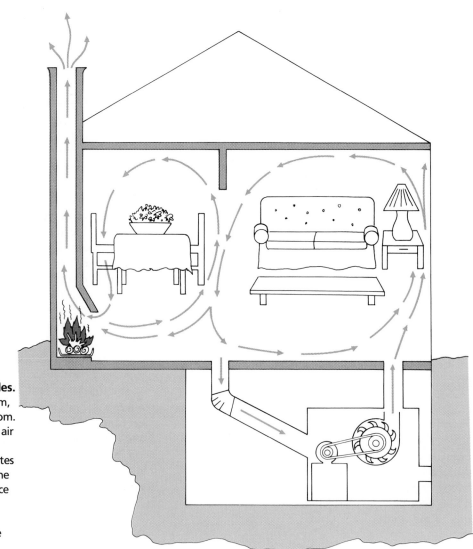

Figure 5.9 Convection cycles.
In a forced-air heating system, warm air is blown into a room. The warm air rises, the cold air descends, and a convection cycle is set up, which promotes heat distribution. Some of the cold air returns to the furnace for heating. Notice how a great deal of heat energy is lost up the chimney of a fire place.

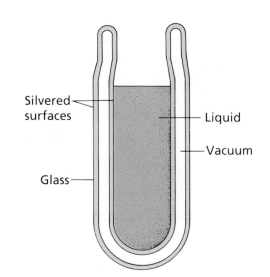

Silvered surfaces

Glass

Liquid

Vacuum

Figure 5.10 A vacuum bottle.
In a vacuum Thermos bottle, glass is a poor conductor, the partial vacuum between the glass walls prevents convection cycles therein, and the silvered surfaces prevents radiation losses by reflection. These design elements serve to keep hot foods hot and cold foods cold.

cannot be transferred from one glass wall to the other by convection. Finally, the inner surfaces of the glass bottle are silvered to prevent heat transfer by radiation. Thus hot coffee or a cold drink in the bottle remains hot or cold for some time.

5.6 Phases of Matter

As we saw in Section 5.3, the addition of heat can cause a substance to change phase. The three common phases of matter are the solid phase, the liquid phase, and the gaseous phase. All substances exist in each of the three phases of matter at some temperature and pressure. At normal room temperature and atmospheric pressure a substance will be in one of the three phases. For instance, at room temperature oxygen is a gas, water a liquid, and copper a solid.

The principal distinguishing features of solids, liquids, and gases can be understood if we look at the various phases from a molecular point of view. Most substances are made up of very small particles called molecules. A molecule is the smallest division of a substance that still retains the properties of the substance.

In a pure **solid** the molecules are usually arranged in a specified three-dimensional manner. This orderly arrangement of molecules is called a **lattice.** Figure 5.11 shows an example of a lattice in only two dimensions. The molecules (represented in the figure by small circles) are bound to each other by electric forces.

Upon heating, the molecules gain kinetic energy and vibrate about their positions in the lattice. The more heat that is added, the stronger the vibrations become. The molecules move farther apart, and as shown diagrammatically by Fig. 5.11, the solid expands.

When the melting point of a solid is reached, the heat breaks apart the bonds that hold the molecules in place (the heat of fusion, 80 kcal/kg for water). As bonds break, holes are produced in the lattice, and nearby molecules can move toward the holes. As more and more holes are produced, the lattice becomes significantly distorted.

Figure 5.12 shows an arrangement of the molecules in a liquid. There are many holes in the liquid, which has only a slight lattice structure. A molecule can easily move to a new spot, because there are so many holes. A **liquid** is an arrangement of molecules with many holes, so that the molecules are free to move and assume the shape of the container. Upon heating of a liquid, the individual molecules gain kinetic energy, and even more holes are produced. The result is that the liquid expands.

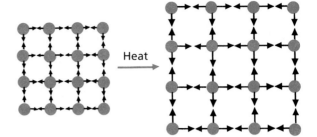

Figure 5.11 Crystalline lattice.
A schematic diagram of a crystal lattice in two dimensions. Heating causes the molecules to vibrate with greater amplitudes in the lattice, thus increasing the volume of the solid.

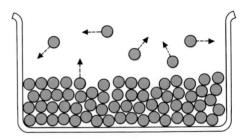

Figure 5.12 Liquid molecules.
A diagram illustrating the arrangement of molecules (small circles) in a liquid. The molecules are packed closely together and form only a slight lattice structure. Some molecules may acquire enough energy to break free of the liquid. This is called vaporization.

When the boiling point is reached, the heat energy is sufficient to break the molecules completely apart from each other. The heat of vaporization is the heat per kilogram necessary to free the molecules completely from each other. Because the electric forces holding the different molecules together are quite strong, the heat of vaporization is fairly large. When the molecules are completely free from each other, the gaseous phase is reached.

A **gas** is made up of molecules that exert little or no force on one another except when they collide. The distance between molecules in a gas is quite large compared with the size of the molecules. The molecules in a gas are moving rapidly. As a result, a volume of gas has no definite shape but assumes the size and shape of its container.

Continued heating of a gas causes the molecules to move faster and faster. Eventually, the molecules and atoms are ripped apart by collisions with one another. Inside hot stars, such as our Sun, atoms and molecules do not exist, and another phase of matter, called a **plasma,** (no relationship to blood plasma) occurs. A plasma is a gas of electrically charged particles. On Earth we have plasmas in gas discharge tubes, such as fluorescent and neon lamps.

We have seen that when a substance is in a single phase, heating increases the kinetic energy of the molecules of which it is composed. When a substance is changing phases, heating supplies the energy to overcome the attractive forces holding the different molecules together. Since the temperature rises only when the substance is not changing phase, we conclude that *temperature can be defined as a measure of the average kinetic energy of the molecules.*

5.7 The Gas Laws and Kinetic Theory

We can look at the gaseous phase of matter to learn about thermodynamic processes and molecular kinetic theory. Molecular theory views the molecules of a gas to be moving at high speeds. They collide with each other *and* the walls of the container (Fig. 5.13a). Each collision with a wall exerts a tiny force on the wall. However, the frequent collisions of billions of gas molecules exert a fairly steady average force, or pressure, on the wall. **Pressure** is defined as the force per unit area ($p = F/A$). The SI units of pressure are N/m^2, which is called a pascal (Pa).

The pressure of a given quantity of gas depends on its volume and temperature. Suppose the temperature of a gas is kept constant and its volume is decreased by one-half (Fig. 5.13b). Such a process, in which the temperature remains constant, is called an **isothermal process.** As shown in Fig. 5.13, the pressure increases— and, in fact, doubles. Because the gas occupies half the space, it makes twice as many collisions with the container walls, and hence it exerts twice the pressure. This relationship between pressure and volume (V) may be expressed as

$$p_1 V_1 = p_2 V_2 \qquad (5.10)$$

Note: if $V_2 = V_1/2$, then $p_2 = (V_1/V_2)p_1 = 2p_1$. This expression is known as **Boyle's law,** after Robert Boyle (1627–1691), an English chemist who discovered it. The law applies for all gases at normal pressures.

Using kinetic theory, one can show that the product

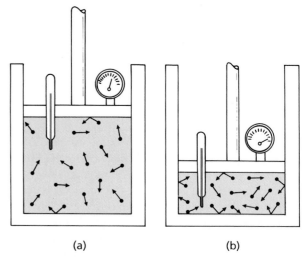

of pV is proportional to the average kinetic energy of the gas molecules:

$$pV \propto \text{average kinetic energy of gas molecules}$$

The greater the kinetic energy of the molecules, the faster they move and the greater number of collisions they have with the walls. An increased pressure results. Also, the greater the molecular energy, the greater is the ability of the molecules to do work and to increase the volume of the container (if not rigid).

The temperature (T) of a gas is a relative indication of the internal energy of a gas, so we may write

$$pV \propto T$$

For dilute gases $pV/T = $ constant, so

$$\frac{p_1 V_1}{T_1} = \frac{p_2 V_2}{T_2} \qquad (5.11)$$

where the subscripts indicate the values at different times. This relationship is known as the **perfect, or ideal, gas law.** Note that the temperature in this equation is the absolute or Kelvin temperature.

An ideal or perfect gas is a theoretical system in which there are only collision interactions between the gas molecules. The perfect gas law holds for real dilute gases over normal temperature ranges. The reason a gas is theoretically perfect or ideal will be discussed shortly. Let's first examine the perfect gas law in terms of the

Figure 5.13 Isothermal (constant-temperature) process. If the volume of a quantity of gas is decreased by one-half while keeping the temperature constant, the pressure of the gas doubles.

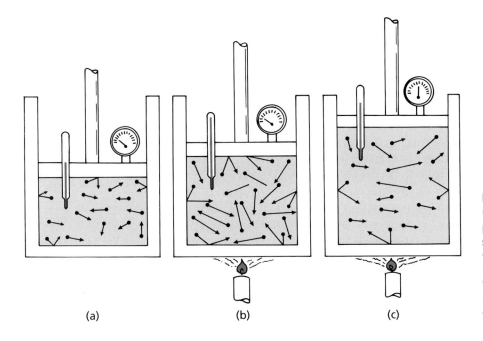

Figure 5.14 Gas processes.
(a) → (b) Isobaric process. The pressure of the gas is held constant, but the temperature and volume change in the process. (a) → (c) General process. When heat is added to a gas, its temperature increases as well as its pressure and volume.

(a)   (b)   (c)

examples illustrated in Fig. 5.14. The first is an **isobaric process,** one in which the pressure is constant. The other variables, temperature and volume, change in the process.

EXAMPLE 6

In an isobaric, or constant-pressure, process a quantity of gas in a cylinder initially at a volume of 0.25 m³ and a temperature of 20°C is heated to a temperature of 70°C. What is the final volume of the gas? See Figs. 5.14a and 5.14b.

Solution Since the process is isobaric, $p_1 = p_2$, and by the gas law,

$$\frac{V_1}{T_1} = \frac{V_2}{T_2} \qquad (5.12)$$

But don't forget that the temperatures are absolute temperatures, so we change to Kelvin units:

$$T_1 = T_c + 273 = 20 + 273 = 293 \text{ K}$$

and $\quad T_2 = T_c + 273 = 70 + 273 = 343 \text{ K}$

Then $\quad V_2 = \left(\dfrac{T_2}{T_1}\right) V_1 = \left(\dfrac{343 \text{ K}}{293 \text{ K}}\right)(0.25 \text{ m}^3) = 0.29 \text{ m}^3$

The relationship in Eq. 5.12 is known as **Charles' law,** after the French physicist who discovered it in the eighteenth century.

In general, when heat is added to a perfect gas and its temperature increases, both its pressure and volume increase (as in Fig. 5.14a directly to 5.14c). In this case Eq. 5.11 applies.

EXAMPLE 7

An atmospheric balloon, when launched, has a volume of 4000 m³, a pressure of 1.0×10^5 N/m², and a temperature of 27°C. What is the balloon's volume when it reaches an altitude where its pressure is 2.0×10^4 N/m² and the temperature is -33°C?

Solution We use the perfect gas law, with $T_1 = 27$°C $+ 273 = 300$ K and $T_2 = -33$°C $+ 273 = 240$ K, and solve for V_2.

$$
\begin{aligned}
V_2 &= \left(\frac{p_1}{p_2}\right)\left(\frac{T_2}{T_1}\right) V_1 \\
&= \left(\frac{1.0 \times 10^5 \text{ N/m}^2}{2.0 \times 10^4 \text{ N/m}^2}\right)\left(\frac{240 \text{ K}}{300 \text{ K}}\right)(4000 \text{ m}^3) \\
&= 16{,}000 \text{ m}^3
\end{aligned}
$$

Thus we see that the balloon expands to four times its original volume.

The perfect gas law allows another *iso-* process. When a quantity of gas is in a rigid container, its volume doesn't change with a change in temperature and pressure. This process is called an *isovolumetric process* or, more commonly, an **isometric process.** In this case

$V_1 = V_2$, and Eq. 5.11 becomes

$$\frac{p_1}{T_1} = \frac{p_2}{T_2} \qquad (5.13)$$

If the ratio p/T is the same (constant) at any time, we may write $p/T = k$, where k is a constant.

Hence for an isometric process $p = kT$. Because the pressure varies directly with temperature, we have another way to measure temperature: with a constant-volume gas thermometer (Fig. 5.15a). Dilute, real gases follow the $p = kT$ relationship. So once the gas thermometer is calibrated on a p-versus-T graph, the temperature may be determined directly from the pressure reading of the gas thermometer.

Of course, at very low temperature real gases liquify, so the temperature range of a gas thermometer is limited. However, if the straight line of the p-versus-T graph is extrapolated to zero pressure, we obtain a value for zero or absolute temperature (Fig. 5.15b). This value turns out to be $-273°C$, which is taken to be 0 K on the Kelvin scale.

According to the kinetic theory, a gas is made up of a very large number of molecules moving at very high speeds (on the average, 1000 mi/h). The molecules continually collide with each other (about 1 billion collisions each second!), as well as bombard nearby or confining surfaces. When they collide with a surface, a force is exerted, and the macroscopic average effect is expressed as pressure, or force per unit area.

The absolute temperature of a gas is directly proportional to the average kinetic energy of its molecules. In a real gas there are intermolecular forces, and hence some of the internal energy of the gas is potential energy. However, in a perfect gas there is only interaction by molecular collision, and so absolute temperature is a relative measure of the total internal energy. (A perfect gas is perfect in the sense that it would remain a gas at any temperature, because there are no intermolecular forces to become dominant and liquify the gas at very low temperatures when the kinetic energy of the molecules becomes small.)

Because the kinetic energy of a perfect gas is its internal energy, it follows that when the temperature of a perfect or ideal gas is doubled, its internal energy is doubled. For example, to double the internal energy of a perfect gas at 200 K, the temperature of the gas would have to be raised to 400 K (with constant volume, so no work would be done; recall the first law). This relation is only true on the absolute-temperature scale. What would it mean to double the temperature of a gas at 0°C or 0°F?

EXAMPLE 8

A quantity of perfect gas in a rigid container at room temperature (20°C) is heated so that the internal energy of the gas is doubled. What is the final temperature of the gas, in degrees Celsius?

Solution The internal energy of the gas is doubled when the *absolute* temperature is doubled, so the initial absolute temperature is

$$T_1 = T_C + 273 = 20° + 273 = 293 \text{ K}$$

Then doubling this temperature gives

$$T_2 = 2 \times T_1 = 2 \times 293 \text{ K} = 586 \text{ K}$$

Converting back to Celsius yields

$$T_C = T_K - 273 = 586 - 273 = 313°C$$

Figure 5.15 Constant-volume gas thermometer.
(a) The volume of a quantity of gas in a rigid container is constant and the pressure is directly proportional to its temperature ($p = kT$). (b) By plotting the pressure versus temperature, the temperature may then be determined directly from the pressure reading of the gas thermometer. At low temperatures, gases do not follow straight-line relationships, but by extending the line to the axis (0 pressure), a value may be obtained for the absolute zero of temperature.

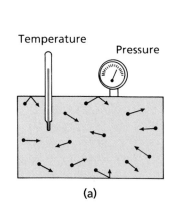

(a)

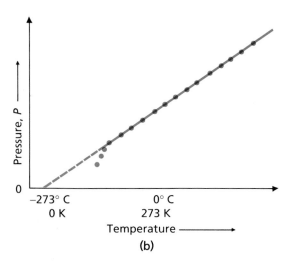

(b)

Keep in mind that not all of the molecules of a gas are moving at the same speed. For this reason, we define absolute temperature as a measure of the *average* kinetic energy of the molecules. Some molecules are moving faster and some are moving slower than the average, and the speeds change with the transfer of energy on collision. A liquid-in-glass thermometer measures the temperature of a gas or other substance by measuring the energy transferred to the thermometer bulb by molecular collisions.

When water (or any liquid) evaporates, the evaporation is due to the escape of more energetic molecules. Evaporation may be enhanced by heating the water, and at the boiling point all the added energy goes into changing the phase rather than the temperature. Pockets of energetic molecules form gas bubbles. When the pressure due to the molecular activity in the bubbles is great enough, they rise and break the surface of the liquid. We then say the liquid is boiling.

Evaporation is a relatively slow phase change of a liquid to a gas, and it is a major cooling mechanism of our bodies. When hot, we perspire, and the evaporation of perspiration is a cooling effect, because energy is lost. This cooling effect is quite noticeable on the bare skin when one gets out of a bath or shower or rubs down with alcohol.

The comforting evaporation of perspiration is promoted by moving air. When we are sweaty and standing in front of a blowing electric fan, we often say that the air is "cool." But the air is the same temperature as the other air in the room. The motion of the air promotes evaporation by carrying away molecules (and energy), and there is a cooling effect.

On the other hand, air can hold only so much moisture at a given temperature. The amount of moisture in the air is commonly expressed in terms of relative humidity (Chapter 20). When it is quite humid, there is little evaporation of perspiration and we feel hot.

Learning Objectives

After reading and studying this chapter, you should be able to do the following without referring to the text:

1. Explain the difference between temperature and heat.
2. Explain how a temperature scale is constructed with reference to the three major scales.
3. Convert from one temperature scale to another, using the appropriate formulas.
4. Explain the units of heat and the food Calorie.
5. Compute the amount of heat necessary to melt ice, boil water, or change the temperature of water, knowing how much mass is involved.
6. State and explain the three laws of thermodynamics.
7. Distinguish between heat engines and heat pumps, and compute the ideal efficiency of a heat engine.
8. Describe three methods of heat transfer, and give an example of each.
9. Discuss the molecular structure of the solid, liquid, and gas phases of matter.
10. Use the perfect gas law and explain the properties of gases in terms of molecular kinetic theory.

Important Words and Terms

temperature	heat	heat engine	solid
thermometer	calorie	thermal efficiency	lattice
thermal expansion	kilocalorie	second law of thermodynamics	liquid
ice point	Btu	third law of thermodynamics	gas
steam point	specific heat	heat pump	pressure
Fahrenheit scale	latent heat of fusion	entropy	isothermal process
degree Fahrenheit	melting point	conduction	Boyle's law
Celsius scale	latent heat of vaporization	thermal conductivity	perfect gas law
degree Celsius	boiling point	thermal insulators	isobaric process
Kelvin scale	sublimation	convection	Charles' law
kelvin	first law of thermodynamics	radiation	isometric process

Questions

Temperature

1. Why is temperature a *relative* measurement?

2. On what property does temperature measurement commonly depend, and how is it applied?

3. What are the ice points and steam points on three temperature scales: Celsius, Fahrenheit, and Kelvin?

Heat

4. Explain what is meant by the statement, "Heat is energy in transit."

5. What is the difference between a calorie and a Calorie?

6. What is the British system unit of heat, and how does it compare in magnitude with metric units?

7. Why do bridges and sidewalks have cracks built into them?

Specific Heat and Latent Heat

8. What is *specific* about specific heat?

9. Water is a common heat storage medium in solar heating. Why?

10. Why does it take a long time for lakes to freeze when the air temperature is below freezing?

11. When does the addition of heat *not* result in a temperature increase?

12. Compare the units of specific heat and latent heat, and explain any differences.

Thermodynamics

13. What do the first law and second law of thermodynamics tell you?

14. Distinguish between thermal efficiency and ideal efficiency.

15. Why is approximately two-thirds of the heat generated by a coal-fired or nuclear power plant wasted? What happens to this waste heat?

16. By expending 1 J of energy in doing work, a home heat pump can provide 2 J of energy to a room. Explain why this process does not violate the conservation of energy.

17. In terms of entropy, why do occupied rooms tend to get messy? What happens when a room is cleaned up?

18. Compare and explain the following statements in terms of the laws of thermodynamics: "Energy can be neither created nor destroyed." "Entropy can be created but not destroyed."

Heat Transfer

19. What are several examples of good thermal conductors and good thermal insulators? In general, what makes a substance one or the other?

20. Why does a vinyl tile floor feel colder than a rug, even though they are both at the same temperature?

21. Explain how coolers and Thermos bottles keep things hot and cold.

22. Is an open fireplace an efficient way of heating a room? Explain.

Phases of Matter

23. Give descriptions of a solid, a liquid, and a gas in terms of shape and volume.

24. What determines the phase of a substance?

The Gas Laws and Kinetic Theory

25. In terms of kinetic energy and potential energy of molecules, distinguish between heat and temperature.

26. Why does the pressure inside an auto tire increase during long, continuous driving?

27. In terms of kinetic theory, explain (a) why a basketball stays inflated and (b) what happens when an inflated balloon is placed in the freezer compartment of a refrigerator.

28. What is the purpose of perspiring when we get hot?

Exercises

Temperature

1. Normal body temperature is 98.6°F. What is the equivalent temperature on the Celsius scale?

   Answer: 37°C

2. Normal room temperature is about 68°F. What is the equivalent temperature on the Celsius scale? On the Kelvin scale?

3. The temperature of outer space is estimated to be 3 K. What is the equivalent temperature on the (a) Celsius scale and (b) Fahrenheit scale? *Answer:* (b) −454°F

4. While in Europe, a tourist hears on the radio that the temperature that day will have a high of 15°C. What is this temperature on the Fahrenheit scale?

5. Show that absolute zero is equivalent to a temperature of about −460°F.

Heat

6. A skier comes down a 10-m high slope, reaching the bottom at a speed of 10 m/s. If the skier's mass is 60 kg, how much heat is produced if all the lost energy goes into heat?

7. A college student produces about 100 kcal of heat per hour during an average day. How many watts is this?
    Answer: 116 W

8. How many kilocalories of energy does a 1200-W hair dryer produce each second?

9. A piece of pie contains 100 Cal of food energy. How would the piece of pie be rated in Btu?
    Answer: 400 Btu

Specific Heat and Latent Heat

10. How much energy is required to raise the temperature of 1 L of water from 20° to 30°C?   *Answer:* 10 kcal

11. (a) How much heat does it take to warm 1.0 kg of water from room temperature (20°C) to the boiling point?
    (b) At 8¢ per kWh, how much does it cost to heat 1.0 kg of water for instant coffee?
    Answer: (b) 0.7¢ (less than 1¢)

12. Equal amounts of heat are added to equal masses of aluminum and copper at the same temperature. Which metal will have a higher final temperature, and how many times greater than the other's final temperature will this be?

13. Twenty joules of work are expended in stirring 400 g of water. If the initial temperature of the water was 25°C, what will be the final temperature of the water? (Assume all the work went into heating the water.)
    Answer: 25.01°C

14. In making iced tea, you cool 1.6 kg of hot tea at 100°C to 0°C. What is the minimum amount of ice necessary to make iced tea?   *Answer:* 2.0 kg

15. How much heat is required to change 500 g of ice at −10°C to water at 20°C?

16. A quantity of steam (200 g) at 110°C is cooled to 90°C with a phase change in the process. How much heat was removed?

Thermodynamics

17. If a heat engine has a thermal efficiency of 40%, how much work is obtained from 50 kcal of heat input?
    Answer: 20 kcal (84 kJ)

18. If a cyclic heat engine has a heat input of 10 kcal of heat energy per cycle and a work output of 2.0×10^4 J, what is the engine's thermal efficiency?

19. Researchers have proposed using the temperature difference between the top and lower portions of the ocean to run a heat engine power plant. If the surface temperature of the ocean water is at 15°C and the lower depth is at 4°C, what is the maximum possible theoretical efficiency?

20. A proposed nuclear electrical generation plant would have $T_h = 350$°C and $T_c = 100$°C, with an efficiency of 42%. Would you support this proposal and buy stock in the electric company?   *Answer:* No; ideal eff = 40%

21. A coal-fired power plant has operating temperatures of $T_h = 320$°C and $T_c = 100$°C. What can you say about the efficiency of the power plant?

The Gas Laws and Kinetic Theory

22. In an isothermal process the volume of a gas increases from 0.25 to 0.50 m^3. How is the pressure affected?
    Answer: decreased by one-half

23. An auto tire contains a volume of air at 22°C and a pressure of 200 kPa. If the volume remains constant and the pressure increases to 250 kPa because of an increase in temperature, what is the new temperature, in degrees Celsius?
    Answer: 96°C

24. A balloon is blown up to a pressure of 29 lb/in^2 so that it occupies a volume of 900 cm^3 at a temperature of 17°C. The next day the temperature of the room is up to 27°C, and the balloon has expanded to 1000 cm^3. What is the new pressure of the balloon?

25. A sample of air occupies a volume of 10 m^3 at a temperature of 0°C and a pressure of 100 kPa. Determine the volume of the air sample at 40°C and 400 kPa.
    Answer: 2.9 m^3

26. A container is filled with helium gas at 22°C. If heat is applied so that the internal energy of the gas doubles, what is the Celsius temperature of the gas?

27. An *adiabatic process* is one in which no heat is added or removed from a system. If 200 J of work is done on a perfect gas in compressing it, what is the change in the internal energy of the system?   *Answer:* 200 J increase

Waves

SINCE WE BEGAN OUR study of energy, much has been said about the forms of energy, the relation of energy to work, and energy conservation—with many questions raised and answered. Other interesting questions remain, however. For example, how is energy changed from one form to another? How is energy transferred to the ball in a baseball game? How do our bodies obtain energy? How is energy transferred to a train, a car, a spaceship, a particle of any size? How do we get energy from the Sun?

A partial answer to these questions has been found in terms of particle collisions and wave motion, two important ways for transferring energy (see Fig. 6.1). If a particle applies a force to another particle through a distance, then a transfer of energy has taken place through particle collision; one particle will gain energy, and the other will lose energy. If we are to believe the conservation of energy, and we know of no instance where the principle has failed, then the increase of energy of one particle will be equal to the decrease of energy of the other, assuming we neglect heat losses. In all cases, we find a transfer of energy takes place whenever two or more particles collide with one another.

When matter is disturbed, energy emanates from the disturbance; this propagation of energy is known as **wave motion.** For example, a stone dropped on the surface of a pond of water will disturb the water, and energy will be transferred outward from the disturbance as wave motion. Only energy is transferred, not matter (the water), as can be noted by observing a floating fishing bobber that goes up and down with the water.

A similar situation can occur in a solid. For example, during an earthquake a disturbance takes place because of a slippage or other cause, and this disturbance is transmitted to all parts of the Earth as wave motion. Again, this disturbance is a transfer of energy, not matter.

The transfer of energy takes place with or without a medium. Sound waves in the air and waves upon stretched strings and steel wires are examples of energy transmissions that require a medium. The neighboring particles of the media react upon one another to transfer the disturbance. Electromagnetic waves, including radio, infrared radiation, visible light, and X-rays, can be transferred without a medium. We say these disturbances are radiated through space.

The data from which we learn about the planets, the Sun, and other stars come to us by means of light. Later, when quantum mechanics is discussed, we shall study the dual nature of light. Strangely enough, light, which we consider to be a wave motion, has some of the properties of matter. This topic will be discussed in Chapter 9.

Our eyes and ears are two wave-detecting devices that serve to link us to our environment. Because a study

Figure 6.1 Energy transfer.
Some examples of transferring energy.

◄ Water waves in a breaking surf.

of wave motion seems relevant to an understanding of our physical environment, a knowledge of wave motion is essential to understanding many scientific principles.

6.1 Wave Properties

The disturbance generating a wave motion is usually periodic; that is, the disturbance is repeated again and again at regular intervals. The plucking of a guitar string or the blowing of a whistle sets up periodic waves (Fig. 6.2). However, the disturbance, and the resulting wave motion, does not necessarily have to be periodic. It may be a simple pulse or a shock wave, such as the sound originating from a book hitting the floor or from a jet plane passing through the sound barrier.

A disturbance in an elastic body will set up wave motion. For example, in Fig. 6.3 a stretched spring is attached between two fixed points. When several coils at one end of the spring are compressed and released, the disturbance is propagated along the length of the coiled spring. That is, the disturbance moves with the velocity vector parallel to the direction in which the

particles were displaced. The velocity is known as **wave velocity.** When the particle displacement and the wave motion are in the same direction, the wave is called a **longitudinal wave.** Sound waves are of this type.

The term **transverse wave** is used to denote wave motion in which the individual particles are displaced perpendicular to the direction of the wave velocity vector. Figure 6.4 is an illustration of a transverse wave in a stretched cord. The cord is disturbed from its equilibrium position by moving the end of the cord up and down. Moving the cord end from side to side—or in any direction—will also produce a transverse wave. The transfer of energy by transverse waves can take place in the absence of a medium. All electromagnetic radiation is of this type of wave motion.

The directions of the spring and cord "particles" relative to the directions of the wave velocities can easily be seen by tying a small piece of ribbon or cloth to the spring and the cord. The piece of cloth will oscillate similar to the way the particles of the medium oscillate.

The study of wave motion has resulted in certain terms that are used to explain the action of all waves (Fig. 6.5). Wave velocity describes the direction and

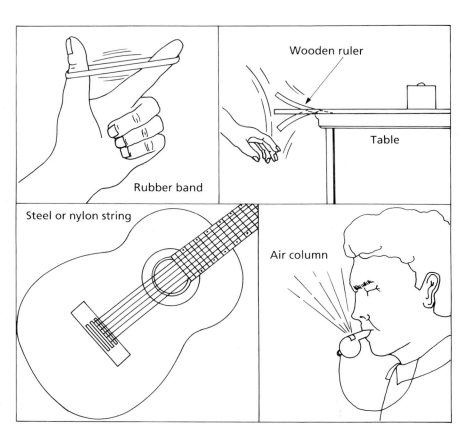

Figure 6.2 Wave motion.
Some common methods used to generate waves.

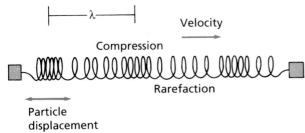

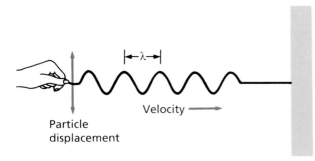

Figure 6.3 Longitudinal wave.
In a longitudinal wave, as illustrated here in a stretched spring, the wave velocity (vector) or direction of wave propagation is parallel to the (spring) particle displacement.

Figure 6.4 Transverse wave.
In a transverse wave, as illustrated here in a stretched cord, the wave velocity (vector) or direction of wave propagation is perpendicular to the (cord) particle displacement.

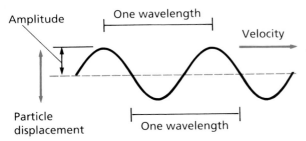

Figure 6.5 Wave description.
Some terms used to describe wave characteristics. See text for descriptions.

magnitude of the wave motion. The wave speed depends in general on the properties of the medium. The wave **frequency** is the number of oscillations that occur during a given period of time, usually 1 s. Frequency is expressed in cycles per second. This unit is given the name hertz. One **hertz** (Hz) is one cycle per second. For example,

if five complete wave crests pass a given spot in one second, the frequency is five cycles per second, or five hertz.

The **period** of a wave is the time it takes for one complete wave oscillation. If five crests pass by a given point in one second, one crest or complete cycle passes in $\frac{1}{5}$ of a second, and the period is $\frac{1}{5}$ second. The frequency is 5 cycles per second. From this example we see that

$$\text{frequency} = \frac{1}{\text{period}}$$

or
$$f = \frac{1}{T} \tag{6.1}$$

The wavelength of a wave is measured in units of length. One **wavelength** is the distance from any point on a wave to the corresponding point on the adjacent wave as shown in Fig. 6.5.

A simple relation between wave speed, wavelength, and period (or frequency) exists. Because speed is distance divided by time, we may write

$$v = \frac{\lambda}{T}$$

or by Eq. 6.1,

$$v = \lambda f \tag{6.2}$$

where v = wave speed measured in meters per second or other units,
λ = wavelength, measured in length units,
T = period of the wave, usually measured in seconds,
f = frequency of the wave, measured in cycles per second or hertz.

EXAMPLE 1

Consider sound waves with a speed of 344 m/s and a frequency of (a) 20 Hz and (b) 20,000 Hz. Find the wavelength of each of these sound waves.

Solution   We rearrange Eq. 6.2 and solve

$$v = \lambda f$$

or
$$\lambda = \frac{v}{f}$$

(a)
$$\lambda = \frac{344 \text{ m/s}}{20 \text{ Hz}} = 17 \text{ m}$$

(b)
$$\lambda = \frac{344 \text{ m/s}}{20,000 \text{ Hz}} = 0.017 \text{ m}$$

The frequencies of 20 and 20,000 Hz given in Example 1 define the general range of audible sound wave frequencies. Thus the wavelengths of sound cover the range from about 1.7 cm for the highest-frequency sound we can hear up to about 17 m for the lowest-frequency sound we can hear. In British units sound waves range from wavelengths of approximately $\frac{1}{2}$ in up to about 50 ft.

The **amplitude** of a wave refers to the maximum displacement of any part of the wave (or wave particle) from its equilibrium position. The amplitude of the wave does not affect the wave speed. The energy transmitted by a wave is related to the square of its amplitude.

6.2 Electromagnetic Waves

When charged particles such as electrons vibrate, energy is radiated away from them in the form of **electromagnetic waves.** Electromagnetic waves consist of vibrating electric and magnetic fields, which will be more thoroughly studied in Chapter 8. They are vector fields. In electromagnetic waves they radiate outward at the speed of light, which is 3×10^8 m/s in vacuum.

A drawing of an electromagnetic wave is shown in Fig. 6.6. The wave is traveling in the x direction. The electric and magnetic field vectors are at angles of 90° to one another, and the velocity vector of the wave is at an angle of 90° to both of the field vectors.

Charged particles are accelerated in many different ways to produce electromagnetic waves of various frequencies. Waves with low frequencies, or long wavelengths, are known as *radio waves* and are produced primarily by causing electrons to oscillate, or vibrate, in an antenna. The frequency of oscillation is controlled by the physical dimensions and other properties of the driving circuit.

The production of electromagnetic waves with frequencies greater than radio waves is accomplished by molecular excitation. In such cases radiation occurs from the collision of molecules in hot gases and solids. Because the molecules carry charged particles that are greatly accelerated as the molecules vibrate, the particles will radiate electromagnetic waves ranging from 10^{10} Hz to about 4.0×10^{14} Hz. This portion of the electromagnetic spectrum includes the *microwave* and *infrared* regions (see Fig. 6.7).

As the temperature of gases and solids is increased to higher and higher values, the atoms composing the molecules become more excited, and electromagnetic radiation in the *visible* and *ultraviolet* regions of the spec-

trum is emitted. Notice that the small portion of the spectrum visible to the human eye (Fig. 6.7) lies between the infrared and ultraviolet regions.

Still more energy applied to the atom will generate waves of higher frequencies, called *X-rays*, which range from 3×10^{17} to 3×10^{19} Hz. If sufficient energy is applied to the atom to disturb the nucleus, radiation known as *gamma rays* is emitted.

The term *light* is commonly used for electromagnetic radiations in or near the visible region; for example, we say ultraviolet light. Only the frequency (or wavelength) distinguishes visible electromagnetic radiation from the other portions of the spectrum. Our human eyes are only sensitive to certain frequencies or wavelengths, but other instruments can detect other portions of the spectrum. For example, a radio receiver can detect radio waves.

Radio waves are not sound waves. They are electromagnetic waves that are detected and then amplified by the radio frequency circuits of the radio receiver. The radio frequency signal is then demodulated—that is, the audio signal is separated from the radio frequency carrier. The audio signal is amplified, then applied to the speaker system that produces sound waves.

Electromagnetic radiation consists of transverse waves. These waves can travel through a vacuum. For instance, radiation from the Sun travels through the

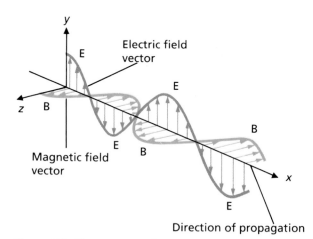

Figure 6.6 Electromagnetic wave.
An illustration of the vector components of an electromagnetic wave. The wave consists of two force fields (electric and magnetic) perpendicular to each other and perpendicular to the direction of wave propagation (velocity vector).

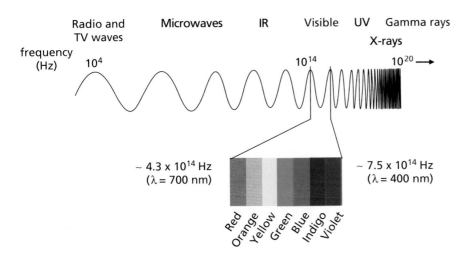

Figure 6.7 Electromagnetic spectrum.
Different frequency (or wavelength) regions are given different names. Notice how the visible spectrum forms only a very small part of the EM spectrum.

vacuum of space before arriving at the Earth. All electromagnetic waves travel at the same speed in a vacuum. This speed is called the speed of light. The **speed of light** in a vacuum is designated by the letter c and has a value of

$$c = 3.0 \times 10^8 \text{ m/s}$$

or

$$c = 186{,}000 \text{ mi/s}$$

(To a good approximation, this value is also the speed of light in air.)

We can use the formula $c = \lambda f$ to find the wavelength of light or any electromagnetic radiation in a vacuum. For example, let's calculate the wavelength of a typical AM radio wave.

EXAMPLE 2

What is the wavelength of the radio waves produced by a station with an assigned frequency of 600 kHz?

Solution Because *kilo-* means one thousand, we have

$$f = (600 \text{ kHz})(10^3 \text{ Hz/kHz})$$
$$= 600 \times 10^3 \text{ Hz} = 6.0 \times 10^5 \text{ Hz}$$

Then with $c = \lambda f$ we get

$$\lambda = \frac{c}{f} = \frac{3.0 \times 10^8 \text{ m/s}}{6.0 \times 10^5 \text{ Hz}}$$
$$= 500 \text{ m}$$

Hence we see that the wavelengths of AM radio waves are quite long. FM radio waves have shorter wavelengths because the frequencies are higher (in the megahertz, MHz, range). Visible light with frequencies on the order of 10^{14} Hz has relatively short wavelengths, as can be seen by the approximation

$$\lambda = \frac{c}{f} \approx \frac{10^8 \text{ m/s}}{10^{14} \text{ Hz}} = 10^{-6} \text{ m}$$

So the wavelength of visible light is on the order of one-millionth of a meter. To avoid using negative powers of ten, we commonly express the wavelengths in smaller units. Two units are the nanometer (nm, with $1 \text{ nm} = 10^{-9} \text{ m}$) and the angstrom (Å, with $1 \text{ Å} = 10^{-10}$ m, named after a Swedish scientist). Using the values given in Fig. 6.7, you should be able to show that the wavelength range for the visible region is between 4×10^{-7} and 7×10^{-7} m. This range corresponds to a range of 400 to 700 nm, or 4000 to 7000 Å.

For visible light, different frequencies are perceived by the eye as different colors, and the brightness depends on the energy of the wave.

6.3 **Sound Waves**

Technically, **sound** is defined as the propagation of longitudinal waves through matter. Sound waves involve particle displacement in any kind of matter—solid, liquid, or gas.

We are most familiar with sound waves in air, which affect our sense of hearing. However, sound also travels in liquids and solids. When you are swimming underwater and someone clicks two rocks together, you can

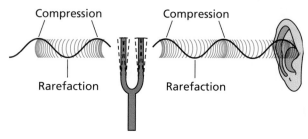

Compression   Compression

Rarefaction   Rarefaction

Figure 6.8 Sound waves.
Sound waves consist of a series of compressions (high-pressure regions) and rarefactions (low-pressure regions) as illustrated here for a vibrating tuning fork. Note how these regions can be described by a waveform.

hear this disturbance. Also, we can hear sound through thin (but solid) walls.

The wave motion of sound depends on the elasticity of the medium. A longitudinal disturbance produces varying pressures and stresses in the medium. For example, consider a vibrating tuning fork, as illustrated in Fig. 6.8. As an end of the fork moves outward, it compresses the air in front of it, and a **compression** is propagated outward. When the fork end moves back, it leaves a region of decreased air pressure and density called a **rarefaction.** With the continual vibration a series of high- and low-pressure regions travel outward, forming a longitudinal sound wave. The waveform may be displayed electronically on an oscilloscope, as shown in Fig. 6.9.

Sound waves may have different frequencies and so form a spectrum similar to the electromagnetic spectrum (Fig. 6.10). However, the **sound spectrum** has much lower frequencies and is much simpler, with only three frequency regions. These regions are based on the *audible range of human hearing,* which is about 20 Hz to 20 kHz (20,000 Hz) and defines the *audible region* of the spectrum. Below it is the *infrasonic region,* and above it is the *ultrasonic region.* (Note the analogy to infrared and ultraviolet light.) The sound spectrum has an upper limit of about a billion hertz (or 1 GHz, gigahertz) because of the elastic limitations of materials.

The infrasonic region of the sound spectrum is of little interest, but the audible region is of immense importance to us in terms of hearing. Sound is sometimes defined as those disturbances perceived by the human ear. As can be seen from Fig. 6.10, this definition would omit a majority of the sound spectrum. Indeed, ultrasound, which we cannot hear, has many practical applications, some of which will be discussed shortly.

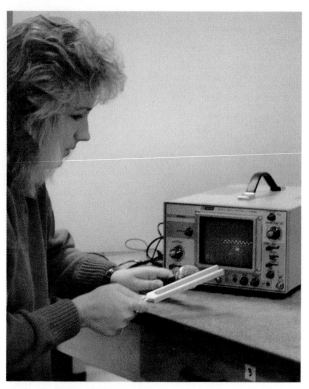

Figure 6.9 Waveform.
The waveform of the tone from a tuning fork can be displayed on an oscilloscope by using a microphone to convert the sound wave to an electrical signal.

QUESTION

If a tree falls in the forest and there is no one there to hear it, is there sound?

Answer This is an old question you may have heard before. Now you know the answer, which is, "It's a matter of definition." There is always sound in terms of longitudinal waves propagating from the disturbance of the falling tree. However, if you define sound as the sensation perceived by the human ear, then you might say there was no sound (heard).

We hear sound because the propagating disturbance causes the eardrum to vibrate, and sensations are transmitted to the auditory nerve through the fluid and bones of the inner ear. The characteristics associated with human hearing are physiological and can differ from their physical counterparts. For example, **loudness** is a relative term. One sound may be louder than another, and as

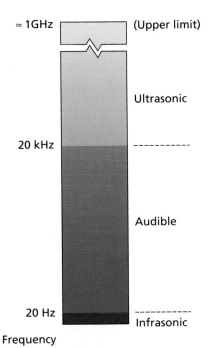

Figure 6.10 Sound spectrum.
The sound spectrum consists of three regions: the infrasonic region ($f < 20$ Hz), the audible region (20 Hz $< f < 20$ kHz), and the ultrasonic region ($f > 20$ kHz).

given as so many joules per second (J/s) through a square meter (m²). But recall that a joule per second is a watt (W), so intensity has the units of W/m².

The loudness or intensity of sound decreases the farther one is from the source. As the sound is propagated outward, it is "spread" over a greater area and so has less energy per area. This characteristic is illustrated for a point source in Fig. 6.11. In this case the intensity is inversely proportional to the square of the distance from the source ($I \propto 1/r^2$); that is, it is an inverse-square relationship. This relationship is analogous to painting a larger room with the same amount of paint (energy). The paint must be spread thinner and so is less "intense."

The minimum sound intensity that can be detected by the human ear (called the *threshold of hearing*) is about 10^{-12} W/m². At a much greater intensity of about 1 W/m² sound becomes painful to the ear. Because of the wide range, intensity is commonly measured on a logarithmic scale, which is more convenient. The sound-intensity level is measured on a decibel (dB) scale, as illustrated in Fig. 6.12.

A **decibel** is one-tenth of a bel (B), a unit named in honor of Alexander Graham Bell, the inventor of the telephone. Because the decibel scale is not linear with intensity, when the sound intensity is doubled, the dB level is not doubled. Instead, the intensity level increases by only 3 dB. That is, a sound with an intensity level of 63 dB has twice the intensity of a sound with an intensity level of 60 dB.

Comparisons are conveniently made on the dB scale in terms of decibel differences and factors of 10:

you might guess, this quality is associated with the energy of the wave. The measurable physical quantity is **intensity** (*I*), which is the rate of energy transfer through a given area. For example, intensity may be

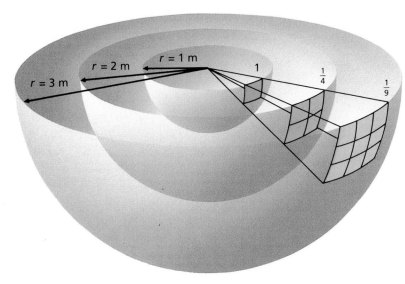

Figure 6.11 Sound intensity.
An illustration of the inverse-square law for a point source ($I \propto 1/r^2$). Notice that when the distance doubles, e.g., from *r* to 2*r*, the intensity decreases to one-fourth, because the sound must pass through four times the area.

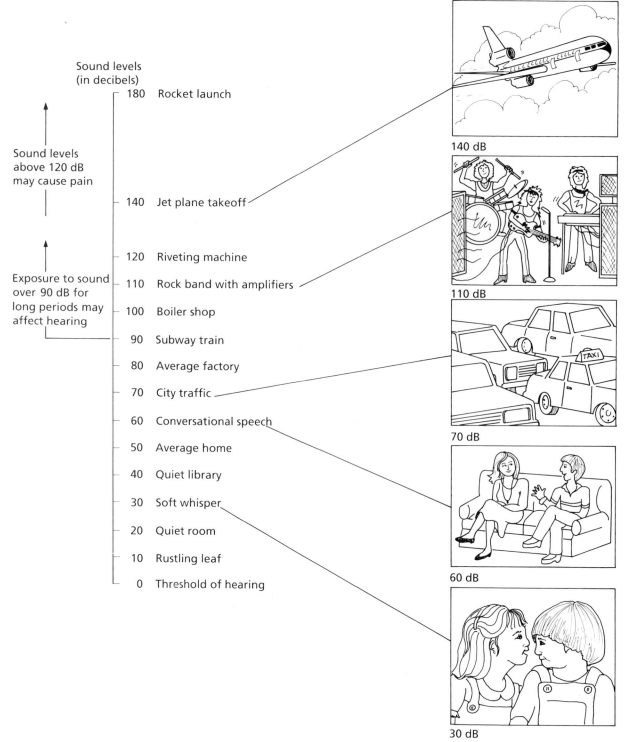

Sound levels
(in decibels)

180	Rocket launch	
140	Jet plane takeoff	
120	Riveting machine	
110	Rock band with amplifiers	
100	Boiler shop	
90	Subway train	
80	Average factory	
70	City traffic	
60	Conversational speech	
50	Average home	
40	Quiet library	
30	Soft whisper	
20	Quiet room	
10	Rustling leaf	
0	Threshold of hearing	

Sound levels above 120 dB may cause pain

Exposure to sound over 90 dB for long periods may affect hearing

140 dB

110 dB

70 dB

60 dB

30 dB

Figure 6.12 Sound-intensity levels.
The decibel scale of sound-intensity levels with sounds from various sources.

An increase of 10 dB increases the sound intensity by a factor of 10.

An increase of 20 dB increases the sound intensity by a factor of 100.

An increase of 30 dB increases the sound intensity by a factor of 1000; and so on.

Loudness is related to intensity, but loudness is subjective and estimates differ from person to person. Also, the ear does not respond equally to all frequencies. For example, two sounds with different frequencies, but with the same intensity level, are judged by the ear to have different loudnesses.

The frequency of a sound wave may be physically measured, whereas **pitch** is the *perceived* highness or lowness of a sound. For example, a soprano has a high-pitched voice as compared with a baritone. Pitch is related to frequency. But if a sound with a single frequency is heard at two intensity levels, nearly all listeners will agree that the more intense sound has lower pitch.

Ultrasound is the term used for sound waves with frequencies greater than 20,000 Hz. These waves cannot be detected by the human ear, but the frequency range for other animals includes ultrasound frequencies. For example, dogs can hear ultrasound, and ultrasonic whistles used to call dogs don't disturb humans. Ultrasonic whistles on cars alert deer to oncoming traffic so that they won't leap across the road in front of cars.

An important use of ultrasound is in examining parts of the body. Thus ultrasound is an alternative to X rays, which may be harmful. The ultrasonic waves allow different materials such as tissue and bone to be "seen" or distinguished by bouncing waves off the object examined. The waves are detected, analyzed, and stored in a computer. An *echogram*, such as the one of an unborn fetus shown in Fig. 6.14, is then reconstructed. X rays might harm the fetus and cause birth defects, but ultrasonic waves have less energetic vibrations and have given no evidence of harming a fetus.

Ultrasound can also be used as a cleaning technique. Minute foreign particles can be removed from objects

HIGHLIGHT

Noise Exposure Limits

Sounds with intensities of 120 dB and higher can be painfully loud to the ear. Brief exposures to even higher sound intensity levels can rupture eardrums and cause permanent hearing loss. However, long exposure to relatively lower sound (noise) levels can also cause hearing problems. (Noise is defined as unwanted sound.) Such exposures may be an occupational hazard, and in some jobs ear protectors must be worn (Fig. 6.13). You may have experienced a temporary hearing loss after being exposed to a loud band for a long time or a loud bang for a short time.

Federal standards now set permissible noise exposure limits for occupational loudness. These limits are listed in Table 6.1. Notice that a person can work on a subway train (90 dB, Fig.

Figure 6.13 Sound-intensity safety. An airport cargo worker wears ear protectors to prevent ear damage from the high sound-intensity levels of jet-plane engines.

6.12) for 8 h, but a person should only play in (or listen to) an amplified rock band (110 dB) continuously for $\frac{1}{2}$ h.

Table 6.1 Permissible Noise Exposure Limits

Maximum Duration per Day (h)	Sound Level Intensity (dB)
8	90
6	92
4	95
3	97
2	100
$1\frac{1}{2}$	102
1	105
$\frac{1}{2}$	110
$\frac{1}{4}$ or less	115

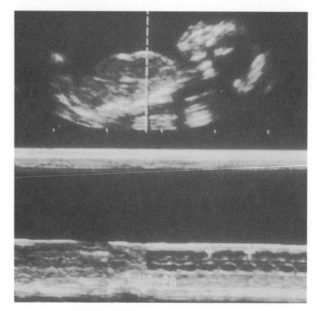

Figure 6.14 Echogram.
A fetal echogram, or sonogram, in which the outline of the baby's face can be seen in the top right yellow region. There is no evidence that ultrasound scans can harm a fetus, as X-rays can.

placed in a liquid bath through which ultrasound is passed. The wavelength of ultrasound is on the same order of magnitude cf the particle size, and the wave vibrations can get into small crevices and "scrub" particles free. Thus ultrasound is especially useful in cleaning objects with hard-to-reach recesses, such as rings and other jewelry. Ultrasonic cleaning baths for false teeth are also commercially available.

The speed of sound in a particular medium depends on the makeup of the material. In general, the **speed of sound** in air at 20°C is

$$v_{sound} = 344 \text{ m/s (770 mi/h)}$$

or approximately $\frac{1}{5}$ mi/s.

The speed of sound varies directly with temperature over the normal temperature range. That is, as the temperature increases, the speed of sound increases, and vice versa. Note that the speed of sound in air is much less than the speed of light.

The relatively slow speed of sound in air may be observed at a baseball game. A spectator may see a batter hit the ball but hear the "crack" of the bat slightly later if he or she is sitting far from home plate. Similarly,

one may see the smoke or flash from a fired rifle but hear the report later, because the sound comes to an observer much slower than the visual signal, which travels at the speed of light.

In general, as the density of the medium increases, the speed of sound increases. The speed of sound is about 4 times faster in water than in air and, in general, about 15 times faster in solids.

Using the speed of sound and the frequency, we can easily compute the wavelength of a sound wave.

EXAMPLE 3

What is the wavelength of a sound wave in air with a frequency of (a) 2200 Hz and (b) 22 MHz?

Solution   Using the relation

$$\lambda = \frac{v_{sound}}{f}$$

we have, with $v_{sound} = 344$ m/s:
(a) For $f = 2200$ Hz, which is in the audible range,

$$\lambda = \frac{344 \text{ m/s}}{2200 \text{ Hz}} = 0.16 \text{ m}$$

This wavelength is about $\frac{1}{2}$ ft.
(b) For $f = 22$ MHz $= 22 \times 10^6$ Hz, which is in the ultrasonic region,

$$\lambda = \frac{344 \text{ m/s}}{22 \times 10^6 \text{ Hz}} = 16 \times 10^{-6} \text{ m}$$
$$= 0.000016 \text{ m} = 16 \ \mu\text{m}$$

or 16 micrometers. Hence the wavelength of ultrasound is on the order of particle size and can be used in cleaning baths, as described earlier.

6.4 Standing Waves and Resonance

Most of us have shaken one end of a stretched cord or rope and have observed wave patterns that seem to "stand" along the rope when we shake the cord just right. We refer to these waveforms as **standing waves,** which are caused by the interference of waves traveling down and back along the rope. When two waves meet, they interfere, and the combined waveform of the superimposed waves is the sum of the waveforms or particle displacements of the medium.

Consider waves traveling in a rope in opposite directions, as illustrated in Fig. 6.15a. These waves may

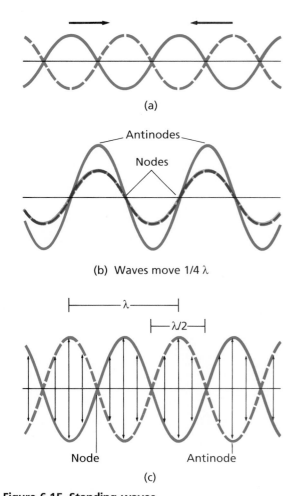

(a)

(b) Waves move 1/4 λ

(c)

Figure 6.15 Standing waves.
Waves traveling in opposite directions on a stretched rope continually interfere so as to produce a standing wave with nodes and antinodes as illustrated here.

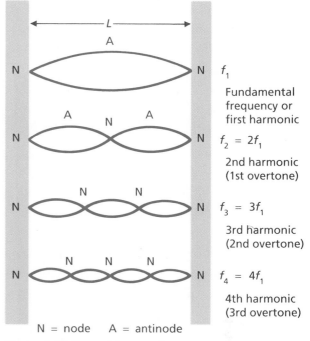

N = node   A = antinode

Figure 6.16 Natural frequencies.
(a) An illustration of the series of allowed vibrations, or standing waves, for a stretched string fixed at each end. Notice that only half-wavelength segments can "fit in" to the string because of the node conditions at each end.

result from shaking one end of the rope, with reflections returning from the fixed end. Notice that as the waves move, the crests and troughs periodically add and subtract. The net result is a standing waveform in the rope. The zero points, or the points that remain stationary, are called **nodes.** The points of maximum displacement are called **antinodes** (Fig. 6.15b). Notice that the distance between two nodes (or two antinodes) is one-half the wavelength ($\lambda/2$) of the standing wave (Fig. 6.15c). That is, each standing wave "loop" is half a wavelength.

Suppose a vibrating string of length L were fixed at both ends (Fig. 6.16). In this situation there is a node at each end of the string. As a result, only particular

numbers of wave loops can "fit in" the string, i.e, $L = \lambda/2$, $L = 2(\lambda/2)$, $L = 3(\lambda/2)$, etc. Hence there is a series of possible wavelengths, given generally by

$$\lambda_n = \frac{2L}{n} \quad n = 1, 2, 3, 4, \ldots \quad (6.3)$$

For example, for $n = 1$, we have $\lambda_1 = 2L$, or $L = \lambda_1/2$, which corresponds to the first case in Fig. 6.16.

The frequencies of the standing waves may be determined by using Eq. 6.1,

$$f = \frac{v}{\lambda}$$

and substituting for the general wavelengths (Eq. 6.3). We have

$$f_n = \frac{nv}{2L} \quad n = 1, 2, 3, 4, \ldots \quad (6.4)$$

These frequencies are referred to as the characteristic or *natural frequencies* of the stretched string. The lowest

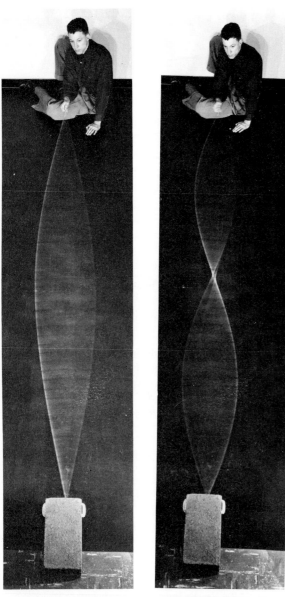

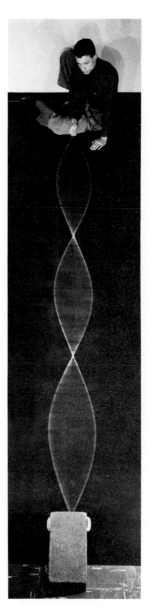

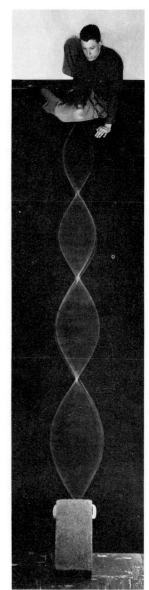

(b) Actual standing waves in a stretched rubber string.
Figure 6.16 (*cont.*)

frequency uses $n = 1$ and is

$$f_1 = \frac{v}{2L} \qquad (6.5)$$

which is called the **fundamental frequency** or the **first harmonic.** The higher frequencies, $f_2, f_3, \ldots,$ are multiples of the fundamental frequency, e.g., $f_2 = 2f_1,$ and are called the **second harmonic, third harmonic,** and so on. Taken together, the harmonics higher than the first harmonic are called **overtones.**

EXAMPLE 4

If the frequency of the first harmonic of a stretched string is 440 Hz, what are the frequencies of the second harmonic (first overtone) and the third harmonic (second overtone)?

Solution To solve this problem, we use Eq. 6.4 with $n = 1, 2$, and 3. The first harmonic, or fundamental frequency, uses $n = 1$, and $f_1 = v/2L = 440$ Hz. Notice that Eq. 6.4 may be written as

$$f_n = \frac{nv}{2L} = nf_1 = n \times 440 \text{ Hz}$$

where $n = 1, 2, 3, 4, \ldots$. Thus

$$f_1 = 1 \times 440 \text{ Hz} = 440 \text{ Hz} \qquad \text{(first harmonic)}$$
$$f_2 = 2 \times 440 \text{ Hz} = 880 \text{ Hz} \qquad \text{(second harmonic)}$$
$$f_3 = 3 \times 440 \text{ Hz} = 1320 \text{ Hz} \quad \text{(third harmonic)}$$

When a stretched string or an object is acted upon by a periodic driving force with a frequency equal to one of the natural frequencies, the oscillations have large amplitudes. This phenomenon is called **resonance,** and in this case there is maximum energy transfer to the system.

A common example of driving a system in resonance is pushing a swing. A swing is essentially a pendulum with only one natural frequency, which depends on the rope length. When a swing is pushed periodically with a period of $T = 1/f$, energy is transferred to the swing and its amplitude gets larger (higher swings). If the swing is not pushed at its natural frequency, the pushing force may be applied as the swing approaches or after it has reached its maximum amplitude and is swinging back. In either case, the swing is not driven in resonance.

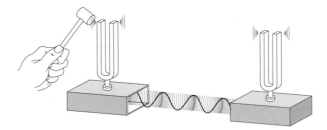

Figure 6.17 Resonance.
(a) When one tuning fork is activated, the other tuning fork of the same frequency will be driven in resonance and start to vibrate.

A stretched string, as discussed above, has many natural frequencies, instead of just one, as a pendulum has. When the string is shaken, standing waves are set up. But when the driving frequency corresponds to one of the natural frequencies, the amplitude at the antinodes will be larger because of resonance.

There are many examples of resonance. The structure of the throat and nasal cavities gives the human voice a particular tone due to resonances. Tuning forks of the same frequency can be made to resonate, as illustrated in Fig. 6.17a. A steel bridge or any elastic structure is capable of vibrating at natural frequencies, sometimes with dire consequences (Fig. 6.17b). People marching in columns across bridges are told to "break step" and not march at a periodic cadence, which might

Figure 6.17 (*cont.*)
(b) Unwanted resonance. The famous Tacoma Narrows bridge, which collapsed after the wind drove the bridge into resonance vibrations.

correspond to a natural frequency of the bridge and result in resonance and large oscillations that could cause structural damage.

There is also *electrical* resonance. When we tune a radio or TV receiver, turning the dial adjusts the natural frequency of an alternating electric circuit to the assigned broadcast frequency of a particular station. There is then maximum energy transfer to the circuit for this frequency, and other frequencies or stations are excluded.

Musical instruments use standing waves and resonance to produce different tones. Standing waves are formed on strings fixed at both ends on stringed instruments such as the guitar, violin, and piano. When a stringed instrument is tuned, a string is tightened or loosened, which adjusts the tension and the wave speed in the string. This adjustment changes the frequency or pitch, because the length of the string is fixed ($\lambda f = v$). A vibrating string does not produce a great disturbance in air, but the body of a stringed instrument such as a violin acts as a sounding board and amplifies the sound. Thus the body of such an instrument acts as a resonance cavity, and sound comes out through holes in the top surface.

Similarly, standing waves are set up in wind instruments in air columns. Organ pipes have fixed lengths similar to fixed strings, so only a certain number of wavelengths can be fitted in. (Determining the frequencies for standing waves in organ pipes is left as an exercise.) However, the length of an air column and the frequency or tone can be varied in some instruments, such as a trombone or trumpet, by varying the length of the column.

A musical note is designated by a particular frequency, and a musical scale is a series of these notes. Musical-scale intervals are based on pleasing or consonant sounds when the frequencies of two notes form a whole-number ratio. The simplest scale in Western music is illustrated in Fig. 6.18. The international standard frequency of 440 Hz for the note A_4 is the reference frequency for this scale. Starting with a C note on a piano and playing successively higher notes on the white keys (to the right) gives the familiar "do-re-me-fa-sol-la-ti-do." This is the C-major scale.

Notice in Fig. 6.18 that the frequency doubles from C_4 to C_5 (from 264 Hz to 528 Hz). This interval is called an **octave** (Latin meaning "eight"); it has eight steps or notes. A piano range covers seven octaves, with C_4 of the middle octave being "middle C." The frequency of each note—D, E, etc.—doubles with each octave interval. The black keys of the piano help divide an octave into more equal intervals, which generally sound more pleasing.

The *quality* (or timbre) of a sound depends on the waveform or the number of harmonics or overtones present. The dominant frequency of a tone is the fundamental frequency, but the overtones distinguish it from another tone with the same frequency by giving it a different quality. For example, you can sing the same note as a famous singer, but a different combination of overtones gives the singer's voice a different quality,

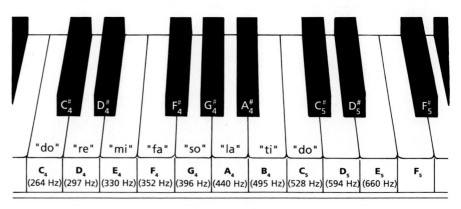

Figure 6.18 Musical scale. A simple musical scale that uses 440 Hz (A_4) as a reference frequency.

perhaps a pleasing "richness." It is the quality of our voices that gives them different sounds.

6.5 The Doppler Effect

When we watch a race and a racing car with a loud engine approaches, we hear a higher-than-usual sound frequency. When the car passes by, the frequency suddenly shifts lower and a low-pitched "whoom" sound is heard. As shown in Fig. 6.19, the waves are "spread out" behind the moving source. Similar frequency

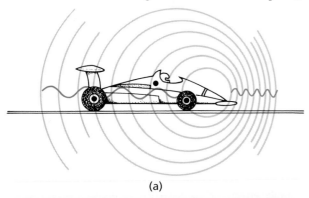

(a)

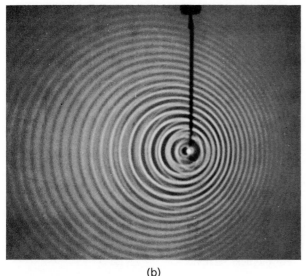

(b)

Figure 6.19 Doppler effect.
Above: Because of the motion of the source, illustrated here as a car, sound waves are "bunched up" in front and "spread out" in back. This results in shorter wavelengths, or an increased frequency, in front of the source and longer wavelengths, or a lower frequency, behind the source. *Below:* The Doppler effect in water waves in a ripple tank. The source or disturbance is moving to the right.

changes may be heard when a large truck passes by. The apparent change in the frequency of the source is called the **Doppler effect.**[*]

The reason for the observed change in frequency (and wavelength) of a sound is illustrated in Fig. 6.19. As a moving sound source approaches an observer, the waves are "bunched up" in front of the source. With the waves closer together (shorter wavelength) an observer perceives a higher frequency. Behind the source the waves are spread out, and with an increase in wavelength a lower frequency is heard ($f = v/\lambda$). If the source is stationary and the observer moves toward and passes the source, the shifts in frequency are also observed. Hence the Doppler effect depends on the *relative* motion of the source and the observer.

Waves propagate outward in front of a source as long as the speed of sound is greater than the speed of the source (Fig. 6.20). However, as the speed of the source approaches the speed of sound in a medium, the waves begin to bunch up. When the speed of the source exceeds the speed of sound in the medium, a V-shaped bow wave is formed. This wave is readily observed for a motor boat traveling faster than the wave speed in water.

In air, when a jet aircraft travels at a supersonic speed (a speed greater than the speed of sound in air), the bow wave is in the form of a shock wave that trails out and downward from the aircraft. When this high-pressure, compressed wave front passes over an observer, he or she hears a sonic boom. But the high-pressure bow wave travels with the supersonic aircraft and does not occur only at the instant the aircraft "breaks the sound barrier" (first exceeds the speed of sound).

The Doppler effect is a general effect that occurs for all kinds of waves, such as water waves, sound waves, and light (electromagnetic) waves. In the Doppler effect for visible light the frequency is shifted toward the blue end of the spectrum when the light source (such as a star) is approaching. (Blue light has a shorter wavelength, or higher frequency.) In this case we say a Doppler *blue shift* has occurred. When a stellar light source moves away from us and the frequency is shifted toward the red (longer wavelength) end of the spectrum, we say that a Doppler **red shift** has occurred. The magnitude of the frequency shift is related to the speed of the

---

[*] After Christian Doppler (1803–1853), an Austrian physicist who first described the effect.

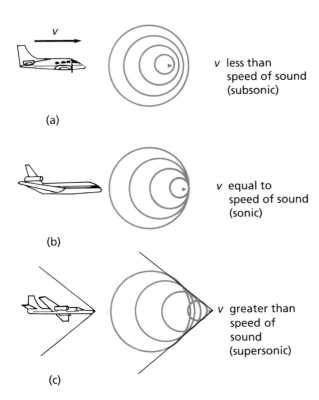

(a)

v less than
speed of sound
(subsonic)

(b)

v equal to
speed of sound
(sonic)

(c)

v greater than
speed of
sound
(supersonic)

Figure 6.20 Bow wave and sonic boom.
Just as a moving boat forms a bow wave in water, a moving aircraft forms a bow wave in air. The sound waves "bunch up" in front of an airplane for increasing subsonic speeds [(a) to (b)]. A plane traveling at supersonic speeds forms a high-pressure shock wave in the air that is heard as a sonic boom when it passes over an observer.

source. As we will see in Chapter 19, light from galaxies shows red shifts, which indicates that they are moving away from us and that the universe is expanding. The degree or magnitude of the red shift is related to the relative recession speed of a galaxy, and this relationship allows us to determine how far the receding galaxies are from our own.

As we have seen, the Doppler shift of an electromagnetic wave can be used to determine the speed of an astronomical object. It can also be used here on Earth. For instance, the radar (radio waves) that police use in determining the speed of moving vehicles utilizes the Doppler effect.

Learning Objectives

After reading and studying this chapter, you should be able to do the following without referring to the text:

1. Define the two main types of waves, give examples of each, and list five properties of each wave.

2. List the various regions of the electromagnetic spectrum in terms of increasing frequency.

3. List the regions of the sound spectrum, and compare the wavelengths and speeds of audible sound and visible light.

4. State the wave properties that give rise to pitch, loudness, quality, color, and brightness.

5. Distinguish frequency from pitch, and loudness from intensity, in reference to sound waves.

6. Define *natural frequencies*, and explain how they are determined for a stretched string.

7. Describe what produces standing waves in a stretched string, and describe the effect of resonance.

8. Explain how different musical instruments produce different notes, and describe how a musical scale is formed.

9. Explain what causes a sonic boom.

10. Discuss how we know that distant galaxies are moving away from us.

11. Define and explain the important words and terms listed in the next section.

Important Words and Terms

wave motion	wavelength	sound spectrum	standing waves	overtones
wave velocity	amplitude	loudness	nodes	resonance
longitudinal wave	electromagnetic waves	intensity	antinodes	octave
transverse wave	speed of light	decibel	fundamental frequency	Doppler effect
frequency	sound	pitch	first harmonic	red shift
hertz	compression	ultrasound	second harmonic	
period	rarefaction	speed of sound	third harmonic	

Questions

Wave Properties

1. What is the difference between a longitudinal wave and a transverse wave? Give an example of each.

2. What are the SI units of (a) wavelength, (b) frequency, and (c) period?

3. Does the equation $v = \lambda f$ give the correct units for speed? Explain.

4. What determines the amplitude of a wave?

Electromagnetic Waves

5. List, in order of increasing frequency, the various types of electromagnetic waves.

6. Which end of the visible spectrum has the longer wavelength?

7. Are radio waves sound waves? Explain.

8. What is the order of magnitude of the wavelength of visible light? How does the wavelength of visible light compare with that of audible sound?

9. What wave property distinguishes between (a) different colors, (b) different brightnesses, and (c) different radio stations?

Sound Waves

10. What makes a sound wave audible?

11. What is the chief wave property that distinguishes between sound (a) pitches, (b) intensities, and (c) quality?

12. If an astronaut on the moon dropped a hammer, would there be sound? Explain.

13. Can ultrasound be heard? Give some uses or applications of ultrasound.

14. Why does the music from a marching band in a spread-out formation on a football field sometimes sound discordant?

15. Does doubling the decibels of a sound level double the intensity? Explain.

Standing Waves and Resonance

16. Why can only half wavelengths be fitted into a vibrating stretched string?

17. How many harmonics does a stretched string have? How many does a simple pendulum have?

18. What is the effect if a system is driving in resonance? Is a particular frequency required?

19. What determines the pitch of a string on a violin? How does a violinist get a variety of notes from one string?

20. How is a musical scale constructed? What is an octave?

The Doppler Effect

21. How is the wavelength of sound affected when (a) a source moves toward a stationary observer and (b) an observer moves away from a stationary source?

22. What would be the situation for a sound (a) "blue shift" and (b) "red shift"?

23. Compare the crack of a whip with a sonic boom.

24. Radar and sonar are based on similar principles. Sonar (which stands for *sound na*vigation ranging) uses ultrasound, and radar (which stands for *radio de*tecting *a*nd ranging) uses radio waves. Explain the principles of detecting and ranging in these applications.

Exercises

Wave Properties

1. A periodic wave has a period of 3.0 s. What is the wave frequency?

2. A sound wave has a frequency of 200 Hz. What is the distance between crests or compressions of the wave?

3. Waves moving on a lake have a speed of 2.0 m/s, with a distance of 5.0 m between adjacent crests.
   (a) Determine the frequency of the waves.
   (b) Find the period of the wave motion.
   Answer: (a) 0.40 Hz

4. The speed of sound in water is 1530 m/s. What is the wavelength of a 2000-Hz sound wave?
   Answer: 0.765 m

Electromagnetic Waves

5. Compute the wavelength of radio waves emitted by an AM radio station operating at 1340 kHz.
   Answer: 2.2×10^2 m

6. What is the frequency of blue light that has a wavelength of 420 nm? *Answer:* 7.1×10^{14} Hz

7. Compute the wavelength, in nm and Å, of an X-ray with a frequency of 10^{18} Hz. *Answer:* 30 nm, or 3 Å

8. How many orders of magnitude greater is the frequency of visible light than the frequency of audible sound? Use the upper limit of each range.

Sound Waves

9. What are the wavelength limits of the audible range of the sound spectrum?

10. Compute the wavelength in air of ultrasound with a frequency of 60,000 Hz if the speed of sound is 344 m/s.
    Answer: 5.7×10^{-3} m

11. During a thunderstorm 4 s elapses between observing a lightning flash and hearing the resulting thunder. Approximately how far away was the lightning flash?
    Answer: 0.8 mi

12. A subway train has a sound intensity level of 90 dB, and a rock band has a sound intensity of about 110 dB. How many times greater is the sound intensity of the band than the subway train?
    Answer: 100 times

13. In Table 6.1 the maximum exposure limit decreases from 6 to 4 h when the sound intensity level goes from 92 to 95 dB. By what factor does the intensity increase for this dB increase?

14. A rock band with an intensity level of 123 dB turns its speakers down so that the intensity is $\frac{1}{100}$ of this value. What is the new intensity level, in dB?

15. A speaker is playing with an output of 80 dB. If the volume is turned up so that the output intensity is 10,000 times greater, what will be the new intensity level?
    Answer: 120 dB

16. A new jackhammer has only $\frac{1}{10}$ the sound intensity of an old hammer that was rated at 118 dB. What is the sound intensity level of the new hammer?

Standing Waves and Resonance

17. What is the fundamental frequency of a stretched string 1.0 m long if the wave speed in the string is 240 m/s? What are the frequencies of the first and second harmonics?

18. If the frequency of the first harmonic of a stretched string is 256 Hz, what are the frequencies of the first and second overtones?
    Answer: 512 Hz and 768 Hz

19. The frequency of the third harmonic of a stretched string is 660 Hz. What is the frequency of the fourth harmonic of the string?

20. Standing waves may be set up in air columns in organ pipes, as illustrated in Fig. 6.21.
    (a) Show that the characteristic frequencies for an open organ pipe (open at both ends, with an antinode at each end) are given by $f_n = nv/2L$, where $n = 1, 2, 3, 4, \ldots$.
    (b) Show that the characteristic frequencies for a closed organ pipe (closed at one end with a node at that end) are given by $f_m = mv/4L$, where $m = 1, 3, 5, \ldots$

21. Use the results of Exercise 20. What is the frequency of the (a) first harmonic and (b) fifth harmonic for a closed organ pipe with a length of 2.0 m when the temperature is 20°C?
    Answer: (b) 215 Hz

22. Use the result of Exercise 20. What is the length of an open organ pipe that has a fundamental frequency of 440 Hz?

Figure 6.21 Organ pipes.
Pipe organs use resonance to produce different notes.

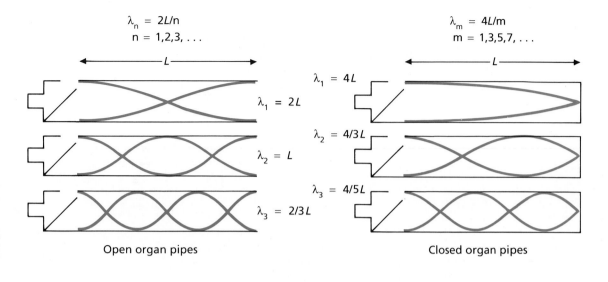

$$\lambda_n = 2L/n$$
$$n = 1,2,3, \ldots$$

$$\lambda_m = 4L/m$$
$$m = 1,3,5,7, \ldots$$

L

$\lambda_1 = 2L$

$\lambda_2 = L$

$\lambda_3 = 2/3L$

$\lambda_1 = 4L$

$\lambda_2 = 4/3L$

$\lambda_3 = 4/5L$

L

Open organ pipes

Closed organ pipes

(b)

Figure 6.21 (*cont.*)
An illustration of the modes of vibration of air columns in open and closed organ pipes of fixed lengths. An open organ pipe has antinodes at each end, while a closed organ pipe has a node at the closed end of the pipe. The characteristic frequencies of a particular pipe depend on its length. Different pipe lengths have different fundamental frequencies or dominant tones.

CHAPTER 7

Wave Effects

THE EFFECTS OF waves—particularly sound and light waves—are all around us. We are aware of many of these effects, but they are such a part of our experience that we take them for granted and rarely try to analyze them. For instance, if you speak loudly to someone in another room, you know the person hears you. Sound waves travel around corners; however, visible light waves do not. You can be heard in the next room but not seen.

Similarly, when we look up at the sky, we see a blue sky, white clouds, and sometimes even a rainbow. These are natural phenomena that owe their description and understanding to the effects of electromagnetic waves interacting with air and water droplets in the atmosphere. We are well aware of these phenomena, but we rarely stop to consider the wave effects that cause them.

Mirrors and lenses also are based on the wave effects of reflection and refraction. We will discuss these two effects and then describe the basic principles of lenses and mirrors. This discussion will lead to an understanding of many common optical devices such as the human eye, slide projectors, and eyeglasses.

There are many wave effects. In fact, two of them—resonance and the Doppler effect—have already been discussed. In this chapter we will discuss most of the other important wave phenomena that affect us all the time.

7.1 Reflection

Waves travel through space in a straight line and will continue to do so unless forced to deviate from their original direction. A change in direction takes place when light strikes and rebounds from a surface or the boundary between two media. A change in direction by this method is called **reflection.**

◄ The setting Sun looks flattened because of refraction.

Reflection may be thought of as light "bouncing" off a surface. However, it is really much more complicated and involves the absorption and emission of complex atomic vibrations of the reflecting medium. To describe reflection in a simple manner, we consider the reflection of light rays. A **ray** is a straight line that represents the motion of light. A beam of light may be represented by a group of parallel rays.

A light ray is reflected from a surface in a particular way. As illustrated in Fig. 7.1, the angles of the incident and reflected rays are measured relative to the *normal*, a line perpendicular to the reflecting surface. The **law of reflection** states:

> **The angle of reflection θ_r is equal to the angle of incidence θ_i.**

Also, the reflected and incident rays are in the same plane.

The reflection from very smooth or mirror surfaces is called **regular reflection** (Fig. 7.2). In regular reflection incident parallel rays are parallel on reflection. However, rays reflected from relatively rough surfaces are not parallel; this is called **diffuse reflection.** The reflection

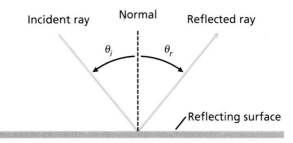

Incident ray   Normal   Reflected ray

θ_i θ_r

Reflecting surface

Figure 7.1 Law of reflection.
The angle of incidence (θ_i) equals the angle of reflection (θ_r) relative to the normal or a line perpendicular to the reflecting surface. The rays and the normal also lie in a plane.

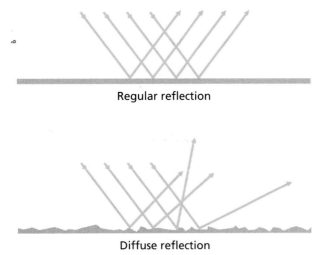

Regular reflection

Diffuse reflection

Figure 7.2 Reflection.
A rough surface produces a diffuse reflection. A smooth (mirror) surface produces a regular reflection.

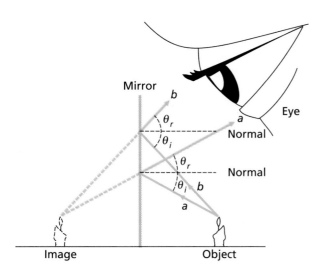

Figure 7.3 Ray diagram.
By tracing the reflected rays, mirror images are located where the rays intersect or appear to intersect behind the mirror.

from the pages of this book is diffuse reflection. The law of reflection applies to each type of reflection, but the rough surface causes the light rays to be reflected in different directions.

Rays may be used to determine the image formed by a mirror. A *ray diagram* for determining the apparent location of an image formed by a plane mirror is shown

in Fig. 7.3. The image is located by drawing two rays emitted by the object and applying the law of reflection. Where the rays intersect or appear to intersect locates the image. Notice for a plane mirror that the image is located behind or "inside" the mirror at the same distance the object is in front of the mirror.

Figure 7.4 shows a ray diagram for the light rays involved when a person sees a complete or head-to-toe image. Applying the law of reflection reveals that one can see one's total image in a plane mirror that is only half of one's height. Also, the person's distance from the mirror is not a factor.

It is the reflection of light that allows us to see things. Look around you. What you see in general is light reflected from walls, ceiling, floor, and other objects. Of course, there must be one or more sources of light present. These sources are generally lamps or the Sun. If you are in a dark room, then there is no reflected light, and the room is black. At night in a lighted room a transparent windowpane serves quite well as a mirror. Yet during the day we see through it, because transmitted light during the day masks the reflected light. We often see beautiful reflections in nature, as shown in Fig. 7.5. (Is the picture really right-side-up? Turn the book over and see.)

Figure 7.4 Complete figure.
For a person to see his or her complete figure in a plane mirror, the height of the mirror must be one-half of the height of the person as can be easily shown by ray tracing.

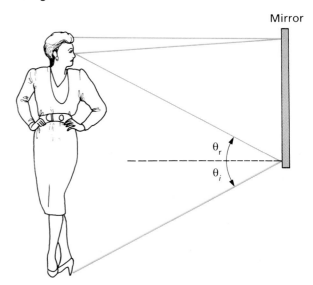

Figure 7.5 Natural reflection.
Beautiful reflections, such as the one on the still surface of this mountain lake, are often seen in nature.

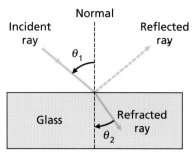

Figure 7.7 Refraction.
When light enters a transparent medium, it is deviated, or refracted, from its original path. As illustrated here for light passing from air into a denser medium such as glass, the rays are refracted or "bent" toward the normal ($\theta_2 < \theta_1$—the angle of refraction is less than the angle of incidence). Some light is also reflected at the surface.

7.2 Refraction and Dispersion

Refraction

When light strikes a transparent medium, some light is reflected and some is transmitted. This is illustrated in Fig. 7.6 for a beam of light incident to the surface of a body of water. On investigation, we find that the transmitted light has changed direction, because the speed of light changes in going from one medium to another. The deviation of light from its original path because of a speed change is called **refraction.** You have probably observed refraction effects for an object in a glass of water. For example, a fork or pencil in a glass of water will appear to be displaced and perhaps severed. (Try it.)

The directions of the incident and refracted rays are expressed in terms of the angle of incidence θ_1 and the angle of refraction θ_2, which are measured relative to a normal (line perpendicular) to the surface boundary of the medium (see Fig. 7.7). The different speeds in the different media are expressed in terms of the ratio of the speeds. This ratio is known as the **index of refraction** n:

$$\text{index of refraction} = \frac{\text{speed of light in vacuum}}{\text{speed of light in medium}} \quad (7.1)$$

or

$$n = \frac{c}{c_m}$$

The index of refraction is a pure number, because c and c_m are measured in the same units. The indices of refraction of some common substances are given in Table 7.1. Notice that the index for air is close to that for a vacuum.

Figure 7.6 Refraction in action.
A beam of light is refracted—i.e., its direction is changed—on entering the water in the glass tank.

Table 7.1 Indices of Refraction of Some Common Substances

Substance	n
Water	1.33
Crown glass	1.52
Diamond	2.42
Air (0°C, 1 atm)	1.00029
Vacuum	1.00000

When light passes obliquely ($\theta_1 > 0$) into a denser medium—for example, from air into water or glass—the light rays are refracted or "bent" toward the normal ($\theta_2 < \theta_1$). It is the slowing of the light that causes this deviation. Complex processes are involved; but intuitively, we might expect the passage of light by atomic absorption and emission through the denser medium to take longer. For example, the speed of light in water is about 75% of that in air or a vacuum.

EXAMPLE 1

What is the speed of light in water?

Solution From Table 7.1 the index of refraction for water is $n = 1.33$. Then from Eq. 7.1 we have

$$c_m = \frac{c}{n} = \frac{3.0 \times 10^8 \text{ m/s}}{1.33} = 2.3 \times 10^8 \text{ m/s}$$

Thus c_m is about 75% of c, as can be seen by the ratio

$$\frac{c_m}{c} = \frac{1}{n} = \frac{1}{1.33} = 0.75 \; (= 75\%)$$

To help understand how light is bent or refracted when it passes into another medium, consider a band marching across a field and entering a wet, muddy region obliquely (at an angle), as illustrated in Fig. 7.8a. As the marchers enter the muddy region, they keep marching at the same frequency (cadence). But slipping in the muddy ground, they don't cover as much ground and are slowed down.

The marchers in the same row on solid ground continue on with the same stride, and as a result, the direction of the marching column is changed as it enters the muddy region. This change in direction with change in marching speed is also seen when a marching band turns a corner and the inner members mark time.

We may think of wave fronts as analogous to marching rows (Fig. 7.8b). In the case of light, the wave frequency (cadence) remains the same, but the wave speed and the wavelength change ($c_m = \lambda_m f$).

When light goes from a denser medium into a less dense medium—for example, from water into air—the ray is refracted, or bent away from the normal. We may see this refraction by tracing, in reverse, the ray of light going from glass into air in Fig. 7.6. This type of refraction is shown in Fig. 7.9. But notice that an interesting thing happens as the angle of incidence becomes larger. The refracted ray is bent farther from the normal, and at a particular critical angle θ_c the refracted ray is along

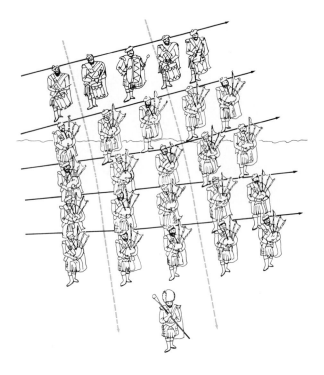

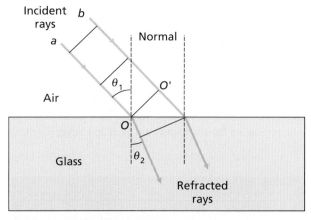

Figure 7.8 Refraction analogy.
(a) *Above:* Marching obliquely into a muddy field causes a band to change direction. The cadence or frequency remains the same, but the marchers in the mud slip and travel shorter distance (shorter wavelengths). This is analogous to the refraction of a wavefront (b) (*below*).

the boundary of the two media. For angles greater than θ_c the light is reflected and none is refracted. This phenomenon is called **total internal reflection.** An illustration of reflection, refraction, and total internal

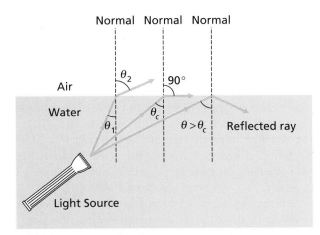

Figure 7.9 Internal reflection.
When light goes from a denser medium into a less dense medium, such as illustrated here for water into air, it is refracted away from the normal. At a certain critical angle, θ_c, the angle of refraction is 90°. For incidence above the critical angle, the light is internally reflected.

Figure 7.10 Reflection, refraction, and total internal reflection.
Light enters from the left and is partially reflected and refracted by the triangular piece of glass. The lower two beams are totally internally reflected and emerge from the lower surface of the glass.

reflection is shown in Fig. 7.10. With total internal reflection a prism can be used as a mirror.

Internal reflection enhances the brilliance of cut diamonds. A diamond is cut so that most of the light entering the diamond is internally reflected. The light then emerges from only certain portions with a brilliance far exceeding that of the incoming light. This effect creates the diamond's beautiful sparkle.

Another example of total internal reflection occurs when a fountain of water is illuminated from below. The light is totally reflected within the streams of water, providing a spectacular effect. Similarly, light can travel along transparent plastic tubes called "light pipes." When the incident angle for light in the tube is greater than the critical angle, the light undergoes a series of internal reflections down the tube.

Light can also travel along thin fibers, and bundles of such fibers are used in the relatively new field of *fiber optics* (Fig. 7.11). You have probably seen fiber optics used in decorative lamps. An important use of the flexible fiber bundle is to pipe light to hard-to-reach places. Light may also be transmitted down one set of fibers and reflected back through another so that an image of the illuminated area may be seen. This illuminated area may be a person's stomach or heart in medical applications or some recessed region that cannot normally be seen.

Figure 7.11 Fiber optics.
A fiber-optic bundle held between a person's fingers. Notice how the ends of the fibers are lit due to the transmission of light by multiple internal reflections.

Fiber optics is also used in telephone communications. In this application electronic signals in wires are replaced by light (optional) signals in fibers.

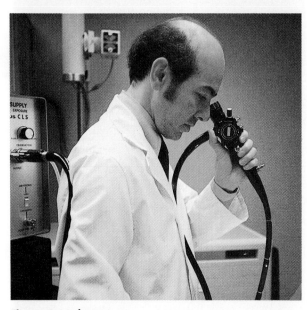

Figure 7.11 (*cont.*)
Above: A fiber-optics application in the form of a decorative lamp. *Below:* Another fiber-optics application. A doctor looks through an endoscope used for digestive tract operations. Light can be reflected down some fibers and back through others, allowing a view of otherwise inaccessible places.

Dispersion

The angle of refraction of light depends somewhat on the light's frequency or wavelength. The fact that different frequencies are refracted at slightly different angles is called **dispersion.** When white light (electromagnetic radiation containing all wavelengths visible to the human eye) passes through a glass prism, as illustrated in Fig. 7.12, the light rays are refracted on entering the glass and refracted again on leaving the glass. Because the speed changes on entering (decreases) and on leaving (increases), the directions will change accordingly.

The amount of refraction is a function of the wavelength, with the shortest wavelengths being deviated from their path by the greatest amount. Violet light has a shorter wavelength than red light. A prism used in this fashion produces the visible color spectrum, and we say the white light has been dispersed. A diamond is said to have "fire" because of colorful dispersion, in addition to having brilliance due to internal reflection.

The **spectroscope** is an instrument that separates a source of light into its respective frequencies, or wavelengths. Figure 7.13a shows the major parts of the instrument. Light from a source falls upon the slit, a very narrow opening used to reduce blurring of the spectral lines by preventing overlaps of the spectrum from other parts of the incoming beam of light. The light beam is then collimated (made parallel) and directed as a parallel beam upon the prism, which separates the light into its respective wavelengths.

The light is then focused and directed upon a detecting device. The detecting instrument is usually the eye for a spectroscope, which is used simply to view the color spectrum. When a scale for measuring the frequency or wavelength of a component is included, we call the instrument a *spectrometer.* (Fig. 7.13b) [A diffraction grating (Section 7.3), which produces sharper lines between colors, is commonly used instead of a prism.] Also, if the light intensity of the different components is recorded in some manner, such as on a photographic plate or a plotter, we commonly call the instrument a *spectrograph.*

Astronomers, chemists, physicists, and other scientists use these instruments to study the universe. Every substance, when sufficiently heated, gives off characteristic frequencies. These frequencies can be detected, their spectra studied, and the substance identified.

The study of spectra is called *spectroscopy.* The study of the universe with these instruments has given us more basic information about ourselves and the universe than any other instrument.

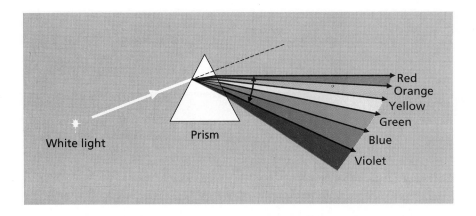

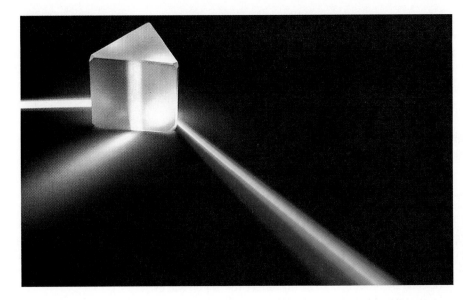

Figure 7.12 Dispersion.
Above: White light is dispersed into a spectrum of colors by a prism because the shorter wavelengths toward the blue end of the spectrum are refracted more than the longer wavelengths toward the red end.
Left: An actual spectrum produced by dispersion in a prism.

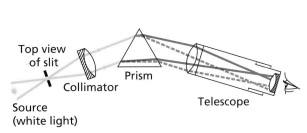

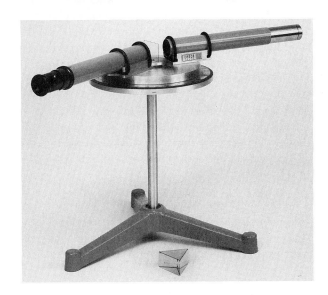

Figure 7.13 Spectroscope and spectrometer.
(a) *Above:* A spectroscope is an instrument used to separate light into its respective component frequencies or colors. (b) *Right:* Equipped with a scale to measure the frequency, the instrument becomes a spectrometer.

Atomic Hydrogen

Sodium Lamp

Figure 7.13 (*cont.*)
(c) Examples of line spectra that can be observed with a spectroscope.

When we observe a photograph taken by a spectrograph of a known or unknown source of electromagnetic radiation, we obtain an image of the slit that is illuminated by the source. The images, representing definite wavelengths, appear as bright vertical lines (see Fig. 7.13c).

7.3 Diffraction, Interference, and Polarization

Diffraction

Waves will also deviate from a straight-line path when they pass through a narrow slit. The effect also occurs when light from a fairly bright source passes the edge of an opaque object. Close observation will show a shadow at the edges surrounded by narrow bands, or fringes. This phenomenon can easily be seen by holding two fingers close together to form a narrow slit and observing a bright light from a distance. If a fluorescent lamp is available, position the fingers parallel to the length of the lamp and try it. In all these cases, the deviation of light waves is referred to as **diffraction.**

The diffraction of a water wave as it passes through a small slit is shown in Fig. 7.14. Note how the waves bend around the slit as they pass through. All waves—sound, light, and so on—show this type of bending as they go through relatively small slits. Sound is heard around corners due sometimes to diffraction as well as to reflection.

In our study of reflection and refraction we located a mirror image by drawing straight lines; that is, we solved the problem by using geometric methods. The study of optics using these methods is known as **geometric optics.** Diffraction, however, cannot be explained by using geometric optics. We must return to the theory of wave motion to give a satisfactory explanation of the effects observed. A study of optics from this point of view is known as **physical optics.**

In terms of physical or wave optics, the key ratio to a quantitative understanding of diffraction is

$$\frac{\text{wavelength of wave}}{\text{size of opening or obstacle causing diffraction}}$$

or
$$\frac{\lambda}{d}$$

If λ/d is much less than 1,—that is, λ is much less than d—very little diffraction occurs. However, if λ/d is approximately 1 or greater (λ is greater than d), diffraction effects are easily observed. Figure 7.14 is a good example of this effect for water waves.

We are well aware that sound waves bend around everyday objects while light waves do not. For instance, if we hold a newspaper in front of our face and speak, the sound of our voice can be heard, but our face cannot be seen. The phenomenon of diffraction and the ratio of λ/d help to explain why.

We know that audible sound waves have wavelengths of centimeters to meters, while visible light waves have wavelengths of around 10^{-6} m. Ordinary objects have dimensions d of centimeters to meters. Thus we get

$$\left(\frac{\lambda}{d}\right)_{\substack{\text{visible} \\ \text{light}}} \approx 10^{-6} \ll 1$$

$$\left(\frac{\lambda}{d}\right)_{\substack{\text{audible} \\ \text{sound}}} \geq 1$$

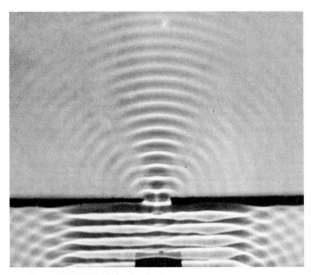

Figure 7.14 Diffraction.
Plane water waves are diffracted as they pass through a slit. Notice how the waves bend or curve around the corners of the slit.

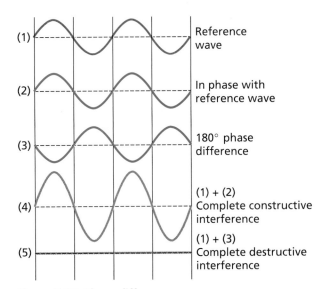

Figure 7.15 Phase differences.
Waves (1) and (2) are in phase. If the waves interfere, complete constructive interference occurs. Wave (1) and (3) are 180° out of phase. If these waves interfere, complete destructive interference occurs.

From these statements we see that diffraction readily occurs for audible sound but not for visible light.

When you sit in a lecture room, a movie theater, or a concert hall, the sound will easily diffract around the people in front of you, but the light will not. This effect occurs because sound wavelengths are so long and light wavelengths are so short compared with the size of people.

Radio waves are electromagnetic waves of very long wavelength, some being hundreds of meters, so (λ/d) for radio waves is much greater than 1, and radio waves are easily diffracted around houses, trees, and so on.

Interference

Interference effects are defined as changes in wave motion produced by phase and amplitude relations of two or more waves. The amplitude of a wave has been defined (Section 6.1) as the maximum displacement of the disturbance from its midposition. Two or more waves may reinforce one another when they are changing in the same direction, or they may cancel one another when varying in opposite directions. When two or more waves have their displacement at all times in the same direction, they are said to be *in phase* with one another. If their displacements at all times are in opposite directions, they

are *completely out of phase* with one another. Figure 7.15 illustrates the concept of phase.

When interference occurs between waves, the two waves must remain exactly in phase for **constructive interference** (when waves reinforce each other) and 180° out of phase for **destructive interference** (when they completely cancel each other). For *complete* destructive interference, the two waves must have the same amplitude. They must also have the same wavelength (Fig. 7.15).

The colorful displays seen in oil films and soap bubbles can be explained by interference. Consider light waves incident on a thin film of oil on the surface of water or on a wet road. Part of the light is reflected at the air-oil surface and part is transmitted. Part of the light in the oil film is then reflected at the oil-water surface (see Fig. 7.16).

The two reflected waves may be in phase, totally out of phase, or somewhere in between. In Fig. 7.16 the waves are shown in phase, but this result will occur only for certain angles of observation, wavelengths of light (colors), and thicknesses of oil film. At certain angles and oil thicknesses only one wavelength of light shows constructive interference. The other visible wavelengths interfere destructively. Thus these wavelengths are transmitted and not reflected.

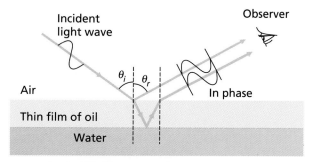

Figure 7.16 Thin-film interference.
When the reflected rays from the bottom and top surfaces of the oil film are in phase, constructive interference occurs and an observer sees only one color of light for a certain angle and film thickness. If the reflected waves are out of phase, destructive interference occurs, which means the light is transmitted at the bottom of the film surface rather than reflected. Because the film thickness varies, a colorful display is seen for the different wavelengths of light.

Hence different wavelengths interfere constructively for different oil-film thicknesses, and an array of colors is seen. In soap bubbles, where the thickness of the soap film moves and changes with time, so does the array of colors.

Diffraction can also give rise to interference. This interference can arise from the bending of light around the corners of a single slit (see the regions of constructive and destructive interference in Fig. 7.14), but an instructive technique employs two narrow double slits that can be considered point sources, as illustrated in Fig. 7.17. When the slits are illuminated with monochromatic light, the diffracted light through the slits spreads out and interferes constructively and destructively at different points where crest meets crest and crest meets trough, respectively. By placing a screen a distance from the slits, an observer can see an interference pattern of alternate bright and dark fringes.

This experiment was done in 1801 by the English scientist Thomas Young. It demonstrated the wave nature of light, and Young was able to compute the wavelength of light from the geometry of the experiment.

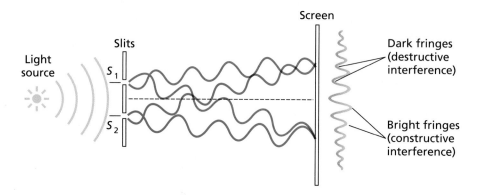

Figure 7.17 Double-slit interference.
Above: Light through two narrow slits which act as point sources interferes, giving rise to regions of constructive interference, or bright fringes, and regions of destructive interference, or dark fringes. *Right:* An actual double-slit interference pattern.

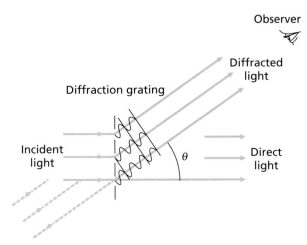

Figure 7.18 Diffraction-grating interference pattern.

Direction from which diffracted light appears to be coming

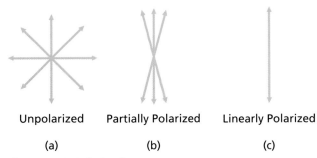

Figure 7.19 Polarization.
(a) When the electric field vectors are randomly oriented as viewed along the direction of propagation, the light is unpolarized. (b) With preferential orientation, the light is partially polarized. (c) If the field vectors lie in a plane, the light is linearly polarized.

This principle may be extended. A **diffraction grating** consists of many narrow, parallel slits spaced very closely together. There may be 10,000 lines per centimeter or more on a diffraction grating. Lines may be ruled on glass, but more commonly a thin film of aluminum is deposited on a glass plate. Then some of the metal is removed to make regularly spaced, parallel lines. The interference patterns produced by diffraction gratings are very sharp compared with the patterns for only two slits (see Fig. 7.18).

Diffraction gratings are used to separate the components or wavelengths of light from various sources, such as stars and other light sources. Different elements and compounds, when heated and made incandescent, give off certain characteristic spectra. From the spectra the composition of a light source can be analyzed. For example, the gaseous element helium was first identified in the solar spectrum—hence the name helium (from *helios*, the Greek word for the Sun).

Polarization

Light waves are transverse electromagnetic waves with the electric and magnetic field vectors oscillating perpendicular to the direction of propagation (see Fig. 6.6). The atoms of a light source generally emit light waves that are randomly oriented, and a beam of light has transverse field vectors in all directions. When we view a beam of light from the front, the transverse field vectors may be represented as shown in Fig. 7.19a. In the figure the electric field vectors are in a plane perpendicular to

the direction of propagation. Such light is said to be **unpolarized**—that is, the field vectors are randomly oriented. **Polarization** refers to the preferential orientation of the field vectors. If there is some partial preferential orientation of the field vectors, the light is **partially polarized** (Fig. 7.19b). If the vectors are in a single plane, the light is **linearly polarized** (Fig. 7.19c).

A light wave may be polarized by several means. A common method is the use of Polaroid film. Polaroid film has a polarization direction associated with the long molecular chains of the polymer film. The polarizer allows only the components in a plane to pass, as illustrated in Fig. 7.20a. The other field vectors are absorbed and do not pass through the polarizer.

The human eye cannot detect polarized light, so an analyzer, perhaps another polarizing sheet, is needed. If a second polarizer is placed in front of the first polarizer, as illustrated in Fig. 7.20b, then little (theoretically no) light is transmitted and the sheets appear dark. When the polarization directions of the sheets are at 90°, the Polaroid films are said to be "crossed" (Fig. 7.20c). The polarization of light is experimental proof that light is a transverse wave. Longitudinal waves, such as sound, cannot be polarized.

A common application of polarization is in Polaroid sunglasses. The lenses of these glasses are polarizing sheets oriented so that the polarization direction is vertical. When sunlight is reflected from a surface, such as water or a road, the light is partially polarized in the horizontal direction. The reflections increase the intensity, which an observer sees as glare. Polarizing

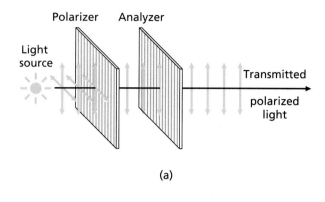

(a)

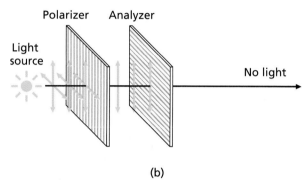

(b)

(c)

Figure 7.20 Polarized light.
(a) Light is polarized when it passes through the polarizer and also passes through the analyzer if it is similarly oriented. (b) When the polarization direction of the analyzer is perpendicular to that of the polarizer ("crossed Polaroids"), little or no light is transmitted. (c) The actual conditions of (a) and (b).

sunglasses allow only the vertical component of the light to pass, and the horizontal component is blocked out, which reduces the glare (Fig. 7.21).

Figure 7.21 Polarizing sunglasses.
Above: Light reflected from surfaces is generally polarized in the horizontal direction. By orienting the polarizing direction of sunglasses vertically, the horizontal component is blocked, thereby reducing the glare. *Below:* Old-time glare-reduction advertisement from the Polaroid Corporation Archives. Notice the vintage of the cars and the price of the glasses.

7.4 Spherical Mirrors

Spherical surfaces can be used to make practical mirrors. The geometry of a spherical mirror is shown in Fig. 7.22. A spherical mirror is a section of a sphere of radius *R*. A line drawn through the center of curvature *C* perpendicular to the mirror surface is called the *principal*

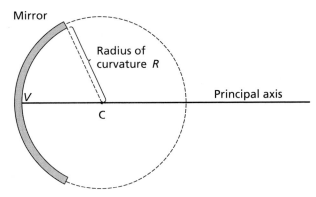

Figure 7.22 Spherical mirror geometry.
A spherical mirror is a section of a sphere with a center of curvature C. The focal point f is midway between C and the vertex V. The radius of curvature R then is twice the focal length (distance between V and f, and also commonly designated f). Hence, $R = 2f$ or $f = R/2$.

axis. The point where the principal axis meets the mirror surface is called the *vertex* (V in the figure).

Another important point in spherical mirror geometry is the *focal point f.* The distance from the vertex to the focal point is called the **focal length.** (What is "focal" about the focal point and focal length will become evident shortly.) For a spherical mirror the focal length is one-half the value of the radius of curvature of the spherical surface. Expressed in symbols, the focal length is

$$f = \frac{R}{2} \qquad (7.2)$$

where f = the focal length,

R = the radius of curvature.

The inside surface of a spherical section is said to be *concave* (as though looking into a recess or cave); and if light is reflected from this surface, we have a **concave mirror.** A concave mirror is commonly called a **converging mirror,** for the reason illustrated in Fig. 7.23a. Light rays parallel to the principal axis converge and pass through the focal point. Off-axis parallel rays converge in the focal plane.

Similarly, the outside surface of a spherical section is said to be *convex;* and when light is reflected from this surface, we have a **convex mirror.** A convex mirror is commonly called a **diverging mirror.** Parallel rays along the principal axis are reflected so that they *appear* to diverge from the focal point (Fig. 7.23b). Considering

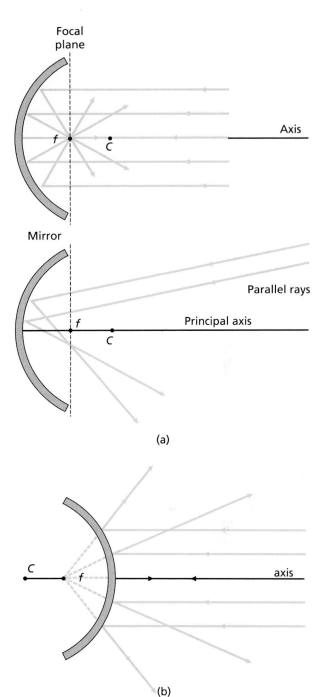

(a)

(b)

Figure 7.23 Spherical mirrors.
(a) Rays parallel to the principal axis of a concave spherical mirror converge at the focal point. Rays not parallel to the axis converge in the focal plane. Extended images are formed under these conditions. (b) Rays parallel to the axis of a convex, or diverging, spherical mirror are reflected so as to appear to diverge from the focal point inside the mirror.

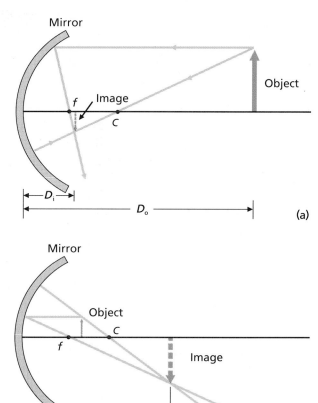

Figure 7.23 (cont.)
(c) The divergent property is used to give an expanded field of view, as shown here. (Consider the reverse ray tracing in Fig. 7.23b.)

reverse ray tracing, light rays coming to the mirror from the surroundings are made parallel, and an expanded field of view is seen in the diverging mirror. Diverging mirrors are used to monitor store aisles and to give truck drivers a wide rear view of traffic (Fig. 7.23c).

The images formed by spherical mirrors may be found by using ray diagrams. We commonly use an arrow as the object, and we determine the location and size of the image by drawing two rays:

1. Draw a ray parallel to the principal axis that is reflected through the focal point.

2. Draw a ray through the center of curvature C that is perpendicular to the mirror surface and is reflected back along the incident path.

The intersection of these rays locates the position of the image.

These rays are shown in Fig. 7.24 for an object at various positions in front of a concave mirror. The characteristics of an image are described as being real or virtual, upright (erect) or inverted, and larger or smaller than the object. A **real image** is one for which the light rays converge so that an image can be formed on a screen. A **virtual image** is one for which the light rays diverge and cannot be formed on a screen. Both real and virtual images are formed by a concave (converging) spherical mirror.

Notice that the real images are formed in front of the mirror where a screen could be positioned. Virtual images are formed behind the mirrors where the light

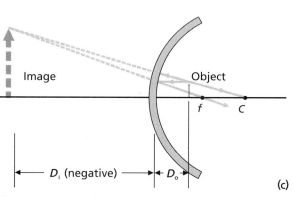

Figure 7.24 Ray diagrams.
(a) The ray diagram for an object beyond the center of curvature C for a concave spherical mirror showing where the image is formed. (b) Ray diagram for a concave spherical mirror for an object between f and C. The images are real and could be seen on a screen placed at the image distances in these cases. Notice how the image moves out and gets bigger as the object moves toward the mirror. (c) Ray diagram for a concave spherical mirror with the object inside the focal point. In this case, the image is virtual and formed and is behind, or "inside," the mirror.

rays *appear* to converge. For a converging mirror, a virtual image results when the object is inside the focal point, and a virtual image always results for a diverging mirror, wherever the object is located. (What type of image is formed by a plane mirror?)

Our ray diagrams have to be drawn to scale if we are to find the locations and sizes of the images. However, the location of the image may be found more quickly by using the **spherical mirror equation:**

$$\frac{1}{D_o} + \frac{1}{D_i} = \frac{1}{f} \tag{7.3}$$

where D_o = object distance from the vertex of the mirror,

D_i = image distance from the vertex of the mirror,

f = focal length.

The values of the object distance D_o and the focal length f are considered to be positive for a concave mirror. (For a convex mirror f is given a negative value.) Then if D_i is positive, the image is real or formed in front of the mirror. If D_i is negative, the image is virtual and behind the mirror.

The relative size of the image compared with that of the object is given by the **magnification** M:

$$M = -\frac{D_i}{D_o} \tag{7.4}$$

The minus sign in the equation is used to tell whether the image is upright or inverted. For example, if D_i is positive, M will be negative, which indicates that the image is inverted. But if M is positive (negative D_i), then the image is upright. This sign convention is illustrated in the following examples. Even when we use the mirror equation, initially making a quick ray diagram sketch is helpful, so that the general image characteristics are known in advance.

EXAMPLE 2

An object is placed 20 cm from a concave mirror with a focal length of 5 cm.
(a) At what distance from the mirror will the image be formed?
(b) What are the image characteristics?

Solution (a) We are given $D_o = 20$ cm and $f = 5$ cm. Then using Eq. 7.3 in the form

$$\frac{1}{D_i} = \frac{1}{f_o} - \frac{1}{D_o}$$

we obtain

$$\frac{1}{D_i} = \frac{1}{5} - \frac{1}{20}$$

$$= \frac{4}{20} - \frac{1}{20} = \frac{3}{20}$$

or $$D_i = \frac{20}{3} \text{ cm} = 6.7 \text{ cm}$$

(The centimeter units were omitted in the computation of the fractions for clarity.)
(b) Because D_i is positive, we know that the image is real; that is, it is formed 6.7 cm *in front* of the mirror. To find the other characteristics, we use the magnification equation:

$$M = -\frac{D_i}{D_o} = -\frac{\frac{20}{3} \text{ cm}}{20 \text{ cm}} = -\frac{1}{3}$$

Thus because M is negative, the image is inverted. The numerical value of M indicates that the image is smaller than the object. A magnification (factor) of $\frac{1}{3}$ indicates that the image is one-third as tall as the object. For example, if the object were 30 cm tall, the image would be 10 cm tall. Note that this example generally corresponds to the ray diagram in Fig. 7.24a.

EXAMPLE 3

Consider a concave mirror with a radius of curvature of 50 cm. An object is located 20 cm from the mirror.
(a) Where will its image be?
(b) Will the image be upright or inverted, larger or smaller?

Solution (a) We are given $R = 50$ cm, so $f = R/2 = (50 \text{ cm})/2 = 25$ cm, and $D_o = 20$ cm. A convenient form of Eq. 7.3 is obtained by solving directly for D_i. The result is

$$D_i = \frac{D_o f}{D_o - f} \tag{7.3}$$

Then $$D_i = \frac{(20 \text{ cm})(25 \text{ cm})}{20 \text{ cm} - 25 \text{ cm}}$$

$$= -100 \text{ cm} = -33.3 \text{ cm}$$

The minus sign indicates that the image is virtual, and it is located 100 cm behind or "inside" the mirror.
(b) Using the magnification equation, we get

$$M = -\frac{D_i}{D_o} = -\frac{(-100 \text{ cm})}{20 \text{ cm}} = +5$$

Because M is positive, the image is upright and five times larger than the object. This example corresponds to the ray diagram in Fig. 7.24c with the object inside the focal point. A virtual, upright image is always formed for a concave mirror in this case.

Example 3 corresponds to the case of a makeup mirror, which is a concave mirror of relatively long focal length. Makeup mirrors give moderate magnification so that facial features can be better seen (Fig. 7.25).

Convex mirrors may also be treated by ray diagrams and the spherical mirror equation. In this case the rays of a ray diagram are drawn by using the law of reflection, but they are extended through the focal point and the center of curvature inside or behind the mirror, as shown in Fig. 7.26. A virtual image is formed where these extended rays intersect. As we may gather from the figure, even though the object distance may vary, the image of a convex mirror is always virtual, upright, and smaller than the object (see Fig. 7.23b).

When we use the spherical mirror equation for convex mirrors, everything is the same as before, *except* that the focal length of a convex mirror is taken to be negative. The following example illustrates the use of the mirror equation for this case.

EXAMPLE 4

What is the magnification of an image in a convex mirror for an object 6 m from the mirror if the focal length of the mirror is 2 m? (Consider an image in a store mirror, as in Fig. 7.23c.)

Solution Using the previous form of Eq. 7.3 with

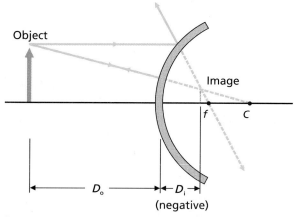

Figure 7.26 Diverging-mirror ray diagram.
The ray diagram for a convex spherical mirror with an object in front of the mirror. A convex mirror always forms a virtual image.

$f = -2$ m and $D_o = 6$ m, we get

$$D_i = \frac{D_o f}{D_o - f} = \frac{(6 \text{ m})(-2 \text{ m})}{6 \text{ m} - (-2 \text{ m})} = -\frac{12 \text{ m}}{8} = -1.5 \text{ m}$$

which tells that the image is virtual (which we already know). Then

$$M = -\frac{D_i}{D_o} = -\frac{(-1.5 \text{ m})}{6 \text{ m}} = +\frac{1}{4}$$

and the image is smaller than the object—only one-fourth as tall. The positive value of M indicates that the image is upright.

7.5 **Lenses**

A lens consists of material such as a transparent piece of glass or plastic that refracts light waves to give an image of an object. Lenses are extremely useful and are found in eyeglasses, telescopes, magnifying glasses, cameras, and many other optical devices.

In general, there are two main classes of lenses. A **converging lens** or **convex lens** is thicker at the center than at the edge. A **diverging lens** or **concave lens** is thicker at the edges. These two classes and some of the possible shapes for each are illustrated in Fig. 7.27.

Light passing through a lens is refracted twice—once at each surface. The lenses most commonly used are known as thin lenses. Thus when constructing ray diagrams, we can neglect the thickness of the lens, and we can assume that the two surfaces that causes refraction are in essentially the same plane.

Figure 7.25 Magnification.
Concave makeup mirrors give moderate magnification so facial features can be better seen.

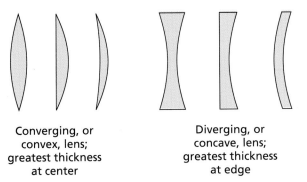

Converging, or convex, lens; greatest thickness at center

Diverging, or concave, lens; greatest thickness at edge

Figure 7.27 Lenses.
Different types of converging and diverging lenses.

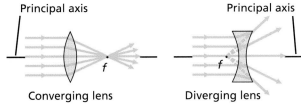

Principal axis

Principal axis

Converging lens

Diverging lens

Figure 7.28 Lens focal points.
For a converging spherical lens, rays parallel to the principal axis and incident on the lens converge at a focal point on the opposite side of the lens. Rays parallel to the axis of a diverging lens appear to diverge from a focal point on the incident side of the lens.

The principal axis for a lens goes through the center of the lens and is labeled in Fig. 7.28. Rays coming in parallel to the principal axis are refracted toward the principal axis by a converging lens. For a converging lens the rays are focused at point f, the focal point. For a diverging lens the rays are refracted away from the principal axis and appear to emanate from the focal point.

We can show how lenses refract light to form images by a graphic procedure similar to what we did with mirrors. In this procedure two rays are drawn from the tip of the object.

1. The first ray travels straight through the center of the lens and is not bent.

2. The second ray travels parallel to the principal axis and then is bent by the lens along a line drawn through a focal point of the lens.

The image of the tip of the arrow occurs where these two rays meet.

Examples of this procedure are shown in Fig. 7.29. Notice that only the focal points for the respective

Figure 7.29 Ray diagrams.
(a) *Above:* Ray diagram for a converging lens with the object outside of the focal point. (b) *Center:* A real image is formed on the opposite side of the lens that could be viewed on a screen as shown in the picture with a candle object. Notice that the image on the screen is inverted. (c) *Below:* Ray diagram for a diverging lens. A virtual image is formed on the object side of the lens.

surfaces are shown, which are all that are needed. The lenses do have radii of curvature, but for spherical lenses $f \neq R/2$, in contrast to $f = R/2$ for spherical mirrors. The characteristics of the images formed by a converging or convex lens (Fig. 7.29a) change just the way those of a converging mirror change as the object is brought toward the mirror from a distance. Beyond the focal point an inverted, real image is formed, which becomes larger as the object approaches the focal point. The magnification becomes greater than 1 when the object distance is less than $2f$. Once inside the focal point of a converging mirror, an object always forms a virtual image. For lenses a *real image* is formed on the opposite side of a lens from the object and can be seen on a screen (Fig. 7.29b). A *virtual image* is formed on the object side of the lens.

For a diverging or concave lens the image is always upright and smaller than the object. When looking through a concave lens, we see images as shown in Fig. 7.30. Possibly you have seen a flat plastic lens like this one. They are sold commercially and sometimes used on the back of vans to show a good rear view of traffic (similar to a diverging side mirror). The lens is flat on one side and has circular grooves on the other. The grooves are made as illustrated in Fig. 7.31. Such a lens is called a *Fresnel lens,* after Augustin Fresnel (1788–1827), a French physicist who realized that all of the refraction of a lens takes place at its surface, so the interior part is not really needed. A similar grooving

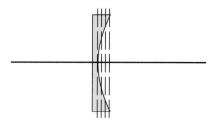

Ordinary concave lens

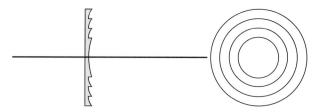

Flat grooved concave lens   Head on view of grooves

Figure 7.31 Fresnel lens.
A flat, grooved concave lens is made with the same curvature as an ordinary concave lens.

process can be used to make a flat-grooved converging, or plano-convex, lens.

In working out lens problems, we usually want to know the image distance, which is the distance between the lens and an image (see Fig. 7.29). The object distance is labeled D_o, and the image distance is labeled D_i. In the one-lens situations we will meet D_o will always be positive.

The image distance D_i is positive if the image is on the opposite side of the lens from the object. Conversely, D_i is negative if the image is on the same side as the object. We have the latter case in Fig. 7.29b, so D_i is labeled "negative."

The distance between the focal point and the lens is the focal length f. The focal length f is taken to be positive for a convex lens and negative for a concave lens.

Thin-lens problems can be solved analytically with a single equation as long as the proper signs are used. The **thin-lens equation** is

$$\frac{1}{D_o} + \frac{1}{D_i} = \frac{1}{f} \tag{7.5}$$

Note that the thin-lens equation and the spherical mirror equation (Eq. 7.3) are mathematically identical.

When D_i is positive, the image will be a real image; that is, it can be brought to a focus on a screen. When

Figure 7.30 Diverging lens.
Like a diverging mirror, a diverging lens gives an expanded field of view.

D_i is negative, the image will be virtual and can be viewed only by looking through the lens.

The **magnification** of a lens is given by the same formula we had for mirrors:

$$M = -\frac{D_i}{D_o} \qquad (7.6)$$

The following examples illustrate the use of the lens equation.

EXAMPLE 5

An object is placed 6 cm in front of a convex lens with a focal length of 10 cm. Calculate the image position.

Solution Substituting the values into the lens equation (Eq. 7.5), we get

$$\frac{1}{D_i} = \frac{1}{f} - \frac{1}{D_o}$$

$$\frac{1}{D_i} = \frac{1}{10} - \frac{1}{6} = \frac{6}{60} - \frac{10}{60} = -\frac{4}{60} \text{ cm}$$

or $\qquad D_i = -15 \text{ cm}$

Because D_i is negative, the image is virtual and located on the same side of the lens as the object.

Such a convex lens acts like a simple magnifying glass, as illustrated in the ray diagram in Fig. 7.32. The object to be viewed and magnified must be located inside the focal point of the lens. The magnification is given by

$$M = -\frac{D_i}{D_o} = -\frac{(-15 \text{ cm})}{6 \text{ cm}} = +2.5$$

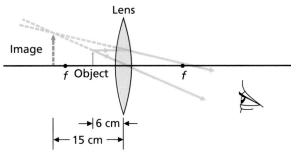

Figure 7.32 Simple magnifying glass.
An object inside the focal point of a converging lens forms an enlarged virtual image that is viewed by the eye and the object appears magnified.

EXAMPLE 6

A concave lens with a focal length of -80 cm is used to view an object 120 cm from the lens.

(a) Where will the image be located?
(b) What are its characteristics?

Solution (a) This situation is similar to the one for the concave lens ray diagram shown in Fig. 7.29b. We are given $D_o = 120$ cm and $f = -80$ cm. Then using the alternative form of Eq. 7.5, we get

$$D_i = \frac{D_o f}{D_o - f} = \frac{(120 \text{ cm})(-80 \text{ cm})}{120 \text{ cm} - (-80 \text{ cm})} = -48 \text{ cm}$$

Because D_i is negative, we know the image is virtual and on the same side of the lens as the object.
(b) Using the magnification equation, we get

$$M = -\frac{D_i}{D_o} = -\frac{(-48 \text{ cm})}{120 \text{ cm}} = +0.4$$

Hence the image is upright (positive) and only 0.4 times as tall as the object.

A slide projector consists of a convex lens and a light source to make the slide bright. The slide is placed in a position that is slightly farther from the lens than the focal length. Because the image will be inverted, the slide is placed upside down, as indicated in Fig. 7.33. The image is brought to focus on a screen placed a distance D_i from the lens. As shown in Fig. 7.33, the image will be much larger than the slide itself.

EXAMPLE 7

A slide projector has a convex lens with a focal length of 20 cm. The slide is placed upside down 21 cm from the lens. How far away should the screen be from the slide projector's lens so that the slide is in focus?

Solution To solve this problem, we recognize that the slide will be in focus when the screen is placed at the

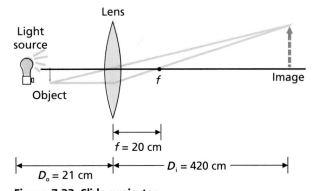

Figure 7.33 Slide projector.
A diagram of the principle of a slide projector, which produces a magnified real image that can be viewed on a screen. Notice the object (slide) is placed upside down so the image is upright on the screen.

distance D_i from the lens. We use the lens equation to compute D_i with $D_o = 21$ cm and $f = 20$ cm.

$$D_i = \frac{D_o f}{D_o - f} = \frac{(21\text{ cm})(20\text{ cm})}{21\text{ cm} - 20\text{ cm}} = 420\text{ cm}$$

Hence the image is real and will be sharply formed on a screen placed 4.2 m away from the projector lens. Also,

$$M = -\frac{D_o}{D_i} = -\frac{420\text{ cm}}{21\text{ cm}} = -20$$

Because M is negative, the image will be inverted. So the image is not upside down on the screen; for this reason we put slides in upside down in a projector. The image will be 20 times bigger than the slide dimensions in each direction. That is, a 2-in $\times$ 2-in slide will appear on a screen 4.2 m away as a 40-in $\times$ 40-in image.

The human eye contains a convex lens, along with other refractive media in which most of the light refraction occurs. Even so, we can learn a great deal about the optics of the eye by considering only the focusing action of the lens of the eye. As illustrated in Fig. 7.34, the lens focuses the light entering the eye on the *retina*. The photoreceptors of the retina, called *rods* and *cones*, are connected to the optic nerve, which sends signals to the brain. (The rods are more sensitive than the cones and are responsible for light and dark "twilight" vision; the cones are responsible for separating or distinguishing wavelengths and color.)

Because the distance between the lens and the retina does not vary, we have a situation in which D_i is a constant. Because D_o varies for different objects, the focal length of the lens of the eye must vary. The lens of the eye is called the *crystalline lens*, and it is made of glassy fibers. By action of the attached ciliary muscles the shape and the focal length of the lens is varied as it is made thinner and thicker. Through variation of the focal length the distances at which the eye can see objects clearly range from a few centimeters to infinity.

Notice in Fig. 7.34 that an image focused on the retina in normal vision is upside down. Why, then, don't we see things upside down? For some reason the brain interprets things as right-side-up and they are visualized as being properly oriented.

Defects of vision occur when the image does not fall on the retina. If the image falls in front of the retina, as in Fig. 7.35a, the eye is said to be **nearsighted.** As the object moves closer, the image distance moves back to the retina. Hence near objects can be seen clearly by a nearsighted person. Glasses consisting of diverging lenses will correct this defect.

The opposite effect occurs when the image falls behind the retina (see Fig. 7.35b). This effect causes the eye to be **farsighted,** and distant objects are seen clearly. Glasses consisting of converging lenses will correct this defect.

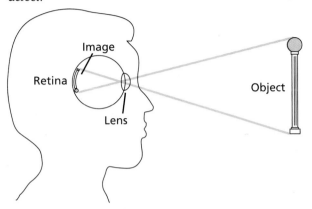

Figure 7.34 The human eye.
The human eye has a convex lens, along with other refractive media, which focuses light on the retina at the back of the eye.

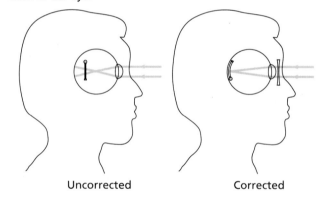

Figure 7.35 Vision defects.
(a) Nearsightedness is corrected by wearing glasses with a diverging lens.

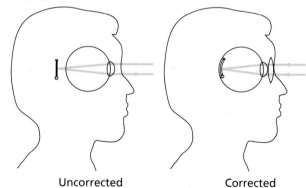

(b) Farsightedness is corrected by wearing glasses with a converging lens.

Learning Objectives

After reading and studying this chapter, you should be able to do the following without referring to the text:

1. State the law of reflection, and distinguish between regular and diffuse reflections.

2. Draw ray diagrams illustrating and explain the principles of refraction and total internal reflection.

3. Explain what is meant by the dispersion of light, and explain why it occurs.

4. Discuss spectroscopy and some of the instruments used in this study.

5. Distinguish between geometrical optics and physical optics.

6. Discuss the difference between sound waves and visible light waves with regard to diffraction and polarization.

7. Explain what causes the colors when light is reflected from a thin film of oil on water.

8. Explain how Polaroid sunglasses work.

9. Use diagrams to describe the images seen in a concave mirror when the object is (a) close to the mirror and (b) far from the mirror.

10. Use a diagram to describe the images seen in a convex mirror.

11. Draw a diagram showing how a slide projector with a convex lens works.

12. Use a diagram to show how a concave lens works, and explain how a flat plastic concave lens is made.

13. Explain how the human eye works, and explain how farsightedness and nearsightedness can be corrected.

14. Define and explain the important words and terms listed in the next section.

Important Words and Terms

reflection	physical optics	diverging mirror
ray	constructive interference	real image
law of reflection	destructive interference	virtual image
regular reflection	diffraction grating	spherical mirror equation
diffuse reflections	unpolarized light	magnification
refraction	polarization	converging lens
index of refraction	partially polarized light	convex lens
total internal reflection	linearly polarized light	diverging lens
dispersion	focal length	concave lens
spectroscope	concave mirror	thin-lens equation
diffraction	converging mirror	nearsighted
geometric optics	convex mirror	farsighted

Questions

Reflection

1. If a light ray strikes a flat mirror surface at an angle of 45° to the normal, at what angle will it be reflected? Explain the principle involved.

2. What is the difference between regular and diffuse reflections? Does the law of reflection apply in both cases? Explain.

3. Why can you see your reflection in a windowpane at night but not during the day?

Refraction and Dispersion

4. What is refraction, and what causes it?

5. Explain how the refraction of light on entering a denser medium is analogous to a situation for a marching band.

6. How is light bent or refracted on entering a (a) denser medium and (b) less dense medium?

7. Draw a diagram to show how internal reflection occurs.

8. What causes dispersion?

9. Explain why diamonds have brilliance and "fire."

Diffraction, Interference, and Polarization

10. Why do sound waves bend around everyday objects, while the bending of light rays is not generally observed?

11. Why do you see colors when looking at a film of oil on water?

12. What is a diffraction grating, and what does it do?

13. What are the differences between unpolarized, partially polarized, and linearly polarized light?

14. How do Polaroid sunglasses work?

Spherical Mirrors

15. Distinguish between real images and virtual images for spherical mirrors.

16. What are the relationships between the center of curvature, the focal point, and the vertex for spherical mirrors?

17. Explain when real and virtual images are formed by a (a) concave mirror and (b) convex mirror.

Lenses

18. Distinguish between real images and virtual images for convex and concave lenses.

19. Explain when real and virtual images are formed by a (a) convex lens and (b) concave lens.

20. Describe how a flat, converging Fresnel lens is made.

21. How does a magnifying glass magnify?

22. Why are slides put in a slide projector upside down, and what is done in focusing a slide image on a screen?

23. How is the human eye able to focus on objects at various distances so quickly?

24. What are (a) nearsightedness and (b) farsightedness, and how are these conditions corrected?

Exercises

Reflection

1. Light is incident on a flat mirror surface at an angle of 30° relative to the surface. What is the angle of reflection?

2. Show that for a person to see his or her complete (head-to-toe) image in a plane mirror, the mirror must have a length of at least one-half of the person's height (see Fig. 7.4). Does the person's distance from the mirror make a difference? Explain.

3. How much longer must the minimum length of a plane mirror be for a 6-ft 4-in man to see his complete head-to-toe image than for a 5-ft 2-in woman?   *Answer:*   7 in

Refraction

4. From the equation for the index of refraction, determine the speed of light in a diamond.

5. The speed of light in a particular type of glass is measured to be 1.6×10^8 m/s. What is the index of refraction of the glass?

6. What percentage of the speed of light in a vacuum is the speed of light in crown glass?

Diffraction

7. What will be the waveform resulting from the interference of a reference wave and a similar wave 90° out of phase. See Fig. 7.15 for reference. (Make a sketch.)

8. A slit opening in a water tank is 0.50 cm wide. Determine above what wavelength the diffraction of water waves on the surface of the water passing through the slit will be easily observed.

9. Let the dimension of an obstacle be 1.0 m.
   (a) Find λ/d for visible light with a wavelength of 600 nm.
   (b) Find λ/d for 2000-Hz sound waves.
   (c) Find λ/d for 1200-kHz radio waves.
   (d) Which of these waves are easily diffracted around the 1-m obstacle?   *Answer:*   (d) waves in (c)

Spherical Mirrors

10. Sketch ray diagrams for a concave mirror for objects at (a) $D_o > R$, (b) $R > D_o > f$, and (c) $D_o < f$.

11. An object 3 cm tall is placed 50 cm from a concave spherical mirror with a focal length of 20 cm.
    (a) Sketch a ray diagram of this situation.
    (b) Analytically determine the image distance, the image characteristics, and the height of the image.
    Answer: $D_i = 33.3$ cm
    $h_i = 2$ cm

12. A woman with a 6-cm-high forehead holds a makeup mirror with a radius of curvature of 120 cm a distance of 20 cm from her face. (a) Sketch a ray diagram of this situation. (b) Determine analytically the height of the forehead image the woman sees in the mirror.
    Answer: 9 cm

13. An object 5 cm tall is placed 10 cm from a convex spherical mirror with a focal length of 3.0 m. Determine the height and characteristics of the image.

14. Find the magnification of the image of an object located at (a) R and (b) f of a concave mirror.
    Answer: (a) $M = 1$

Lenses

15. Sketch ray diagrams for a spherical convex lens for objects at (a) $D_o > 2f$, (b) $2f > D_o > f$, and (c) $D_o < f$.

16. A 3-cm-tall object is placed 30 cm in front of a convex lens with a focal length of 20 cm.
    (a) Sketch a ray diagram of the situation.
    (b) Determine the image distance, characteristics, and height.   *Answer:*   (b) $D_i = 60$ cm, $h_i = 6$ cm

17. A simple magnifying glass has a focal length of 8 cm. An object is placed 6 cm in front of it. What is the magnification of the image?   *Answer:* $M = 4$

18. A slide projector has a convex lens with a focal length of 8.0 cm. A slide is placed 8.2 cm from the lens.
    (a) How far from the lens should a screen be placed so that the image of the slide is in focus?
    (b) If the slide film is 1 in × 1 in, what size screen is needed to see the total slide area?

19. A concave lens has a focal length of 20 cm. A 3-cm-tall object is placed 10 cm from the lens.

    (a) Sketch a ray diagram of the situation.
    (b) Determine the image distance, characteristics, and height, using the thin-lens equation.

20. Find the magnification of the image of an object located at a distance of (a) $2f$ and (b) f from a convex lens.   *Answer:*   (b) $M = \infty$

21. A particular human eye has a focal length of 2.4 to 2.2 cm and an image distance of 2.4 cm. What is the closest an object can be brought to the eye and still be in focus? (This distance is called the *near point* of the eye.)   *Answer:*   26.4 cm

22. The human eye can focus clearly on objects from a distance of about 25 cm (the near point of the eye) to infinity. If the distance from the crystalline lens to the retina is 2.5 cm, what is the range of the focal length of the lens, assuming the thin-lens equation applies to the eye?

CHAPTER **8**

Electricity and Magnetism

> Like charges repel, and unlike charges attract each other, with a force that varies inversely with the square of the distance between them. . . . Frictional forces, wind forces, chemical bonds, viscosity, magnetism, the forces that make the wheels of industry go round—all these are nothing but Coulomb's law. . . .
>
> —J. R. Zacharias

A Brief History

THE MAGNETIC PROPERTIES of the mineral lodestone (magnetite, Fe_3O_4) were known to the Greeks as early as 600 B.C. Thales of Miletus (640–546 B.C.), an early Greek mathematician and astronomer, was aware of the properties of attraction and repulsion of lodestone with similar pieces of lodestone. He also knew of an electrostatic effect called the amber effect, that is, the attraction of bits of straw to an amber rod that had been rubbed with wool.

The word *magnet* seems to have been derived from Magnesia, a province in Asia Minor, where the Greeks first discovered lodestone. The Chinese were perhaps the first to use the lodestone as a compass both on land and sea. Early records indicate that ships sailing between Canton, China, and Sumatra as early as A.D. 1000 were navigated by the use of the magnetic compass.

In the thirteenth century a Frenchman, Petrus P. de Maricount, described the magnetic compass in some detail and applied the term *pole* to the regions on the compass where the fields of influence were the strongest. The north-seeking pole he called N, and the south-seeking pole he called S. The attraction of unlike poles, the repulsion of like poles, and the formation of new unlike poles when a magnet was broken into two pieces were also described by de Maricount.

In 1600 Dr. William Gilbert (1540–1603), court physician to Queen Elizabeth, published his book on magnetism, *De Magnete*. The book contained a summary of information then known about electricity and magnetism, plus experiments carried out by Gilbert. These experiments included information on the dip (the angle the Earth's magnetic field makes with the Earth's surface) and declination (the angle the compass needle deviates from the geographical north) of the compass, the loss of magnetism by a magnet when heated, and experiments with a sphere-shaped magnet, which led him to the conclusion that the Earth acts like a huge magnet.

Gilbert was also aware of the amber effect. We now know that this effect is due to repulsion and attraction of electric charges. Gilbert carried out many experiments on the amber effect with an instrument he called a versorium (from the Latin *verso*, "to turn around"). The versorium was nothing more than a slender, arrow-type, nonconducting material balanced on a pivot point so as to give a high degree of sensitivity to the force of attraction when an amber rod or other substances were placed in its vicinity. With his versorium he discovered that many substances possess the amber effect. Gilbert is responsible for the word *electron*, which is very familiar today. He classified those substances possessing the amber effect as "electrics" (from the Greek *electron*, "amber").

Another major study of electricity and magnetism took place in the eighteenth century, when Charles Coulomb (1736–1806) established the inverse-square law of attraction and repulsion between electrostatic charges and verified the same law to hold for magnetic poles.

Electric and magnetic phenomena were studied in detail during the 1800s. In 1820 Hans Christian Oersted (1777–1851) found that a compass is deflected by a current-carrying wire. In 1831 Michael Faraday (1791–1867) and Joseph Henry (1797–1878) independently found that a magnet inserted into a coil of wire would induce an electric current.

The laws of electricity and magnetism discovered by Coulomb, Oersted, Faraday, Henry, and others were studied in detail by James Clerk Maxwell (1831–1879), a Scottish physicist. Maxwell wondered why the physical laws were not symmetric when expressed in mathematical form. By applying the concept of symmetry, he discovered an additional law, which completed the equations of electromagnetism in 1865.

Maxwell studied these equations and found that, according to the equations, light is made up of electromagnetic waves. Radio waves were also predicted at a much lower frequency than visible light waves. The theoretical fact that electromagnetic waves of radio frequency are possible led to experiments in which radio waves were generated and detected. This discovery

---

◀ **Electricity in nature: a summer lightning storm.**

opened up a new era of wireless communication and eventually brought us commercial radio and television and a host of other devices. Many of these practical developments will be described in this chapter.

8.1 Electric Charge

Electric charge is a fundamental quantity. The property of electric charge is associated with certain subatomic particles, and experimental evidence leads us to the conclusion that there are two types of charge. They are designated as *positive* (+) and *negative* (−) for distinction. All matter, according to modern theory, is made up of small particles called *atoms*, which are composed in part of negatively charged particles called **electrons,** positively charged particles called **protons,** and neutral particles that carry no electric charge, called **neutrons.** Table 8.1 summarizes the fundamental properties of these atomic particles.

As the table indicates, all three particles have a certain mass, but only the electrons and protons possess electric charges. The magnitudes of the electric charge on the electron and the proton are equal, but their natures are different, as expressed by the plus and minus signs. When we have the same number of electrons and protons, the *total* charge is zero (same number of positive and negative charges of equal magnitude), and we have an electrically *neutral* situation.

Also, scientists believe that the electronic charge is the fundamental or smallest unit of free charge in nature. Evidence indicates that subatomic particles called **quarks** have fractional electronic charges—for example, charges of $\pm\frac{2}{3}$ and $\pm\frac{1}{3}$ of the electronic charge of 1.6×10^{-19} coulomb. According to the quark theory, quarks combine to form protons and neutrons. However, quarks are believed to exist only in the nucleus of the atom and not in a free state. Therefore, the fractional-charge quarks are not a consideration in our study of electricity.

The unit of electric charge is called the **coulomb** (C), named after Charles Coulomb, the French scientist who studied electrical effects. Electric charge is usually designated by the letter q. A $+q$ indicates that an object has an excess number of positive charges, or more protons than electrons. A $-q$ indicates an excess of negative charge, or more electrons than protons.

When charge "flows," or is in motion, we say we have an electric current. **Current** is defined as the time rate of flow of electric charge. It is measured in units of amperes, named after André Ampère (1775−1836), another early French investigator of electricity. One **ampere** (A) is equal to a flow of one coulomb of charge per second.

In symbol notation we relate electric current, commonly designated by the letter I, and the electric charge by the formula

$$I = \frac{q}{t} \tag{8.1}$$

where I = electric current, measured in amperes,
q = electric charge flowing past a given point, measured in coulombs,
t = time for the charge to move past the point, measured in seconds.

From Eq. 8.1 we can write

$$q = I \times t \tag{8.2}$$

or 1 coulomb = 1 ampere × 1 second.

This equation is important because the experimental quantity used as the fundamental unit is the ampere. To be precise, the *coulomb* is defined as the amount of charge that flows past a given point in one second when the current is one ampere.

Scientists have long known that like charges repel and unlike charges attract. This statement is sometimes called the **law of charges.** Charles Coulomb first quantified this attraction and repulsion in what is now known

Table 8.1 Some Properties of Atomic Particles

Particle	Symbol	Mass	Charge
Electron	e^-	9.11×10^{-31} kg	-1.6×10^{-19} C
Proton	p^+	1.67×10^{-27} kg	$+1.6 \times 10^{-19}$ C
Neutron	n	1.67×10^{-27} kg	0

as Coulomb's law for electric charges. **Coulomb's law** states:

> **The force of attraction or repulsion between two charged bodies is directly proportional to the product of the two charges and inversely proportional to the square of the distance between them.**

This law can be written as

$$F = \frac{kq_1q_2}{r^2} \qquad (8.3)$$

where F = force of attraction or repulsion,
q_1 = magnitude of first charge,
q_2 = magnitude of second charge,
r = distance between charges.

The k is a proportionality constant with a value of

$$k = 9 \times 10^9 \ \frac{\text{newton} \times \text{meter}^2}{\text{coulomb}^2}$$

Coulomb's law is similar in form to Newton's law of universal gravitation (Chapter 3, $F = Gm_1m_2/r^2$). The forces are both dependent on the square of the separation distance. One obvious difference between the two laws is that Coulomb's law depends on charge, while Newton's law depends on mass.

There are two other important differences. One is that Coulomb's law can give rise to either an attractive or a repulsive force, depending on whether the two charges are different or the same. (This is how we know there are two types of electric charges.) The force of gravitation, on the other hand, is *always* attractive.

The other important difference is that the electric forces are comparatively much stronger than the gravitational forces. An electron and a proton are attracted to each other both electrically and gravitationally. However, when we deal with atoms, the gravitational forces can be ignored; we consider only the electric forces of attraction and repulsion.

An object with an excess of electrons is said to be negatively charged, and an object with a deficiency of electrons is said to be positively charged. A negative charge can be placed on a rubber rod by stroking the rod with fur. (Electrons are transferred from the fur to the rod by friction.) In Fig. 8.1 a rubber rod that has been rubbed with fur is shown suspended by a thin thread that allows the rod to swing freely. When a similar rubber rod that has been stroked with fur is brought close to the suspended rod, it will swing away from the second rubber rod; that is, the charged rods repel one another.

The same procedure using two glass rods that have been stroked by silk will show similar results (Fig. 8.2). Here electrons are transferred from the rod to the silk.

Although the experiments show repulsion in both cases, the charge on the glass rods is different from the charge on the rubber rods. As shown in Fig. 8.3, the

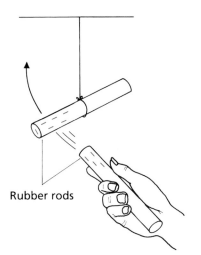

Figure 8.1 Repulsive electric force. Two negatively charged bodies repel one another.

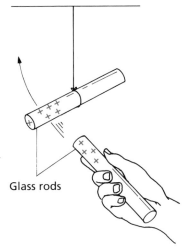

Figure 8.2 Repulsive electric force. Two positively charged bodies repel one another.

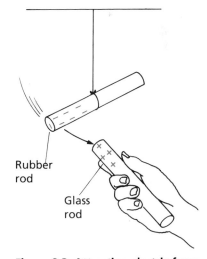

Figure 8.3 Attractive electric force. A negatively charged body and a positively charged body attract one another.

A Surface of a plate coated with a photo-conductive metal is electrically charged as it passes under wires.

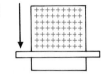

B Original document is projected through a lens. Plus marks represent latent image retaining positive charge. Charge drained in areas exposed to light.

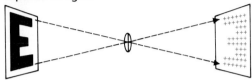

C Negatively charged powder (toner or "dry ink") is applied to the latent image, which now becomes visible.

D Positively charged paper attracts dry ink from plate, froming direct positive image.

E Image is fused into the surface of the paper or other material by heat for permanency.

Figure 8.4 Electrostatic copying.
The steps involved in making an electrostatic photocopy. See text for description.

charges on the stroked rubber and glass rods attract one another. The charge placed on the glass rod is positive; on the rubber rod the charge is negative.

The important principle of electrostatics—that unlike charges attract—is the basis of photocopying machines such as a Xerox copier. In these machines the material to be copied is placed face down on a transparent glass plate, and a copy is produced. When a page is copied, light is reflected off the page onto a positively charged cylinder or plate made of selenium. Selenium is a photoconductor, and charge leaks away from the areas struck by the light, leaving an electrostatic "image" of the dark regions or print of the copy page (Fig. 8.4).

Next, negatively charged black ink powder is dusted on the cylinder, sticking to the charged part of the selenium cylinder. A positively charged piece of paper is then passed over the cylinder, and the ink is attracted to the paper. Then the ink is fused onto the paper by heating the paper. The paper copy is then ejected.

Although the inner workings of the machine are fairly complicated, the main principle is simply the attraction of unlike charges. Perhaps you have noticed the static charge on the photocopies when they are ejected from the machine.

From Coulomb's law we see that as two charges get closer together, the force of attraction or repulsion increases. This important effect can give rise to nonzero electric forces, as illustrated in Fig. 8.5. When a negatively charged rubber comb is brought near a small piece of paper, the charges in the paper molecules are acted upon by electrical forces—positive charges attracted, negative charges repelled—and there is an effective separation of charge. The molecules are then said to be *polarized* with definite regions of charge.

Because the positive-charge regions are closer to the comb than the negative-charge regions, the attractive forces are stronger than the repulsive forces. Thus there is a net attraction between the comb and the piece of paper. Small bits of paper may be picked up by the comb, which indicates that the attractive electrical force is stronger than the paper's gravitational weight force. Keep in mind, however, that overall the paper is uncharged; only molecular regions are charged. This condition is termed *charging by induction.* Now you know why a balloon will stick to a ceiling or wall after being rubbed on a person's hair or clothing. The balloon is charged by the frictional rubbing through a transfer of charge. When the balloon is placed on a wall, the charge induces regions of charge in the molecules of the wall material, attracting the balloon to the wall (Fig. 8.5b).

Another demonstration of the electrical force is shown in Fig. 8.6. When a charged rubber rod is brought close to a thin stream of water, the water is attracted toward the rod and the stream is bent. Water molecules have a permanent separation of charge or regions of different charges. They are called *polar molecules.*

Because atoms are fundamental parts of matter, it follows that electrons and protons are always present in any material. The electrons, which are located in the outer part of the atom, move within the material with varying degrees of freedom. The protons, located in the inner part of the atom, are much heavier and are not free to move. In metals, such as copper, silver, and

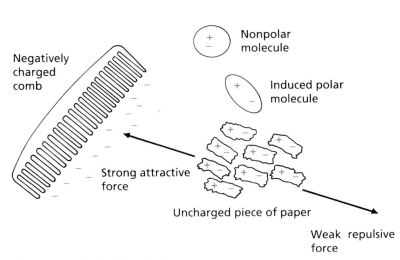

Figure 8.5 Polarization of charge.
Left: When a negatively charged rubber comb is brought near small pieces of paper, the molecules are polarized with definite regions of charge which give rise to a net attractive force. As a result, the bits of paper are attracted and adhere to the comb. *Right:* Charged balloons cling to the ceiling and wall because of the attractive forces resulting from molecular polarization.

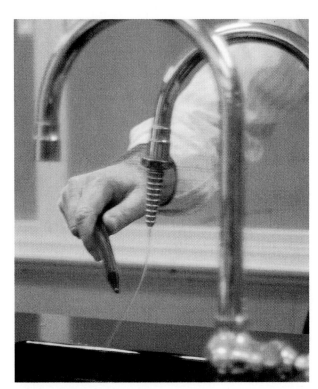

Figure 8.6 Bending water.
A charged rod brought close to a small stream of water will bend the stream because of the polarization of the water molecules.

aluminum, some of the electrons can move easily. Such metals are called **conductors.** In materials such as wood, paper, cork, glass, and plastics, the electrons cannot move freely. These materials are called **insulators.** Materials that are neither good conductors nor good insulators are called *semiconductors.*

8.2 Electricity

The effects produced by moving charges give rise to what we generally call **electricity.** For charges to move, they must be acted upon by positive or negative charges.

Consider the situation shown in Fig. 8.7. We start out with some unseparated charges and then begin to separate them. It takes very little work to pull the first negative charge to the left and the first positive charge to the right. When the next negative charge is moved to the left, it is repelled by the negative charge already there, so more work is needed. Similarly, it takes more work to move the second positive charge to the right. As we separate more and more charges, it takes more and more work.

Because work is done in separating the charges, we have **electric potential energy.** If we insert a negative charge between the separated charges, it will move to the right, as shown in Fig. 8.7; that is, it will acquire a velocity and kinetic energy.

Unseparated charges   Separated charges

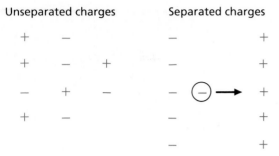

Figure 8.7 Electrical potential energy.
Work must be done in separating positive and negative charges. When separated, the charges have electric potential energy and would move if free to do so.

Instead of speaking of electric potential energy, we usually speak of a potential difference or voltage. The voltage is defined as the amount of work it would take to move a charge between two points, divided by the value of the charge. That is, **voltage** is the work per unit charge, or the electrical potential energy per unit charge. That is,

$$\text{voltage} = \frac{\text{work}}{\text{charge}}$$

or
$$V = \frac{W}{q} \qquad (8.4)$$

The **volt** (V) is the unit of voltage and is equal to one joule per coulomb. Voltage is caused by a separation of charge. Once the charges are separated, a current can be set up, and we get electricity.

When we have a current, it meets with some opposition because of collision within the conducting material. This opposition to the flow of charge is called **resistance.** The unit of resistance is the **ohm** (Ω, Greek letter omega). There is a simple relationship between voltage, current, and resistance that was formulated by Georg Ohm (1787–1854), a German physicist, which applies to many materials. This law is called **Ohm's law,** and in equation form it is written as

$$\text{voltage} = \text{current} \times \text{resistance}$$

or
$$V = IR \qquad (8.5)$$

From this equation we see that one ohm is a volt per ampere.

An example of a very simple electric circuit is shown in Figure 8.8, along with a circuit diagram. The battery provides the voltage through chemical activity (chemical energy). A conducting wire is placed between the positive and negative terminals, and electrons in the wire can move away from the negative terminal of the battery toward the positive terminal. If we put a light bulb along the path of the electrons, it provides resistance and will light up (conversion of electrical energy to radiant energy).

The switch in the circuit allows the path of the electrons to be closed or open. When the switch is open, there is not a complete path or circuit for charge to flow through, and there is no current. When the switch is closed, the circuit is completed, current flows, and the bulb will light up.

When current exists in a circuit, work is done to overcome the resistance of the circuit and power is expended. Recall that one definition of power is

$$P = \frac{W}{t}$$

From Eq. 8.4, $W = qV$, and substituting for W, we get

$$P = \frac{q}{t} V$$

We now recognize that $q/t = I$ to get

$$P = IV \qquad (8.6)$$

If we use Ohm's law for V, we get $P = I(IR)$, or

$$P = I^2R \qquad (8.7)$$

The power that is dissipated in an electric circuit is frequently in the form of heat. This heat is called *joule heat* or "I^2R losses," as given by Eq. 8.7. This heating effect is used in electric stoves, heaters, cooking ranges, hair dryers, and so on. Hair dryers have heating coils of low resistance so as to get a large current for large I^2R losses. When a light bulb lights up, much of the power goes to produce heat as well as light. The units of power are watts, and light bulbs are rated in watts (see Fig. 8.9).

EXAMPLE 1

Find the current and resistance of a 60-W, 120-V light bulb in operation.

Solution   Using Eq. 8.6, we get the current given by

$$I = \frac{P}{V}$$

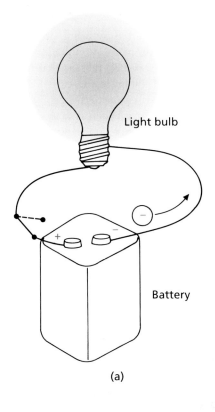

Light bulb

Battery

(a)

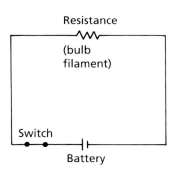

Resistance

(bulb filament)

Switch

Battery

(b)

Figure 8.8 Electric circuit.
(a) A simple electric circuit in which the battery supplies the voltage and the light bulb the resistance. When the switch is closed, electrons flow from the negative terminal of the battery toward the positive terminal. Electrical energy is expended in heating the bulb filament. (b) The circuit diagram for the circuit.

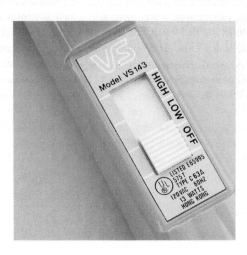

Figure 8.9 Wattage (power) ratings.
Left: A 60-W light bulb dissipates 60 J of electrical energy each second. *Right:* A curling iron uses 13 W at 120 V. Given the wattage and voltage ratings, one can find the current drawn by an element, $I = P/V$.

or $$I = \frac{60 \text{ W}}{120 \text{ V}} = 0.50 \text{ A}$$

We can then use Eq. 8.5 to get the resistance.

$$R = \frac{V}{I}$$

or $$R = \frac{120 \text{ V}}{0.50 \text{ A}} = 240 \text{ } \Omega$$

An electric company delivers electric energy. Our electric utility bills are computed for the number of kilowatt-hours (kWh) used. This is a unit of energy, because energy is used to do work and

$$W = P \times t$$

or $$(E = P \times t)$$ (8.8)

where E is the energy used.

Hence we pay an electric power company for energy, not power.

EXAMPLE 2

How much does it cost, at $0.08 per kWh, to leave a 100-W light bulb on for a month?

Solution   We use Eq. 8.8 to figure the total energy in kWh used in one month. The number of hours in a 30-day month is 24 h × 30 = 720 h. So the total energy consumed, from Eq. 8.8, is

$$E = P \times t$$
$$= 100 \text{ W} \times 720 \text{ h}$$
$$= 72{,}000 \text{ Wh}$$
$$= 72 \text{ kWh}$$

The cost is then

$$72 \text{ kWh} \times \$0.08/\text{kWh} = \$5.76$$

Electric Current and Circuits

There are two principal forms of electric current. In a battery circuit, such as in Fig. 8.8, the electron charge flow or current is always in one direction, from the negative terminal to the positive terminal. This type of current is called **direct current**, or **dc.** We use direct current in battery-powered devices such as flashlights, radios, and automobiles.

The other common type of current is **alternating current**, or **ac,** which is produced by constantly changing the voltage from positive to negative to positive and so on. (Although the usage is redundant, we commonly say ac current and ac voltage.) Alternating current is produced by electric companies and is used in the home. The frequency of changing from positive to negative voltages is usually at the rate of 60 cycles per second (cps), or 60 Hz (see Fig. 8.9). The average voltage varies between 110 and 120 V, and household ac voltage is commonly listed as 110 V, 115 V, or 120 V. The formulas for Ohm's law (Eq. 8.5) and power (Eqs. 8.6 and 8.7) apply to both dc and ac circuits containing only resistances.

One of the principal reasons we use alternating current in our homes and businesses is that the voltages of alternating currents can easily be increased or decreased by means of devices called *transformers*, which are explained in more detail in Section 8.4. In the transmission of electrical power, transformers help to minimize the energy lost to joule heating in the transmission lines by reducing the current.

Notice that Eq. 8.6 ($P = IV$) tells us that the same amount of power can be transmitted by using either a high voltage and a low current, or a low voltage and a high current. Equation 8.7 ($P = I^2R$) tells us that for a given resistance we should minimize the current to minimize the joule heat or I^2R losses due to the resistance of the wires of the transmission lines. When a transformer is used to increase the voltage ("stepped up"), the current is decreased, because by the conservation of energy the energy per time (power) is constant ($P = IV$).

Thus an electric company transmits energy at very high voltages, which are obtained by the use of transformers. Voltages may be as high as 765,000 V. The current is reduced by the same factor as the voltage is increased, so the I^2R losses in the transmission wires are greatly reduced.

This high voltage is then stepped down at area distribution stations and substations and again at the home by using transformers (Fig. 8.10), until the voltage is approximately 220 to 240 V. (The principle of a transformer will be discussed in Section 8.4 after some electromagnetic principles have been learned.)

Because electricity is transmitted at very high voltages, insulators are used to keep the wires away from the supporting towers and away from each other (Fig. 8.11). If the wires were in close proximity, the high voltage could cause an ionization of the air, which would provide a conductive path and leakage losses.

Once the electricity enters the home or business, it is used in circuits to power (energize) appliances. Plugging the appliances, lamps, and other electrical items into a wall outlet places them in a circuit. There are two basic ways of connecting elements in a circuit: in *series* or in *parallel.* An example of a series circuit is illustrated in Fig. 8.12. Notice that the lamps are conveniently represented as resistances in the circuit diagram. In a **series circuit** the same current passes through all of the resistances, and the total resistance of the circuit is simply the sum of the individual resistances. That is, the total voltage is given by

$$V = IR_1 + IR_2 + IR_3 + \cdots$$
$$= I(R_1 + R_2 + R_3 + \cdots)$$

where V = voltage of the voltage source,
   I = current passing through the various resistors,
   R_1 = first resistance,
   R_2 = second resistance, etc.

And if we write $V = IR_s$, where R_s is the total circuit

Figure 8.10 Stepping down.
After being transmitted over power lines at high voltages (and low currents) to reduce the I^2R losses, the voltage is stepped down by transformers at distribution substations (left) and by utility-pole transformers (right) to deliver 220–240 V electricity.

(series) resistance, then

$$R_s = R_1 + R_2 + R_3 + \cdots \qquad (8.9)$$

What this equation means is that the resistances in series can be replaced with a single resistance R_s, and the same current will flow and the same power will be dissipated. For example, $P = I^2R_s$ is the power used in the whole circuit. (The resistances of the connecting wires are considered negligible.)

The example of lamps or resistances in series in Fig. 8.12 could as easily have been a string of Christmas tree lights connected in series. Christmas tree lights were once connected in this manner. When a bulb burned out, the whole string of lights went out, because there was no longer a completed path for current flow and the circuit was "open." Having a burned-out or broken bulb is like opening a switch in the circuit to turn off the lights. However, in most strings of lights purchased today, one light can burn out and the others will remain lit. Why? Today's lights have a parallel connection, which will be explained in our discussion of parallel circuits.

Figure 8.11 High voltage.
Transmission towers support high-voltage transmission lines. Insulators keep the wires away from the towers, and interline separators keep the wires from touching each other.

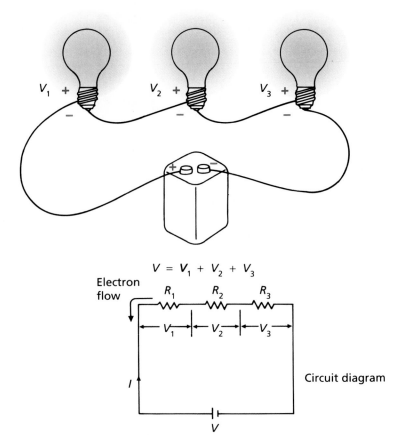

Figure 8.12 Series circuit.
The light bulbs are connected in series, so the same current flows through each of the bulbs.

The other basic type of simple circuit is called a parallel circuit. As illustrated in Fig. 8.13, in a **parallel circuit** the voltage across each resistance is the same, but the current through each resistance may vary. Notice that the current from the voltage source divides at the junction where all the resistances are connected together. This arrangement is analogous to liquid flow in a large pipe coming to a junction where it divides into several smaller pipes. Because there is no build up of charge at the junction, the charge entering the junction must equal the charge leaving the junction (conservation of charge), and we may write

$$I = I_1 + I_2 + I_3 + \cdots$$

Using Ohm's law ($I = V/R$), we may write, for the different resistances,

$$I = \frac{V}{R_1} + \frac{V}{R_2} + \frac{V}{R_3} + \cdots$$

The voltage V is the same for each R, because the voltage source is effectively connected "across" each resistance and each gets the same voltage effect.

Writing Ohm's law in terms of the *total* resistance of the parallel circuit R_p, we have

$$I = \frac{V}{R_p}$$

By comparison with the previous equation, we get

$$\frac{1}{R_p} = \frac{1}{R_1} + \frac{1}{R_2} + \frac{1}{R_3} + \cdots \qquad (8.10)$$

For a circuit with only two resistances in parallel, this equation can be conveniently written as

$$R_p = \frac{R_1 R_2}{R_1 + R_2} \qquad (8.11)$$

EXAMPLE 3

Three resistors have values of $R_1 = 6\ \Omega$, $R_2 = 6\ \Omega$, and $R_3 = 3\ \Omega$. What is the total resistance of the resistors

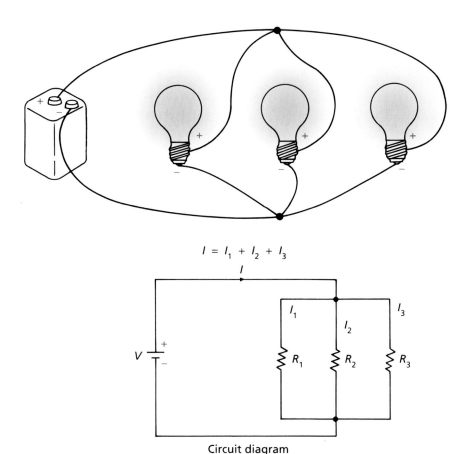

$$I = I_1 + I_2 + I_3$$

Circuit diagram

Figure 8.13 Parallel circuit.
The light bulbs are connected in parallel, so the current from the battery divides at the junction. The amount of current that flows in each branch is determined by the relative resistances in the branches; the greatest current flows through the path of least resistance.

when connected in parallel, and how much current will be drawn from a 12-V battery if it is connected to the circuit?

Solution Let's first combine R_1 and R_2 into a single equivalent resistance, using Eq. 8.11:

$$R_{p_1} = \frac{R_1 R_2}{R_1 + R_2} = \frac{(6\ \Omega)(6\ \Omega)}{6\ \Omega + 6\ \Omega} = 3\ \Omega$$

Hence an equivalent circuit is a resistance R_{p_1} connected in parallel with R_3. Apply Eq. 8.11 to these parallel resistances to find the total resistance:

$$R_p = \frac{R_{p_1} R_3}{R_{p_1} + R_3} = \frac{(3\ \Omega)(3\ \Omega)}{3\ \Omega + 3\ \Omega} = 1.5\ \Omega$$

The same current will be drawn from the source for a 1.5-Ω resistor as for the three resistances in parallel.

The problem can also be solved by using Eq. 8.10. In this case R is found by using the lowest common denominator for the fractions:

$$\frac{1}{R_p} = \frac{1}{R_1} + \frac{1}{R_2} + \frac{1}{R_3} = \frac{1}{6} + \frac{1}{6} + \frac{1}{3}$$

$$= \frac{1}{6} + \frac{1}{6} + \frac{2}{6} = \frac{4}{6\ \Omega}$$

or

$$R_p = \frac{6\ \Omega}{4} = 1.5\ \Omega$$

The current drawn from the source is then given by Ohm's law, using R_p:

$$I = \frac{V}{R_p} = \frac{12\ V}{1.5\ \Omega} = 8.0\ A$$

An interesting fact about resistances connected in parallel is that *the total resistance is always less than the smallest parallel resistance.* Such is the case in Example 3. (Try to find a parallel circuit that proves otherwise.)

Home appliances are wired in parallel (Fig. 8.14). There are two major advantages to the parallel circuit.

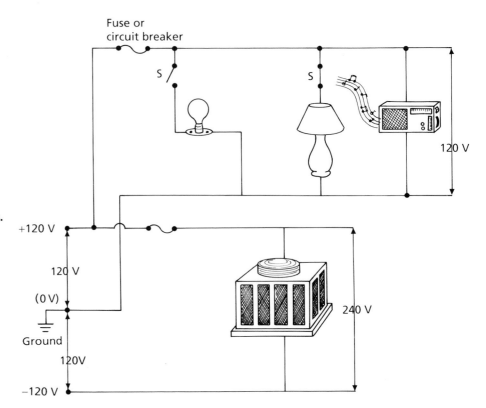

Figure 8.14 Household circuits.
As illustrated here, household circuits are wired in parallel, normally at 120 V. Because they are thus independent, any particular circuit element can operate when others in the same circuit do not. For large appliances, such as a central air conditioner, the connection is made from the +120 V to the −120 V potential leads to give a voltage difference of 240 V.

1. The same voltage (110–120 V) is available through-out the house. This arrangement makes it much easier to design the appliances.*

2. If one appliance fails to operate, the others are not affected. In a series circuit, if one fails, none of the others will operate.

QUESTION

Consider the Christmas lights that remain on when one lamp burns out. How are they connected?

ANSWER The bulbs could be wired in parallel. As shown in the household circuit in Fig. 8.14, if a lamp bulb burns out, the other components in the parallel circuit continue to operate. However, the total resistance of a string of lights wired in parallel would be small, and an undesirably and dangerously large current would flow in the circuit. Also, parallel wiring would require a lot of additional wire, which is expensive. For cost-effectiveness

each lamp is wired in parallel with a resistor, as illustrated in Fig. 8.15. When the filament of a bulb burns out in this case, the resistor provides a path for the current, and all of the other bulbs remain lit.

Electrical Safety

Electrical safety for both people and property is an important consideration in using electricity. For example, in household circuits, as more and more appliances are turned on, more and more current flows and the wires get hotter and hotter. The fuse shown in the circuit diagram in Fig. 8.14 is a safety device that prevents the wires from getting too hot. When too much current flows, the fuse gets so hot that it melts and acts as an open switch. The damaged fuse prevents any current from flowing until the fuse is replaced.

A circuit breaker has the same function as a fuse. When the current gets too high, the circuit breaker triggers a switch to open, halting any current from flowing. When the trouble is corrected, the circuit breaker can be pushed in to close the switch, and current can again flow in the circuit.

Switches, fuses, and circuit breakers are always

---

* The voltage for large appliances is 220–240 V, which is available by connecting across the two incoming potentials, as shown in Fig. 8.14. This potential is similar to a height difference between two positions for gravitational potential energy.

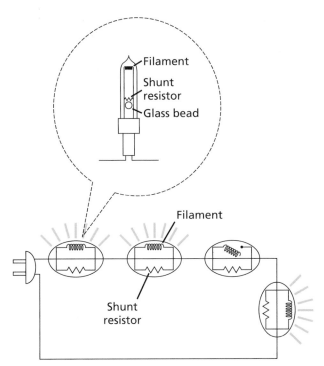

Figure 8.15 Shunt resistors.
With a small shunt resistor wired in parallel with each bulb filament, the filament of any individual bulb can burn out, but there is still a complete circuit so that the other filaments continue to glow.

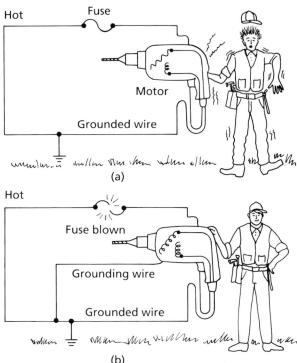

(a)

(b)

Figure 8.16 Electrical safety.
(a) *Above:* Without a dedicated ground wire, the casing of an appliance may be at a high potential without blowing a fuse or tripping a circuit breaker. If someone touches the casing, a dangerous shock can result. (b) *Below:* By grounding the casing with a dedicated ground wire through a third prong on the plug, the fuse would be blown and the casing would be at zero potential.

Figure 8.17 Electrical plugs.
(a) *Left:* A three-prong plug and socket. The third, round prong is a dedicated grounding wire used for safety purposes. (b) *Right:* A two-prong polarized plug and socket. Note that one blade or prong is larger than the other. The ground or neutral (zero potential) side of the line is wired to the large-prong side of the socket. This distinction or polarization permits paths to ground for safety connections.

placed in the hot or high-voltage side of the line so that when they are opened, current flow is interrupted. However, fuses and circuit breakers may not always give protection from electrical shock. A hot wire inside an appliance or power tool may break loose and come into contact with the housing or casing, putting it at a high voltage. The fuse does not blow if there is not a large current flow, such as when being shorted to ground (zero potential). Should the casing be conductive and a person touch it, as illustrated in Fig. 8.16a, a path is provided to ground and the person receives a shock.

This condition is prevented by *grounding* the casing, as shown in Fig. 8.16b. Then if a hot wire touches the casing, a current flows and the fuse blows. This grounding process is the purpose of the three-prong plugs found on many electrical tools and appliances (Fig. 8.17a).

A *polarized* plug is shown in Figure 8.17b. You have probably noticed that some plugs have one blade larger than the other and will only fit into a wall outlet in one way. Polarized plugs are an older type of safety feature. Being polarized or directional, one side of the plug is always connected to the ground side of the line. The casing of an appliance can be connected to ground by this means, with a similar effect as the three-wire system. However, a dedicated grounding wire is better, because the polarized system depends on several components of the circuit being wired or connected properly.

An electric shock can be very dangerous, and touching exposed electric wires should always be avoided. Many people are killed every year through electric shocks. The danger is proportional to the amount of electric current that goes through the body. The amount of current going through the body is given by Ohm's law as

$$I = \frac{V}{R_{body}} \qquad (8.11)$$

where R_{body} = resistance of the body.

A current of 0.001 A can be felt as a shock, and a current as low as 0.05 A can be fatal. A current of 0.1 A is nearly always fatal.

The amount of current, as indicated in Eq. 8.11, is very dependent on the body's resistance. The body's resistance varies considerably, mainly due to the dryness of the skin. Because our bodies are mostly water, the skin resistance makes up most of the body's resistance.

A dry body can have a resistance as high as 500,000 Ω, and the current from a 110-V source will be 0.00022 A. The danger occurs when the skin is moist or wet. Then the resistance of the body can go as low as 100 Ω, and the current will rise to 1.1 A. Injuries and death from shocks usually occur when the skin is wet. Therefore, appliances such as radios should not be used near a bathtub. Should a plugged-in radio happen to fall into the bathtub, then the whole tub, including the person in it, may be plugged into 110 V.

8.3 Magnetism

One of the first things we notice in examining a bar magnet is that it has two regions of magnetic concentration: poles at the ends of the magnet. We call these two poles north seeking, N, and south seeking, S, because a compass needle was one of the earliest applications of magnetism. The N pole of a magnet, such as a compass, points north, and the S pole points south.

A basic principle of magnetic force, called the **law of poles,** has been mentioned already in the introduction to this chapter.

Like poles repel and unlike poles attract.

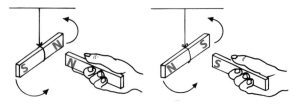

Like poles repel

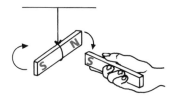

Unlike poles attract

Figure 8.18 Law of poles.
(a) *Above:* Like poles repel and unlike poles attract. (b) *Below:* The adjacent poles of the magnets in the photo must be like poles. Why?

That is, N and S poles, N–S, attract, and N–N or S–S poles repel each other (Fig. 8.18a). The strength of the attraction or repulsion depends on the strength of the magnetic poles. Also, in a manner similar to Coulomb's law, the strength of the magnetic force is inversely proportional to the distance between the magnetic poles. Figure 8.18b shows some-toy magnets that seem to defy gravity due to their magnetic repulsion.

Every magnet produces a force on every other magnet. In order to discuss these force effects, we introduce

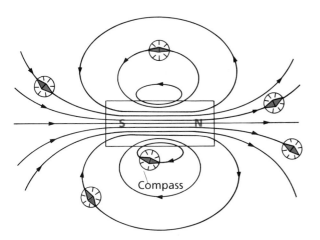
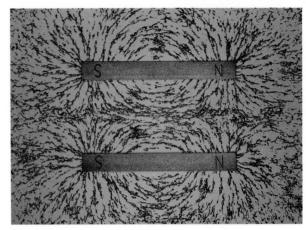

Figure 8.19 Magnetic field.
(a) *Left:* Magnetic field lines may be found by using a small compass. The N pole of the compass points in the direction of the field at any point. (b) *Right:* Iron filings become induced magnets and conveniently outline the pattern of the magnetic field.

the concept of a magnetic field. A **magnetic field** is a set of imaginary lines that indicates the direction a small compass needle would point if it were placed at a particular spot. Figure 8.19a shows the magnetic field lines around a simple bar magnet. The arrows in the field lines indicate the direction in which a compass N pole would point. The closer together the field lines, the stronger the magnetic force is on the imaginary compass.

Magnetic field patterns are frequently generated by using small iron filings. The iron filings are magnetized and act like small compasses. The magnetic field produced in this way using two pieces of naturally magnetic lodestone (Fe_3O_4) is shown in Fig. 8.19b.

The field concept also applies to the electric force. An **electric field** is a set of imaginary lines that indicates the direction a small positive charge would move if it were placed at a particular spot. The electric fields are indicated in Fig. 8.20 for a single positive charge, a single negative charge, and a positive and negative charge together. This last configuration is called an *electric dipole* because there are two poles involved. A common magnet such as that shown in Fig. 8.19 would be called a *magnetic dipole* because it also has two poles. Note the similarity and difference between the fields for an electric dipole and a magnetic dipole.

Two facts about magnetic and electric fields are evident from Figs. 8.19 and 8.20. First, field lines never cross. If they did, it would indicate that at the point of crossing the compass pointed in two directions or the positive charge would move in two directions. Second,

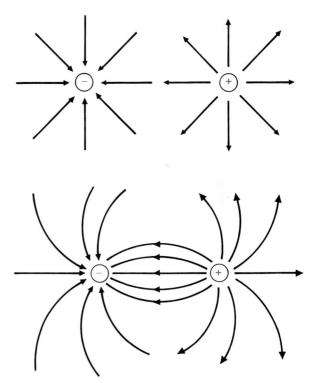

Figure 8.20 Electric field.
Electric field lines are in the direction a positive test charge would feel a force. Hence, for single negative and positive charges, the field lines are directed as shown. For an electric dipole, which is a positive and negative charge together, the field lines begin at the positive charge and end at the negative charge.

electric fields begin and end on electric charges, whereas magnetic field lines are continuous. Magnetic field lines have no beginning or end. All magnets occur with two poles, i.e., as *di*poles.

Electric charge can be thought of as isolated single poles or *monopoles*. A **magnetic monopole** would consist of a single N or S pole without the other. There is nothing to prevent a magnetic monopole from existing, but so far none has been found. The discovery of a magnetic monopole would be an important fundamental development.

Electric and magnetic fields have directions associated with them, so they are *vector quantities*. Electromagnetic waves, discussed in Chapter 6, are caused by electric and magnetic fields that vary with time.

Electricity and magnetism are discussed together in this chapter because they are linked. In fact, the source of magnetism can be understood as moving and "spinning" electric charges. Hans Oersted, a Danish physicist, first discovered in 1820 that a compass needle was deflected by an electric current—carrying wire. To see this effect, one must have a fairly strong electric current.

Different configurations of current-carrying wires give different magnetic field configurations. Some of them are shown in Fig. 8.21. A straight wire produces a field in a circular pattern around the wire. A single loop of wire acts like a small bar magnet, and a coil of wire creates a field similar to a bar magnet.

Individual electrons in an atom have a magnetic field due to what is best understood as a spinning motion. This motion causes all electrons to act like magnetic dipoles or very small bar magnets. The electrons also go in orbit around the atom's nucleus. This effect is similar to a single loop of wire and again creates a magnetic field similar to a small bar magnet.

Thus there are magnetic effects resulting from an electron's movement and its spinning motion. Most atoms have many electrons, and the magnetic effects of all these electrons usually cancel out. So most materials are not magnetic or only slightly magnetic. In some atoms, however, the magnetic effect can be strong.

Materials that are highly magnetic are called **ferromagnetic.** These materials include the elements iron, nickel, cobalt, several rare-earth elements, and certain alloys of these and other elements. In ferromagnetic materials many atoms combine their magnetic fields in the same direction, which gives rise to magnetic **domains,** or local regions of alignment. A single magnetic domain is very small and acts like a tiny bar magnet.

In iron the domains can be aligned or nonaligned. A piece of iron with the domains randomly oriented is

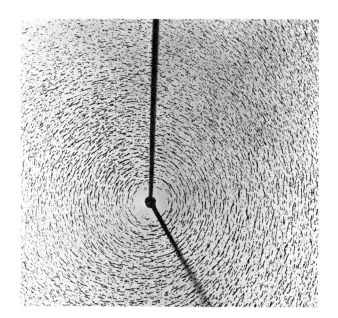

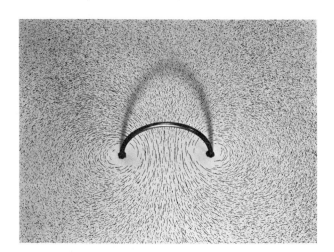

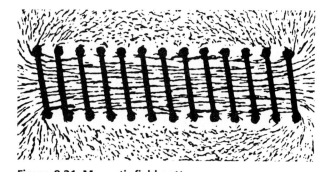

Figure 8.21 Magnetic-field patterns.
Iron-filing patterns near current-carrying wires outline the magnetic fields for (from top) a long straight wire, a single wire loop, and a coiled wire.

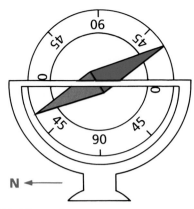

Figure 8.26 Dip angle.
The magnetic dip or the angle the Earth's magnetic field makes relative to the horizontal is measured with a dip needle.

Honeybees also are believed to use the direction of the Earth's magnetic field as well as the Sun's direction to communicate the location of a good food source.

8.4 Electromagnetism

The interaction of electrical and magnetic effects is known as **electromagnetism.** Electromagnetism is one of the most important concepts in physical science, and much of our current technology is directly related to this crucial interaction. We can summarize the two basic principles of electromagnetism as follows:

1. Moving electric charges create magnetic fields.
2. A magnetic field may deflect a moving charge.

The first idea was mentioned in Section 8.3 when we discussed the source of magnetism. One of the most important applications of this principle is called an electromagnet. An **electromagnet** consists of a current-carrying coil of insulated wire wrapped around a piece of soft iron (Fig. 8.27). When the current is turned on, a magnetic field is created inside the coil (see Fig. 8.21, top picture). The soft iron is magnetized by this field and makes the magnetic field about 2000 times stronger. When the current is turned off, the iron loses most of its magnetism.

The advantages of the electromagnet are many. First, it can be switched on and off. Second, the strength of the electromagnet is controlled by the amount of current flowing in the wire. Third, the poles of the magnetic field can be reversed by reversing the current.

Electromagnets are found in a variety of devices and appliances, such as doorbells and telephones, and

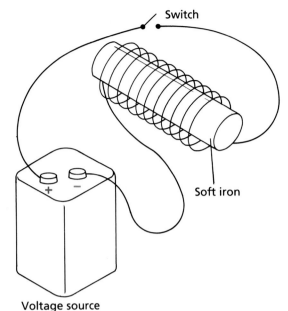

Voltage source

Figure 8.27 Electromagnet.
An electromagnet consists of a coil of wire wrapped around a piece of soft iron. When the switch is closed, charge flows in the wire and magnetizes the iron, creating a magnet. When the switch is open, no charge flows and the iron is not magnetized.

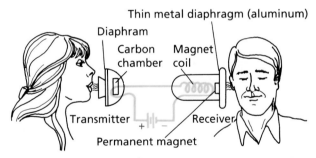

Figure 8.28 Telephone.
A simplified diagram of a (one-directional) telephone circuit. Sound waves are converted into varying electrical pulses in a transmitter. The pulses travel along the phone lines to the receiver where they are converted back to sound waves by the actions they have on an electromag net that drives a diaphragm.

in devices used for moving magnetic metals. The mechanism in a telephone receiver will serve as an example of one use. A simplified diagram of a telephone circuit is shown in Fig. 8.28.

When a telephone number is dialed, a circuit is completed with another telephone's bell. When the ring-ing phone is lifted, the circuit between the speakers and receivers of the two telephones is completed. A telephone conversation consists of converting the sound waves into a varying electric current. This varying current travels along the wire to the other telephone's receiver where it is converted back to sound waves.

The transmitter of a phone contains a diaphragm that vibrates due to the sound waves that are spoken. The diaphragm vibrates against a chamber that contains carbon granules. As the diaphragm vibrates, the pressure on the carbon granules varies, which causes more or less electric resistance in the circuit. The resistance is low when the granules are pressed together and increases as they spread apart. Because of Ohm's law, this varying resistance causes a varying electric current to be transmitted.

At the receiver end there is an electromagnet and a permanent magnet, which is attached to a disk. The varying electric current gives the electromagnet a varying magnetic strength. The electromagnet attracts the permanent magnet with a variable force. The changing force on the permanent magnet and disk causes the receiver disk to vibrate. This vibration sets up sound waves that closely resemble the original sound waves, and the voice is heard at the other telephone.

Magnetic Force and Moving Electrical Charge

The second basic principle of electromagnetism has been stated in a qualitative way: A magnetic field will deflect moving charges. Quantitatively, we want to understand how the moving charges are deflected. The formula for the magnitude of the force on a moving charge due to a magnetic field is given by

$$F_{mag} = qv \times B \qquad (8.12)$$

where F_{mag} = force on a moving charged particle due to the magnetic field B,

q = charge on the particle,

v = velocity of the particle,

B = magnetic field.

If the velocity is zero, there is no force on the particles. The force F_{mag} is a *vector quantity*, as are all forces. The direction of F_{mag} takes some practice to understand, and it is perpendicular to both the velocity vector and the direction of the magnetic field. In addition, in order for the force to exist, the velocity vector must have some component perpendicular to the magnetic field. If the

velocity vector and magnetic field vector are parallel or antiparallel, there will be no force.

The best way to understand this deflecting force is by looking at some examples. Figure 8.29 shows a beam of electrons traveling from left to right. The magnetic field direction is vertical. The force on the electrons is then out from the paper or toward the viewer. Two photographs of an electron beam are shown in Fig. 8.30. In the upper photo the beam is undeflected. In the lower photo the nearly horizontal magnetic field of the bar magnet causes the beam to be deflected downward.

The electrons in conducting wires also experience effects caused by magnetic fields. In Fig. 8.31a a non-conducting wire is shown in a magnetic field. Because the electrons are not moving, there is no force on the wire. In Fig. 8.31b the electrons in the wire are moving to the right. This situation is similar to the one in Fig. 8.29, except that now the force causes the whole wire to be forced out of the paper toward the viewer. This discussion demonstrates the general principle that a current-carrying wire placed in a magnetic field will experience a force.

Motors and Generators

A **motor** is a device that converts electrical energy into mechanical energy. The basic principle involved is that a current-carrying wire in a magnetic field experiences

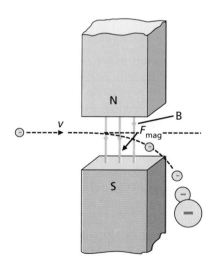

Figure 8.29 Magnetic deflection.
Electrons in a vertical magnetic field B, as shown, experience a force F_{mag} that deflects them out of the page. See text for explanation.

Figure 8.30 Magnetic force on moving charges.
(a) *Above:* The presence of a beam of electrons in a discharge tube is made evident by a fluorescent strip in the tube that allows the beam path to be seen. (b) *Below:* The magnetic field of a magnet gives rise to a force on the electrons and the beam is deflected.

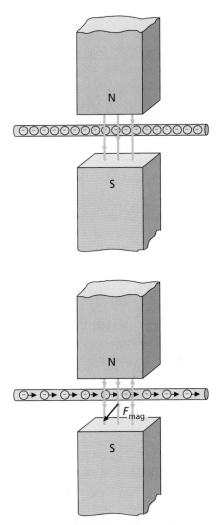

Figure 8.31 Magnetic field and current.
(a) *Above:* A stationary wire with no current does not experience a force. (b) *Below:* A current-carrying wire in a magnetic field experiences a force. With the electron current flow to the right, the magnetic force F_{mag} would be out of the page, as in the case illustrated here. This force on a current-carrying wire is the basic principle of the electric motor.

a force, as in Fig. 8.31b. If the wire is a pivoted loop, the force gives rise to a torque, and the loop rotates. Motors generally have many windings on a rotating armature to enhance the effect. The rotating loops cause a connected shaft to rotate, which is used to do mechanical work. The conversion of electric energy to mechanical energy is enhanced by many loops of wire and stronger magnetic fields.

One might ask if the reverse is possible. That is, is the conversion of mechanical energy into electrical energy possible? Indeed it is, and this principle is the basis of electrical generation. As illustrated in Fig. 8.32, suppose that an applied (mechanical) force sets a wire in a magnetic field in motion. The electrons in the wire are charges moving in a magnetic field and hence experience a magnetic force (F_{mag}) to the right. The electrons move in the conductor, and an electrical current is set up. Note that a current is generated without any batteries, plugs, or other external voltage sources. This is the basic principle used in an electrical generator.

A **generator** converts mechanical work or energy into electrical energy. The processes are more complex than described here, but the effect is the same. Generators are used in power plants to convert other forms of energy into electrical energy. Falling water and pressurized steam from burning coal or fuel oil are two sources of energy used to turn turbines that supply the mechanical energy in the generation process. The electricity is carried to homes and businesses, where it is either converted back into mechanical energy to do work or converted into heat energy.

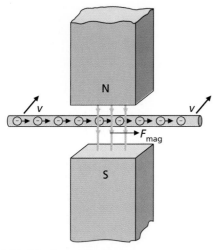

Figure 8.32 Motion and current.
If a wire is moved perpendicularly to a magnetic field as illustrated here, the magnetic force causes the electrons in the wire to move, setting up a current in the wire. This is the basic principle of a generator.

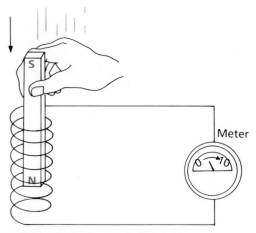

Figure 8.33 Electromagnetic induction.
An illustration of Faraday's experiment showing electromagnetic induction. The reading on the meter indicates a current in the circuit as the magnet is moved through the coil of wire.

In the formula for the force due to a magnetic field (Eq. 8.12), the velocity v is the relative velocity between the charged particles and the magnetic field. So far, all of our examples have been with stationary magnetic fields. In 1831 Michael Faraday, an English scientist, did a famous experiment in which he moved the magnet rather than the wire in order to create or induce an electric current. A diagram of this experiment is shown in Fig. 8.33.

We discussed in Section 8.2 how **transformers** were used to increase or decrease voltages associated with alternating current. Transformers are a good example of the Faraday effect of a current (or voltage) induced by a time-varying magnetic field.

A transformer is illustrated in Fig. 8.34. Alternating current creates a varying magnetic field in the iron core. The changing magnetic field in the core also passes through the secondary coil, where it induces a current.

The voltage and current in the secondary coil are different from those in the primary coil. The amount of voltage is proportional to the ratio of the number of turns in the coils. By putting more turns in the secondary coil than in the primary, we can step up the voltage. Similarly, we can decrease the voltage by putting fewer turns in the secondary than in the primary. The relationship for the voltage change is

$$V_2 = \left(\frac{N_2}{N_1}\right) V_1 \qquad (8.13)$$

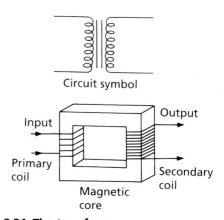

Figure 8.34 The transformer.
The circuit symbol for and the basic features of a transformer. It consists of two coils wrapped on a piece of iron. Alternating current in the primary coil creates a varying magnetic field, which is concentrated by the iron through the secondary coil. This in turn induces a voltage and a current in the secondary coil.

where V_2 = secondary voltage,
V_1 = primary voltage,
N_1 = number of turns in primary coil,
N_2 = number of turns in secondary coil.

EXAMPLE 4

A transformer has 500 windings in its primary coil and 25 in its secondary coil. If the primary voltage is 4400 V, find the secondary voltage.

Solution   We use Eq. 8.13 to get

$$V_2 = \left(\frac{25}{500}\right)(4400 \text{ V}) = 220 \text{ V}$$

This result is typical of a step-down transformer on a utility pole near residences.

If the voltage is stepped up by a transformer, the current is stepped down, and vice versa. The factor of the change in the current is just given by the inverse of the turn ratio in Eq. 8.13. For example, in Example 4 the current would be stepped up by a factor of $\frac{500}{25} = 20$.

8.5 Electronics

We live in an age of high technology. Television, computers, and calculators are just three examples of recent developments of the electronics era of today. Electronics is the branch of physics and engineering that deals with the emission and control of electrons. It can take place in an evacuated tube, such as a television picture tube, or in solid semiconductor materials that are the basis of solid-state electronics.

A television picture tube is an example of a **cathode ray tube, CRT**. In Fig. 8.35 the basic features of a CRT are shown. The electron gun is a heated filament that ejects electrons. The electrons are accelerated by a voltage and pass through a narrow opening, forming an electron beam. The beam of electrons is then deflected by either electric or magnetic fields. In our example (Fig. 8.35) there are electromagnets. The input current to the electromagnets determines the strength of the magnetic fields and thereby regulates the deflection of the beam.

The deflected electron beam strikes a phosphorescent screen. The intensity of the electron beam controls the brightness of the spot where the electron beam strikes the screen. The beam sweeps across 700 dots to form a line and then goes back to form a new line. It produces every other line and then goes back and fills in the alternate lines to prevent a noticeable flicker. Thirty pictures are made every second to create the illusion of movement on the screen. The input signal to the beam is a variable current that controls the intensity of the electron beam and hence the brightness of each dot on the screen. The input signal is received from the antenna, which gets it from a TV transmitter.

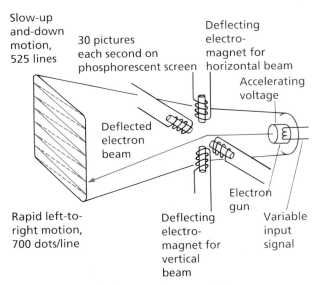

Figure 8.35 Cathode ray tube and television.
A simplified diagram of a cathode ray tube (CRT) such as those used in television sets. The electron gun ejects electrons, which are accelerated by a voltage. The electron beam is deflected by electromagnets mounted above and below and on each side of the path of the beam, which cause the beam to sweep across and up and down the screen. The variable input controls the intensity of the electron beam and hence the brightness of each dot on the fluorescent screen. Thirty slightly different pictures made each second give the illusion of a moving image.

A color TV generally has three electron guns, and each dot on the screen consists of a triad of red, green, and blue dots. Varying the intensities of the three beams gives each dot a different color, so that a color picture is produced.

Two basic electronic components used in TV sets and other electronics devices need to be mentioned. One is the diode; the other is the transistor. A **diode** is a device that allows current to flow in only one direction. It has a negligible resistance in one direction and a practically infinite resistance in the reverse direction. Diodes have many uses; for instance, they can be used to change alternating current into direct current.

A **transistor** is a solid-state *triode*, that is, an element with three components. The third component provides a means of controlling the transistor current and varying the voltage. This process allows for the amplification, or gain, of an input signal. Solid-state diodes and transistors offer great advantages over the older vacuum tubes, which they have all but replaced. Major advantages are in size, power consumption, and material economy (Fig. 8.36). Consider the now-common transistor radio, which can be quite small. You may remember or have seen pictures of older vacuum tube radios; they were quite large. Also, the vacuum tubes had glowing heating elements that consumed a great deal of power compared with a comparable solid-state component—a difference of watts versus milliwatts.

Using diodes, transistors, and resistors, many types of electronics circuits, called integrated circuits, can be made. They are now made in miniature by techniques such as depositing conducting metal onto layers of material made primarily from silicon, a good semiconductor. The silicon layers can be used to make diodes, transistors, and resistors. There is plenty of silicon available, since sand is mainly silicon dioxide.

Techniques have been developed to make the integrated circuits smaller and smaller. Integrated circuits made in this way are commonly called "chips," and they can be quite complex even if quite small. A chip is shown in Fig. 8.37.

Microchips are used in a variety of applications (Fig. 8.38). Chips are used in hand calculators and miniature

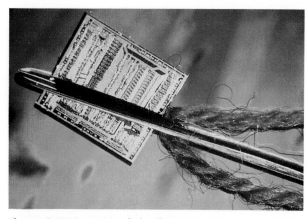

Figure 8.37 Integrated circuits.
A microprocessor integrated circuit, or "chip," held in the eye of a threaded needle.

Figure 8.38 Microchips in action.
Microchips are used in hand-held calculators (above) and miniature TV sets, and in personal computers (below).

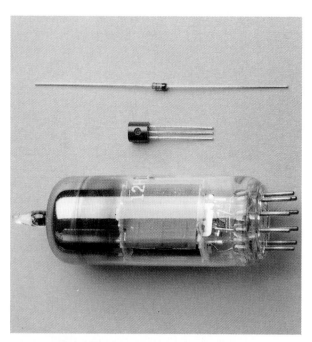

Figure 8.36 Solid-state diode and transistor.
A diode (two leads) and a transistor (three leads) are shown with a vacuum tube. Solid-state diodes and transistors have all but replaced vacuum tubes in many applications.

TV sets. Also, a personal computer uses one or more chips to form the logic circuits, and a CRT allows information to be displayed for viewing by the operator.

As our knowledge of electronics grows, computers are becoming smaller and more powerful, and this trend will undoubtedly continue in the future.

Learning Objectives

After reading and studying this chapter, you should be able to do the following without referring to the text:

1. Give the units of electric charge, current, voltage, resistance, energy, and power.

2. State the basic force law for (a) charges and (b) magnetic poles.

3. Explain briefly why some materials are conductors and some are insulators.

4. Explain why electrons, rather than protons, move in conductors such as metal wires.

5. Explain the relationship of voltage, current, and resistance in terms of Ohm's law.

6. Explain the difference between ac and dc, and tell why ac is used in homes.

7. Explain what transformers do and how they work.

8. Explain the difference and draw diagrams for series and parallel circuits, and explain why home appliances are wired in parallel rather than series.

9. Explain why a shock from a 110-V ac line may or may not be fatal.

10. Briefly give an atomic description of what causes iron to be magnetic.

11. Briefly describe how an electromagnet works, and give some applications.

12. Explain the basic idea behind the operation of an electric generator.

13. Explain briefly the basic idea behind the operation of an electric motor.

14. Explain how images are produced on TV screens.

15. State a use for a (a) diode, (b) transistor, and (c) chip, and give the advantages of these solid-state devices over vacuum tubes.

16. Define and explain the important words and terms listed in the next section.

Important Words and Terms

electrons	insulators	series circuit	magnetic dip
protons	electricity	parallel circuit	electromagnetism
neutrons	electric potential energy	law of poles	electromagnet
quarks	voltage	magnetic field	motor
coulomb	volt	electric field	generator
current	resistance	magnetic monopole	transformers
ampere	ohm	ferromagnetic	cathode ray tube, CRT
law of charges	Ohm's law	domains	diode
Coulomb's law	direct current, dc	Curie temperature	transistor
conductors	alternating current, ac	magnetic declination	

Questions

Electric Charge

1. Describe the three basic subatomic particles.

2. What is a quark?

3. What is an electric current, and what is the unit of current?

4. Describe any similarities between Newton's law of gravitation and Faraday's law for electrical force. Describe any differences.

5. Explain how a charged rubber comb attracts bits of paper and how a charged balloon sticks to a wall or ceiling.

6. Why are some materials good conductors and others are not?

Electricity

7. Distinguish between electric potential energy and voltage.

8. Explain Ohm's law, and give the unit of measurement for each term.

9. How does electrical power depend on (a) current and voltage and (b) current and resistance?

10. Distinguish between alternating current and direct current.

11. What is a major reason for using ac current in homes and businesses?

12. What would happen if electric power were transmitted from the generating plant to your home at 120 V?

13. Why are home appliances wired in parallel rather than in series? Are today's Christmas tree lights wired in series or parallel?

14. Discuss the safety features of (a) fuses, (b) circuit breakers, (c) three-pronged plugs, and (d) polarized plugs.

15. What causes electrical shocks, and what precautions should be taken to avoid them?

Magnetism

16. Compare (a) the law of charges and the law of poles, (b) electric and magnetic fields, and (c) electric and magnetic dipoles.

17. What happens if a bar magnet is cut in half? Can it be continually cut in half to finally get two magnetic monopoles? Explain.

18. What is a ferromagnetic material, and what are some examples?

19. Why does a permanent magnet attract a piece of iron? What would happen if the piece of iron were above the Curie temperature?

20. Distinguish between magnetic dip and declination.

Electromagnetism

21. How does an electromagnet work?

22. How do telephone transmitters and receivers work?

23. Describe the basic principle of an electric motor.

24. Describe the basic principle of an electric generator.

25. Distinguish between a motor and a generator.

26. What is the principle of a transformer, and how are transformers used?

Electronics

27. How are images produced on a TV screen?

28. Distinguish between a diode and a transistor.

29. Explain what is meant by an integrated circuit and a chip, and give some examples of applications.

Exercises

Electric Charge

1. How many electrons make up one coulomb of charge?
   Answer: 6.25×10^{18}

2. A charge of 2.0 C passes by a point in a conductor in 4.0 s. How much current flows in the conductor?

3. A current of 1.5 A flows in a conductor for 6.0 s. How much charge passes by a given point in the conductor during this time?

4. There is a net passage of 4.8×10^{18} electrons by a point in a wire conductor in 0.25 s. What is the current in the wire?
   Answer: 3.1 A

5. What happens to the force between an electron and a proton if the distance between them is tripled?

6. What are the forces on two charges of $+0.50$ and $+2.0$ C, respectively, if they are separated by a distance of 3.0 m?

7. Use Eq. 8.3 to find the force of attraction between a proton and an electron that are 5.0×10^{-11} m apart (the arrangement in a hydrogen atom).

Electricity

8. To separate a 0.25-C charge from another charge, 30 J of work is done. What is the voltage between the charges? How is this voltage related to energy?

9. If an electrical component with a resistance of 40 Ω is connected to a 120-V source, how much current will flow through the component?
   Answer: 3.0 A

10. What battery voltage is necessary to supply 0.50 A of current to a circuit with a resistance of 3.0 Ω?

11. (a) How much current is drawn by a flashlight using batteries that add to 3 V to produce a power output of 0.5 W?
    (b) What is the resistance of the flashlight bulb?
    Answer: (b) 18 Ω

12. How much does it cost to run a 1500-W hair dryer 30 min each day for one month at 8¢ per kWh?
    Answer: $1.80

13. Show that electric power is given by the expression $P = V^2/R$.

14. A refrigerator motor using 1000 W runs one-third of the time. How much does the electricity cost each month at $0.08 per kWh?

15. A 100-W light bulb is turned on.
    (a) How much current flows through the bulb?
    (b) What is the resistance of the bulb?
    (c) How much energy is used each second?
    (Assume a voltage of 120 V.)

16. How much power is used by the appliance shown in Fig. 8.9?

17. In Fig. 8.14 the voltage is 110 V.
    (a) What is the current in a light bulb that has a resistance of 100 Ω?
    (b) What is the current in a toaster with a resistance of 10 Ω?
    (c) What is the total current if only the light bulb and toaster are turned on? *Answer:* (c) 12.1 A

18. A portable electric heater has six resistances of 1.5 Ω, each wired in series. The applied voltage is 120 V.
    (a) Compute the current flow in the heater.
    (b) Compute the power output of the heater.

19. The heating element of an iron operates on 110 V at a current of 10 A.
    (a) What is the resistance of the iron?
    (b) What is the power dissipated by the iron?

20. Three resistors with values of 20, 30, and 40 Ω, respectively, are connected in series and hooked to a 90-V battery.
    (a) How much current flows in the circuit?
    (b) How much power is expended in the circuit?

21. A Christmas tree light bulb filament and its resistor in parallel have resistances of 0.35 and 0.50 Ω, respectively. What is the total resistance of the bulb?

22. (a) What is the total resistance in Fig. 8.13 if the three resistances in parallel are $R_1 = 60\ \Omega$, $R_2 = 30\ \Omega$, and $R_3 = 20\ \Omega$? *Answer:* (a) 10 Ω
    (b) Use Ohm's law and the resistance found in part (a) to compute the total current if $V = 110$ V.
    Answer: (b) 11 A

23. A person with dry skin has a body resistance of 50,000 Ω.
    (a) If the person touches the terminals of a 2-V battery, what is the current through the body?
    (b) If the person's body is wet with perspiration and the body resistance drops to 2000 Ω, what is the current?

Magnetism

24. Sketch the magnetic fields between (a) a north pole and a south pole, and (b) two north poles.

25. Using Fig. 8.25, find the magnetic declination where you live. You must live in the coterminous United States to use the figure.)

Electromagnetism

26. A transformer has 300 windings on its secondary and 50 windings on its primary. The primary is connected to a 12-V source.
    (a) What is the voltage output of the secondary?
    (b) How much current can the secondary supply?

27. A power company runs current along 765,000-V transmission lines. This voltage is changed by a transformer that has 900 turns in the primary coil and 150 turns in the secondary coil. What is the secondary voltage produced?

28. A transformer with 1000 turns in the primary coil has to decrease the voltage from 4400 to 220 V for home use. How many turns should be in the secondary coil?
    Answer: 50

CHAPTER **9**

Atomic Physics

> And who loves Nature more
> Than he, whose painful art
> Has taught and skilled his heart
> To read her skill and lore?
> —Robert Bridges

I N THE LATTER PART of the nineteenth century scientists thought physics was in fairly good order. The principles of mechanics, wave motion, sound, and optics were reasonably well understood. Electricity and magnetism had been combined into electromagnetism, and light had been shown to be a wave (electromagnetic wave). There were some rough edges, but it seemed that only a few refinements were needed.

However, as scientists probed deeper into the microscopic and submicroscopic world of the atom, they observed strange things—strange in the sense that the phenomena could not be explained by classical principles. These discoveries were unsettling, and as time and investigations went on, scientists became aware that not everything about nature was known. In fact, they had to use new approaches in the description of nature that seemed radical at that time.

The development of physics since about 1900 is commonly termed *modern physics.* Classical or Newtonian physics was generally concerned with the macrocosm—the universe, movement of the planets, description of observable phenomena, etc. —, whereas modern physics generally considers the microcosm—the atom and its structure. Chapters 9 and 10 will be concerned with some topics of modern physics. Chapter 9 will look chiefly at atomic physics, which deals with the atom as a whole and its electronic structure. Chapter 10 will go to the heart of the matter and look at the central core or nucleus of the atom—nuclear physics.

9.1 The Dual Nature of Light

Even before the turn of the century scientists knew that light of all frequencies was emitted by the atoms of an incandescent solid, such as the filament of a light bulb. As the temperature is increased, more radiation is emitted and the component of maximum intensity is shifted to a higher frequency. As a result, a very hot solid appears

to go from a dull red to a bluish white (Fig. 9.1). This outcome is expected, because the hotter the solid, the greater the electron vibrations in the atoms and the higher the frequency of the emitted radiation.

However, according to classical wave theory, the intensity or energy of the emitted radiation should be proportional to the square of the frequency ($I \propto f^4$). This discrepancy was termed the ultraviolet catastrophe: "ultraviolet" because the difficulty occurred at high frequencies beyond the violet end of the visible spectrum, and "catastrophe" because it predicted the energy intensity to be very much greater than was actually observed.

The dilemma was resolved in 1900 by German physicist Max Planck (pronounced "Plahnk") (Fig. 9.2). He introduced a radical new idea that explained the observed thermal radiation intensity. In doing so, Planck took the first step toward a new theory of physics called quantum physics. Classically, an electron oscillator may vibrate with any frequency or have any energy up to some maximum value. But **Planck's hypothesis** states that the energy was *quantized,* or that the oscillators could have only discrete or certain amounts of energy. Moreover, the energy of an oscillator depended on its frequency according to the relationship

$$E = hf \tag{9.1}$$

where h is constant, called *Planck's constant*, and is 6.63×10^{-34} J·s (a very small number).

Thus Planck introduced the idea of a quantum, a discrete amount of energy. This concept was in radical contrast to the classical idea that an electron oscillator, like a mass oscillating on a spring, could have continuously different energies.

In 1905 Albert Einstein used Planck's hypothesis to describe light in terms of particles rather than waves. He did so to explain what is called the **photoelectric effect.** In this phenomenon, as scientists had observed in the latter part of the nineteenth century, electrons are emitted when certain metallic materials are exposed to light. This direct conversion of light (radiant energy)

◀ **A laser making a star and halo.**

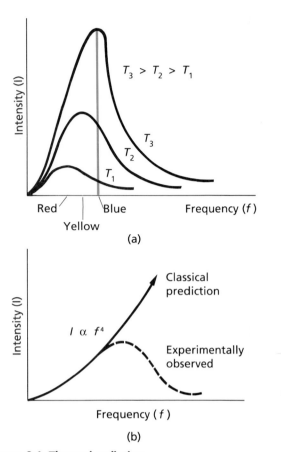

Figure 9.1 Thermal radiation.
(a) As the temperature of an incandescent solid increases, the component of maximum intensity shifts to a higher frequency. (b) Classically, it is predicted that the intensity should be proportional to f^4, which indicates a much greater energy than actually observed. *Above right:* The radiation component of maximum intensity of a hot solid determines its color, as shown here by hot steel coming out of a furnace.

into electrical energy now forms the basis of photocells used in photographic light meters and solar energy applications (Fig. 9.3).

Some aspects of the photoelectric effect could not be explained by classical theory. For example, the amount of energy needed to free an electron from a photomaterial could be calculated. But according to classical theory, where light is a wave with a continuous flow of energy, it would take an appreciable time for an electron to be emitted. However, as we know, current flows from photocells almost immediately on being exposed to light.

Applying Planck's hypothesis, Einstein assumed that light or any electromagnetic radiation was quantized. He termed the quanta (plural) of light **photons**, which were "packets" or "particles" of energy. Planck's rela-

tionship, with the frequency f taken to be the frequency of the light, was used. That is, a quantum of light contained an amount of energy of

$$E = hf$$

or
$$\frac{\text{energy of}}{\text{a photon}} = \frac{\text{Planck's}}{\text{constant}} \times \frac{\text{frequency}}{\text{of light}}$$

Because the frequency of light is related to the speed of light c and its wavelength λ by $f = c/\lambda$, we have

$$E = hf = \frac{hc}{\lambda} \tag{9.2}$$

According to this equation, the shorter the wavelength (or higher frequency) of light, the greater is the energy of its photons. For example, photons of blue light

Figure 9.2 Max Planck (1858–1947).
While a professor of physics at the University of Berlin in 1900, Planck proposed that the energy of thermal oscillators existed in only discrete amounts, or quanta, thus introducing the idea of quantum physics. The important small constant h that appears over and over again in quantum physics is called Planck's constant. Planck was awarded the Nobel Prize in physics in 1918 for his contributions to quantum theory.

Figure 9.3 Photoelectric effect in action.
Radiant (solar) energy is converted directly into electrical energy in the world's largest array of solar panels at Mount Lagune Air Force Station in California.

have more energy than photons of longer-wavelength red light.

By considering light to be composed of photons or quanta of energy, Einstein was able to successfully explain the photoelectric effect. (He received a Nobel Prize for this work, not for his more famous theory of relativity.) The classical time delay to get enough energy to free an electron is not a problem with quanta of energy. The quanta or packets of energy provide a given amount of energy all at once. An analogy of wave and quantum energy is shown in Fig. 9.4.

The concept of light being composed of discrete packets or photons is confusing to many people. How can light be composed of particles when it shows wave phenomena such as diffraction, interference, and polarization? But as just mentioned, Einstein's photoelectric theory, in which light is considered to be made up of photons, satisfied the scientific method by being

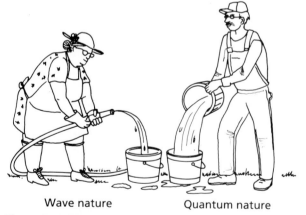

Wave nature   Quantum nature

Figure 9.4 Wave and quantum analogy.
A wave supplies a continuous flow of energy, somewhat analogous to a stream of water. A stream of water would take an appreciable time to fill a bucket. A quantum supplies energy in a packet or bundle, analogous to a bucket of water. Dumping the water from the bucket fills the other almost instantaneously. This is analogous to a quantum supplying the energy in the photoelectric effect.

experimentally valid. We have a confused situation. Is light a wave or a particle? The answer to this is couched in the term the **dual nature of light**, which simply means that light acts sometimes like a wave and sometimes like a particle.

9.2 The Bohr Theory of the Hydrogen Atom

The quantum theory played an important role in the development of a simplified model of the atom. In the late 1800s a great deal of experimental work was being done with gas-discharge tubes—for example, mercury-vapor and neon tubes. When the light was analyzed with a spectrometer, discrete or line spectra were observed instead of the continuous spectra observed from incandescent sources such as filament lamps. That is, only spectral lines of certain frequencies or wavelengths were found (Fig. 9.5). Scientists did not understand why only certain wavelengths of light were emitted by the atoms in various excited gases.

An explanation of the spectral lines observed for hydrogen was put forth in 1913 by the Danish physicist Niels Bohr (Fig. 9.6). In 1911, the British physicist Lord Ernest Rutherford had shown that the protons of an atom were in a nucleus or central core in the atom. Being the simplest atom, hydrogen has only one proton in its nucleus and one associated electron. Bohr's theory assumed that the electron of the hydrogen atom revolved about the nuclear proton in a circular orbit, much the same as a planet orbits the Sun or a satellite orbits the Earth, but with the electrical force instead of the gravitational force supplying the centripetal force. Furthermore, the hydrogen electron could be in only certain *discrete* orbits of given radii. (This assumption is not true for gravitationally maintained satellites. For example, an Earth satellite may have any orbital radius if given the proper tangential velocity.)

The possible electron orbits were characterized by whole-number values, $n = 1, 2, 3, 4, \ldots$, and n is called the **principal quantum number** (Fig. 9.7). The *quantized* circular orbits in the Bohr theory were a result of an application of quantum theory to the motion of the hydrogen electron. Bohr reasoned that a *discrete* line

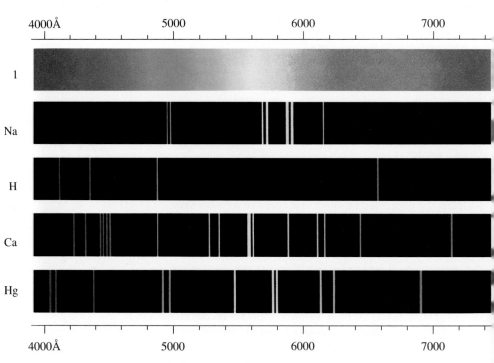

A continuous spectrum of white light is shown above the line spectra of light emitted from excited atoms in sodium (Na), hydrogen (H), calcium (Ca), and mercury (Hg) discharge tubes. Notice how each line spectrum is different of characteristic for a given element.

Figure 9.5 Continuous and discrete line spectra.

Figure 9.6 Niels Bohr (1885–1962).
Bohr was one of the foremost scientists of the twentieth century. His initial application of quantum theory to the hydrogen atom, for which he was awarded the Nobel Prize in 1922, led to the development of our present-day understanding of atomic structure and spectra. Bohr's subsequent work in nuclear theory played an important part in the understanding of nuclear fission. He was forced to flee his native Denmark during World War II because of his part in the anti-Nazi resistance movement. He came to the United States where he was involved in the atomic bomb project.

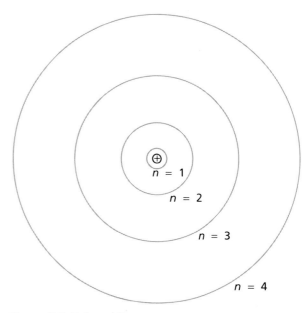

Figure 9.7 Bohr orbits.
The Bohr theory predicts only certain discrete orbits for the hydrogen electron. Notice that the orbits are unevenly spaced.

spectrum resulted from a *discrete* amount or quantum of a parameter.

However, there was still a classical problem with Bohr's theory. According to classical theory, an accelerating electron radiates electromagnetic energy. An electron in a circular orbit has centripetal acceleration and hence should radiate energy. This loss of energy would cause the electron to spiral into the nucleus, similar to an Earth satellite in a decaying orbit. But this event doesn't happen in the atom, and Bohr made a very nonclassical assumption in his theory. He postulated that the hydrogen electron does not radiate energy when in a bound, discrete orbit but does so only when it makes a "quantum jump" or transition from one discrete orbit to another.

The allowed orbits of the hydrogen atom are commonly expressed in terms of *energy states* or *levels*, with

each state corresponding to a particular orbit (Fig. 9.8). Keep in mind that a particle (or satellite) in a circular orbit with a particular radius has a particular energy. We characterize the energy levels as being states in a **potential well.** Like an object in a hole or well, energy is required to lift it to a higher level; and if the top of the well is the zero reference point, the energy levels in the well can have negative values.

In a hydrogen atom the electron is normally at the bottom of the well, or in the **ground state** ($n = 1$), and must be given energy, or "excited," to raise it up in the well to a higher energy level. The states above the ground state, with $n > 1$ ($n = 2, 3, 4, \ldots$), are called **excited states.** The levels in the energy well resemble the rungs of a ladder. Just as a person going up and down a ladder must do so in discrete steps on the ladder rungs, so must a hydrogen electron be excited (or de-excited) by discrete amounts. Notice, however, that the energy "rungs" of the hydrogen atom are not evenly spaced. Also, if enough energy is applied to excite the electron to the top of the well, the electron is no longer bound to the nucleus, and the atom is ionized.

A mathematical development of Bohr's theory is given in Appendix V. The important results are the predictions of the radii of the orbits and the energies of

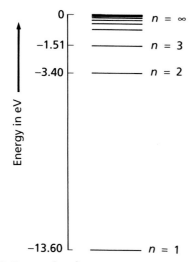

Figure 9.8 Energy levels.
The energy-level diagram for the hydrogen atom. Each Bohr orbit has a particular energy value, or level. The lowest level, $n = 1$, is called the ground state, and the higher levels are called excited states.

Table 9.1 Allowed Values of the Hydrogen Electron's Radius and Energy for Low Values of n

n	r_n	E_n
1	0.53 Å	-13.60 eV
2	2.12 Å	-3.40 eV
3	4.77 Å	-1.51 eV
4	8.48 Å	-0.85 eV
5	13.25 Å	-0.54 eV
$\vdots$	$\vdots$	$\vdots$
$\to \infty$	$\to \infty$	$\to 0$

the orbits. The radius of a particular orbit is given by

$$r_n = 0.53n^2 \text{ Å} \qquad (9.3)$$

where n is the principal quantum number of an orbit, $n = 1, 2, 3, \ldots$,
  r is measured in angstrom units $(1 \text{ Å} = 10^{-10} \text{ m})$.

For $n = 1$ we have $r_1 = 0.53$ Å; for $n = 2$ the orbit radius is $r_2 = (0.53)(2)^2 = (0.53)(4) = 2.12$ Å; and so on. A list of several of the allowed values of r is given in Table 9.1. Notice that the radii are not evenly spaced, as was depicted in Fig. 9.7.

The total energy of an electron in one of the allowed orbits is given by

$$E_n = \frac{-13.6}{n^2} \text{ eV} \qquad (9.4)$$

where n is the principal quantum number of the orbit
  E is measured in electron volts (eV)

An **electron volt** is the amount of energy an electron acquires when it is accelerated through an electric potential of one volt. The eV is a common unit of energy in atomic and nuclear physics, and 1 eV is equal to 1.6×10^{-19} J.

Equation 9.4 predicts that the energy of the electron in a hydrogen atom has only discrete or allowed values. For example,

$$E_1 = \frac{-13.6}{(1)^2} \text{ eV} = -13.6 \text{ eV}$$

$$E_3 = \frac{-13.6}{(3)^2} \text{ eV} = \frac{-13.6}{9} \text{ eV} = -1.51 \text{ eV}$$

Other electron energy values are given in Table 9.1 for low values of n, or orbits near the nucleus. These values correspond to the energy levels in the diagram in Fig. 9.8. The minus sign, indicating a negative energy value, shows that the electron is in a potential well. With $n = \infty$ the electron is at the top of the well and $E = 0$.

But how did the theory stand up to experiment? Recall that Bohr was trying to explain discrete line spectra. According to Bohr's theory, an electron can make transitions only between the allowed orbits or energy levels. In these transitions the total energy must be conserved. If the electron is initially in an excited state, it may lose energy by changing to a less excited (lower) state. In this case the electron's energy loss is carried away by a photon—a quantum of light. The total energy *before* must equal the total energy *after* the process, so we have

$$E_{n_i} = E_{n_f} + E_{\text{photon}}$$

where E_{n_i} = initial energy of electron,
  E_{n_f} = final energy of electron,
  E_{photon} = energy of photon given off.

Thus we have

$$E_{\text{photon}} = E_{n_i} - E_{n_f}$$

A schematic diagram of the process of photon emission is shown in Fig. 9.9a; Figure 9.9b shows the reverse process of photon absorption. The absorption of energy causes the atom to become further excited.

The transitions for photon emission are shown on an energy-level diagram in Fig. 9.10. The electron may "jump" down one or more energy levels in becoming de-excited. In this situation discrete amounts of energy are given off. Recall that $E_{\text{photon}} = hf$, so we may write the preceding energy conservation equation as

$$hf = E_{n_i} - E_{n_f}$$

And with $f = c/\lambda$ we have

$$\frac{hf}{\lambda} = E_{n_i} - E_{n_f}$$

or

$$\lambda = \frac{hc}{E_{n_i} - E_{n_f}}$$

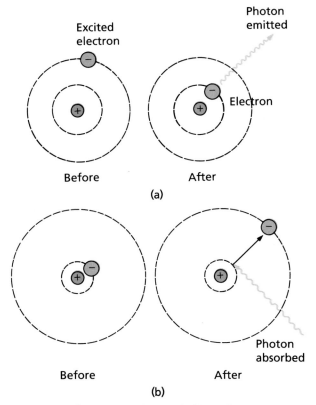

Before   **After**

(a)

Before   **After**

(b)

Figure 9.9 Photon emission and absorption.
(a) When an electron in an excited hydrogen atom makes a transition to a lower energy level, the atom loses energy by emitting a photon. (b) When an atom absorbs a photon, the electron is excited into a higher energy level.

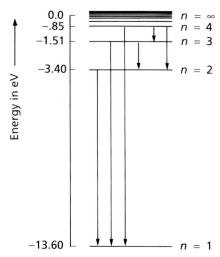

Figure 9.10 Hydrogen atom transitions.
Some possible transitions for the hydrogen electron in returning to the ground state. The electron may make one or more "jumps" in the de-excitation process. The energy of the photon is proportional to the difference in the energy levels, so the lengths of the transition arrows give an indication of the relative energies of the emitted photons.

If we express E_{n_i} and E_{n_f} in eV and h and c in appropriate units, we have

$$\lambda = \frac{12{,}400}{E_{n_i} - E_{n_f}} \, \overset{\circ}{A} = \frac{12{,}400}{\Delta E} \, \overset{\circ}{A} \qquad (9.5)$$

To find the wavelength of the light in angstroms, simply divide 12,400 by the difference between the initial and final energies (in eV) of the electron. The current trend is to express wavelength in nanometers (nm). In this case Eq. 9.5 becomes

$$\lambda = \frac{1240}{\Delta E} \, \text{nm} \qquad (9.6)$$

EXAMPLE 1

What is the wavelength emitted when a hydrogen electron makes a transition from the $n = 3$ to the $n = 2$ energy level?

Solution   We have $n_i = 3$ and $n_f = 2$. First find ΔE:

$$E = E_{n_i} - E_{n_f} = -1.51 \text{ eV} - (-3.40 \text{ eV}) = 1.89 \text{ eV}$$

where   the energy values were taken from Table 9.1 (or Fig. 9.10).

Then

$$\lambda = \frac{12{,}400}{\Delta E}\,\text{Å} = \frac{12{,}400}{1.89}\,\text{Å} = 6560\,\text{Å} \;(656.0\text{ nm})$$

which is the wavelength of red light.

EXAMPLE 2

Light with a blue-green color ($\lambda = 486$ nm) is observed to be emitted by hydrogen atoms. What are the quantum numbers of the initial and final energy levels of the electrons?

Solution   With $\lambda = 486$ nm we have

$$\lambda = 486\text{ nm} = \frac{1240}{\Delta E}$$

Solving for ΔE gives

$$\Delta E = E_{n_i} - E_{n_f} = \frac{1240}{486} = 2.55\text{ eV}$$

The problem now is to find a transition of 2.55 eV in the hydrogen atom. That is, we need to find two energy levels that differ by 2.55 eV. On inspection of the energy-level diagram in Fig. 9.10, we see that n_f cannot be the ground state ($n = 1$), because this value would give too large an energy difference. Similarly, the final state cannot be $n = 3$ or greater, because this value would give too *small* an energy difference. Hence we must have $n_f = 2$. And $n_i = 4$ seems to give about the right energy difference. Check to see whether it works: $\Delta E = E_4 - E_2 = -0.85\text{ eV} - (-3.40\text{ eV}) = 2.55\text{ eV}$, so $n_f = 4$ and $n_i = 2$ are the correct (and only) energy levels that allow for photon emission of this wavelength.

The Bohr theory predicts that the hydrogen atom will emit light with discrete wavelengths. When the wavelengths for the allowed transitions were computed and compared with the wavelengths of the lines of the hydrogen spectrum, researchers found that the theory agreed with the experimental data (see Fig. 9.11). Transitions to a particular final state form a series of transitions. These series were named in honor of early spectroscopists who discovered or experimented with these wavelengths of light. For example, the series of lines in the visible spectrum, which corresponds to the transitions to a final state of $n = 2$, is called the Balmer series.

Thus quantum theory and the quantum nature of light scored another success. As you might imagine, the energy-level arrangements for atoms, other than hydro-

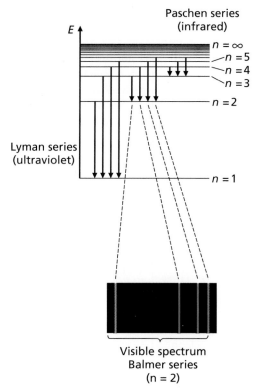

Figure 9.11 Spectral lines.
The transitions between discrete energy levels give rise to discrete spectral lines. Transitions to a particular state form a series; e.g., transitions to $n = 2$ is called the Balmer series, and the spectral lines are in the visible region.

gen, with more than one electron are quite complex. Even so, the line spectra of various atoms are indicative of their energy-level spacings and provide characteristic "fingerprints" by which atoms and molecules may be spectroscopically identified.

9.3 Quantum-Physics Applications

Much of modern physics and chemistry is based on the study of the energy levels of various systems. When light is emitted, scientists study the *emission spectrum* to learn about the energy levels of the system. Some chemists do research in the field of molecular spectroscopy—the study of the energy levels of molecules and their associated spectra. Molecules can have quantized energy levels owing to molecular vibrations, rotations, or excited atoms. Of course, different molecules have quite different spectra.

An *absorption spectrum* is obtained when light passes through a gas or liquid. Many of the wavelengths pass through, but photons that have the proper energy to excite the molecules to higher energy levels are absorbed. The absorption of a particular wavelength can be total or partial, depending on how much absorptive material is present, and other factors. An absorption spectrum appears as dark lines (absorbed wavelengths) on a bright background.

The various gases in the atmosphere absorb light of particular wavelengths. The main absorbing gases are carbon dioxide (CO_2), water (H_2O), and ozone (O_3). The absorption properties of these gases, as well as the atmosphere as a whole, will be discussed in Chapter 20.

The water molecule has some rotational energy levels spaced very closely together. The energy differences are such that microwaves are absorbed. Microwave photons have relatively low energies, with wavelengths of about 0.05 to 30 cm. The principle of the *microwave oven* is based on microwave absorption by water molecules. The water molecules in food absorb microwave radiation, thereby heating and cooking the food. The interior metal sides of the oven reflect the radiation and remain cool.

Because it is the water content that is crucial in microwave heating, materials such as paper plates and ceramic or glass dishes do not get hot immediately in a microwave oven. However, they often become warm after contact with the hot food. Some people think that microwaves penetrate the food and heat it throughout. This is *not* the case. On the average, microwaves penetrate only a few centimeters before being completely absorbed. The interior of a large mass of food is heated by conduction. For this reason microwave oven users are advised to let foods sit for a time after microwaving. Otherwise, the outside of the food may be quite hot, while the center remains cold.

As a safety precaution, microwave ovens should not be operated with the door open. Human tissue contains water and can be cooked as easily as food.

Another device based on energy levels is the laser. The development of the laser was a great success for the modern approach to science. Scientific discoveries have often been made accidentally like this, and even though some of these discoveries were put to practical use, no one fully understood how or why they worked. The discovery of X-rays is a good example. Similarly, early inventors applied a trial-and-error approach until they discovered something that would work. Edison's invention of the incandescent lamp is a good example. However, the laser was first developed "on paper" and then built with the expectation that it would work as predicted.

The word **laser** is an acronym for *light amplification by stimulated emission of radiation.*[*] The amplification of light provides a very intense beam. The key is the stimulated emission. Ordinarily, when an atom is excited, it returns to its ground state in a short time. This process is called *spontaneous emission.* In the process there is one photon in and one out (Fig. 9.12). However, an excited atom can be *stimulated* to emit a photon. In a **stimulated emission** process an excited atom is struck by a photon of the same energy of the allowed transition, and two

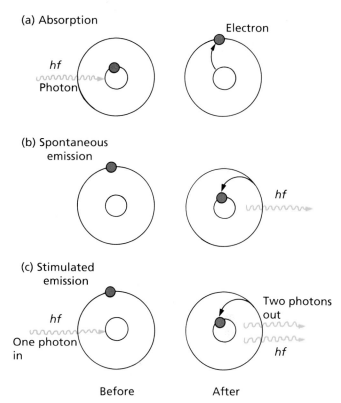

Figure 9.12 Spontaneous and stimulated emissions. When an atom absorbs a photon and becomes excited (a), it may spontaneously return to its ground state in a short time with the emission of a photon (b). If, however, the excited atom is struck by a photon with the same energy as the absorbed photon, the atom is stimulated to emit a photon (c).

---

[*] The principle of the laser was first developed for microwave frequencies, and the first device was called a *maser* (*m* = microwave).

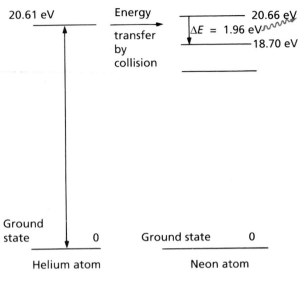

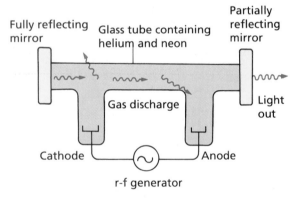

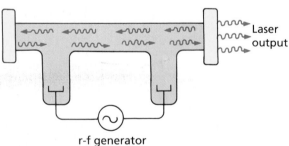

Figure 9.13 Laser action.
(a) *Above:* The red light of a He-Ne laser beam passing through a transparent cell. (b) *Below:* A schematic diagram of a He-Ne laser tube. Stimulated emissions and reflections from the end mirrors set up an intense beam along the axis of the tube. (c) *Above right:* The energy-level diagram for those levels involved in the lasing action of the helium-neon laser. See text for description.

photons are given off (one in, two out—amplification). Of course, in this process we do not get something for nothing. Energy is needed to initially excite the atom.

One of the common lasers used in the laboratory is the helium-neon gas laser (Fig. 9.13a). Let's focus on it. The gas mixture is about 85 percent helium and 15 percent neon. The helium and neon atoms are placed in an excited state by a radio frequency (rf) voltage applied across the laser tube (Fig. 9.13b). Normally, the excited atom would spontaneously emit a photon almost immediately. However, some states are said to be *metastable*, meaning that the electron stays in such a state for a longer period of time (but still only a fraction of a second). A particular energy level of the helium atom is a metastable state. As illustrated in the energy-level diagram in Fig. 9.13c, the neon atom also has an energy level at about the same energy. Hence there is time for a helium atom to collide with an unexcited neon atom and transfer energy to it before emitting a photon spontaneously. The excited neon atom will decay after a short time to another level, and the photon emitted in the transition has a wavelength that is in the visible red region of the spectrum.

The amplification of the light emitted by the neon atoms is accomplished by reflections from mirrors placed at each end of the laser tube. Reflecting back and forth in the tube, the photons cause stimulated emissions, and an intense beam of light builds up along the direction of the tube axis. Part of the beam emerges through one of the end mirrors, which is partially silvered. Because the light is amplified and very intense, a laser beam should never be directed toward a person's eyes (or reflected into the eyes). The direct viewing of a laser beam can cause serious eye damage.

The emitted light beam has some relatively unique properties. Light from common sources, such as an incandescent lamp, is said to be *incoherent*. That is, tne waves have no particular relationship to each other. In such light sources the excitation occurs randomly, and atoms emit randomly at different frequencies (different transitions). As a result, an incoherent beam is "chaotic" (Fig. 9.14a). For example, if you threw a handful of gravel into a pond, the resulting waves would be incoherent.

The light waves from a laser, on the other hand, have the same wavelength, phase, and direction. Such light is said to be **coherent light** (Fig. 9.14b). Because a laser beam is so directional, there is very little spreading of the beam. This feature has permitted us to reflect a pulsed laser beam back to Earth from a mirror placed on the moon by astronauts. Such experiments are used to accurately measure the distance between the Earth and the moon, so that small fluctuations in the moon's orbit can be studied.

Lasers and laser light are used in an increasing number of applications. For instance, long-distance communications use laser beams in space and in optical fibers.

Lasers are used in the medical field as diagnostic and surgical tools. The intense heat produced by the laser light over a small area can drill very small holes in metals and can weld metal machine parts. Laser "scissors" are used to cut cloth in the garment industry. There are applications in surveying, weapons systems, chemical processing, photography, and holography (the process of making three-dimensional images). Laser printers are used in computer printouts.

A common application of the laser is in compact disc players. A laser "needle" is used to read the information (sound) stored on a disc in small dot patterns. The dots produce reflection patterns that are read by photocells and converted to electronic signals. The laser in this case is a small solid-state semiconductor laser. Another very common laser application is found in the supermarket (Fig. 9.15). You have probably noticed the pink glow of He-Ne laser light in a supermarket checkout line that uses an optical scanner for reading the product codes on items. Each item has a particular code that identifies it to a computer that contains the programmed price of the item.

Another example of quantum phenomena is X-rays, which are used widely in medical and industrial fields. **X-rays** were discovered accidentally in 1895 by German physicist Wilhelm Roentgen while he was working with

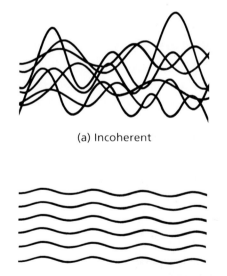

(a) Incoherent

(b) Coherent

Figure 9.14 Incoherent and coherent light.
(a) In incoherent light the waves have no particular relationship to each other and are random, or "chaotic". (b) Waves of coherent light, on the other hand, have the same wavelength, phase, and direction.

Figure 9.15 Optical scanning.
A laser scanner being used in a supermarket. The light reflected from the light areas of the code bar gives an identifying reflection pattern that is detected by the scanner, and this information is sent to a computer that supplies the price of the item.

gas-discharge tubes. He noticed that a piece of fluorescent paper was glowing, apparently from having been exposed to some unknown radiation being emitted from a discharge tube. He called it X-radiation.

In a modern X-ray tube electrons are accelerated through a large electrical voltage toward a metal target (Fig. 9.16a). When the electrons strike the target, they interact with the electrons in the atoms of the target material, and the electrical repulsion decelerates the incident electrons. The result is the emission of high-frequency X-ray quanta. X-rays are called *bremsstrahlung* ("braking rays") in German.

An X-ray spectrum is shown in Fig. 9.16(b). Notice that there is a *cutoff wavelength* λ_o below which no X-rays are emitted for a given tube voltage. Also, for very large tube voltages there are intense "spikes" or spectral lines. These spikes are characteristic of the target material

and are called *characteristic X-rays*. Both features of the X-ray spectrum can be explained by quantum considerations.

The low cutoff wavelength corresponds to the quantum of the highest frequency or energy ($E = hf = hc/\lambda$). Such quanta result when an incident electron is stopped completely and gives up all of its energy. There can be no quanta of greater frequency or lower wavelength, because the maximum energy has been given up. The cutoff wavelength may be made smaller by increasing the tube voltage so as to give the electrons more energy.

In the atoms of a target material there are many electrons. The electrons in the lower energy levels near the nuclear protons are shielded from the incident electrons by the atomic electrons in outer orbits, and so there is little interaction with the inner electrons.

At large tube voltages the incident electrons have sufficient energy to occasionally eject an electron from an inner orbit. This ejection leaves a vacancy that may be filled by an electron from a nearby outer orbit, and this leaves a vacancy in that orbit. The transitions in filling these vacancies in the inner orbits give rise to the characteristic spectral lines, similar to line spectra produced by transitions in the hydrogen atom.

9.4 Matter Waves and Quantum Mechanics

As part of the dual nature of light, what was thought to be a wave sometimes acts as a "particle." But can the reverse be true? That is, can particles have a wave nature? This question was considered by the French physicist Louis de Broglie (Fig. 9.17), who in 1925 postulated that matter, as well as light, has properties of both waves and particles. According to the hypothesis of de Broglie, any moving particle has a wave associated with it whose wavelength is given by

$$\lambda = \frac{h}{mv} \tag{9.7}$$

where λ = wavelength of the moving particle,
   m = mass of the moving particle,
   v = magnitude of the velocity of the particle,
   h = Planck's constant.

The waves associated with moving particles are called **matter waves** or **de Broglie waves**.

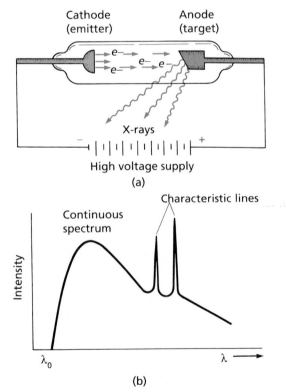

(b)

Figure 9.16 X-rays.
(a) X-rays are produced in a tube in which electrons from the cathode are accelerated toward the anode. On interacting with the atoms of the anode material, they are slowed down and emit energy in the form of X-rays.
(b) An X-ray spectrum. See text for description.

Figure 9.17 Louis de Broglie (1892–1987).
Prince Louis de Broglie, a French nobleman, studied medieval history at the Sorbonne. He enlisted in the French army and was a radioman in World War I. The experience created an interest in physics, and he presented his concept of matter waves in 1925 in a doctoral thesis. It was met with much skepticism at first. When it was shown that electrons can produce diffraction patterns, however, de Broglie's theory was taken more seriously. He was awarded the Nobel Prize for physics in 1929.

EXAMPLE 3

What are the de Broglie wavelengths for (a) an electron moving at 7.3×10^5 m/s and (b) a 1000-kg car traveling at 90 km/h (25 m/s, or about 55 mi/h)?

Solution (a) Using Eq. 9.7, we have

$$\lambda = \frac{h}{mv}$$

$$= \frac{6.63 \times 10^{-34} \text{ J·s}}{(9.11 \times 10^{-31} \text{ kg})(7.3 \times 10^5 \text{ m/s})}$$

$$= 10 \times 10^{-10} \text{ m} = 10 \text{ Å} = 1.0 \text{ nm}$$

(b) Similarly, for the 1000-kg car we get

$$\lambda = \frac{h}{mv} = \frac{6.63 \times 10^{-34} \text{ J·s}}{(1000 \text{ kg})(25 \text{ m/s})}$$

$$= 2.7 \times 10^{-38} \text{ m}$$

Because a wave is generally detected by the interaction with an object of about the same size as the wavelength, we see why the matter waves of common moving objects are not evident.

De Broglie's hypothesis was met with skepticism at first, but it was experimentally verified in 1927 by G. Davisson and L. H. Germer in the United States. They showed that a beam of electrons exhibited a diffraction pattern. Because diffraction is a wave phenomenon, a beam of electrons must have wavelike properties. For appreciable slit diffraction a wave must pass through a slit with a width approximately the size of the wavelength. Visible light has wavelengths on the order of 400 to 700 nm, and slits with widths of these sizes can be made quite easily. Electrons, as shown in Example 3, may have wavelengths on the order of 1.0 nm. Slits of this width cannot be made.

However, nature has provided us with suitably small slits in the form of crystal lattices. The atoms in these crystals are arranged in rows (or some other orderly arrangement), and the rows of atoms make natural "slits." Davisson and Germer bombarded nickel crystals with electrons and obtained a diffraction pattern on a photographic plate. Diffraction patterns made by X-rays (electromagnetic radiation of very short wavelength) and an electron beam incident on a thin aluminum foil are shown in Fig. 9.18. The similarity in the diffraction patterns from the electromagnetic waves and electron particles is evident. Electron diffraction demonstrates the "dual nature of matter."

An electron microscope uses the concept of matter waves. A beam of electrons, rather than a beam of light, is used to "view" an object. In one technique the beam of electrons bounces off a surface. The beam spot is scanned across the specimen by means of deflecting coils, much as is done in a television tube. Surface irregularities cause directional variations in intensity of the reflected electron beam that gives a contrasted image.

The amount of fuzziness of an image due to diffraction effects is proportional to the wavelength that is used. A typical beam wavelength in an electron microscope is 1.0 nm. This length is very short compared with the wavelength of visible light at about 500 nm.

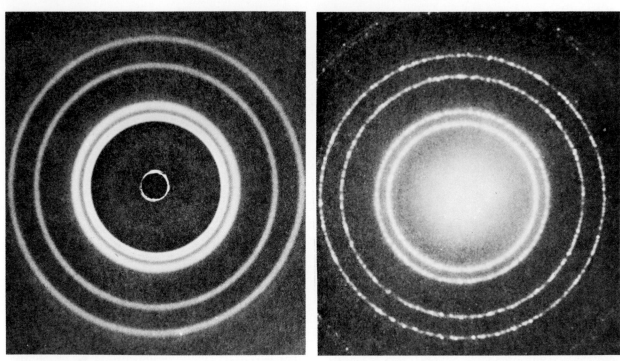

Figure 9.18 Diffraction patterns.
Two photographs showing diffraction patterns produced by X-rays and electrons. The X-ray pattern (left) can be explained using a wave description of X-rays. The electron pattern (right) can be explained using de Broglie or matter waves.

Hence finer detail or greater resolution can be achieved with an electron microscope, as well as greater magnification. Some electron photo micrographs are shown in Fig. 9.19.

Another big step toward understanding the nature of atoms and nuclei was taken in 1926 when Erwin Schrödinger, an Austrian physicist (Fig. 9.20), presented a widely applicable mathematical equation that gave new meaning to matter waves. This equation, known as the **Schrödinger equation,** is written in its simple form as

$$(E_k + E_p)\psi = E\psi$$

where E_k, E_p, and E are the kinetic energy, potential energy, and total energy, respectively.

The term ψ (Greek letter psi) is called the **wave function** and represents the wave associated with a particle. The Schrödinger equation is a sort of formulation of the conservation of energy.

The detailed form of the Schrödinger equation is quite complex, and the equation is difficult to solve. One simple case that was solved almost immediately was that of the hydrogen atom. The possible energy levels were found to be exactly the same as those Bohr had obtained in 1913. But there are additional results from the solution to the Schrödinger equation that involve the wave function ψ.

At first, scientists were not sure how ψ should be interpreted. They finally concluded that ψ^2 (the wave function squared) represented the *probability* that the hydrogen electron would be at a certain distance from the nucleus. In Bohr's theory the electron can be in circular orbits with discrete radii given by $r = 0.53n^2$ Å. A plot of ψ^2 versus r for the hydrogen electron is shown in Fig. 9.21a. It shows that the most *probable* radius for the hydrogen electron is one with $r = 0.53$ Å, which is the ground state for the hydrogen atom. The electron *could* be found at other radii but with lower probability (i.e., it is less likely).

As a result of the dual natures of waves and particles, a new kind of physics based on the synthesis of wave and quantum ideas was born in the 1920s and 1930s.

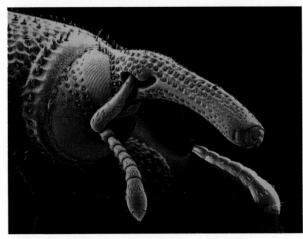

Figure 9.19 Electron micrographs.
Scanning electron micrographs of a diamond stylus in the grooves of a phonograph record and a maize weevil showing one of its compound eyes.

Figure 9.20 Erwin Schrödinger (1887–1961).
Born in Austria, he served in World War I and later became a professor of physics in Germany. In 1925, Schrödinger published a paper in which moving particles were treated mathematically as waves, making him one of the founders of quantum mechanics. He left Germany in 1933 when the Nazis came to power, and was later a professor of physics at the University of Dublin.

It was called **quantum mechanics** and replaced the classical-mechanics view that everything moved according to *exact* laws of nature with the concept of probability. For example, the quantum mechanics analysis of the hydrogen atom using the Schrödinger equation predicts that the electron would *most probably* be in the discrete orbits predicted by the Bohr theory. This idea gave rise to the idea of an "electron cloud" around the nucleus, with the cloud density reflecting the probability that the electron could be in that region (Fig. 9.21b).

There is another important aspect of quantum mechanics. According to classical mechanics, there is no limit to the accuracy of a measurement. The accuracy can be continually improved by refinement of the measurement instrument and/or procedure, to the point at which there may be no uncertainty in the measurement. This philosophy resulted in a *deterministic* view of nature. For example, if you know the position and velocity of a particle *exactly* at a particular time, you can determine where it will be in the future and where it was in the past (assuming no future or past unknown forces).

However, quantum theory predicts otherwise and sets limits on measurement accuracies. This idea was developed by the German physicist Werner Heisenberg (Fig. 9.22) and is called Heisenberg's uncertainty principle, which can be stated as:

> **It is impossible to simultaneously know a particle's exact position and velocity.**

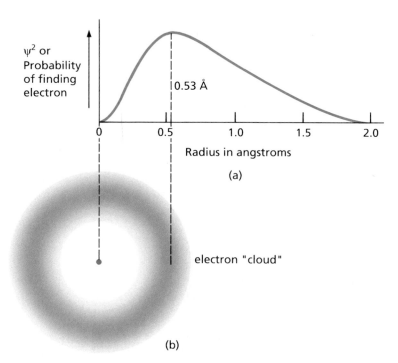

ψ^2 or Probability of finding electron

0.53 Å

0   0.5   1.0   1.5   2.0

Radius in angstroms

(a)

electron "cloud"

(b)

Figure 9.21 ψ^2 probability.
(a) *Above:* The square of the wavefunction (ψ^2) gives the probability of finding the electron at a particular radius. As shown here, the radius with the greatest probability of containing the electron is 0.53 Å, which corresponds to the first Bohr radius. (b) *Below:* The probability of finding the electron at other radii gives rise to the concept of an electron "cloud," or probability density.

Figure 9.22 Werner Heisenberg (1901–1976).
At the age of 23, Heisenberg wrote a paper explaining quantum mechanics. His theory was entirely different from Schrödinger's, yet it produced the same results. Schrödinger later wrote a paper showing that his and Heisenberg's theories were mathematically identical. Heisenberg's name is chiefly associated with the uncertainty principle, which he conceived as a result of his quantum theory. During World War II, Heisenberg was in charge of the German nuclear energy program.

This concept is often illustrated with a simple example. Suppose you want to measure the position and momentum (i.e., velocity) of an electron, as illustrated in Fig. 9.23. If you are to see the electron and determine its location, at *least* one photon must bounce off the electron and come to your eye. In the collision process, some of the photon's energy and momentum are transferred to the electron. (This situation is analogous to a classical collision of particles or billard balls. A collision involves a transfer of momentum and energy.)

After the collision, the electron recoils. Hence, in the process of trying to locate the position very accurately, which means that the uncertainty in position Δx is very small, a rather large uncertainty is caused in knowing the electron's velocity or momentum, because it recoiled at some angle to its original track. (Recall that $p = mv$, so $\Delta p = m \, \Delta v$).

In the case of light the position of an electron can be measured, at best, to an accuracy of about the wavelength λ of the light, that is, $\Delta x \approx \lambda$. The photon "particle" has a momentum of $p = h/\lambda$. (From the de Broglie hypothesis $\lambda = h/mv = h/p$.) Similarly, the momentum of the electron will be uncertain by at least an amount $\Delta p \approx h/\lambda$.

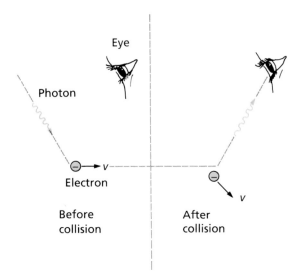

Eye

Photon

v

Electron

v

Before
collision

After
collision

Figure 9.23 Uncertainty.
Imagine trying to accurately determine the location of an electron with a single photon, which must strike the electron and come to the detector. The electron recoils after the collision, which introduces a great degree of uncertainty in knowing the electron's velocity.

The total uncertainty is the product of the individual uncertainties, and

$$(\Delta p)(\Delta x) = (m\,\Delta v)(\Delta x) \approx \left(\frac{h}{\lambda}\right)\lambda = h$$

and

$$m\,\Delta v\,\Delta x \approx h \qquad (9.8)$$

where m = the mass of the object being measured,
Δv = the minimum uncertainty in velocity,
Δx = the minimum uncertainty in position,
h = Planck's constant (6.63×10^{-34} J·s).

Thus Heisenberg's uncertainty principle states that the product of the *minimum* uncertainties are on the order of Planck's constant. Equation 9.8 gives the minimum uncertainties—that is, the best we can ever hope to measure *simultaneously*. The more accurate we measure the position (make Δx smaller), the greater the uncertainty in the velocity, $m\,\Delta v \approx h/\Delta x$, and vice versa. If we could measure the exact location of a particle ($\Delta x \to 0$), then we would have no idea about its velocity ($\Delta v \to \infty$).

EXAMPLE 4

If the uncertainty in the velocity of an electron is 1.4×10^7 m/s, what is the approximate minimum uncertainty in its position?

Solution   The mass of an electron is $m = 9.11 \times 10^{-31}$ kg, and from Heisenberg's uncertainty principle (Eq. 9.8) we have

$$\Delta x \approx \frac{h}{m\,\Delta v}$$

$$\approx \frac{6.63 \times 10^{-34}\ \text{J·s}}{(9.11 \times 10^{-31}\ \text{kg})(1.4 \times 10^7\ \text{m/s})}$$

$$\approx 0.52 \times 10^{-10}\ \text{m} = 0.52\ \text{Å}$$

The uncertainty in the velocity is quite large, which makes the uncertainty in the position of the electron relatively small. However, notice that the uncertainty in the position is on the order of the Bohr radius for the hydrogen electron in the ground state.

9.5 Atomic Quantum Numbers

When the Schrödinger equation is solved for the hydrogen atom, several quantum numbers result. One is n, the principal quantum number, which we discussed in the Bohr theory. However, several other such numbers arise, each with limitations. They are designated by the letters l, m, and m_s and are called the orbital, magnetic, and spin quantum numbers, respectively.

The **orbital quantum number l** is associated with the value of the angular momentum of the electron. (Recall that the angular momentum of a particle going in a circle is calculated by multiplying its mass times its velocity times the radius of the circle.) The quantum number l can have the values $0, 1, 2, 3, \ldots, (n - 1)$. The last term means that l must always be less than the value of the principal quantum number n for a given energy level.

EXAMPLE 5

What are the possible values of l for a hydrogen electron in a state designated by the principal quantum number (a) $n = 1$ and (b) $n = 3$?

Solution   (a) For $n = 1$ the only value of l for the electron is $l = 0$, because l must be less than n.
(b) For $n = 3$, l is less than n for $l = 0, 1$, and 2, so there are three possible l values.

The values of l are commonly designated by letters. This designation comes from historical spectroscopic notation. The letters $s, p, d, f, g, \ldots$, correspond to the values $l = 0, 1, 2, 3, 4, \ldots$, respectively. (The first four letters originally stood for the spectroscopic terms *sharp, principal, diffuse,* and *fundamental.*) That is,

$l = 0$ is designated by the letter *s*.

$l = 1$ is designated by the letter *p*.

$l = 2$ is designated by the letter *d*.

$l = 3$ is designated by the letter *f*.

$l = 4$ is designated by the letter *g*.

After *f* and *g* the letters continue alphabetically for higher values of *l*. However, our discussion will be limited to the first few *l* values.

A third quantum number is the **magnetic quantum number** *m*. This quantum number is associated with the orientation of the angular momentum and the region of space the electron occupies. The value of *m* can be any positive or negative integer from -1 to $+1$, including 0. Hence there can be $2l + 1$ values of *m* for a given *l*. For example, if $l = 2$, there are five possible values of *m*: $2l + 1 = 2(2) + 1 = 5$.

EXAMPLE 6

What are the possible *m* values for (a) $l = 0$ and (b) $l = 2$?

Solution (a) For $l = 0$ there can be only one value of *m*, that is, $m = 0$. Notice that $2l + 1 = 2(0) + 1 = 1$. (b) For $l = 2$, *m* can have the values $-2, -1, 0, +1,$ and $+2$. Notice how the values are just the integer numbers in the series $-l, \ldots, 0, \ldots, +l$.

The fourth and last quantum number is the **spin magnetic quantum number** m_s. This number is strictly a quantum mechanic effect and has no classical analog. However, the orbiting hydrogen electron is sometimes likened to the Earth revolving about the Sun. Like the rotating or spinning Earth, the electron "spins." Spin can be either clockwise or counterclockwise. The quantum number m_s is arbitrarily designated either $+\frac{1}{2}$ or $-\frac{1}{2}$ to indicate the two quantum mechanical spins. No matter what the values of the other three quantum numbers (n, l, and m), the value of m_s is always either $+\frac{1}{2}$ or $-\frac{1}{2}$, with each of these values occurring for each value of *m*. For example, there is a $\pm\frac{1}{2}$ m_s pair for $m = 0$, a $\pm\frac{1}{2}$ m_s pair for $m = 1$, etc.

The four quantum numbers for atomic electrons are summarized in Table 9.2. For the hydrogen atom only the principal quantum number *n* determines the energy of the electron. However, for other elements, those with two or more electrons, the quantum number *l* also is associated with the energy of each electron. The quantum numbers *m* and m_s have nothing to do with the energy of atomic electrons.

9.6 Multielectron Atoms and the Periodic Table

Because the quantum numbers *n* and *l* are in general associated with the energy levels of atoms, they are important in determining the properties of atoms. The energy levels and the number of electrons occupying them gives us insight into the chemical properties of atoms. Therefore, we will take a look at the atomic energy levels before we get to the chapter on chemistry (Chapter 11).

Table 9.2
Quantum Numbers for an Atomic Electron

Designation	Meaning	Possible Values
n	Principal quantum number associated with effective volume and energy	Any positive integer: 1, 2, 3, . . .
l	Orbital quantum number associated with angular momentum and energy	Any positive integer less than n: $l = 0, 1, 2, 3, \ldots, (n - 1)$
m	Magnetic quantum number associated with orientation of angular momentum	From $-l$ to $+l$ in integer steps, including 0
m_s	Spin magnetic quantum number; quantum mechanical property	$+\frac{1}{2}$ or $-\frac{1}{2}$ for each value of m

Because only n and l are associated with the values of the energy of the electrons in atoms other than hydrogen (n only), each atomic energy level is labeled by using these two quantum numbers. The common notation is to write n as a number, followed by the letter that stands for the value of l. For example, $1s$ means the energy level for $n = 1$ and $l = 0$; $3d$ designates the energy level with $n = 3$ and $l = 2$; and $4p$ means the level with $n = 4$ and $l = 1$.

For each energy level there are various sets of all four quantum numbers possible, because m and m_s have different values at each level.

EXAMPLE 7

What are the possible sets of quantum numbers up to and including the 3*d* level?

Solution The 3 indicates the principal quantum number ($n = 3$), and the *d* means $l = 2$. For this particular l level we can have $m = -2, -1, 0, +1$, and $+2$; and m_s can be $+\frac{1}{2}$ or $-\frac{1}{2}$ for each value of m. Hence there are 10 different sets of quantum numbers for the *d* level. This set is illustrated in Fig. 9.24, along with the other possible sets of quantum numbers with smaller n or l values.

Notice in Fig. 9.24 that the *total* number of possible sets of quantum numbers for a given n is $2n^2$. Also, the number of possible sets for any given l is $2(2l + 1)$.

The energy levels for the hydrogen atom run sequentially upward. The situation for other multielectron atoms is not so simple. The numerical sequence has numbers out of order. For example, as illustrated in Fig. 9.25, the $4s$ level is below the $3d$ level, because multielectron atoms have more than one electron orbiting the nucleus and there are additional electrical forces between the electrons. Electrons in outer orbits are also "shielded" from the attractive force of the nucleus by the electrons in inner orbits. Solving this problem is extremely complicated, and a simple, more detailed explanation of why the levels are so ordered cannot be given.

However, there is a convenient way to remember the order: Write down the levels in a triangle, as shown

		Values for m and m_s	Number of different sets
$n = 1$	$l = 0$	0 ∧ +1/2 −1/2	2
$n = 2$	$l = 0$	0 ∧	2
	$l = 1$	−1, 0, +1 ∧ ∧ ∧	6
$n = 3$	$l = 0$	0 ∧	2
	$l = 1$	−1, 0, +1 ∧ ∧ ∧	6
	$l = 2$	−2, −1, 0, +1, +2 ∧ ∧ ∧ ∧ ∧	10
$n = 4$	$l = 0$	0 ∧	2
	$l = 1$	−1, 0, +1 ∧ ∧ ∧	6
	$l = 2$	−2, −1, 0, +1, +2 ∧ ∧ ∧ ∧ ∧	10
	$l = 3$	−3, −2, −1, 0, +1, +2, +3 ∧ ∧ ∧ ∧ ∧ ∧ ∧	14

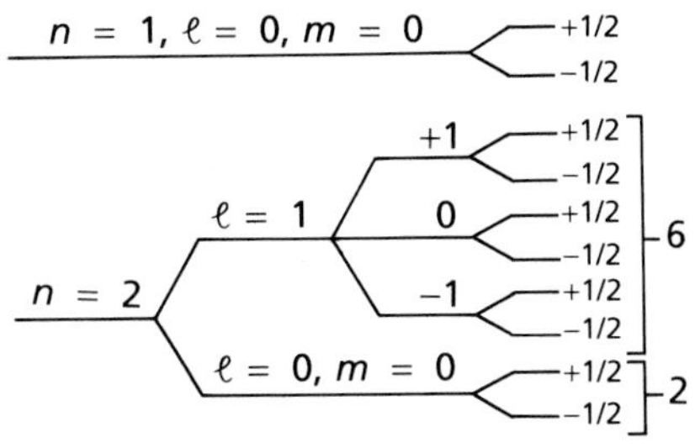

Figure 9.24 Sets of quantum numbers.
A tabulation and display of the possible quantum numbers for a given n and the number of different sets or combinations. (In the table the m_s values of $+\frac{1}{2}$ and $-\frac{1}{2}$ are shown for $n = 1$ only. They are represented by the wedge symbol for other n quantum numbers.

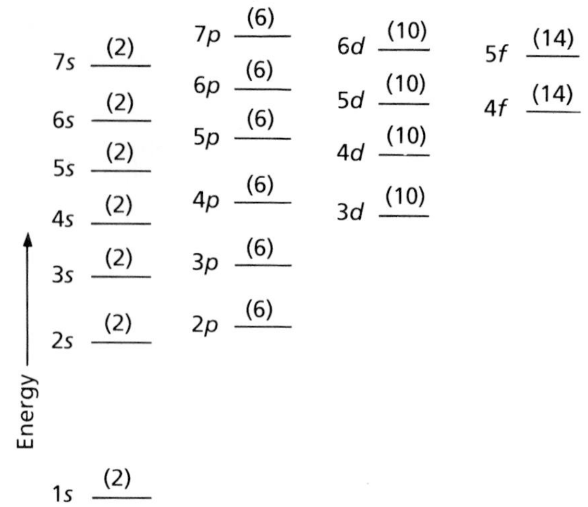

7s $\underline{\quad(2)\quad}$	7p $\overline{\quad(6)\quad}$	6d $\underline{\quad(10)\quad}$	5f $\underline{\quad(14)\quad}$
6s $\underline{\quad(2)\quad}$	6p $\overline{\quad(6)\quad}$	5d $\underline{\quad(10)\quad}$	4f $\underline{\quad(14)\quad}$
5s $\underline{\quad(2)\quad}$	5p $\overline{\quad(6)\quad}$	4d $\underline{\quad(10)\quad}$	
4s $\underline{\quad(2)\quad}$	4p $\overline{\quad(6)\quad}$	3d $\underline{\quad(10)\quad}$	
3s $\underline{\quad(2)\quad}$	3p $\overline{\quad(6)\quad}$		
2s $\underline{\quad(2)\quad}$	2p $\overline{\quad(6)\quad}$		
1s $\underline{\quad(2)\quad}$			

Energy ↑

Figure 9.25 Multi-electron-atom energy levels.
A typical diagram for a multi-electron atom (not to scale). In a multi-electron atom, the energy levels depend on the *n* and *l* quantum numbers. The number of different sets of quantum numbers for each level is shown in parentheses.

Figure 9.26 Energy-level ordering.
A mnemonic diagram for the ordering of energy levels in multi-electron atoms. Writing the energy levels in an array as shown, the order is from upper right to lower left along successively lower diagonal lines.

in Fig. 9.26. Then draw diagonal lines going from the upper right-hand to the lower left-hand portion, as shown. For the correct order, start with the top diagonal, go in the direction shown, and then go on to the next diagonal.

The ground state of an atom is the combination of energy levels with the lowest total energy. For the

hydrogen atom the ground state is the 1s level. Almost all naturally occurring atoms are in the ground state. If an atom is not in the ground state, its electrons will give off energy in the form of photons and fall to the ground state. To understand the more complicated, multielectron atoms, we must know how the electrons distribute themselves in the ground-state energy levels. The **Pauli exclusion principle**, set forth in 1928 by the German physicist Wolfgang Pauli, indicates how this distribution occurs. The Pauli exclusion principle is:

> **No two electrons can have the same set of quantum numbers.**

Hence each different *set* of quantum numbers (n, l, m, m_s) corresponds to a different energy state that can be occupied by only one electron. Then by filling up the lower energy levels with one electron for a particular set of quantum numbers, we can "build up" the various electron configurations for the atoms. Energy-level diagrams are illustrated in Fig. 9.27 for the 3 electrons in the lithium (Li) atom, for the 10 electrons in the neon (Ne) atom, and for the 11 electrons in the sodium (Na) atom. (Compare the number of electrons in each level with the number of sets of quantum numbers in Fig. 9.24.)

Rather than draw diagrams each time we want to represent the electron arrangement in an atom, we can use a shorthand notation called the **electron configuration**. When we write the electron configuration, we give the levels in order of increasing energy, and we designate the number of electrons in each level by a superscript. For example, $2p^5$ means that there are 5 electrons in the $2p$ state ($n = 2$, $l = 1$). The electron configurations for the atoms shown in Fig. 9.27 are as follows:

Li (3 electrons)	$1s^2\ 2s^1$
Ne (10 electrons)	$1s^2\ 2s^2\ 2p^6$
Na (11 electrons)	$1s^2\ 2s^2\ 2p^6\ 3s^1$

Notice that the superscripts add to the number of electrons in an atom.

A multielectron atom can be thought of in terms of the Bohr model with electrons orbiting the nucleus. As in the hydrogen atom, the lower the value of n, the closer the orbit is to the nucleus. All orbits or energy levels with the same n value are said to be in an **electron shell,** the $n = 1$ shell, $n = 2$ shell, etc.

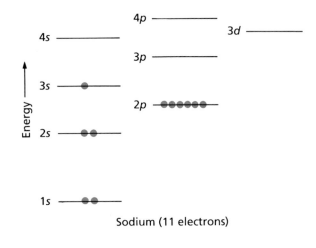

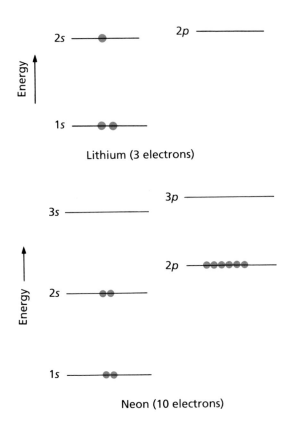

Figure 9.27 Electrons and energy levels.
The energy-level diagrams with electron occupancy for the ground states of lithium, neon, and sodium. The Pauli exclusion principle determines how many electrons can be in a particular energy level. (Energy-level spacing is not to scale.)

Electrons with the same n value are in the same shell. Electrons with the same values of n and l are said to be in the same **electron subshell.** For example, sodium has 6 electrons in the 2p subshell, as indicated in the preceding electron configuration for sodium.

Notice in Fig. 9.27 that the spaces between the energy levels are not equal. For instance, there is a large gap between the 1s and 2s levels. In general, there are large gaps between the s levels and the levels below them (except the 1s level). The gaps are smaller between the other levels, such as between 4s–3d and 3d–4p. The energy levels (such as 4s, 3d, and 4p) between the large gaps have approximately the same energy.

A set of energy levels that have about the same energy is called an **electron period.** For example, the energy gaps between periods are indicated by the vertical lines drawn between energy levels in the following electron configuration:

$$1s^2|2s^22p^6|4s^23d^{10}4p^6|5s^24d^{10}5p^6| \text{ etc.}$$

These periods form the basis of the chemical periodic table, which will be studied in Chapter 11. The table was developed empirically in the 1800s, because the structure of the atom was not known at that time. The elements were arranged in horizontal rows, and when the chemical properties of a next element were similar to a previous element, that element was placed below the previous element. So all the elements in a vertical column had similar properties. This periodic recurrence of similar chemical properties is related to the electron periods.

If the electron configurations are placed on the periodic table for the atom of each element, we find a very interesting occurrence (Fig. 9.28). Notice that, in general, the outer subshell configurations of the atoms in a vertical column are the same or similar. Look at Li and Na in the far left column. The electron configurations for these atoms were written previously. They both have a single electron in the outermost s subshell. In fact, all the atoms in this column do. Hence we may conclude that the similarity of the chemical properties of these elements is due to a similar outer electron configuration. Keep this idea in mind when you begin the study of chemistry in Chapter 11.

PERIODS

s

Period	1	2																	13	14	15	16	17	18

Main table (s, d, p blocks)

Period 1:
- 1 **H** $1s^1$
- 2 **He** $1s^2$

Period 2:
- 3 **Li** $2s^1$
- 4 **Be** $2s^2$
- 5 **B** $2s^22p^1$
- 6 **C** $2s^22p^2$
- 7 **N** $2s^22p^3$
- 8 **O** $2s^22p^4$
- 9 **F** $2s^22p^5$
- 10 **Ne** $2s^22p^6$

Period 3:
- 11 **Na** $3s^1$
- 12 **Mg** $3s^2$
- 13 **Al** $3s^23p^1$
- 14 **Si** $3s^23p^2$
- 15 **P** $3s^23p^3$
- 16 **S** $3s^23p^4$
- 17 **Cl** $3s^23p^5$
- 18 **Ar** $3s^23p^6$

Period 4:
- 19 **K** $4s^1$
- 20 **Ca** $4s^2$
- 21 **Sc** $4s^23d^1$
- 22 **Ti** $4s^23d^2$
- 23 **V** $4s^23d^3$
- 24 **Cr** $4s^13d^5$
- 25 **Mn** $4s^23d^5$
- 26 **Fe** $4s^23d^6$
- 27 **Co** $4s^23d^7$
- 28 **Ni** $4s^23d^8$
- 29 **Cu** $4s^13d^{10}$
- 30 **Zn** $4s^23d^{10}$
- 31 **Ga** $4s^24p^1$
- 32 **Ge** $4s^24p^2$
- 33 **As** $4s^24p^3$
- 34 **Se** $4s^24p^4$
- 35 **Br** $4s^24p^5$
- 36 **Kr** $4s^24p^6$

Period 5:
- 37 **Rb** $5s^1$
- 38 **Sr** $5s^2$
- 39 **Y** $5s^24d^1$
- 40 **Zr** $5s^24d^2$
- 41 **Nb** $5s^14d^4$
- 42 **Mo** $5s^14d^5$
- 43 **Tc** $5s^14d^6$
- 44 **Ru** $5s^14d^7$
- 45 **Rh** $5s^14d^8$
- 46 **Pd** $4d^{10}$
- 47 **Ag** $5s^14d^{10}$
- 48 **Cd** $5s^24d^{10}$
- 49 **In** $5s^25p^1$
- 50 **Sn** $5s^25p^2$
- 51 **Sb** $5s^25p^3$
- 52 **Te** $5s^25p^4$
- 53 **I** $5s^25p^5$
- 54 **Xe** $5s^25p^6$

Period 6:
- 55 **Cs** $6s^1$
- 56 **Ba** $6s^2$
- 57 **La*** $6s^25d^1$
- 72 **Hf** $6s^25d^2$
- 73 **Ta** $6s^25d^3$
- 74 **W** $6s^25d^4$
- 75 **Re** $6s^25d^5$
- 76 **Os** $6s^25d^6$
- 77 **Ir** $6s^25d^7$
- 78 **Pt** $6s^15d^9$
- 79 **Au** $6s^15d^{10}$
- 80 **Hg** $6s^25d^{10}$
- 81 **Tl** $6s^26p^1$
- 82 **Pb** $6s^26p^2$
- 83 **Bi** $6s^26p^3$
- 84 **Po** $6s^26p^4$
- 85 **At** $6s^26p^5$
- 86 **Rn** $6s^26p^6$

Period 7:
- 87 **Fr** $7s^1$
- 88 **Ra** $7s^2$
- 89 **Ac**** $7s^26d^1$
- 104 **Unq** $7s^26d^2$
- 105 **Unp** $7s^26d^3$
- 106 **Unh** $7s^26d^4$
- 107 **Uns** $7s^26d^5$
- 108 **Uno** $7s^26d^6$
- 109 **Une** $7s^26d^7$

*Lanthanides

58 **Ce**	59 **Pr**	60 **Nd**	61 **Pm**	62 **Sm**	63 **Eu**	64 **Gd**	65 **Tb**	66 **Dy**	67 **Ho**	68 **Er**	69 **Tm**	70 **Yb**	71 **Lu**
$6s^24f^15d^1$	$6s^24f^35d^0$	$6s^24f^45d^0$	$6s^24f^55d^0$	$6s^24f^65d^0$	$6s^24f^75d^0$	$6s^24f^75d^1$	$6s^24f^95d^0$	$6s^24f^{10}5d^0$	$6s^24f^{11}5d^0$	$6s^24f^{12}5d^0$	$6s^24f^{13}5d^0$	$6s^24f^{14}5d^0$	$6s^24f^{14}5d^1$

**Actinides

90 **Th**	91 **Pa**	92 **U**	93 **Np**	94 **Pu**	95 **Am**	96 **Cm**	97 **Bk**	98 **Cf**	99 **Es**	100 **Fm**	101 **Md**	102 **No**	103 **Lr**
$7s^25f^06d^2$	$7s^25f^26d^1$	$7s^25f^36d^1$	$7s^25f^46d^1$	$7s^25f^66d^0$	$7s^25f^76d^0$	$7s^25f^76d^1$	$7s^25f^96d^0$	$7s^25f^{10}6d^0$	$7s^25f^{11}6d^0$	$7s^25f^{12}6d^0$	$7s^25f^{13}6d^0$	$7s^25f^{14}6d^0$	$7s^25f^{14}6d^1$

184

Figure 9.28 Electron periods.
A table of electron periods with the electron configuration (or the latter part of it) for each atom.

Learning Objectives

After reading and studying this chapter, you should be able to do the following without referring to the text:

1. State Planck's hypothesis, and explain why it was so radical.

2. Explain what is meant by the dual nature of light.

3. Describe the photoelectric effect, and explain the quantum theory that describes it.

4. Describe quantitatively the Bohr theory of the hydrogen atom.

5. Explain what is illustrated in an energy level diagram, and explain why the energies are negative.

6. Explain how microwave ovens heat foods and not the oven itself.

7. State what the term *laser* stands for, how a laser works, and what is unique about laser light.

8. Explain how X-rays are produced.

9. State de Broglie's hypothesis concerning matter waves, and give an example calculation.

10. Give an example or proof that matter waves exist.

11. Explain the meaning of the wave function (ψ) in Schrödinger's equation.

12. State and explain the implications of Heisenberg's uncertainty principle.

13. State and explain Pauli's exclusion principle.

14. Draw ground state energy level diagrams for some elements.

15. Write the electron configurations for some elements.

16. Explain how electron periods are associated with the periodic table and how the outer subshell structure of elements with similar chemical properties compare.

17. Define and explain the other important words and terms listed in the next section.

Important Words and Terms

Planck's hypothesis
quantum
photoelectric effect
photons
dual nature of light
Bohr theory
principal quantum number
potential well
ground state
excited states

electron volt
laser
stimulated emission
coherent light
x-rays
matter waves, de Broglie waves
Schrödinger equation
wave function
quantum mechanics
Heisenberg's uncertainty principle

orbital quantum number l
magnetic quantum number m
spin magnetic quantum number m_s
Pauli exclusion principle
electron configuration
electron shell
electron subshell
electron period

Questions

The Dual Nature of Light

1. What was Planck's hypothesis, and why was it so different?

2. Is light a wave or a particle?

3. What proof can you give that light is a wave? a particle?

The Bohr Theory of the Hydrogen Atom

4. Some hydrogen atoms have one proton and an electrically neutral particle called a neutron in the nucleus. How would a neutron affect the Bohr theory?

5. According to the Bohr theory, what is the approximate size of a hydrogen atom in the ground state?

6. Why are the energies of the hydrogen electron all negative?

7. How did Bohr address the problem that an orbiting accelerating electron should emit radiation?

8. How does the Bohr theory explain the discrete line spectrum of hydrogen?

Quantum-Physics Applications

9. Why does a microwave oven heat a potato but not a ceramic plate?

10. Why are microwave ovens constructed in such a way that they will not operate when the door is open?

11. What does the word *laser* stand for, and what are the principle and the function of a laser?

12. What is unique about light from a laser source, and why should you never look directly into a laser beam?

13. Why are X-rays called "braking rays," and why does increasing the tube voltage shift the cutoff wavelength of an X-ray spectrum to a shorter wavelength?

Matter Waves and Quantum Mechanics

14. What is a matter wave?

15. How was a beam of electrons shown to have wavelike properties?

16. What is quantum mechanics?

17. How does the Heisenberg uncertainty principle change the classical deterministic view of the universe?

Atomic Quantum Numbers

18. State the four quantum numbers for an electron in a hydrogen atom, and explain the association of each.

19. What do the letters *s, p, d,* and *f* mean?

20. What determines the electron energy in a (a) hydrogen atom and (b) multielectron atom?

Multielectron Atoms and the Periodic Table

21. Explain the Pauli exclusion principle.

22. Why are the possible sets of atomic quantum numbers so important?

23. Distinguish between electron configuration, electron shell, electron subshell, and electron period.

24. Why is a 3*f* energy level impossible?

25. What is the basis of the periodic table of elements, and what do the elements in a vertical column have in common?

Exercises

The Dual Nature of Light

1. The human eye is most sensitive to yellow-green light with a wavelength of approximately 550 nm.
   (a) Determine the frequency of this electromagnetic radiation.
   (b) What is the energy of a photon of this light in joules and in electron volts?

2. What is the difference in the energies of a photon of red light and a photon of blue light?

3. Light is measured to have photons with an energy of 1.4 eV. Is this visible light? Give proof of your answer.

The Bohr Theory of the Hydrogen Atom

4. What is the radius of the electron orbit of the hydrogen atom for each of the following principal quantum numbers?
   (a) $n = 2$ (b) $n = 6$ (c) $n = 10$
   Answer: (c) 53 Å

5. What is the energy of the electron of a hydrogen atom for each of the orbits designated by the following principal quantum numbers?
   (a) $n = 2$ (b) $n = 6$ (c) $n = 10$
   Answer: (c) −0.136 eV

6. What is the ionization energy, in eV, for a hydrogen atom if the electron is in an orbit with a principal quantum number of (a) $n = 2$, (b) $n = 6$, and (c) $n = 10$? (The *ionization energy* is the amount of energy that would have to be given to the atom to ionize, or remove the electron from, the proton.)   *Answer:*   (c) +0.136 eV

7. What is the energy of the photons, in eV, emitted in the following transitions of an electron in a hydrogen atom?
   (a) $n = 4$ to $n = 2$
   (b) $n = 6$ to $n = 2$
   (c) $n = 3$ to $n = 1$
   (d) $n = 4$ to $n = 3$ *Answer:*   (b) 3.02 eV

8. (a) What are the wavelengths corresponding to the transitions in Exercise 7?
   (b) Which wavelengths of part (a) are in the visible region?
   (c) Which are in the ultraviolet and infrared regions?
   Partial answer: (b) $n = 6$ to $n = 2$. $\lambda = 4100$ Å, visible

9. A hydrogen atom absorbs a photon of wavelength 656 nm.
   (a) How much energy did the atom absorb?
   (b) What were the initial and final states of the electron in the atom?   *Answer:*   (b) $n = 2$ to $n = 3$

10. What is the shortest wavelength present in the hydrogen spectrum?

Quantum Physics Applications

11. (a) Determine the energy of a microwave photon with a wavelength of 1.0 cm.
    (b) How many photons are needed to produce 1 cal of heat?   *Answer:*   (a) 2.0×10^{-23} J

12. Calculate the wavelength of the radiation emitted when an electron in an excited neon atom falls from a 20.66-eV energy level to a 18.70-eV energy level.

13. There is spontaneous emission from the 18.70-eV energy level to the ground state in a neon atom. Determine the wavelength of the emitted photon.

Matter Waves and Quantum Mechanics

14. The Bohr radius for the ground state is 0.529 Å.
    (a) What is the circumference of the electron orbit?
    (b) Set this circumference equal to the de Broglie wavelength of the electron. Use the de Broglie equation to determine how fast the electron is moving in its orbit. Compare your answer with the speed of the electron according to the Bohr theory—that is, $v = 2.19 \times 10^6$ m/s. ($m_e = 9.11 \times 10^{-31}$ kg)

    Answer: (b) same

15. Calculate the de Broglie wavelength of a 0.5-kg ball moving with a constant velocity of 26 m/s (approximately 60 mi/h).

16. What is the de Broglie wavelength for the Earth moving in its orbit? (*A reminder:* $m_E = 6.0 \times 10^{24}$ kg and $v_E = 3.0 \times 10^4$ m/s.)

17. What is the minimum uncertainty in the velocity of an electron if the uncertainty in its position is 10^{-4} m?

18. Compute the minimum uncertainty in the velocity of a 0.5-kg ball when the uncertainty in position is 10^{-4} m.

19. The minimum uncertainty in the momentum of an electron is on the order of Planck's constant. How well can the electron be located?

Atomic Quantum Numbers and Multielectron Atoms

20. What are the possible l values for (a) $n = 3$ and (b) $n = 5$?

21. What are the numbers of possible electron states for the electron shells in Exercise 20?   *Answer:* (b) 50

22. What are the possible values of the quantum number m for (a) $l = 2$ and (b) $l = 3$?

23. (a) List the possible sets of quantum numbers for an electron in the $4d$ subshell.
    (b) How many electrons can occupy this subshell?

24. Consider the following electron configuration:

$$1s^2 2s^2 2p^6 3s^2 3p^6 4s^2 3d^9 4p^1$$

    (a) To which atom does this configuration correspond?
    (b) Is the atom in the ground state?
    Answer: (a) Zn (b) no

25. How many electrons are in the outermost shell of the atom described in Exercise 24?

26. Write the electron configuration for each of the following: (a) Al, (b) Cl, and (c) K.

27. Draw the ground state energy level diagrams for each of the following elements: (a) Al, (b) Cl, and (c) K.

28. Draw schematic diagrams of electrons orbiting the nucleus in subshells for each of the following elements: (a) Al, (b) Cl, and (c) K.

CHAPTER 10

Nuclear Physics

> Children of yesterday, heirs of tomorrow
> What are you weaving? Labor and Sorrow?
> Look to your looms again. Faster and faster
> Fly the great shuttles prepared by the Master.
>
> —Mary A. Lathbury

INSTEIN'S EQUATION ($E = mc^2$), which relates mass to energy, has been known since 1905. But its importance to the human race was not realized until 1939. In December of 1938 the German scientists Otto Hahn and Fritz Strassman showed that, under appropriate conditions, uranium atoms would split into two approximately equal parts with the accompanying release of large amounts of energy.

Hahn conveyed the startling discovery to Lise Meitner, his co-worker of twenty-five years. Miss Meitner, forced to flee Germany in 1938, and her physicist nephew, Otto Frisch, then worked out the theory of the splitting of the atom during Christmas vacation in Sweden. Frisch communicated their results to the great Danish physicist Neils Bohr (whose atomic theory was covered in Chapter 9). Bohr was just setting out for a scientific meeting in Washington, D.C., and so he conveyed to America the startling news Frisch had told him.

In 1939 the Italian physicist Enrico Fermi (Fig. 10.1) first proposed that through the splitting of the atom large amounts of mass energy could be released for useful work. Soon after, because of the world political climate, the first censorship of scientific information took place. In 1945 the awesome results of atom splitting became known when the first atomic bomb was exploded above a New Mexico desert. The world has never been the same since.

10.1 The Atomic Nucleus

All matter encountered in day-to-day living is made up of atoms. An atom is a basic building block of matter. It is composed of negatively charged particles, called electrons, which surround a positively charged nucleus. The **nucleus** is the central core of the atom. It consists of **protons**, which are positively charged particles, and

◀ **The containment building and cooling towers of a nuclear reactor.**

neutrons, which have no electrical charge. The protons and neutrons residing in a nucleus are collectively called **nucleons**. The basic properties of protons, neutrons, and electrons are given in Table 10.1.

The nucleus forms a very small core of the atom, having a diameter of about 10^{-14} m. In contrast, the

Figure 10.1 Enrico Fermi (1901–1954).
Fermi was one of the few scientists capable of both great theoretical and experimental work (shown here with Niels Bohr on the right in a 1931 photo taken on the Appian Way in Italy). After winning the 1938 Nobel Prize in physics, he persuaded the Italian dictator Mussolini to let him take his family with him to Stockholm for the presentation ceremony. He never returned to Italy, but sailed instead to America. Thus it was that an "enemy alien" was in charge of the research that brought about the first controlled nuclear chain reaction in 1942, and later the atomic bomb.

Table 10.1 Major Constituents of an Atom

Particle	Mass	Charge	Comments
Electron	0.91×10^{-30} kg	-1.6×10^{-19} coulomb	Goes in orbit around nucleus
Proton	1673×10^{-30} kg	$+1.6 \times 10^{-19}$ coulomb	1840 times more massive than electron
Neutron	1675×10^{-30} kg	None	Slightly more massive than proton

electrons may be thought of as traveling in orbits that are about 10^{-10} m from the nucleus (see Section 9.2). These small dimensions are difficult for us to comprehend, but note that 10^{-10} m is 10,000 times larger than 10^{-14} m. Thus the diameter of an atom is approximately 10,000 times the diameter of a nucleus.

A schematic diagram of a helium atom is shown in Fig. 10.2. It is not drawn to scale because if the nucleus were drawn an inch in diameter, the electrons would have to be drawn about one-fifth of a mile away. Nuclei are extremely small, even compared with atoms.

Between the electrons and the nucleus there is nothing. Most of the atom's volume consists of empty space. It is a sobering thought to realize that most of matter consists of a void. If nuclei could be packed together, the resulting matter would be about 10^{12} times as dense as lead. This density can occur in dense stars but not in atoms on Earth.

In terms of size, electrons determine the size of atoms. In terms of mass, it is the nucleons that are important, as a study of Table 10.1 shows. The nucleons contribute over 99.9% of the mass of an atom.

In previous chapters we have studied two forces of nature—the gravitational force and the electromagnetic force. The electromagnetic force is the only important force on the electrons in an atom. The electromagnetic force between a proton and an electron in an atom ($r = 10^{-10}$ m) is 10^{39} times greater than the corresponding gravitational force. It is the electromagnetic force that is responsible for the structure of atoms, molecules, and, hence, matter in general. That paper tears, rocks are hard, grass is green, and pencils write results from the action of the electromagnetic force. In fact, about 95% of life's experiences are governed by this one force.

In order for an atom to be neutral (have a total charge of zero), the number of electrons and protons must be the same. Atoms are distinguished by the number of charged particles (protons) in the nucleus. An **element** is a substance that has the same number of protons in all of its atoms. As an example, each atom of the element oxygen has eight protons in its nucleus. Each neutral oxygen atom has eight electrons in motion about its nucleus, but it is the number of protons in the nucleus that determines the "species" of the atom.

An electron may be removed or added to give an *ion*. For example, if an electron is removed from an oxygen atom, we have an *oxygen* ion (O^+) with 8 nuclear protons and 7 electrons. If the number of electrons were used to determine the species of this atom with a missing electron, we would say it was nitrogen (7 protons and 7 electrons), which is not the case. The species of an atom cannot be changed with the removal or addition of electrons. However, it can be changed by altering the number of protons in nuclear processes, as we will learn shortly.

In a nucleus the protons are packed closely together. According to Coulomb's law, like charges will repel each other, and the protons will not remain together. In fact, according to Coulomb's law, the closer the protons, the greater the repulsive force is. Because the protons are closer than 10^{-14} m to each other, the repulsive electric forces are huge.

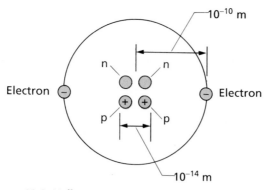

Figure 10.2 Helium atom.
A schematic diagram of a helium atom (not to scale). The nucleus is made up of two protons and two neutrons (designated p and n, respectively). The two electrons orbit about the nucleus.

For the protons to remain together, we need another force—a strong nuclear force. This strong nuclear force is not as easy to understand as the electric force. The **strong nuclear force** is very strongly attractive at short range and is reduced to zero at longer range. This nuclear force is strong enough to hold many nuclei together. However, not all combinations of nuclei will remain together. The question of which combinations stay together and which ones fly apart is discussed in Section 10.2.

The exact formula for the strong nuclear force is not known. It is too complicated to have a simple formula. However, two things about the strong nuclear force are well known. For distances between nuclear particles much greater than the diameter of a nucleus ($r \approx 10^{-14}$ m), the force is zero. For the very short nuclear distances, it is very strongly attractive. So we can write its *approximate* formula as

$$F_{nucl} = \text{very strongly attractive,}$$
$$\text{if } r \text{ is less than about } 10^{-14} \text{ m} \qquad (10.1)$$

$$F_{nucl} = 0 \text{ if } r \text{ is greater than about } 10^{-14} \text{ m}$$

where r = distance between two particles inside the nucleus.

A **stable nucleus** is one that will not separate of its own accord into two or more groups of particles. An **unstable nucleus** will spontaneously fly apart because the repulsive forces inside the nucleus are greater than the attractive forces. Nuclear stability will be discussed in Section 10.2.

Nuclei with many protons experience both strong electric repulsive forces from all the other protons and strongly attractive nuclear forces. A typical large nucleus is shown in Fig. 10.3. A proton on the edge of the nucleus is attracted only by the six or seven nearest nucleons. Because the strong nuclear force is a short-range force, only the nearby nucleons contribute to the attractive force. There is no attraction from all the other nucleons.

The repulsive electrical force between the protons in the nucleus is long range ($F = kp^2/r^2$) and acts between any two protons, no matter how far they are apart. As more and more protons are added to the nucleus, the electric repulsive forces increase while the nuclear attractive forces remain the same. When there are more than 83 protons in the nucleus, the electric forces of repulsion overcome the nuclear attraction, and the nucleus splits apart (usually into two unequal parts) of its own accord.

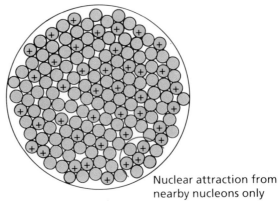

Nuclear attraction from nearby nucleons only

Electric repulsion from all protons

Figure 10.3 Multi-nucleon nucleus.
The protons on the edge of the nucleus, as illustrated here, are attracted by only six or so nearby nucleons, but they are repelled by all the other protons. When the number of protons exceeds 83, the electrical repulsion of the other protons is stronger than the nucleon attraction and the nucleus is unstable.

The **atomic number** (or proton number) of an atom is the number of protons in the nucleus of the atom.* The **neutron number** is, as the term indicates, the number of neutrons in the nucleus. The **mass number** is the number of protons plus neutrons in the nucleus. The properties of the nucleus itself depend not only on the number of protons it has but also on the number of neutrons in it.

Nuclear **isotopes** are nuclei that have the same number of protons but different numbers of neutrons. As such, they are nuclei of the same species of atom. As an example, the nuclei of isotopes of carbon, all of which have six protons, may have six, seven, or eight neutrons.

Isotopes of a particular element are easily represented and distinguished using nuclear notation. To illustrate this notation, consider the general nucleus representation shown in Fig. 10.4a. The X represents the chemical symbol of the element—for example, C for

---

* The *atomic number* is an older chemical designation. It is the number on the periodic table that determines the type of atom. In nuclear physics *proton number* is sometimes preferred as being more descriptive. The atomic number really refers to the number of electrons in an atom. For a neutral atom, this number is equal to the number of protons.

Mass number
(p + n)

X

Proton number
(p)

Neutron number
(n)

(a) Nuclear notation

$^{12}_{6}C_{6}$ $^{13}_{6}C_{7}$ $^{14}_{6}C_{8}$

Carbon–12   Carbon–13   Carbon–14

(b) Isotopes of carbon

$^{1}_{1}H_{0}$ $^{2}_{1}H_{1}$ $^{3}_{1}H_{2}$

Hydrogen   Deuterium   Tritium

(c) Isotopes of hydrogen

Figure 10.4 Nuclear notation and isotopes.
(a) The notation used in designating a particular nuclear species. (b) Isotopes of carbon all have 6 protons, but a different number of neutrons, as shown here for three isotopes of the family. There are others. (c) The isotopes of hydrogen all have one proton. There are only three isotopes of hydrogen and they are given special names.

carbon.* The left superscript is the *mass number*, the total number of protons and neutrons in the nucleus. The left subscript is the *atomic* or *proton number*, the number of protons in the nucleus. (The atomic number really refers to the number of electrons in an atom. For a neutral atom, this number is equal to the number of protons.) The right subscript is the *neutron number*, the number of neutrons in the nucleus. Notice that the proton number and the neutron number add up to the mass number. It is common to omit the neutron number from the notation, because it is readily determined by subtracting the proton number from the mass number. The proton number and the chemical symbol in effect specify the same thing. (Why?)

---
* The chemical symbols for all the elements are given in the periodic table in Chapter 11.

Three isotopes of carbon are shown in Fig. 10.4b represented by this notation. Note that they all have six protons and hence are carbon nuclei. But because they have different numbers of neutrons in their nucleii, they are isotopes of carbon—sort of like members of the same element family, all Joneses or Smiths, but with distinguishing characteristics and names. Isotopes of the same element are named according to their mass numbers. For example, the carbon isotopes in Fig. 10.4 are called carbon-12 (C-12), carbon-13 (C-13), and carbon-14 (C-14).

The isotopes of hydrogen are listed in Fig. 10-4c. These isotopes are given specific names. The most common isotope of hydrogen ($^{1}_{1}H_{0}$)is called ordinary hydrogen, or simply hydrogen. The isotope $^{2}_{1}H_{1}$ is called a deuteron (deuterium in atomic form), and $^{3}_{1}H_{2}$ is called a triton (tritium).

10.2 **Nuclear Stability**

The nuclear interactions that give rise to nuclear stability are quite complicated. However, by looking at some of the general properties of stable nuclei, we can obtain some criteria of nuclear stability. One criterion has already been mentioned with regard to the nuclear force. There are no stable nuclei with proton numbers greater than 83. As pointed out in the previous section, the repulsive electric forces resulting from the protons in the nucleus apparently overcome the attractive nuclear forces between nucleons when the nuclei have more than 83 protons. Therefore, bismuth (Bi), with 83 protons, has the most massive stable nucleus, the nucleus of the isotope $^{209}_{83}Bi$.

Another observation can be made by considering the nucleon populations, or the numbers of protons and neutrons, in stable nuclei. For low mass numbers of about 40 or less, most stable nuclei have about the same number of protons and neutrons. Examples of such nuclei are $^{4}_{2}He_{2}$, $^{12}_{6}C_{6}$, $^{23}_{11}Na_{12}$, and $^{27}_{13}Al_{14}$. For stable nuclei with mass numbers greater than 40, the number of neutrons exceeds the number of protons, and increasingly so as the mass number increases. Examples of these heavier stable nuclei are $^{62}_{28}Ni_{34}$, $^{114}_{50}Sn_{64}$, and $^{208}_{82}Pb_{126}$. These trends in nucleon populations may be seen in a plot of neutron number versus proton number for stable nuclei, as shown in Fig. 10.5. Notice how the stable nuclei lie above the line on which the number of protons and neutrons are equal. Such nuclei presumably have extra neutrons, so that the attractive nuclear forces between

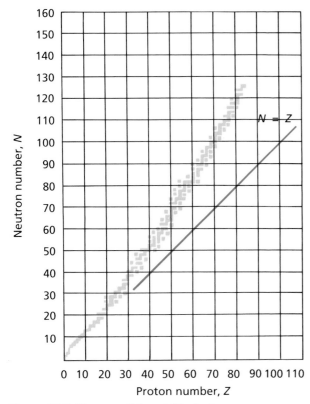

Figure 10.5 N versus Z for stable isotopes.
As may be seen from the graph, the stable heavier nuclei lie above the N-Z line, or have more neutrons than protons. This gives rise to a greater attractive nuclear force for stability.

Table 10.2 The Pairing Effect in Stabilizing Nuclei

Proton Number	Neutron Number	Number of Stable Nuclei
1. Even	Even	165
2. $\begin{cases} \text{Even} \\ \text{Odd} \end{cases}$	$\left.\begin{matrix} \text{Odd} \\ \text{Even} \end{matrix}\right\}$	105
3. Odd	Odd	4

themselves—that is, protons tend to pair up and neutrons tend to pair up to give stable even-even combinations. When there are odd numbers of nucleons, there is, for the most part, an instability associated with the pairing up of the last proton and neutron. The instability is less if there is only one odd nucleon (odd-even or even-odd combination). The stability associated with the even-even combinations of nucleons is called the **pairing effect**, and it gives a qualitative criterion for stability. For example, we would expect the isotope $^{23}_{11}\text{Na}_{12}$ to be stable but not $^{22}_{11}\text{Na}_{11}$, and this is actually the case.

In summary, the general criteria for nuclear stability are:

1. All nuclei with proton numbers greater than 83 are unstable.

2. (a) Most even-even nuclei are stable.
   (b) Many even-odd and odd-even nuclei are stable.
   (c) Only four odd-odd nuclei are stable.

3. (a) Stable nuclei with mass numbers less than 40 have approximately the same number of protons and neutrons.
   (b) Stable nuclei with mass numbers greater than 40 have more neutrons than protons.

EXAMPLE 1 _____

Is the isotope $^{38}_{16}\text{S}$ expected to be stable?

Solution To get a general idea whether a nucleus is stable, we apply each of the above criteria to the nucleus. If *any* of the conditions are not satisfied, then we can expect the nucleus to be unstable. If it satisfies all the general criteria for nuclear stability, a nucleus would be expected to be stable. We now apply the criteria to the nucleus $^{38}_{16}\text{S}$:

1. Satisfied. With a proton number less than 83, this criterion does not indicate any instability. Isotopes

nucleons will be greater than the repulsive electric forces between the numerous protons. However, notice in Fig. 10.5 that the stable nuclei stop at proton number 83.

Another fact about stable nuclei that is not evident from Fig. 10.5 is that many such nuclei have even numbers of both protons and neutrons. Also, very few stable nuclei have odd numbers of both protons and neutrons. A survey of the stable isotopes shows that 165 nuclei have an even-even combination. There are 105 nuclei that are odd-even or even-odd with respect to proton and neutron numbers. Interestingly, there are only four stable nuclear isotopes that have odd numbers of both protons and neutrons. These isotopes are the lowest odd-proton-number nuclei, $^{2}_{1}\text{H}_{1}$, $^{6}_{3}\text{Li}_{3}$, $^{10}_{5}\text{Bi}_{5}$, and $^{14}_{7}\text{N}_{7}$. See Table 10.2.

These even and odd combinations indicate that protons and neutrons in stable nuclei seem to pair up among with proton numbers greater than 83 are immediately ruled out of being stable.

2. Satisfied. In complete notation, $^{38}_{16}S_{22}$, this sulfur isotope is seen to be an even-even nucleus.

3. Not satisfied. The mass number is less than 40, but the proton number (16) and the neutron number (22) are not approximately equal.

Hence S-38 is expected to be unstable (and it actually is).

Albert Einstein introduced his special theory of relativity in 1905. One of the results of Einstein's assumptions was that mass is a form of energy. If it were possible to convert a very small amount of mass completely into energy of other forms, the total energy of other forms produced would be tremendous. Einstein's famous equation relating mass and energy is

$$E = mc^2 \qquad (10.2)$$

where E = energy,
         m = mass,
         c = speed of light in a vacuum.

For example, a mass of 1 kg has an equivalent energy of $E = mc^2 = (1 \text{ kg})(3 \times 10^8 \text{ m/s})^2 = 9 \times 10^{16}$ J. This is a fantastically large amount of energy, as the following example illustrates.

EXAMPLE 2

A typical nuclear power plant (see Fig. 10.6) produces 3600 megawatts (MW) of power in the form of heat energy per second, of which about 1000 MW becomes usable electric power. How much (uranium) mass is converted to heat energy each day?

Figure 10.6 Nuclear power plant.
Two domed containment buildings are shown, along with two water-cooling towers.

Solution To solve this problem, we use Eq. 10.2, realizing that 3600 MW = 36×10^8 W is 36×10^8 J/s. So every second, 36×10^8 J of energy is produced from mass conversion. The amount of mass converted each second is then

$$m = \frac{E}{c^2} = \frac{36 \times 10^8 \text{ J}}{9 \times 10^{16} \text{ J/kg}}$$
$$= 4 \times 10^{-8} \text{ kg}$$

So 4×10^{-8} kg of mass is converted to energy each second. Multiplying by the number of seconds in 24 h gives the daily mass conversion:

(4 × 10^{-8} kg/s)(24 h/day)(3600 s/h)

$$= 3.5 \times 10^{-3} \text{ kg/day}$$

Thus about 3.5 *grams* of mass are converted into energy each day in a typical nuclear power plant.

Note, however, that although mass is a form of energy, we cannot arbitrarily convert mass to energy. This conversion occurs primarily in nuclear processes, as we will learn later in this chapter.

The units of mass and energy commonly used in nuclear physics are quite different from those used in preceding chapters. Because protons and neutrons are so small, they are usually measured in very small units of mass called atomic mass units (amu). An **amu** is defined as $\frac{1}{12}$ the mass of the ^{12}C isotope. That is,

$$\text{mass of } ^{12}\text{C} = 12.0000 \text{ amu} \qquad (10.3)$$

One amu is equal to 1.66×10^{-27} kg. The masses of nuclear particles in terms of these units are

$$\text{mass of proton} = 1.00783 \text{ amu}$$

$$\text{mass of neutron} = 1.00867 \text{ amu}$$

Each nuclear isotope has a specific mass. In nature, a given element usually occurs as several different isotopes. Hence the **atomic weight** of an element that appears on the periodic table is the average mass (weight) of a naturally occurring element. For example, the most common isotope of carbon is C-12, which makes up about 98.89 percent of the carbon found in nature. There are also C-13 and C-14 isotopes that contribute to the atomic weight. Thus when you see the atomic weight for carbon on the periodic table, you will find that it is given as slightly greater than 12.000, actually 12.011.

The unit of energy commonly used in nuclear physics is the MeV, which is 1 million electron volts. Recall

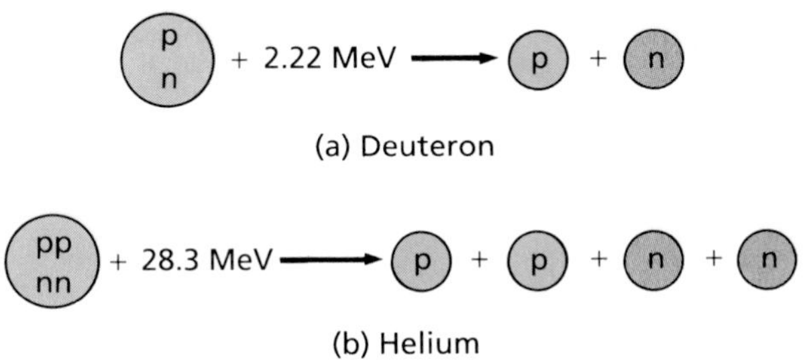

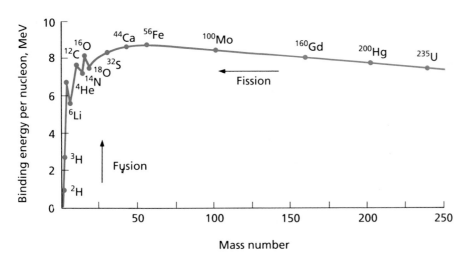

Figure 10.7 Binding energy.
(a) An illustration of the binding energy of a deuteron. It takes 2.22 MeV to separate the deuteron into a free proton and neutron. (b) It takes 28.3 MeV to separate the nucleons of a helium nucleus, which indicates that on the average, its nucleons are more tightly bound than those of the deuteron.

that an *electron volt* is the amount of kinetic energy an electron (or proton) acquires when it is accelerated through an electric potential of one volt. One electron volt (eV) is equal to 1.6×10^{-19} joule. If 1 amu of mass could be converted into energy, 931 MeV would be obtained. So we have the relation

$$(1 \text{ amu})c^2 = 931 \text{ MeV} \qquad (10.4)$$

The concept of *binding energy* plays an important role in the discussion of nuclear stability. The **total binding energy** of a nucleus is the energy necessary to separate a nucleus into protons and neutrons that are free to move independently of one another. This number

gives an indication of how tightly the nucleons in a nucleus are bound together, and hence it is a relative measure of nuclear stability. This concept is illustrated in Fig. 10.7. It takes 2.22 MeV to separate the proton and neutron of a deuteron, while 28.3 MeV is required to separate the nucleons of a helium nucleus.

To get a more meaningful idea of the cohesion of the nucleons, we consider the **binding energy per nucleon.** This number is simply the total binding energy divided by the number of nucleons. For example, for the deuteron the binding energy per nucleon is 2.22 MeV/2 = 1.11 MeV. Similarly, the binding energy per nucleon for the helium nucleus with four nucleons is 28.3 MeV/4 = 7.08 MeV. Hence the nucleons of a helium nucleus are more tightly bound, on the average, than those of a deuteron.

It is instructive to determine the binding energy per nucleon for various stable nuclei and plot these values versus mass number, as shown in Fig. 10.8. Notice that the curve has a maximum in the vicinity of iron (Fe). With a large binding energy per nucleon, iron is one of the most stable nuclei. The curve also gives us an idea of ways we might convert mass into energy. Suppose that, in effect, we could turn the arrows around in Fig. 10.7. For example, if we could take a proton and a neutron and put them together to form a deuteron, energy would be released. Similarly, if we could take two deuterons and put them together to form a helium nucleus, energy would be released. This energy comes from the conversion of mass into energy. If the masses of two protons and two neutrons are added, we would

Figure 10.8 E_b **versus mass number.**
The curve indicates that the most stable nuclei occur in the vicinity of iron (Fe). Notice that if two light nuclei could be "fused" together, a more stable nucleus would result. Also, if a heavy nucleus could be "split" apart or fissioned, the resulting lighter nuclei would be higher on the binding energy curve and be more stable.

find that the helium nucleus with two protons and two neutrons has less mass.

Notice in Fig. 10.8 that helium is higher on the curve than deuterium (^{2}H). Putting together two deuterons would then form a more stable nucleus with the release of energy. Processes of this sort, in which light nuclei come together to form more stable nuclei with the release of energy, are called *fusion reactions.* Notice in Fig. 10.8 that if a large nucleus such as U-235 could be split apart into two lighter nuclei, the lighter nuclei would be higher on the curve and energy would be released. These processes are called *fission reactions.* Both fission and fusion reactions are important in the production of nuclear energy, as we will see in Section 10.5.

10.3 Radioactive Decay

So far we have limited our discussion to stable nuclei. What about nuclei that are unstable—that is, those that tend to fly apart of their own accord? Nuclei that disintegrate of their own accord are called **radioactive nuclei.**

Radioactive decay was discovered in 1896 by Henri Becquerel, who found that uranium compounds spontaneously emit some very penetrating radiation. In 1898 the husband-and-wife team of Pierre and Marie Curie announced the discovery of two new radioactive elements, radium and polonium. (See the chapter Highlight.)

We now know that there are four common ways by which a radioactive nucleus can disintegrate: alpha decay, beta decay, gamma decay, and fission. In all of these radioactive decay processes energy is given off. This energy is usually in the form of kinetic energy of the products of the decay process.

Alpha decay is the disintegration of a nucleus into an alpha particle ($^4_2$He$_2$ nucleus) and the nucleus of another element. An example is

$$^{232}_{90}\text{Th}_{142} \longrightarrow {}^{228}_{88}\text{Ra}_{140} + {}^4_2\text{He}_2$$

In radioactive decay the original nucleus is sometimes

HIGHLIGHT

Madame Curie

Marie Sklodowska (1867–1934) was born in Warsaw, Poland. As a student, she became involved in a revolutionary organization, and because of these activities, she left the country and went to Paris. There she earned a degree in science. In 1895 she married Pierre Curie, a physicist well known for his work on crystals and magnetism. (See Figure 10.9.)

Marie started research on radioactivity, and Pierre joined her in this work. In 1898 they announced the discovery of two new radioactive elements that they had isolated, radium and polonium (the latter named after Marie Curie's native country). They had painstakingly isolated 10 mg of radium and a smaller amount of polonium from 8 tons of uranium pitchblende ore.

The Curies were awarded the Nobel Prize in physics in 1903 for their work in radioactivity. The prize was shared with Henri Becquerel, who had discovered the radioactive properties

Figure 10.9 Marie and Pierre Curie.

of uranium in 1896. Pierre Curie was killed in a horse-drawn carriage accident in 1906, and Marie Curie (commonly known as Madame Curie) was appointed to his post at the Sorbonne.

Madame Curie was awarded the Nobel Prize in chemistry in 1911 for her work on radium. She was the first person to win two Nobel Prizes.

The remainder of her career was spent in establishing laboratories for research on radioactivity and the use of radium in the treatment of cancer. Madame Curie died in 1934 from leukemia (cancer of the blood), which may have been caused by overexposure to radioactivity. She died one year before the Curies' daughter, Irene Joliot-Curie, and her husband Frederic Joliot were awarded the Nobel Prize in chemistry.

called the *parent nucleus* and the resulting nucleus the *daughter nucleus.*

Note that the sums of the proton numbers and of the mass numbers on both sides of the arrow in the equation are equal, that is, $90 = 88 + 2$ for the proton numbers and $232 = 228 + 4$ for the mass numbers. This result is a general rule for all nuclear reactions and involves the *conservation of charge* and the *conservation of mass (energy)*, respectively. In general, the neutron numbers do not have to be equal on both sides of the arrow, although they frequently are.

Alpha decay is a very common type of decay for elements with a proton number greater than 82. The heavy nucleus that results from the decay (^{228}Ra in our example) decays again. A chain of decays results until a stable nucleus such as $^{208}_{82}\text{Pb}_{126}$ is formed.

EXAMPLE 3

The nucleus $^{238}_{92}\text{U}$ decays by alpha decay. What are the resulting products of this decay?

Solution Because alpha decay implies the emission of a nucleus of ^4_2He, the other product nucleus must have a proton number of $92 - 2 = 90$ and a mass number of $238 - 4 = 234$. Thus, the decay is

$$^{238}_{92}\text{U} \longrightarrow {}^{234}_{90}\text{Th} + {}^4_2\text{He}$$

Beta decay is the disintegration of a nucleus into a beta particle (electron) and the nucleus of another element. An example is

$$^{14}_{6}\text{C}_8 \longrightarrow {}^{14}_{7}\text{N}_7 + {}^{\ 0}_{-1}\text{e}$$

In this decay we have represented an electron as $^{\ 0}_{-1}\text{e}$. Because its electric charge is opposite that of a proton, it has a charge number of -1. Also, because its mass is negligible compared with that of a proton (see Table 10.1), its mass number is zero.

The proton numbers and mass numbers on both sides of the arrow are seen to be equal in our example of beta decay. However, the neutron number in this beta decay example has gone from 8 to 7. What apparently happens in beta decay is that a neutron is changed into a proton and an electron. The proton remains in the nucleus and the electron is emitted.

Gamma decay is slightly different from either alpha or beta decay. In **gamma decay** the proton number, mass number, and neutron number remain the same, but

the nucleus decreases its energy by emitting a "particle" or quantum of electromagnetic energy called a **gamma ray.** The gamma ray has energy but no mass or charge. A typical example of gamma decay is

$$^{204}_{82}\text{Pb}^*_{122} \longrightarrow {}^{204}_{82}\text{Pb}_{122} + \gamma$$

In this type of decay the symbol γ (Greek letter gamma) represents the particle of electromagnetic radiation that is emitted by the nucleus. The asterisk implies that the $^{204}\text{Pb}^*$ nucleus is in an excited state, analogous to an atom being in an excited state with an electron in a higher energy level (Chapter 9). When the nucleus de-excites, a gamma particle is emitted and the nucleus is in the "ground state," so to speak. Gamma rays are different from alpha and beta particles in that gamma rays have no charge or mass associated with them.

In all radioactive processes energy is released. This energy can be destructive if it impinges on living tissue. In alpha and beta decay the energy is in the form of kinetic energy of either alpha or beta particles. In gamma radiation it is in the form of high-energy electromagnetic radiation. In some nuclear reactions the most common form is the kinetic energy of emitted neutrons.

Alpha and beta particles ($^4_2\text{He}_2$, $^{\ 0}_{-1}\text{e}$) are charged particles and can be easily stopped by materials such as paper and wood. The alpha and beta particles interact with charged particles within the materials, following Coulomb's law of electrical attraction and repulsion. Gamma rays and neutrons are not charged particles and are difficult to stop. Thick shielding is needed to protect living things from these kinds of radiation. This shielding is usually in the form of either lead or concrete for gamma rays and either water or concrete for neutrons.

Some unstable nuclei take a long time to decay; others decay very rapidly. The quantity that describes the rate of decay is called the half-life. The **half-life** of a given radioactive substance is the time it takes for half of the radioactive nuclei to decay. The half-lives of typical radioactive isotopes range from millionths of a second to billions of years. The concept of half-life is often misunderstood. An example best illustrates this concept. Consider the reaction

$$^{90}_{38}\text{Sr} \longrightarrow {}^{90}_{39}\text{Y} + {}^{\ 0}_{-1}\text{e}$$

This reaction is a typical beta decay, and the half-life is 28 years. Thus if in 1990 you had 200 μg of ^{90}Sr, then 28 years later (2018) you would have half as much, that is, 100 μg of ^{90}Sr. Then 28 years later, in 2046, you would have half again as much, or 50 μg of ^{90}Sr, and so on. The following table lists the process:

Year	Amount of ^{90}Sr
1990	200 μg
2018	100 μg
2046	50 μg
2074	25 μg
2102	12.5 μg
⋮	⋮

If we plot the amount of the sample of ^{90}Sr remaining versus the time, we get the graph shown in Fig. 10.10.

Some isotopes have half-lives of billions of years or more. Consequently, they can still be found occurring naturally. For example, the half-life of ^{238}U is 4.5×10^9 years. Since the age of the Earth is believed to be approximately 4.5×10^9 years also, we see that about half of the uranium present at the time the Earth formed should still be here today. That is, if a rock contained 2 kg of uranium when the Earth formed 4.5×10^9 years ago, 1 kg would still be present in the rock. The other kilogram of ^{238}U would have decayed into other elements. The uranium still present could be mined and processed.

When the naturally occurring isotope ^{238}U decays, it emits an alpha particle, and a ^{234}Th nucleus remains. The ^{234}Th then decays, and the series of decays continues until stable ^{206}Pb is formed. The ^{238}U radioactive series

produces many additional unstable isotopes. Because of the long half-life of ^{238}U, these radioactive isotopes occur naturally.

Radioactive isotopes can also occur naturally from the atmospheric bombardment of high-energy particles from outer space. These high-energy particles hit stable nuclei, and the collisions produce radioactive isotopes. For example, ^{14}C is formed in the atmosphere in this manner.

Some of the naturally occurring radioactive isotopes are given in Table 10.3. Many of them, such as ^{14}C, exist in our bodies at all times. Little is known about the effects of this low-radiation dosage to which humans have been exposed throughout history.

Radioactive isotopes have many uses. One important use is found in the dating of very old artifacts. All organic matter (all matter that was once alive, such as wood, coal, bones, and parchment) contains carbon. Natural carbon occurs in three isotopes, ^{12}C, ^{13}C, and ^{14}C. The first two are stable, while ^{14}C decays by beta emission into ^{14}N with a half-life of 5730 years.

In the upper atmosphere ^{14}C is formed as a result of bombardment by cosmic rays from outer space. The ^{14}C decays at about the same rate at which it was created, so the percentage of carbon that is ^{14}C remains relatively constant. Organic materials acquire the ^{14}C in the form of carbon dioxide, and in living things there is enough

Table 10.3 Half-Life of Some Naturally Occurring Radioactive Nuclei

Isotope	Half-Life
^{8}Be	2×10^{-16} second
^{214}Po	1.6×10^{-4} second
^{26}Al	7 seconds
^{34}Cl	33 minutes
^{231}Th	26 hours
^{222}Rn	4 days
^{22}Na	2.6 years
^{3}H	12.5 years
^{14}C	5.73×10^3 years
^{235}U	0.7×10^8 years
^{238}U	4.5×10^9 years
^{40}K	1.4×10^9 years
^{87}Rb	6×10^{10} years
^{115}In	6×10^{14} years
^{124}Sn	1.7×10^{17} years

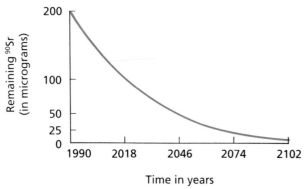

Figure 10.10 Radioactive decay.
A graph of the radioactivity of strontium-90 versus time. We assume an initial sample of 200 μg of Strontium-90 beta, which decays with a half-life of 28 years.

^{14}C to give about 16 beta emissions per minute per gram of carbon.

Once the organic matter dies, it ceases to take in the ^{14}C from the atmosphere, and the amount of ^{14}C that it contains slowly decays. By measuring the beta decay rate in old materials, scientists can determine the age of the material, because the ^{14}C half-life is known. This process is called *radioactive, or C-14, dating.*

EXAMPLE 4

What is the age of an old bone that emits 4 beta particles per minute per gram of carbon?

Solution   Living material has a rate of 16 beta emissions per minute per gram. After one half-life, or 5730 years, this rate would be decreased to 8, and after two half-lives, or 11,460 years, this rate would be cut to 4 beta particles per minute per gram of carbon. Thus the bone is about 11,460 years old. (See Table 10.4.)

Using similar methods, archeologists are able to date things as old as 40,000 years (see Fig. 10.11). Using isotopes with much longer half-lives enables geologists to measure the ages of meteorites from outer space or very old rock. In this manner the approximate age of the Earth has been determined to be approximately 4.5 billion years.

The activity of a radioactive source is commonly measured in a unit called the curie. One **curie** (Ci) is arbitrarily defined as 3.7×10^{10} disintegrations per second.* This number is approximately the number of disintegrations given off by one gram of ^{226}Ra. The curie, as defined, is a fairly large unit. Therefore, the millicurie and the microcurie are commonly used in the laboratory.

Large radioactive doses can be harmful to living

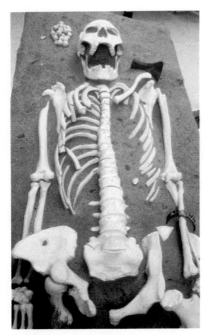

Figure 10.11 Radioactive dating.
Carbon-14 dating is often used to date human skeletons found at archeological sites.

cells. They can cause mutation in genes, and they are particularly harmful to cells of the fetus. For this reason the federal government maintains rigid safety standards for radioactive processes.

Radioactive isotopes have many uses in medicine, agriculture, and biology. In medicine they are often used as tracers. The movement of isotopes can easily be traced in plant and animal systems, using radiation-detecting instruments such as Geiger counters. Tracers therefore can be used to monitor the flow of blood in biological systems, to study the metabolism of plants and animals, and to determine the structure of complex molecules. An example of tracers used in cancer research is shown in Fig. 10.12.

As another example, the radioactive isotope of iodine, ^{131}I, is used in a measurement connected with the thyroid gland. The patient is administered a prescribed amount of the isotope, and like regular iodine in the diet, it is absorbed by the thyroid gland. The radioactive isotope allows doctors to monitor the iodine intake of the thyroid because the radioactive ^{131}I can be traced as it is released into the bloodstream in the form of protein-bound iodine.

Radioisotopes can also be used to treat diseased

Table 10.4 Decay of ^{14}C

Time Since Death	Beta emissions/min/g of C
0 years	16
5,730 years	8
11,460 years	4
17,190 years	2
22,920 years	1

* In the SI system one disintegration per second is called a becquerel (Bq).

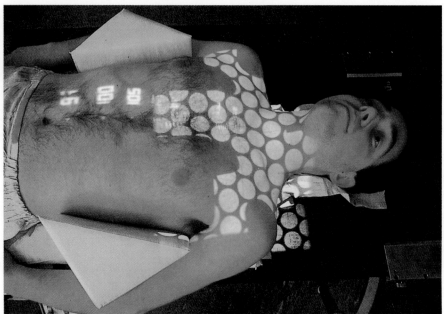

Figure 10.12 Radiation therapy. Radiotherapy on a linear accelerator is here being used to treat Hodgkin's disease, a cancer of the lymphatic system. The illuminated discs over the patient's chest indicate the areas which are to receive radiation. The pattern is defined by an arrangement of lead blocks suspended below the head of the accelerator, which shields the lungs from excessive radiation.

cells. Like healthy cells, diseased cells can be damaged or destroyed by nuclear radiation. By focusing an intense beam of radiation on diseased tissue, such as a cancerous tumor, its cells can be destroyed and its growth impaired or stopped.

10.4 Nuclear Reactions

Radioactive nuclei, through alpha and beta decay, spontaneously change, or transmute, into the nuclei of other elements, with the emission of charged particles. Scientists wondered, of course, whether the reverse process was possible; that is, could a particle be added to a nucleus to change it into that of another element? The answer was yes.

Lord Rutherford, a British scientist, produced the first induced nuclear reaction in 1919 by bombarding nitrogen (N-14) gas with alpha particles from a radioactive source. Particles coming from the gas were identified as protons. Rutherford reasoned that an alpha particle colliding with a nitrogen nucleus can occasionally knock out a proton. The result is an *artificial* transmutation of a nitrogen nucleus into an oxygen nucleus. In nuclear notation the reaction is

$$\underset{(4.00260)}{{}^{4}_{2}\text{He}} + \underset{(14.00307)}{{}^{14}_{7}\text{N}} \longrightarrow \underset{(16.99913)}{{}^{17}_{8}\text{O}} + \underset{(1.00783)\ \text{amu}}{{}^{1}_{1}\text{H}}$$

where the masses of the nuclei and particles are given in parentheses. (The reason for listing the masses will be discussed shortly.) Notice that the conservation of nucleons and the conservation of charge holds in this reaction, as in all nuclear reactions.

The reaction in Rutherford's experiment was discovered almost by accident, because it took place so infrequently. But think of the implications of its discovery: One element had been changed into another! This was the age-old dream of the alchemists, although their main concern was to change common metals, such as mercury and lead, into gold. Such reactions can now be done. Large machines called *particle accelerators* use electric fields to accelerate charged particles to very high energies (Fig. 10.13). The energetic particles are used to bombard nuclei and initiate nuclear reactions. Different reactions require different energies. One nuclear reaction that is initiated when protons (${}^{1}_{1}\text{H}$) strike mercury nuclei is

$$ {}^{1}_{1}\text{H} + {}^{200}_{80}\text{Hg} \longrightarrow {}^{197}_{79}\text{Au} + {}^{4}_{2}\text{He}$$

Isotope Au-197 is the only stable isotope of gold. However, making gold by this process is not very economical. It costs more to produce the gold than it is worth.

Figure 10.13 Particle accelerator.
The Fermi National Accelerator Laboratory at Batavia, Illinois. The large circle is the main accelerator with a 6.4-km (4-mi) circumference. A series of magnets bend and focus a proton beam that travels around the circle 50,000 times a second. With each round trip more energy is added by a radio-frequency system, producing highly energetic particles used in nuclear research.

Neutrons produced in nuclear reactions can in turn be used to induce nuclear reactions. Neutrons are especially effective as nuclear "bullets" in initiating reactions because they have no electric charge. They do not experience repulsive electrical interactions with nuclear protons, as would alpha-particle or proton projectiles. As a result, they are more likely to penetrate the nucleus and induce a reaction.

Nuclear reactions are said to be either endoergic or exoergic, depending on the mass and energy involved in a process. In an **endoergic reaction** energy is absorbed. That is, the kinetic energy of the reacting particles is at least partially converted into mass. A scientist can see this result by comparing the masses of the particles and nuclei on each side of a nuclear equation. For example, if you add the masses on each side of the equation for Rutherford's reaction given above, you will find that the combined masses of the products (right side of equation) are greater than that of the reactants (left side of equation). The difference is 0.00129 amu. Using the relationship $E = mc^2$, that is, 931 MeV is equivalent to 1 amu, we have

$$(0.00129 \text{ amu})(931 \text{ MeV/amu}) = 1.20 \text{ MeV}$$

Hence 1.20 MeV of kinetic energy was converted into mass.

In an **exoergic reaction** energy is released. That is, mass is converted into energy. Because we are interested in sources of energy, this reaction is very important. An example is

$$_1^2\text{H} + {}_1^3\text{H} \longrightarrow {}_2^4\text{He} + {}_0^1\text{n}$$
$$(2.01410) \quad (3.01605) \quad\quad (4.00260) \quad (1.00867) \text{ amu}$$

(The symbol ${}_0^1\text{n}$ represents a neutron.)

If we compare the combined masses of the reactants and products, we find that mass is lost in the reaction. This mass is converted into energy and carried away by the products as kinetic energy.

EXAMPLE 5

How much energy is released in the previous reaction?

Solution Totaling the masses, we have

${}_1^3\text{H} = 3.01605$	${}_2^4\text{He} = 4.00260$
${}_1^2\text{H} = \underline{2.01410}$	${}_0^1\text{n} = \underline{1.00867}$
5.03015 amu	5.01127 amu

and there is a mass difference of $\Delta m = 0.01888$ amu.

Converting this mass difference to its equivalent energy, we get

$$(0.01888 \text{ amu})(931 \text{ MeV/amu}) = 17.6 \text{ MeV}$$

So 17.6 MeV is released in this exoergic reaction.

As we will see in the next section, such exoergic nuclear reactions are important energy sources.

10.5 **Fission and Fusion**

Fission is the process in which a large nucleus is "split" into two intermediate nuclei of approximately equal size, with the emission of neutrons and the release of a relatively large amount of energy. In the fission process mass is converted into energy. As an example, consider the fission decay of $^{236}_{92}U_{144}$. If ^{235}U is bombarded with low-energy neutrons, $^{236}_{92}U$ is formed. The ^{236}U then fissions, or splits, into smaller nuclei. These two reactions may be represented as

$$^{235}_{92}U + ^1_0n \longrightarrow ^{236}U$$

$$^{236}_{92}U \xrightarrow{\text{Fission}} \text{two intermediate nuclei}$$
$$+ \text{ neutrons} + \text{energy}$$

A typical fission decay of ^{236}U is (see Figure. 10.14a)

$$^{236}_{92}U_{144} \longrightarrow ^{140}_{54}Xe_{86} + ^{94}_{38}Sr_{56} + 2(^1_0n)$$

This reaction is just one of various possible fission decay reactions of U-236. Another is

$$^{236}_{92}U \longrightarrow ^{132}_{50}Sn_{82} + ^{101}_{42}Mo_{59} + 3(^1_0n)$$

EXAMPLE 6

Complete the following fission reaction:

$$^{236}_{92}U \longrightarrow ^{88}_{36}Kr + ^{144}_{56}Ba + \underline{\qquad}$$

Solution The proton numbers are balanced, since $92 = 36 + 56$. The mass number on the left is 236, and the sum of the mass numbers on the right is $88 + 144 = 232$. Hence if the mass numbers (and the neutron numbers, which you may write out explicitly) are to balance, there must be 4 extra neutrons on the right side of the equation. The reaction is then

$$^{236}_{92}U \longrightarrow ^{88}_{36}K + ^{144}_{56}Ba + 4(^1_0n)$$

Note that the mass numbers are now balanced.

Here are three important features of nuclear fission:

1. The intermediate product nuclei are almost always radioactive. (This feature gives rise to nuclear-waste-disposal problems, which we will discuss later.)

2. Relatively large amounts of energy are produced in fission reactions, typically on the order of 200 MeV per reaction.

3. Of particular interest is that neutrons are released. Neutron bombardment of U-235 starts the fission process. Hence with the release of neutrons, there

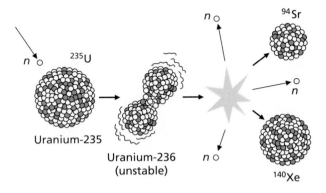

Uranium-235

Uranium-236 (unstable)

(a) Fission

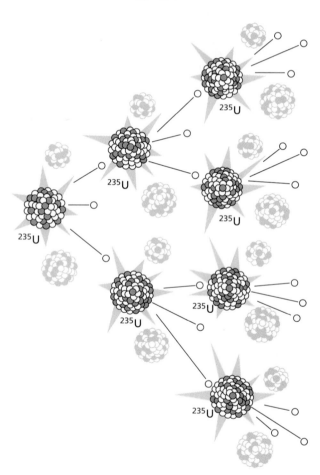

(b) Chain reaction

Figure 10.14 Fission chain reaction.
(a) In a fission reaction, such as that for uranium-235, a neutron is absorbed and the nucleus splits apart into two lighter nuclei with the emission of two or more neutrons. (b) If the emitted neutrons cause other fission reactions, the process grows in a "chain" of events.

is the possibility of further fissioning, or a sustained fission reaction in the form of a chain reaction.

In a **chain reaction,** one initial reaction triggers a series of subsequent reactions. In the case of fission, one neutron hitting a nucleus of ^{235}U forms ^{236}U, which fissions to give off two or more neutrons. These neutrons then hit other ^{235}U nuclei and fission occurs again and similarly again. A schematic diagram of a chain-reaction process is illustrated in Fig. 10.14b. Each time a nucleus fissions, energy is released; and as the chain grows, the energy output increases.

Of course, the process of energy production by fission is not as simple as just described. For a chain reaction to proceed or occur, there must be a sufficient amount of fissionable material (^{235}U) present. Otherwise, neutrons would escape from the sample before reaction with a nucleus, and the chain reaction would not grow. The chain would be "broken," so to speak. The minimum amount of fissionable material necessary to sustain a chain reaction is called the **critical mass.**

Natural uranium is composed of 99.3% U-238 and only 0.7% fissionable U-235 isotope. So that more fissionable U-235 nuclei are present in a sample, the U-235 is concentrated, or "enriched." Enriched uranium used in nuclear reactors for the production of electricity is on the order of 3% to 5% U-235. Weapons-grade material may be enriched to 99% or more. This percentage provides many fissionable nuclei for a large release of energy. In a bomb a critical mass of highly enriched material must be held together for a short time to get an explosive release of energy. (Otherwise, the energy release would cause the sample to fly apart, and there would not be a critical mass.) Segments of the fissionable material in an "atomic" bomb or "nuclear" device are separated before explosion so that there is not a critical mass for the chain reaction. A chemical explosive is used to bring the segments together into an interlocking configuration that holds them together long enough for a large fraction of the material to undergo fission. The result is an explosive release of energy.

A bomb is an example of *uncontrolled* fission, which is very undesirable. A nuclear reactor is an example of *controlled* fission, in which we control the growth of the chain reaction and the release of energy.

A **nuclear reactor** operates on the principle of a controlled chain reaction. The basic design of a fission nuclear reactor is shown in Fig. 10.15. Enriched uranium fuel pellets in tubes form fuel rods in the reactor core, in which the fission takes place. The chain reaction and the energy output is controlled by means of *control rods.* These rods are inserted into the core to slow the reaction and are retracted to allow it to grow for more energy output. Certain materials, such as boron or cadmium, that absorb neutrons are used as control rod materials. The control rods are adjusted so that only a certain amount of neutrons are absorbed, ensuring that the chain reaction proceeds and energy is released at a fairly steady rate. On the average, this control requires that one neutron from each fission event initiate only one additional fission process. If more energy is needed, the rods are withdrawn further. When fully inserted into the core, the control rods absorb enough neutrons to stop the chain reaction, and the reactor is shut down.

A reactor is basically an energy source, and the heat energy is removed by a coolant flowing through the core. In U.S. reactors the coolant is most commonly water. The water flowing through the fuel assemblies not only acts as a coolant—heat transfer agent, which is used to produce steam to generate electricity but also acts as a moderator. The U-235 nuclei react best with "slow" neutrons. The neutrons emitted from the fission reactions are relatively "fast" neutrons, with energies that are not best suited for U-235 fission. The fast neutrons are slowed down, or "moderated," by energy losses through collisions. The neutrons transfer energy to the water molecules in collision processes; and after only a few collisions the neutrons are slowed down to the point that they react easily with the U-235 nuclei.

In addition to U-235, the other fissionable isotope of importance is $^{239}_{94}$Pu. This plutonium isotope is produced by the conversion of ^{238}U into ^{239}Pu. Fast neutrons induce this reaction. As a result, Pu-239 is made in nuclear reactors, because all of the neutrons are not moderated or slowed down. As a fissionable material, the Pu-239 extends the refueling time of a reactor. In a **breeder reactor** this process is promoted, and more fissionable material (Pu-239) is produced than consumed (U-235). The isotope Pu-239 can be chemically separated from the fission by-products and used as the fuel in an ordinary nuclear reactor or in weapons. Its use in weapons is a major concern, because Pu-239 can be obtained from regular reactors.

Another important class of exoergic nuclear reactions is the fusion reaction. **Fusion** is the process in which light nuclei are "fused" together to form heavier nuclei. An example of a fusion reaction was given in the previous section, and another is

$$^2_1H + {}^2_1H \longrightarrow {}^3_1H + {}^1_1H$$

In this reaction two deuterons fuse to form a triton, with

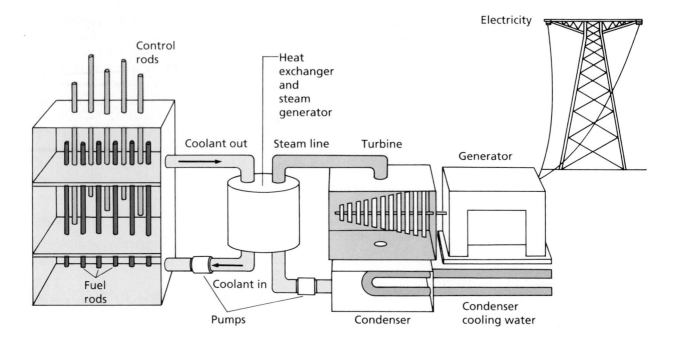

Figure 10.15 Nuclear reactor.
Above: The elements of a nuclear reactor and a schematic diagram showing how a reactor is used in the generation of electricity. *Below:* The core of a research reactor at Sandia National Laboratory seen looking down through shielding water. The characteristic blue glow is called Cerenkov radiation, which is light produced by rapidly moving charged particles.

the release of energy. This example is termed a D-D (deuteron-deuteron) reaction. The fusion reaction discussed in the previous section is a D-T reaction.

In the case of fusion there is no critical mass or size, because there is no chain reaction to maintain. However, there are problems in practical fusion-energy production. One major problem is how to produce a self-sustaining reaction. The repulsive force between positively charged nuclei opposes fusing. With only a single proton, this force is minimal for hydrogen. Even so, initiating hydrogen fusion reactions is a problem. If the temperature of hydrogen gas is raised, the kinetic energy of the molecules increases. Hence attaining a sufficiently high temperature is all that is necessary for fusion to occur. However, this fusion process requires temperatures on the order of millions of degrees.

At such high temperatures the hydrogen atoms are stripped of their electrons, and a gas of charged particles results. This gas is called a **plasma.** Plasmas have such special properties that scientists refer to them as the "fourth phase of matter."

Large amounts of fusion energy have been released in an uncontrolled manner—in a hydrogen (H) bomb. In the H-bomb a nuclear fission bomb is used to trigger or supply the energy needed to initiate the fusion reaction (Fig. 10.16). Because of the high temperatures required to initiate fusion reactions, they are sometimes called *thermonuclear reactions.*

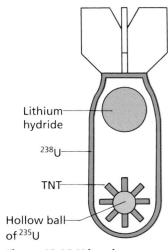

Lithium
hydride

^{238}U

TNT

Hollow ball
of ^{235}U

Figure 10.16 H-bomb.
Left: To detonate a hydrogen bomb, the TNT is exploded, forcing the ^{235}U together
to get a critical mass and an "atomic" explosion. The fusionable material is deuterium
in the lithium hydride (LiH), and in the heated plasma D-D reactions occur. Also, neutrons from the fission bomb react with the lithium to give tritium (^{3}H), and D-T fusion
reactions also take place. The bomb is surrounded with ^{238}U, which tops off the explosion with a fission reaction. *Right:* The result of the preceding.

Controlled fusion might be accomplished by adding
fuel in small amounts to a fusion reactor. However, fusion
reactors are not yet practical. Major problems arise in
the confinement of the ultrahot plasma *and* the confinement of sufficient energy to initiate the reaction. No
material can be used to hold the reaction. Tungsten, the
material with the highest melting point, melts at around
3370°C. Also, heat is readily transferred from the confinement region at high fusion temperatures.

The confinement problem is being approached by
two methods. One method is *magnetic confinement.* Because the plasma is a gas of charged particles, it can be
controlled and manipulated with electric and magnetic
fields. In magnetic confinement magnetic fields are used
to confine the plasma in a "magnetic bottle" or "ring"
(Fig. 10.17). Electric fields produce currents that heat and
raise the temperature of the plasma.

The other approach is *inertial confinement.* In this
technique laser, electron, or ion beams would be used
to initiate fusion. Hydrogen fuel pellets would be
dropped into a reaction chamber (Fig. 10.18). High-
energy pulses would cause the pellet to "implode", resulting in compression and high temperatures. If the
pellet stayed intact for a sufficient time, fusion might be
initiated.

Neither approach has yet been successful. Problems

Figure 10.17 Magnetic confinement.
A magnetic confinement vehicle or tokamak for fusion
research at the Princeton Plasma Physics Laboratory. The
tokamak at the center has a magnetic confinement
"ring" for the plasma. Beam injectors are seen around
the periphery.

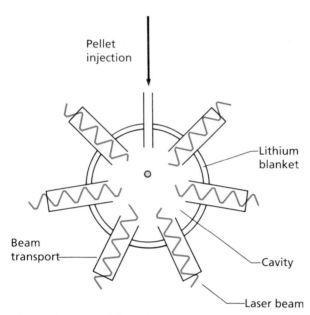

Figure 10.18 Inertial confinement.
An illustration of a reactor cavity for inertial confinement.
Fuel pellets dropped into the cavity are imploded by laser
light (or a particle beam), inducing a fusion miniexplosion.
Neutrons react with the lithium blanket to produce tri-
tium, which can be recovered and used as fuel.

arise with plasma temperature and confinement time for
magnetic confinement, and lasers and particle beams are
not yet powerful enough to produce fusion in inertial
confinement. Practical energy production from fusion is
not expected until well into the twenty-first century.

Even so, fusion is the promising energy source of
the future, because it does not have the problems associated with fission. The by-products of fusion are radioactive but have relatively short lifetimes compared
with those of fission, which gives rise to problems with
nuclear waste. Also, with a continuous-fission chain reaction, there is the possibility of a nuclear accident, as
occurred in 1979 at the Three Mile Island (TMI) nuclear
plant in Pennsylvania and in 1986 with the Russian
reactor in Chernobyl (Fig. 10.19). The term *meltdown* is
commonly used in speaking of such accidents. If heat
energy is not continuously removed from the core of a
fission reactor, the fuel rods may fuse. In that event the
reaction could not be controlled with the control rods,
and energy would no longer be removed by the coolant
flowing between the fuel rods. The fissioning mass would
become extremely hot, "melt down" through the floor
of the containment vessel, and enter the environment.

There was a partial meltdown at TMI with a slight
fusing of the fuel rods, but little radioactive material
escaped into the environment. What did escape was
primarily radioactive gases. Chernobyl, however, did
experience a meltdown, explosions, and fire. This particular type of reactor used carbon blocks for a moderator. Gas explosions caused the carbon to catch fire;
radioactive material escaped with the smoke, and radioactive fallout spread over Europe. Many deaths occurred
in the immediate region.

Figure 10.19 Chernobyl.
The site of a nuclear accident that resulted in meltdown
and the escape of large amounts of radioactivity into
the environment. A protective wall is being constructed
around the damaged Unit 4.

QUESTION

If control were lost in a nuclear reactor, would it explode like a bomb?

Answer No. A nuclear reactor cannot explode like an "atomic" bomb with the characteristic mushroom cloud.

The enriched uranium used in reactors is 3% to 5% U-235, whereas in nuclear weapons it is 90% or more. In a bomb the critical mass must be held together to achieve a nuclear explosion. In a reactor the material is not enriched enough to explode. The major concern for a reactor is a core meltdown, which may be accompanied by *chemical* explosions but not by nuclear explosions.

Learning Objectives

After reading and studying this chapter, you should be able to do the following without referring to the text:

1. Describe the general structure of an atom, including a comparison of nuclear and atomic mass and size.
2. Explain why a nucleus doesn't fly apart because of the repulsive electrical forces between the protons.
3. Explain why all elements with more than 83 protons are unstable.
4. Give three general criteria for nuclear stability.
5. Explain radioactivity and describe the three general modes of decay.
6. Explain how old bones and artifacts can be dated.
7. Draw a diagram of and describe the operation of a nuclear reactor in general terms.
8. Calculate the energy released or absorbed in a nuclear reaction.
9. Distinguish between fission and fusion.
10. Explain why there is no practical fusion reactor, and describe the approaches toward controlled fusion.
11. Describe the major possible dangers associated with nuclear reactors.
12. Define and explain the other important words and terms listed in the following section.

Important Words and Terms

nucleus	neutron number	alpha decay	chain reaction
protons	mass number	beta decay	critical mass
neutrons	isotopes	gamma decay	nuclear reactor
nucleons	pairing effect	gamma ray	breeder reactor
element	amu	half-life	fusion
strong nuclear force	atomic weight	curie	plasma
stable nucleus	total binding energy	endoergic reaction	
unstable nucleus	binding energy per nucleon	exoergic reaction	
atomic number	radioactive nuclei	fission	

Questions

The Atomic Nucleus

1. How much of the mass of an atom is contained in the nucleus?
2. How does the diameter of an atom compare with that of its nucleus?
3. What property of atoms specifies an element?
4. Why don't nuclei fly apart because of the repulsive electrical forces between protons?
5. Why are nuclei with more than 83 protons unstable?

Nuclear Stability

6. How do the numbers of protons and neutrons in nuclei compare?
7. What is the pairing effect?
8. What is an electron volt?
9. Does a stable nucleus have more mass than the free nucleons that compose it? Explain.
10. Discuss two ways mass can be converted into energy.

Radioactive Decay

11. Compare the physical properties of alpha, beta, and gamma radiation in terms of charge, mass, and penetrating ability.

12. If uranium is radioactive, why does it occur naturally on Earth?

13. What is meant by radiocarbon dating, and how is it done?

14. Give some useful applications of radioactive isotopes.

Nuclear Reactions

15. What two quantities are conserved in all nuclear reactions?

16. How are nuclear reactions induced?

17. If the products of nuclear reaction have more mass than the reactants, what type of reaction is it?

18. Is mass converted to energy in an endoergic reaction? Explain.

Fission and Fusion

19. Define the following terms: (a) chain reaction, (b) critical mass, (c) moderator, (d) control rods, and (e) meltdown.

20. How is energy produced in a nuclear reactor?

21. What is a breeder reactor?

22. Do we have fusion reactors? If not, why not?

23. What is a plasma, and how is it confined?

Exercises

The Atomic Nucleus

1. Write the names of each of the following isotopes. Then give the proton number, neutron number, and mass number for each.
   (a) ^6_3X (b) $^{10}_5X$ (c) $^{12}_6X$ (d) $^{34}_{16}X$
   (e) $^{90}_{38}X$ (f) $^{235}_{92}X$ (g) $^{239}_{93}X$ (h) $^{240}_{94}X$
   Answer: (e) strontium 38, 52, 90

2. Write the nuclear symbols for (a) the three isotopes of hydrogen, and (b) the isotopes of oxygen from O-15 through O-18.

Nuclear Stability

3. Only two isotopes of $_{19}K$ (potassium) are stable. Pick which two they are from the following list:
   $^{38}K, \ ^{39}K, \ ^{40}K, \ ^{41}K, \ ^{42}K$

4. Only two isotopes of antimony (Sb) are stable. Pick the two stable isotopes from the following list:
   $^{120}Sb, \ ^{121}Sb, \ ^{122}Sb, \ ^{123}Sb, \ ^{124}Sb$

5. Apply the three general criteria for nuclear stability and predict whether the following isotopes are unstable or stable:
   (a) ^{12}C (b) ^{14}N (c) ^{16}O
   (d) ^{18}F (e) ^{90}Y (f) ^{96}Sn
   (g) ^{22}N (h) ^{249}Cf (i) ^{226}Ra
   Answer: (e) odd-odd, so unstable
   (i) more than 83 protons, so unstable

6. State whether you would expect the following isotopes to be stable or unstable, and give your reasons.
   (a) ^{20}C (b) ^{100}Sn (c) ^6Li
   (d) ^9Be (e) ^{22}Na (f) ^{23}Na
   (g) ^{50}V (h) ^{209}Bi (i) ^{209}Po
   (j) ^{44}Ca (k) ^2H (l) ^{14}C

7. What is the equivalent mass energy of a carbon 12 atom?

8. How much mass is converted to energy in one day in a submarine's nuclear reactor that generates 1 MW of power?   *Answer:* 1 mg

Radioactive Decay

9. Complete the following nuclear decay schemes, and state whether the process is alpha decay, beta decay, gamma decay, or fission.
   (a) $^{47}_{21}Sc \longrightarrow \, ^{47}_{22}Ti \, + \, \underline{\quad}$
   (b) $^{210}_{84}Po^* \longrightarrow \, ^{210}_{84}Po \, + \, \underline{\quad}$
   (c) $^{210}_{84}Po \longrightarrow \, ^{206}_{82}Pb \, + \, \underline{\quad}$
   (d) $^{240}_{94}Pu \longrightarrow \, ^{97}_{38}Sr \, + \, ^{139}_{56}Ba \, + \, \underline{\quad}$
   (e) $^{229}_{90}Th \longrightarrow \, ^4_2He \, + \, \underline{\quad}$
   (f) $^{47}_{21}Sc^* \longrightarrow \, ^{47}_{21}Sc \, + \, \underline{\quad}$
   (g) $^{237}_{93}Np \longrightarrow \, ^0_{-1}e \, + \, \underline{\quad}$
   Answer: (c) ^4_2He, alpha decay

10. Bismuth 214 may decay by emitting an alpha or a beta particle. Write the symbol notation for the daughter nucleus in each case.

11. Thorium 229 ($^{229}_{90}Th$) undergoes alpha decay.
    (a) What is the daughter nucleus of this process?
    (b) This daughter nucleus undergoes beta decay. What is the "granddaughter" nucleus?

12. Write the nuclear equation expressing (a) the beta decay of $^{60}_{27}Co$ and (b) the alpha decay of $^{226}_{88}Ra$.

13. The half-life of ^{131}I is 8 days. If a thyroid cancer patient is given 16 μg of ^{131}I, how many micrograms will still be in his thyroid after 24 days?

14. A skeleton of a prehistoric woman is found to yield 2 beta rays per minute per gram of carbon. About how old is the skeleton?   *Answer:* 17,190 years

15. The activity of a given sample of radioactive material at 9 A.M. was 8 microcuries (μCi). At 12 noon the activity is measured as 2 μCi.
    (a) Determine the half-life of the material.
    (b) How many disintegrations were given from the sample at 12 noon?

Nuclear Reactions

16. Complete the following nuclear reaction equations by balancing proton numbers and mass numbers.
    (a) $^{27}_{13}\text{Al} + ^4_2\text{He} \longrightarrow ^{30}_{15}\text{P} + $ _____
    (b) $^{48}_{20}\text{Ca} + ^1_1\text{H} \longrightarrow ^2_1\text{H} + $ _____
    (c) $^{23}_{11}\text{Na} + \gamma \longrightarrow $ _____
    (d) $^{236}_{92}\text{U} \longrightarrow ^{131}_{53}\text{I} + 3^1_0\text{n} + $ _____
    (e) $^{20}_{10}\text{Ne} + ^{16}_8\text{O} \longrightarrow ^{12}_6\text{C} + $ _____
    (f) $^6_3\text{Li} + ^6_3\text{Li} \longrightarrow ^7_3\text{Li} + $ _____

    Answer: (d) $^{102}_{39}\text{Y}$

17. Complete the following nuclear reactions.
    (a) $^6_3\text{Li} + ^1_1\text{H} \longrightarrow ^3_2\text{He} + $ _____
    (b) $^{27}_{13}\text{Al} + ^1_0\text{n} \longrightarrow ^1_1\text{H} + $ _____

    (c) $^{58}_{28}\text{Ni} + ^2_1\text{H} \longrightarrow ^{59}_{28}\text{Ni} + $ _____
    (d) $^{235}_{92}\text{U} + ^1_0\text{n} \longrightarrow ^{138}_{54}\text{Xe} + 5(^1_0\text{n}) + $ _____

18. Given the following masses, calculate the amount of energy released in the fusion reactions below.

    $$^1_0\text{n} = 1.00867 \text{ amu} \qquad ^3_1\text{H} = 3.01605 \text{ amu}$$

    $$^1_1\text{H} = 1.00783 \text{ amu} \qquad ^3_2\text{He} = 3.01603 \text{ amu}$$

    $$^2_1\text{H} = 2.01410 \text{ amu} \qquad ^4_2\text{He} = 4.00260 \text{ amu}$$

    (a) $^2_1\text{H} + ^2_1\text{H} \longrightarrow ^3_1\text{H} + ^1_1\text{H} + $ energy
    (b) $^3_1\text{H} + ^2_1\text{H} \longrightarrow ^4_2\text{He} + ^1_0\text{n} + $ energy

    Answer: (a) 4.02 MeV

19. Are the following reactions endoergic or exoergic?
    (a) $\quad ^{16}\text{O} \quad + \quad ^1_0\text{n} \quad \longrightarrow \quad ^{13}_6\text{C} \quad + \quad ^4_2\text{He}$
    $\quad\quad$ (15.99491) $\quad$ (1.00867) $\quad\quad$ (13.00335) $\quad\quad$ (4.00260) amu
    (b) $\quad ^3_2\text{He} \quad + \quad ^1_0\text{n} \quad \longrightarrow \quad ^2_1\text{H} \quad + \quad ^2_1\text{H}$
    $\quad\quad$ (3.01603) $\quad$ (1.00867) $\quad\quad$ (2.01410) $\quad\quad$ (2.01410) amu

CHAPTER 11

The Periodic Table

> Just as without knowing the cause of gravitation it is possible to make use of the law of gravitation, so for the aims of chemistry it is possible to take advantage of the laws discovered by chemistry without being able to explain their causes.
>
> —Dmitri Mendeleev

THE OBJECTS IN our environment are composed of matter. **Matter** may be defined as anything that exists in time, occupies space, and has mass. The elements are the different kinds of atoms that make up the matter in our everyday environment. The physical, chemical, and combining properties of the various elements affect us constantly. To cite two examples, the element oxygen is necessary to sustain human life, while the diamond in an engagement ring is pure carbon. With only a few exceptions, the 88 naturally occurring elements in our environment (although chemists have varying opinions regarding the number, depending on how they define the phase "naturally occurring") either singularly or in combination, are the components of all matter. The science of chemistry deals with the composition, structure, properties of matter and the transformations that this matter undergoes.

The modern understanding of matter, based on the electron structure of atoms, is relatively recent, but chemistry had its beginnings early in recorded history. Egyptian hieroglyphs that date back to 3400 B.C. show wine presses. Wine making, of course, required a chemical fermentation process. By 2000 B.C. the Egyptians and Mesopotamians possessed the knowledge required to produce and work the metals gold, silver, lead, copper, iron, and bronze—an alloy of copper and tin. Dyes were discovered during this period, and methods of fixing them were known.

In the fourth century B.C. the Greek philosopher Aristotle developed the false idea that all matter was composed of four elements: earth, air, fire, and water. As the Greek culture faded, the pseudoscience of alchemy was born. Its main goals were to change base metals into gold and to find the "elixir of life," which could restore an aging human body to its youthful vigor. Alchemy was a mixture of magic and experimentation, and it was practiced in China as well as the West. Although it was practiced by many charlatans, its finest practitioners were dedicated experimentalists, who contributed to the real science of chemistry. The goals of the alchemists were never achieved, but by the time the scientific method was born in the sixteenth century, much had been discovered about the properties of matter.

In 1869 Dmitri Mendeleev in Russia and Julius Meyer in Germany, working independently, published a classification of the known elements showing a relationship between the magnitude of the atomic weights and the properties of elements. This relationship led to the conclusion that the properties of elements are periodic functions of their atomic weights. In 1914 Henry Moseley, an English physicist, showed that atomic numbers rather than atomic weights are fundamental in determining the properties of elements, thus laying the foundation for the modern periodic table. In this chapter we shall learn the properties of elements and how to use the periodic table to determine many of these properties.

11.1 Elements

In the very earliest of civilizations, nine elements were isolated: gold, silver, lead, copper, tin, iron, carbon, sulfur, and mercury. The alchemists added five more: antimony, arsenic, bismuth, phosphorus, and zinc. As the list grew, a precise definition of an element was needed. The concept of an element as we know it today was formulated by Robert Boyle (Fig. 11.1), an English chemist, in 1661. In his book *The Skeptical Chemist*, Boyle proposed that the name "element" be applied to those simple bodies that could not be separated into components by chemical methods. In 1789 the French scientist Antoine Lavoisier (1743–1794) (Fig. 11.2) expanded Boyle's definition and applied the term *element* to the matter that remained after complete chemical decomposition had occurred.

Lavoisier, who is sometimes called the father of chemistry, introduced quantitative methods into chemistry and discovered the role of oxygen in combustion. In 1789 his list of elements included 23 that we recognize today as elements. The modern definition of the word **element** is a substance in which all the atoms have the same atomic number. Our modern listing of the elements

---

◄ **The element gold, the most malleable and ductile metal.**

Figure 11.1 Robert Boyle.
This engraving shows Boyle with Denis Papin on the development of the pneumatic machine.

Figure 11.2 Antoine Lavoisier performing an experiment with oxygen.

has grown to 109, with more expected to be created artificially in the laboratory.

In order to designate the different elements, we use a symbol notation. The notation used today was first conceived by the Swedish chemist Jöns Jacob Berzelius. He used the first one or two letters of the Latin name for each element. Thus sodium is designated Na for "natrium," silver is Ag for "argentum," tin is Sn for "stannum," and so forth.

Since Berzelius' time, most elements have been symbolized by the first one or two letters of the English name. Examples are C for carbon, O for oxygen, H for hydrogen, and Ca for calcium. Note that the first letter is always capitalized and the second is lowercase. This symbol notation of the elements has proved to be quite useful.

In Table 11.1 each element is listed in alphabetical order, along with its symbol, atomic number, and atomic weight. There are 103 elements listed in this table.

A newly discovered element is named by the person or persons making the discovery, but the name is not officially recognized until the International Union of Pure and Applied Chemistry (IUPAC) gives its approval. The IUPAC has recommended using the name unnilquadium (abbreviated Unq) for element 104, and it has recommended similar names for elements 105, 106, 107, 108, and 109. Figure 11.3 gives the abbreviated names for these elements. The names are based on Latin words for

the atomic numbers (*nil* for zero, *un* for one, *bi* for two, *tri* for three, *quad* for four, etc.).

When the atoms of two or more elements react chemically, they often join to form what chemists call **molecules** of a compound. The **chemical formula** for the compound is written by putting the element symbols adjacent to each other and using subscripts to designate the number of atoms of each element in the compound. For instance, water is written H_2O. This formula means that in a single molecule of the compound water there are two atoms of hydrogen and one atom of oxygen. If more than one molecule is to be designated, a number, called a coefficient, is placed in front of the molecule. For instance, 3 H_2O means there are three molecules of water, each having two atoms of hydrogen and one atom of oxygen.

Recall (Section 10.1) that the atomic number (proton number) of an element refers to the number of protons in the nucleus of an atom. The atomic number is also equal to the number of electrons in the neutral atom. We will see shortly, however, that sometimes electrons are gained or lost, so that the atom is no longer neutral. It is the number of protons, not the number of electrons, that determines the element. Each element can have several isotopes. **Isotopes** are forms of the same element whose nuclei have the same number of protons but varying numbers of neutrons. For example, carbon atoms all have 6 protons but may have 6, 7, or 8 neutrons.

Table 11.1 Alphabetical List of the Elements

Element	Symbol	Atomic Number	Atomic Mass	Mass Number*	Element	Symbol	Atomic Number	Atomic Mass	Mass Number*
Actinium	Ac	89	(227)		Mercury	Hg	80	200.6	202
Aluminum	Al	13	26.98	29	Molybdenum	Mo	42	95.94	98
Americium	Am	95	(243)		Neodymium	Nd	60	144.2	142
Antimony	Sb	51	121.8	121	Neon	Ne	10	20.18	20
Argon	Ar	18	39.95	40	Neptunium	Np	93	(237)	
Arsenic	As	33	74.92	75	Nickel	Ni	28	58.70	58
Astatine	At	85	(210)		Niobium	Nb	41	92.91	93
Barium	Ba	56	137.3	138	Nitrogen	N	7	14.01	14
Berkelium	Bk	97	(247)		Nobelium	No	102	(259)	
Beryllium	Be	4	9.012	9	Osmium	Os	76	190.2	192
Bismuth	Bi	83	209.0	209	Oxygen	O	8	16.00	16
Boron	B	5	10.81	11	Palladium	Pd	46	106.4	106
Bromine	Br	35	79.90	79	Phosphorus	P	15	30.97	31
Cadmium	Cd	48	112.4	114	Platinum	Pt	78	195.1	195
Calcium	Ca	20	40.08	40	Plutonium	Pu	94	(244)	
Californium	Cf	98	(251)		Polonium	Po	84	(209)	
Carbon	C	6	12.01	12	Potassium	K	19	39.10	39
Cerium	Ce	58	140.1	140	Praseodymium	Pr	59	140.9	141
Cesium	Cs	55	132.9	133	Promethium	Pm	61	(145)	
Chlorine	Cl	17	35.45	35	Protactinium	Pa	91	(231)	
Chromium	Cr	24	52.00	52	Radium	Ra	88	226.0	
Cobalt	Co	27	58.93	59	Radon	Rn	86	(222)	
Copper	Cu	29	63.55	63	Rhenium	Re	75	186.2	187
Curium	Cm	96	(247)		Rhodium	Rh	45	102.9	103
Dysprosium	Dy	66	162.5	164	Rubidium	Rb	37	85.47	85
Einsteinium	Es	99	(252)		Ruthenium	Ru	44	101.1	102
Erbium	Er	68	167.3	166	Samarium	Sm	62	150.4	152
Europium	Eu	63	152.0	153	Scandium	Sc	21	44.96	45
Fermium	Fm	100	(257)		Selenium	Se	34	78.96	80
Fluorine	F	9	19.00	19	Silicon	Si	14	28.09	28
Francium	Fr	87	(223)		Silver	Ag	47	107.9	107
Gadolinium	Gd	64	157.3	158	Sodium	Na	11	22.99	23
Gallium	Ga	31	69.72	69	Strontium	Sr	38	87.62	88
Germanium	Ge	32	72.59	74	Sulfur	S	16	32.06	32
Gold	Au	79	197.0	197	Tantalum	Ta	73	180.9	181
Hafnium	Hf	72	178.5	180	Technetium	Tc	43	(98)	
Helium	He	2	4.003	4	Tellurium	Te	52	127.6	130
Holmium	Ho	67	164.9	165	Terbium	Tb	65	158.9	159
Hydrogen	H	1	1.008	1	Thallium	Tl	81	204.4	205
Indium	In	49	114.8	115	Thorium	Th	90	232.0	
Iodine	I	53	126.9	127	Thulium	Tm	69	168.9	169
Iridium	Ir	77	192.2	193	Tin	Sn	50	118.7	120
Iron	Fe	26	55.85	56	Titanium	Ti	22	47.90	48
Krypton	Kr	36	83.80	84	Tungsten	W	74	183.9	184
Lanthanum	La	57	138.9	139	Uranium	U	92	238.0	
Lawrencium	Lr	103	(260)		Vanadium	V	23	50.94	51
Lead	Pb	82	207.2	208	Xenon	Xe	54	131.3	132
Lithium	Li	3	6.941	7	Ytterbium	Yb	70	173.0	174
Lutetium	Lu	71	175.0	175	Yttrium	Y	39	88.91	89
Magnesium	Mg	12	24.31	24	Zinc	Zn	30	65.38	64
Manganese	Mn	25	54.94	55	Zirconium	Zr	40	91.22	90
Mendelevium	Md	101	(258)						

Note: The atomic masses are based on $^{12}C = 12.0000$. If the element does not occur naturally, the mass of the longest-lived isotope is given in parentheses.

* The mass number stated is for the most abundant stable isotope that is found naturally in our environment.

The Periodic Table of the Elements

PERIODS	1A (1)	2A (2)	3B (3)	4B (4)	5B (5)	6B (6)	7B (7)	8B (8)	8B (9)	8B (10)	1B (11)	2B (12)	3A (13)	4A (14)	5A (15)	6A (16)	7A (17)	8A (18)
1	1 H 1.008																	2 He 4.003
2	3 Li 6.941	4 Be 9.012											5 B 10.81	6 C 12.01	7 N 14.01	8 O 16.00	9 F 19.00	10 Ne 20.18
3	11 Na 22.99	12 Mg 24.31											13 Al 26.98	14 Si 28.09	15 P 30.97	16 S 32.06	17 Cl 35.45	18 Ar 39.95
4	19 K 39.10	20 Ca 40.08	21 Sc 44.96	22 Ti 47.88	23 V 50.94	24 Cr 52.00	25 Mn 54.94	26 Fe 55.85	27 Co 58.93	28 Ni 58.69	29 Cu 63.55	30 Zn 65.38	31 Ga 69.72	32 Ge 72.59	33 As 74.92	34 Se 78.96	35 Br 79.90	36 Kr 83.80
5	37 Rb 85.47	38 Sr 87.62	39 Y 88.91	40 Zr 91.22	41 Nb 92.91	42 Mo 95.94	43 Tc (98)	44 Ru 101.1	45 Rh 102.9	46 Pd 106.4	47 Ag 107.9	48 Cd 112.4	49 In 114.8	50 Sn 118.7	51 Sb 121.8	52 Te 127.6	53 I 126.9	54 Xe 131.3
6	55 Cs 132.9	56 Ba 137.3	57–71*	72 Hf 178.5	73 Ta 180.9	74 W 183.9	75 Re 186.2	76 Os 190.2	77 Ir 192.2	78 Pt 195.1	79 Au 197.0	80 Hg 200.6	81 Tl 204.4	82 Pb 207.2	83 Bi 209.0	84 Po (209)	85 At (210)	86 Rn (222)
7	87 Fr (223)	88 Ra 226	89–103†	104 Unq (261)	105 Unp (262)	106 Unh (263)	107 Uns (262)	108 Uno (265)	109 Une (266)									

s — REPRESENTATIVE ELEMENTS
d — TRANSITION ELEMENTS
p — REPRESENTATIVE ELEMENTS
NOBLE GASES

INNER TRANSITION ELEMENTS (f)

Metals ← | → Nonmetals

*Lanthanides	57 La* 138.9	58 Ce 140.1	59 Pr 140.9	60 Nd 144.2	61 Pm (145)	62 Sm 150.4	63 Eu 152.0	64 Gd 157.3	65 Tb 158.9	66 Dy 162.5	67 Ho 164.9	68 Er 167.3	69 Tm 168.9	70 Yb 173.0	71 Lu 175.0
†Actinides	89 Ac† (227)	90 Th 232.0	91 Pa (231)	92 U 238.0	93 Np (237)	94 Pu (244)	95 Am (243)	96 Cm (247)	97 Bk (247)	98 Cf (251)	99 Es (252)	100 Fm (257)	101 Md (258)	102 No (259)	103 Lr (260)

Atomic numbers are shown above the symbols; atomic mass below. A value (of atomic mass) given in parentheses denotes the mass number of the isotope having the longest known half-life.

Figure 11.2 The Periodic Table of the Elements

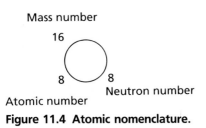

Mass number

Atomic number

Neutron number

Figure 11.4 Atomic nomenclature.

Table 11.2 Relative Abundances of Elements in the Earth's Crust

Element	Approx. Percentage (weight)
Oxygen (O)	46.5
Silicon (Si)	27.5
Aluminum (Al)	8.1
Iron (Fe)	5.3
Calcium (Ca)	4.0
Magnesium (Mg)	2.7
Sodium (Na)	2.4
Potassium (K)	1.9
All Others	1.6
Total	100.0

The symbol notation, plus numbers indicating the atomic number, mass number, and neutron number for the oxygen 16 isotope, is shown in Fig. 11.4. The isotope's symbol, atomic number, and atomic mass are obtained from the periodic table of elements (see Fig. 11.3). Additional information concerning the isotope is known, because the proton number is the same as the atomic number, and in a neutral atom the electron number equals the proton number. The neutron number is determined by subtracting the atomic number from the mass number.

The atomic mass of an element (called the atomic weight by chemists, because the mass is usually determined by weighing, which is done by comparing masses) is usually given under the symbol of the element in the periodic table (see Fig. 11.3). This number, which has no units, is equal to the average mass (in amu) of an atom of the element.

The mass number of an isotope is the integer (whole number) that approximates the isotopic mass. For example, the isotopic mass of hydrogen 1 is 1.007825. This number, to the nearest integer, gives the mass number of hydrogen 1. Thus the mass number of hydrogen 1 is 1. The mass number of the most abundant stable isotope of each element found naturally in our environment is given in Table 11.1.

The element hydrogen is found in nature as three isotopes (see Fig. 11.5). The three isotopes have similar chemical properties, because they all have the same electron structure. They differ in physical properties and can be separated in a mass spectrograph because they have different masses. Ordinary hydrogen ($^{1}_{1}H$) is called protium, hydrogen 2 ($^{2}_{1}H$) is called deuterium, and hydrogen

3 ($^{3}_{1}H$) is called tritium. In a given sample of hydrogen about one atom in 5000 is deuterium and about one atom in 10,000,000 is tritium. Heavy water (D_2O) is composed of two atoms of deuterium and one atom of oxygen. Protium and deuterium are stable atoms, whereas tritium is radioactive, with a half-life of 12.4 years.

The atomic weight of each element is the average weight of the isotopes contained in the naturally occurring element. Atomic weights today are based on the ^{12}C isotope, which is given a *relative* atomic weight of exactly 12.0000. The reason naturally occurring carbon has an atomic weight slightly greater than 12.0000 is that there is some ^{13}C and ^{14}C, as well as ^{12}C, in naturally occurring carbon.

Only a few of the elements exist in large amounts in the Earth's crust. About 75% of the weight of the Earth's crust is composed of only two elements—oxygen and silicon. Aluminum and iron, two important metals, are quite abundant also, but many elements exist naturally in only minute quantities. Several, of course, do not occur at all in nature and can only be made artificially. Table 11.2 gives the approximate percentage by weight of the more abundant elements in the Earth's crust.

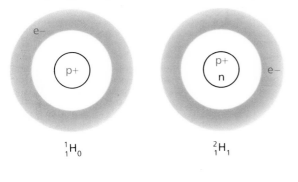

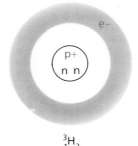

$^{1}_{1}H_{0}$

$^{2}_{1}H_{1}$

$^{3}_{1}H_{2}$

Figure 11.5
The three isotopes of hydrogen: ordinary hydrogen (protium), deuterium, and tritium.

PERIODS	s	s	d	d	d	d	d	d	d	d	d	d	p	p	p	p	p	p / s
1	1 **H** $1s^1$																	2 **He** $1s^2$
2	3 **Li** $2s^1$	4 **Be** $2s^2$											5 **B** $2s^2 2p^1$	6 **C** $2s^2 2p^2$	7 **N** $2s^2 2p^3$	8 **O** $2s^2 2p^4$	9 **F** $2s^2 2p^5$	10 **Ne** $2s^2 2p^6$
3	11 **Na** $3s^1$	12 **Mg** $3s^2$											13 **Al** $3s^2 3p^1$	14 **Si** $3s^2 3p^2$	15 **P** $3s^2 3p^3$	16 **S** $3s^2 3p^4$	17 **Cl** $3s^2 3p^5$	18 **Ar** $3s^2 3p^6$
4	19 **K** $4s^1$	20 **Ca** $4s^2$	21 **Sc** $4s^2 3d^1$	22 **Ti** $4s^2 3d^2$	23 **V** $4s^2 3d^3$	24 **Cr** $4s^1 3d^5$	25 **Mn** $4s^2 3d^5$	26 **Fe** $4s^2 3d^6$	27 **Co** $4s^2 3d^7$	28 **Ni** $4s^2 3d^8$	29 **Cu** $4s^1 3d^{10}$	30 **Zn** $4s^2 3d^{10}$	31 **Ga** $4s^2 4p^1$	32 **Ge** $4s^2 4p^2$	33 **As** $4s^2 4p^3$	34 **Se** $4s^2 4p^4$	35 **Br** $4s^2 4p^5$	36 **Kr** $4s^2 4p^6$
5	37 **Rb** $5s^1$	38 **Sr** $5s^2$	39 **Y** $5s^2 4d^1$	40 **Zr** $5s^2 4d^2$	41 **Nb** $5s^1 4d^4$	42 **Mo** $5s^1 4d^5$	43 **Tc** $5s^1 4d^6$	44 **Ru** $5s^1 4d^7$	45 **Rh** $5s^1 4d^8$	46 **Pd** $4d^{10}$	47 **Ag** $5s^1 4d^{10}$	48 **Cd** $5s^2 4d^{10}$	49 **In** $5s^2 5p^1$	50 **Sn** $5s^2 5p^2$	51 **Sb** $5s^2 5p^3$	52 **Te** $5s^2 5p^4$	53 **I** $5s^2 5p^5$	54 **Xe** $5s^2 5p^6$
6	55 **Cs** $6s^1$	56 **Ba** $6s^2$	57 **La*** $6s^2 5d^1$	72 **Hf** $6s^2 5d^2$	73 **Ta** $6s^2 5d^3$	74 **W** $6s^2 5d^4$	75 **Re** $6s^2 5d^5$	76 **Os** $6s^2 5d^6$	77 **Ir** $6s^2 5d^7$	78 **Pt** $6s^1 5d^9$	79 **Au** $6s^1 5d^{10}$	80 **Hg** $6s^2 5d^{10}$	81 **Tl** $6s^2 6p^1$	82 **Pb** $6s^2 6p^2$	83 **Bi** $6s^2 6p^3$	84 **Po** $6s^2 6p^4$	85 **At** $6s^2 6p^5$	86 **Rn** $6s^2 6p^6$
7	87 **Fr** $7s^1$	88 **Ra** $7s^2$	89 **Ac**** $7s^2 6d^1$	104 **Unq** $7s^2 6d^2$	105 **Unp** $7s^2 6d^3$	106 **Unh** $7s^2 6d^4$	107 **Uns** $7s^2 6d^5$	108 **Uno** $7s^2 6d^6$	109 **Une** $7s^2 6d^7$									

	f	f	f	f	f	f	f	f	f	f	f	f	f	f
***Lanthanides**	58 **Ce** $6s^2 4f^1 5d^1$	59 **Pr** $6s^2 4f^3 5d^0$	60 **Nd** $6s^2 4f^4 5d^0$	61 **Pm** $6s^2 4f^5 5d^0$	62 **Sm** $6s^2 4f^6 5d^0$	63 **Eu** $6s^2 4f^7 5d^0$	64 **Gd** $6s^2 4f^7 5d^1$	65 **Tb** $6s^2 4f^9 5d^0$	66 **Dy** $6s^2 4f^{10} 5d^0$	67 **Ho** $6s^2 4f^{11} 5d^0$	68 **Er** $6s^2 4f^{12} 5d^0$	69 **Tm** $6s^2 4f^{13} 5d^0$	70 **Yb** $6s^2 4f^{14} 5d^0$	71 **Lu** $6s^2 4f^{14} 5d^1$
****Actinides**	90 **Th** $7s^2 5f^0 d^2$	91 **Pa** $7s^2 5f^2 6d^1$	92 **U** $7s^2 5f^3 6d^1$	93 **Np** $7s^2 5f^4 6d^1$	94 **Pu** $7s^2 5f^6 6d^0$	95 **Am** $7s^2 5f^7 6d^0$	96 **Cm** $7s^2 5f^7 6d^1$	97 **Bk** $7s^2 5f^9 6d^0$	98 **Cf** $7s^2 5f^{10} 6d^0$	99 **Es** $7s^2 5f^{11} 6d^0$	100 **Fm** $7s^2 5f^{12} 6d^0$	101 **Md** $7s^2 5f^{13} 6d^0$	102 **No** $7s^2 5f^{14} 6d^0$	103 **Lr** $7s^2 5f^{14} 6d^1$

Figure 11.6
The periodic table illustrating the four broad categories of elements plus the electron configuration of the highest occupied electron level for each element.

11.2 The Periodic Table

By 1870 a total of 65 elements had been discovered. Using the methods of many dedicated chemists, researchers identified the atomic weight and other properties of many elements. However, except for a few sketchy attempts, a system of classifying the elements had not been found. It remained for the great Russian chemist Dmitri Mendeleev to formulate a satisfactory classification scheme. Mendeleev surveyed all data on the known elements and was able to arrange them in a table. A modern version of his table, called the periodic table of the elements, is shown in Fig. 11.3 and on the inside front cover of this book.

HIGHLIGHT

History of the Periodic Table

The story of the periodic table began with Johann Dobereiner, a German chemist, in 1829, when he reported a relationship between the properties of certain elements and their atomic weights. Dobereiner arranged a few elements with similar properties into groups of three, which he called triads. But the arrangement failed when he attempted to apply the model to other known elements.

Thirty-five years later, in 1864, an English chemist, John Newlands, noted a relationship between the atomic weights and the chemical properties of elements. He observed that when the known elements were arranged according to increasing atomic weights, every eighth element had similar properties. Newlands called the relationship the law of octaves, comparing the elements to the notes in an octave of music. Although his model was successful with some groups of elements, his model failed when he attempted to fit all known elements into his law of octaves.

The first detailed and useful periodic table of the chemical elements, in order of increasing atomic weights, was worked out and published in 1869 by Dmitri Mendeleev ("men-deh-lay-eff") (1834–1907), a Russian chemist. The modern periodic table of today resembles the horizontal rows and vertical columns in his original table. (Julius Meyer, a German chemist and physi-

Figure 11.7
Dmitri Mendeleev, a Russian chemist, worked out the first useful periodic table of the chemical elements.

cian, published evidence for the periodic classification of the elements about this same time, but his work did not contain the details published by Mendeleev.)

Mendeleev's table was very useful because of the importance he placed on the periodic appearance of physical and chemical properties at regular intervals and because of his prediction of the properties and positions of unknown elements in the table. He even left vacancies in his table for them. When chemists later discovered these elements, such as gallium, scandium, and germanium, they found that they had the properties Mendeleev had predicted, and his periodic table was universally accepted.

Mendeleev realized there were a few positions in his table where the increasing atomic weight and the properties of the element did not coincide. For example, on the basis of atomic weight, iodine should come before tellurium; but iodine's properties indicate it should *follow* tellurium.

Mendeleev did not know, and was unable to explain, why the chemical properties occurred at regular intervals with increasing atomic weight. But his table provided useful information concerning the chemical elements.

We now know that the periodic law is a function of the atomic number of an element, not the atomic weight, and that the chemical properties of an element depend on the electron structure. The atomic weight depends upon the total number of neutrons located in the nucleus of the atom, and each isotope of an element has a different number of neutrons. Different neutron numbers mean different atomic weights, but the atomic number is the same for each element.

The modern periodic table is a very valuable aid, and it is universally used by the scientific community.

The **periodic table** arranges the elements in rows according to their electron periods, with all the elements in the same column having similar properties. Mendeleev's original table was based on atomic weights, but H. G. J. Moseley, an English physicist, showed in 1913 that properties of the elements are determined by their atomic numbers. The modern statement of the **periodic law** is as follows:

The properties of elements are periodic functions of the atomic number.

The periodic table is most easily understood in terms of the electron configurations of the elements. The atomic number of each element corresponds to the number of electrons in a neutral atom of that element. The element's position in the table depends on the energy level of its last several electrons. For example, the electron configuration of sulfur, atomic number 16, is

$$1s^2 2s^2 2p^6 3s^2 3p^4$$

The last 4 electrons are in the $3p$ level in the third period. Thus sulfur occupies the fourth position in the $3p$ level in the third row of the periodic table. In each box of the table the element's atomic number, symbol, atomic weight (or most stable isotope, if the element is not found in nature), and number of electrons in the outermost shells are given. Thus for sulfur there are six electrons in the outer, unfilled shell; two are in the $3s$ subshell, and four are in the $3p$ subshell.

The rows of the periodic table are called **periods,** and they correspond to the electron periods given in Fig. 11.6. The obvious grouping of columns into the first 2 on the left side, the middle 10, and the last 6 on the right side, correspond to the s, d, and p energy levels, which can contain a maximum of 2, 10, and 6 electrons, respectively. Two groups of 14 elements, each corresponding to the f levels, which hold a maximum of 14 electrons, are placed separately at the bottom of the table.

11.3 Classification of Elements

By examining the periodic table, we can classify the elements into four broad categories:

1. *Elements in which the s and p subshells are completely filled with electrons.* Each of these elements has electrons completely filling up all the subshells in a period. They are helium, with 2 electrons in the $1s$ level, and neon, argon, krypton, xenon, and radon,

each of which has 8 electrons in the outer shell— 2 in the s subshell and 6 in the p subshell. These elements are in the column at the far right labeled 8A and normally occur as gases. Only rarely do they react with other elements. They are called the **noble gases.**

2. *Elements in which an s or p subshell is incomplete.* The s and p subshells together contain a maximum of 8 electrons. Groups of elements that have their last electrons in one of these subshells are called **representative elements.** They are labeled according to the number of electrons in their s and p subshells. Elements in columns labeled 1A, 1B, 2A, 2B, 3A, 4A, 5A, 6A, and 7A are included in this category. The numeral indicates the number of s and p electrons in the outer shell. Columns 1B and 2B are distinguished from 1A and 2A as follows: Elements in columns 1B and 2B have filled d subshells in the last period, while elements in columns 1A and 2A have empty d subshells in the last period.

3. *Elements in which the last electrons fill the d subshell in the period.* These elements are placed in the middle columns of the periodic table. The columns are labeled 3B, 4B, 5B, 6B, 7B, and 8B. The numeral designates the number of s and d electrons in the period. (The exceptions to this rule are the three columns labeled 8B. These elements have from 8 to 10 electrons in the s and d levels.) The elements in this category are called **transition elements.**

4. *Elements in which the last electrons are in the f subshell.* These elements are placed in two rows below the main table. They are in the **lanthanide,** or **rare earth, series:** cerium (Ce) through lutetium (Lu), and the **actinide series:** thorium (Th) to lawrencium (Lr). These two groups of elements make up the **inner transition elements.**

Figure 11.6 illustrates, in color, the four broad categories of elements in the periodic table and the electron configuration of the highest occupied electron level for each element. Note that for some elements the electron configuration does not adhere to the mnemonic diagram (Fig. 9.26) for the ordering of electron levels. For example, chromium, copper, silver, and gold do not adhere to the mnemonic diagram. Evidently, the configuration shown for these elements is a more stable arrangement.

The chemical properties of elements are determined by the number of electrons in the **valence shell**—the outermost shell (or shells in some higher–atomic number atoms) that has valence electrons.

A representative element has valence electrons in the outermost shell (*s* and *p* electrons) (see Figs. 11.3 and 11.6). For a transition element the valence shell contains electrons in the outermost shell and *d* electrons in the next inner shell. For an inner transition element the valence shell contains electrons in the outermost shell, *d* electrons in the first inner shell, and *f* electrons in the second inner shell.

Elements tend to gain, lose, or share valence electrons so that they will have a filled outer electron shell. Except for helium, elements that have one or two electrons in the *s* level tend to lose these electrons easily, so that there will be no electrons in the outer shell. Elements that have six or seven *s* and *p* level electrons tend to acquire electrons to fill up the *s* and *p* subshells. Elements with 3, 4, or 5 valence electrons either gain or lose electrons to get 0 or 8 in the outer shell.

One of the earliest classifications of elements was into metals and nonmetals. This classification was done originally on the basis of certain distinctive properties. Our modern definition is that a **metal** is an element that tends to lose its valence electrons. A **nonmetal** is an element that tends to gain electrons to complete an outer shell.

The large majority of elements are metals. The reason there are so many metals can be readily seen in terms of electron structure. When the last electrons in the electron configuration occupy *d* and *f* levels, the electrons go into inner shells. The elements in which this happens (those in the transition series and inner transition series) thus have only one or two *s* electrons in their outermost shells. Therefore, all of these elements tend to lose their valence electrons, so they are metals. Only the elements on the far right of the periodic table can be nonmetals. The actual dividing line between metals and nonmetals cuts through the periodic table like a staircase (see Fig. 11.3).

One of the first ways in which metals and nonmetals were distinguished was by their chemical properties. When an element combines with oxygen, the resulting compound is called an **oxide.** Many oxides of metals dissolve in water to form aqueous solutions of bases. (Bases will be defined and discussed in Chapter 14.) In general, the metallic character of the elements increases as you go down and to the left of the periodic chart. The nonmetallic properties increase upward and to the right. Thus francium is the most metallic element and fluorine the most nonmetallic. Certain chemical and physical properties are shown in Table 11.3. Some elements display properties of both metals and nonmetals. Elements that exhibit this hybrid behavior are called **semi-metals,** or **metalloids.** The periodic arrangement of the

Table 11.3 Some Chemical and Physical Properties of Metal and Nonmetals

Chemical Properties	
Metals	*Nonmetals*
1. Have one to five electrons in their outer energy shell	1. Have four to eight electrons in their outer energy shell
2. Lose their outer electrons easily	2. Gain electrons readily
3. Form hydroxides that are basic	3. Form hydroxides that are acidic
4. Are good reducing agents	4. Are good oxidizing agents
5. Have lower electronegativities	5. Have higher electronegativities

Physical Properties	
Metals	*Nonmetals*
1. Good conductors of heat and electricity	1. Poor conductors of heat and electricity
2. Malleable—capable of being beaten into thin sheets	2. Brittle—if a solid
3. Ductile—capable of being stretched into wire	3. Nonductile
4. Possess metallic luster	4. Do not possess metallic luster
5. Opaque as a thin sheet (translucent if extremely thin)	5. Transparent as a thin sheet
6. Solids at room temperature (the one exception is mercury, Hg)	6. Solids, liquids, or gases at room temperature

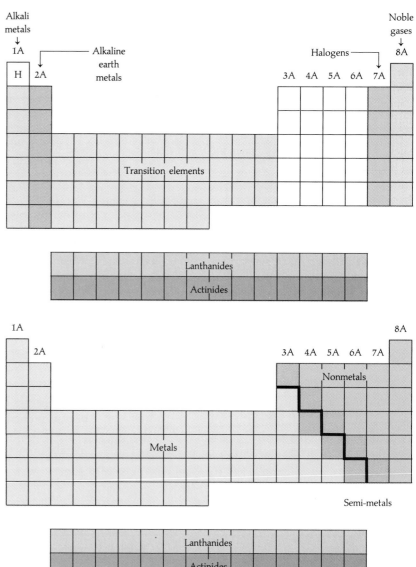

Figure 11.8
Diagram illustrating specific divisions and groups of elements in the periodic table.

metals, nonmetals, and semimetals is illustrated with color in Fig. 11.8. See Fig. 11.9 for elements belonging to each classification.

The elements can also be classified according to whether they are solids, liquids, or gases at room temperature and atmospheric pressure. Only two, bromine and mercury, occur as liquids. Eleven occur as gases: hydrogen, nitrogen, oxygen, fluorine, chlorine, and the six noble gases. All the rest are solids.

Figure 11.9 Three representative elements (right): metal (copper), nonmetal (sulfur), and semimetal (arsenic).

11.4 Periodic Characteristics

In this section the periodic characteristics of a few properties of the elements will be discussed. One is the size of the atoms that compose the elements. Figure 11.10 illustrates a comparison of the atomic radii of some atoms. The radius of an atom is defined as half the distance between the nuclei in a molecule composed of two like atoms. The radius is defined in this manner because the radius is obtained by measuring the distance between atoms in a chemical compound. Generally speaking, the atomic radii decrease from left to right in a period. Notice the large radius of each alkali metal (column 1A) with respect to the other elements of that period. Alkali metal atoms have large radii because the outer electron is loosely bound to the nucleus.

In a family of elements the atomic radii increase from top to bottom. Each successive element of the family or group has an additional energy level containing electrons. As the charge on the nucleus increases (more protons), the outer electrons are bound more firmly, thus decreasing the atomic radius across the rows. This sequence continues until the outer shell of an element is filled. The radius will then jump to a much higher value, because the next element has an electron in a new outer shell.

When an atom gains or loses electrons, it acquires a net electric charge. The net electric charge is the number of protons minus the number of electrons. For example, for 8 protons and 10 electrons the net electric charge

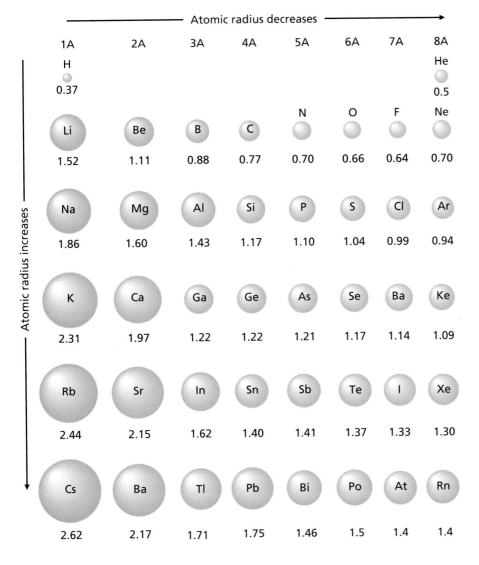

Figure 11.10 The radii of a few atoms, given in angstoms
One Å = 10⁻¹⁰ m. To obtain the radius in nanometers (nm), move the decimal point one place to the left.

is minus 2, written 2—. An atom or group of atoms with a net electric charge is called an **ion.** Magnesium, atomic number 12, normally has 12 protons and 12 electrons for a net electric charge of zero. When it loses its two outer (valence) electrons, it has 12 protons and 10 electrons, for a net charge of 2+.

The net charge on the atom is sometimes called the **valence.** The valence also can be thought of as the number of electrons an atom can give up (or acquire) to achieve a filled outer shell. Magnesium has a valence of 2+. Recall that magnesium has two valence electrons also. The valence of positive ions is frequently equal to the number of valence electrons. The negative charge of an ion can also be an indication of valence. Consider oxygen, with 8 protons and 10 electrons, for a net charge of 2—. Thus oxygen has a valence of 2—. Note that when an element gains or loses electrons, it does not lose its chemical identity. It is the atomic number, not the number of electrons, that determines the identity of the element.

The amount of energy necessary to remove one electron from an atom is called its **ionization potential.** The first ionization potential is the energy required to remove the most loosely held electron. The second ionization potential is the energy to remove a second electron, and so on. The periodic nature of the first ionization potential is shown for the first 90 elements in Fig. 11.11.

Notice that the alkali metals have the lowest first ionization potentials. Their last electron is in an outer shell, and it does not take much energy to remove the electron. Elements located to the right of the alkali metals in the table have more electrons added in the same shell when protons are added to the nucleus. The added protons bind the electrons more and more strongly until the shell is completely filled. In general, the ionization potential increases from left to right in the periodic table. There are a few exceptions. The noble gases have the highest ionization potentials.

11.5 Groups of Elements

A row of the periodic table is called a period. The elements in a column of the table are said to be in the same **group,** or **family,** of elements. All the elements in a group have similar electron configurations; therefore, they have similar properties. In particular, if one element in a group reacts with a given substance, then all the elements in the group will react in an analogous manner with that substance. The molecular formulas of the compounds produced will also be similar. In this section we will discuss several of these groups of elements. See Fig. 11.8, which illustrates, in color, the different groups.

The Noble Gases

Each of the noble gases contains complete periods of electrons; and except for helium, they all contain eight electrons in their outermost shells. They are **monatomic;** that is, they exist as single atoms in nature. This group of elements almost never reacts with other elements to form compounds. We can conclude, then, that electron configurations with completed periods, or eight electrons in the outer shell, are quite stable. This conclusion is of great importance when one considers the stability of other atoms. All atoms, except those of hydrogen and helium, tend to accept, donate, or share electrons with other atoms so as to have an octet of electrons (eight electrons) in their outer shell.

The noble gases were once called inert gases because they did not react. However, in 1962 chemists caused Xenon to react with fluorine to form a stable compound. Hence the terminology *inert gases* no longer applies. The noble gases are sometimes called the **rare gases** because of their scarcity. Helium is obtained from natural gas, and radon is a radioactive by-product of radium's decay; but the other noble gases are found only in the air. Their chief use is in "neon" signs. Neon signs contain minute amounts of various rare gases in a sealed glass vacuum tube. When an electric current is passed through the tube, the gases glow. The particular color of the glow depends on the mixture of rare gases used. Figure 11.12 shows the red-orange light from the element neon.

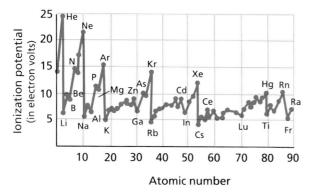

Figure 11.11
Graph showing the relationship between the first ionization potential and the atomic number for the first 90 elements in the periodic table.

Figure 11.12
Mixtures of neon and other gases emit bright colors when excited by an electric discharge.

Figure 11.13
Potassium is a soft metallic element that is extremely active in water, as shown by this photograph. It is stored in oil to keep moisture away from the metal.

Because the noble gases have similar electron configurations, many of their physical and chemical properties are similar. Table 11.4 summarizes some of their physical properties and shows that the melting points, boiling points, and atomic radii increase with increasing atomic number. However, the important fact here is that they have similar melting points and boiling points, particularly when compared with other atoms in the periodic table.

The Alkali Metals

The elements in the far left column of the periodic table, except for hydrogen, are called the **alkali metals.** Hydrogen is classified as a nonmetal. It is the only element that can achieve a closed outer shell by either gaining or losing an electron. The electron configurations of the alkali metals all end with one s electron in the outer shell. Thus they have one valence electron. The alkali metals tend to lose this outer electron quite easily so that a stable, closed-shell electron configuration will be reached. Thus they react readily with other elements.

The elements sodium and potassium (see Fig. 11.13) are abundant in the Earth's crust, but lithium, rubidium, and cesium are rare. Francium, which is radioactive, is very rare. The most common compound containing an alkali metal is table salt, NaCl. Other alkali metal compounds known to the ancients were potash (potassium carbonate, K_2CO_3) and washing soda (sodium carbonate, Na_2CO_3).

Table 11.4 Physical Properties of the Noble Gases

Element	Symbol	Atomic Number	Atomic Weight	Melting Point (°C)	Boiling Point (°C)	Atomic Radius (Å)
Helium	He	2	4.00260	−272	−268.9	0.93
Neon	Ne	10	20.179	−248.6	−246.1	1.12
Argon	Ar	18	39.948	−189.3	−186	1.54
Krypton	Kr	36	83.80	−157.2	−153.4	1.69
Xenon	Xe	54	131.30	−111.0	−108.1	1.90
Radon	Rn	86	(222)	−71	−62	2.2

The formulas and names of some common stable compounds of sodium are given in the list.

NaCl	Sodium chloride (common table salt)
NaOH	Sodium hydroxide (caustic soda)
Na_2CO_3	Sodium carbonate (washing soda)
$NaHCO_3$	Sodium hydrogen carbonate (baking soda; bicarbonate of soda)
Na_2O	Sodium oxide

By knowing these compounds of sodium and the fact that all the elements in a group produce similar compounds, we automatically know a great deal about the compounds of the other alkali metals. Thus we expect potassium oxide to have the molecular formula K_2O. We do not expect it to be KO, KO_2, KO_3, K_2O_5, and so on. Similarly, lithium carbonate should be, and is, Li_2CO_3.

Some of the physical properties of the alkali metals are given in Table 11.5. The concept of groups can be used to estimate the melting and boiling points (at present unknown) of the radioactive element francium. We expect all the alkali metals to have similar properties, so we would guess from Table 11.5 that francium's melting and boiling points should be about 20° and 680°C, respectively. These guesses, of course, are not exact but should be close.

The Halogens

The elements of the periodic table in the column next to the far right are called the **halogens**. See Figs. 11.8 and 11.14. They have seven electrons in the outer s and p subshells and need only one more electron to achieve a closed-shell electron configuration. Thus they tend to react quite strongly with other elements and so are not found free in nature.

Figure 11.14 Three halogens: chlorine, bromine, and iodine.

The reactivity of the halogens decreases as you go down the periodic table. Generally, the halogens are all classified as nonmetals. At room temperature, fluorine and chlorine occur as gases, bromine as a liquid, and iodine and astatine as solids. The type of iodine found in a medicine cabinet is not the pure solid element. It is a tincture—solid iodine dissolved in denatured alcohol. Some of the basic properties of the halogens are given in Table 11.6.

Fluorine is the most reactive of all the elements. It corrodes even platinum, a metal that withstands most other chemicals. In a stream of fluorine gas, wood, rubber, and even water burst into flame. Fluorine is a pale-yellow,

Table 11.5 Physical Properties of the Alkali Metals

Element	Symbol	Atomic Number	Atomic Weight	Melting Point (°C)	Boiling Point (°C)	Atomic Radius (Å)
Lithium	Li	3	6.941	180	1326	1.23
Sodium	Na	11	22.9898	97.5	889	1.57
Potassium	K	19	39.102	63.4	757	2.03
Rubidium	Rb	37	85.4678	38.8	679	2.16
Cesium	Cs	55	132.9055	28.7	690	2.35
Francium	Fr	87	(223)	—	—	2.7

Table 11.6 Physical Properties of the Halogens

Element	Symbol	Atomic Number	Atomic Weight	Melting Point (°C)	Boiling Point (°C)	Atomic Radius (Å)	Physical State
Fluorine	F	9	18.9984	−218	−188	0.72	Gas
Chlorine	Cl	17	35.453	−101	−34	0.99	Gas
Bromine	Br	35	79.90	−7.3	58.8	1.14	Liquid
Iodine	I	53	126.9045	114	184	1.33	Solid
Astatine	At	85	(210)	—	—	1.40	Solid

highly poisonous gas. It was responsible for the deaths of several very able chemists before it was finally isolated. Chlorine is also a poisonous gas. It is used as a disinfectant in swimming pools and as a purifying agent in public water supplies. About one part chlorine is used per one million parts water. Bromine is a foul-smelling, poisonous liquid. It is also used as a disinfectant. Iodine is a blue-black solid. Most table salt is now "iodized"; i.e., NaI is added, to supplement the human diet, because an iodine deficiency causes thyroid problems.

Some formulas and compound names for the halogens are:

NaI	Sodium iodide
$AlCl_3$	Aluminum chloride
SnF_4	Stannic fluoride
PCl_5	Phosphorus pentachloride
HBr	Hydrogen bromide
$CaCl_2$	Calcium chloride

By analogy with these formulas, we can write the correct formulas of many other stable compounds—for example, $NaBr$, AlF_3, and $SnCl_4$.

The Alkaline Earths

The **alkaline earths** are found in column 2A of the periodic table. They contain two valence electrons and are thus metals. They are quite active chemically, but they are not as reactive as the alkali metals. They also have higher melting points and are generally harder and stronger than their neighbors in column 1A. Some of the properties of alkaline earths are given in Table 11.7.

Calcium is used by the bodies of mammals in the formation of bones and teeth. When radioactive strontium-90 is ingested, it goes into bone marrow because of its similarity to calcium. If enough strontium-90 is ingested, it can destroy the bone marrow or perhaps cause cancer. Magnesium is used in making lightweight metal alloys. Beryllium has a similar use. Some typical alkaline earth compounds are:

$CaCO_3$	Calcium carbonate
SrO	Strontium oxide
$Mg(OH)_2$	Magnesium hydroxide
$BaSO_4$	Barium sulfate
$BeCl_2$	Beryllium chloride

Table 11.7 Physical Properties of the Alkaline Earths

Element	Symbol	Atomic Number	Atomic Weight	Melting Point (°C)	Boiling Point (°C)	Atomic Radius (Å)
Beryllium	Be	4	9.01218	1283	1500	0.89
Magnesium	Mg	12	24.305	650	1120	1.36
Calcium	Ca	20	40.08	850	1490	1.74
Strontium	Sr	38	87.62	790	1348	1.91
Barium	Ba	56	137.34	704	1638	1.98
Radium	Ra	88	(226)	700	1500	—

Calcium carbonate is a mineral found in nature in many forms. The most common form is limestone, a sedimentary rock formed from seashells. Limestone is used for building roads and to neutralize acid soils. The beautiful formations of stalactites and stalagmites found in underground caverns are composed of calcium carbonate.

There are other groups of elements which, because of similar group characteristics, can be classified as families. We have considered only four as examples. As mentioned previously, one element, hydrogen, is unique. It has one valence electron and lacks one valence electron. It acts sometimes like an alkali metal, forming HCl, H_2S, etc. (similar to $NaCl$, Na_2S), and sometimes like a halogen, forming NaH, CaH_2, etc. (compare with $NaCl$, $CaCl_2$). It is the lightest of the elements. At room temperature it is a colorless, odorless, highly flammable gas. It has a melting point of $-259°C$ and a boiling point of $-252.7°C$.

Learning Objectives

After reading and studying this chapter, you should be able to do the following with the use of the periodic table:

1. Write the symbol notation plus numbers indicating the atomic number, mass number, and neutron number (as illustrated in Fig. 11.4) for any isotope.

2. Name the symbol for each element given in the periodic table; specify the element's atomic number, atomic weight, atomic mass, proton number, neutron number, electron number, group number, period, and valence; and specify the element as a metal, semimetal, or nonmetal.

3. Name the four main classifications of elements, and specify the elements in each classification.

4. Write the electron configuration for the highest occupied electron level for any element.

5. State the phase of each element when the element is at normal pressure and temperature.

6. State the relative size of one atom with respect to another, using the text.

7. Arrange a given number of atoms in order of increasing ionization potential, using the text.

8. Define and explain the important words and terms in the next section without referring to the text.

Important Words and Terms

matter	periods	valence shell	ionization potential
element	noble gases	metal	group, family
molecules	representative elements	nonmetal	monatomic
chemical formula	transition elements	oxide	alkali metals
isotopes	lanthanide series, rare earth series	semimetal, metalloid	halogens
periodic table	actinide series	ion	alkaline earths
periodic law	inner transition elements	valence	

Questions

Elements

1. Distinguish between an atom and an element.

2. Explain why there are many more different kinds of atoms when we know of only 109 different kinds of elements.

3. What was the view held by the Greek philosopher Aristotle concerning the composition of matter?

4. State the differences between deuterium and ordinary hydrogen.

5. Distinguish between atomic number and atomic weight.

6. On which isotope are all atomic weights based?

The Periodic Table

7. What is the modern basis for the periodic law?

8. A search is on for the element with atomic number 114. Which element will it most resemble?

9. State the group number of each of the following elements:
   (a) scandium   (b) lead   (c) niobium
   (d) hahnium,   (e) uranium

10. State the period of each of the following elements:
    (a) helium   (b) zinc   (c) radium
    (d) europium   (e) plutonium

Classification of Elements

12. Define the terms *metal*, *semimetal*, and *nonmetal*.

13. State some chemical properties that distinguish metals from nonmetals.

14. State some physical properties that distinguish metals from nonmetals.

15. What are the two most abundant metals?

16. What determines the chemical properties of an element?

17. List the elements that are gases and liquids at room temperature and atmospheric pressure.

11. Looking at the periodic table, we notice that there are four major groupings of columns. There are 2 columns on the left, 10 in the middle, 6 on the right, and 14 below. What causes these groupings of 2, 10, 6, and 14 columns?

Periodic Characteristics

18. Distinguish between an atom and an ion.

19. Distinguish between valence and valence shell.

20. How would you expect the ionization potential and atomic radius of francium to compare with the other alkali metals?

Groups of Elements

21. Explain why the alkali metals have larger radii than the elements with one or two, more or less, protons.

22. List several elements that do not react readily. List several elements that are highly reactive.

Exercises

Elements

1. Write the symbol notation plus numbers indicating the atomic number, mass number, and neutron number for the following isotopes: sodium 23, carbon 14, calcium 40, iodine 131, and uranium 238.
   Answer: for carbon 14, $^{14}_{6}C_8$

2. What are the two most abundant elements in the Earth's atmosphere?

The Periodic Table

3. Using the periodic table, determine the atomic number, proton number, atomic weight, mass number, neutron number, electron number, and number of valence electrons of each of the following neutral elements, and state whether each is a metal, a nonmetal, or a noble gas.

   (a) lithium   (b) neon   (c) calcium
   (d) chlorine   (e) niobium   (f) gold
   (g) holmium   (h) uranium
   Answer: (c) 20, 20, 40.08, 40, 20, 2, metal

4. Classify the following elements as noble gas, representative element, transition element, or inner transition element:

   (a) krypton   (b) iron   (c) iodine
   (d) strontium   (e) hydrogen   (f) helium
   (g) nitrogen   (h) mercury   (i) uranium

5. Determine the most common valence of each of the elements in Exercise 4.

6. Write the electronic configuration for the highest occupied electron level for the following elements (refer to Fig. 11.6): sodium, copper, selenium, iodine, and uranium.
   Answer: for copper, $4s^1 3d^{10}$

7. Write the electronic configurations for the three coinage metals—copper, silver, and gold. Use the mnemonic diagram in Chapter 9 and the periodic table (Figure 11.6).

Classification of Elements

8. List two elements in which the *s* and *p* subshells are completely filled with electrons.

9. List two elements in which the last electrons fill the *d* subshell.

10. List two elements in which the *s* or *p* subshell is incomplete.

11. List two elements in which the last electrons are in the *f* subshell.

12. Arrange the elements cesium, lithium, potassium, rubidium, and sodium according to increasing atomic radii.

13. Arrange the elements argon, helium, krypton, and neon according to decreasing ionization potential.
    Answer: He, Ne, Ar, Kr

14. Which atom has the largest radius in each of the following pairs of elements: sodium or potassium, carbon or silicon, polonium or selenium, argon or xenon?

15. Arrange the elements cobalt, helium, cesium, tin, and chlorine in order of increasing atomic radii.
    Answer: He, Cl, Co, Sn, Cs

16. Arrange the elements helium, magnesium, rubidium, zinc, and xenon in order of increasing ionization potential.

17. What is the difference in the value for the ionization potentials (in eV) of (a) neon and lithium and (b) argon and sodium?   *Answer:* (a) $22 - 6 = 16$ eV

Groups of Elements

18. What are the melting and boiling points of the following elements: helium, sodium, chlorine, and calcium?

CHAPTER 12

Compounds, Molecules, and Ions

> **Happy the man who, studying nature's laws,**
> **Through known effects can trace the secret cause.**
>
> —Virgil

THERE ARE 109 known elements. In Chapter 11 they were arranged in order of increasing atomic number to form the periodic table. With the use of the table we learned many of the properties of these known elements.

Matter found in our environment is usually in the form of mixtures. A **mixture** is a nonchemical, or physical, combination of two or more substances of varying proportions. When we look around, we see many examples of mixtures, such as ocean water, soil, concrete, plants, and animals. Our environment is composed of millions of compounds and mixtures. In this chapter we lay the foundations for compound formation and study the properties of some of the compounds that make up our world.

A **compound** is a substance composed of two or more elements chemically combined in a definite proportion. As you now know, an element is made up of only one kind of atom. Some compounds are made up of ions, others of molecules. We stated in Chapter 11 that an ion is an atom or a chemical combination of atoms with a net electric charge. A **molecule** is an uncharged particle of an element or a compound. The properties of many ions and molecules play an important role in our environment. Some molecules, such as the nitrogen oxides, sulfur dioxide, carbon dioxide, and carbon monoxide, pollute the air we breathe. Others play a vital role in our being alive. For instance, we will see in Chapter 20 that the ozone, carbon dioxide, and water molecules in the air absorb radiant energy, thus keeping the Earth at a liveable temperature. Similarly, the various minerals and rocks that we find in our Earth obtain their properties from those of the various molecules and ions that constitute them.

The formation of ions and molecules results from the electromagnetic forces between all the various electrons and the nuclei of the atoms involved. It is a very complicated problem, because there are so many forces to consider. Modern chemists use large computers and the techniques of quantum mechanics to try to understand how molecules and ions form. Although there is still a lot to be understood, chemists are now able to produce new compounds having almost any desired property. We begin our study with some of the basic principles of compound formation and the properties of some of the different types of compounds.

12.1 Principles of Compound Formation

The noble gases, with their outer shells filled with electrons, are said to have closed-shell electron configurations. The noble gases are monatomic, and it is extremely difficult to make them combine with any of the other elements. This observation suggests that closed-shell configurations result in extremely stable substances. Another important fact to remember in compound formation (see Section 11.5) is that elements in the same group or family form similar compounds. Because all of the elements in the same family have an equal number of valence electrons, we conclude that it is the outer electrons that are important in compound formation. We now have a reason for the two basic assumptions made to explain compound formation:

1. The only electrons that take part in compound formation are those in the outer shell.
2. Compounds are formed when each atom in the molecule or ion achieves a closed-shell electron configuration.

Both of these assumptions, however, are violated at various times in compound formation. In the transition and inner transition elements some of the inner d or f shell electrons sometimes take part in compound formation. This reaction violates the first assumption. There are also exceptions to the second assumption. Occasionally, a compound is formed in which one or more of the atoms do not have a closed-shell configuration.

◀ **Photomicrograph of NaCl (table salt).**
The photo is in false color.

However, because most compounds do obey our basic assumptions, we will use them and neglect the few exceptions that occur.

Except for elements in the first period (hydrogen and helium), there can be a maximum of eight electrons in the outer shell of elements in every period. In order for an atom to achieve a noble gas (closed-shell) configuration, it must have either eight or zero valence electrons. Atoms achieve eight or zero valence electrons by gaining, losing, or sharing electrons with other atoms. The concept that atoms tend to have eight electrons in their outer shell when forming molecules or ions is called the **octet rule.**

Hydrogen, of course, is the most important exception to this rule. It must have two electrons in its outer shell when it combines with other atoms. Other exceptions to the octet rule occur with elements in the two transition groups. However, in the remaining chapters we shall be dealing almost exclusively with the representative elements, for which the octet rule holds, in general.

There are two ways in which individual atoms can achieve a closed-shell electron configuration. They can transfer electrons or they can share electrons. In the transfer of electrons one or more elements lose some or all of their outer electrons, and another one or more elements gain these same electrons to achieve closed-shell configurations. Compounds formed by this electron transfer process are called **ionic compounds.** Because these compounds in the liquid phase conduct electricity, they are sometimes referred to as **electrovalent compounds.** The majority of compounds are formed when electrons are shared between different atoms. Compounds formed by the electron-sharing process are called **covalent compounds.** Now let us examine these two kinds of compounds.

12.2 **Ionic Compounds**

Compounds formed when electrons are lost or gained are referred to as ionic compounds. This name is used because when the electrons are lost or gained by the various atoms, ions are formed. Thus an **ion** is an atom or group of atoms that has a net positive or negative electric charge.

The change in energy when an atom gains an electron, forming a negative ion, is called **electron affinity.** The nonmetals have high electron affinities, and the metals have low electron affinities. Thus only the non-

metals tend to form negative ions by gaining electrons. An atom that needs only one or two electrons to fill its outer shell can easily acquire the electrons from atoms that have low ionization energy—i.e., atoms with only one or two electrons in the outer shell, such as those in groups 1A and 2A of the periodic table.

To illustrate, let us consider how common table salt, NaCl, is formed. Figure 12.1a illustrates the loss of an electron from a neutral sodium atom to form a plus-one sodium ion. The neutral atom has the same number of electrons and photons. Thus the net electric charge on the atom is zero. After the loss of an electron the number of protons is one more than the number of electrons, leaving a net charge on the sodium ion of plus one. Similarly, Fig. 12.1b illustrates the gain of an electron by a neutral chlorine atom to form a negative-one chlorine ion.

Note the difference in the sizes of the atoms and their ions. The sodium atom has a radius of 1.86 Å, but the ion has a radius of 0.95 Å. The chlorine atom has a radius of 0.99 Å, and the ion has a radius of 1.81 Å.

When chlorine, a green gas (see Fig. 11.14) composed of Cl_2 molecules, is united with the metal sodium, they react chemically to form the ionic compound sodium chloride. Electrons are lost by the sodium atoms and gained by the chlorine atoms to form ions, which are bonded by the attractive electric force between the positive and negative ions. Figure 12.2 shows a model of sodium chloride molecules, illustrating how the atoms of sodium and chlorine are arranged.

The formation of sodium chloride can also be illustrated by using what is known as the electron dot notation. With this notation the nucleus and inner electrons of an element are represented by the element symbol and the valence electrons by dots. The sodium and chlorine atoms of Fig. 12.1a are represented in the electron dot notation as

$$\text{Na}\cdot \qquad :\ddot{\text{Cl}}:$$

The sodium and chlorine ions of Fig. 12.1b are represented as

$$\text{Na}^+ \qquad :\ddot{\text{Cl}}:^-$$

$$+1 \qquad\qquad -1$$
$$\text{charge} \qquad \text{charge}$$

The charge of the ion is represented by the number of plus or minus signs. An ion with two negative charges is sometimes designated by a double minus sign ($=$)

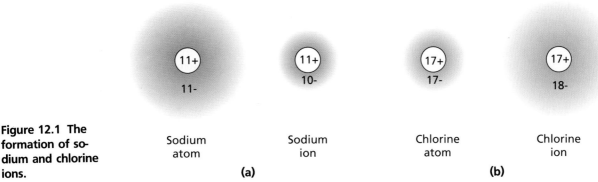

Figure 12.1 The formation of sodium and chlorine ions.

Sodium atom

Sodium ion

Chlorine atom

Chlorine ion

(a)

(b)

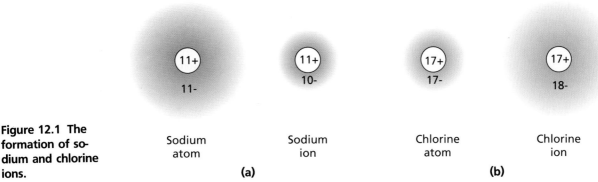

= Cl⁻   = Na⁺

Figure 12.2 Schematic diagram of the sodium chloride crystal.
The sodium ions are the smaller spheres with the plus charge indicated. The chlorine ions have one more shell, so they are drawn as bigger spheres with the minus charge indicated.

but most often by $2-$. Consider the following examples:

$$Mg^{++} \quad \text{or} \quad Mg^{2+}$$

$$:\overset{..}{O}:^{=} \quad \text{or} \quad :\overset{..}{O}:^{2-}$$

The charge on most ions of the representative elements can easily be determined as follows: The positive charge is equal to the number of valence electrons. The negative charge is eight minus the number of valence electrons. For example, aluminum, in column 3A of the periodic table, has three valence electrons. The aluminum

ion thus has a charge of $3+$. Sulfur, with six valence electrons, tends to fill its outer shell. Its ion has a negative charge of 8 minus 6, or a charge of $2-$. Some of the main group elements plus their ions are given in Table 12.1, along with an illustration of their electron dot notation. We have not included elements in group 4A of the periodic table because these elements share electrons and thus form covalent rather than ionic compounds.

In the formation of simple ionic compounds one element loses its electrons and the other element gains them, resulting in ions. The ions are then held together by the electrical attraction between them. The formation of NaCl can be represented as

$$Na\cdot + :\overset{..}{\underset{..}{Cl}}: \longrightarrow Na \quad :\overset{..}{\underset{..}{Cl}}:$$

$$\begin{array}{cc} +1 & -1 \\ \text{charge} & \text{charge} \end{array}$$

In this example the sodium atom lost one electron and the chlorine atom gained one. Thus we see that for every ion of sodium there is one ion of chlorine. This one-to-one correspondence is necessary to balance the charges on the ions.

Another example of an ionic compound is calcium oxide, CaO. Its formation is represented

$$Ca: + :\overset{..}{\underset{..}{O}}: \longrightarrow Ca \quad :\overset{..}{\underset{..}{O}}:$$

$$\begin{array}{cc} +2 & -2 \\ \text{charge} & \text{charge} \end{array}$$

In the formation of this compound the calcium atom has lost two electrons and the oxygen atom has gained two. Thus to get a neutral compound, we need one ion of each element.

Now consider what happens with calcium and chlorine. Calcium has two electrons to lose, but each atom

Table 12.1 Electron Dot Notation and Common Ions for Some Main Group Elements

1A		2A		3A		5A		6A		7A	
Atom	Ion	Atom	Ion	Atom	Ion	Atom	Ion	Atom	Ion	Atom	Ion
H·	H$^+$										
Li·	Li$^+$	Be:	Be^{2+}							:F̈:	:F̈:$^-$
Na·	Na$^+$	Mg:	Mg^{2+}	Al:	Al^{3+}	·N̈:	:N̈:$^{3-}$	:O:	:Ö:$^{2-}$	:C̈l:	:C̈l:$^-$
K·	K$^+$	Ca:	Ca^{2+}	Ga:	Ga^{3+}	·P̈:	:P̈:$^{3-}$	:S:	:S̈:$^{2-}$	:B̈r:	:B̈r:$^-$

of chlorine can gain only one. There must be two atoms of chlorine to accept both of the electrons and still maintain a neutral charge on the compound. We have

$$Ca: + :\ddot{C}l: + :\ddot{C}l: \longrightarrow Ca \quad :\ddot{C}l: \quad :\ddot{C}l:$$

$$\begin{matrix} +2 & -1 & -1 \\ \text{charge} & \text{charge} & \text{charge} \end{matrix}$$

All of the atoms in the compound have closed shells, and the total charge is zero. The molecular formula of calcium chloride is $CaCl_2$.

We are now beginning to see how molecular formulas arise. The numbers of atoms of the various elements involved in the compound are determined by the requirements that the total charge must be zero and that all the atoms are to have closed-shell electron configurations. As another example, consider sodium and sulfur. Each sodium atom loses one electron and each sulfur atom gains two, so there must be two sodium atoms for every sulfur atom.

$$Na· + Na· + :\ddot{S}: \longrightarrow Na \quad Na \quad :\ddot{S}:$$

$$\begin{matrix} +1 & +1 & -2 \\ \text{charge} & \text{charge} & \text{charge} \end{matrix}$$

The formula for the compound formed, sodium sulfide, is thus seen to be Na_2S.

12.3 Properties of Ionic Compounds

Ionic compounds are formed by the electrical attraction of the ions. These compounds generally have higher melting and boiling points than covalent compounds. Ionic compounds occur in the solid phase as crystals. A

crystal is an orderly arrangement of atoms. The NaCl crystal shown in Fig. 12.2 is an example of a cubic crystal; there are many other possible crystal arrangements. The cubic nature of the NaCl crystal can be seen by looking closely at a grain of ordinary table salt.

When we write the formula NaCl, it means that in the compound sodium chloride there is one ion of Na$^+$ to one ion of Cl$^-$. On a microscopic scale it is impossible to associate any one ion of Na$^+$ with a particular ion of Cl$^-$. However, there will always be a one-to-one correspondence between the Na$^+$ and Cl$^-$ ions. Similarly, the formula $CaCl_2$ means that in the compound calcium chloride there are two chlorine ions for every calcium ion.

One of the more important properties of ionic compounds is their behavior when an electric current is passed through the compound. This behavior is illustrated as follows: If a light bulb is connected in series with a battery, it lights up. Electrons flow from the negative terminal of the battery, through the light bulb, and back to the positive terminal. If the wires are cut, the electrons cannot flow (no electric current), and the light bulb will not light.

Now, if the cut wires are inserted into an ionic compound, such as NaCl in water, the bulb lights. That is, ionic compounds, in the liquid phase, conduct an electric current. Figure 12.3 is a diagram illustrating how the conduction takes place, and Fig. 12.4 is a photograph that demonstrates the effect.

The explanation of how ionic compounds conduct an electric current is rather complex. The electrons do not flow through the liquid. The metal conductors (copper wires) placed in the liquid are charged positively and negatively because they are connected to the positive and negative terminals of the battery.

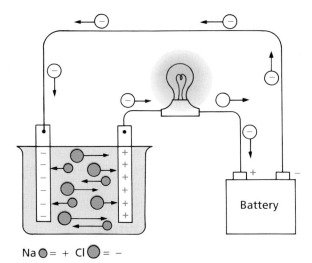

Na $\bigcirc$ = + Cl $\bigcirc$ = −

Figure 12.3 Diagram illustrating how melted NaCl conducts electricity.
The lighted bulb indicates the flow of an electric current. The electric current is achieved because of the reactions of the ions at the electrodes.

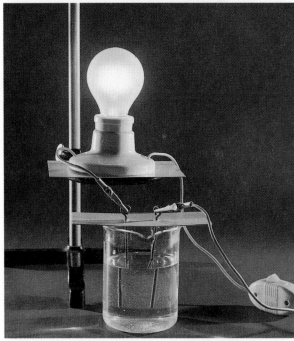

These metal conductors are called electrodes. The positively charged electrode is called the **anode;** the negatively charged electrode is called the **cathode.** The positively charged sodium ions are attracted to the cathode, while the negatively charged chlorine ions are drawn toward the anode. For this reason a positively charged ion is called a **cation,** and a negatively charged ion is called an **anion.** When a sodium ion reaches the cathode, an electron is taken from the cathode, and the sodium ion becomes a sodium atom. The reaction that takes place at an electrode is called a **half-reaction.** The half-reaction at the cathode is represented by

$$Na^+ + e^- \longrightarrow Na$$

The net effect of this process is to take electrons from the cathode. At the anode the negatively charged chlorine ions give up an electron to the positively charged anode. Then two chlorine atoms form a covalent chlorine gas molecule, which appear as bubbles on the anode. The half-reaction is

$$2\ Cl^- \longrightarrow Cl_2 + 2\ e^-$$

This half-reaction donates electrons to the anode. The electric current is thus achieved because of the reaction of the ions at the electrodes, and the light bulb then lights up.

Figure 12.4 Photographs demonstrating the conduction of an electric charge by an ionic compound.
Pure water is used in the apparatus in the top photo and no conduction of the electric charge is taking place. The light bulb is not on. A small amount of salt (NaCl) has been added to the beaker of water in the bottom photo. The light bulb is on, indicating a flow of electric charge.

To summarize, at the cathode electrons are removed; at the anode they are added. The net result is an electron flow, or current. Eventually, all the sodium and chlorine ions react. At this point the current ceases to flow, and the light goes out. An electric current will also be conducted when molten NaCl is used. Most ionic compounds dissolve in water, and the solutions will conduct an electric current.

The half-reaction that occurs at the anode is an example of oxidation. **Oxidation** is the process in which electrons are lost. At the cathode reduction takes place. **Reduction** is the process in which electrons are gained. The word *oxidation* comes from the word *oxygen*, because it is the material that frequently causes oxidation. For example, when pure iron (Fe) is exposed to the air, the oxygen atoms combine with the iron to form rust, which for our purpose can be designated by Fe_2O_3. The oxygen atoms gain electrons, so they will have a filled shell and are said to be reduced. The iron atoms lose electrons; they are oxidized.

The total reaction in which electrons are gained and lost is called an **oxidation-reduction reaction.** If we add the two half-reactions in our example, we get

$$2\,Na^+ + 2\,e^- \longrightarrow 2\,Na$$
$$2\,Cl^- \longrightarrow Cl_2 + 2\,e^-$$
$$\overline{2\,Na^+ + 2\,Cl^- + 2\,e^- \longrightarrow 2\,Na + Cl_2 + 2\,e^-}$$

We have multiplied the half-reaction at the cathode by 2 before comparing it with the anode half-reaction. We multiply by 2 to balance the number of electrons gained and lost. The two electrons on both sides of our oxidation-reduction reaction cancel, and we get

$$2\,NaCl \longrightarrow 2\,Na + Cl_2$$

as our final reaction. By sending an electric current into an aqueous solution of NaCl, sodium and chlorine ions can be converted into pure sodium metal and chlorine gas. In general, the production of an oxidation-reduction reaction by means of an electric current is called **electrolysis.** The electrolysis of sodium and many other metals is an important commercial process that produces the metals in a pure form.

Electrolysis is also the means by which electroplating is accomplished. The material to be plated is used as the cathode in electrolysis, and a melted ionic compound or a solution of the ionic compound in water is the source of the metal to be plated.

So far, we have discussed several properties of ionic compounds. Their high binding energies, produced by the strong electrostatic attractions between ions, causes them to have high melting points. They do not exist as single molecules, but the individual ions occur in a ratio given by the molecular formula. They do not conduct electricity in the solid form but do in the liquid form. Finally, most ionic compounds dissolve in water, and in solution they conduct electricity. The mechanism by which this dissolving process occurs will be discussed after we have learned something about the structure of the water molecule.

12.4 Covalent Compounds

In the formation of ionic compounds electrons are transferred; in covalent compounds electrons are shared. Hydrogen gas, H_2, is the simplest example of a covalent molecule. It contains two atoms in each molecule. Molecules containing two atoms are called *diatomic*. When two hydrogen atoms are brought together, there is an attraction between opposite electrons and protons and repulsion between the two electrons and between the two protons. When the positively charged nuclei get about an angstrom apart, they begin to repel each other. The hydrogen molecule consists of a total system of two nuclei and two electrons. Once the atoms are close together, the two electrons no longer orbit around individual nuclei. They both orbit around both of the protons, tracing out a circular path around them.

Of course, as was previously mentioned, the concept of electrons in orbits is somewhat naive. The quantum mechanical description is based on the probability that an electron will occupy any given spot around the nuclei. In the hydrogen molecule both electrons are shared equally by both nuclei, and it is this sharing of electrons that tends to hold the atoms together. There is said to be a **covalent bond** between the two atoms. Contrast this situation to an **ionic bond,** which is formed when two oppositely charged ions attract each other.

In the electron-dot notation the separated hydrogen atoms are written as

$$H\cdot \qquad H\cdot$$

and the H_2 molecule is written as

$$H\!:\!H$$

The two dots between the hydrogen atoms indicate that these electrons are being shared. The bond can also be represented by a dash,

$$H\text{—}H$$

which indicates a pair of shared electrons.

Not all atoms will share electrons. For instance, the helium atom, with a closed $1s$ shell, will not form a stable molecule with another helium atom. Two helium atoms repel each other, but two hydrogen atoms attract. The molecule H_2 is stable, but He_2 is not. It takes complicated quantum mechanical calculations to determine which molecules will be stable and which will not. However, our basic assumption serves as a good guide. Stable covalent molecules are formed when the atoms share electrons in such a way as to give all atoms a closed outer-shell configuration.

Let us consider more examples of covalent bonding. Two chlorine atoms in the electron-dot notation are given by

$$:\ddot{Cl}\cdot \qquad \cdot\ddot{Cl}:$$

Each chlorine atom needs one electron to have a closed shell. If each shares its unpaired electron, we get the following Cl_2 molecule:

$$:\ddot{Cl}\!:\!\ddot{Cl}: \quad \text{or} \quad :\ddot{Cl}\text{—}\ddot{Cl}:$$

In the Cl_2 molecule each chlorine atom has six electrons plus two shared electrons, giving it a closed outer shell.

Covalent bonds can also be formed between unlike atoms. Consider the molecule HCl. The hydrogen and chlorine atoms have the dot structures

$$\dot{H} \qquad \cdot\ddot{Cl}:$$

If each shares one electron, they both have closed outer shells:

$$H\!:\!\ddot{Cl}: \quad \text{or} \quad H\text{—}\ddot{Cl}:$$

Sometimes, more than one electron is shared by each atom. Consider how carbon dioxide, CO_2, is formed. In the dot notation two oxygen atoms and one carbon atom look like this:

$$:\ddot{O}: \qquad \cdot\dot{C}\cdot \qquad :\ddot{O}:$$

In order for us to get a stable molecule, we need to have eight electrons around each atom. If only two electrons are shared between each oxygen atom and the carbon atom, we get the following structure:

$$:\ddot{O}\!:\!\ddot{C}\!:\!\ddot{O}: \qquad \text{(unstable)}$$

This structure is unstable because, although the carbon atom has eight electrons around it, each oxygen atom has only six electrons around it. For a stable molecular structure there must be four electrons shared between the carbon atom and each oxygen atom. The correct dot and dash structure of the CO_2 molecule is

$$:\ddot{O}\!:\!:\!C\!:\!:\!\ddot{O}: \quad \text{or} \quad :\ddot{O}\!=\!C\!=\!\ddot{O}:$$

An ordinary single covalent bond consists of two shared electrons. A sharing of four electrons produces a double bond, and a sharing of six electrons produces a triple bond. Nitrogen gas, N_2, is an example of a triple bond. Two nitrogen atoms may be represented as

$$\cdot\ddot{N}\cdot \qquad :\dot{N}\cdot$$

To satisfy the octet rule each nitrogen atom must share three electrons. The molecule N_2 looks like

$$:N\!:\!:\!:\!N: \quad \text{or} \quad :N\!\equiv\!N:$$

Another example of the triple bond is carbon monoxide, CO. Carbon and oxygen in dot notation look like

$$\cdot\dot{C}\cdot \qquad :\ddot{O}:$$

The carbon molecule needs four more electrons and the oxygen molecule needs two to satisfy the octet rule. The only way the octet rule can be satisfied is for the oxygen atom to share four of its electrons with the carbon atom and for the carbon atom to share two of its electrons with the oxygen atom. Altogether, six electrons are shared:

$$:C\!:\!:\!:\!O: \quad \text{or} \quad :C\!\equiv\!O:$$

One of the most common molecules containing three atoms is water, H_2O. The individual atoms are, of course,

$$H\cdot \qquad :\ddot{O}: \qquad \cdot H$$

When they combine, each hydrogen atom shares its electron with the oxygen atom, which shares two of its electrons:

$$H\!:\!\ddot{O}\!:\!H \quad \text{or} \quad H\text{—}\ddot{O}\text{—}H$$

The electron dot notation is used merely to indicate how the electrons are shared among atoms, not to indicate the spatial arrangement of the atoms. We could also have written H_2O as

$$H:\ddot{O}: \quad \text{or} \quad :\ddot{O}-H \quad \text{or} \quad \overset{\cdot\cdot}{\underset{H}{\ddot{O}}}\diagdown_H$$

The actual molecule looks more like these last three. The two O:H bonds form an angle of about 105° with each other.

The compound ammonia, NH_3, is a covalent combination of one nitrogen atom and three hydrogen atoms. Each hydrogen atom shares its electron with the nitrogen atom as follows:

$$H:\ddot{N}:H \quad \text{or} \quad :N-H$$

Methane gas, CH_4, is an example of a simple molecule with five atoms. Its structure can be represented by

$$H:\ddot{C}:H \quad \text{or} \quad H-C-H$$

More complicated electron dot structures arise when elements other than hydrogen are involved. The halogens combine analogously to hydrogen. Consider CH_3Cl. Its structure is just like CH_4, with a chlorine atom replacing a hydrogen atom:

$$H:\ddot{C}:\ddot{Cl}: \quad \text{or} \quad H-C-Cl$$

Other compounds with halogens are NH_2Cl, NI_3, $CHBrCl_2$. Their dot structures are

$$H:\ddot{N}:H \qquad :\ddot{I}:\ddot{N}:\ddot{I}: \qquad :\ddot{Cl}:\ddot{C}:H$$

Note that in each of the preceding structures the total number of electrons for the molecule is equal to the total number of valence electrons in the atoms making

up the molecule. Also note that each atom has a closed shell of electrons around it.

Sometimes, a molecule must have single, double, and even triple bonds simultaneously in order to satisfy the octet rule. The molecules C_2H_3Br and C_2HI are examples:

$$H:\ddot{C}::\ddot{C}:\ddot{Br}: \quad \text{or} \quad \overset{H}{\diagdown}C=C\diagup^{\ddot{Br}:}$$

$$H:C:::C:\ddot{I}: \quad \text{or} \quad H-C\equiv C-\ddot{I}:$$

In these structures the carbon atoms are bonded with double and triple bonds, while the other elements are held to the carbon atoms by single bonds.

Frequently, atoms combine with a net charge. Such combinations are called *complex ions*, or polyatomic ions. They are also known as radicals. The atoms within the complex ion are covalently bonded. The whole aggregation then behaves like an ion in forming compounds. There is usually a strong bond between atoms in complex ions, so it is difficult to break them up. In chemical reactions they frequently act as a single unit. Some important examples of complex ions are given in Table 12.2.

The structure of the polyatomic ions of known charge can be determined by writing down the free atoms in their dot notation, adding or subtracting electrons depending on the charge on the ion, and then following the rules of compound formation. The structures of NO_3^- and PO_4^{3-} are two common and important examples:

$$\left[:O::N:\ddot{O}:\right]^- \qquad \left[:\ddot{O}:\ddot{P}:\ddot{O}:\right]^{3-}$$

Table 12.2 Some Important Complex Ions

Ammonium	NH_4^+	Carbonate	CO_3^{2-}
Hydroxide	OH^-	Sulfate	SO_4^{2-}
Nitrate	NO_3^-	Sulfite	SO_3^{2-}
Cyanide	CN^-	Phosphate	PO_4^{3-}
Nitrite	NO_2^-		

Some compounds contain both ionic and covalent bonds. Sodium hydroxide, NaOH, is an example. Its structure is

$$Na^+[:\overset{\cdot\cdot}{\underset{\cdot\cdot}{O}}:H]^-$$

The hydroxide ion, $[OH]^-$, is covalently bonded together as a unit. In sodium hydroxide the hydroxide unit and sodium are bound by an ionic bond. Like most ionic compounds, sodium hydroxide dissolves in water. Other examples of ionic compounds formed by complex ions are silver nitrate, $Ag^+[NO_3]^-$, and barium sulfate, $Ba^{2+}[SO_4]^{2-}$

So far, we have classified compounds as being either ionic or covalent. In ionic bonding the electrons are transferred from one element or group of elements to another. In covalent bonding the electrons are shared. Actually, most bonds are intermediate between these two extremes. For most bonds the electrons are unequally shared between two atoms, because the atoms are unlike. Whenever two unlike atoms share electrons, one will have a greater attraction for the shared electrons than the other.

The tendency to attract shared electrons is called **electronegativity,** the measure of the ability of an atom to attract electrons in the presence of another atom. Consider the polar molecule HCl. The chlorine atom needs an electron to fill its outer shell. The hydrogen atom also needs an electron to fill its outer shell; but if it loses its electron, it will have a closed shell. The chlorine atom has a higher electronegativity than the hydrogen atom. Electronegativity increases from left to right within a period. The result is that although the electrons are shared between the two atoms, they tend to spend more time at the chlorine end than at the hydrogen end of the molecule. We get a molecule that looks like

$$H^+:\overset{\cdot\cdot}{\underset{\cdot\cdot}{Cl}}:^-$$

The + and − indicate that the chlorine end of the molecule tends to be negatively charged and the hydrogen end tends to be positively charged. A bond like this one, in which the electrons are unequally shared, is called a **polar bond.** All bonds between unlike atoms are polar bonds. Bonds between like atoms have a perfect sharing and are called **nonpolar bonds.**

Sometimes, a molecule has several polar bonds arranged symmetrically. An example is methane, CH_4. Its

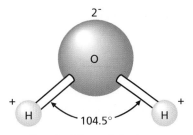

Figure 12.5 Ball-and-stick model of a water molecule, showing its polar character.

structure is

$$\begin{array}{c} H^+ \\ | \\ H^+ - \overset{\displaystyle C^{4-}}{\underset{\displaystyle |}{}} - H^+ \\ H^+ \end{array}$$

Methane has four polar bonds, but they are arranged so that no one side or end of the molecule has a more positive or negative charge than any other side or end. Because of this symmetry, it is a **nonpolar molecule.** Most molecules have one end or side positively charged and one end or side negatively charged. Such molecules are called **polar molecules.** The HCl molecule is an example of a polar molecule.

Another example of a polar molecule is water. Its structure is

$$\overset{\displaystyle \overset{\cdot\cdot}{\underset{\cdot\cdot}{O}}{}^{2-}}{H^+ \qquad H^+}$$

In the water molecule we see that the oxygen end is negatively charged, and the hydrogen ends are positively charged. To emphasize the polar property of the water molecule, we sometimes represent it as in Fig. 12.5. In ionic solutions the negative side of the water molecule attracts positive ions, while the positive ends attract negative ions.

12.5 Properties of Covalent Compounds

Because of the nature of the bonding involved, covalent compounds have quite different properties from those of ionic compounds. Although the covalent bond is strong within a molecule, the various molecules in the compound only weakly attract each other. Their binding energies in the solid and liquid forms are therefore not very high, and their melting points and boiling points are low compared with those for ionic compounds. For instance, carbon tetrachloride, CCl_4, melts at $-23°C$; the ionic compound NaCl melts at 800°C. Many covalent compounds occur as liquids or gases at room temperature.

Many molecules are bonded by polar bonds. The more polar the bond, the more unequal the electron sharing is. An ionic bond is actually a very polar bond. Polar molecules are partially covalent and partially ionic. In general, with the exception of the hydrogen atom, combinations of atoms from the far left and far right (ignoring the noble gases) of the periodic table form ionic compounds—for example, NaCl. Also, compounds formed from complex ions tend to be ionic. Compounds tend to be covalent if they are formed with elements from the same or adjacent columns of the periodic table, or if they are formed from any two nonmetallic atoms, such as hydrogen and nitrogen. In addition, all compounds formed with carbon are covalent.

Covalent compounds occur in the gas phase as individual molecules with a specific molecular formula. A molecule of gaseous CCl_4 consists of one carbon atom and four chlorine atoms. Ionic compounds have electric charges on the ions, and an ionic gas molecule might have any number of atoms in the ratio given by the molecular formula.

In liquid form covalent compounds do not conduct electricity well. Benzene, a nonpolar covalent compound, does not conduct electricity at all. Pure water, a polar covalent compound, conducts electricity only very slightly. Ordinary tap water or groundwater is far from pure. It usually has many ions dissolved in it, and therefore, tap water is an excellent conductor of electricity. Thus, if a lightning bolt were to hit a lake or swimming pool, the occupants would certainly be killed by the electric shock.

Water is a polar compound. When a solid ionic compound, such as table salt, is put into water, the ionic compound usually dissolves. What occurs when NaCl dissolves in water can be explained with the aid of Fig. 12.6. This figure shows that the positively charged ends of the water molecules are attracted to the Cl^- ions, and the negatively charged ends of the water molecules are attracted to the Na^+ ions. The water molecules surrounding the ions tend to decrease the attraction of the Na^+ and Cl^- ions for one another, which dissolves the salt crystal. Ions surrounded by water molecules are called *solvated ions.* The dissolving process is called *solvation.*

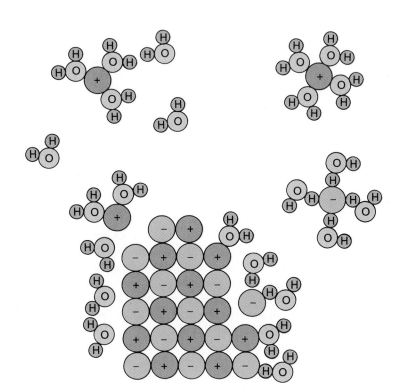

Figure 12.6 A two-dimensional drawing illustrating how sodium chloride (NaCl) is dissolved in water (HOH.)
The purple circles depict positive sodium ions, and the green circles depict negative chlorine ions. See the text for an explanation.

12.6 Oxidation Number

A useful concept in writing chemical formulas and equations is that of the oxidation number, or oxidation state. The **oxidation number,** or **oxidation state,** is essentially a measure of the electric charge that an atom in a molecule would have if it were ionically bound. When we assign oxidation numbers, the atom with the greater electron affinity is given a negative oxidation number. The atom with the lesser electron affinity is given a positive oxidation number.

Here are some general rules for assigning oxidation numbers:

1. The oxidation state of any free or uncombined element is zero.

2. The sum of the oxidation numbers of all atoms in the formula for a compound is zero.

3. The sum of the oxidation numbers of all atoms in an ion is equal to the charge on the ion.

4. Elements in group 1A always have the oxidation number $+1$, except hydrogen, which is -1 for hydrides of active metals.

5. Elements in group 2A have an oxidation number $+2$ in their compounds.

6. The maximum positive oxidation number for any group of representative elements is equal to the number of the group.

7. The most negative oxidation number of a group of the representative elements is equal to 8 minus the group number.

8. Metallic elements usually show only positive oxidation numbers.

9. The most important oxidation number for oxygen is -2.

10. The oxidation number for the halogens is usually -1.

Using these rules, we can determine the correct formula for a stable compound. Also, the oxidation numbers of other elements in a neutral molecule or charged ion can be determined. The sum of the oxidation numbers of all the atoms involved in a molecule or ion must be equal to the charge on the molecule or ion. In the most basic sense, the concept of oxidation number is just a means of keeping track of the various charges involved in molecules and ions.

Some examples follow.

EXAMPLE 1

Is BaCl the correct formula for barium chloride?

Solution Barium (Ba) is in Group 2A and has an oxidation number of $+2$. Chlorine (Cl) must take on an oxidation number of -2 for the compound to have a total oxidation number of zero. Chlorine is in Group 7A and can only have one negative value, which is -1. Therefore, BaCl is not a stable compound and does not exist. For a total oxidation number of zero for the compound two chlorine atoms must be used in the formula. Thus the formula is $BaCl_2$.

EXAMPLE 2

Does the compound Fe_2O_3 exist?

Solution If the sum of the oxidation numbers add to zero, then the answer is yes. Oxygen (O) has an oxidation number of -2. The total oxidation number for the three oxygen atoms is -2×3, or -6. Iron (Fe) has a positive, variable oxidation number and can take on a value of $+3$; and $+3 \times 2 = +6$. Finally, $+6 + (-6) = 0$. Therefore, the compound does exist.

EXAMPLE 3

What is the oxidation number of Cl in HCl?

Solution Because H has an oxidation number of $+1$, Cl has an oxidation number of -1.

EXAMPLE 4

What is the oxidation number of iron in Fe_2O_3?

Solution Oxygen has an oxidation number of -2. The total negative charge on the molecule is $3 \times (-2) = -6$. Because the molecule is neutral, there must be six positive charges. There are two iron atoms, so each must have $+3$. For the total charge on the molecule we get $2 \times (+3) + 3 \times (-2) = 0$.

EXAMPLE 5

What is the oxidation number of phosphorus in the $(PO_4)^{3-}$ ion?

Solution The oxidation number of oxygen is -2. There are four oxygen atoms, so there are $4 \times (-2) = -8$ negative charges. The total charge is -3, so phosphorus must have an oxidation number of $+5$. We get $+5 + [4 \times (-2)] = -3$.

EXAMPLE 6

What is the correct formula for calcium phosphate?

Solution The phosphate ion, $(PO_4)^{3-}$, has an oxidation number of -3; the calcium ion has an oxidation number of $+2$. The total charge on the molecule must be zero. To have an equal number of positive and negative charges, we need three calcium ions and two phosphate ions. Then we have $3 \times (+2) + 2 \times (-3) = 0$. The formula is then $Ca_3(PO_4)_2$.

Many elements have several possible oxidation states. In $FeCl_3$ iron has an oxidation state of $+3$; in $FeCl_2$, $+2$. Tin has an oxidation number of $+2$ and $+4$ in $SnCl_2$ and $SnCl_4$, respectively. Chlorine usually has an oxidation number of -1, as in the preceding examples. However, it has an oxidation number of $+3$ in $NaClO_2$, $+5$ in $NaClO_3$, and $+7$ in $NaClO_4$.

12.7 Naming Compounds

The naming of simple compounds in straightforward. Compounds with only two elements in them are called *binary compounds.*

Binary compounds with one oxidation state are named by giving the name of the more metallic element, followed by the name of the more nonmetallic element but with the suffix *-ide.* Examples are as follows:

NaCl	Sodium chloride
NaH	Sodium hydride
CaI_2	Calcium iodide
HCl	Hydrogen chloride
KBr	Potassium bromide
Ba_3P_2	Barium phosphide

If there is a complex ion involved, it is given its usual name without a new ending. Some examples follow.

NH_4Cl	Ammonium chloride
$(NH_4)_2SO_4$	Ammonium sulfate
H_3PO_4	Hydrogen phosphate
$CaCO_3$	Calcium carbonate

When nonmetallic elements exist in more than one oxidation state in a binary compound with another element, the compounds are distinguished by using the prefixes *mono-* ("one"), *di-* ("two"), *tri-* ("three"), *tetra-* ("four"), *penta-* ("five"), *hexa-* ("six"), *hepta-* ("seven"), and *octa-* ("eight"). The prefixes designate the number of atoms of the element that occur in the molecule. The prefix *mono-* is sometimes omitted.

CO	Carbon monoxide
CO_2	Carbon dioxide
NO_2	Nitrogen dioxide
N_2O_4	Dinitrogen tetraoxide
N_2O_5	Dinitrogen pentaoxide
SO_2	Sulfur dioxide
SO_3	Sulfur trioxide

If the metallic element has two possible oxidation numbers, they are sometimes distinguished by Roman numerals indicating the oxidation number.

$FeCl_2$	Iron(II) chloride
$FeCl_3$	Iron(III) chloride
Hg_2O	Mercury(I) oxide
HgO	Mercury(II) oxide

Under an older system of nomenclature the lower oxidation state has the ending *-ous,* and the higher oxidation state has the ending *-ic.* Under this older system the four compounds above have the names ferrous chloride, ferric chloride, mercurous oxide, and mercuric oxide.

12.8 Carbon and Some Simple Organic Compounds

One of the most important classes of compounds is the **organic compounds,** covalent compounds that contain the element carbon. Scientists originally believed that organic compounds (compounds derived from living matter) were exclusively of plant or animal origin— hence the name; but this belief has proved to be false, because many of the so-called organic compounds can be made in the laboratory from minerals (inorganic compounds). The study of carbon compounds is known as organic chemistry.

The carbon atom has four electrons in its outer shell. This electron structure gives it the ability to form long-chain and ring-shaped molecules.

There are over 6 million carbon compounds, and others are being added to the list constantly. Many of these carbon compounds are found in things we use daily, such as food, fuels, drugs, detergents, perfumes, and synthetic fibers. Carbon is the essential element of life, because it is the basic constituent of the complex molecules of proteins, fats, and carbohydrates, plus the nucleic acids. In this section we will examine the element carbon and some of its simpler compounds.

The element carbon exists in two crystalline struc-

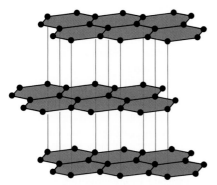

Figure 12.7 Crystalline structure of graphite.
Each carbon atom is bonded to three other carbon atoms in the same plane. Atoms in one plane are weakly bonded to atoms in adjoining planes.

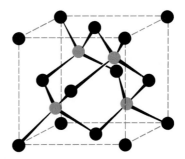

Figure 12.8 The crystalline structure of diamond.
Note that each carbon atom is bonded to four others, spatially arranged at the corners of a regular tetrahedron.

Figure 12.9 Two forms of carbon: graphite and diamond.

tures, graphite and diamond. A third form, called amorphous carbon, has the same crystalline structure as graphite. Coke and charcoal are examples of amorphous carbon. Graphite is soft, black, and slippery and is a good conductor of electricity. Diamond is very hard (the hardest substance known, with perhaps the exception of a substance called boron carbide), colorless, and transparent, and it will not conduct an electric current. Because both diamond and graphite are composed of carbon atoms, the difference between graphite and diamond must be in the bonding or crystalline structure.

The carbon atoms in graphite are arranged in hexagonal rings. Each carbon atom is bonded to three other carbon atoms that lie in the same plane, and thus each carbon atom forms a part of three hexagons (Fig. 12.7). The hexagons extend in all directions within the formed plane. The remaining electron of the carbon atom is relatively free and shifts from one bond to another in an endless fashion. Thus there is a free valence electron that gives graphite its ability to conduct an electric current. The attractive forces between planes are very weak and allow the planes to slide easily over one another, which accounts for the slippery feeling of graphite. Graphite is used in many lubricants because of this property, but it is probably best known as the black substance used in making "lead" pencils.

The crystalline structure of diamond is quite different from that of graphite. The four valence electrons of carbon atoms in diamond are arranged with those of four neighboring carbon atoms to form a tetrahedral structure, as shown in Fig. 12.8. The carbon atoms at the vertices are in turn bonded to four other carbon atoms. Thus the entire structure is one big crystal, which is extremely strong in all directions. Besides being very

hard, diamond has a high melting point. When heated to 1000°C in the absence of air, diamond will change to graphite. (See Fig. 12.9.)

Compounds containing only carbon and hydrogen are known as **hydrocarbons.** There are various groups or series of hydrocarbon compounds. For example, the **alkanes** are hydrocarbons that have a composition that satisfies the general formula

$$C_nH_{2n+2}$$

where n = the number of carbon atoms,
$2n + 2$ = the number of hydrogen atoms,
$n = 1, 2, 3, \ldots$

The formula says that the number of hydrogen atoms present in the compound is twice the number of carbon atoms plus two.

Methane (CH_4) is the first member of the alkane series. Ethane (C_2H_6) is the second member, propane (C_3H_8) the third, and butane (C_4H_{10}) the fourth. Table 12.3 lists some other members of the series. The prefix n in front of the name in Table 12.3 stands for the word

Table 12.3 Some Compounds of the Alkane Series
of Hydrocarbons

CH_4	Methane
C_2H_6	Ethane
C_3H_8	Propane
C_4H_{10}	n-Butane
C_5H_{12}	n-Pentane
C_6H_{14}	n-Hexane
C_7H_{16}	n-Heptane
C_8H_{18}	n-Octane
C_9H_{20}	n-Nonane
$C_{10}H_{22}$	n-Decane
$C_{11}H_{24}$	n-Undecane
$C_{12}H_{26}$	n-Dodecane
$C_{15}H_{32}$	n-Pentadecane
$C_{20}H_{42}$	n-Eicosane

normal. Normal hydrocarbons are those whose molecules are all straight chains, with no carbon-hydrogen groups branching off the chain.

The hydrocarbon's structure is easy to visualize when we write the **structural formula** (a graphic representation of the way the atoms are connected to one another) instead of its molecular formula. For example, the structural formula for methane (CH_4) is written

$$H-\overset{\displaystyle H}{\underset{\displaystyle H}{\overset{|}{\underset{|}{C}}}}-H$$

The structural formula is similar to the electron-dot notation. Each dash corresponds to two shared electrons and so represents a covalent bond. The structural formulas of ethane and propane are

Ethane   Propane

The alkanes make up many well-known petroleum products. Methane is the principal constituent in natural gas used for cooking stoves. Gasoline contains many of the alkanes from pentane to decane ($n = 5$ to $n = 10$). Kerosene contains the alkanes with $n = 10$ to 16. The alkanes with higher values of n make up other products such as diesel fuel ($n = 12$ to 20), fuel oil ($n = 14$ to 22), lubricating oil ($n = 20$ to 30), petroleum jelly

($n = 22$ to 40), and paraffin wax ($n = 25$ to 50). The alkanes are also used as starting materials for many other products such as paints, plastics, drugs, detergents, insecticides, and cosmetics.

The alkanes are highly combustible. They react with the oxygen in air to form heat, carbon dioxide, and steam. If the combustion is not complete, carbon in the form of black soot and carbon monoxide is formed. The alkanes are colorless. Recall that paraffin wax, kerosene, and "white" gasoline used in camp stoves all lack color. Most gasolines used in automobiles, however, are colored, because a little dye has been added to indicate that it contains tetraethyl lead, a deadly poison. The alkanes are nonpolar compounds. The alkane molecules thus are not attracted to the polar molecules of water, and so solvation does not occur to any appreciable extent.

When two hydrogen atoms are removed from a molecule of an alkane, two electrons of the carbon atoms become unpaired. For the octet rule to be satisfied, a double bond must be formed. Hydrocarbons with a double bond are called **alkenes.** For example,

Ethane   Ethene
(or ethylene)

The general formula for the alkene series is C_nH_{2n}. The series begins with C_2H_4 (ethene, or ethylene), shown by the structural formula above. Some of the normal alkenes are listed in Table 12.4. A number, instead of n (normal), preceding the name indicates the presence of

Table 12.4 Some Compounds of the Alkene Series
of Hydrocarbons

CH_2CH_2	Ethene (ethylene)
CH_3CHCH_2	Propene (propylene)
$C_2H_5CHCH_2$	1-Butene
$C_3H_7CHCH_2$	1-Pentene
$C_3H_7CHCHCH_3$	2-Hexene
$C_5H_{11}CHCH_2$	1-Heptene
$C_6H_{13}CHCH_2$	1-Octene
$C_7H_{15}CHCH_2$	1-Nonene
$C_{10}H_{20}$	1-Decene
$C_{15}H_{30}$	1-Pentadecene
$C_{20}H_{40}$	1-Eicosene

a double bond and its position. The carbons in the chain are numbered starting at the end—which end doesn't matter. For example, 1-pentene and 2-pentene have the following structural formulas:

$$H-\underset{\underset{H}{|}}{\overset{\overset{H}{|}}{C}}-\underset{\underset{H}{|}}{\overset{\overset{H}{|}}{C}}-\underset{\underset{H}{|}}{\overset{\overset{H}{|}}{C}}-\overset{\overset{H}{|}}{C}=\overset{\overset{H}{|}}{C}-H$$

1-Pentene

$$H-\underset{\underset{H}{|}}{\overset{\overset{H}{|}}{C}}-\underset{\underset{H}{|}}{\overset{\overset{H}{|}}{C}}-\overset{\overset{H}{|}}{C}=\overset{\overset{H}{|}}{C}-\underset{\underset{H}{|}}{\overset{\overset{H}{|}}{C}}-H$$

2-Pentene

The alkanes are all saturated hydrocarbons. A **saturated hydrocarbon** contains only single bonds, and its hydrogen content is at a maximum. The alkenes are known as **unsaturated hydrocarbons,** because they can take on hydrogen atoms to form saturated hydrocarbons or atoms of substances other than hydrogen to form derivatives.

When a second pair of hydrogen atoms is removed from the carbon atoms that have a double bond, a triple bond is formed. The general formula for this series, **alkyne,** is C_nH_{2n-2}, and the simplest member is ethyne (or acetylene), $HC{\equiv}CH$ (see Fig. 12.10). The members of this series are more reactive than those of the alkene series. Some members of the alkyne series are listed in Table 12.5.

The general formula C_nH_{2n}, which gives the structure of the alkenes, also gives the structure for a saturated hydrocarbon series called the *cycloalkanes.* The prefix *cyclo-* indicates the ring structure, and the ending *-ane* indicates the saturated condition of the chain. The carbon

Figure 12.10 An iron worker cutting steel with an oxyacetylene torch.

atoms of this series are arranged in a ring, with two hydrogen atoms connected to each carbon atom in the ring. These compounds are called the **cyclic hydrocarbons.** The first member of the series is cyclopropane, which has the structural formula

$$\underset{\underset{\displaystyle H}{|}}{\overset{\displaystyle\overset{H}{\diagdown}\ \overset{H}{\diagup}}{C}}$$

Cyclopropane

Various numbers of double bonds can also be introduced into cyclic structures. One of the most important cyclic hydrocarbons is benzene (C_6H_6). It has the structural formula

Benzene ring

Table 12.5 Some Compounds of the Alkyne Series of Hydrocarbons

$HC{\equiv}CH$	Ethyne (acetylene)
$CH_3C{\equiv}CH$	Propyne (methylacetylene)
$C_2H_5C{\equiv}CH$	1-Butyne
$C_3H_7C{\equiv}CH$	1-Pentyne
$C_3H_7C{\equiv}CCH_3$	2-Hexyne
$C_5H_{11}C{\equiv}CH$	1-Heptyne
$C_6H_{13}C{\equiv}CH$	1-Octyne
$C_7H_{15}C{\equiv}CH$	1-Nonyne
$C_{10}H_{18}$	1-Decyne

Because the molecule takes the structure of a ring, it is known as the **benzene ring.** The molecule has the ability to switch back and forth between the two structures shown. This property is called *resonance* and makes the benzene molecule more stable than the regular hydrocarbons. Sometimes, the benzene structure is written

Benzene ring

Benzene is obtained from coal tar, a by-product of soft coal. When other atoms or groups of atoms are substituted for the hydrogen atoms in the benzene ring, a vast number of different compounds can be produced. These compounds include such things as perfumes, explosives, drugs, solvents, insecticides, lacquers, and a host of others (Fig. 12.11).

So far, we have been naively drawing structural formulas as if molecules had only two dimensions. The seemingly obvious principle that molecules exist in three dimensions was not realized until 1874, some ten to twenty years after the concepts of ring and chain molecules were discovered. In 1874 Jacobus van't Hoff in Holland and Joseph Le Bel in France independently reached the conclusion that many unexplained chemical mysteries could be explained if molecules were viewed in three dimensions.

The molecule CH_4 is actually a tetrahedron with a carbon atom in the center. Its three-dimensional structure looks like this:

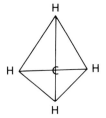

This tetrahedral arrangement is the basic geometric shape of organic molecules.

The way chains are built up in three dimensions is illustrated by normal butane (*n*-butane):

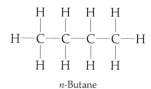

n-Butane

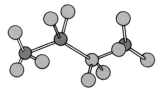

n-Butane

There are three hydrogen atoms on the two end carbon atoms and two hydrogen atoms attached to the middle two carbon atoms.

However, another arrangement of the atoms of the butane molecule is possible. It is an isomer of *n*-butane. **Isomers** are molecules with the same molecular formula but a different structure or arrangement of atoms. They can exist wherever the structural buildup does not violate the octet rule. Isobutane also has 4 carbon atoms and 10 hydrogen atoms, but its structural formula is

Figure 12.11 Chemists making perfume.

Isobutane's structural formula is different from n-butane's, because it has three hydrogen atoms attached to three of the carbon atoms, while the fourth carbon atom has only one hydrogen atom bonded to it. The different structures can be seen from the two-dimensional views of the molecules and from the three-dimensional, stick-and-ball models. Although they have the same molecular formula and the same composition, these compounds have different physical and chemical properties.

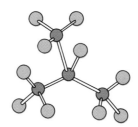

Isobutane

Isobutane

Learning Objectives

After reading and studying this chapter, you should be able to do the following without referring to the text:

1. State the two basic assumptions made to explain compound formation.

2. State the octet rule.

3. Describe ionic bonding, state the properties of ionic compounds, and give a few examples of ionic compounds.

4. Describe covalent bonding, state the properties of covalent compounds, and give a few examples of covalent compounds.

5. Distinguish between an oxidation and a reduction process, and give examples of each.

6. Distinguish between polar and nonpolar molecules, and give some examples of each.

7. Define *oxidation number*, and state the general rules for assigning oxidation numbers.

8. State the general rules for naming compounds.

9. Describe organic compounds, and state the general formula for some compounds known as hydrocarbons.

10. Define or explain the important words and terms listed in the next section.

Important Words and Terms

mixture	anion	oxidation number
compound	half-reaction	oxidation state
molecule	oxidation	organic compounds
octet rule	reduction	hydrocarbons
ionic compounds	oxidation-reduction reaction	alkanes
electrovalent compounds	electrolysis	structural formula
covalent compounds	covalent bond	alkenes
ion	ionic bond	saturated hydrocarbon
electron affinity	electronegativity	unsaturated hydrocarbon
crystal	polar bond	alkynes
anode	nonpolar bonds	cyclic hydrocarbons
cathode	nonpolar molecule	benzene ring
cation	polar molecules	isomers

Questions

Introduction

1. Distinguish between a compound and a mixture. Give an example of each.

2. Distinguish between (a) a compound and a molecule and (b) between a compound and an ion. Give an example of each.

Principles of Compound Formation

3. Which electrons of an atom take part in the formation of compounds?

4. What is the electron configuration for the most stable state of an atom or ion that takes part in the formation of a compound?

5. During compound formation, how is the number of valence electrons in an atom related to its tendency to lose or gain electrons?

6. Distinguish between ionic and covalent bonding.

7. State the octet rule.

Ionic Compounds

8. Define the term *ion,* and give an example.

9. Give an example of a compound not composed of molecules.

10. What is the meaning of electron affinity?

Properties of Ionic Compounds and Covalent Compounds

11. What force is responsible for molecular formation?

12. Distinguish between an anion and a cation.

13. Compare the physical properties of ionic compounds with those of covalent compounds.

14. A compound has a boiling point of $-10°C$. Is the compound ionic or covalent? Why?

15. A compound melts at a very high temperature of $1000°C$. Would you expect it to be ionic or covalent? Why?

16. Explain how an ionic liquid conducts an electric charge.

Covalent Compounds

17. Why is hydrogen a diatomic gas while helium is monatomic?

18. Where on the periodic table do you find elements with (a) high electronegativity and (b) low electronegativity?

Oxidation Number

19. Explain the meaning of oxidation number.

20. Compare the oxidation number with the number of electrons in the outermost shell of a neutral atom.

Carbon and Some Simple Organic Compounds

21. Why are single bonds sometimes formed, while at other times double or triple bonds are formed?

22. List some important alkanes and their uses.

23. Write the general formula for (a) the alkanes and (b) the alkynes.

24. What are isomers?

Exercises

Ionic and Covalent Compounds

1. Draw the electron dot notation for the following atoms and their ions:
   (a) Mg   (b) S   (c) Br   (d) K
   (e) Hg   (f) Ni   (g) Al
   Answer: (a) Mg: and Mg^{2+}

2. (a) Specify the following as either ionic or covalent substances:
   CO_2, NaCl, CaF_2, NH_4^+, $(NO_2)^-$, N_2H_3Cl
   (b) Draw the electron dot formulas for the above compounds and ions.

3. Given the following sets of atoms, state which will form ionic bonds and which will form covalent bonds. State the reason for your answer.
   (a) two atoms of oxygen
   (b) silicon and chlorine
   (c) magnesium and oxygen
   (d) sodium and bromine
   (e) hydrogen and phosphorus
   Answer: (b) covalent; silicon has four electrons in outer shell, which it prefers to share

4. Specify the following molecules as either polar or nonpolar:
   H_2O, CO, HCl, CH_4, O_2, NH_3 *Answer:* NH_3, polar

Oxidation Number

5. Find the oxidation numbers of all elements in the following formulas:
   (a) H_2SO_4
   (b) N_2
   (c) CO
   (d) CO_2
   (e) SO_3^{2-}
   (f) SO_4^{2-} *Answer:*
   (g) CrO_4^{2-}
   (h) $Cr_2O_7^{2-}$ (g) O, 2 −; Cr, 6 +

6. Determine whether stable compounds with the following formulas exist. Justify your answer.
   (a) FeO
   (b) KCl_2
   (c) Mg_2O_3
   (d) N_2O_5

Naming Compounds

7. Write formulas for the following compounds.
   (a) barium nitrate
   (b) aluminum sulfate
   (c) vanadium(II) oxide
   (d) vanadium(III) oxide
   (e) calcium phosphate
   (f) ammonium nitrite
   (g) calcium iodide
   (h) sodium sulfite
      Answer: (e) $Ca_3(PO_4)_2$

8. What are the names of the following compounds?
   (a) CuO
   (b) Cu_2O
   (c) Na_2SO_4
   (d) C_6H_6
   (e) C_8H_{18}

Carbon and Some Simple Organic Compounds

9. Draw structural formulas for the following organic compounds.
   (a) benzene
   (b) propane
   (c) propene
   (d) propyne

10. Draw structural formulas for three isomers of pentane.

11. Write the formula for each compound that you formed in Exercise 3.   *Answer:*   (b) $SiCl_4$

CHAPTER 13

Some Chemical Principles

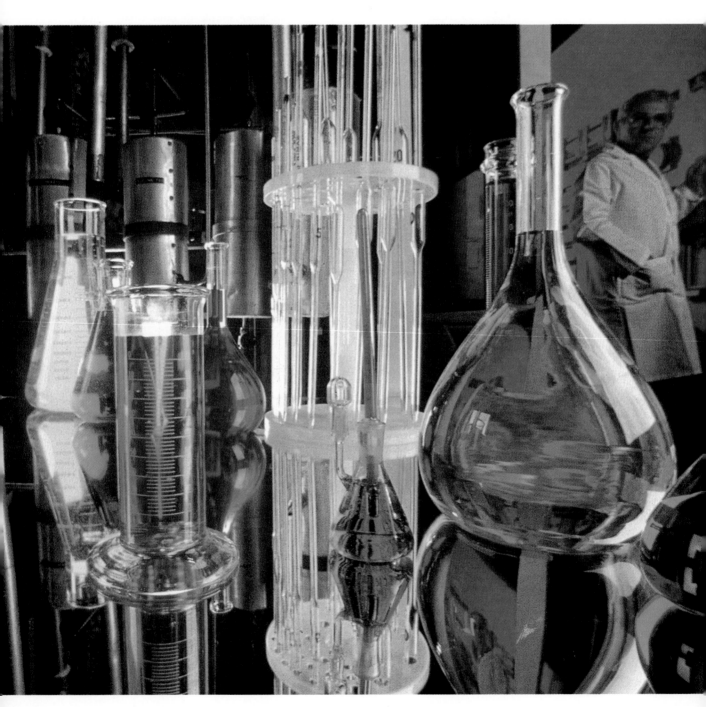

The chemists are a strange class of mortals, impelled by an almost insane impulse to seek their pleasure among smoke and vapor, soot and flame, poisons and poverty; yet among all these evils I seem to live so sweetly that I may die if I would change places with the Persian King.

—Johann Joachim Becher

I N THIS CHAPTER some of the basic principles of chemistry will be discussed. Most of them were discovered before 1850 by chemists such as Becher (1635–1682) who earnestly pursued their research in small labs that they financed themselves. Historically, these laws and principles were known before Mendeleev constructed his periodic table in 1869. It was long after 1869, of course, that the structure of atoms and molecules was understood. In fact, the principles we shall study in this chapter helped to establish our modern understanding of molecular structure. When these principles were first discovered, the reasons behind them were unknown. The explanation for them lies in our present understanding of atoms and molecules discussed in the previous three chapters.

Many of these very old laws concerning proportions and mixtures are important to us today. For instance, if the exhaust of a car is emitting black smoke, we know that an incorrect mixture of air and gasoline vapor is being burned in the engine. Such exhaust gases are one of the chief pollutants of our atmosphere. Mixtures are also important to all cooks. If a recipe is not followed closely, the wrong proportions of ingredients will yield an unpalatable dish. The chemists of Becher's day have left us a legacy that we would be wise to guard carefully.

13.1 Types of Matter

Matter is classified in several different ways. One of the ways, which was mentioned previously, is the classification of matter into four phases: solid, liquid, gas, and plasma. In the solid phase matter is rigid and has a definite shape. It expands or contracts comparatively little with pressure and temperature changes. In the liquid phase matter assumes the shape of the container and maintains a horizontal surface because of the effect of gravity. Like solids, liquids do not expand or contract much when pressure and temperature are varied. In the gas phase

◄ **A research-and-development chemical laboratory for chemical processes.**

matter fills the container completely and uniformly. Gravitational effects are small because the individual gas molecules have such a small mass. Gases expand and contract quite easily when pressure and temperature are varied. In fact, the gas laws tell us that the volume of a gas is inversely proportional to its pressure and directly proportional to its temperature.

Matter that is of uniform composition throughout is described as **homogeneous.** That is, any sample will be like any other sample. Matter that is of nonuniform composition is described as **heterogeneous.** As examples, a well-stirred cup of coffee with cream and sugar is a homogeneous mixture, while a pizza is a heterogeneous mixture.

A **pure substance** is a homogeneous sample of matter, all specimens of which have identical properties and composition. A **compound** is a substance composed of two or more elements chemically combined in a definite proportion. That is, a compound is composed of only one kind of molecule. All compounds are substances, but not all substances are compounds. Pure elements are substances, but they are not compounds. For example, pure hydrogen gas is a substance, but it is not a compound; steam is both a substance and a compound.

A sample of matter composed of two or more substances in varying amounts that are not chemically combined is called a **mixture.** Mixtures can be homogeneous or heterogeneous. A **solution** is a homogeneous mixture. From this definition a solution need not be a liquid; it may also be a solid or a gas. Salt water is a liquid solution. Metal alloys are examples of solid solutions. Air is an example of a gaseous solution. Fig. 13.1 summarizes these different types of matter.

The characteristics of a substance are known as its properties. There are two basic types of properties, physical and chemical. **Physical properties** are those that do not involve a change in the chemical composition of the substance. Among these properties are density, hardness, color, melting point, boiling point, electrical conductivity, thermal conductivity, and specific heat. These properties may change with a change in pressure

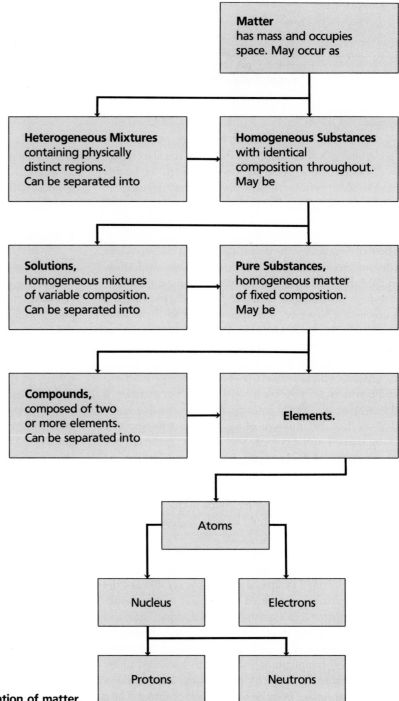

Figure 13.1
The classification of matter.

or temperature. Changes that do not alter the chemical composition of the substance are called **physical changes.** The processes of freezing or boiling water are good examples of physical changes brought about by changing the temperature. Other examples of physical changes are dissolving table salt in water, heating a piece of metal, and evaporating water.

The properties involved in the transformation of one substance into another are known as **chemical properties.** When wood burns, oxygen in the air unites with the different substances in the wood to form new substances. When iron corrodes, it combines with oxygen and water to form a new substance commonly known as rust. These examples illustrate chemical properties of wood and iron, respectively. Changes that result in the formation of new substances are known as **chemical changes.** The fermenting of wine, burning of gasoline, souring of milk, discharging of a battery, and exploding of gun powder are all examples of chemical changes. All chemical changes involve a production or absorption of energy.

13.2 Early Chemical Laws

We have seen in previous chapters that mass is a form of energy. Mass and energy are related by Einstein's formula, $E = mc^2$. Total energy, including mass energy, is always conserved in any process. In chemical reactions energy is produced or absorbed in the form of heat, kinetic energy, or potential energy. There is, then, a corresponding gain or loss in mass energy, but the total energy does not change. However, the change in mass in a chemical reaction is so infinitesimal that it cannot be detected. If the total mass involved in a chemical reaction is precisely weighed before and after the reaction takes place, no change can be detected. This principle is known as the **law of conservation of mass** and is stated as follows:

> **There is no detectable change in the total mass during a chemical process.**

This law was formulated in 1774 by Antoine Lavoisier. Lavoisier had developed balances that could measure $\frac{1}{100}$ the weight of a drop of water. He put tin in a vessel, sealed it, and carefully weighed it. The vessel was then heated so that the tin reacted with the air in the vessel to form a new compound. Lavoisier next reweighed the sealed container and found the weight to be identical with the original weight. Because the container had been sealed, no air could go in or out. When Lavoisier opened the vessel, air rushed in. This result showed that the tin had reacted with the air inside, or something in the air, to form the new compound. This experiment of Lavoisier's was one of the first great quantitative chemical experiments. Now, two centuries later, Lavoisier's law of conservation of mass is still almost exactly true.

The French chemist J. L. Proust (1755–1826) was one of the first to propose, in the late 1790s, that elements combine with one another in a definite ratio by weight. Proust's proposal is now known as the law of definite composition or the **law of definite proportions.** The law states:

> **Different samples of a pure compound always contain the same elements in the same proportion by weight.**

For instance,

$$9 \text{ g } H_2O = 8 \text{ g oxygen and } 1 \text{ g hydrogen}$$

$$36 \text{ g } H_2O = 32 \text{ g oxygen and } 4 \text{ g hydrogen}$$

In both cases the ratio by weight of oxygen to hydrogen is 8 to 1. This ratio results because each molecule of H_2O is composed of one atom of oxygen (atomic weight 16) and two atoms of hydrogen (atomic weight 1, total weight of hydrogen = 2×1 or 2). Thus each molecule of H_2O is composed of 16 parts oxygen by weight and 2 parts hydrogen by weight, or a ratio of 8 to 1. Because a pure covalent compound is simply a very large collection of identical molecules, the proportion by weight of any element in a covalent compound will be the proportion by weight that it occupies in an individual molecule of that covalent compound.

The fractional amount of carbon (C) and oxygen (O) in CO_2 can be found as follows:

$$\text{fractional amount of carbon in } CO_2 = \frac{\text{atomic weight of C}}{\text{molecular weight of } CO_2}$$

$$= \frac{12}{44} = \frac{3}{11}$$

$$\text{fractional amount of oxygen in } CO_2 = \frac{\text{molecular weight of } O_2}{\text{molecular weight of } CO_2}$$

$$= \frac{32}{44} = \frac{8}{11}$$

EXAMPLE 1

How many grams of carbon and oxygen are in 55 g of carbon dioxide (CO_2)?

Solution   Because $\frac{3}{11}$ of every sample of CO_2 must be carbon and $\frac{8}{11}$ of every sample of CO_2 must be oxygen, the amount of carbon is $\frac{3}{11} \times 55$ g = 15 g, and the amount of oxygen is $\frac{8}{11} \times 55$ g = 40 g. In both the 11-g and 55-g samples of CO_2 the ratio is 3 parts carbon to 8 parts oxygen.

For ionic compounds, a similar submicroscopic description is easily given. Because the ions occur in a definite proportion, the weight of the elements in the compound are in the same proportion by weight as they are in the smallest combination of ions that gives the formula of the compound.

EXAMPLE 2

If 25 g of calcium carbonate ($CaCO_3$) consists of 10 g calcium, 3 g carbon, and 12 g oxygen, how many grams of each element are contained in 100 g of $CaCO_3$?

Solution Using the law of definite proportions, we know that $\frac{10}{25}$ of every sample must be calcium, $\frac{3}{25}$ carbon and $\frac{12}{25}$ oxygen. Thus in 100 g of $CaCO_3$ we have

$$\text{amount of calcium} = \left(\frac{10}{25}\right) \times 100 \text{ g} = 40 \text{ g}$$

$$\text{amount of carbon} = \left(\frac{3}{25}\right) \times 100 \text{ g} = 12 \text{ g}$$

$$\text{amount of oxygen} = \left(\frac{12}{25}\right) \times 100 \text{ g} = 48 \text{ g}$$

Sometimes, the elements that are to be combined to form a compound are not in the correct proportions. In this situation one of the elements will be completely used and the one in excess will be partially used in the correct proportion, with some of it left over.

EXAMPLE 3

How many grams of CO_2 can be made from 36 g of carbon and 40 g of oxygen?

Solution The ratio of carbon to oxygen in CO_2 is 3:8. So 8 g of oxygen is needed for every 3 g of carbon. If we divide 8 g into the 40 g of oxygen available, we get 5. If we divide 3 g of carbon into the 36 g available, we get 12. Because 12 is greater than 5, the carbon is in excess. We have 5×8 g ($= 40$ g) of oxygen available. This amount will combine with 5×3 g ($= 15$ g) of the carbon to make 55 g of CO_2. Then 21 g of carbon is left over. In summary,

$$\text{amount of oxygen available} = 40 \text{ g}$$

$$\text{amount of carbon available} = 36 \text{ g}$$

amount of carbon needed
(to combine with 40 g of $= 15$ g
oxygen to produce CO_2)

amount of CO_2 produced $= 40 \text{ g} + 15 \text{ g}$
$= 55 \text{ g}$

amount of carbon left over $= 21 \text{ g}$

In this example it is clear that the law of conservation of mass has been obeyed. We began with a total of 76 g (40 g + 36 g) and we ended with a total of 76 g (55 g + 21 g).

Our modern science of chemistry is based upon postulates proposed and published by John Dalton (1766–1844) in 1808 (Fig. 13.2). Dalton proposed that (1) elements are composed of small indivisible particles called atoms that are identical for that element but are different for atoms of other elements, and (2) chemical combination is the bonding of a definite number of atoms of each of the combining elements to form one molecule of the formed compound.

Dalton's atomic theory has provided us with data concerning the properties of atoms and compounds, plus the mass, energy, and volume relationships of reactions between atoms. His theory also leads to the law of multiple proportions. This law describes a relationship that exists for different compounds that are formed from the same elements. Carbon monoxide, CO, and carbon dioxide, CO_2, are examples. The **law of multiple proportions** states:

> **Whenever two elements combine to form a series of compounds, there is a ratio of small whole numbers between the various amounts of one element and constant amount of another element that it can be combined with.**

Some examples will clarify the statement of this law. First, consider CO and CO_2. These two compounds are both composed of the same two elements, carbon and oxygen. The two amounts by weight of oxygen that combine with the same weight of carbon are in a ratio of small whole numbers and so conform to the law. In CO, 16 g of oxygen will combine with 12 g of carbon; in CO_2, 32 g of oxygen will combine with 12 g of carbon. The numbers 16 and 32 are in the ratio 1:2, which is a ratio of small whole numbers, as predicted by the law of multiple proportions.

We can see from our understanding of molecules why this law must be true. In a single molecule of CO there is one oxygen atom; in a single molecule of CO_2 there are two oxygen atoms. This ratio of 1:2 is then

13.3 Atomic and Molecular Weights

John Dalton was the first to study systematically the weight relationships among the various elements. These weight relationships are given in the periodic table as atomic weights. In the present-day table the weight relationships are based on an arbitrary scale that assigns the ^{12}C isotope the value 12.0000. All other naturally occurring elements are given a weight value relative to that of the ^{12}C isotope (see Table 11.1).

The atomic weights for most elements have been determined to several decimal places. For purposes of simplicity, however, we shall round these values off to the nearest integer or half integer. For example, we shall consider the atomic weights of hydrogen, carbon, oxygen, and chlorine to be approximately 1, 12, 16, and 35.5, respectively.

Because these are only relative weights, they do not have units. For atomic weights to have units, the concept of gram atomic weight is necessary. A **gram atomic weight** is defined as the weight of an atom of an element relative to ^{12}C, expressed in grams. The gram atomic weights of hydrogen and carbon are then 1 and 12 g, respectively. The number of gram atomic weights of a given sample of an element is found by dividing the mass of the sample by one gram atomic weight. As an example, in 20 g of carbon there are $\frac{20}{12} = 1.67$ gram atomic weights.

One gram atomic weight of any element contains the same number of atoms (see Fig. 13.3). This number is called one mole of atoms and has the value of 6.02×10^{23}. One **mole** of atoms is defined as the number of atoms in 12 g of carbon 12.

The **formula weight** of a compound is the sum of the atomic weights given in the formula of the compound. The formula weight of $CaCO_3$ is thus $40 + 12 + (3 \times 16) = 100$. The formula weight is a relative quantity like the atomic weight. The **gram formula weight** is the formula weight of the compound expressed in grams. The gram formula weights of $CaCO_3$ and H_2O are 100 and 18 g, respectively. A gram formula weight is also referred to as a mole. In 200 g of $CaCO_3$ there are 2 moles.

Figure 13.2 John Dalton (1766–1844).
Born the son of a poor weaver in the village of Eaglesfield, England, Dalton was a child prodigy who opened his own school at the age of 12 and became a professor at the University of Manchester in his mid-twenties. His greatest contribution to science was his atomic theory of matter. He took the word *atom* from the Greek word *atomos,* meaning "indivisible." Dalton remained a bachelor all his life, because he "never had time to marry."

reflected in the amounts of oxygen that will combine with equal amounts of carbon to form the two different compounds.

A good example of the multiple-proportion law is found in the compounds of nitrogen and oxygen. The amounts of oxygen that unite with 1 g of nitrogen are 0.571, 1.142, 2.284, and 2.885. If each of these numbers is divided by the smallest (0.571), the results are 1, 2, 3, 4, and 5. All the ratios of the amounts of oxygen that combine with 1 g of nitrogen are in small whole numbers. That is, ratios such as 1.738:1 or 1:1.23 do not occur. In the early 1800s discovery of the law of multiple proportions was an important step toward the molecular understanding of chemical compounds.

EXAMPLE 4

How many moles are there in (a) 9 g of H_2O and (b) 65 g of NaCl?

Figure 13.3 One mole each of three elements.
One mole of copper is 63.55 g. One mole of mercury is 200.6 g. One mole of lead is 207.2 g. Each sample contains Avogadro's number—6.02×10^{23} atoms.

Solution   (a) The gram formula weight of H_2O is 18 g, so in 9 g there are 9 g/18 g = 0.5 mole. (b) The gram formula weight of NaCl is 23 g + 35.5 g = 58.5 g. Thus in 65 g there are 65 g/58.5 g = 1.11 moles.

In covalent compounds the formula given represents a discrete molecule of the compound. Therefore, the terms **molecular weight** and **gram molecular weight** are often used for covalent compounds instead of formula weight and gram formula weight. However, in ionic compounds distinct molecules do not exist in the crystal or in solution. Thus the terms *molecular weight* and *gram molecular weight* do not apply to ionic compounds, and so *formula weight* and *gram formula weight* are the terms used.

The **percentage composition** of an element in a compound is the percentage by weight of each element in the compound. The percentage composition can be found by computing the total weight of the element in the compound, then dividing the total weight by the formula weight of the compound. The resulting answer must then be multiplied by 100 to get a percentage.

EXAMPLE 5_____

What is the percentage composition of the elements in table salt, NaCl? In sulfuric acid, H_2SO_4?

Solution   First. find the formula weight. The formula weight of NaCl is 23 + 35.5 = 58.5. The percentage

composition is then

$$Na = \frac{23}{58.5} \times 100 = 39.3\%$$

$$Cl = \frac{35.5}{58.5} \times 100 = 60.7\%$$

The formula weight of sulfuric acid is $(2 \times 1) + (1 \times 32) + (4 \times 16) = 2 + 32 + 64 = 98$. The percentage composition is then

$$H = \frac{2}{98} \times 100 = 2.04\%$$

$$S = \frac{32}{98} \times 100 = 32.65\%$$

$$O = \frac{64}{98} \times 100 = 65.31\%$$

13.4 Molecular Volumes

In 1808 the French chemist Joseph-Louis Gay-Lussac performed some experiments on gases that had a profound effect on our modern understanding of atoms and molecules. Gay-Lussac mixed different gases together to form new gases. As he did this, he took pains to measure the volumes of the initial and final gases. He found that when he mixed 2 volumes of hydrogen gas with 1 volume of oxygen gas, he obtained 2 volumes of steam. Similarly, when he mixed 1 volume of nitrogen gas with 3 volumes of hydrogen gas, he got 2 volumes of ammonia. In all the experiments he measured the gas volumes at the same temperature and pressure. From these and many other experiments he developed what is called **Gay-Lussac's law of combining volumes:**

> **When gases combine to form new gaseous compounds, all at the same temperature and pressure, the volumes of the initial and final gases are in the ratio of small whole numbers.**

The explanation of Gay-Lussac's law of combining volumes was first proposed by the Italian chemist Amedeo Avogadro in 1811. Avogadro explained Gay-Lussac's experiments by making two assumptions. His first assumption, which has since been confirmed many times, is now known as **Avogadro's law:**

> **Equal volumes of gases at the same temperature and pressure contain equal numbers of molecules.**

Avogadro's second assumption (which is also correct) was that sometimes the molecules of a gaseous element consist of two atoms. That is, hydrogen gas, nitrogen gas, oxygen gas, and many others all have two atoms in a molecule. They are diatomic gases. Of course, the noble gases are all monatomic. We have seen previously that molecules are formed in this way because each atom needs to have eight electrons to fill its outer shell.

Avogadro's two assumptions form the basis for our modern understanding of Gay-Lussac's experiments. Figure 13.5 shows schematically what takes place when 2 volumes of hydrogen and 1 volume of oxygen unite to form 2 volumes of steam. According to Avogadro's law, equal numbers of molecules are shown in each volume of gas. Two of the hydrogen atoms combine with one oxygen atom to form one steam molecule.

Figure 13.6 is an explanation of a similar situation in which nitrogen and hydrogen combine to form ammonia gas. Because experimentally we find that 3 volumes of hydrogen combine with 1 volume of nitrogen to form 2 volumes of ammonia gas, we can deduce

HIGHLIGHT

The Mole

The International System of Units (SI) defines the mole as the amount of substance that contains as many elementary entities as there are atoms in 0.012 kg of carbon-12.

To the chemist, the mole is a general term referring to 6.02×10^{23} chemical units. For example, a mole of atoms is 6.02×10^{23} atoms. A mole of molecules is 6.02×10^{23} molecules. A mole of ions is 6.02×10^{23} ions. And a mole of electrons is 6.02×10^{23} electrons.

To the physical science student, a dozen eggs is 12 eggs. Similarly, a mole of eggs is 6.02×10^{23} eggs. A gross of entities is 144 entities. Similarly, a mole of entities is 6.02×10^{23} entities.

The number 6.02×10^{23} is called Avogadro's number, in honor of Amedeo Avogadro (1776–1856), (Fig. 13.4) an Italian professor of physics who, in 1811, formulated what is known today as Avogadro's law. This law states that equal volumes of gases at the same temperature and pressure contain equal numbers of molecules. Avogadro did not know the number, but he knew it must be very large.

How can we comprehend an extremely large number such as Avogadro's number? For comparison, 6.02×10^{23} is more than 10 times greater than the estimated number of all the stars in the entire universe. Suppose we want to count the number of atoms in 1 mole (0.012 kg) of carbon-12. How long will it take the entire population (about 5 billion people) of the planet Earth to

Figure 13.4 Amedeo Avogadro (1776–1856).

count 6.02×10^{23} atoms? If each person counted, uninterrupted, one atom each second, it would take about 4 million years to count all the atoms.

How do we count the number of particles in 1 mole? The unit for measuring electric charge is the coulomb (C), and it takes 96,485 C (called one faraday) to liberate 1 mole of singly charged ions from compounds. For example, by electrolysis (chemical change accomplished by an electric current), one faraday will liberate 1 mole of chlorine from sodium chloride (NaCl). The electron is an elementary subatomic particle that has a negative charge of 1.602×10^{-19} C. The flow of electrons liberates the ions in the process of electrolysis. Thus we have a method to calculate Avogadro's number (N), as follows:

$$N = \frac{96,485 \text{ C/faraday}}{1.602 \times 10^{-19} \text{ C/electron}}$$
$$= 6.02 \times 10^{23} \text{ electrons per faraday}$$

this value is the number of electrons in a mole of electrons, or the number of atoms in a mole of atoms, or the number of molecules in a mole of molecules.

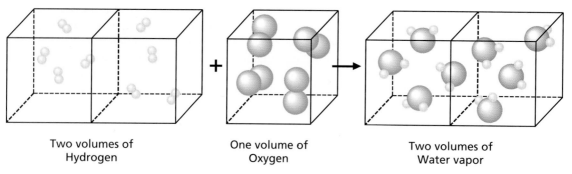

Two volumes of Hydrogen + One volume of Oxygen → Two volumes of Water vapor

Figure 13.5 An illustration of Avogadro's explanation of Gay-Lussac's experiment.
There are four molecules in each volume. This satisfies Avogadro's law.

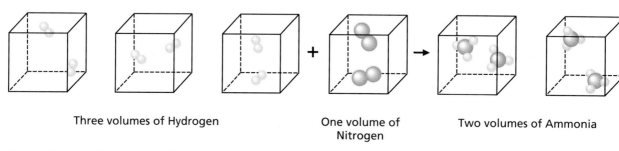

Three volumes of Hydrogen + One volume of Nitrogen → Two volumes of Ammonia

Figure 13.6 An illustration of Avogadro's explanation of the formation of ammonia.
There are two molecules in each volume. This satisfies Avogadro's law.

the molecular formula for ammonia. To deduce it, we use, Avogadro's law, along with his assumption that H_2 and N_2 are the correct molecular formulas for hydrogen and nitrogen gas. If we then put 2 molecules of each gas into each volume, we find that the molecular formula of ammonia can only be NH_3.

Equal numbers of gas molecules occupy equal volumes at the same temperature and pressure. However, the weights of different molecules are different, and equal volumes do not, of course, weigh the same. For instance, an oxygen molecule is about 16 times as heavy as a hydrogen molecule. Thus one volume of oxygen gas will weigh about 16 times the weight of an equal volume of hydrogen gas. The volume unit used in chemistry is a liter. A **liter** (L) is equal to 1000 cm³, or roughly equivalent in volume to a quart.

One liter of oxygen gas weighs 1.43 g at standard conditions of temperature and pressure, **STP.** Standard conditions are 0°C and sea level atmospheric pressure. Throughout the rest of the book, when a volume of gas is used, the temperature and pressure will be assumed to be at the standard conditions unless otherwise spec-

ified. One liter of hydrogen gas weighs about $\frac{1}{16}$ of the weight of a liter of oxygen, or about 0.09 g. The weights of other gases can be found by using the ratio of their molecular weights to that of oxygen.

EXAMPLE 6

What is the weight of 1 L of nitrogen gas?

Solution   The molecular weight of N_2 is 28. The molecular weight of O_2 is 32. The weight of a volume of any gas is proportional to its molecular weight.

$$\frac{\text{wt. of 1 L } (N_2)}{\text{mol. weight } N_2} = \frac{\text{wt. of 1 L } O_2}{\text{mol. weight } O_2}$$

$$\frac{\text{wt. of 1 L } (N_2)}{28} = \frac{1.43 \text{ g}}{32}$$

or

$$\text{wt. of 1 L } N_2 = \frac{28}{32} \times 1.43 \text{ g} = 1.25 \text{ g}$$

One mole of oxygen gas weighs 32 g. Because 1 L of oxygen gas weighs 1.43 g, the volume occupied by

32 g is given a simple ratio:

$$\frac{\text{vol. of 32 g}}{32 \text{ g}} = \frac{\text{vol. of 1.43 g}}{1.43 \text{ g}}$$

or

$$\frac{\text{vol. of 1 mole } O_2}{32 \text{ g}} = \frac{1 \text{ L}}{1.43 \text{ g}}$$

or

$$\text{vol. of 1 mole } O_2 = \frac{32 \text{ g}}{1.43 \text{ g}} \times 1 \text{ L}$$

$$= 22.4 \text{ L}$$

Similarly, we find that 1 mole of H_2 equals about $\frac{2}{0.09} \times 1 \text{ L} = 22.2 \text{ L}$. If we had used a more exact figure for the molecular weight of hydrogen (2.016 instead of 2), we would have obtained 22.4 L again. In fact, 1 mole of any gas occupies 22.4 L. The quantity 22.4 L is known as the **gram molecular volume.** It is the volume of 1 mole of any gas at standard conditions.

EXAMPLE 7

What volume will 27 g of steam occupy at STP?

Solution The molecular formula of steam is H_2O. The gram molecular weight is $1 \times 2 + 16 \times 1 = 18$ g. So there are $\frac{27}{18} = 1.5$ moles of steam in 27 g of steam. And 1.5 moles of any gas occupies $1.5 \times 22.4 \text{ L} = 33.6 \text{ L}$.

EXAMPLE 8

What is the weight of 1 L of nitrogen gas at STP?

Solution This problem has been worked in a previous example, but now we will do it in a simpler fashion. First, we find how many moles there are in 1 L of nitrogen gas (or any gas):

$$\text{number of moles} = \frac{1 \text{ L}}{22.4 \text{ L/mole}} = 0.0446 \text{ mole}$$

The weight of 0.0446 mole of nitrogen is $0.0446 \times$ gram molecular weight of nitrogen gas, or $0.0446 \text{ mole} \times 28$ g/mole $= 1.25$ g.

We see from the examples that it is important to find the number of moles of the gas. The volume can then be found by multiplying the number of moles by 22.4 L:

$$\text{volume of gas} = \text{no. of moles} \times 22.4 \text{ L}$$

If the weight of the gas is desired, it can be found by multiplying the number of moles by the gram mo-

lecular weight of the substance. This formula works for liquids and solids as well as gases.

$$\text{weight} = \text{number of moles}$$
$$\times \text{gram formula weight of substance}$$

13.5 Avogadro's Number

Avogadro's law states that equal volumes of gases at the same temperature and pressure contain equal numbers of molecules. The number of molecules in one mole of a gas or liquid or solid is called **Avogadro's number.** Avogadro's number can be found by several different methods to be 6.02×10^{23} molecules per mole. Thus in 22.4 L of any gas at standard conditions, there will be 6.02×10^{23} molecules of that gas. The number of molecules in any covalent compound can be easily found, once the number of moles of the covalent compound is known. The formula is

$$\text{no. of molecules} = \text{no. of moles}$$
$$\times \text{Avogadro's number}$$

EXAMPLE 9

How many molecules are there in 6 g of water?

Solution In 6 g of water there is $\frac{6}{18} = \frac{1}{3}$ mole. The number of molecules is then

$$\frac{1}{3} \times 6.02 \times 10^{23} = 2.0 \times 10^{23} \text{ molecules}$$

in 6 g of water.

Again, we see that if we can find the number of moles, we can find the quantity we are looking for, in this case the number of molecules.

In order to work problems concerning volumes of gases, weights of substances, or the number of molecules in a covalent compound, we first need to find the number of moles from information given in the problem. The number of moles can be found from any of three different equations. They are

$$\text{number of moles} = \frac{\text{weight in grams}}{\text{gram formula weight (or gram atomic weight if a substance is monatomic)}}$$

$$\text{number of moles} = \frac{\text{volume of gas in liters}}{22.4 \text{ L}}$$

$$\text{number of moles} = \frac{\text{number of molecules}}{6.02 \times 10^{23}}$$

The correct equation to use depends on the information given in the problem.

EXAMPLE 10

How many molecules are there in 11.2 L of nitrogen gas at STP?

Solution First, find the number of moles. Then, find the number of molecules.

$$\text{number of moles} = \frac{11.2 \text{ L}}{22.4 \text{ L/mole}} = 0.5 \text{ mole}$$

$$\begin{aligned}\text{number of molecules} &= 0.5 \text{ mole} \times 6.02 \\ &\quad \times 10^{23} \text{ molecules/mole} \\ &= 3.01 \times 10^{23} \text{ molecules}\end{aligned}$$

Avogadro's number also enables us to get an approximate idea of the size of molecules. For instance, water has a density of 1 g/cm^3. Using this information and Avogadro's number, we can find the approximate size of a water molecule. First, we find the number of molecules in 1 cm^3 of water. There is $\frac{1}{18}$ mole in 1 g of water and $(\frac{1}{18}) \times 6.02 \times 10^{23}$ molecules = 0.33×10^{23} molecules in 1 g, or 1 cm^3, of water. Next, we assume that the molecules have equal volumes in the liquid. The volume occupied by one molecule is then the total volume (1 cm^3) divided by the number of molecules in that volume:

$$\text{vol. of one water molecule} = \frac{1 \text{ cm}^3}{0.33 \times 10^{23}}$$

$$= 3 \times 10^{-23} \text{ cm}^3$$

If we then assume that each H_2O molecule is a sphere (which we know it is not; it looks more like a triangle), we can find the radius of that sphere from

$$\tfrac{4}{3}\pi r^3 = 3 \times 10^{-23} \text{ cm}^3$$

We then calculate the radius of the sphere:

$$r^3 = \frac{3}{4\pi} \times 3 \times 10^{-23} \text{ cm}^3$$

$$= 0.72 \times 10^{-23} \text{ cm}^3$$

$$= 7.2 \times 10^{-24} \text{ cm}^3$$

Take the cube root to find r:

$$r = \sqrt[3]{7.2} \times 10^{-8} \text{ cm}$$

The cube root of 8 is 2, so the cube root of 7.2 is about 2 and the radius of a water molecule is about 2×10^{-8} cm, or about 2 Å. This value seems reasonable, because we already know that the Bohr radius of the hydrogen atom is 0.529 Å. We would expect the water molecule to be a few times larger than a single hydrogen atom.

13.6 Solutions

Solutions are homogeneous mixtures. They can be solid, liquid, or gaseous solutions. Metal alloys and air have previously been cited as examples of solid and gaseous solutions. The most common solutions, however, are liquids. The substance in excess in a solution is called the **solvent.** The substance dissolved is the **solute.** Solutions in which water is the solvent are called **aqueous solutions.** A solution in which the solute is present in only a small amount is called a **dilute solution.** If the solute is present in large amounts, the solution is **concentrated solution.** When the maximum amount of solute possible is dissolved in the solvent, the solution is called a **saturated solution.**

The concentration, or amount of solute dissolved, is frequently expressed in terms of the molar concentration. The **molar concentration,** or **molarity,** is the number of moles of solute per liter of solution. Thus a one molar solution, written 1.0 M, has one gram formula weight of solute dissolved in one liter of solution. In general,

$$\text{Molarity} = \frac{\text{number of moles of solute}}{\text{number of liters of solution}}$$

Note that it is the *number of liters of solution,* not the number of liters of solvent, that is used. For example, a 1.0 M solution of cane sugar ($C_{12}H_{22}O_{11}$) is made by dissolving 1 mole (346 g) of cane sugar in water so that the final solution is 1.0 L (see Fig. 13.7). One liter of water is not added to the mole of solute; only enough water is added to make the volume of the solution 1.0 L.

The concentration of a solution can be any value up to the saturation limit. The molarity desired can be obtained in any number of ways, depending on the amount of solution desired. For example, a 2.0 M solution can be made by dissolving 2 moles of solute in water to make a liter of solution or, if we wish to make a smaller quantity, by dissolving 1 mole of solute in 0.5 L of solution. The concentration is the same either way.

EXAMPLE 11

What is the molarity of 40 g of salt dissolved in water to give 0.80 L of solution?

Figure 13.7 A 1.0 M sucrose solution.

Solution First, find the number of moles of salt. Then, find the molarity.

$$\text{number of moles} = \frac{40 \text{ g}}{58.5 \text{ g/mole}} = 0.68 \text{ mole}$$

$$\text{Molarity} = \frac{0.68 \text{ mole}}{0.80 \text{ L}} = 0.85 \text{ M}$$

EXAMPLE 12

How many grams of H_2SO_4 are present in 0.60 L of a 3.0M solution of sulfuric acid?

Solution First, find the number of moles; then, convert the answer to number of grams:

$$\text{Molarity} = \frac{\text{number of moles}}{\text{liter of solution}}$$

or

$$\text{number of moles} = \text{molarity} \times \text{liter of solution}$$
$$= 3.0 \text{ moles/L} \times 0.60 \text{ L}$$
$$= 1.8 \text{ moles}$$

Now, find the gram formula weight of H_2SO_4:

$$\text{grams } H_2SO_4 \text{ in 1 mole} = (2 \times 1) + (1 \times 32) + (4 \times 16)$$
$$= 98 \text{ g}$$

$$\text{grams in 1.8 moles} = 1.8 \text{ mole} \times 98 \text{ g/mole}$$
$$= 176 \text{ g}$$

Remembering that Avogadro's number (6.02×10^{23}) is the number of molecules in a mole of any substance, we can find the number of molecules or ions in any concentration of solution. For example, a single liter of a 2.0M concentration of table sugar, a covalent compound, contains $2 \times 6.02 \times 10^{23} = 12.04 \times 10^{23}$ sugar molecules. If we have an ionic compound such as NaCl, we can compute the number of ions in the solution. A 0.5-L quantity of a 3.0M NaCl solution contains $2 \times 3.0 \times 0.5 \times 6.02 \times 10^{23} = 18.06 \times 10^{23}$ ions. The factor 2 comes about because NaCl contains two ions, Na^+ and Cl^-.

When a solute is added to a solvent, it dissolves and, if thoroughly stirred, distributes itself uniformly throughout the solvent. The distribution of molecules or ions is the same at the top and bottom of the solution. As more solute is added, the solution becomes more and more concentrated; more and more solute is at the top and at the bottom. Finally, after a given quantity has been added, no more solute dissolves, and we say the solution is saturated.

The **solubility** of a given solute is the amount of solute that will dissolve in a specified volume of solvent (at a given temperature) to produce a saturated solution. The solubility depends on the temperature of the solution. If the temperature is raised, the solubility of solids almost always increases. For this reason hot water dissolves more of the solute than cold water.

If a gas such as CO_2 is being dissolved in water, the solubility increase is directly proportional to the pressure. This principle is used in the manufacture of soft drinks. The carbon dioxide is first forced into the beverage at high pressure. Then the beverage is bottled and tightly capped to maintain the pressure on the CO_2. Once the bottle is opened, the pressure inside the bottle is reduced to normal atmospheric pressure, and the CO_2 starts escaping from the liquid (see Fig. 13.8). If the bottle is allowed to remain open for some time, most of the CO_2 escapes, and the drink loses its zingy taste.

The solubility of gases in most liquids decreases with increasing temperature. This effect is also demonstrated with soft drinks. If the beverage is allowed to warm before opening the bottle, the solubility of the

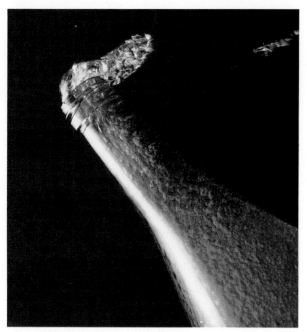

Figure 13.8
Dissolved carbon dioxide (CO_2) escapes from a carbonated beverage when pressure on the liquid surface is lowered.

CO_2 decreases, and when the bottle is opened, the CO_2 may escape so fast that the beverage shoots out of the bottle.

When saturated solutions of solids are prepared at high temperatures and then cooled, the solubility drops, and the excess solid crystallizes (forms crystals) and separates from the solution. However, if there are no crystals of the solid already present in the saturated solution, crystallization may not take place if the saturated solution is carefully cooled. The solution will then contain a larger amount of solute than the solubility of the solute dictates.

Solutions that contain more than the normal maximum amount of dissolved solute are said to be **supersaturated solutions.** Such solutions are unstable. The introduction of a "seed" crystal will cause the excess solute to crystallize. As discussed later in Chapter 21, this is the basic principle behind cloud seeding to cause rain. If the air is supersaturated with water vapor, the introduction of certain types of crystals into the clouds greatly increases the probability that the water vapor will form raindrops. Under certain conditions the carbon dioxide solution comprising a soft drink also may be supersaturated. Vigorously shaking the solution causes the excess gas to be liberated.

Whenever a solid is dissolved in water, the freezing point of the water solution is lowered and the boiling point is raised. These changes are related to the number of molecules or ions of solute present in the solution. Ordinary tap water boils at a slightly higher temperature and freezes at a slightly lower temperature than pure water, because of the presence of impurities. Similarly, the salt in ocean water causes the freezing point of salt water to be lower than fresh water.

A well-known application of this principle of lowering the freezing point is the use of antifreeze in car radiators. The antifreeze in use today is usually ethylene glycol. The more antifreeze that is present, the more the freezing point is lowered. See Fig. 13.9.

As we stated previously, water is a polar molecule. Water dissolves ionic compounds or polar covalent compounds because of its electrical attraction for such substances. However, when a nonpolar covalent compound such as carbon tetrachloride or gasoline is combined with water, it is not attracted by the water molecules. The water molecules attract only each other, and the

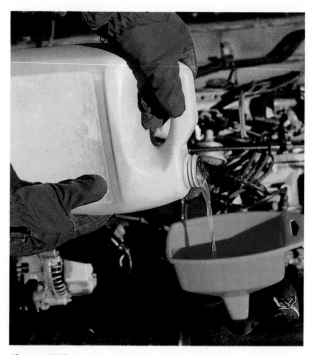

Figure 13.9
An antifreeze added to water in an automobile radiator produces a solution with a freezing point low enough to prevent freezing during cold weather.

nonpolar molecules are forced away from the water. For this reason nonpolar compounds will not dissolve in water.

Nonpolar compounds will dissolve each other, however, but they will not dissolve ionic or polar covalent compounds. As a general rule, "like dissolves like." If you wish to dissolve stains made by a nonpolar compound such as grease, use nonpolar solvents such as gasoline and benzene. For a polar compound, such as sugar, water should be used.

Learning Objectives

After reading and studying this chapter, you should be able to do the following without referring to the text:

1. Describe the various types of matter, and give examples of each.
2. Distinguish between physical and chemical properties of matter.
3. State and explain the following laws:
   (a) conservation of mass
   (b) definite proportions
   (c) multiple proportions
   (d) Gay-Lussac's law of combining volumes
   (e) Avogadro's law
4. Distinguish between gram atomic weight and gram formula weight.

5. Distinguish between gram molecular weight and molecular weight.
6. State the physical characteristics required for standard conditions.
7. State the numerical value of Avogadro's number.
8. Distinguish between solvent and solute.
9. Describe the principal characteristics of solutions.
10. Distinguish between molar and molar solution.
11. Define or explain the important words and terms listed in the next section.

Important Words and Terms

homogeneous	law of multiple proportions	gram molecular volume
heterogeneous	gram atomic weight	Avogadro's number
pure substance	mole	solvent
compound	formula weight	solute
mixture	gram formula weight	aqueous solutions
solution	molecular weight	dilute solution
physical properties	gram molecular weight	concentrated solution
physical changes	percentage composition	saturated solution
chemical properties	Gay-Lussac's law of combining volumes	molar concentration, molarity
chemical changes	Avogadro's law	solubility
law of conservation of mass	liter	supersaturated solutions
law of definite proportions	STP	

Questions

Types of Matter

1. Describe the physical properties that distinguish solids, liquids, and gases.
2. Identify each of the following as a physical or chemical change:
   (a) melting of ice
   (b) burning of a match
   (c) fermenting of wine
   (d) dissolving of sugar in water
   (e) rusting of steel
   (f) magnetizing of a sewing needle

3. Distinguish between homogeneous and heterogeneous matter.

4. Identify three pure substances, and list some of the properties used to identify them as pure substances.

5. Give some examples of solutions that are not in the liquid phase.

Early Chemical Laws

6. State the law of conservation of mass.

7. Is the law of conservation of mass exactly true or only approximately true?

8. What is the molecular explanation of the law of definite proportions?

9. What is the molecular explanation of the law of multiple proportions?

Atomic and Molecular Weights

10. Distinguish between gram atomic weight and gram molecular weight. Give an example of each.

11. Distinguish between gram atomic weight and gram formula weight.

12. State the difference between gram molecular weight and molecular weight.

13. How is 1 mole defined?

Molecular Volumes

14. State Gay-Lussac's law of combining volumes.

15. What were Avogadro's assumptions that explained Gay-Lussac's law of combining volumes?

16. How is it possible to mix one volume of a gas with three volumes of another gas and end up with only two volumes?

17. Why is the volume of a gas occupied by one mole always the same, no matter what the gas (assuming standard conditions)?

18. What is the meaning of the term *standard conditions*?

Avogadro's Number

19. State Avogadro's law.

20. Define Avogadro's number. State its numerical value.

Solutions

21. Describe the following terms as applied to solutions: solute, solvent, aqueous, dilute, concentrated, saturated, supersaturated.

22. Define a 1.0M solution. Give an example.

23. Why do soft drinks contain bubbles? What are the bubbles?

24. Why does an opened soft drink go flat after a period of time?

25. Why is antifreeze added to the water in a car's radiator during the winter?

26. On what type of compounds is carbon tetrachloride a useful cleaning agent? Explain your answer.

Exercises

Atomic and Molecular Weights

1. Calculate the formula weight of the following compounds.
   (a) water (H_2O)
   (b) carbon dioxide (CO_2)
   (c) methane (CH_4)
   (d) benzene (C_6H_6)

2. Calculate the formula weight of the following compounds.
   (a) sodium chloride (NaCl)
   (b) calcium carbonate ($CaCO_3$)
   (c) hydrochloric acid (HCl)
   (d) sodium bicarbonate ($NaHCO_3$)

3. Determine the percentage composition of the following compounds.
   (a) water (H_2O)
   (b) sugar ($C_{12}H_{22}O_{11}$)
   (c) table salt (NaCl)
   (d) aspirin ($CH_3COOC_6H_4COOH$)

4. Determine the percentage composition of the following compounds.
   (a) carbon monoxide (CO)
   (b) potassium chlorate ($KClO_3$)
   (c) manganese dioxide (MnO_2)
   (d) iron rust (Fe_2O_3)

5. Calculate the number of grams of sodium chloride (NaCl) that can be formed from 100 g of sodium and 105 g of chlorine. What element is not completely used? How much is not used?

6. Calculate the number of grams of carbon dioxide (CO_2) that can be formed from 36 g of carbon and 100 g of oxygen. What element is not completely used? How much is not used?

7. Calculate the number of grams of sulfur that can be produced from 128 g of sulfur dioxide (SO_2).

8. Calculate the number of grams of lime (CaO) and carbon dioxide (CO_2) that can be produced from 200 g of calcium carbonate ($CaCO_3$).

Molecular Volumes

9. Determine the weight of 1 L of argon gas.

10. Compute the volume at standard conditions that will be occupied by the following gases.
    (a) 40 g of argon
    (b) 150 g of xenon
    (c) 48 g of oxygen
    (d) 14 g of carbon dioxide   *Answer:* (c) 33.6 L

Avogadro's Number

11. Find the number of moles present in each of the following:
    (a) 56 g of carbon monoxide
    (b) 30 L of carbon monoxide gas at standard conditions
    (c) 3.01×10^{23} molecules of carbon monoxide

12. Determine the number of molecules contained in the following quantities:
    (a) 1 g of water
    (b) 44.8 L of carbon monoxide gas
    (c) 4.0 moles of sugar   *Answer:* (a) 0.33×10^{23}

Solutions

13. How many grams of H_3PO_4 are present in 0.5 L of a 2.0 M solution of phosphoric acid?

14. Find the molarity of the following solutions.
    (a) 2.5 moles of sugar in 1 L of solution
    (b) 49 g of H_2SO_4 in 2 L of solution
    (c) 12 g of HCl in 0.7 L of solution

          Answer: (c) 0.47 M

CHAPTER 14

Chemical Reactions

> **There's nothing constant in the universe,**
> **All ebb and flow, and every shape that's born**
> **Bears in its womb the seed of change.**
>
> —Ovid, *Metamorphoses*

CHEMICAL CHANGE IN nature takes place continuously. Green plants absorb carbon dioxide from the atmosphere and, with energy from the Sun and chlorophyll as a catalyst, react with water from the soil to form sugar and oxygen. This complex chemical reaction is called *photosynthesis.* The animal life on our planet ingests the sugar and inhales the oxygen to obtain energy. Chemical change in the body then returns water and carbon dioxide to the environment. This interaction between plants and animals and their environment constitutes an ecosystem.

Our environment is composed of atoms that combine with one another to form molecules, which in turn react with other atoms or molecules to produce the many substances we need and use. In the production of new products energy is released or absorbed, depending on the nature of the chemical reaction. There are countless chemical reactions, but only a few will be studied here.

Something of the nature of chemical reactions can be learned by studying examples of two specific types of reactions: oxidation-reduction and acid-base reactions. Everyone is acquainted with a match burning in air and the rusting of iron. These are examples of a chemical reaction known as an oxidation-reduction reaction. Most of us are also acquainted with a sour or acid stomach and the substances used to neutralize it. Neutralization is an example of an acid-base reaction.

We will see in Chapters 15, 20, and 24 that chemical reactions are important in the chemistry of large molecules, in atmospheric processes, and in geology. The present chapter discusses the basic chemistry involved in chemical reactions—in particular, oxidation-reduction and acid-base reactions.

14.1 Basic Concepts

When two or more substances are brought together, they may react chemically to form a new substance. For example, hydrogen and oxygen react to form water. The reaction can be written, in words, as follows:

$$\text{hydrogen} + \text{oxygen} \longrightarrow \text{water}$$

This word equation is read, "hydrogen plus oxygen reacts to give water." The arrow $\rightarrow$ indicates the direction of the reaction and has the meaning of "reacts to give" or "yields."

Instead of using words, we can write a symbol equation to indicate a chemical reaction. Consider a representative equation in which substances A and B react to form substances C and D. This equation is written as

$$\boxed{A} \ + \ \boxed{B} \ \longrightarrow \ \boxed{C} \ + \ \boxed{D}$$

$$\text{Reactants} \qquad\qquad\qquad \text{Products}$$

Such an expression is called a *chemical equation.* It is read, "Substances A and B react to yield the substances C and D." The initial substances A and B are called the **reactants,** and the final substances C and D are called the **products.**

In any chemical reaction three things take place:

1. The reactants disappear or are diminished.

2. Different substances appear as products. These products have chemical and physical properties that are significantly different from the original reactants.

3. Energy is either released or absorbed. This energy may take the form of heat, light, electricity, etc.

The energy characteristics of chemical reactions will be discussed more fully in Section 14.2.

Now let us suppose that we turn our original reaction around, so that when C and D are combined, A and B are formed as the products. We get

$$\boxed{C} \ + \ \boxed{D} \ \longrightarrow \ \boxed{A} \ + \ \boxed{B}$$

If we start with A and B, we get C and D; if we start with C and D, we get A and B. A reaction of this type is called a **reversible reaction.** The reversible reaction can be written as

$$\boxed{A} \ + \ \boxed{B} \ \rightleftharpoons \ \boxed{C} \ + \ \boxed{D}$$

◄ The rapid combustion of fuel with oxygen starts the Space Shuttle on the way to Earth orbit.

The double arrows mean that the reaction may proceed in either direction.

Equations written with formulas provide more information about the chemical reaction than word or symbol equations. For example, the following reversible reaction involves iron, steam, and hydrogen gas. The equation is

$$3 \text{ Fe} + 4 \text{ H}_2\text{O} \rightleftharpoons \text{Fe}_3\text{O}_4 + 4 \text{ H}_2$$

If steam is passed over hot iron, the reaction proceeds from left to right. If hydrogen gas is passed over heated Fe_3O_4, the reaction goes from right to left.

In a reversible reaction such as

$$A + B \rightleftharpoons C + D$$

all four substances (A, B, C, and D) are present in the vessel containing the substances. On a molecular scale, molecules of A and B are combining to form molecules of C and D; and at the same time molecules of C and D are combining to form molecules of A and B. When the two competing reactions are occurring at the same rate, we say that the reactions have reached a state of equilibrium. **Equilibrium** is a dynamic process in which there is a fixed number of molecules of all reactants and products present. However, individual molecules are continuously changing back and forth, because of interactions with each other.

The amounts of A, B, C, and D present at equilibrium depend on many things. Among them are temperature, pressure, and the excess amounts of reactants and products present. As an example, consider again the reaction

$$3 \text{ Fe} + 4 \text{ H}_2\text{O} \longrightarrow \text{Fe}_3\text{O}_4 + 4 \text{ H}_2$$

and the reverse reaction

$$\text{Fe}_3\text{O}_4 + 4 \text{ H}_2 \longrightarrow 3 \text{ Fe} + 4 \text{ H}_2\text{O}$$

In both of these reactions a stream of gas is passed over a hot solid. If steam is passed over hot iron, the hydrogen is swept out by the steam, and the first reaction dominates. If hydrogen gas is passed over Fe_3O_4, the steam is swept out by the excess hydrogen, and the second reaction occurs. This example demonstrates just one of many methods by which a reversible reaction may be shifted from right to left or left to right.

Every reaction is to some extent reversible. That is, there will always be some of every kind of atom or molecule present that can possibly occur in a reaction. For example, consider the reaction

$$2 \text{ HI} \longrightarrow \text{H}_2 + \text{I}_2$$

This reaction says that hydrogen iodide decomposes into hydrogen and iodine. However, in any sample of hydrogen and iodine there will be some hydrogen iodide, even though it may be a very small amount. To show explicitly that this is the case, we write the reaction with a small arrow going from right to left:

$$2 \text{ HI} \rightleftharpoons \text{H}_2 + \text{I}_2$$

Writing the reaction this way emphasizes the reversibility, however slight, of the reaction. In the discussions that follow, though, we will omit the small arrow from right to left. It will be understood that at equilibrium there are always some molecules of all the various reactants and products in the reaction.

Chemical equations are written for a chemical reaction, but until the equation is balanced, it cannot express a chemical equality. The equation is balanced by using the law of conservation of mass. That is, in a chemical reaction atoms cannot be created or destroyed. Thus there must be the same number of atoms on each side of the arrow in any equation for a chemical reaction. As an example, consider the "roasting" of zinc ore (ZnS). In this process sulfide ore is heated in air in order to convert the zinc sulfide to an oxide from which pure zinc can be extracted. The chemical equation for the roasting of the ore zinc sulfide is

$$2 \text{ ZnS} + 3 \text{ O}_2 \longrightarrow 2 \text{ ZnO} + 2 \text{ SO}_2$$

The numbers in front of the chemical formulas in this equation are called *coefficients.* They indicate the relative amounts of each substance that participate in the reaction. For instance, in the above reaction, for every 3 moles of oxygen gas reacting with every 2 moles of zinc sulfide, the result will be 2 moles of zinc oxide and 2 moles of sulfur dioxide. An inspection of the equation will show that it is indeed balanced. A tabulation of the number of atoms on both sides of the equation shows

$$\left.\begin{array}{l}\text{2 atoms of zinc} \\ \text{2 atoms of sulfur} \\ \text{6 atoms of oxygen}\end{array}\right\} \longrightarrow \left\{\begin{array}{l}\text{2 atoms of zinc} \\ \text{2 atoms of sulfur} \\ \text{6 atoms of oxygen}\end{array}\right.$$

$$\text{Reactants} \qquad\qquad \text{Products}$$

This equality of atoms occurs because atoms cannot be created or destroyed in chemical reactions.

When we balance chemical equations, the subscripts on the various elements represent the number of atoms in each molecule, and these numbers cannot be changed. The numerical coefficients to be found must be such that there are an equal number of atoms on both sides of the arrow. As an example, we consider balancing the equation for the decomposition of potassium chlorate ($KClO_3$) when heated. (This is a common method of producing oxygen in the laboratory.)

EXAMPLE 1

Balance the equation:

$$KClO_3 \longrightarrow KCl + O_2$$

Solution   To balance the equation, we inspect it to see which elements have a different number of atoms on the two sides of the equation. We see that the potassium and chlorine atoms are balanced already. Only the oxygen is not. Of course, once we adjust the equation to balance the oxygen atoms, we may unbalance the potassium and chlorine atoms.

To get the oxygen atoms equal on both sides, we must put a 2 in front of the $KClO_3$ (potassium chlorate) and a 3 in front of the O_2, giving

$$2 KClO_3 \longrightarrow KCl + 3 O_2$$

The oxygen is now balanced, but the potassium and chlorine are unbalanced. However, we can balance them by putting a 2 in front of the KCl. Finally, we have the balanced equation

$$2 KClO_3 \longrightarrow 2 KCl + 3 O_2$$

EXAMPLE 2

Balance the equation (decomposition of water into hydrogen and oxygen):

$$\text{water} \longrightarrow \text{hydrogen} + \text{oxygen}$$

or

$$H_2O \longrightarrow H_2 + O_2$$

Both hydrogen and oxygen exist as diatomic molecules in the free state.

Solution   The equation is balanced as written except for the oxygen atoms. To balance the oxygen, we place the coefficient 2 in front of the water molecule:

$$2 H_2O \longrightarrow H_2 + O_2$$

This step balances the oxygen; that is, the oxygen atoms are equal on both sides of the yield sign. But now the hydrogen is not balanced. To balance the hydrogen, we place the coefficient 2 in front of the hydrogen molecule. The equation, now balanced, reads

$$2 H_2O \longrightarrow 2 H_2 + O_2$$

This equation illustrates the diatomic nature of hydrogen and oxygen, because the correct formula for the water molecule is H_2O.

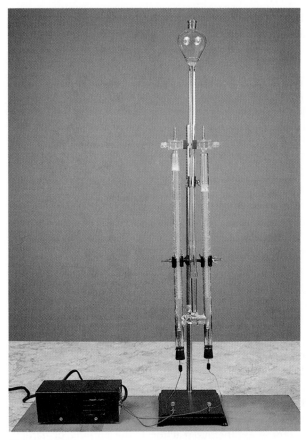

Figure 14.1
The apparatus used to decompose water (H_2O) into hydrogen and oxygen gas by means of an electric current.

The process of electrolysis, which is the separation of a compound into simpler substances by means of a direct electric current, was discussed in Section 12.3. Figure 14.1 shows an apparatus used to decompose water into hydrogen gas and oxygen gas by means of an electric current.

In addition to balancing the number of atoms on both sides of an equation, we must also balance the electric charge, because electric charge cannot be created or destroyed. Consider the equation for the ionization of sulfuric acid.

$$H_2SO_4 \longrightarrow 2 H^+ + SO_4{}^{2-}$$

In this equation the charges on each side of the arrow add up to zero. The number of atoms on both sides of this equation is also the same, as it must be.

Now let us balance a more difficult equation.

EXAMPLE 3

Balance the equation:

$$Pb + PbO_2 + H^+ + SO_4^{2-} \longrightarrow PbSO_4 + H_2O$$

Solution This reaction occurs in a lead storage battery when it is being used. To balance the equation, we first inspect it to see what is not balanced. The lead is not balanced, so we put a 2 in front of the $PbSO_4$ on the right. To balance the sulfur, we put a 2 in front of the SO_4^{2-} on the left. We then have

$$Pb + PbO_2 + H^+ + 2 SO_4^{2-} \longrightarrow 2 PbSO_4 + H_2O$$

Now we need a 4 in front of the H^+ to get a total charge of zero on the left (4 + charges and 4 − charges) and a 2 in front of the H_2O on the right to balance the equation. The equation is correctly written as

$$Pb + PbO_2 + 4 H^+ + 2 SO_4^{2-} \longrightarrow 2 PbSO_4 + 2 H_2O$$

14.2 Energy and Rate of Reaction

All chemical reactions involve a change in energy. The energy is either released or absorbed in the form of heat, light, electrical energy, sound, etc. If energy is released in a chemical reaction, the reaction is called an **exothermic reaction.** If the energy is absorbed, the reaction is called an **endothermic reaction.** An example of a common exothermic reaction is the burning of natural gas, which is composed primarily of methane:

$$CH_4 + 2 O_2 \longrightarrow CO_2 + 2 H_2O + energy$$

This reaction occurs when a gas stove is lighted (see Fig. 14.2). An example of an endothermic reaction is the

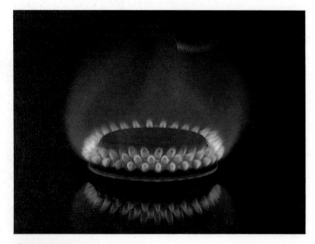

Figure 14.2
A gas flame is an exothermic reaction.

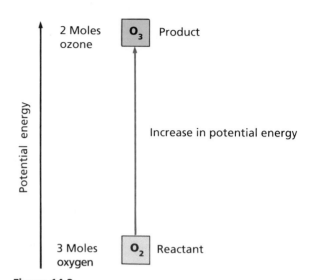

Figure 14.3
Energy is absorbed in the reaction, because the ozone has a higher potential energy than the oxygen. This is an endothermic reaction.

production of ozone, the triatomic molecule of oxygen:

$$3 O_2 + energy \longrightarrow 2 O_3$$

This reaction occurs in the upper atmosphere, where the energy is provided by the ultraviolet radiation from the Sun. It also occurs near electric discharges. Ozone has a pungent odor, which is readily detected even in very small amounts. It is the unusual odor one notices when near a toy electric train that is operating and its motor is producing sparks.

In the two preceding equations we have explicitly included the word *energy* to show the exothermic or endothermic nature of the reaction. This step is usually not done, however.

The energy associated with a chemical reaction is related to the bonding energies between the atoms that form the molecules. The process is complex, because during a chemical reaction some chemical bonds are broken while others are formed, and it is difficult to determine the energies associated with each. In general, energy is absorbed when bonds are broken and given off when they are formed. The ozone reaction is illustrated in Fig. 14.3. Energy must be supplied to the diatomic oxygen molecule in order to break the bonds. Thus energy is absorbed and the reaction is endothermic.

When methane burns in air, energy is released. The potential energy of the products is less than the potential energy of the reactants. This reaction is illustrated in Fig. 14.4.

A chemical reaction may occur when the reactants are brought together; that is, when two or more molecules or ions collide with one another, they may react chemically, but *only if the kinetic energy of the molecules is above a minimum value.* In other words, a certain amount of energy is needed to start a chemical reaction.

For example, we are all aware of the necessity of striking a match before it ignites. In this example we must contribute some energy—through friction—to initiate the chemical reaction of a match burning. Another example is the burning of methane gas. A flame is needed to ignite the methane, because the CH and O_2 bonds must be broken initially. Once the stove is lighted, the net energy released suffices to break the bonds of still more CH_4 and O_2 molecules, and the reaction proceeds continuously, giving off energy in the form of heat and light.

The energy necessary to get a chemical reaction started is called the **activation energy.** It is the kinetic energy colliding molecules must possess in order to react chemically. However, once the activation energy is supplied, an exothermic reaction can release more kinetic energy than was supplied. This concept is illustrated on a different scale in Fig. 14.5. In the figure a boulder is on top of a cliff. If enough energy is supplied to raise the boulder to the edge of the cliff, the boulder will roll down the cliff, releasing great amounts of energy. A

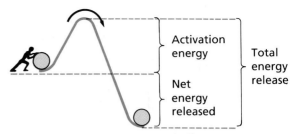

Figure 14.5 A graphic illustration of the energetics of an exothermic chemical reaction.
If enough energy is supplied to activate the reaction (lifting the boulder), the total energy released (the falling boulder) will equal the activation energy plus the net energy released.

similar situation exists for an exothermic reaction. Once the activation energy is supplied, the formation of new molecular bonds can release more kinetic energy than was absorbed in breaking the original bonds. In endothermic reactions more kinetic energy is absorbed to break molecular bonds than is generated when new bonds are formed. In this case there is a net loss of kinetic energy. This loss is usually perceived on a large scale as a "loss of heat."

When an exothermic chemical reaction takes place so very rapidly that there is a large change in the volume of the gases and an almost instant liberation of the energy of the reaction, an explosion occurs. Some of the energy in an explosion is given off in the form of sound energy. A reaction that proceeds more slowly than an explosion and yet is still quite rapid is termed *combustion.* In combustion, heat and light are produced in the form of fire. Common examples of combustion are the burning of natural gas, wood, paper, or coal in air. When gasoline in the form of heptane is burned, the combustion reaction is

$$C_7H_{16} + 11\ O_2 \longrightarrow 7\ CO_2 + 8\ H_2O$$

In the automobile engine there is insufficient oxygen to completely burn all of the carbon, and the reaction is more commonly

$$C_7H_{16} + 9\ O_2 \longrightarrow 4\ CO_2 + 2\ CO + C + 8\ H_2O$$

In this reaction the poisonous gas carbon monoxide, CO, is formed, along with the element carbon, which is black. The gases coming from an automobile tail pipe consist of significant amounts of poisonous carbon monoxide. For this reason running an automobile engine in a closed garage is very dangerous. The carbon monoxide

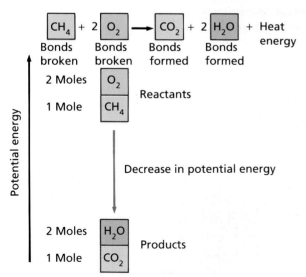

Figure 14.4
Energy is released in the reaction, because the products have less potential energy. This is an exothermic reaction.

accumulating from the exhaust can be fatal. The black color of some exhaust gases indicates the presence of large amounts of carbon, which shows that incomplete oxidation is occurring.

The rate of a reaction depends on several factors. One of the most important factors affecting the rate of reaction is the temperature, because the average kinetic energy of the molecules is directly proportional to the temperature. In many cases the rate of reaction is significantly altered by even a slight change in temperature. In fact, as a general rule of thumb, a 10°C increase in temperature leads to a doubling of the reaction rate. This rule is sometimes referred to as the **ten-degree rule.**

The boundary of an atom or molecule is defined by its outermost orbiting electrons. These outer electrons are repelled by the electrons of neighboring atoms or molecules, and thus the particles may not react unless they possess enough energy to overcome this tendency. The kinetic energy possessed by any one atom or molecule may or may not be great enough for reaction. Molecules and atoms that have low activation energies readily react because a greater number of the collisions are effective. If the temperature of the reactants is raised, the kinetic energy of the molecules will be increased, and the reaction rate will increase.

In some reactions the energy of activation of the molecules or atoms can be partially altered by adding a substance called a catalyst. A **catalyst** is a substance that speeds or slows the rate of reaction but is not itself consumed in the reaction. Catalysts have no effect on the equilibrium mixture of a reaction. They only lessen the amount of time required to reach equilibrium conditions.

Some catalysts act by providing a surface on which the reactants are adsorbed and concentrated. Others act as "carriers." That is, they unite with a reactant to form an intermediate substance that takes part in the chemical process and then decomposes to release the catalyst. The use of nitrogen oxide (NO) to produce SO_3 in the manufacture of sulfuric acid is an example of a catalyst that acts as a carrier. The reaction is

$$2\ SO_2 + O_2 \longrightarrow 2\ SO_3 \quad \text{(slow reaction)}$$

$$\left. \begin{array}{l} 2\ NO + O_2 \longrightarrow 2\ NO_2 \\ SO_2 + NO_2 \longrightarrow SO_3 + NO \end{array} \right\} \text{(fast reactions)}$$

The NO combines with oxygen to form NO_2. The NO_2 then reacts with SO_2 to form SO_3, releasing NO, which can be used again.

Another example of the use of a catalyst occurs in the decomposition of potassium chlorate ($KClO_3$). A common method for preparing oxygen in the laboratory is the decomposition of potassium chlorate by heating. When heated to 400°C, the white crystalline solid breaks down, producing potassium chloride and releasing oxygen. The reaction can be written as

$$2\ KClO_3 + \text{heat} \longrightarrow 2\ KCl + 3\ O_2$$

The process, however, is slow and heat requirements are high. If a small amount of manganese dioxide (MnO_2) is mixed with the $KClO_3$, the reaction takes place rapidly at 250°C. The manganese dioxide does not enter into the reaction but acts only as a catalyst.

The way in which many catalysts work is not known. A substance that acts as a catalyst for one reaction may or may not act as a catalyst for another. Catalysts are used extensively in manufacturing processes. They also play an important part in biochemical processes. The human body has many thousands of catalysts called **enzymes,** which act to control various physiological reactions. Scientists also believe that vitamins and hormones are catalysts.

14.3 Oxidation-Reduction Reactions

The element oxygen was discovered in 1774 by Joseph Priestley (1733–1804), an English clergyman, who referred to the element as "air." Priestley made the discovery by heating red rust of mercury [mercury(II) oxide] in a closed glass flask with a burning glass (lens). The reaction Priestley observed was

$$2\ HgO \longrightarrow 2\ Hg + O_2$$

In this reaction the oxide has decomposed, but it is much more common for oxygen to combine with another substance to form the oxide. We have seen that combustion is a reaction that frequently involves oxygen. When oxygen combines with another substance we call the reaction *oxidation.* For instance, the equation for the oxidation of carbon to form carbon dioxide is

$$C + O_2 \longrightarrow CO_2$$

Another example of oxidation is the souring of wines, that is, the changing of alcohol to acetic acid. The reaction can be written as

$$\underset{\text{Ethyl alcohol}}{CH_3CH_2OH} + O_2 \longrightarrow \underset{\text{Acetic acid}}{CH_3COOH} + H_2O$$

Acetic acid is the principal component of vinegar.

When oxygen is removed from a compound, the reaction is called *reduction.* For example, when the ore hematite, iron(III) oxide, is heated with coke (carbon) in a blast furnace, reduction occurs, the iron oxide being reduced to iron. This reaction is extremely important in the steel-making industry. It can be written as

$$2 \ Fe_2O_3 + 3 \ C \longrightarrow 4 \ Fe + 3 \ CO_2$$

We observe from the equation that oxygen is removed from the iron oxide. We say that the iron has been reduced. At the same time we see that oxygen has reacted with the carbon to form carbon dioxide; that is, the carbon has been oxidized. Thus the two reactions, oxidation and reduction, always occur together. A reaction in which both oxidation and reduction occur is called an **oxidation-reduction reaction.**

Oxidation-reduction reactions do not always involve oxygen. So that the term will apply to a larger number of reactions, oxidation-reduction reactions are also stated in terms of electrons gained or lost by atoms. By definition, the loss of electrons by an atom or ion is called **oxidation.** The gain of electrons by an atom or ion is called **reduction.**

When an atom or ion gives up electrons, it is **oxidized,** and when an atom or ion gains electrons, it is **reduced.** For example, when copper (Cu) combines with oxygen to form copper(II) oxide, the copper atoms lose electrons and are thus said to be oxidized. The oxygen atoms gain electrons and are reduced. Thus we have

$$2 \ Cu + O_2 \longrightarrow 2(Cu^{2+}O^{2-})$$

Because all the electrons lost by atoms or ions must be gained by other atoms or ions, it follows that oxidation and reduction occur at the same time and at the same rate. The atom that is oxidized is the **reducing agent;** the atom that is reduced is called the **oxidizing**

HIGHLIGHT

Glass

Glass is classified as a supercooled liquid that possesses the physical property of very high resistance to flow. Any material meeting this criteria may be considered to be a glassy solid rather than a crystalline or true solid. Thus glass is a solid in which the atoms, ions, or molecules do not take on ordered positions. This type of substance is commonly called an *amorphous* solid.

Glass is produced when silica, SiO_2, is heated above its melting point (about 1700°C) and cooled rapidly. The common varieties of glass (such as that used in windows and clear bottles) are made by heating mixtures of Na_2CO_3 and $CaCO_3$ with SiO_2. The reactions that occur are

$$Na_2CO_3 \quad + SiO_2 \longrightarrow$$
Sodium carbonate   Silica

$$Na_2SiO_3 \quad + \quad CO_2$$
Sodium silicate   Carbon dioxide

$$CaCO_3 \quad + SiO_2 \longrightarrow$$
Calcium carbonate   Silica

$$CaSio_3 \quad + \quad CO_2$$
Calcium silicate   Carbon dioxide

A mixture of Na_2SiO_3 and $CaSiO_3$ is the basic substance in ordinary glass. Pyrex glass is ordinary glass with boron oxide, B_2O_3, added to it. The addition of B_2O_3 gives the glass a low coefficient of expansion, which enables it to withstand changes in temperature without breaking. This glass is used for beakers, test tubes, and other items in the chemistry laboratory.

When potassium oxide, K_2O, is added, a very hard glass is produced, which can be abraded for eyeglasses and contact lenses. If some of the calcium is replaced by lead, a high-density and high-index-of-refraction glass is obtained. It is called flint glass, and it is used in making cut glass items.

If a colored glass is desired, an appropriate substance is added during the manufacturing process. For example, cobalt oxide is added for blue glass, manganese oxide for violet glass, gold or selenium for red glass (see Fig. 14.6), uranium oxide for yellow glass, and iridium for black glass.

Figure 14.6 Glass blowing—a bubble of molten glass.

agent. In the above example the metal (copper) reduced the oxygen; therefore, it was the reducing agent. Likewise, the nonmetal (oxygen) caused the oxidation to occur, so it was the oxidizing agent. Generally speaking, metals are reducing agents and nonmetals are oxidizing agents.

As we just mentioned, according to the electron concept, the element oxygen does not have to take part in an oxidation-reduction reaction. If we place a shiny iron nail in a copper sulfate solution, the iron nail becomes coated with a film of copper, and we find iron ions in the solution. This process is an oxidation-reduction reaction. The copper has replaced the iron on the nail. The iron atoms lose electrons and are oxidized into ferrous (2+) ions. The copper ions in solution gain electrons and are reduced to copper atoms. The reaction is shown in the following equation:

$$Fe + Cu^{2+}SO_4^{2-} \longrightarrow Fe^{2+}SO_4^{2-} + Cu$$

or
$$Fe + Cu^{2+} \longrightarrow Fe^{2+} + Cu$$

In the reaction of copper sulfate with iron,

$$CuSO_4 + Fe \longrightarrow Cu + Fe^{2+}SO_4^{2-}$$

the iron displaces the copper. Such reactions are called **displacement reactions.** Thus we see that some metals give up their valence electrons while others gain electrons. The activity of a metal—that is, its propensity to lose electrons—is due to its electron configuration. When metals compete for electrons, the more active metal yields to the less active. Thus iron atoms, being more active than copper atoms, lose electrons to the copper ions and become iron ions. The copper ions, on the other hand, gain electrons and are converted to neutral copper atoms.

The relative activity of any metal can be discovered by placing the metal in an ionic solution containing another metal and observing whether the metal replaces the one in solution. If it does, it is more active, because it has given electrons to the metal in solution. If this process is carried out for all metals, an order-of-activity series can be obtained. Such a series is called the **electromotive series** (Fig. 14.7).

There are thousands of important oxidation-reduction reactions. We will mention only a few more examples. In the commercial production of metals the metals must be separated from other elements present in the metals' ores. Frequently, the ores are sulfides or carbonates. The first step is to "roast" the ores. Roasting consists of heating the ores to very high temperatures

Element	Ion	Most Active
Lithium	Li → Li$^+$	
Rubidium	Rb → Rb$^+$	
Potassium	K → K$^+$	
Cesium	Cs → Cs$^+$	
Barium	Ba → Ba^{2+}	
Strontium	Sr → Sr^{2+}	
Calcium	Ca → Ca^{2+}	
Sodium	Na → Na$^+$	
Magnesium	Mg → Mg^{2+}	
Beryllium	Be → Be^{2+}	
Aluminum	Al → Al^{3+}	
Manganese	Mn → Mn^{2+}	
Zinc	Zn → Zn^{2+}	
Chromium	Cr → Cr^{3+}	
Gallium	Ga → Ga^{3+}	
Iron	Fe → Fe^{2+}	
Cadmium	Cd → Cd^{2+}	
Thallium	Tl → Tl$^+$	
Cobalt	Co → Co^{2+}	
Nickel	Ni → Ni^{2+}	
Tin	Sn → Sn^{2+}	
Lead	Pb → Pb^{2+}	
Hydrogen	H → H$^+$	
Copper	Cu → Cu^{2+}	
Mercury	Hg → Hg^{2+}	
Silver	Ag → Ag$^+$	
Gold	Au → Au$^+$	Least Active

Figure 14.7 The electromotive, or activity, series. The higher the element is in the series, the more active the element. For example, Fe is more active than Cu, and Zn is more active than Fe.

in the presence of air. Some typical reactions are

$$2\ ZnS + 3\ O_2 \longrightarrow 2\ ZnO + 2\ SO_2$$

$$PbCO_3 \longrightarrow PbO + CO_2$$

The roasting process converts the metal to its oxide. The next step is to reduce the oxide to the pure metal. Carbon and carbon monoxide are frequently used in this step. For example,

$$ZnO + C \longrightarrow Zn + CO$$

$$Fe_2O_3 + 3\ CO \longrightarrow 2\ Fe + 3\ CO_2$$

These reactions are extremely important in the commercial manufacturing of metals.

14.4 Electrochemical Reactions

As we have seen in previous chapters, the properties of elements and compounds can be explained in terms of their electron structures. Microscopically, chemistry is intimately intertwined with electric charges. In this section we concern ourselves with several important chemical reactions that relate to the transfer of electric charges.

Substances that dissolve in water to give solutions that will conduct an electric current are called **electrolytes.** In Section 12.3 we discussed how table salt, NaCl, in water conducts an electric current. All ionic compounds are electrolytes. Acids and bases are also electrolytes. They also ionize in water and conduct an electric current.

Also in Section 12.3 we discussed electrolysis and identified the cathode and the anode. In electrolysis the **cathode** is the electrode at which electrons enter the electrolytic solution. The **anode** is the electrode where electrons are withdrawn from the electrolytic solution. In electrochemistry the important reactions occur at the anode and the cathode.

Electrochemical cells are extremely important. The dry cell is used in flashlights, transistor radios, and other devices to provide an electrical potential. The common dry cell is shown schematically in Fig. 14.8. The dry cell is basically a zinc can with a carbon rod imbedded in a paste of ammonium chloride, NH_4Cl. A porous separator is inserted between the zinc can and the NH_4Cl. Some MnO_2 surrounds the carbon rod also. When the cell is connected, electrons flow externally from the zinc can to the carbon rod. The chemical reactions occurring at the cathode and the anode depend on the size of the current required from the cell. When small currents are drawn, the following reactions probably occur:

$$Zn + 4 NH_3 \longrightarrow Zn(NH_3)_4^{2+} + 2 e^- \quad \text{(anode)}$$

$$2 MnO_2 + 2 NH_4^+ + 2 e^- \longrightarrow$$
$$Mn_2O_3 + 2 NH_3 + H_2O \quad \text{(cathode)}$$

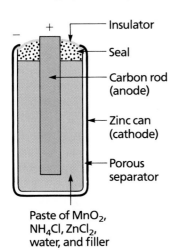

Figure 14.8 Cross-section of a dry cell, showing the important parts.

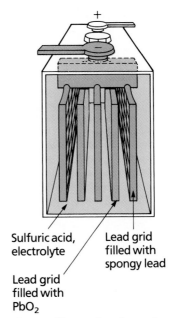

Figure 14.9 Diagram illustrating the major components of a lead storage battery.

The dry cell is called an irreversible cell, because the application of an external potential will not recharge it. An example of a reversible cell is the lead storage battery used in automobiles.

The lead storage battery consists of a set of lead grids in a dilute sulfuric acid solution (Fig. 14.9). One grid, or plate, contains spongy lead, the other lead dioxide, PbO_2. These plates are mounted in the solution of sulfuric acid, H_2SO_4, which acts as an electrolyte by ionizing in water to yield H^+ and SO_4^{2-} ions.

The lead grid acts as the anode, and the reaction during discharge is

$$Pb + SO_4^{2-} \longrightarrow PbSO_4 + 2 e^- \quad \text{(anode)}$$

The PbO_2 acts as the cathode. The reaction during discharge is

$$PbO_2 + 4H^+ + SO_4^{2-} + 2 e^- \longrightarrow$$
$$PbSO_4 + 2 H_2O \quad \text{(cathode)}$$

If an external electrical potential is applied to a discharged storage battery, it can be recharged. The reactions that occur during recharge are just the reverse of those that occur during discharge. During discharge the sulfuric acid is consumed. Conversely, it is regenerated by the charging process. Because the presence of H_2SO_4 indicates a charged state, the condition of a lead storage battery may be determined by measuring the

Figure 14.10 Sears Incredicell Battery.

Figure 14.11 Rusting—a destructive chemical reaction.

density of the sulfuric acid solution. The higher the density, the more the battery is charged. When the auto engine is running, the battery is being charged (provided the alternator is working properly). The battery is discharging when the car is being started. A set of lead plates, such as shown in Fig. 14.9, generates a potential of 2.05 V. Thus, a 12-V battery contains six sets of plates, while a 6-V battery contains three sets of plates.

The maintenance-free battery is similar to the lead storage battery in operation but requires no maintenance over its full life. Three major factors qualify a battery as maintenance-free: (1) Terminal corrosion has been reduced to a point where it is no longer a factor in limiting battery life; (2) no additional water is required during the life of the battery; and (3) the battery should have a long shelf life. These characteristics have been achieved with the Sears DieHard Incredicell battery shown in Fig. 14.10.

The single most destructive chemical reaction that occurs in the United States is the rusting, or corrosion, of iron (see Fig. 14.11). In fact, about one out of every four workers in the steel industry is working to replace iron products that have been ruined by rusting. The corrosion of iron is brought about by the presence of both oxygen and water. Iron will not rust in dry air. The explanation of corrosion involves electrochemistry, as we shall see.

When iron is in contact with water containing an electrolyte, the iron, being very active, tends to give up its electrons according to the reaction

$$Fe \longrightarrow Fe^{2+} + 2\,e^- \quad \text{(anode)}$$

This reaction can be considered a reaction at an anode. Usually, one part of the iron will be more active than other parts. At the less active part of the iron, H^+ ions

react to form hydrogen gas, which collects on the iron surface:

$$2\,H^+ + 2\,e^- \longrightarrow H_2 \quad \text{(cathode)}$$

The accumulation of hydrogen gas on the iron surface tends to inhibit the reaction. However, dissolved oxygen in the water reacts with the H_2 to form water:

$$2\,H_2 + O_2 \longrightarrow 2\,H_2O$$

The iron ions react with the OH^- ions present in the water and then with oxygen to form $Fe_2O_3 \cdot 2\,H_2O$, which is iron rust:

$$Fe^{2+} + 2\,OH^- \longrightarrow Fe(OH)_2$$

$$4\,Fe(OH)_2 + O_2 \longrightarrow 2(Fe_2O_3 \cdot 2\,H_2O)$$
Iron rust

The pollutant sulfur dioxide, SO_2, which is emitted during the burning of coal, will combine with water to form dilute H_2SO_3. This, then, supplies H^+ ions for the cathode reaction in the rusting process. Thus SO_2 pollution is partially responsible for the destruction of billions of dollars worth of structural materials each year.

Rusting can be prevented by coating the iron surface with a material such as paint or a ceramic enamel, like that used in sinks, stoves, or refrigerators. Some alloys of iron, such as stainless steel [a mixture of iron, chromium (15%) and nickel (8%)], are corrosion resistant.

14.5 Acids and Bases

The acid-base concept originated with the beginning of chemistry and is an important classification of chemical compounds. The word *acid* is derived from the Latin *acidus*, which means "sour." The sour taste is one of the properties of acids, but the student should never actually taste an acid in the laboratory. An **acid** is a substance that, when dissolved in water, has the following properties:

1. Conducts electricity.
2. Changes the color of litmus dye from blue to red.
3. Tastes sour. (The student should *never* taste any acid.)
4. Reacts with a base to neutralize its properties.
5. Reacts with metals (e.g., Zn, Mg) to liberate hydrogen gas.

A **base** is a substance that, when dissolved in water, has the following properties:

1. Conducts electricity.
2. Changes the color of litmus dye from red to blue.
3. Tastes bitter and feels slippery (as with acids, the student should *never* taste a base).
4. Reacts with an acid to neutralize the acid's properties.

One of the first theories formulated to explain the acid-base concept was put forth by Svante Arrhenius (1859–1927), a Swedish chemist who proposed in 1887 that the characteristic properties of aqueous solutions of acids were due to the hydrogen ion (H^+), and the characteristic properties of aqueous solutions of bases were due to the hydroxide ion (OH^-).

According to the **Arrhenius acid-base concept,** when a substance like hydrogen chloride (HCl) is added to water, some HCl molecules dissociate into H^+ ions and Cl^- ions, which exist in the solution in equilibrium with the HCl molecules that are not dissociated. The acidic properties of HCl are due to the H^+ ions. Similarly, when a substance such as sodium hydroxide (NaOH) is added to water, partial dissociation also occurs, and the Na^+ ions and OH^- ions exist in equilibrium with NaOH molecules that are not dissociated. The basic properties of NaOH are due to the OH^- ions.

However, a substance need not have H^+ or OH^- ions in order to have acid and base properties. For example, ammonia, NH_3, and sodium carbonate, Na_2CO_3, contain no OH^- ions, but in water solutions these compounds are bases. In water the reactions are

$$NH_3 + H_2O \longrightarrow NH_4^+ + OH^-$$
$$Na_2CO_3 + H_2O \longrightarrow HCO_3^- + OH^- + 2\,Na^+$$

The OH^- ions come from the water, so aqueous so-

Figure 14.12 A common base.

lutions of NH_3 and Na_2CO_3 have the same properties as other bases.

A base common to many households is sodium bicarbonate, $NaHCO_3$, known as baking soda. Its solution is weakly basic, the reaction being

$$NaHCO_3 + H_2O \longrightarrow H_2CO_3 + OH^- + Na^+$$

Acids act upon the carbonate ion of baking soda to give off carbon dioxide, a gas:

$$NaHCO_3 + H^+\text{(from an acid)} \longrightarrow$$
$$Na^+ + H_2O + CO_2$$

This reaction is involved in the leavening process in baking. Baking powders contain baking soda plus an acidic substance such as $KHC_4H_4O_6$ (cream of tartar). When this combination is dry, no reaction occurs; but when water is added, CO_2 is given off. Some common bases are shown in Fig. 14.12.

The Arrhenius theory, as we noted, is restricted to aqueous solutions containing H^+ or OH^- ions. Another theory of acids and bases was conceived independently in 1923 by J. N. Brönsted (1879–1947), a Danish physical chemist, and T. M. Lowry (1874–1936), an English chemist, and developed by others since then.

The Brönsted-Lowry theory defines an acid as a substance that acts as a proton donor and a base as a substance that acts as a proton acceptor.

According to this theory, reactions involving the transfer of a proton are known as **acid-base reactions.**

Because a hydrogen ion (H^+) is in reality a proton, a proton donor must be a substance that has hydrogen atoms. For example, when hydrogen chloride (HCl) is

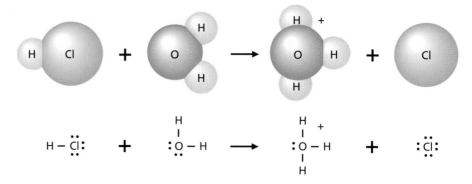

Figure 14.13
When hydrogen chloride (HCl) is dissolved in water (H_2O), the hydrogen ion (H^+) is transferred to the water molecule, forming the hydronium ion (H_3O^+), and the chlorine ion (Cl^-).

placed in water, the hydrogen ion (H^+) is transferred from the HCl to the water (H_2O) molecule, as shown by the following equation:

$$HCl + H_2O \longrightarrow H_3O^+ + Cl^-$$

This reaction proceeds almost entirely from left to right. The H_3O^+ ion, called the **hydronium ion**, and the Cl^- ion are formed in this reaction. In solution the H^+ ions of acids are joined to water molecules to give hydronium ions, H_3O^+ (see Fig. 14.13).

Sulfuric acid (H_2SO_4) has two hydrogen atoms per molecule. When this acid molecule is placed in a water solution, dissociation occurs as shown:

$$H_2SO_4 + H_2O \longrightarrow \underset{\substack{\text{Hydronium} \\ \text{ion}}}{H_3O^+} + \underset{\substack{\text{Hydrosulfate} \\ \text{ion}}}{HSO_4^-}$$

The hydrosulfate ion, in turn, can act like an acid, as its proton is transferable to a water molecule:

$$HSO_4^- + H_2O \longrightarrow H_3O^+ + \underset{\text{Sulfate ion}}{SO_4^{2-}}$$

Acids are classified as strong or weak. A strong acid is one that ionizes almost completely in solution. The high propensity of an acid to ionize, and thus be classified as strong, is due to a weak binding energy of the H^+ ion. The acids HCl, HNO_3, and H_2SO_4 have weakly bound H^+ ions and therefore are strong acids. A weak acid is one that does not ionize to any great extent; that is, only a small number of the acid's protons unite with H_2O to form the H_3O^+ ion. Weak acids are formed from compounds that strongly bind the H^+ ion.

The Brönsted-Lowry concept holds that there must be a base that corresponds to every acid; that is, every acid, when it donates or gives up a proton (H^+ ion), becomes a base; and every base that accepts a proton becomes an acid. This relation can be expressed as

$$\text{acid} + \text{base} \longrightarrow \text{base} + \text{acid}$$

or, using HCl as an example, as

$$HCl + H_2O \longrightarrow Cl^- + H_3O^+$$

and

$$H_3O^+ + Cl^- \longrightarrow H_2O + HCl$$

According to the Brönsted-Lowry concept, acids and bases are classified as strong or weak according to their

Table 14.1 List of Common Acids and Their Conjugate Bases

	Acid Name	Reaction	Conjugate Base	
Strong	Hydrochloric	$HCl \longrightarrow H^+ + Cl^-$	Chloride ion	Weak
	Nitric	$HNO_3 \longrightarrow H^+ + NO_3^-$	Nitrate ion	
	Sulfuric	$H_2SO_4 \longrightarrow H^+ + HSO_4^-$	Bisulfate ion	
	Phosphoric	$H_3PO_4 \longrightarrow H^+ + H_2PO_4^-$	Dihydrogen phosphate ion	Bases
Acids	Lactic	$HC_3H_5O_3 \longrightarrow H^+ + C_3H_5O_3^-$	Lactate ion	
	Acetic	$HC_2H_3O_2 \longrightarrow H^+ + C_2H_3O_2^-$	Acetate ion	
	Carbonic	$H_2CO_3 \longrightarrow H^+ + HCO_3^-$	Bicarbonate ion	
	Ammonium ion	$NH_4^+ \longrightarrow H^+ + NH_3$	Ammonia	
Weak	Water	$H_2O \longrightarrow H^+ + OH^-$	Hydroxide ion	Strong

Table 14.2 Some Common Acids Found in Foods

Acid	Formula	Food
Acetic	$HC_2H_3O_2$	Vinegar
Citric	$H_3C_6H_5O_7$	Citrus fruits
Carbonic	H_2CO_3	Soft drinks
Lactic	$HC_3H_5O_3$	Sour milk
Oxalic	$H_2C_2O_4$	Rhubarb
Tartaric	$H_2C_4H_4O_6$	Grapes

Figure 14.14
Grapes contain tartaric acid.

ability to donate or accept protons. Hydrochloric acid (HCl) is a strong acid; that is, the HCl molecule gives up its proton very readily. Accordingly, the chloride ion is a weak base; that is, it does not accept a proton readily to form a HCl molecule. Table 14.1 lists a few common acids and the base produced when the acids donate their protons.

The base produced from an acid is called the **conjugate base,** and an acid produced from a base by the addition of a proton is called the **conjugate acid.** It follows from what has been said that the stronger the acid, the weaker its conjugate base; and the stronger the base, the weaker its conjugate acid.

Acids are very important industrially. Sulfuric acid is one of the most important chemical products in the United States. It is used in the refining of petroleum, the "pickling" of steel, and the manufacturing of fertilizers and numerous other products. A dilute solution of hydrochloric acid is present in the human stomach and is used in the digestion of our foods. There are persons who must take a daily supplement of this acid because it is lacking in their bodies. Many acids can be found in our foods; a few examples are given in Table 14.2 and Fig. 14.14.

An important property of an acid is the disappearance of its characteristic properties when brought into contact with a base and vice versa. This mixing of acid and base has to be in the proper proportions, however, for the properties of each to disappear. When the proper proportions are mixed and the characteristics of both the acid and the base disappear, we say that neutralization has taken place. **Neutralization** is the mutual disappearance of the H^+ ions that are responsible for the characteristic properties of the acid and the OH^- ions that are responsible for the characteristic properties of the base. The other ions associated with the acid and the base undergo no chemical change and remain as ions in the aqueous solution or appear as a crystalline solid if the water is evaporated. For example, if hydrochloric

acid (HCl) is added to potassium hydroxide (KOH) in the proper proportions, neutralization takes place as shown by the following equation:

$$H^+ + Cl^- + K^+ + OH^- \longrightarrow K^+ + Cl^- + HOH$$

This equation represents the typical reaction of an acid with a hydroxide base. Note that the metal ion (K^+) and the nonmetal ion (Cl^-) appear as ions on both sides of the equation, indicating that they do not take part in any chemical reaction; whereas the H^+ ion combines with the OH^- ion to form a molecule of water. The water molecule is usually written H_2O, but it was written HOH in this equation to show that the water molecule is formed by the proton donated by the acid and the hydroxide ion donated by the base. If equal numbers of H^+ ions and OH^- ions are mixed together, they will combine to form water molecules, and the acid and base properties of the original substances will disappear.

The following examples illustrate the principles involved for neutralization to occur.

EXAMPLE 4

How many milliliters of 1.0 M HCl are needed to neutralize 10 mL of 1.0 M KOH? (Refer to Section 13.6 on molarity if necessary.)

Solution   The number of H^+ ions must equal the number of OH^- ions for neutralization to take place. Because the acid and the base are both 1.0 M and the ion concentration is the same—that is, there is one H^+ ion for each HCl molecule and one OH^- ion for each KOH molecule—then the volumes must be equal. Thus 10 mL of HCl is needed to neutralize 10 mL of NaOH. They combine equally to form water (HOH), and the acid and base characteristics disappear.

EXAMPLE 5

How many milliliters of 1.0 M H_2SO_4 are needed to neutralize 10 mL of 2.0 M NaOH?

Solution Again, the number of H^+ ions must equal the number of OH^- ions for neutralization to take place. Because the acid is 1.0 M and the base 2.0 M, the acid is twice as strong as the base. But the H^+ ion concentration for the acid is two, and the OH^- ion concentration for the base is one. Thus 10 mL of acid is needed to neutralize 10 mL of the base.

More involved neutralization problems with odd values of molarity can be solved by using the following equation:

$$M_a V_a C_a = M_b V_b C_b$$

where M_a is the molarity of the acid,
M_b is the molarity of the base,
V_a is the volume of the acid,
V_b is the volume of the base,
C_a is the concentration of H^+ ions per molecule of acid,
C_b is the concentration of OH^- ions per molecule of base.

EXAMPLE 6

How many milliliters of 4.8 M phosphoric acid (H_3PO_4) are needed to neutralize 10 mL of 1.6 M calcium hydroxide [$Ca(OH)_2$]?

Solution   4.8 M $\times V_a \times 3 = 1.6$ M $\times$ 10 mL $\times 2$

$$V_a = \frac{1.6 \times 10 \times 2}{4.8 \times 3} = 2.22 \text{ mL}$$

The positive and negative ions that remain after the H^+ ions have combined with the OH^- ions constitute a **salt.** If the water of the solution is removed by evaporation, the remaining solid is the salt in crystalline form. All salts are ionic compounds consisting, usually, of a positive metallic ion and a negative nonmetallic ion.

To summarize, when an acid and a base are mixed, they produce water plus a salt. This reaction is shown by the following equations:

HCl   +   NaOH $\longrightarrow$ NaCl   +   HOH

acid   +   base $\longrightarrow$ salt   +   water

HNO_3 +   KOH $\longrightarrow$ KNO_3 +   HOH

acid   +   base $\longrightarrow$ salt   +   water

The most common salt is table salt (NaCl). It is frequently called simply salt, but there are many other salts according to the strict chemical definition.

All salts are ionic compounds; most salts are crystalline solids at ordinary temperatures. Some salts have a salty taste; others taste sweet, bitter, or sour or have no taste. Aqueous salt solutions may have either acidic, basic, or neutral characteristics. In addition to being formed when neutralization occurs, salts are produced by the direct union of a metal and a nonmetal by electron transfer and also by reactions between metallic and nonmetallic oxides.

Some salts occur in our environment in the **hydrated** form; that is, they contain one or more molecules of water bonded with the salt. The salt and water are combined according to the law of definite proportions. As examples, sodium thiosulfate, $Na_2S_2O_3 \cdot H_2O$ (hypo), magnesium sulfate, $MgSO_4 \cdot 7 H_2O$ (epsom salt), and sodium carbonate, $Na_2CO_3 \cdot 10 H_2O$ (washing soda), always have 1, 7, and 10 molecules, respectively, of water bonded to the molecule of salt. **Anhydrous salts** are those that do not have water bonded to the molecule of salt. Sodium chloride, NaCl (table salt), is an example.

The partial neutralization of an acid or a base produces what are known as acid salts or basic salts. A common acid salt is potassium acid tartrate, $KHC_4H_4O_6$ (cream of tartar); a common basic salt is lead carbonate, $Pb_3CO_3(OH)_4$ (white lead).

14.6 Acids and Bases in Solution

Pure water has the unusual property of acting as either an acid or a base because it dissociates to a small extent, producing both hydrogen and hydroxide ions. That it does slightly dissociate is shown by the fact that pure water conducts a very feeble electric current. The ionization of pure water is shown in the following equation:

$$H_2O + H_2O \longrightarrow H_3O^+ + OH^-$$

Thus we see that water can act as an acid or a base, according to the definitions given previously. In the formation of H^+ (or H_3O^+) and OH^- ions, one molecule acts as a proton donor and the other as a proton acceptor, as stated by the Brönsted-Lowry theory.

The dissociation of water to produce H^+ and OH^- ions is very slight. The concentration is only one ten-millionth of a mole of H^+ and OH^- ions in one liter of water at 25°C; expressed in powers of 10, the concentration is 1×10^{-7} mole/L. This concentration of H^+ and OH^- ions is used to define a neutral solution. Any solution containing 1×10^{-7} mole/L of both H^+ ions and OH^- ions at 25°C is defined as a **neutral solution.**

Table 14.3 The pH Concept

pH	Concentration of H^+ Ions		Acidic or Basic Property
−1	1.0×10^{1}	mole/L	Very acidic
1	1.0×10^{-1}	mole/L	↑
4	1.0×10^{-4}	mole/L	
6	1.0×10^{-6}	mole/L	Acidic
7	1.0×10^{-7}	mole/L	Neutral
8	1.0×10^{-8}	mole/L	Basic
10	1.0×10^{-10}	mole/L	↓
15	1.0×10^{-15}	mole/L	Very basic

An **acid solution** is defined as one in which the H^+ ion concentration is *greater* than 1×10^{-7} mole/L. An **alkaline solution** is one in which the OH^- ion concentration is greater than 1×10^{-7} mole/L. We can also define an alkaline solution as one in which the H^+ ion concentration is *less* than 1×10^{-7} mole/L. In all cases, the temperature is assumed to be 25°C.

The concentration of the H^+ ions is a measure of the acidity or basicity of a solution. The concentration can be expressed in powers of 10, as stated above, but this practice has proved to be rather awkward, and a method known as the pH method has been devised for expressing the H^+ ion concentration. Stated in simple terms, the **pH** is the exponent of the negative power to which 10 is raised when powers of 10 are used to express the concentration. For example, a solution containing H^+ ions with a concentration of 0.0000001 (1×10^{-7}) mole/L has a pH of 7. A concentration of 1×10^{-11} has a pH of 11.

A neutral solution has a pH of 7. A pH less than 7 indicates an acid solution, and a pH greater than 7 indicates a basic solution. A solution with a pH of 5 is said to be slightly acidic, and one with a pH of 9 is slightly basic. Soda water (carbonic acid) has a pH of 4.5, blood 7.4, sea water 8.5, and milk of magnesia 10.5. Solutions in which the pH is 7 or more points from the neutral position of 7 are strongly acidic or basic. The concept of pH is illustrated in Table 14.3.

The pH of substances found in our environment varies. Most body fluids have a normal pH range, and a continued deviation from the normal range usually indicates some disorder in a body function. Thus the pH value can be used as a means of diagnosis. Table 14.4 gives an approximate pH of some common substances. Actual values range on each side of the value given in the table.

Table 14.4 Approximate pH of Some Common Substances

Substance	pH	Substance	pH
Battery acid	0.0	Bread	5.5
Stomach acid	1.2	Potatoes	5.8
Lemons	2.3	Coffee	6.0
Vinegar	2.8	Rainwater	6.2
Soft drinks	3.0	Corn	6.3
Apples	3.1	Milk (cow)	6.5
Grapefruit	3.1	Pure water	7.0
Wines	3.2	Blood (human)	7.4
Oranges	3.5	Eggs	7.8
Tomatoes	4.2	Sea water	8.5
Beer	4.5	Clorox	9.0
Bananas	4.6	Milk of magnesia	10.5
Carrots	5.0	Oven cleaner	13.0

Learning Objectives

After reading and studying this chapter, you should be able to do the following without referring to the text:

1. Describe a chemical reaction, and distinguish between the reactants and the products.

2. Explain chemical equilibrium, and give an example.

3. Distinguish between exothermic and endothermic reactions, and give an example of each.

4. Explain how temperature affects the rate of a chemical reaction.

5. Describe chemical reaction rates in the presence of a catalyst.

6. Explain oxidation-reduction reactions, and give an example.

7. Explain the electromotive series.

8. Define and give an example of (a) an acid, (b) a base, (c) a salt, and (d) a neutralization process.

9. Distinguish between the Arrhenius acid-base concept and the Brönsted-Lowry theory of acids and bases.

10. Define and explain the important words and terms listed in the next section.

Important Words and Terms

reactants	oxidized	acid-base reactions
products	reduced	hydronium ion
reversible reaction	reducing agent	conjugate base
equilibrium	oxidizing agent	conjugate acid
exothermic reaction	displacement reactions	neutralization
endothermic reaction	electromotive series	salt
activation energy	electrolytes	hydrated
ten-degree rule	cathode	anhydrous salts
catalyst	anode	neutral solution
enzymes	acid	acid solution
oxidation-reduction reaction	base	alkaline solution
oxidation	Arrhenius acid-base concept	pH
reduction	Brönsted-Lowry theory	

Questions

Basic Concepts

1. What is a chemical reaction?

2. What three things occur in every chemical reaction?

3. Explain the concept of dynamic equilibrium.

Energy and Rate of Reaction

4. List the factors that influence the rate of a chemical reaction.

5. Give an example of (a) a fast chemical reaction and (b) a slow reaction.

6. Explain why chemical reactions proceed faster as the temperature is increased.

7. Why do chemical reactions proceed at a faster rate when the concentrations of the reactants are increased?

8. What is the origin of the energy released in the chemical reaction?

9. Distinguish between exothermic and endothermic reactions.

10. How does a catalyst affect a reaction's equilibrium mixture?

11. How is the activation energy related to the total energy released by a chemical reaction?

12. Describe the role of a catalyst in a chemical reaction.

13. The human body converts sugar into carbon dioxide and water at body temperature, 98.6°F or 37°C. Why are much higher temperatures required for the same conversion in the laboratory?

Oxidation-Reduction Reactions

14. Which would be easier to oxidize, sodium or chlorine? Explain.

15. Which would be easier to reduce, sodium or chlorine? Explain.

16. Which is the better oxidizing agent, nitrogen or fluorine? Why?

17. Why must one keep wines sealed or covered?

18. Why are the precious metals gold and silver found free in nature, while the metals iron and magnesium are found in ores?

19. Identify some uses of the electromotive series.

Electrochemical Reactions

20. Explain the difference between a reversible and an irreversible electrochemical cell.

21. Why is an automobile battery so heavy?

22. What is the most destructive chemical reaction in terms of dollars?

23. What two substances are necessary for iron to rust?

Acids and Bases

24. Describe two theories that define an acid and a base.

25. In respect to the Arrhenius concept of acids, bases, and salts, what is meant by neutralization?

26. Distinguish between a hydrogen ion and a hydronium ion.

Acids and Bases in Solution

27. Explain how water can be both an acid and a base.

28. How is the strength of an acid or base expressed?

Glass

29. What is an amorphous solid?

30. Describe the structure of glass.

31. What is the composition of ordinary glass?

32. How are different colors of glass made?

Exercises

Basic Concepts

1. Balance the following chemical equations:
   (a) When an acid reacts with a metal, H_2 gas is given off:
   $$HCl + Fe \longrightarrow FeCl_2 + H_2$$

   (b) Nitric acid may be prepared by the following reaction:
   $$H_2O + NO_2 \longrightarrow HNO_3 + NO$$
   Answer: (b) $H_2O + 3\ NO_2 \longrightarrow 2\ HNO_3 + NO$

2. Balance the following chemical equations:
   (a) When octane gasoline is burned completely, the reaction is
   $$C_8H_{18} + O_2 \longrightarrow CO_2 + H_2O$$

   (b) Borax is a good water-softening agent. One of the important reactions of the borate ion in water is
   $$B_4O_7{}^{2-} + H_2O \longrightarrow H_3BO_3 + OH^-$$

Oxidation-Reduction Reactions

3. State which element or ion is oxidized and which reduced in the following reactions:
   (a) The Haber process for manufacturing ammonia is
   $$N_2 + 3\ H_2 \longrightarrow 2\ NH_3$$

   (b) An important reaction in steel production is
   $$Fe_2O_3 + 3\ CO \longrightarrow 2\ Fe + 3\ CO_2$$
   Answer: (b) Iron is reduced; carbon is oxidized.

4. State which element or ion is oxidized and which reduced in the following reactions:
   (a) Sulfur in coal burns to form the pollutant sulfur dioxide:
   $$S + O_2 \longrightarrow SO_2$$

   (b) Hydrofluoric acid is so strong it can etch glass. It is formed as follows:
   $$H_2 + F_2 \longrightarrow 2\ HF$$

5. Referring to the electromotive series, use an arrow to indicate the direction of the following displacement reactions:
   (a) $Na + K^+ \qquad Na^+ + K$
   (b) $Ni + Cu^{2+} \qquad Ni^{2+} + Cu$
   Answer: (a) $Na + K^+ \longleftarrow Na^+ + K$

6. Referring to the electromotive series, use an arrow to indicate the direction of the following displacement reactions:
   (a) $Sn + Fe^{2+} \qquad Sn^{2+} + Fe$
   (b) $Pb + Fe^{2+} \qquad Pb^{2+} + Fe$

Acids and Bases

7. One molar nitric acid (HNO_3) is added to 1.0 M potassium hydroxide (KOH) to form a salt plus water, as shown by the following equation:
   $$HNO_3 + KOH \longrightarrow \underset{salt}{\rule{1cm}{0.4pt}} + HOH$$

   (a) Write the formula for the salt.
   (b) Balance the equation.

8. One molar phosphoric acid (H_3PO_4) is added to 1.0 M sodium hydroxide (NaOH) to form a salt plus water, as shown by the following equation:
   $$H_3PO_4 + NaOH \longrightarrow \underset{salt}{\rule{1cm}{0.4pt}} + HOH$$

   (a) Write the formula for the salt.
   (b) Balance the equation.

9. One molar sulfuric acid (H_2SO_4) is added to 2.0 M potassium hydroxide (KOH) to form a salt plus water, as shown by the following equation:
   $$H_2SO_4 + KOH \longrightarrow \underset{salt}{\rule{1cm}{0.4pt}} + H_2O$$

   (a) Write the formula for the salt.
   (b) Balance the equation.
   (c) Calculate the volume of H_2SO_4, in cubic centimeters, needed to neutralize 10 mL of KOH.

10. One molar hydrochloric acid (HCl) is added to 1.0 M calcium hydroxide [$Ca(OH)_2$] to form a salt plus water, as shown by the following equation:
    $$HCl + Ca(OH)_2 \longrightarrow \underset{salt}{\rule{1cm}{0.4pt}} + HOH$$

    (a) Write the formula for the salt.
    (b) Balance the equation.
    (c) Calculate the volume of HCl, in cubic centimeters, needed to neutralize 10 mL of $Ca(OH)_2$.

Acids and Bases in Solution

11. State whether the following solutions are acidic or basic.
    (a) 10 g of H_2SO_4 dissolved in 1 L of water
    (b) Blood
    (c) A solution with a pH of 10
    (d) A solution with 1×10^{-9} mole/L of OH^- ion concentration

12. State whether the following solutions are acidic or basic.
    (a) 10 g of $Mg(OH)_2$ dissolved in 1 L of water
    (b) Pure water
    (c) A solution with a pH of 7
    (d) A solution with 1×10^{-5} mole/L of H^+ ion concentration

13. Determine the hydrogen ion concentration in a solution with a pH of 8.

14. Determine the pH of a solution that has a hydrogen ion concentration of 1×10^{-4} mole/L.

CHAPTER 15

Complex Molecules

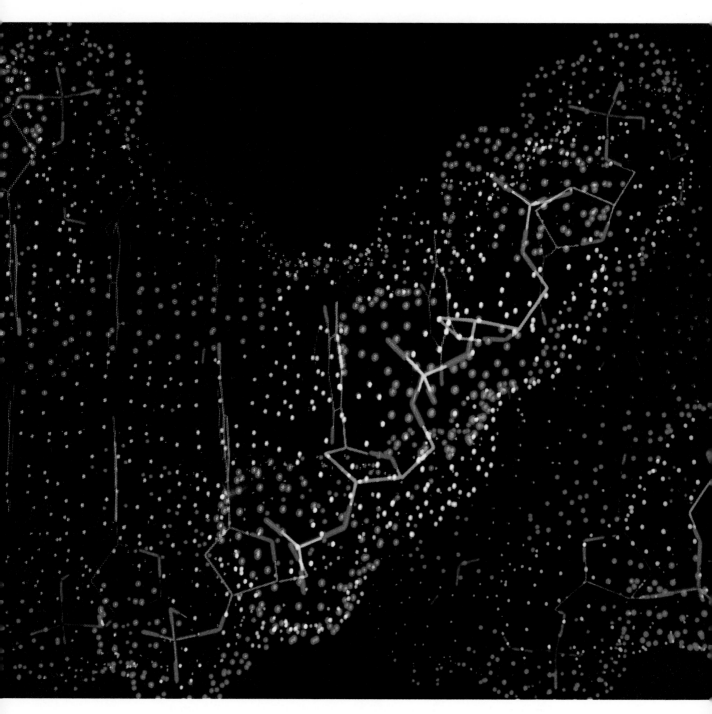

I am part of all you see
In Nature: part of all you feel: / I am the impact of the bee
Upon the blossom; in the tree / I am the sap—that shall reveal
The leaf the bloom—that flows and flutes
Up from the darkness through its roots.   —Madison Cawein

PREVIOUSLY, WE LEARNED that atoms of the chemical elements are nature's building blocks. In Chapter 12 we discovered that compounds are formed by combining the atoms of various elements. Many naturally occurring compounds are made up of very complex molecules. Because the fundamental elements are known, scientists hoped that these complex molecules could be analyzed and understood. It is true that a great degree of success has been achieved in analyzing those compounds whose molecules contain only a few atoms. However, these small molecules comprise only the inanimate portion of nature such as gases, water, and minerals. The chemical structure of the living portion of nature is grossly complex. **Complex molecules** may be either an aggregate of atoms or repeating units of smaller molecules chemically bonded together, thus giving rise to great molecular weights and chainlike structures.

A simple example of building molecules from repeating units is the hydrocarbon series discussed in Chapter 12. Specifically, the fundamental repeating unit, or **monomer,** of the alkane series can be structurally written as

$$\left[\begin{array}{c} H \\ | \\ -C- \\ | \\ H \end{array} \right]_n$$

where it is understood that other fundamental units are attached to the open bonds, and the series is terminated by hydrogen atoms.

The subscript n, which indicates the fundamental unit, may be repeated n times, and the resulting chainlike structure is called a **polymer.** For example, if $n = 3$, the resulting compound is propane (see Section 12.8). This simple example is meant to introduce the concept of repeating units; remember that polymers, whether oc-

curring naturally or prepared synthetically, may have repetition numbers (n) that may be in the thousands or hundreds of thousands. In addition, the fundamental unit may itself be quite complex, and the chain configurations and variations may differ greatly.

Nature has been in the business of making large complex molecules for a long time. All living and growing matter is composed of these molecules, some being so complex that they are still not completely understood. The complexity of nature's molecules should come as no surprise, because they compose the structure of life itself. The newest frontier of science is concerned with determining the structures and interactions of the complex molecules of which living things are made. Society someday may be forced to make some fundamental decisions regarding how far it will allow science to go in solving the mysteries of life.

15.1 Common Organic Compounds

The number and complexity of the organic compounds in nature can, in large part, be attributed to the ability of the carbon atom to form single, double, and even triple bonds with itself and other elements. There are over 6 million organic compounds, and their identification, classification, and structural determination has been—and still is—the task of workers in the field of organic chemistry. Only a few of the more common compounds will be presented here, most of which should be familiar to the student because of their use and occurrence in everyday life.

Consider the following list of classes of compounds. A specific example from each class is given to show how the complexity of molecules can develop, starting with just a simple hydrocarbon.

$$\begin{array}{ccc} H & H \\ | & | \\ H-C-C-H \\ | & | \\ H & H \end{array}$$

Hydrocarbon (ethane)

◄ **A computer-graphic model showing the double-helix structure of DNA (deoxyribonucleic acid).**

1. *Alkyl halides*

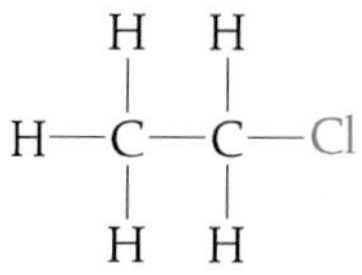

Ethyl chloride

2. *Alcohols*

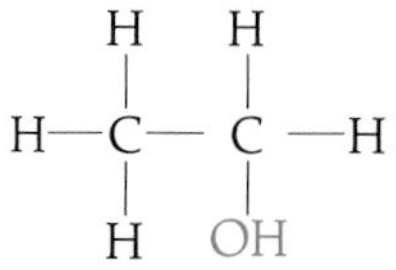

Ethyl alcohol

3. *Aldehydes*

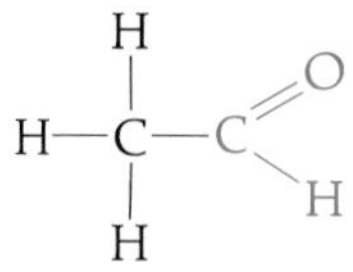

Acetaldehyde

4. *Acids (organic)*

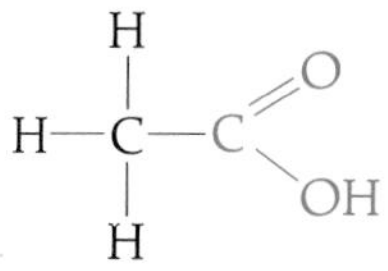

Acetic acid

5. *Esters*

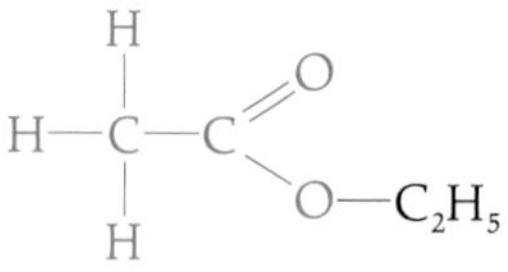

Ethyl acetate

The colored portions of the formulas indicate the organic functional group that characterizes the general physical and chemical properties of these compounds.

Alkyl Halides

Alkyl groups are alkanes minus one hydrogen atom. They have the general formula C_nH_{2n+1}. An **alkyl halide** is formed by replacing the hydrogen atom(s) in hydrocarbons with halogens—chlorine, bromine, fluorine, and iodine. Ethyl chloride, the example given in the list, is used as a local anesthetic and is the product of the reaction of ethane with chlorine:

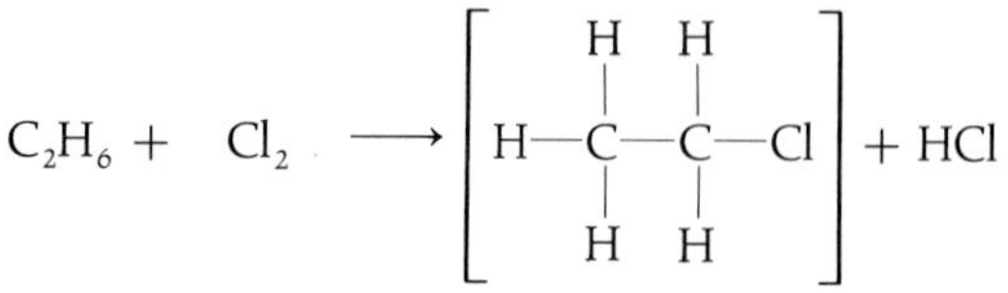

Ethane   Chlorine   Ethyl chloride

The halogen may replace more than one of the hydrogen atoms; in fact, they may all be replaced, as the following series shows:

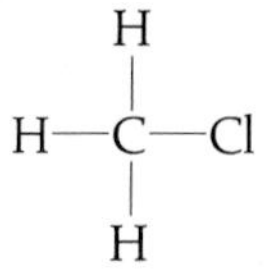

Methyl
chloride 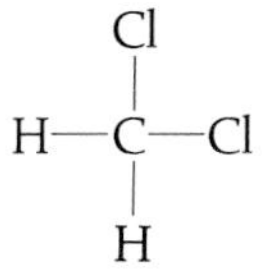   Dichloromethane

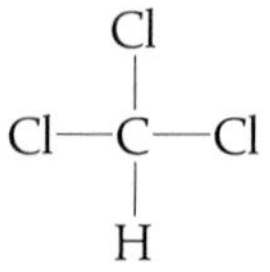

Trichloromethane   Tetrachloromethane
(chloroform)   (carbon tetrachloride)

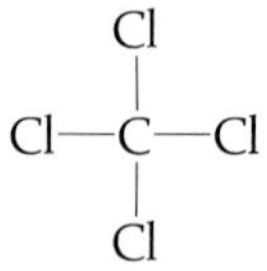

Similar compounds can be formed with bromine and iodine.

Carbon tetrachloride is considered a cancer-causing agent, and its use has been greatly reduced. It is no longer used in fire extinguishers or in fabric cleaners.

Alcohols

Alcohols are organic compounds that contain one or more OH groups that have been substituted for one or more hydrogen atoms. Ethyl alcohol (C_2H_5OH) is probably the most important alcohol known. It is made from sugars by the action of yeast in the process of fermentation:

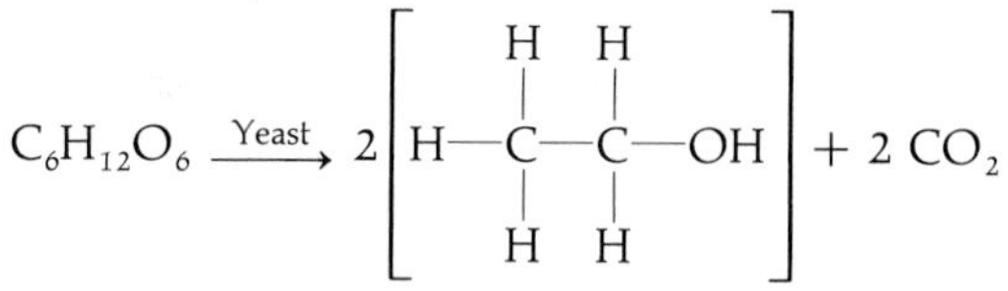

Ethyl alcohol

or synthetically from ethylene and water:

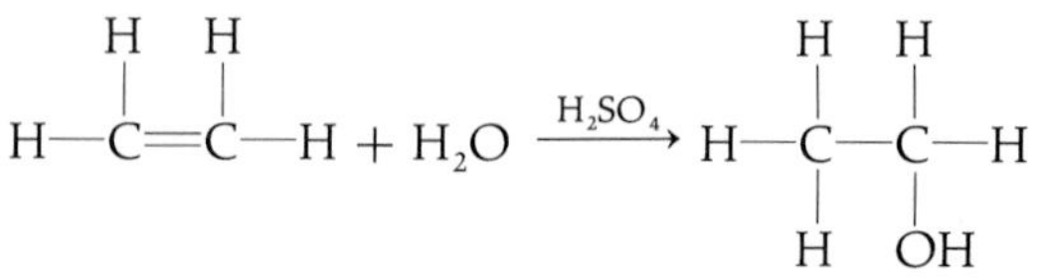

Ethylene   Ethyl alcohol

Winemaking is based on the fermentation of grape juice (see Fig. 15.1).

Figure 15.1 A wine cellar.
Wine-making is based on the fermentation of grape juice.

Ethyl alcohol is a colorless liquid that mixes with water in all proportions. It is the least toxic of all alcohols and is used in alcoholic beverages. Ethyl alcohol is also used as a solvent and in the production of many substances, including perfumes, dyes, varnishes, antifreeze, and ethyl ether.

Alcohols are characterized by the —OH, or hydroxyl, group; hence they are the organic equivalent of the inorganic bases. Many alcohols exist, some with one (—OH) group, others with two or more (—OH) groups. Ethylene glycol is an example of an alcohol with two hydroxyl groups:

$$H-\underset{\underset{\displaystyle H-\underset{\displaystyle H}{C}-OH}{|}}{\overset{\displaystyle H}{\underset{|}{C}}}-OH$$

Ethylene glycol

Ethylene glycol is widely used as an antifreeze in automobiles.

Aldehydes

Aldehydes are characterized by the $-C{\overset{\displaystyle O}{\underset{\displaystyle H}{}}}$ group and are formed when alcohols react with oxygen (are oxidized). When ethyl alcohol is oxidized, acetaldehyde is formed:

$$2\left[H-\underset{\underset{\displaystyle H}{|}}{\overset{\overset{\displaystyle H}{|}}{C}}-\underset{\underset{\displaystyle OH}{|}}{\overset{\overset{\displaystyle H}{|}}{C}}-H \right] + O_2 \longrightarrow$$

Ethyl alcohol

$$2\left[H-\underset{\underset{\displaystyle H}{|}}{\overset{\overset{\displaystyle H}{|}}{C}}-C{\overset{\displaystyle O}{\underset{\displaystyle H}{}}} \right] + 2\ H_2O$$

Acetaldehyde

A more common aldehyde, formaldehyde, is prepared similarly from methyl alcohol (CH_3OH). Formaldehyde is used as a disinfectant and tissue preservative.

Organic Acids

The further action of oxygen on aldehydes produces a group of compounds known as **organic,** or **carboxylic, acids,** which are characterized by the molecular arrangement $-C{\overset{\displaystyle O}{\underset{\displaystyle OH}{}}}$ called the **carboxyl** group. There are many of these carboxylic acids. Acetic acid, whose dilute natural form is vinegar, is formed by the following reaction:

$$2\left[H-\underset{\underset{\displaystyle H}{|}}{\overset{\overset{\displaystyle H}{|}}{C}}-C{\overset{\displaystyle O}{\underset{\displaystyle H}{}}} \right] + O_2 \longrightarrow 2\left[H-\underset{\underset{\displaystyle H}{|}}{\overset{\overset{\displaystyle H}{|}}{C}}-C{\overset{\displaystyle O}{\underset{\displaystyle OH}{}}} \right]$$

Acetaldehyde   Acetic acid

The simplest carboxylic acid, formic acid, is common in insects and is the cause of painful discomfort from insect bites or bee stings (see Fig. 15.2). It is prepared by the oxidation of formaldehyde:

$$H-C{\overset{\displaystyle H}{\underset{\displaystyle O}{}}} + oxygen \longrightarrow H-C{\overset{\displaystyle OH}{\underset{\displaystyle O}{}}}$$

Formaldehyde   Formic acid

Figure 15.2 A honeybee collecting pollen.

Esters

When carboxylic acid reacts with an alcohol, an ester is formed. For example,

Acetic acid   Ethyl alcohol

Ethyl acetate
(an ester)

An **ester** is a compound that conforms to the general formula

$$R\!-\!\overset{\overset{\displaystyle O}{\|}}{C}\!-\!O\!-\!R'$$

where R and R' are any alkyl groups. Both R and R' may be the same group, but they are usually different. Esters are analogous to salts. Inorganic acids and bases react to form salts. Organic acids and alcohols react to form esters.

Esters possess very distinct and pleasant odors. The fragrance of many flowers and the pleasant taste of ripe fruits are due to esters. For example, bananas contain the ester amyl acetate,

$$CH_3\overset{\overset{\displaystyle O}{\|}}{C}\!-\!O(CH_2)_4CH_3$$

Amyl acetate

and oranges contain octyl acetate,

$$CH_3\overset{\overset{\displaystyle O}{\|}}{C}\!-\!O(CH_2)_7CH_3$$

Octyl acetate

Another ester is acetylsalicylic acid, commonly known as aspirin and called by some the wonder drug:

Acetylsalicylic acid

Aspirin is used for the treatment of colds, headaches, and other minor aches and pains. It is also used in the treatment of arthritis and rheumatic fever and as protection against heart attacks.

The above group of compounds may appear to be rather complex; however, the examples given are selected for their simplicity. For very complex molecules we need only look as far as ourselves or other living matter.

15.2 Biochemistry

Biochemistry is the study of the chemistry of living cells, and the molecules that make up these living cells are the most complicated matter known. The molecules of living matter are extremely large and complex. These molecules (molecular weights greater than 1 million) contain tens of thousands of atoms connected in three-

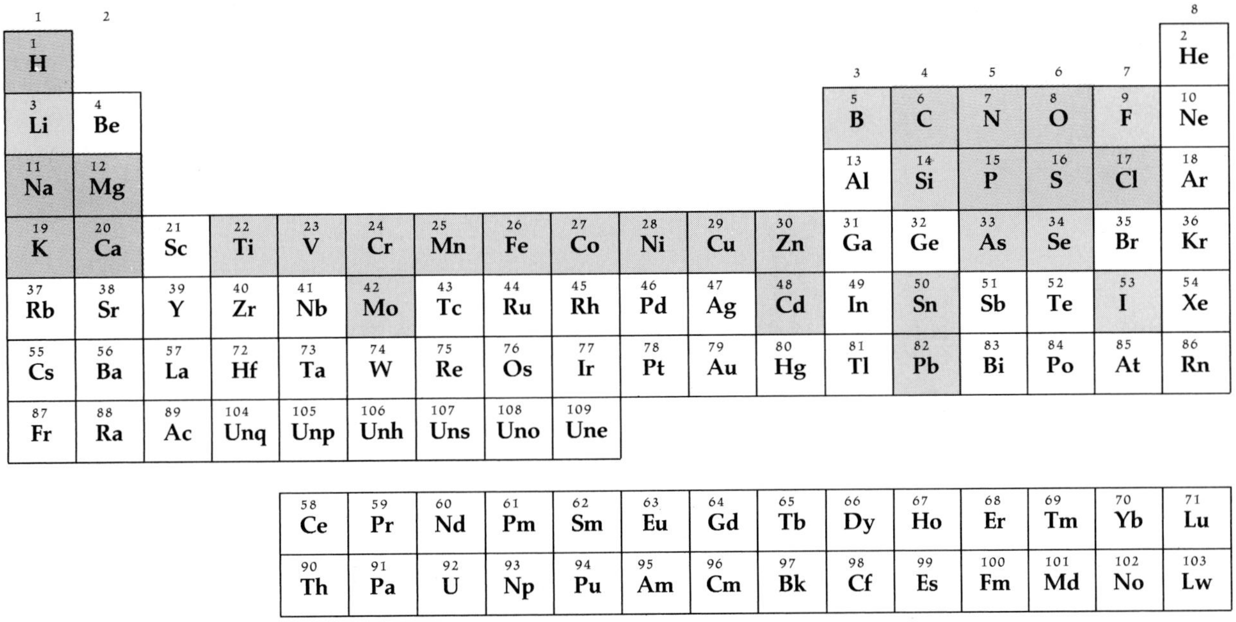

Figure 15.3 Periodic Table illustrating the elements essential for living cells.
The elements shown in red are the more abundant ones found in the human body.
Those shown in blue are the trace elements.

dimensional patterns that are complex in structure and very difficult to analyze.

The elements that are essential for living cells are shown in Fig. 15.3. The more abundant elements are shown in red, and the trace elements are shown in blue. Table 15.1 gives the percentages by mass and the function of each of the essential elements found in the human body.

Common organic compounds in living matter are carbohydrates, proteins, fats (lipids), nucleic acids, and vitamins. This section begins with the composition and structure of carbohydrates.

Carbohydrates

Carbohydrates are compounds composed of carbon, hydrogen, and oxygen with the hydrogen-to-oxygen ratio usually 2 to 1, the same as in the water molecules. The general formula for the carbohydrates is $C_n(H_2O)_x$. Thus the name *carbohydrates* was given to those substances known as hydrates of carbon, although the nature of these compounds is not what the name implies. The most important carbohydrates are sugars, starches, and cellulose. The simplest sugars, glucose ($C_6H_{12}O_6$) and

fructose ($C_6H_{12}O_6$), are isomers. Their structural formulas are

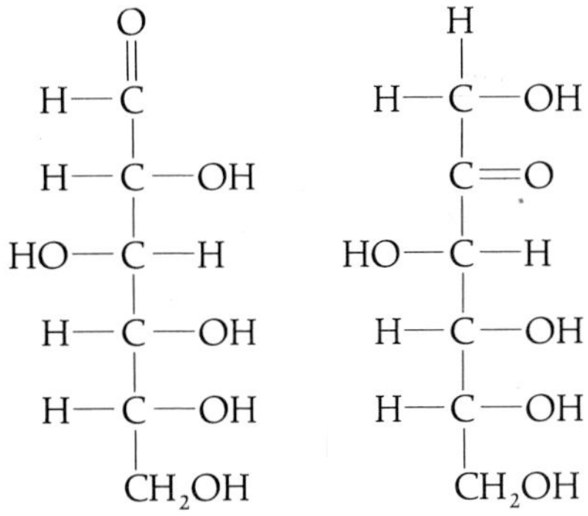

Glucose, also known as dextrose and grape sugar, is found in sweet fruits, such as grapes and figs, and in flowers and honey. It is present in the blood to the extent of 0.1% but is present in much greater amounts in persons suffering from diabetes. Glucose is formed in plants by the action of sunlight and chlorophyll on carbon dioxide (CO_2) from the air and water (H_2O) from the soil. The chemical reaction in simple terms is

$$6\ CO_2 + 6\ H_2O + energy \longrightarrow C_6H_{12}O_6 + 6\ O_2$$

Glucose

Table 15.1 Percentage by Mass of the Essential Elements and Some of Their Functions

Element	Percent by Mass in the Human Body	Function
Oxygen	65	Component of water and many organic compounds
Carbon	18	Component of all organic compounds
Hydrogen	10	Component of water and many inorganic and organic compounds
Nitrogen	3	Component of both inorganic and organic compounds
Calcium	1.5	Major component of bone; essential to some enzymes and to muscle action
Phosphorus	1.2	Essential in cellular synthesis and energy transfer
Potassium	0.2	Cation in intracellular fluid
Chlorine	0.2	Anion inside and outside the cells
Sulfur	0.2	Component of proteins and some other organic compounds
Sodium	0.1	Cation in extracellular fluid
Magnesium	0.05	Essential to some enzymes
Iron	<0.05	In hemoglobin, myoglobin, and other proteins
Zinc	<0.05	Essential to many enzymes
Cobalt	<0.05	Found in vitamin B_{12}
Copper	<0.05	Essential to several enzymes
Iodine	<0.05	Essential to thyroid hormones
Selenium	<0.01	Essential to some enzymes
Fluorine	<0.01	In teeth and bones
Nickel	<0.01	Essential to some enzymes
Molybdenum	<0.01	Essential to some enzymes
Silicon	<0.01	Found in connective tissue
Chromium	<0.01	Essential in carbohydrate metabolism
Vanadium, tin, manganese, bromine	<0.01	Exact function not known

with energy being absorbed in the process; that is, the energy coming from the sunlight is stored as chemical energy in the plant.

Fructose, also called fruit sugar, the sweetest of all sugars, is found in fruits and honey (Fig. 15.4). Combined with glucose, it forms sucrose, also known as common sugar or cane sugar. The reaction is written as follows:

$$C_6H_{12}O_6 + C_6H_{12}O_6 \longrightarrow C_{12}H_{22}O_{11} + H_2O$$

Fructose   Glucose   Sucrose

Sucrose has two important isomers, maltose or malt sugar and lactose or milk sugar, which are sweet, crystalline solids soluble in water. Lactose is present in milk (about 4%) and is an essential ingredient in food for infants.

Starch has the general formula $(C_6H_{10}O_5)_n$, where n may take on values up to 3000. Thus starch is a polymer (many monomers associated together) that consists of long chains of glucose units. Starch, a noncrystalline substance, is formed by plants in seeds, tubers, and fruits. Starch as a food is present in potatoes, grains (corn and rice), and vegetables. In the digestion process the starches are converted to glucose. Glycogen (animal starch), a smaller and more highly branched polymer of glucose, is stored in the liver and muscles of animals as a reserve food supply that is easily converted to energy.

The energy in foods is given in terms of the energy they can deliver when oxidized to their final products. These final products are carbon dioxide, water, and nitrogenous compounds that are given off as waste prod-

Figure 15.4 Fruit—a source of fructose.

ucts. For example, when glucose is oxidized, 674 kilocalories (kcal) of energy per mole are provided. Because there is 180 g in one mole of glucose, 1 g of glucose provides 3.74 kcal of energy. This energy is used to do work and provide body heat in living organisms.

Cellulose has the same general formula, $(C_6H_{10}O_5)_n$, as starch, but its structure is different; thus it has different properties. Cellulose is the main component of the cell walls of plants and receives its name for this reason. It is the most abundant organic substance found in our environment. Cellulose cannot be digested by humans, because our digestive system does not contain the necessary enzymes to break the linkages in the molecular chain. The digestive track of termites and some animals (cows, goats, deer, and others) does have the necessary enzymes to break the linkages and obtain nutrition from cellulose. There are many commercial uses of cellulose, such as rayon, cellophane, explosives, and paper.

Proteins

Proteins are one of the essential components of living cells. Proteins are composed of carbon, oxygen, nitrogen (approximately 16% in all proteins), and hydrogen. Some also contain small amounts of sulfur, phosphorus, iron, and copper. The molecular structure of proteins is complex, but the basic chemical units are amino acids. An *amino acid* is an organic compound that contains an amine group and an acid group. The **amino (functional) group** has the structural formula

When an amino group is combined with a hydrocarbon group, an amine is formed. Similarly, when the carboxyl functional group

is combined with a hydrocarbon group, an acid is formed. Thus an amino acid contains both of these groups.

There are over twenty known amino acids; eight of them are essential in the human diet. The simplest amino acid is glycine, whose structural formula is

When two molecules of glycine combine, a molecule of water is eliminated. Two molecules of glycine can be shown as

Because a water molecule is given off, an amide linkage is formed, and the resulting dipeptide is

Amide linkage

This process can be repeated by linking more glycine molecules to the protein whose structural formula appears above. When it is repeated again and again, long-chain proteins of very high molecular weight are formed. The molecular weight of proteins ranges from a few thousand up to millions for the most complex. The approximate weights of a few examples are insulin, 6000; hemoglobin, 65,000; fibrinogen, 450,000; myosin, 850,000; and hemocyamine, 9 million.

The structure of proteins is three dimensional. The primary structure relates to the number and the sequence of the amino acids that are held together by peptide bonds. The secondary structure relates to the regularly recurring arrangement of the amino acid chain held to-

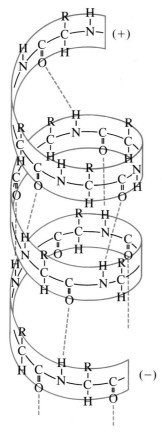

Figure 15.5 The alpha-helix structure of a polypeptide chain.
The hydrogen bonds are illustrated by the blue lines.

gether by hydrogen bonding between an oxygen atom in the carboxyl group of an amino acid and a hydrogen atom connected to a nitrogen atom of a different amino acid. One of the most common polypeptides protein chains is the alpha-helix, which is illustrated in Fig. 15.5.

Lipids

Lipid is a general term that includes such substances as fats, oils, and waxes.

Fats are used in the diets of humans. In the digestive process the fats are broken down into glycerol and acids, which are absorbed into the bloodstream and oxidized to produce energy that may be used immediately or stored for future use. Fats are also used by the body as an insulator to prevent loss of heat.

Fats are esters composed of glycerol $[C_3H_5(OH)_3$,

an alcohol] and organic acids known as fatty acids. Fatty acids are acids that are present in any organic acid. A typical fatty acid is stearic acid, $C_{17}H_{35}COOH$ (a component of beef fat). When stearic acid is combined with glycerol, glyceryl stearate (a fat) is obtained. The reaction is

$$3\left[C_{17}H_{35}-\overset{\overset{\displaystyle O}{\|}}{C}-OH\right] + \begin{array}{c} H-\overset{\overset{\displaystyle H}{|}}{C}-OH \\ H-\overset{|}{C}-OH \\ H-\overset{|}{C}-OH \\ \overset{|}{H} \end{array} \longrightarrow$$

Stearic acid   Glycerol

$$\begin{array}{c} H \\ | \\ C_{17}H_{35}COO-\overset{}{C}-H \\ C_{17}H_{35}COO-\overset{|}{C}-H + 3\,H_2O \\ C_{17}H_{35}COO-\overset{|}{C}-H \\ | \\ H \end{array}$$

Glyceryl stearate (a fat)

The stearate structure is a large-chain hydrocarbon containing 18 carbon atoms; it is written in short form here to save space.

Liquid fats, such as vegetable oils, are composed of hydrocarbon chains with double bonds between some of the carbon atoms; that is, they are unsaturated. These oils can be changed to solid fats by a process called **hydrogenation.** In this process hydrogen is added to the carbon atoms that have the double bond; thus the hydrocarbon chains become saturated. An example of hydrogenation is the changing of ethylene (C_2H_4), which contains a double bond between the carbon atoms, to ethane (C_2H_6), which has a single bond between the carbon atoms. The reaction is

$$\begin{array}{c} H \\ \diagdown \\ H \end{array} C=C \begin{array}{c} H \\ \diagup \\ H \end{array} + H_2 \longrightarrow H-\overset{\overset{\displaystyle H}{|}}{\underset{\underset{\displaystyle H}{|}}{C}}-\overset{\overset{\displaystyle H}{|}}{\underset{\underset{\displaystyle H}{|}}{C}}-H$$

Ethylene   Ethane

When cottonseed oil (a liquid) is hydrogenated, margarine (a solid) is obtained. The reaction is

$$\left[\begin{array}{l} CH_3(CH_2)_7CH{=}CH(CH_2)_7COOCH_2 \\ CH_3(CH_2)_7CH{=}CH(CH_2)_7COOCH \\ CH_3(CH_2)_7CH{=}CH(CH_2)_7COOCH_2 \end{array}\right] + 3\,H_2 \longrightarrow$$

Cottonseed oil (a liquid)

$$\left[\begin{array}{l} CH_3(CH_2)_{16}COOCH_2 \\ CH_3(CH_2)_{16}COOCH \\ CH_3(CH_2)_{16}COOCH_2 \end{array}\right]$$

Margarine (a solid)

Thus liquid fats (oils) are esters of unsaturated acids, and solid fats are esters of saturated acids.

Waxes are esters derived from long-chain mono-hydric alcohols (an alcohol with only one OH group per molecule) and fatty acids. Beeswax from the honeycomb of the honeybee contains myricyl palmitate ($C_{15}H_{31}COOC_{31}H_{63}$) and is used in pharmaceutical preparations. Carnauba wax, which comes from the Brazilian wax palm, is used in commercial waxes for wood and metal surfaces. Chinese wax, which is secreted by an insect, is used in medicine.

The term **detergent** means cleansing agent and commonly refers to a soap substitute; but ordinary soap, which is a cleansing agent, is also a detergent (see Fig. 15.6). Ordinary soap is made by combining fats or oils with alkalis such as sodium or potassium hydroxide. When fats are treated with sodium hydroxide, glycerol and sodium salts of the fatty acid, sodium palmitate, sodium stearate, and sodium oleate are formed. A typical soap is sodium stearate ($C_{17}H_{35}COO^-Na^+$), whose structural formula is

$$H{-}\underset{\underset{\displaystyle H}{|}}{\overset{\overset{\displaystyle H}{|}}{C}}{-}{-}{-}(C_{15}H_{30}){-}{-}{-}\underset{\underset{\displaystyle H}{|}}{\overset{\overset{\displaystyle H}{|}}{C}}{-}C\overset{\displaystyle O}{\underset{\displaystyle O^-}{}}\ Na^+$$

Soaps have a cleansing action because the carboxyl end of the soap ion is water soluble and the hydrocarbon end is oil soluble. Thus the hydrocarbon end can dissolve fats, and the carboxyl or ion end surrounds the dirt particles and forms a film around them so that they can be washed away. Hence through a combination of emulsifying action and particle suspension, soaps serve as detergents.

Figure 15.6 Detergent—a cleansing agent.

Although soaps have a cleansing action, they have the disadvantage of forming precipitates when used in acidic solutions or with hard water, which contains calcium, magnesium, and iron ions.

The modern detergents, which are soap substitutes, contain a long hydrocarbon chain that is nonpolar (for example, $C_{12}H_{25}{-}$) and a polar group such as sodium sulfate ($-OSO_3Na$). This detergent has the structural formula

$$H{-}\underset{\underset{\displaystyle H}{|}}{\overset{\overset{\displaystyle H}{|}}{C}}{-}{-}{-}(CH_2)_{10}{-}{-}{-}\underset{\underset{\displaystyle H}{|}}{\overset{\overset{\displaystyle H}{|}}{C}}{-}O{-}\underset{\underset{\displaystyle O}{|}}{\overset{\overset{\displaystyle O}{|}}{S}}{-}O^-Na^+$$

Sodium lauryl sulfate

Synthetic detergents such as sodium lauryl sulfate do not form precipitates with the calcium, magnesium, and iron ions. Therefore, they are effective cleansing agents in hard water.

Nucleic Acids

Nucleic acids are very-high-molecular-weight polymers present in living cells. Scientists generally believe that they control the processes of heredity and take an active role in building proteins from amino acids. There are two types of nucleic acids: *deoxyribonucleic acid*, **DNA,** which is located primarily in the nucleus of the cell, and *ribonucleic acid*, **RNA,** which is located in the cytoplasm, the outer structure of the cell. Figure 15.7 is a diagram

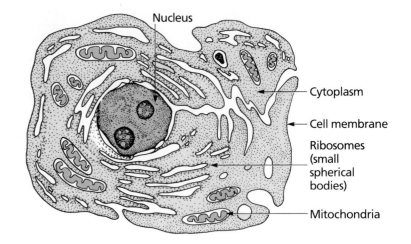

Figure 15.7 Diagram of a typical living cell.
Every living cell contains a nucleus with cytoplasm surrounding it. The cell membrane encloses everything. The whole mass of living matter is called protoplasm.

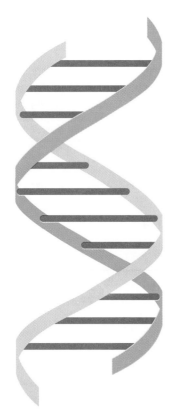

Figure 15.8 Diagram illustrating the DNA molecule.
The DNA molecule has the structure of a double-stranded helix. The spiral frames are formed from sugars and phosphates. The connecting links are composed of nitrogen compounds.

of a living cell with some of its components. DNA has the molecular structure of a double-stranded helix (see Fig. 15.8). It is composed of simple sugars (deoxyribose), phosphates, and nitrogen compounds, of which there are four: adenine, thymine, cytosine, and guanine, commonly labeled A, T, C, and G, respectively. The sugars and phosphates form the spiral frames of the double-stranded helix, and the nitrogen bases form the connecting links, with adenine always joined with thymine and cytosine always joined with guanine to form links between the spiral frames. The molecules vary in length, but they may be several inches long and contain billions of cross-linkages. The DNA molecular weight is greater than 1 million. RNA has a similar helical structure, except that the linkage is different and the molecular weight is smaller, ranging from 20,000 to 1 million.

The biological significance of DNA and how it affects heredity is opening up an exciting new field of study, because DNA is found in every living cell (plant and animal) in our environment. This topic will be discussed in Section 15.3.

Vitamins

Vitamins are organic substances that are needed in minute amounts to perform specific functions for normal growth and nutritional needs of the human body. The word *vitamine* was coined in 1912 by researchers who thought the substances were amines that were vital for good health. Since then the *e* has been dropped, because all vitamins are not amines.

Vitamin A was the first vitamin to be discovered and identified. Elmer V. McCollum, an American scientist, identified vitamin A in 1912 through an experiment in nutrition with rats. He noted that rats grew and remained healthy when fed a fat-source diet of butter and egg yolk, but their health declined and they eventually died when lard or olive oil was substituted for the butter. This observation led McCollum to believe that some mysterious substance was present in the butter or egg yolk or both. Through a series of experiments he extracted this substance, a yellow-colored oil, from the butter. He called it "fat-soluble A," which today is known as vitamin A.

Vitamin A is an alcohol, with the molecular formula $C_{20}H_{29}OH$ and the structural formula

It is found in eggs, milk, butter, and green and yellow vegetables. It increases the resistance to infections and promotes good night vision.

There are many other important vitamins (see Fig. 15.9), some whose functions are well known and others whose importance is not yet understood. Table 15.2 lists some of the important vitamins and their functions.

Figure 15.9 Vitamin C crystal.

Table 15.2 Some Common Vitamins and Their Functions

Vitamin	Function
A_1, retinol	Necessary for protein synthesis
A_2, 3-dehydroretinol	Prevents night blindness
B_1, thiamine	Necessary for carbohydrate metabolism; prevents beriberi
B_2, riboflavin	Acts as an oxygen carrier to the cell; prevents lesions of the eyes, mouth, and skin
B_6, pyridoxine	Takes part in the decarboxylation of amino acids; alleviates dermatitis, irritability, and apathy
B_{12}, cobalamin	Synthesizes certain amino acids; prevents pernicious anemia
C, ascorbic acid	Needed in the formation of tissue; prevents scurvy
D_2, calciferol (activated ergosterol)	Increases the absorption of calcium and phosphorus from the small intestines
D_3, activated 7-dehydrocholesterol	Prevents rickets
E, alpha-tocopherol	Antioxidant; no known therapeutic use
K	The antihemorrhaging vitamin

15.3 The Ingredients of Life and the Genetic Code

As we have noted, the molecular structures of living matter are quite complex. However, scientists generally agree that when the Earth was formed, only simple substances such as water vapor (H_2O), hydrogen (H_2), ammonia (NH_3), and methane (CH_4) were present. How, then, did the complex molecules of life form from these simple substances?

The first experimental insight to this question was gained in 1952 when Stanley Miller, a graduate student in chemistry, performed the now famous **Miller-Urey experiment.** In studying the process of forming complex molecules, Miller circulated through a glass container a mixture of water vapor, hydrogen, ammonia, and methane that he thought might be similar to our primordial atmosphere. Some form of energy must be present to initiate a chemical reaction in the mixture, so a sparking device was placed in the container. This sparking could conceivably be likened to lightning flashes in the early atmosphere. As the gases continuously passed through the spark, drops of liquid began to collect on the bottom of the container.

When this liquid was analyzed, it was found to contain several organic compounds that are present in living matter, the most significant of which were four amino acids, the building blocks of proteins. Thus the Miller-Urey experiment demonstrated a possible way in which the compounds of living matter might have been formed originally. A schematic diagram of the apparatus used in the Miller-Urey experiment is shown in Fig. 15.10.

Subsequent experiments yielded as many as 10 amino acids from similar reactions. This synthesis of the amino acids is extremely significant because of their relation to proteins. As we discussed previously, proteins are one of the essential components of living matter. They are produced by living cells and make up approximately 75% of the dry weight of our bodies. Proteins are also the main component of plant-cell interiors. Proteins have a complex molecular structure, being composed of chemical combinations of about 20 amino acids.

The Miller-Urey experiment demonstrated how nature might have prepared the amino acids. The next logical question, ignoring that of the beginning of life itself, might be: How do living cells form the correct protein from the various combinations of amino acids? That is, how does an animal cell generate proteins characteristic of the animal, and a plant cell, proteins characteristic of the plant? In short, how are the characteristics of life reproduced? The answer to this question, of course, involves heredity.

Why offspring resemble their parents has always fascinated us. Geneticists thought for a long time that the characteristic traits were carried in the blood and were passed to a child by mixing of the parental blood. This misconception gave rise to the idea of blood lines, such as royal blood, having bad blood, or being full-blooded for some race. Science, however, has traced the trait-carrying agent to structures called chromosomes within the nucleus of the living cell.

Chromosomes are threadlike structures that split in such a way that when a body cell divides to produce new cells, the two resulting daughter cells have the same number of identical chromosomes as the parent cell. This process of cell division is called **mitosis.** Because of

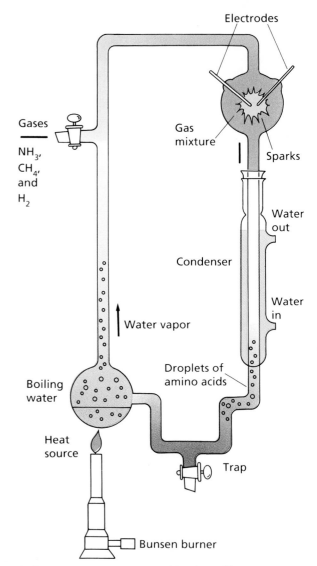

Figure 15.10 Apparatus used in the Miller-Urey experiment.
Steam, ammonia, methane, and hydrogen were introduced into the reaction chamber, and the products cooled and collected in the trap. It was found that 10 amino acids could be made in this way.

mitosis, cells are reproduced with the same characteristics. All mature body cells in the same animal or plant have the same number of chromosomes.

In another type of cell division, called **meiosis,** cells result with just half the number of chromosomes as the original cell. Meiosis occurs in the reproductive organs of sexually reproducing plants and animals and involves the sex cells—sperm in the male and egg in the female. Because the sperm and the egg cells unite in the fertilization process, it is evident that for the cell of the offspring to have the proper number of chromosomes and combination of traits, the new egg and sperm cells must each have half as many chromosomes as the original cell. Thus, as a result of meiosis, each child gets a set of chromosomes from each parent and passes half of these to its own offspring. The difference between mitosis and meiosis is illustrated in Fig. 15.11.

Human egg cells and sperm cells have 23 chromosomes each, but humans have thousands of inherited

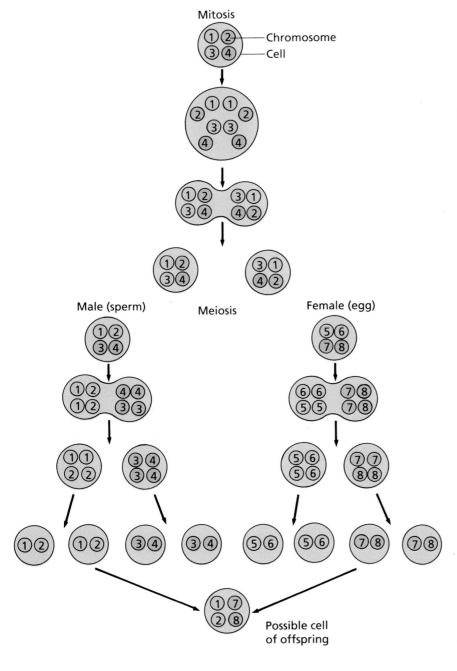

Figure 15.11 Schematic illustration of the processes of mitosis and meiosis.
In mitosis, the chromosomes divide in such a way before the cell divides that each daughter cell has the original number of chromosomes. In meiosis, the immature reproductive cells of both male and female divide so that one-half of the original number of chromosomes is present in the new cell. Thus, the correct number of chromosomes is present in the body cells of the offspring when a male and female cell unite.

traits and characteristics. Therefore, the chromosomes must have a multiple structure to carry all these traits. On a closer look at the threadlike structure of the chromosome, researchers found that what appeared to be a thread was actually made up of small beads hooked together in a necklacelike fashion. Each human chromosome might contain as many as a thousand of the beads, or **genes,** which are the carriers of inherited traits. Thus a cell may inherit two genes for the same characteristic in meiosis, only one of which is **dominant,** or influences the trait. The other is **recessive** and does not influence the characteristics of the offspring. The recessive gene is, however, a part of the cell and may be transferred again in meiosis and become dominant in some later generation. The appearance of a child with red hair in a family whose generations have all had black hair is such an example. Some ancestor of the child must have added a gene for red hair that remained recessive for several generations of the family.

Science has focused much attention on the gene. It consists of the nucleic acid DNA. As noted previously, DNA is composed of sugars, phosphates, and the nitrogen compounds adenine (A), thymine (T), cytosine (C), and guanine (G). The molecule is a spiral ladderlike structure with the sugars and phosphates forming the sides and the nitrogen compounds the rungs (see Fig. 15.8). The nitrogen compounds can form in only certain combinations, A with T and C with G. However, the combined nitrogen compounds may form the rungs of the molecular DNA ladder in any order, such as AT, CG, TA, AT, GC, CG, AT, and so on. Scientists now believe that the arrangement of the nitrogen compound sequences, or **genetic code,** gives each gene its special characteristic. A simple gene may be a small portion of the DNA molecule a few hundred or thousand rungs long. In a single human cell the DNA molecule is thought to be about 3 ft long when uncoiled and contain some 6 billion rungs—a truly complex molecule. J. Watson and F. Crick were awarded the Nobel Prize in 1962 for their work in elucidating the helical structure of the DNA molecule.

According to current theory, when a chromosome of the living cell divides in mitosis, the DNA ladder literally unzips and the two individual parts pick up more raw materials from the nucleus around them. Because an A must combine with a T and a C with a G, the old substances pick up new partners such that two duplicate DNA molecules are formed with the same code and, hence, the same genes. This unzipping and duplicating action is illustrated in Fig. 15.12.

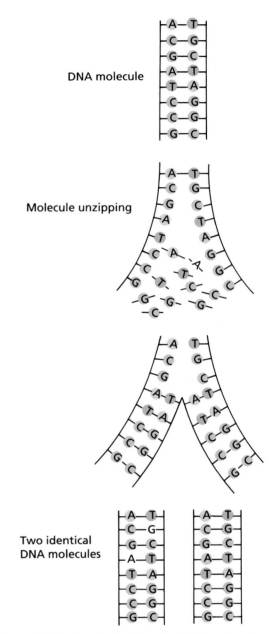

Figure 15.12 The duplication of a DNA molecule. The ladder structure "unzips" in the middle. Then the appropriate nitrogen compounds join each half of the ladder, and two new identical ladders are formed. In reality the ladders are spirals, as shown in Fig. 15.5.

Should an "accident" occur in the reconstruction of the molecule that affects the order of the DNA nitrogen compound pairs, the characteristic of the gene is altered and a **mutation,** or genetic change, may occur. Mutations may arise in a number of ways not fully understood. They may be caused by photons (gamma rays, X-rays, light, etc.) that hit the molecule, or mutations may be caused by chance. According to quantum mechanics, events have only a probability of occurring. That is, there is a high probability the gene will exactly reproduce. Sometimes, by chance, it does not.

To complete the story and return to the original question of how the proteins are reproduced, one must consider another nucleic acid, ribonucleic acid (RNA). This nucleic acid is a single-stranded molecule that is similar in structure to the DNA molecule. The biosynthesis of protein occurs on subcellular structures called ribosomes, which are found in the cytoplasm of the cell. Energy is required to form the peptide bond, and specific enzymes are required to catalyze the chemical reactions that bring about the condensation of the amino acids into the peptide chain. The amino acid must be condensed into a chemical called transfer ribonucleic acid (tRNA). The tRNA orients an amino acid so that it will be brought to the precise adjacent amino acid to form the protein. Information providing the exact sequence of amino acids in the proteins to be synthesized is furnished by the nucleic acid called messenger ribonucleic acid (mRNA). Thus the amino acids are assembled in a predesignated way, and the proteins and cells grow as dictated by the individual's inherited traits contained in the genetic DNA code.

The science of genetics entered a new era in the late 1970s with techniques called recombinant DNA technology, or genetic engineering. This new technology came about primarily because of (1) new instrumentation and techniques for understanding molecular genetics, (2) the use of radioactive tracers that can detect tiny amounts of specific macromolecules, and (3) the use of enzymes known as nucleases that hydrolyze nucleic acids.

Through genetic engineering, scientists can now transfer genes from mammals into bacteria and cause the microbes to produce large amounts of protein that can then be used for the treatment of diseases. Human insulin can now be synthesized by recombining the human insulin gene into a plasmid, a small fragment of DNA found in bacterial cells. Interferon is a protein that is sensitive to certain viruses and interferes with viral replication. Only small amounts of interferon are available from natural sources. Through genetic engineering interferon can be produced in large amounts for the treatment of certain types of cancer and for medical research. A major goal of genetic engineering is the cure of mo-

lecular diseases such as cystic fibrosis, goiter, hemophilia, and sickle-cell anemia, plus many others.

15.4 **Artificial Molecules—Plastics**

Chemists have for a long time tried to duplicate the compounds of nature. Alchemy, or the attempt to transform chemically base metals into gold, was an early example of one such effort. As the science of chemistry progressed and the formulas and basic components of some of the natural compounds became known, chemists were able to synthesize some of these natural compounds by causing the appropriate combinations of elements or compounds to react. During this early trial-and-error development there were, no doubt, as many accidental as deliberate discoveries.

From the attempts to synthesize nature's compounds, synthetics were discovered. A **synthetic** is a material whose molecule has no duplicate in nature. The first synthetic was prepared by Leo Baekeland in 1907. By mixing phenol (C_6H_5OH) and formaldehyde (HCHO), he produced a polymeric combination of these two compounds, $HOCH_2C_6H_4OH$. The polymer is commercially known as Bakelite, a common electrical insulator.

This discovery set off a series of investigations for synthetic materials. Necessity also helped catalyze these endeavors. During World War II many natural supplies, such as rubber, were cut off, thus providing impetus to the search for synthetic products that would take the place of the natural materials that were in short supply. Chemists became aware that substituting different elements or compounds in a complex molecule would change its physical properties. For example, the substitution of a chlorine atom for a hydrogen atom in ethane produces ethyl chloride, which has vastly different properties from ethane. By knowing the general properties of the substituted groups, a chemist can tailor a chemical molecule to satisfy a given requirement.

As a result of this scientific approach, a host of synthetic molecules have been constructed. Probably the best known of these is the group of synthetic polymers commonly called **plastics.** Plastics have become an integral part of our modern life, being used in clothing, shoes, buildings, autos, sports, art, electrical appliances, toothbrushes, toys, and a myriad of other things. (See Fig. 15.13.)

Not only are the properties of plastics affected by their chemical composition, but the alignment of their long molecular chains can also cause differences in their

Figure 15.13 Manufacturing of synthetic plastic.

Monomers of Three Common Polymers

$$\left[\begin{array}{cc} H & H \\ | & | \\ -C-C- \\ | & | \\ H & H \end{array}\right]_n$$

Polyethylene

$$\left[\begin{array}{cc} H & H \\ | & | \\ -C-C- \\ | & | \\ H & CH_3^* \end{array}\right]_n$$

Polypropylene

$$\left[\begin{array}{cc} H & H \\ | & | \\ -C-C- \\ | & | \\ H & \bigcirc^{**} \end{array}\right]_n$$

Polystyrene

qualities. For example, if a thin sheet of polymer is heated to its softening point, then stretched and cooled while held fixed, the molecular chains are aligned along the direction of stretch and the material is said to be oriented, being less flexible than before. Should the material be reheated without being held, the sheet will shrink back to its original unstretched unaligned state. This property is of great use in the packaging industry, where meats or other items are wrapped in stretched material, which is then heated. The result is a skintight polymer wrapping. The next time you have occasion to hold a piece of plastic wrapping, stretch it to align the molecules. If the plastic will not stretch, then the material is probably already oriented. Heating will unalign the molecular chains if it has been stretched while heated. Other chemical "tricks," such as putting short side chains or branches on the main molecular chains, can affect the physical properties of a complex molecule. The long polymeric chains may also be chemically cross-linked, or tied together, to obtain desirable properties.

We will consider a few of the more common plastics, along with their monomer unit, to show how these molecules are constructed. Three polymers of similar structure are listed below, with the monomer of each.

Polyethylene is the simplest of the artificial molecules and may be thought of as a very long alkane; nature's alkanes do not have the thousands of repeating units found in synthetic molecules. By replacing a hydrogen atom in the polyethylene monomer with a methyl group or a phenyl group, one gets polypropylene and polystyrene, respectively. Changing the molecular structure changes the physical properties of the polymers. For example, the melting points of the three polymers are 135°, 175°, and 220°C, respectively.

These polymers have many applications. Because of their chemical inertness, they are used for chemical storage containers as well as many other packaging applications.

Another common polymer of simple structure that has found many uses, especially in coating cooking utensils, is Teflon, a fluorocarbon resin. Its structure is

$$\left[\begin{array}{cc} F & F \\ | & | \\ -C-C- \\ | & | \\ F & F \end{array}\right]_n$$

Teflon

---

* $-CH_3$ is referred to as the methyl group.

** $\bigcirc$ is the symbol for benzene with one hydrogen missing ($-C_6H_5$), which is called a phenyl group.

When the hydrogen atoms in the ethylene monomer are replaced with fluorine atoms, a compound called tetrafluoroethylene is produced. It polymerizes to form the hard, strong, high-melting-point, low-surface-friction, chemically resistant material known by the trade name *Teflon*. When a frying pan is coated with Teflon, foods will not stick. We should also mention that the majority of the polymers are good electrical insulators and are used to great advantage in this application.

Some polymers may be made in fiber or thread form; hence clothing, rugs, and even artificial grass for sporting events are made from plastics. One such polymer is nylon, which has found wide use in hosiery (see Fig. 15.13). Nylon has a somewhat complicated monomer,

Nylon

Other polymers, such as rayon and Dacron, are well-known substitutes for natural fibers.

Technology and the development of polymers have gone hand in hand, each aiding the other. As mentioned, early wartime necessity spurred the development of synthetics. When the supply of natural rubber was cut off, a synthetic rubber called neoprene was developed. The different structures of natural rubber and neoprene are

Rubber (polymer of isoprene)

Neoprene

However, need did not precede discovery in all cases. Our entry into the space age created a great demand for different and unusual materials. But some materials that could stand the rigorous space environment were already available. Two examples are polyethylene terephthalate and polycarbonate, known commercially as Mylar polyester film and Lexan polycarbonate resin, respectively. They are extremely tough materials—the Echo satellites were constructed of thin sheets of Mylar, while the durable Lexan serves as astronauts' helmet visors. They also have everyday applications.

Another group of polymers with applications from bathtub sealer to spaceships is the silicones. These compounds are rubbery and flexible. One such polymer, RTV 615, is elasticlike rubber, transparent, and has a melting point four times that of steel. It has been used as heat shields for space vehicle reentry where the temperatures are in excess of 8000°C. This material transmits virtually no heat, owing to a slow, layer-by-layer decomposition of its molecules that uses excess heat as the latent heat of vaporization.

The list of polymers and their uses could go on and on. However, it should be evident from the few mentioned that artificial molecules are a tribute to our ingenuity.

15.5 Drugs

A **drug** is a compound that may produce a physiological change in human beings or other animals and may become a poison when used in excessive amounts. The molecular composition of drugs is usually quite complex, because the majority of drugs are extracted from natural compounds. Because of the human involvement and the complexity of the human body, the development of new or synthetic drugs is relatively slow compared with the development of other materials such as plastics.

A great degree of caution must be exercised in administering new drugs, and many experiments must be run on laboratory animals in an effort to find any side effects that may be harmful to humans. Occasionally, these effects are latent and emerge only after widespread use, as in the case of thalidomide, the sleeping compound believed to be a cause of deformities in the human fetus. Often there is no conclusive evidence of a compound being harmful to humans, but preventive action is taken anyway, as in the case of the banning of cyclamate sweeteners when research showed that large doses caused cancer in rats.

From our definition of a drug we can see that drugs are useful for purposes other than medicinal, although they are in large part used to correct some body malfunction. One of the oldest drugs known is alcohol. As a by-product of fermented natural beverages, ethyl alcohol has been consumed by humans probably since the beginning of civilization. Some alcohols have severe physiological effects; methyl alcohol, for example, may cause blindness if taken internally. In small amounts ethyl alcohol acts as a depressant, although it is commonly, and falsely, considered to be a stimulant. Acting on the brain's control mechanisms, it can cause loss of muscular control, such as slurred speech, mental disorganization, sleep, and nausea when taken in moderate amounts. In addition, alcohol acts on the parts of the brain that control blood circulation and the kidneys. When alcohol is taken into the body, the capillaries dilate, thus bringing more blood nearer the surface of the body. This effect causes a warm feeling but also a more rapid loss of body heat. Its effect on the kidneys is to depress the control center of the brain that ordinarily slows the kidney's excretion. Addiction to alcohol is also possible after prolonged use. Figure 15.14 shows some common substances containing drugs.

A widely used drug is aspirin. **Aspirin** is the compound acetylsalicylic acid:

Aspirin
(acetylsalicylic acid)

Aspirin is used primarily as an analgesic. An **analgesic** is a drug that relieves pain without dulling consciousness. Aspirin also has the ability to reduce body temperature and is thus effective for reducing fevers. It is estimated that enough acetylsalicylic acid is produced in the United States to supply every person in the United States with 400 aspirin tablets each year.

Relatively recently developed drugs are the antibiotics, or "wonder" drugs. An **antibiotic** is an antibacterial substance that is produced by a living organism and is administered to fight bacterial infections in the body. Before the development of the wonder drugs, death was quite common from serious infections, and their discovery was one of the major milestones in medicine. Probably the best known of these drugs are the penicillins, which are derived from molds of a certain type. There are now many other wonder drugs such as the mycins, which have proved to be more effective than penicillin in many cases.

Another group of drugs called **alkaloids** are complex, nitrogen-containing compounds found in plants. They include cocaine ($C_{17}H_{24}NO_4$), morphine ($C_{17}H_{19}NO_3$), quinine ($C_{20}H_{24}N_2O_2$), nicotine ($C_{10}H_{14}N_2$), and strychnine ($C_{21}H_{22}N_2O_2$). Strychnine is best known as a heart stimulant, although in large amounts it is a poison, as are all drugs.

Tobacco is a source of nicotine and contains from approximately 0.5 to 5% nicotine in its leaves. Because it is a dangerous poison, nicotine finds little use as a drug, but it is an effective insecticide. Quinine was used as an early treatment for malaria but has been replaced by more effective drugs. Morphine is a narcotic found in crude opium, which is extracted from the oriental poppy. It is used to relieve pain and induce sleep. Addiction to this drug may occur with prolonged use.

There has been increasing concern over the misuse of drugs, especially the narcotics. **Heroin,** a derivative of morphine, is a habit-forming drug. Once the habit is formed, the craving for drugs is insatiable; and the user will go to any extreme, including violence, to obtain a certain drug or the money to buy it, thus creating a serious social problem. **Marijuana** comes from the leaves and flowering tops of the Indian hemp plant. Smoking or ingestion of marijuana causes dizziness, hilarity, delusions of grandeur, heightened mental awareness, and mental confusion, depending on the user. Scientists generally agree that marijuana is not habit forming, and its effects are similar to those of alcohol. However, the user becomes apathetic and may, after prolonged and heavy use, suffer permanent loss of mental acuity.

Another group of drugs are the psychedelics, or hallucinogenic drugs, one of which is **LSD** (lysergic acid diethylamide). This dangerous drug causes prolonged stimulation, which may be followed by insanity. Sen-

Figure 15.14 Common substances containing drugs.

sations are amplified, and exaggerations and hallucinations are experienced. The LSD molecule is quite complex (Fig. 15.15). The lower uncolored portion of the formula shows an *indole* ring; such a ring is found in a similar structure called serotonin. **Serotonin** is a compound found in the brain and is believed to transfer signals to the nerve cells. The theory is that because of a structure similar to that of serotonin, LSD interferes with this transfer, and the signals are altered. Thus the drug could change the sensory signals, producing hallucinations and possibly almost any other reaction. Because of the risk of lasting brain disorders (including "flashbacks"), LSD is a drug that poses serious risk to the user.

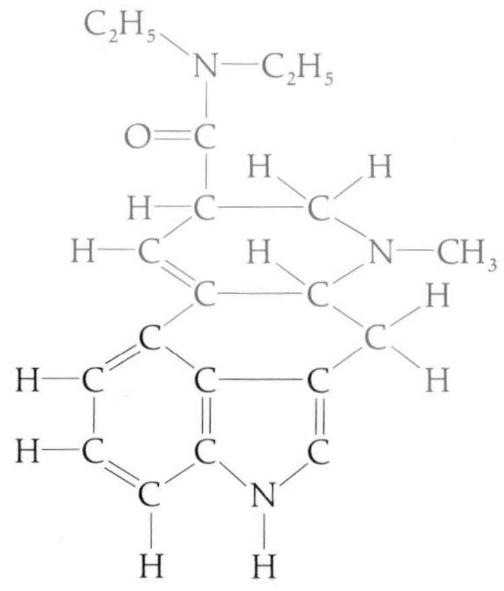

Of a somewhat different social impact are the drugs contained in birth control pills, another success with complex molecules. The drugs used in these pills are synthesized and have proved to be more effective than similar natural compounds. The first of these drugs to be synthesized was **norethynodrel.** Scientists had known for some time that the hormone progesterone would prevent ovulation in the human female about 85% of the time; this hormone, however, is hardly an effective contraceptive. By synthesizing compounds similar in structure to progesterone, researchers found a compound, norethynodrel, that turned out to be virtually 100% effective at one-third the dosage of progesterone when given with a small amount of another hormone, estrogen.

Figure 15.15 LSD molecule.

Learning Objectives

After reading and studying this chapter, you should be able to do the following without referring to the text:

1. Write the organic functional group for a few types of organic compounds.

2. List one or two uses for each of the organic compounds formed by using the functional groups referred to in Learning Objective 1.

3. State the product formed when an organic acid reacts with an alcohol.

4. State and explain the general formula for carbohydrates.

5. Describe the chemical structure of proteins.

6. State the similarities and differences of fats, oils, and waxes.

7. Describe the general structure of the DNA and RNA molecules.

8. Explain the function of DNA and RNA in cell reproduction and the genetic code.

9. Describe the Miller-Urey experiment, and state its significance.

10. Define or explain the important words and terms listed in the next section.

Important Words and Terms

complex molecules	alcohols	ester
monomer	aldehydes	carbohydrates
polymer	organic acids, carboxylic acids	starch
alkyl halide	carboxyl	cellulose

hydrogenation	mutation	mitosis	antibiotic
waxes	synthetic	meiosis	alkaloids
detergent	plastics	genes	heroin
nucleic acids	drug	dominant	marijuana
DNA	proteins	recessive	LSD
RNA	amino group	genetic code	serotonin
vitamins	lipid	aspirin	norethynodrel
Miller-Urey experiment	chromosomes	analgesic	

Questions

Common Organic Compounds

1. Describe the structure of a polymer.

2. What functional group characterizes an alkyl halide? an alcohol? an aldehyde? an acid? an ester?

3. How are aldehydes formed chemically?

4. By what general chemical reaction are alcohols formed?

5. Name the alcohol used in alcoholic beverages. Draw its structural formula.

6. What is the result of oxygen reacting with an aldehyde?

7. Write in words the general equation for the reaction of an organic acid with an alcohol.

Biochemistry

8. What is the meaning of the word *carbohydrate?* Is the nature of carbohydrate compounds what the name implies? Explain.

9. What is the chemical composition of proteins?

10. What is hydrogenation?

11. Give the reason that termites can digest cellulose, the most abundant organic substance, but humans cannot.

The Ingredients of Life and the Genetic Code

12. Distinguish between mitosis and meiosis.

13. How is the DNA molecule constructed?

14. How does the DNA molecule reproduce itself?

15. Explain the function of DNA and RNA in cell reproduction and the genetic code.

Artificial Molecules—Plastics

16. What is the meaning of the term *synthetic?*

17. What is the common name for synthetic polymers?

18. Give the names of three common polymers.

19. What is the trade name for polyethyene terephthalate?

Drugs

20. What is the definition of a drug?

21. Is asprin a drug?

22. What is an analgesic?

23. What is the acetylsalicylic acid content of regular aspirins?

24. Describe the effects of LSD use by humans, and give the probable cause.

Exercises

Common Organic Compounds

1. Considering the monomer unit CH_2, what is the repetition number of each of the following?
   (a) *n*-octane   (b) *n*-eicosane   (c) polyethylene

2. Complete and balance the following equation. Name the compound formed.

$$CH_3COOH + C_2H_5OH \longrightarrow \underline{\quad\quad} + H_2O$$

      Acetic   Ethyl
       acid   alcohol

3. Complete and balance the following equation. Name the compound formed.

$$CH_3OH + O_2 \longrightarrow \underline{\quad\quad} + H_2O$$

      Methyl
      alcohol

4. Complete and balance the following equation.

$$CH_2O + O_2 \longrightarrow \underline{\quad\quad}$$

                  Formic acid

5. Complete and balance the following equation.

$$CO_2 + H_2O + energy \longrightarrow \underline{\hspace{1cm}} + O_2$$
$$\text{Glucose}$$

6. Given the general formulas below, write molecular and structural formulas for the first member of each group. Then name each compound.

$$R\text{—OH} \quad R\overset{\overset{\displaystyle O}{\|}}{\text{—C—H}} \quad R\overset{\overset{\displaystyle O}{\|}}{\text{—C—OH}} \quad R\overset{\overset{\displaystyle O}{\|}}{\text{—C—O—R}}$$

Biochemistry

7. Write the general formula for (a) the alkyl group and (b) a carbohydrate.

8. Show the difference in the structural formulas for glucose and fructose.

9. Determine the percentage of protein in a 100-g sample of food, if the sample contains 5 g of nitrogen.

10. From Table 15.2, prepare a list of the food sources of the important vitamins. Explain the effects of a deficiency of these important vitamins in a person's diet. (Outside source work will be necessary.)

The Solar System

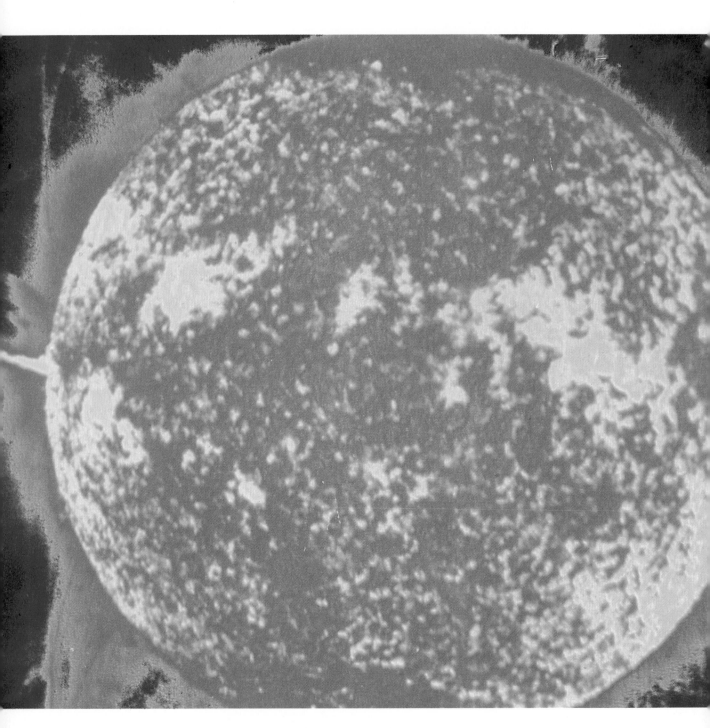

The acquisition of knowledge begins with simple fundamental ideas that expand to broader and more complex concepts.

A S WE GO about our daily lives, we often lose our sense of curiosity and wonder. But on a clear dark night when we gaze toward the heavens, we are struck by the awesome grandeur of the universe. At times like these, we marvel at our seeming insignificance, and we begin to ask ourselves questions. What is the universe? How did it begin, and how will it end? When and where did we originate? Can we possibly know all things and control the forces of the universe?

The science concerned with such questions is astronomy, probably the oldest science. The ancient Babylonians and Greeks were two early civilizations that watched the skies and originated many of the words used today in describing our universe. They were limited in what they could observe by the ability of the unaided eye. With the invention of the telescope and the discovery and development of photography, more accurate measurements of celestial objects were possible.

In 1957 the first artificial satellite was launched by the Soviet Union. Shortly thereafter, the United States began its own space program. The National Aeronautics and Space Administration (NASA) was formed, and astronauts and many artificial satellites were propelled into space. Most of these satellites orbited the Earth, but some were sent to the moon, Mars, Venus, Mercury, Jupiter, Saturn, and Uranus. As these satellites transmitted pictures and data back to the Earth, our knowledge of the solar system exploded. In fact, the well-read college student of today knows more about our solar system than the most distinguished scientist knew in 1960.

As we come to learn more about the moons and planets of our solar system, we also learn more about our own planet Earth. But as we discover more, our sense of curiosity increases, for there are always more and more questions needing answers.

◄ The Sun, showing solar activity.

16.1 The Planet Earth

The Earth is one of nine planets that revolve around the Sun. These nine planets and their four dozen or so known satellites, plus thousands of asteroids and countless comets and meteoroids, make up the local **solar system.** Other similar planetary systems may exist in the universe (see Section 16.6).

Although we are unable to sense directly the motion of our home planet, it is undergoing several motions simultaneously. Two that have major influences on our daily lives will be explained in this section: (1) the daily rotation of the Earth on its axis and (2) the annual revolution of the Earth around the Sun. A third motion, precession, will be discussed in Section 17.5.

The Earth revolves eastward around the Sun and sweeps out a plane called the orbital plane of the Earth. This motion of the Earth produces an apparent annual westward motion of the Sun on the celestial sphere. The apparent path of the Sun on the celestial sphere is called the **ecliptic.** The word *ecliptic* is derived from eclipse, because eclipses of the Sun and moon occur when the moon is on or near the great circle forming the apparent annual path of the Sun.

The Earth is rotating eastward around a central internal axis that is tilted 23.5° from a line that is perpendicular to the orbital plane of the Earth. Later discussion will show why the 23.5° tilt of the axis and the revolving of the Earth around the Sun is the reason for the four seasons we experience annually.

For a study of astronomy one must know the difference between rotating and revolving. A mass is said to be in **rotation** when it spins on an axis. An example is a spinning toy top or a ferris wheel at an amusement park. Revolving, or **revolution,** is the movement of one mass around another. The Earth revolves around the Sun, the moon revolves around the Earth, and electrons revolve around the nucleus of an atom.

The fact that the Earth rotates on its axis was not generally accepted until the nineteenth century. A few scientists had considered the possibility of a rotating Earth, but no definite proof was given to support their beliefs; therefore, their ideas were not acceptable.

In 1851 an experiment demonstrating the rotation of the Earth was performed by J. B. Léon Foucault (1819–1868), a French engineer. He used a 200-ft pendulum that today is called a **Foucault pendulum.** More noticeable results can be seen if the experiment is performed at the north or south pole.

Picture a large single-room building with a ceiling over 200 ft high located at the north pole. See Fig. 16.1. Fastened to the ceiling, precisely above the north pole, is a support hook having very little friction, from which a 200-ft, fine steel wire is attached. Connected to the lower end of the wire is a massive iron ball with a short sharp steel needle attached permanently to its underside. On the floor, under the pendulum, is a layer of fine sand, which can be slightly furrowed by the sharp needle as the pendulum swings back and forth.

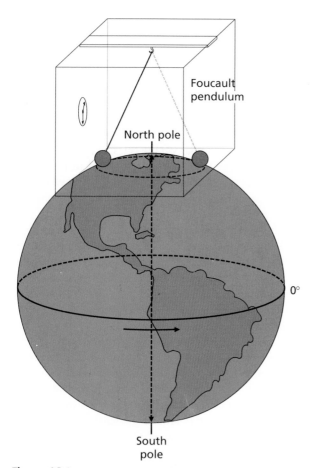

Someone starts the pendulum swinging by displacing it to one side with a strong fine thread, with one end attached to the side of the ball and the other end attached to one wall of the building where a 24-h wall clock is mounted. To prevent any sideways motion, the iron ball is allowed to become motionless before the thread is parted by burning. Extreme care is taken to prevent any lateral external forces from being applied to the upper support point of the 200-ft wire. As the pendulum swings freely back and forth, the sharp needle point traces its path in the layer of fine sand. After a few minutes the plane of the swinging pendulum appears to be rotating clockwise, as shown by the markings in the sand. At the end of an hour the plane of the swinging pendulum has rotated 15° clockwise from its original position. When 6 h have elapsed, the plane of the swinging pendulum appears to have rotated 90° clockwise and is parallel to the wall that holds the 24-h clock. With the passing of each hour the plane of the swinging pendulum appears to rotate another 15° clockwise. At the end of 24 h the plane has made an apparent rotation of 360°

A person who believes in a motionless Earth would argue that the pendulum actually rotated 360°, because one rotation of the swinging pendulum has been observed by anyone stationed in the large room. A different view of the experiment can be obtained if we make the walls of the building of a transparent material such as glass and perform the experiment sometime during the winter months for the Northern Hemisphere. The north pole has 24 h of darkness during these months, and the stars are always visible. When starting the pendulum

Figure 16.1
Diagram showing a Foucault pendulum positioned in a room located at the North Pole of Earth. To an observer in the room, the pendulum will appear to change its plane of swing by 360° every 24 hours.

this time, we take care to place the iron ball in direct line with the stars Alpha and Beta, the pointers in the cup of the Big Dipper. As the minutes pass, we observe, as before, the apparent rotation of the plane of the swinging pendulum in a clockwise direction in reference to the large room and the clock on the wall. We also observe, through the transparent walls of the room, that the pendulum still swings in the same direct line with the stars Alpha and Beta; that is, the pendulum, Alpha, and Beta are in the same plane. There has been no rotation of the pendulum in reference to the fixed stars. No forces have been acting on the pendulum to change its plane of swing. Only the force of gravity has been acting vertically downward to keep it swinging. Therefore, the pendulum does not rotate, but the building and the Earth rotate eastward, or turn counterclockwise, once during the 24-h period. A rotating Earth has a major influence on weather, deflecting winds from their paths and causing cyclones and other cyclic storms.

The Foucault pendulum is an experimental proof of rotation. What kind of an effect can we expect because of the Earth's motion around the Sun?

As the Earth orbits the Sun once each year, the relative positions of nearby stars change with respect to more distant stars. This effect is called parallax. In general, **parallax** is the apparent motion, or shift, that occurs between two fixed objects when the observer changes position. To see parallax for yourself, hold your finger at a fixed position in front of you. Close one eye, move your head from side to side, and notice the apparent motion between your finger and some object out the window. Note also that the apparent motion becomes less as you move your finger farther away. Figure 16.2 is an illustration of the parallax of a nearby star as measured from the Earth in relation to stars that are more distant.

As the Earth revolves around the Sun, there is an apparent shift in the positions of the nearby stars in respect to the stars that are farther away because of the motion of the Earth. Because the stars are at very great distances from the Earth, the parallax angle is very small.

Because the stars are so far away, the parallax of the stars illustrated in Fig. 16.2 cannot be seen with the unaided eye. It was first observed with a telescope in 1838 by Friedrich W. Bessel (1784–1846), a German astronomer and mathematician. The observation of parallax was undisputed proof that the Earth really does go around the Sun. Today, the measurement of the parallax angle is the best method we have of determining the distances to nearby stars.

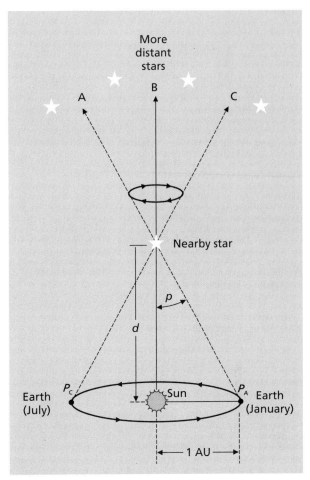

Figure 16.2
Parallax of a star is the apparent displacement of a star that is located fairly close to the Earth, with respect to more distant stars. When the observer is at P_A, the star appears in the direction A. As the Earth revolves counterclockwise, the star appears to be displaced and appears in the direction indicated for different positions of the Earth. Positions P_A and P_C are a few months apart. The angle of parallax p is also shown.

A second proof of the Earth's orbital motion around the Sun is the telescopic observation of a systematic change in the position of all stars annually. The observed effect (called the aberration of starlight) is due to the finite velocity of light and the motion of the Earth around the Sun. The **aberration of starlight** is defined as the apparent displacement in the direction of light coming from a star owing to the orbital motion of the Earth.

If you have driven a car during a snow or rain storm, you have observed the aberration of snowflakes or raindrops. Assume snow is falling vertically and your car is at rest. You will observe snowflakes falling vertically downward. Start the car and drive northward at a slow speed. Now you will observe the snowflakes coming toward the windshield of the car at a slight angle from the vertical. If you increase the speed of the car, you will observe the snowflakes coming toward you at a greater angle with the vertical. Stop the car, and you will observe the snowflakes falling vertically again. If you now travel east, south, west, or any direction, you will observe similar effects. It is the motion of the car that produces the apparent change in the direction of the snowflakes.

The great distance to stars is measured in a unit called the parsec. The unit's name comes from the first three letters of the word *parallax* plus the first three letters of the measuring unit, the second, which is used to measure angle.

A circle contains 360°. A degree is divided into 60 equal divisions, each of which is called a minute; and the minute is further divided into 60 equal divisions, each of which is called a second. Thus one second is an angular measurement equal to $\frac{1}{3600}$ of a circle. One **parsec** is defined as the distance to a star when the star exhibits a parallax of one second. The parsec will be explained in greater detail in the next chapter.

The Earth is an oblate spheroid—flattened at the poles and bulging at the equator. Its shape is due primarily to the rotation of the Earth on its axis. Although the difference in the diameter at the poles and at the equator is about 27 mi, the difference is very small when compared with the total diameter of the Earth, which is about 8000 mi. The ratio of 27 to 8000 is approximately $\frac{1}{300}$, which is a rather small fraction. If the Earth were represented by a basketball, which is approximately 10 in in diameter, the eye would not detect a difference of $\frac{1}{30}$ in in the diameter. The Earth is a more nearly perfect sphere than the average basketball.

The full Earth appears 4 times as large in diameter and more than 60 times as bright as the full moon when viewed from a distance of 240,000 mi. The Earth appears much brighter because the clouds and water areas are much better reflecting surfaces than the dull, dark surface of the moon.

The Earth has the highest density of any planet. The Earth's density is 5.5 g/cm³, or 5.5 times as dense as water.

The Greek scientist Eratosthenes calculated the circumference of the Earth in 250 B.C. He knew that the noon Sun on the first day of summer was directly overhead (on the zenith) at Syene (now Aswan), Egypt. Eratosthenes lived in Alexandria, which was located 500 mi due north of Syene. He had received reports that deep wells at Syene were lighted all the way to the bottom on the first day of summer, which meant that the Sun was directly overhead at Syene. At Alexandria, Eratosthenes discovered that a vertical stick cast a shadow on this same date, which positioned the Sun about $7\frac{1}{2}°$ south of his zenith. Assuming the Earth to be a sphere and the Sun to be such a great distance away that the rays of light would come in parallel to one another, he calculated the diameter of the Earth to be 7850 mi, which is very close to the value of 7918 that we use today.

Figure 16.3 illustrates the principles involved in Eratosthenes' calculations. Because there are 360° in a circle, and the distance between Syene and Alexandria is known, the *circumference* can be calculated by using the

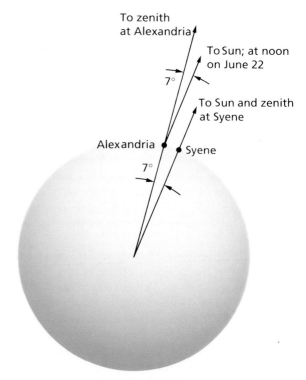

Figure 16.3
The method Eratosthenes used to calculate the size of Earth. The positions of Alexandria and Syene are shown. He calculated the circumference of Earth to be 24,000 mi, which is very close to the value we accept today.

following relationship:

$$\frac{circumference}{360°} = \frac{\text{dist. Syene to Alexandria}}{7.5°}$$

$$circumference = \frac{360° \times 500 \text{ mi}}{7.5°}$$

$$= 48 \times 500 \text{ mi}$$

$$= 24{,}000 \text{ mi}$$

This value is very close to our present value of approximately 24,900 mi.

16.2 The Solar System

The rotating motion of the planet Earth was not an easy concept to prove. In early times most people were convinced that the Earth was motionless and that the Sun, moon, planets, and stars revolved around the Earth, which was considered the center of the universe.

This inadequate theory of the solar system was called the Earth-centered, or **geocentric, theory.** The geocentric theory was supported and extended by Aristotle (384–322 B.C.), who was considered a great philosopher.

There were several important problems with the geocentric theory of Aristotle. Two outstanding faults were (1) a failure to explain the variation of brightness of the planets at different times of the year and (2) a failure to explain the motion of some planets.

In order to solve the difficulties of varying brightness and planetary motion, Claudius Ptolemy of Alexandria (second century A.D.) modified and extended the geocentric theory. In the Ptolemaic model the planets moved in circles within circles, called epicycles and deferents. The Ptolemaic theory was extremely complicated, but it was accepted with few questions for over one thousand two hundred years.

The first philosopher to give serious thought to the fact that the Earth rotates about an axis was Ponticus Heracleides (375–310 B.C.), a member of the Greek Academy. The Greek astronomer Aristarchus of Samos (about 310–230 B.C.) is credited with extending the idea of daily rotation of the Earth and originating the Sun-centered, or **heliocentric, theory,** which placed the Sun at the center of the universe with the Earth, moon, and planets revolving around the Sun in different orbits and at different speeds.

These ideas were not popular and were forgotten, probably because it was difficult for people to believe the ground upon which they lived could be in rapid motion. The theory also disagreed with the philosophic doctrine that the Earth was the center of the universe. Aristarchus also failed to support his theory with mathematical details for use in predicting future planetary positions. Lastly, he failed to satisfy his critics concerning the apparent motion stars would have if the Earth's position changed through large distances, as he claimed, in revolving around the Sun.

We now know the answers to these earlier objections to the heliocentric theory. The Earth's atmosphere rotates and revolves with it, and the acceleration caused by these motions is very small. Thus the movement of the Earth is not immediately obvious. The second objection concerned the fact that parallax was not observed. This concept was discussed in the preceding section, and we now know that parallax occurs, but it cannot be seen with the unaided eye because the stars are so far away.

The heliocentric theory remained buried for 18 centuries. The rebirth of the theory and the development of the solar system as we think of it today began with Nicolaus Copernicus (1473–1543), a Polish astronomer (Fig. 16.4), who viewed Aristarchus' heliocentric theory

Figure 16.4
Nicholas Copernicus, a Polish astronomer who developed mathematical proof of the heliocentric theory.

as absurd when he first studied it. But much dissatisfaction was building against the Ptolemaic, or geocentric, system. This theory, which had been useful for more than a dozen centuries for predicting positions of heavenly bodies, was becoming unreliable. There was doubt about the exact length of the year, and the calendar needed to be revised. Copernicus was invited by the church to express his opinion on the revision, but he refused, saying that the position of the Sun and moon were not known with sufficient accuracy to make a reliable calendar.

Copernicus observed and studied the problem for many years and finally became convinced of the validity of the heliocentric theory. He published his work under the title *On the Revolution of the Celestial Orbs* in 1543, the year of his death. Copernicus had succeeded where Aristarchus had failed, because of mathematical proofs that could be used to predict future positions of the planets. Although the results were no more accurate than those attained with the Ptolemaic method, they were much simpler to use. Because Copernicus held to the idea of uniformly moving concentric spheres, he had to retain the epicycle motions, but they were greatly reduced in number.

There was general public opposition to Copernicus's Sun-centered theory. The same arguments that were used against Aristarchus were again applied against Copernicus, but the scientists calculating the positions of the planets used the heliocentric theory in most cases because of its simplicity. The full acceptance of the theory that the Earth rotates about an axis and is one of the planets that revolve around a fixed Sun had to wait another century, until more accurate predictions could be made of planetary motion and reasons given as to why the planets move as they do.

After the death of Copernicus, the study of astronomy was continued and developed by several men, three of whom made their appearance in the last half of the sixteenth century. Notable among these men was Tycho Brahe (1546–1601), a Danish astronomer who built an observatory on the island of Hven near Copenhagen and spent most of his life observing and studying the stars and planets (Fig. 16.5). Brahe is considered the greatest practical astronomer since the Greeks. His measurements of the planets and stars, all made with the unaided eye (the telescope had not been invented), proved to be more accurate than any previously made. Brahe's data, published in 1603, were edited by his colleague Johannes Kepler (1571–1630), a German mathematician and astronomer who had joined Brahe during

Figure 16.5
Tycho Brahe, a Danish astronomer, is known for his very accurate observations, made with the unaided eye, of the positions of stars and planets. The instrument shown, a large quadrant, was used to make these measurements.

the last year of his life. After Brahe's death, his lifetime of observations were at Kepler's disposal, and they proved very useful in providing the data necessary for the formulation of the laws we know today as Kepler's laws of planetary motion.

Kepler was very interested in the irregular motion of the planet Mars. He spent considerable time and energy before he came to the conclusion that the uniform circular orbit proposed by Copernicus was not a true representation of the observed facts. Perhaps because he was a mathematician, he saw a simple type of geometric figure that would give the correct solution to his calculations while removing all the epicycles used by Copernicus. Kepler's first law, known as the **law of elliptical paths**, states:

> **All planets move in elliptical paths around the Sun with the Sun at one focus of the ellipse.**

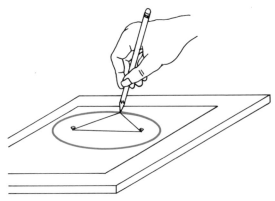

Figure 16.6
An ellipse can be drawn by using two thumb tacks, a
loop of string, a pencil, and a sheet of paper.

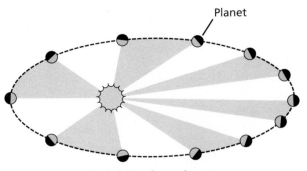

Figure 16.7 Kepler's law of equal areas.
As Earth revolves about the Sun, equal areas are swept
out in equal periods of time.

Note that there is nothing at the other focus of the
ellipse.

An ellipse is a figure that is symmetrical about two
unequal diameters. See Fig. 16.6. An ellipse can be drawn
by using two thumbtacks, a closed piece of string, paper,
and pencil. The points where the two tacks are positioned
are called the foci of the ellipse.

Kepler's first law gives the shape of the orbit but
fails to predict when the planet will be at any position
in the orbit. Kepler, aware of this, set about to find a
solution from the mountain of data he had at his disposal.
After a tremendous amount of work, which was carried
out with no indication that a solution was possible, he
discovered what is now known as Kepler's second law
of planetary motion, or **law of equal areas,** which states:

> **An imaginary line (radius vector) joining a
> planet to the Sun sweeps out equal areas in
> equal periods of time.**

(See Fig. 16.7).

From Fig. 16.7 we can see that the speed of the
revolving body will be greatest when the radius vector
is the least (i.e., when the planet is closest to the Sun),
and its speed the slowest when the planet is farthest
away. Thus a method was provided for determining the
speed, which allows the position of the planet to be
predicted at some future time.

After the publication of his first two laws in 1609,
Kepler began a search for a relationship between the
motions of the different planets and an explanation to
account for these motions. Ten years later he published
De Harmonica Mundi (Harmony of the World), in which

he stated his third law of planetary motion, known as
the **harmonic law:**

> **The ratio of the square of the period and the
> cube of the semimajor axis (one-half the larger
> axis of an ellipse) is the same for all planets.**

This law can be written as

$$\frac{\text{period}^2}{\text{radius}^3} = \text{constant}$$

or algebraically as

$$\frac{T^2}{R^3} = k$$

where T = the period,
R = the radius,
k = a constant that has the same value for all
planets.

Planets go around the Sun in elliptical paths, but
these ellipses are very close to being circles (except for
Pluto). So R in Kepler's third law is approximately the
distance between the planet and the Sun. For the Earth
$T = 365.25$ days (one year) and $R = 93$ million miles,
or 1.5×10^{11} m.

Galileo Galilei (1564–1642), Italian astronomer,
mathematician, and physicist who is usually called Galileo, was one of the greatest scientists of all time (Fig.
16.8). The most important of his many contributions to
science were in the field of mechanics. He originated the
basic ideas for the formulation of Newton's first two
laws of motion, and he founded the modern experimental
approach to scientific knowledge. The motion of bodies,

Figure 16.8
Galileo Galilei, the great Italian scholar, was the first to use the newly invented telescope to observe the planets and stars.

especially the planets, was of prime interest to Galileo. His concepts of motion and the forces that produce motion opened up an entirely new approach to astronomy. In this field he is noted for his contribution to the heliocentric theory of the solar system.

In 1609 Galileo became the first person to observe the moon and planets through a telescope. With the telescope, which he did not invent, he discovered 4 of Jupiter's 16 or more moons, thus proving that the Earth was not the only center of motion in the universe. Equally important was his discovery that the planet Venus went through a change of phase similar to that of the moon, as called for by the heliocentric theory, but contrary to the Ptolemaic theory, which called for a new or crescent phase of Venus at all times.

Galileo also observed the craters of the moon, dark spots on the Sun, star clusters, and individual stars in the Milky Way. He published *Sidereus Nuntius (Starry Messenger)* in 1610, in which he reported his observations with the telescope. He published his major work, *Dialogue of the Two Chief World Systems, Ptolemaic and*

Copernican, in 1632; it was received enthusiastically, but the authority of the church banned the book. Galileo, a Catholic, was forced to renounce his belief in a moving Earth. He was placed under house arrest and forbidden for the rest of his life from any work on astronomy. Thus he had failed to convince his opponents of the advantage of the Sun-centered theory, even though the telescope had presented scientific facts to support it. Kepler did not have such conflicts with the church because he was a German Protestant.

Only the passing of time was to remove most of the prejudice against scientific knowledge and usually allow scientists the freedom to express the facts of nature as they observed them. The works of Copernicus, Kepler, and Galileo were integrated by Newton in 1687 with the publication of the *Principia*.

Sir Isaac Newton (1643–1727), an English physicist regarded by many as the greatest scientist the world has known, formulated the principles of gravitational attraction between bodies and established physical laws determining the magnitude and direction of the forces that cause the planets to move in elliptical orbits in accordance with Kepler's laws. Newton invented calculus and used it to help explain Kepler's first law. Newton also used the law of conservation of angular momentum to explain Kepler's second law.

To explain Kepler's third law, Newton showed that the constant in Kepler's equation was

$$\frac{T^2}{R^3} = \frac{4\pi^2}{Gm_{Sun}}$$

where T = period of a planet,
R = distance between the planet and the Sun,
G = gravitational constant,
m_{Sun} = mass of the Sun.

Newton's explanations of Kepler's laws unified the heliocentric theory of the solar system and brought an end to the confusion of the past. He gave us an ordered system of the Sun and planets satisfactory for the present time.

Today, our solar system is known to consist of one star (the Sun, which contains 99.87% of the material of the system), nine planets (including Earth), about four dozen satellites (our moon is an example), thousands of asteroids (Ceres is the largest, with a diameter of 600 mi), billions of comets, and countless meteoroids. The distribution of the remaining 0.13% of the solar system's mass is shown in Table 16.1. Note that more than half the remaining mass is concentrated in Jupiter.

Table 16.1 The Solar System

Name	Mean Distance from Sun — Million Miles	Mean Distance from Sun — Astron. Units	Titius-Bode Law	Diameter (Miles)	Diameter (Earth = 1)	Mass with Respect to the Earth	Density (g/cm³) (Water = 1)	Period of Revolution	Period of Rotation	Inclination of Axis with the Vertical	Inclination of Orbit with the Ecliptic	Surface Gravity (Earth = 1)	Magnetic Field	Satellites
Sun				864,000		332,000.	1.4		25 days				Yes	9
Mercury	36	0.39	0.4	3,026	0.38	0.055	5.4	88 days	59 days	less than 28°	7°	0.38	Weak	None
Venus	67	0.72	0.7	7,504	0.95	0.82	5.2	225 days	243 days (retrograde)	3°	3.4°	0.91	No	None
Earth	93	1.00	1.0	7,918	1.00	1.00	5.5	365.25 days	24 h	23.5°	0°	1.00	Yes	1
Mars	142	1.52	1.6	4,216	0.53	0.11	3.9	687 days	24.6 h	24°	2°	0.38	Very weak	2
Asteroids	257	2.77	2.8					5 years typical			10° average		No	
Jupiter	483	5.20	5.2	86,000	11.2	318.	1.3	12 years	10 h	3.1°	1.3°	2.53	Yes	16 or more
Saturn	886	9.54	10.0	72,000	9.41	94.3	0.7	29.5 years	10.7 h	26.7°	2.5°	1.07	Yes	17 or more
Uranus	1783	19.19	19.6	33,000	3.98	14.54	1.2	84 years	17.24 h (retrograde)	82°	0.8°	0.92	Yes	15 or more
Neptune	2793	30.07	38.8	31,000	3.81	17.2	1.7	165 years	16 h	29°	1.8°	1.18	Yes	8 or more
Pluto	3666	39.46	77.2	2,000	0.27	0.002	1.7	248 years	6.4 days	?	17°	0.03	?	1

Planets that have orbits smaller than Earth's are classified as inferior planets and those with orbits greater than Earth's, as superior planets. Another method is to classify Mercury, Venus, Earth, and Mars as the inner or **terrestrial planets** because they resemble Earth. Jupiter, Saturn, Uranus, and Neptune are classified as the outer or **Jovian planets** because they resemble Jupiter. (The Roman god Jupiter was also called Jove.) Pluto does not resemble Earth or Jupiter, and many astronomers have suggested that Pluto be classified as an asteroid.

The relative distances of the planets from the Sun are shown in Fig. 16.9. The orbits are all elliptical, but nearly circular, except for that of Pluto. Pluto's orbit actually goes inside Neptune's orbit. The actual position of Pluto is shown for 1990. We see that for several years Neptune will be farther from the Sun than Pluto! Study Fig. 16.9 for a minute and notice how far from the Sun Jupiter is, compared with Mars. Note that the distance from Saturn to Neptune is greater than that from the Sun to Saturn. The distinction between the four inner planets and the five outer planets can also be seen in this illustration.

When viewed from above the solar system (i.e., looking down on the north pole of the Earth), the planets all revolve counterclockwise around the Sun. The planets also spin with a counterclockwise rotation when viewed from above, with the exceptions of Venus and Uranus, which have retrograde rotation—that is, the motion is clockwise, or westward, the opposite of its revolving motion about the Sun.

The relative sizes of the planets are shown in Fig. 16.10. Note the huge size of the outer planets compared with the size of the inner planets. Compare the size of Pluto with the size of the Jovian planets.

The inclinations of the orbits of the planets relative to Earth's orbit are shown in Fig. 16.11. Note that the solar system is contained within a disk shape rather than a spherical shape. Also, note the large inclination of Pluto's orbit.

A simple way to remember the order of some of the planets is to realize that the first letters of the words *Saturn*, *Uranus*, and *Neptune* spell SUN.

A simple method to remember the approximate distance of the planets from the Sun is given by Bode's law. Johann Bode (1747–1826), a German astronomer

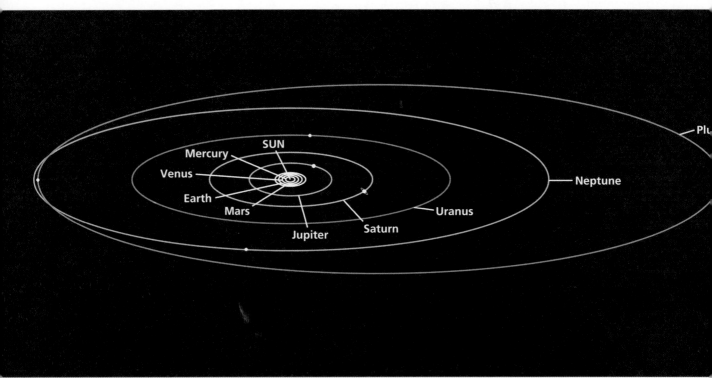

Figure 16.9 The orbits of the planets, drawn to scale.
The actual position of Pluto in 1990 is also shown. The orbits of the planets are all counterclockwise when viewed from above the North Pole of Earth.

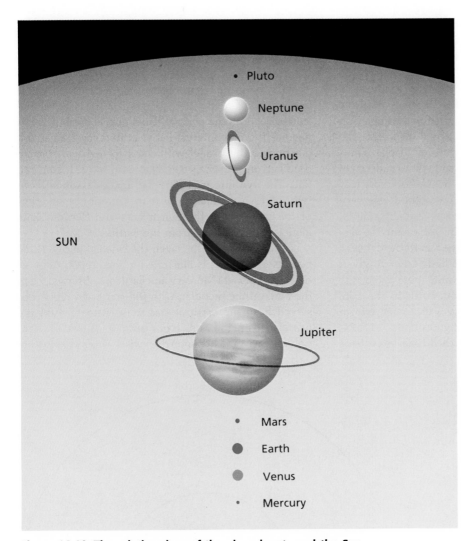

Figure 16.10 The relative sizes of the nine planets and the Sun.

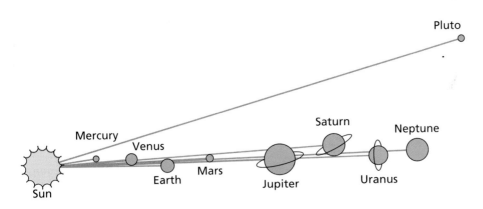

Figure 16.11 The inclination of planetary orbits with the orbital plane of Earth.

and editor of a German astronomical journal, was a strong supporter, but not the discoverer, of the method. Bode's law, in reality, is not a law. That is, it does not represent a physical property of the solar system. The so-called law is a simple method to be followed in calculating the approximate radial distance of some planets from the Sun. See Table 16.1 for a comparison of the results of Bode's law with actual distances. The method was first published by Johann Daniel Titius (1729–1796), a German physicist and mathematician, in 1766. Many astronomers refer to the method more correctly as the **Titius-Bode law.** The method was very useful in the discovery of the asteroids (starlike objects) located between Mars and Jupiter. Values for the distances in astronomical units from the Sun to the different planets, as obtained by the Titius-Bode law, are given in Table 16.1. One **astronomical unit**, abbreviated AU, is equal to the mean distance from Earth to the Sun, which is 9.3×10^7 mi. The law gives the distance from the Sun to the planets when the figures 0, 3, 6, 12, 24, and so on (doubling the number each time, except for

the zero) are each added to 4, and the sum divided by 10. Although the law has no physical interpretation, it does provide an easy method for remembering the distance to most of the planets.

The period of time required for a planet to travel one complete orbital path is referred to as either the sidereal or synodic period. The **sidereal period** is defined as the time interval between two successive conjunctions of the planet with a star (planet and star are together on the same meridian) as observed from the Sun. The **synodic period** is the time interval between two successive conjunctions (either inferior or superior) of the planet with the Sun (planet and Sun on same meridian) as observed from the Earth.

The relationship between the sidereal and synodic periods of a planet is illustrated in Fig. 16.12. At position P_1 the planets Mercury and Earth are observed, from the Sun, on the meridian with the star S. Mercury, observed from the Earth at this same instant, is on the meridian with the Sun. From position P_1 Mercury revolves eastward (counterclockwise) around the Sun

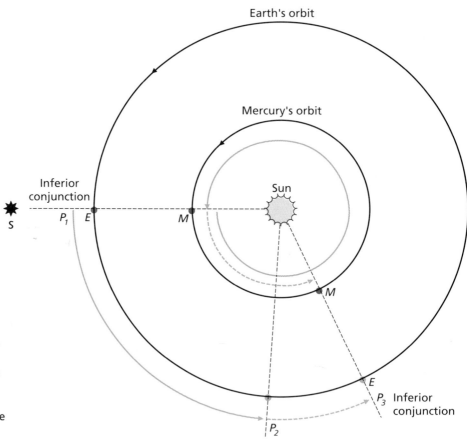

Figure 16.12 The sidereal and synodic periods of the inferior planet Mercury. Mercury and Earth travel the distances shown in color in equal time periods. See the text for an explanation of the diagram.

through 360° back to position P_1. This motion and time period is represented by the solid-color circle in Fig. 16.12. This revolution is the sidereal period for Mercury—the time (88 Earth days) that Mercury requires to make one revolution around the Sun. During this time (88 days) the Earth revolves approximately 87° eastward to position P_2. Mercury continues revolving eastward from position P_1 to position P_3. During this same period the Earth revolves from position P_2 to position P_3. This movement and time period are represented by the broken color lines in Fig. 16.12. At position P_3 an observer on the Earth now again sees Mercury on the meridian with the Sun. The total time for Mercury to revolve from P_1 back to P_1 then to P_3 is the synodic period, equal to 116 Earth days. This is the time it takes the planet Mercury to make one orbit around the Sun as observed from the Earth, or the time from the conjunction at position P_1 to the next conjunction at position P_3. The true period of revolution is the sidereal period.

Opposition is the term used to describe the position of a planet when the planet is 180° from the Sun; that is, the planet is on the opposite side of the Earth from the Sun.

In the next two sections we will discuss the characteristics of the various bodies found in the solar system. Asteroids will be considered in the discussion of the inner planets because most of their orbits are inside Jupiter's. Comets will be considered with the outer planets, because their highly elliptical orbits take them very far from the Sun.

16.3 **The Inner Planets**

Mercury

The planet Mercury is closest to the Sun and has the shortest period of revolution. The early Greeks named it after the speedy messenger of the gods, and it is the fastest moving of the planets because of its position closest to the Sun.

Mercury, at its greatest eastern or western elongation, can be seen only just after sunset or just before sunrise. The elongation (the angular distance between Mercury and the Sun as viewed from the Earth) is only 28°. When Mercury is near eastern elongation, it will appear above the western horizon just after sunset. At western elongation Mercury will be on the eastern horizon shortly before sunrise.

The appearance of Mercury is similar to that of the moon, as can be seen from Fig. 16.13. However, it has a very high density, almost as high as the Earth's. This

Figure 16.13 Photomosaic of Mercury.
This mosaic shows Mercury as it would appear from 50,000 km (31,000 mi) above the surface looking at its southern hemisphere. The planet's south pole is mid-center at the bottom of the photograph. Many craters and bright rays from later impacts dominate the surface, giving it a moonlike appearance.

high density indicates that it probably has an inner core of iron, as does the Earth.

Mercury's rotation period is exactly two-thirds as great as its period of revolution. Thus it rotates exactly three times while circling the Sun twice. This period probably results from tidal gravitational effects from the Sun. As it rotates, the side facing the Sun has temperatures of approximately 700 K, while the dark side is at about 100 K.

Because of Mercury's small size and high daylight surface temperature, the planet should not possess an atmosphere. But the *Mariner 10* mission produced spectral data, operating in the ultraviolet region of the electromagnetic spectrum, that indicate Mercury has an extremely thin atmosphere composed of argon, carbon, helium, nitrogen, and oxygen. These gases produce an atmospheric pressure of only 1/10,000 the atmospheric pressure at the surface of the Earth.

The most perplexing property of Mercury is its weak magnetic field, which was first detected in 1974 and confirmed in 1975. Mercury's magnetic field is about 1% as strong as the Earth's. We believe the Earth's magnetic field is caused by the Earth's rapid rotation, but with Mercury's slow rotation no magnetic field was expected. We will have to wait and see what can explain Mercury's weak magnetic field.

Venus

Venus is our closest planetary neighbor, approaching the Earth at a distance of 26 million miles at inferior conjunction. It is the third brightest object in the sky, exceeded only by the Sun and our moon. Because of its brightness, it was named in honor of the Roman goddess of beauty.

Information concerning Venus has been obtained from the spacecrafts *Venus Pioneer 1*, which orbits the planet; from *Venus Pioneer 2*, which was a multiprobe spacecraft that penetrated Venus' atmosphere; and from the Soviet Venera probes. The *Venus Pioneer 1* orbiter spacecraft and the *Venus Pioneer 2* multiprobe spacecraft are shown in Fig. 16.14.

The relative position of Venus with respect to the Sun and the Earth and the important positions in respect to the Sun as observed from the Earth are shown in Fig. 16.15. When Venus is at superior conjunction, it is in full phase (full illumination of the side facing the Earth) for an observer located on the Earth, but it is not visible to the observer at this time because of the brightness of the Sun. As Venus moves eastward, it appears to the Earth observer as the evening star until it reaches inferior conjunction. The greatest eastern elongation (greatest angular distance from the Sun) occurs 220 days after superior conjunction, but maximum brightness does not occur until Venus has about 39° elongation from the Sun. This occurs about 36 days before and after inferior conjunction.

The synodic period of Venus is 584 Earth days. Viewed from the Earth with a telescope, Venus shows phases. The period of rotation of Venus is 243 Earth days retrograde.

Venus and Earth resemble one another in many ways. They have similar properties, such as average density, mass, size, and surface gravity. But the similarities end there. Venus is covered with a dense atmosphere whose composition is 96% carbon dioxide, some nitrogen (less than 4%), and traces of argon, oxygen, and water vapor. At the surface of Venus the atmospheric pressure is a tremendous 90 atm and the temperature 750 K, or about 900°F. The high temperature is due mainly to the large amount of carbon dioxide in the atmosphere, which produces the "greenhouse effect" (see Section 20.4), so life as we know it cannot exist. Both temperature and pressure decrease with increase in altitude. The temperature at the top of the atmosphere is 220 K.

The surface of Venus can never be seen by an

Figure 16.14
Venus Pioneer 1 (orbiter spacecraft), shown in the upper photo, orbits the planet every 24 hours. *Venus Pioneer 2* multiprobe spacecraft is shown in the lower photo. The probes measured the temperature, pressure, and chemical composition of the atmosphere as a function of altitude. The probes failed to function a short time after impact with the surface of Venus.

observer on the Earth because of dense thick clouds that cover the planet. The clouds are composed mainly of sulfuric acid (H_2SO_4) droplets, along with some water

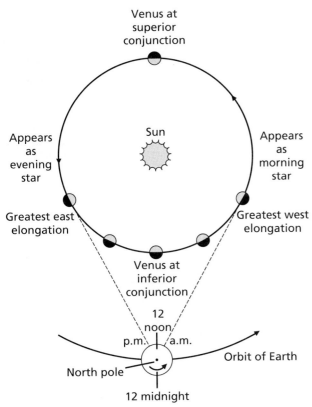

Figure 16.15
The positions Venus must have in order to appear as either our evening or morning star.

Figure 16.16 View of Venus's clouds from the orbiter spacecraft.
The motion of the clouds is from right to left.

droplets. The droplets do not fall out as rain because of the extremely high atmospheric pressure. The clouds occur in four layers, beginning at about 31 km (19 mi) above the surface and extending upward another 37 km (23 mi). The top layer of clouds contains large amounts of yellowish sulfur dust, giving Venus its yellowish or yellow-orange color as viewed from the Earth. See Fig. 16.16. Observation of the top clouds by the orbital spacecraft revealed that Venus's atmosphere makes one rotation every four Earth days in retrograde direction. This rotation is extremely fast compared with the 243 days for rotation of the solid planet.

The topography of Venus consists of highlands, lowlands, and a huge rolling plain, which covers approximately 60% of the planet's surface. (See Fig. 16.17.) The highland regions appear continentlike and rest atop Venus' rolling plain. Ishtar Terra (named after the Babylonian goddess of love), located in the northern hemisphere at about 330° east longitude, is approximately

the same size as Australia. The highland regions are shown in yellow and brown contours and the lower elevations in blue and green.

East of Ishtar Terra is Maxwell Montes (named after James Clerk Maxwell, British theoretical physicist), the highest point on the planet. Maxwell Montes is 35,400 ft above "sea level." Venus has no oceans; the term *sea level* refers to the average radius of the planet.

Southwest of Ishtar the *Pioneer 1* orbiter discovered two very large volcanoes in a region called Beta Regio. The two volcanoes, called Rhea Mons and Theia Mons, have altitudes of almost 20,000 ft (3.8 mi) and bases with diameters of 620 mi.

Figure 16.18 is an artist's conception of Venus' largest highland region, tentatively named Aphrodite Terra after the Greek equivalent of the Roman Venus. The region, about half the size of Africa, is centered at 5° south latitude and 100° east longitude. The huge landmass is about 6000 mi long and 2000 mi wide. Aphrodite Terra consists of eastern and western mountain ranges, separated by a somewhat lower region. The western mountain range rises 23,000 ft above the surrounding

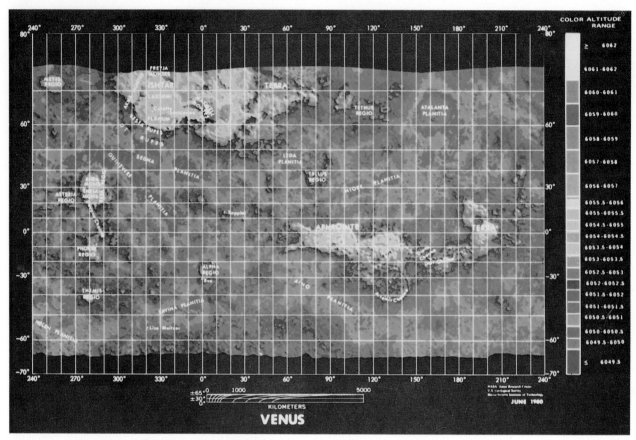

Figure 16.17 A topographic map of Venus's surface, based on radar data taken by *Pioneer 1 Orbiter*.
Most of Venus's surface is covered by plain, shown in blue and green. The highlands are shown in brown and yellow.

Figure 16.18 An artist's conception of Venus's surface.

terrain. The eastern mountains rise 10,000 ft above the surrounding terrain. Radar imaging shows these two mountain ranges to be among the roughest and most broken-up areas on the planet.

Mars

From the Earth, Mars has a reddish color and was named after the bloody Roman god of war. Mars is about one and one-half times as far from the Sun as the Earth. It is tilted on its axis at an angle of 24°, which is very close to the Earth's 23.5° angle of tilt. It rotates once every 24 h 37.4 min, which is very close to a single Earth day. It takes 687 days, or about 23 Earth months, to go around the Sun.

Mars has two small satellites, or moons, named Phobos and Deimos, which are about 20 and 10 mi across, respectively. Both are irregularly shaped and extensively cratered. Like our moon, they keep one side always facing the planet.

The mass of Mars is about one-tenth as large as that of the Earth. Its density is also much less (3.9 g/cm³) than that of Mercury, Venus, and the Earth (5.5 g/cm³). This low density indicates that Mars probably does not have a large iron core in the center like the Earth.

Very little is known about the internal structure of the planet. Mars does have a very weak magnetic field, less than 0.004 times as strong as the Earth's.

Many of the surface features of Mars can be seen in the topographic map of Fig. 16.19. From this map we can observe that Mars has many craters similar to those of the moon. The largest of the so-called basins is located in the southern hemisphere from 50° to 90° east longitude. It is about 1000 mi across. Its smoothness is attributed to dust that has settled out from dust storms. The large volcanic mountain in the upper left of the figure is called *Olympus Mons,* "Mount Olympus." The mountain is very large, towering 15 mi above the plain and having a base with a diameter of 375 mi. The volcano is crowned with a 40-mi-wide crater. The largest volcano on Earth is Mauna Loa on the island of Hawaii. Mauna Loa's base rests on the ocean floor 16,400 ft below the surface of the Pacific Ocean and extends upward another 13,678 ft above the level of the ocean. Thus Mauna Loa is about one-third the size of Mount Olympus.

Three smaller volcanic mountains can be seen to the southeast in Figure 16.19. The large canyon farther southeast is called *Valles Marineris.* See Fig. 16.20 for details. The canyon is approximately 3000 mi long and 20,000 ft deep. Its length is equivalent to the width of the United States. The depth of the canyon is almost

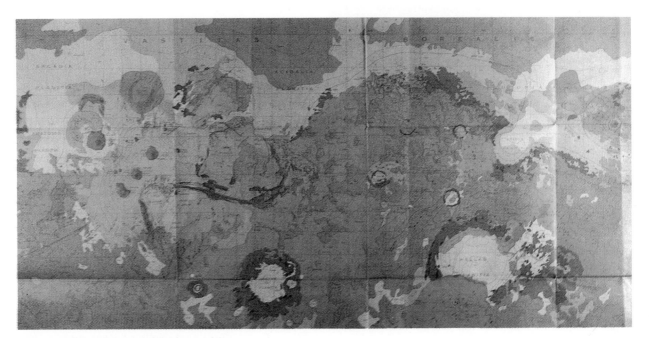

Figure 16.19 A topographic chart of Mars.
Note the many craters, basins, mountains, and canyons.

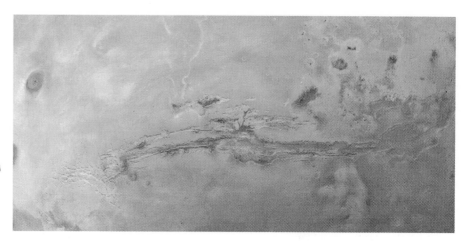

Figure 16.20 An enhanced color mosaic assembled from images taken by the *Viking 1 Orbiter.*
The mosaic shows the great canyon Valles Marineris. The great canyon is 3,000 km (2,500 mi) in length. Geologists believe that Valles Marineris is a fracture in the planet's crust caused by internal forces.

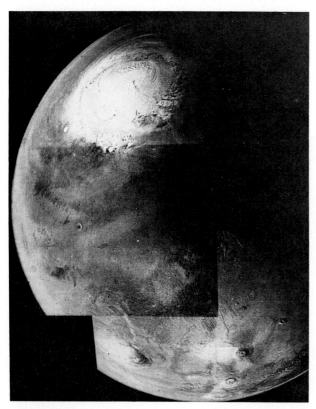

Figure 16.21
This mosaic of three photos shows the northern hemisphere of Mars from the polar cap to a few degrees south of the equator. The huge Martian volcanoes are visible at the bottom.

four times as deep as the Grand Canyon. This tremendous gash in Mars's surface is thought to be a gigantic fracture formed by stress within the planet.

The topographic map shown in Fig. 16.19 does not cover the polar region. Figure 16.21 shows the northern polar cap of Mars. In the winter the polar caps are composed of dry ice (frozen CO_2) and water ice. In the summer the dry ice changes to vapor, leaving behind a residual polar cap of water ice.

The atmosphere of Mars is much more tenuous than that of the Earth. The atmospheric pressure is slightly less than 1% of that on the Earth. The atmosphere is composed of 95% carbon dioxide, 2.7% nitrogen, 1.6% argon, and traces of oxygen, carbon monoxide, krypton, and xenon. Because of the low atmospheric pressure, water cannot exist as a liquid on the Martian surface. Traces of water vapor have been measured in the atmosphere.

Although Mars is exceedingly dry now, there is evidence of water in the past. This evidence is shown in Fig. 16.22, which shows a channel, and Fig. 16.23, which shows teardrop features. Both of these effects are believed to indicate that water once flowed on the surface of Mars.

Our best photographs of the surface of Mars have come from the Viking landers. Figure 16.24 shows the first close-up photograph ever taken of the surface of Mars. These Viking missions have carried out numerous experiments searching for evidence of life on Mars. So far, evidence of life has not been found, but the possibility of life has not been ruled out.

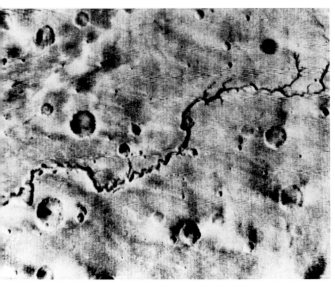

Figure 16.22 Mars channel.
Though not unique on the Martian surface, this meandering "river" is the most convincing piece of evidence that a fluid once flowed over the surface of Mars, draining a large area and eroding a deep channel. The feature is some 575 km (355 mi) long and 5–6 km (3–3.5 mi) wide.

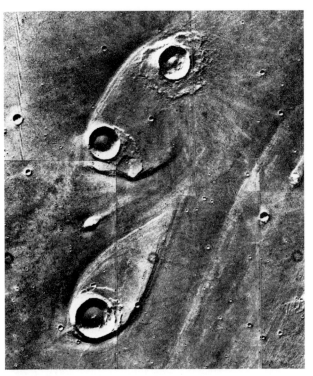

Figure 16.23
These teardrop-shaped features indicate that a fluid flowed on the surface of Mars in the past.

Figure 16.24 Panoramic view taken by *Viking 1* on the surface of Mars.
The horizon features are about 3 km (1.8 mi) away. Patches of bright sand can be discerned among rocks and boulders in the middle distance. The projections on the horizon may represent the rims of distant impact craters.

Asteroids

The Titius-Bode law calls for the next planet beyond Mars to be 2.8 astronomical units from the Sun, but no large planetary body is found at this distance. On January 1, 1801, the first of the many planetary bodies was discovered by Giuseppi Piazzi, an Italian astronomer. This small body is named Ceres after the protecting goddess of Sicily. Ceres is slightly more than 600 mi in diameter, has an orbital period of 4.6 years, and is the largest of more than two thousand objects, which have been named and numbered, orbiting the Sun between Mars and Jupiter. These objects are called **asteroids,** or minor planets. Only Vesta, which is 340 mi in diameter, can be seen with the unaided eye.

The diameters of the known asteroids range from that of Ceres down to only a mile. There are a dozen or more in the 100-to-200-mi range, perhaps a hundred in the 50-to-100-mi range, less than five hundred in the 25-to-50-mi range, and perhaps billions the size of boulders, marbles, and grains of sand. Not all asteroids are spherical in shape. Eros, which can approach the Earth to within 14 million miles, is roughly rod shaped, being 14 mi long and 5 mi thick. Many others are odd and irregular in shape. They all revolve counterclockwise around the Sun, as viewed from above the orbits, or eastward, with an average inclination (degree of incline) to the ecliptic plane of 10°. About a hundred thousand asteroids exist that can be detected with Earth-based telescopes. The total mass of all the asteroids orbiting between Mars and Jupiter is much less than the mass of the Earth's moon. Although most asteroids move in an orbit between Mars and Jupiter, some have orbits that range beyond Saturn or inside the orbit of Mercury.

The **asteroids** are presently believed to be early solar system material that never collected into a single planet. One piece of evidence supporting this view is that there seem to be several different kinds of asteroids. Asteroids at the inner edge of the belt seem to be stony, while the ones farther out are darker, indicating more carbon content. A third group may be mostly composed of iron and nickel.

In 1977 astronomers discovered the first asteriod with an elliptical orbit, which lies mostly between Saturn and Uranus. It is the most remote known asteroid. It has been numbered and named Chiron (kí -ron). Chiron is small, possessing a diameter of less than 200 mi. The asteroid is in a very elliptical orbit with a period of 50 years. Because of its very elliptical orbit, Chiron may eventually collide with a planet or be ejected from the solar system.

16.4 The Outer Planets

The four large outer planets differ in two main respects from the inner planets. Their sizes are much bigger and their densities are much lower (see Table 16.1). Both of these characteristics are generally understood as being due to their greater distances from the Sun and the corresponding lower temperatures in their atmospheres.

When the planets first began to coalesce around 5 billion years ago, the most predominant elements were the two least massive—hydrogen and helium. The heat from the Sun allowed these two elements to escape from the inner planets. That is, the velocities of the molecules of these elements were sufficient to allow them to escape the planets' gravitational pulls. Thus the inner planets were left with mostly rocky cores, giving them a high density. The four large outer planets were much colder, and they retained their hydrogen and helium, which now surround their rocky cores. Thus the four large outer planets consist primarily of hydrogen and helium in various forms, and this composition gives them much lower densities.

Because of their larger mass and greater gravitational pull, the large outer planets also have many more moons than the smaller inner planets do. In fact, we now know that Jupiter, Saturn, and Uranus have large satellite systems with at least 16, 17, and 15 moons, respectively. The four small inner planets have a total of only 3 moons—2 for Mars and 1 for the Earth.

Jupiter

Jupiter, named after the god of the skies because of its brightness and giant size, is the largest planet of the solar system, both in volume and mass. The motion about its axis is faster than that of any other planet, since it takes only 10 h to make one rotation. Jupiter possesses more than half of the total angular momentum of the solar system.

Jupiter's diameter is 11 times as large as that of the Earth, and it has 318 times as much mass. However, its density is only 1.3 g/cm^3. Jupiter consists of a rocky core, a layer of hydrogen in metallic form (because it is at high pressure and temperature), and an outer layer of a liquid mixture of hydrogen and helium. Jupiter's interior structure is shown in Fig. 16.25. Jupiter's atmosphere is composed of about 82% hydrogen and about 17% helium, with a remaining 1% or so of methane, ammonia, water, and several other molecules. The mean surface temperature at the top of the clouds is $-110°C$.

Jupiter has the interesting property of actually giving off twice as much heat as it receives from the Sun. This makes it, in a sense, a star as well as a planet. If Jupiter had about 75 times more mass, nuclear reactions similar to those in the center of the Sun could have started in its interior. Jupiter might be thought of as the smallest star ever observed, but it is usually considered our solar system's largest planet.

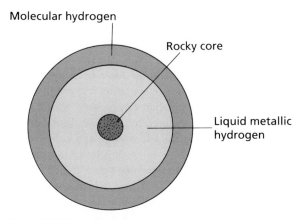

Molecular hydrogen

Rocky core

Liquid metallic hydrogen

Figure 16.25 Jupiter's internal structure.

In Fig. 16.26 the cloud features are easily seen. The clouds of Jupiter show a band structure, and the great red spot stands out. The red spot has an erratic movement and changes color and shape. It sometimes completely disappears. The most recent theory of the great red spot states that it is a huge cyclonic storm similar to a hurricane on Earth but much longer lasting.

Jupiter possesses a tremendous magnetic field. At the top of the atmosphere it is 10 times as strong as the Earth's field. Jupiter's magnetic poles are reversed from those on the Earth, and Jupiter's magnetic poles are 10° from its geographic poles, which are defined by its rotation axis. Its rotation axis is inclined by only a few degrees, so Jupiter does not experience seasonal effects as the Earth and Mars do.

Jupiter has many moons, 16 or more depending on where the line is drawn between a large rock and a small moon. In addition, Jupiter has a very faint ring that is bright enough to be seen from Earth. See Fig. 16.27.

The four largest moons of Jupiter were first discovered by Galileo in 1610. They are sometimes called the Galilean moons of Jupiter. In order of their distances from Jupiter, they are Io, Europa, Ganymede, and Callisto. They were photographed close up by the *Voyager*

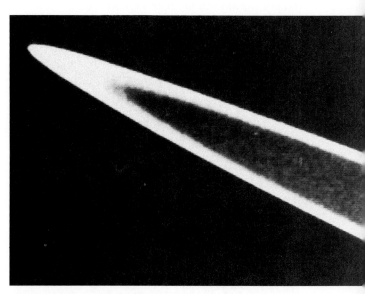

Figure 16.27 Jupiter's ring.
Part of Jupiter's faint ring is shown in this photograph taken by *Voyager 2*. The ring has a distinct outer edge, but the inner edge appears unclear. Seen within the inner edge of the brighter ring is a fainter ring that may extend all the way down to Jupiter's cloud tops.

Figure 16.26 A montage of Jupiter and its four largest moons.
The size of the moons is not to scale. Io is in the left background, bright Europa in the center, Ganymede in the lower left, and Callisto in the lower right.

Figure 16.28 Volcanic explosion on Io.
This photograph was taken by *Voyager 1* from a distance
of approximately 300,000 miles. The solid material has
been thrown upward to an altitude of about 100 miles.
The brightness of the eruption has been enhanced by
computer, but the color (greenish white) is authentic.

1 and *Voyager 2* spacecrafts in 1979. They all have diameters between about 2000 and 3300 mi; Mercury's
diameter is roughly 3000 mi.

One of the most spectacular findings of the Voyager
missions was that Io has many active volcanoes on it.
One of them can be seen in Fig. 16.28. The volcanoes
occur because the gravitational attraction of other nearby
moons, notably Europa, causes Io's orbit to vary so that
it is closer, then farther, from Jupiter. The resulting
changes in Jupiter's gravitational force cause stresses in
the interior rock of Io, and a great deal of frictional heat
is generated, resulting in volcanoes.

Europa is very bright, mainly because its surface is
covered with a mantle of ice approximately 150 mi thick.
Ganymede and Callisto, the two largest satellites, have
cratered surfaces similar to that of the Earth's moon.

Saturn

The most distinctive feature of Saturn is its system of
three prominent rings. The three "classic" rings can be
seen in Fig. 16.29. These rings have been viewed by
Earth-based observers for many years and are the most
spectacular celestial sight that can be seen with a small
telescope. The rings, inclined by 27° to Saturn's orbital
plane, are identified by the letters A, B, and C. The outer
ring, shown in Fig. 16.29, is the A ring, and it is separated
from the bright B ring by the broad dark region known
as the Cassini division. The B ring, which contains the
largest number of particles per unit volume, is the brightest ring. The Cassini division or gap is named in honor
of G. D. Cassini, Italian astronomer, who discovered the
dark region in 1675. The gap is about 3000 mi wide.
The inner C ring has a very low particle density and
appears with less intensity than do rings A and B, because
only a small amount of light is reflected from the smaller
number of particles. The small dark region near the outer
edge of the A ring is known as the Encke division or
as the Keeler gap by some astronomers. Although the
dark regions appear to be without particles, the divisions
or gaps do contain a few tiny particles, which were discovered by the *Voyager 1* spacecraft. The rings, which
are less than 50 m (164 ft) thick, and believed to be
composed of icy particles ranging in size from a few
microns to approximately 10 m (33 ft) in diameter. When

**Figure 16.29 Montage of Saturn and six of its major
satellites.**
Shown are Dione in the forefront; Saturn, Tethys, and Mimas, lower right; Enceladus and Rhea, upper left; and Titan in its distant orbit at the top right.

the distances of the rings are measured in respect to Saturn's center, the outer edge of the C ring is at a radial distance of 57,000 mi. Ring B's outer edge is 73,000 mi from the center of Saturn, and the radial distance of A ring's outer edge is 85,000 mi. Four additional rings have been discovered by *Pioneer 11* and the *Voyager* flybys. The D ring is located inside the C ring and extends from the inner edge of the C ring down to the top of Saturn's clouds. The F, G, and E rings are located beyond the A ring. The F ring, which is very thin and about 60 mi wide, is located next to the A ring. The G ring, which is extremely faint, is beyond the F ring and has a maximum radius of 93,000 mi. The E ring is beyond the G ring. It is also extremely faint and extends outward to more than 150,000 mi.

The *Voyager 1* and *Voyager 2* flights showed the structure of the rings to be very complicated systems of many individual rings. See Fig. 16.30. This highly enhanced color view was assembled from clear, orange, and ultraviolet frames. The photograph, taken from *Voyager 2* at a distance of 5.5 million miles, shows the possible variations in the chemical composition from one part of Saturn's ring system to another. Special computer

Figure 16.30
Possible variations in chemical composition from one part of Saturn's ring system to another are visible in this *Voyager 2* photograph as subtle color variations that can be recorded with special computer-processing techniques. This highly enhanced color view is assembled from clear, orange, and ultraviolet frames obtained at a distance of 5.5 million miles. The C Ring and Cassini Division appear blue in the photo.

processing techniques exaggerate subtle color variations in the photograph.

The structure of Saturn itself is somewhat similar to that of Jupiter. That is, it has a small solid core surrounded by a layer of metallic hydrogen and an outer layer of liquid hydrogen and helium. Saturn's density is only 0.7 g/cm^3, so it would float in water. The temperature of Saturn is very cold, approximately 90 K, or −300°F in the upper atmosphere. This temperature is about 100°F colder than Jupiter. Like Jupiter, Saturn radiates more heat than it gets from the Sun.

Saturn's mass is 95 times that of the Earth, and its diameter is 9 times larger than Earth's. It rotates about once every 11 h. It has a magnetic field that is 1000 times stronger than that of Earth but only 0.05 times as strong as Jupiter's.

Outside the main visible rings lie the 17 or more moons of Saturn. Many of them are quite small, with diameters between about 30 and 100 km (roughly 20 to 60 mi). Most of these 8 small moons were discovered in 1980 by *Voyager 1*. The moons Mimas, Enceladus, Tethys, Dione, Rhea, Hyperion, and Iapetus are similar in two respects. They all have diameters between 385 and 1530 km (240 to 950 mi), and they are all believed to be composed of rock and ice because their densities are between 1 and 2 g/cm^3.

Except for the most distant moon, Phoebe, and possibly Hyperion, all the moons of Saturn rotate once each revolution, similar to our own moon. Thus they always keep the same face toward Saturn. Except for the two most distant moons, Phoebe and Iapetus, all the moons of Saturn move in nearly circular orbits, revolve in the same direction, and lie along the planet's equatorial plane. Phoebe revolves in the opposite direction to that of the other 16 moons, and its orbit is inclined by 150°. Thus Phoebe is probably a moon that was somehow captured by the Saturnian system in the distant past.

The most interesting moon of Saturn is Titan. Titan is the largest moon of Saturn, with a diameter of 5120 km (3180 mi) and a density of 1.9 g/cm^3. It is the only satellite known to have a dense, hazy atmosphere. The main constituent of the atmosphere is nitrogen. Titan's surface cannot be seen, but the surface temperature is about 100 K, or −280°F.

Uranus

Uranus was discovered in 1781 by William Herschel (1738–1822), an English astronomer. The planet's name, chosen in keeping with the tradition of naming planets

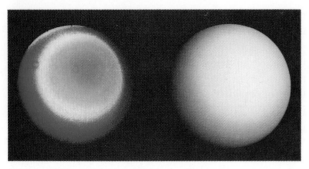

Figure 16.31
These two pictures of Uranus were taken by the narrow-angle camera of *Voyager 2* when the spacecraft was 9.1 × 10⁶ km (5.7 × 10⁶ mi) from the planet. The photo on the right shows Uranus as the human eye would see it from the vantage point of the spacecraft. The photo on the left employs enhanced color. See the text for details.

for the gods of mythology, was first suggested by Johann Bode. Uranus was the father of the Titans and the grandfather of Jupiter.

Figure 16.31 shows two pictures of Uranus taken by *Voyager 2* when the spacecraft was 5.7 million miles from the planet. The right photo has been processed to show Uranus as human eyes would see the planet from the spacecraft. The blue-green color results from the absorption of red light by methane gas in Uranus' atmosphere. The darker shading at the lower left of the disk corresponds to the day-night boundary on the planet. The left photo uses false colors and contrast enhancement to show distinctive details in the polar region of the planet. The false-color picture reveals a dark polar hood surrounded by a series of progressively lighter concentric bands. One possible explanation is the brownish haze or smog concentrated over the polar region of the planet.

The internal structures of Uranus and Neptune are similar but differ from those of Jupiter and Saturn. Uranus and Neptune are much smaller and less massive than Jupiter and Saturn. See Table 16.1 for a comparison of physical properties. Also, their rocky cores are relatively larger compared with their total size. From a study of the physical properties, the internal structure of Uranus is calculated to be in three layers. The inner rocky core, which is about 13,000 km (8100 mi) in diameter, contains about 25% of the planet's mass and is probably composed of iron and silicon. The rocky core is surrounded by a liquid mantle approximately 8,000 km (5000 mi) deep, composed of water, ammonia, and methane ice. The

mantle makes up 65% of the planet's mass. The outer layer, the atmosphere, of the planet is about 11,000 km (6800 mi) thick and is composed of mostly hydrogen plus 12% helium by volume.

Uranus has a ring system that is very thin. Before the *Voyager 2* mission Uranus was known to have nine rings, which were discovered in 1977. All nine rings are shown in false color in Fig. 16.32. The outermost and also the brightest ring, named epsilon, is about 43 km wide. This ring is shown at the top of Fig. 16.32 and appears in a neutral color. Down from epsilon toward the planet, the figure shows the delta, gamma, and eta rings in shades of blue and green; the beta and alpha rings are shown in lighter tones. A final set of three rings known as 4, 5, and 6 are shown in faint off-white tones. The pastel lines seen between the rings are contributed by the process of computer enhancement. A ring newly discovered by *Voyager 2* is called 1985UR1. This ring, which is very thin and hardly visible to *Voyager 2* cameras, is about half-way between epsilon and delta. Ring 1985UR1 is not visible in Fig. 16.32.

The rings of Uranus are composed mainly of boulder-size particles 1 m in diameter and larger, with very few dust size particles present. Because of the lack of dust particles, the rings do not have good reflective properties, like the rings of Saturn, which are filled with tiny particles 1 cm and smaller in size. *Voyager 2* also recorded some very narrow sections of rings. Astron-

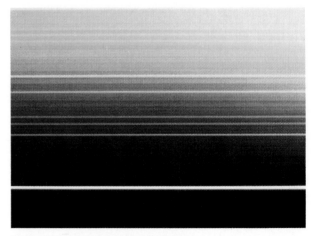

Figure 16.32 The Rings of Uranus.
This enhanced-color view of the rings of Uranus was made from images shot by *Voyager 2* at a distance of 2.59 million miles. The epsilon ring at the top is neutral in color, but there are subtle color differences among the eight other fainter rings.

omers believe that perhaps there are hundreds of partial rings located around Uranus.

Uranus has several other interesting features besides its rings. Its rotation axis is inclined 82° with respect to its orbital motion, which positions the Sun nearly overhead at the north and south poles as Uranus revolves around the Sun. The planet rotates retrograde with a period of approximately 17 h. Uranus has a magnetic field that is about 50 times as strong as the Earth's magnetic field. An unusual feature of the planet's magnetic field is its orientation with the rotational axis of the planet. The magnetic field is tipped 55° to the rotational axis. The clouds of Uranus are deep within its atmosphere and nearly invisible. Thus the planet appears as a bland, almost featureless blue-green disk. The planet's atmospheric temperature, which is fairly uniform over the whole planet, is about −210°C. For some unknown reason, the planet radiates more energy than it receives from the Sun.

The cameras of *Voyager 2* recorded 10 new satellites circling Uranus inside the orbit of Miranda. The first was discovered in December 1985 by *Voyager 2* and named 1985U1. The satellite has a diameter of about 165 km. The other 9 satellites, discovered in 1986, are named 1986U1 through 1986U9. These 9 satellites are very small, with diameters less than 100 km.

Uranus has five major satellites. They are, in order of distance from the planet, Miranda, the smallest and closest, having a diameter of 298 mi and orbiting at a distance of 80,000 mi; Ariel; Umbriel; Titania, the largest, with a diameter of 986 mi; and Oberon, the most distant, orbiting at 362,000 mi. The satellites have densities of about 1.6 g/cm³. This density gives them a composition of an approximate mix of ice (water) and rock.

The surface features of the satellites show that, with the exception of Umbriel, the moons have been tectonically active (see Section 26.3) in the past. Figure 16.33 shows Miranda's surface, with large curvilinear regions of grooves and ridges plus regions that appear chevron shaped. The satellite's surface is pockmarked with craters. The large crater shown in the lower right region of Fig. 16.33 is about 15 mi in diameter. Photos obtained from *Voyager 2* show the surface with very deep valleys and high cliffs ranging in height from 0.3 to 3 mi. The geologic forms on Miranda are the most bizarre forms in the solar system.

The surfaces of Ariel and Titania are pitted with craters having numerous valleys and fault scarps cutting across the highly pitted terrain. Uranus' most distant moon, Oberon, is shown in Fig. 16.34. The picture shows

Figure 16.33
Uranus's innermost large moon, Miranda, is roughly 300 miles in diameter and exhibits a variety of geologic forms—the most bizarre forms in the solar system. Chevron-shaped regions and folded ridges in circular racetrack patterns are visible on the satellite's surface. There are large scarps, or cliffs, ranging up to three miles in height; they are clearly visible in the lower right part of the photo. Next to them is a deep canyon approximately 30 mi wide.

Figure 16.34 Oberon, Uranus's most distant moon.
Note the large mountain, about four miles high, on the lower left edge of the satellite and the very large crater near the center of the picture.

several large impact craters in the planet's icy surface. The large central crater has a high central peak. On the lower left edge of the photo a huge mountain about 4 mi high can be seen.

Neptune

Neptune was discovered in 1846 by John G. Galle (1812–1910), a German astronomer at the Berlin Observatory. Partial credit is also shared by Englishman John Couch Adams and Frenchman U. J. J. Leverrier, two mathematicians. Using Newton's law of gravitation, Adams and Leverrier made calculations that produced information on where to look for a supposed planet that was disturbing the motion of the planet Uranus. The name Neptune was proposed by D. F. Arago, a French physicist who had suggested that Leverrier begin the critical investigation of the planet. He first suggested Leverrier as the name but later withdrew this suggestion because it received little acceptance outside France.

The planet cannot be observed with the unaided eye and appears to have a greenish hue when viewed through a telescope. A close-up photograph of Neptune is shown in Fig. 16.35. The physical makeup of the planet is similar to that of Uranus. Methane and hydrogen have been detected spectrographically, so the planet definitely has a gaseous atmosphere.

Neptune has eight known satellites. Triton, which is larger than our moon, revolves retrograde once every

Figure 16.35 *Voyager 2's* **view of Neptune.**
This false-color image was taken during the fly-by of August 1989. The red ring around the surface is a semi-transparent haze covering the planet.

Figure 16.36 *Voyager 2* **spacecraft.**
The spacecraft weighs 1753 lb, including 249 lb of scientific equipment. The magnetometer boom is 43 ft long, and the high-gain antenna is 12 ft in diameter. The spacecraft shot a series of photographs during its fly-by of Neptune during August 1989.

6 days; it is now believed to be the coldest object in our solar system. The revolving speed is decreasing. Within a few million years Triton will be torn apart by Neptune's gravitational forces, and the planet will develop a ring system similar to those of the other Jovian planets. A faint single ring fragment around Neptune was detected by a group of American and European astronomers during an occultation of the planet on July 22, 1984. The background star dimmed briefly as the faint ring passed between the star and the receiving telescope recording infrared wavelengths. The ring fragment is calculated to be 10 to 15 km wide, about 100 km long, and located some 70,000 to 80,000 km from the planet.

Voyager 2 (see Fig. 16.36) encountered Neptune in August 1989, and astronomers were able to obtain more precise data concerning the planet's faint ring and ring arcs, its eight satellites, its period of rotation, its magnetic field, and its cloud bands.

Another of Neptune's satellites, Nereid, is very small. Its diameter is estimated to be less than 400 miles. Nereid has the most eccentric orbit (0.749) of any satellite in the solar system. The highly elliptical orbit takes the tiny moon from 870,000 to over 6 million miles from Neptune.

Pluto

The planet Pluto, named for the god of outer darkness, is the most distant planet from the Sun. It was discovered by C. W. Tombaugh in 1930 at the Lowell Observatory in Arizona, after a thorough search near the position predicted by theoretical calculations. The planet had been predicted because discrepancies appeared in the orbital motions of Uranus and Neptune. General information concerning Pluto is given in Table 16.1. Because of Pluto's small size and great distance from the Sun, very little is known about the surface features of the planet. Spectroscopic investigations of Pluto indicate the planet is covered with methane ice.

In June 1978 a satellite of Pluto was discovered by James W. Christy. Named Charon, Pluto's moon is about 500 to 600 mi in diameter, making it the largest satellite in relation to its parent planet. Simultaneously, Pluto was found to be much smaller than previously believed. Its diameter is now thought to be about 2000 mi, making it the smallest planet.

The planet is so far away that its temperature is difficult to measure. Also, the diameter, density, and period of rotation given in Table 16.1 are estimated values and may change when more precise data are obtained.

Some astronomers believe that Pluto was once a moon of Neptune, for three reasons:

1. Pluto is much smaller than the other four outer planets (see Fig. 16.10).

2. Pluto does not lie along the planetary disk (see Fig. 16.11).

3. Pluto's orbit is highly elliptical. It actually goes inside the orbit of Neptune (see Fig. 16.9).

Pluto went inside Neptune's orbit in late 1978 and will exit in the year 1999. Thus for more than twenty years Neptune will be the most distant planet from the Sun.

There is one reason to doubt that Pluto was once a moon of Neptune. For Pluto to have been a moon of Neptune, it once must have been very close, and the present orbits suggest that it was *not* close in the past. Detailed studies show that in their present orbits Pluto and Neptune can never come closer than 18 astronomical units. Whether or not Pluto was once a moon of Neptune is thus not at all certain. Some astronomers believe that Pluto and Charon resemble a double asteroid, and Pluto should no longer be classified as a planet.

Planet X

Is there a Planet X? No one knows. There is no persuasive theoretical justification to believe that a planet beyond Pluto does not exist, nor is there positive evidence that a planet beyond Pluto *does* exist. Discrepancies in the orbital motion of the planets Neptune and Uranus, thought to be due to the gravitational influence of some unknown mass, are minor and within the uncertainty of the measurements taken to detect them. Thus there are some who think the unknown mass does not exist and others who think that it does.

Originally, astronomers thought that Pluto was the unknown mass that caused the uncertain discrepancies. But recent calculations of Pluto's mass from the motion of its newly discovered satellite, Charon, indicate that Pluto's mass is too small to produce the discrepancies. Thus the discrepancies, if they are real, must come from another source. U.S. Naval Observatory astronomers have indicated that the gravitational pull from an unknown planet could produce these discrepancies, if the planet possesses a mass between two and five times the mass of the Earth and orbits the Sun at a distance of 50 to 100 astronomical units. The orbit of Planet X should also be highly inclined, similar to Pluto's. The infrared astronomical satellite (IRAS), which circles the Earth in a polar orbit at a height of 560 mi, has detected heat from an object about 50 billion miles away. The search for such a planet will continue, and if it exists, scientists using the new Hubble space telescope may discover it.

Comets

Comets are named from the Latin word *cometes*, which means "long-haired." They are the solar system members that periodically appear in our sky for a few weeks or months, then disappear. A comet occupies a very large volume but has a very small mass; therefore, it must be composed of very fine particles. A **comet** consists of four parts: (1) the **nucleus,** typically a few miles in diameter and composed of rocky or metallic material, solid ices of water, ammonia, methane, and carbon dioxide; (2) the head, or **coma,** which surrounds the nucleus and which can be as much as several hundred miles in diameter, and is formed from the nucleus as it approaches within about 5 astronomical units of the Sun; (3) a long, voluminous, and magnificent **tail,** also formed from the coma by solar winds and radiation, and which can be millions of miles in length; and (4) a spherical cloud of hydrogen, believed to be formed from the dissociation of water molecules in the nucleus, surrounding the coma. The sphere of hydrogen in some comets may have a radius exceeding that of the Sun (Fig. 16.37).

Comets are seen by reflected light from the Sun and the fluorescence of some of the molecules comprising

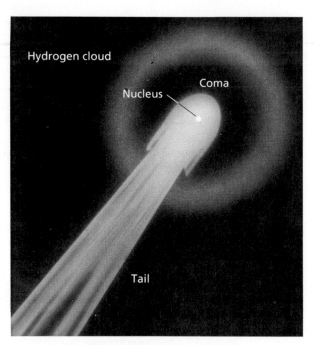

Figure 16.37 The principal parts of a comet.

the comet. As comets approach and move around the Sun, the material in the coma and tail gets larger (Fig. 16.38). This increase in size is evidently caused by the Sun, perhaps by (1) the solar wind, which consists of streams of particles (electrons, protons, and the nuclei of light elements) given off by the Sun and driven outward at speeds of hundreds of miles per second; and (2) radiation pressure generated by the radiant energy given off by the Sun. Scientists believe that only a thin outer shell of the comet's nucleus is heated. As it moves toward its closest approach (perihelion) to the Sun, the increasing solar radiation causes the surface materials to melt and evaporate. The evaporating surface particles form the coma and the long tail, which is driven away from the Sun by the solar wind and radiation pressure. Each time the comet passes near the Sun on its long journey through space, it loses part of its mass. Eventually, the comet loses most of its mass and disappears as a comet, but the rocky or metallic material continues to move around the Sun.

Halley's comet, named after Edmond Halley (1656–1742), a British astronomer, is the brightest and best-

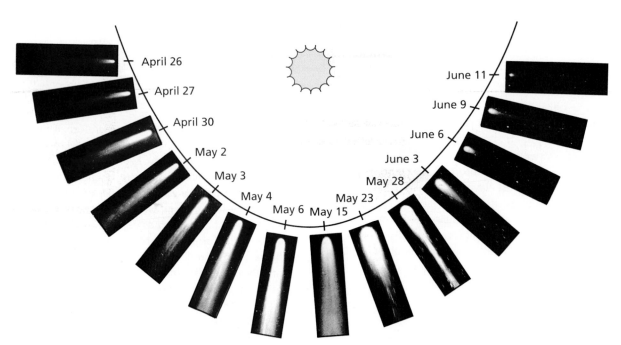

Figure 16.38 Halley's comet.
These fourteen views of Halley's comet were taken between April 26 and June 11, 1910. Note the change in size of the coma and tail.

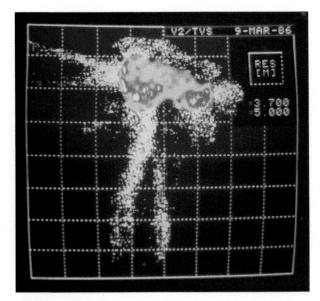

Figure 16.39
This composite image displays the profile of the nucleus of Halley's comet in red, green, and yellow; the jets appear white.

known comet. Halley was the first to suggest and predict the periodic appearance of comets. He observed the comet that bears his name in 1682 and, using Newton's laws of motion, correctly predicted it would return in 76 years. Halley's comet appeared in 1910 and in 1986. The photo in Fig. 16.39 is a composite image showing the profile of comet Halley's nucleus in red, green, and yellow and the jets in white.

The data collected by the spacecrafts that encountered comet Halley confirmed the theory that the comet is a conglomerate of ice and dust. The data indicates that the comet gas is 80% water vapor, 10% to 15% carbon monoxide, and 5% other gases.

The comet's nucleus is a black object shaped like a huge avocado or potato, slightly larger at one end than at the other. The approximate dimensions are 16 km in length, 8 km in one width, and 7.5 km in the other. The calculated volume is between 400 and 500 km^3, and the density is estimated to be between 100 and 400 kg/m^3. Thus the nucleus is composed of very light substances. For comparison, water ice has a density of about 920 kg/m^3.

The nucleus is believed to be composed of dust particles held together mainly by water-ice and some other ices composed of carbon, oxygen, hydrogen, and nitrogen. The surface of the nucleus is irregular, black

as coal, with valleys and hills, and displays relief features spherical in shape, similar to craters. Most of the surface seems to be covered with a dark nonvolatile substance, probably carbon compounds, that forms an insulating crust. The visible surface is fairly cool, having a temperature of about 300 K.

The outgassing of particles producing jet activity appears to take place from small areas of the sunlit surface. Probably less than 10% of the surface is active in producing erupting dust jets. The erupting jets are composed of particles rich in hydrogen, oxygen, carbon, and nitrogen. The largest particles are about the size of a small grain of sand. Water is the major neutral particle detected in the erupting jets and H$_3$O$^+$ the most dominant ion.

The question as to whether the composition of comet Halley is typical for comets remains to be answered, but the lack of internal heating of the comet and external heating only in the vicinity of the Sun may provide scientists with material that is representative of the primitive solar system.

Scientists believe that comets originate and evolve from dirty, icy objects that were part of the primordial debris thrown outward into interstellar space when the solar system was formed. These dirty, icy objects reside in a spherical volume of space that begins beyond the planets and extends outward to an estimated 40,000 to 50,000 astronomical units from the Sun. This far-flung volume of space, which is the reservoir of cometary material, is called the **Oort comet cloud** in honor of Jan Hendrik Oort, the Dutch astronomer who proposed its existence in 1950. The total mass of the Oort cloud is estimated to be 1 to 10 times the mass of the Earth.

The dirty, icy objects that reside within the Oort cloud are perturbed by the weak gravitational force fields of passing stars. These small forces start the icy objects inward on their fall toward the Sun or outward beyond the influence of the Sun's gravitational field. Very few of these cold hard objects become captured comets that come close enough to the Sun to receive sufficient heat to vaporize the icy material. Most pass undetected beyond the orbit of Pluto. Fewer than a thousand comets have been seen and recorded by astronomers, and only one-tenth of these make repeated trips near the Sun.

Where is the outer boundary of the solar system? As we indicated above, the Oort comet cloud has an estimated outer limit of 50,000 astronomical units. Does the Sun's gravitational force field have any influence beyond this limit? The answer is yes, according to some astrophysicists who have made some computer simulations. They have indicated the outer limit for the influence of the Sun's gravitational field is between 80,000 and 100,000 astronomical units. This limit is where the Sun's gravitational force field is balanced by that of the Milky Way. Beyond this boundary the Sun's gravitation force field cannot hold any object.

Meteoroids are interplanetary metallic and stony objects that range in size from a fraction of a millimeter to a few hundred meters. They circle the Sun in elliptical orbits and strike the Earth from all directions with very high velocities. Their high velocity, which is increased by Earth's gravitational force, produces great frictional heating when the meteoroids enter Earth's atmosphere.

A meteoroid is called a **meteor,** or "shooting star," when it enters the Earth's atmosphere and becomes luminous because of the tremendous heat generated by friction with the air. Most meteors are vaporized in the atmosphere, but some larger ones may survive the flight through the atmosphere and strike the Earth's surface. They are then known as **meteorites.**

When a large meteorite strikes Earth's surface, a large hole, called a crater, is created. Figure 16.40 is a photograph of a large meteorite crater near Winslow,

Figure 16.41
The Ahnighto Meteorite being loading aboard ship in-Greenland in 1897. The meteorite has a mass of 36,000 kg and is now on display at the American Museum of Natural History in New York.

Arizona, which scientists estimate to be about 25,000 years old.

The meteoroids are members of the solar system that probably come from the remains of comets and fragments of shattered asteroids. The largest known meteorite has a mass of more than 55,000 kg (120,000 lb) and fell in southwest Africa. The largest known meteorite in the United States, with a mass of about 36,000 kg (79,000 lb) was found near Cape York, Greenland, in 1895 and is on display at the Hayden Planetarium, New York (Fig. 16.41).

Meteorites can be of any size or shape and are classified as metallic or stony. The metallic ones have about 91% iron, 8% nickel, small quantities of cobalt and phosphorus, and traces of many other elements. The stony ones have about 36% oxygen, 26% iron, 18% silicon, and 14% magnesium, with smaller quantities of several other elements. These are average compositions; individual meteorites can vary greatly.

16.5 The Origin of the Solar System

Any theory proposed to explain the origin and development of the solar system must account for the system as it exists now in respect to size, shape, form, substance,

Figure 16.40 The Barringer Meteorite Crater near Winslow, Arizona.
The crater is 1300 m (4264 ft) across, 180 m (590 ft) deep, and its rim 45 m (148 ft) above the surrounding land.

and change. The preceding sections have given a general description of the system in its present state, which, according to our best measurements, has lasted for at least 5 billion years.

Some of the major items that must be accounted for if an acceptable theory of origin is to be obtained are:

1. the origin of the material used to create the system.
2. the forces that were needed to form the system.
3. the size, shape, substance, positions, and motion of the Sun, planets, satellites, asteroids, comets, and meteors.

Presently, there are two types of theories that attempt to explain the origin of the solar system. One group of theories is called **catastrophe theories.** In these theories our solar system arises from a catastrophe such as the near collision of a star with our Sun. This group of theories has several drawbacks:

1. Hot gases drawn from the Sun would not collect to form planets but would be dispersed.
2. The angular momentum of the present planetary system requires an approaching star to be at a minimum distance to account for the angular momentum; but in order for the material to be withdrawn from the Sun, the star would have to approach closer than this minimum distance.

A second group of theories is called **evolutionary theories.** In these theories the solar system forms out of a single rotating cloud of gases owing to the mutual gravitational attraction of all the particles. The main drawback of these theories is that the Sun is now spinning much more slowly than seems reasonable using these theories. Recent advances indicate that there may be a way to explain this apparent discrepancy.

The theory currently in fashion is called the **protoplanet hypothesis** (*proto* is Greek for *"primitive"*). This hypothesis proposes that initially there was a large swirling volume of cold or rarefied gases and dust positioned in space among the stars of the Milky Way. This premise is supported by the fact that today we observe many such nebulae throughout the universe. For some reason, such as a supernova shock wave passing through, the gases began to condense and a larger mass began to form, which produced greater force, causing the entire cloud of gases and dust to shrink (Fig. 16.42).

The collection of particles was slow at first but

became increasingly faster as the central mass became larger. As the particles moved inward, the rotation of the mass had to increase to conserve angular momentum. Because of the rapid turning, the cloud began to flatten and spread out in the equatorial plane. Kepler's third law is a statement that the central part must move faster than the outer parts. This motion set up shearing forces, which, coupled with variations in density, produced the formation of other masses that moved around the large central portion, the protosun, sweeping up more material and forming the protoplanets.

During the early stages of development the space between the protosun and the protoplanets was filled with large amounts of gas and dust, shielding the protoplanets from the protosun, which was beginning to burn and radiate as a star. With the passing of time the space between the protosun and protoplanets became transparent, and the protoplanets' atmospheres were heated and driven off by the pressure of solar radiation. The protoearth was perhaps one thousand times more massive than the planet Earth. The inner protoplanets, receiving more heat than the outer protoplanets, were greatly affected by this heating and lost most of their atmospheres. The outer protoplanets, being at great distances and receiving little solar radiation, were less affected and appear today in a similar protoplanet stage. Planetary satellites are believed to have formed from similar condensations, surrounding their rotating protoplanets, although our moon may have been a smaller protoplanet. The formation of the solar system took place over a 100-million-year period beginning about 5 billion years ago.

The asteroids, which have irregular sizes and shapes, are believed to be the remnants of a planet that never formed. Meteorites are thought to be small asteroids; and by studying the composition of meteorites, we have important clues to the origin of our solar system.

The comets were formed in the outer parts of the nebula, beyond the orbit of Neptune. Because very little material was available at this distance, the comets were very small in size, about 1 mi in diameter, composed of icy ammonia, methane, water, and some silicates. Some of the formed comets have been scattered from their birthplace by disturbances from the planets. Countless others remain in the vast space of their birthplace.

If this improved theory or hypothesis is the true picture of the origin of our solar system, then it is very reasonable to believe there are many, many other systems located throughout the universe, perhaps some with life similar to our own.

Figure 16.42 The formation of the solar system according to the protoplanet hypothesis. The four sketches are not drawn to scale. The top drawing represents a very large volume of space. Interstellar gases and dust particles, with some net angular momentum, collected because of gravitational forces. Most of this collapsed mass formed the Sun. The remainder of the mass formed the planets, which revolve around the Sun in accord with Kepler's laws.

16.6 Other Solar Systems

The search for planets of other stars has been in progress for many years. Recent reports of discoveries of protoplanetary objects around the stars Beta Pictoris and Vega and the detection of a wobble in the motion of the star Van Biesbroeck 8 (VB8), indicating the presence of a substellar object, have aroused new interest in the search. This interest is due to recent advances in technology that increase the possibility of finding substellar objects. The star VB8 is located 21 light-years away in the constellation Ophiuchus. The discovery of a substellar object orbiting VB8 provides evidence that other solar systems may exist in our galaxy.

Detecting extrasolar planets is no easy task. Basically, a star without planets will appear to move in a straight line. A star with planets will have a small wobble superimposed on its straight-line motion due to gravitational effects. Using infrared radiation and a technique called infrared speckle interferometery, scientists at the National Optical Astronomy Observatories in 1983 reported a small wobble in VB8. The detected object, called a "brown dwarf" star by some astronomers and a planet by others, is known as VB8B. The substellar object has a mass about ten times that of Jupiter, a surface temperature of approximately 1400 K, and a diameter nine-tenths that of Jupiter.

Confirmation of VB8B could come when the Hubble Space Optical Telescope is placed in a 300-mi-high orbit by the space shuttle. The optical telescope has a 2.4-m (94-in) mirror, and it will be able to detect objects that are 50 times fainter and resolve objects that are 7 times smaller than any presently made earth-based optical telescope.

Learning Objectives

After reading and studying this chapter, you should be able to do the following without referring to the text:

1. Name the objects (more than a dozen) that make up the solar system.

2. Describe the structure of the solar system.

3. State the two major motions of the planet Earth, and give proof for each motion.

4. List the planets in order of distance from the Sun.

5. List the terrestrial planets and the Jovian planets, and name several ways in which they are different.

6. Describe and differentiate between comets, meteoroids, and asteroids.

7. State briefly the protoplanet hypothesis for the formation of the solar system.

8. Define and explain the important words and terms listed in the following section.

Important Words and Terms

solar system	Kepler's law of equal areas	comet
ecliptic	Kepler's harmonic law	nucleus
rotation	terrestrial planets	coma
revolution	Jovian planets	tail
Foucault pendulum	Titius-Bode law	Oort comet cloud
parallax	astronomical unit	meteoroids
aberration (of starlight)	sidereal period	meteor
parsec	synodic period	meteorites
geocentric theory	conjunctions	catastrophe theories
heliocentric theory	opposition	evolutionary theories
Kepler's law of elliptical paths	asteroids	protoplanet hypothesis

Questions

The Planet Earth

1. State and explain two major motions of the planet Earth.

2. Before 1900, what proof was there that the Earth (a) rotated and (b) revolved?

3. What is the ecliptic?

4. What does the parsec measure? State the definition of one parsec.

5. State why the parallax of a star cannot be seen with the unaided eye.

The Solar System

6. List the major objects that comprise the solar system.

7. Briefly describe the structure of the solar system.

8. What is the explanation of Kepler's second law? (*Hint:* Refer to a conservation law.)

The Inner and Outer Planets

9. What distinguishes the terrestrial planets from the Jovian planets?

10. Which planet is presently the farthest known planet from the Sun? Explain.

11. Which planet has the most elliptical orbit?

12. What is unusual about the satellite Io?

13. What is unique about the satellite Titan?

14. Which planet is the largest? Which satellite of a planet is the largest?

15. Which planet has the greatest orbital velocity? the least?

16. Name the planets that have the highest and lowest surface temperatures.

17. How do the atmospheric pressures on Venus, Earth, and Mars differ?

18. Name the planets that are known to have rings.

19. Compare the physical characteristics of the Earth with those of the other planets of the solar system.

20. How do asteroids, comets, and meteoroids differ?

21. Which planet has the greatest density? The least density?

22. State the reasons why Pluto is thought to have once been a satellite of Neptune. State the opposing reasons.

23. Pluto is considered one of the outer, or Jovian, planets. Compare its physical characteristics with those of the Jovian planets.

24. State the characteristics of Pluto that classify it as (a) a planet and (b) an asteroid.

25. What is the estimated outer limit of the solar system?

The Origin of the Solar System

26. State three major components that must be accounted for in any theory that explains the origin of the solar system.

27. How many years were required for the formation of the solar system?

28. Why do all the planets revolve in the same direction around the Sun?

Other Solar Systems

29. What evidence do astronomers have concerning the possibility of another solar system?

30. How are extrasolar planets detected?

31. What is the diameter of the mirror used in the Hubble Space Optical Telescope?

Exercises

The Solar System

1. Using Newton's formula $T^2/R^3 = 4\pi^2/Gm_{Sun}$, show how the mass of the Sun could be calculated in kilograms (R for the Earth is 1.5×10^{11} m).

2. Use Kepler's second law to show that comets spend most of their time far from the Sun.

3. Show that Kepler's third law holds (approximately) for the Earth and Mars. That is, show that (see Table 16.1)

$$\frac{(365.25)^2}{(93 \times 10^6)^3} = \frac{(687)^2}{(142 \times 10^6)^3}$$

4. Referring to Exercise 3 and Table 16.1, show that Kepler's third law holds (approximately) for the Earth and Venus.

5. Use Kepler's third law to show that the closer a planet is to the Sun, the shorter is its period.

6. Use Kepler's third law to show that the closer a planet is to the Sun, the faster is its speed around the Sun.

7. Determine the distances from the Sun to the terrestrial planets and the asteroids, using the Titius-Bode law. Compare your answers with the actual distances.

8. Determine the distances from the Sun to the Jovian planets, using the Titius-Bode law. Compare your answers with the actual distances.

9. Using the Titius-Bode law, determine the distance from the Sun to Pluto, and compare it with the actual distance.

10. Refer to Table 16.1 and draw a diagram showing how far each inner planet goes around the Sun in 60 days.

11. Refer to Table 16.1 and draw a diagram showing how far each outer planet goes around the Sun in 10 years.

12. List the following distances in order of increasing length: (a) Sun to Earth, (b) Mars to Jupiter, (c) Jupiter to Saturn, (d) Saturn to Uranus.

13. List the planets in order of increasing distance from the Sun.

14. List the planets in order of decreasing size (largest first).

15. List the planets in order of decreasing density (densest first).

16. List the years of the next three appearances of Halley's comet.

CHAPTER 17

Place and Time

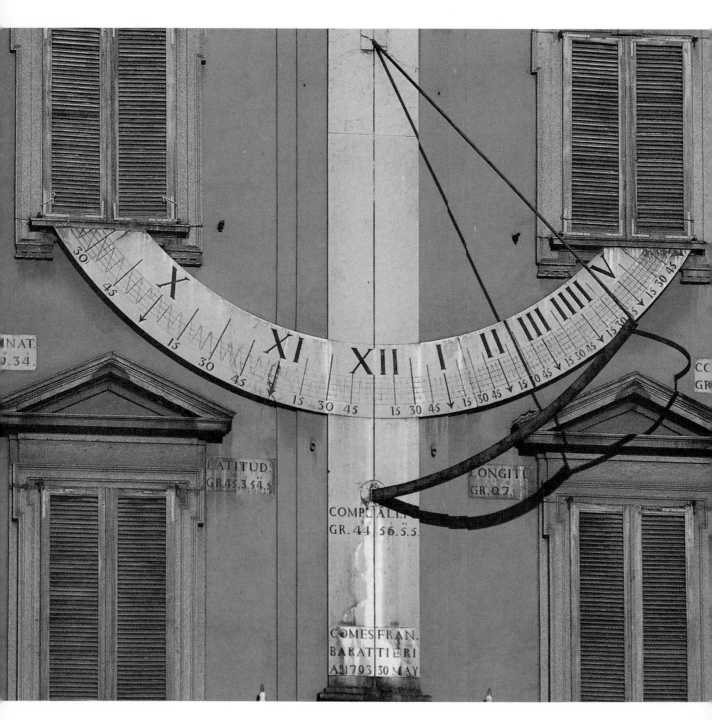

What place have you? / What place have I?

What time have you? / What time have I?

Only here and now / Have you and I.

N PHYSICAL SCIENCE we observe and examine events that take place in our environment. These events occur at different places and at different times. Some occur immediately and are observed, while others occur at great distances and are not observed until a later time. For example, you see a flash of lightning, but you do not hear the sound of thunder until later. A star explodes at some great distance from the planet Earth and years later the event is observed as the radiation reaches the Earth. Thus the events or happenings taking place in our environment are separated in space and time.

Albert Einstein was the first to point out that space and time are related and exist as a single entity. Einstein associated gravitational fields with space and developed the concept of four-dimensional space, which has given us an entirely different concept concerning the reality of our environment. In this chapter we shall study the basic concepts in physical science that refer to our location and the objects of our environment in space and time.

The questions of where and when come up in our daily lives: "The meeting of the next physical science class will be in Heath Hall at 10:00 A.M. on Tuesday." "Sally will spend the Christmas holidays in Florida." "The Space Shuttle was first launched from Cape Canaveral on April 12, 1981." All of these familiar phrases include place and time. Thus on each day of our lives we experience the march of time and the change of position.

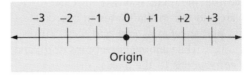

Figure 17.1 A one-dimensional reference system.

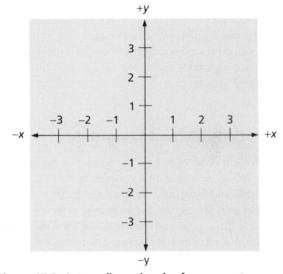

Figure 17.2 A two-dimensional reference system.

17.1 Cartesian Coordinates

The location of an object in our environment requires a reference system that has one or more dimensions. A one-dimensional system is depicted by the number line shown in Fig. 17.1, which illustrates two fundamental features of every coordinate system. A straight line is drawn, which may extend to plus infinity in one direction

and minus infinity in the opposite direction. For the line to represent a coordinate system, an origin must be indicated, and unit length along the line must be expressed. Temperature scales, left-right, above ground–below ground, time past–time future, and profit-loss are all examples of one-dimensional coordinate systems.

A two-dimensional system is shown in Fig. 17.2, in which two number lines are drawn perpendicular to each other and the origin assigned at the point of intersection. The two-dimensional system is called a **Cartesian coordinate system**, in honor of the French philosopher and mathematician René Descartes (1596–1650), the

◄ **A sundial in Italy.**

inventor of coordinate geometry. It is also referred to as a rectangular coordinate system. The horizontal line is normally designated the x-axis and the vertical line, the y-axis. Every position or point in the plane is assigned a pair of coordinates (x and y), which gives the distance from the two lines, or axes. The x number gives the distance from the y-axis, and the y number gives the distance from the x-axis. Many of the cities in the United States are laid out in the Cartesian coordinate system. Usually, one street runs east and west, corresponding to the x-axis, while another street runs north and south, which corresponds to the y-axis.

Our interest in the Cartesian coordinate system results from our desire to determine the location of any position on the surface of the spherical Earth and the location of any objects on the celestial sphere. A spherical surface is a curved surface on which all points are equidistant from a point called the center. The location of any position on a spherical surface can be found by using two reference circles analogous to the coordinate axes mentioned above.

17.2 Latitude and Longitude

The location of an object on the surface of the Earth is accomplished by means of a coordinate system known as latitude and longitude. Because the Earth is turning about an axis, we can use the *geographic poles*, which are defined as those points on the surface of the Earth where the axis projects from the sphere, as reference points.

The *equator*, defined in respect to the poles, is an imaginary line circling the Earth at the surface, halfway between the north and south geographic poles. The equator is a great circle, that is, a circle on the surface of the Earth located in a plane that passes through the center of the Earth. Any such plane would divide the Earth into two equal halves.

The **latitude** of a surface position is defined as the angular measurement in degrees north and south of the equator. The angle is measured from the center of the Earth relative to the equator (Fig. 17.3). Lines of equal latitude are circles drawn around the surface of the sphere parallel to the equator. Any number of such circles can be drawn. The circles become smaller as the distance from the equator becomes greater. These circles are called **parallels,** and when we travel due east or west, we follow a parallel. Latitude has a minimum value of 0° at the equator and a maximum value of 90° north or 90° south at the poles.

Imaginary lines drawn along the surface of the Earth running from the north geographic pole, perpendicular to the equator, to the south geographic pole are known as **meridians.** Meridians are half circles, which are portions of a great circle, because the circle is located in the same plane as the center of the Earth. An infinite number of lines can be drawn as meridians. **Longitude** is defined as the angular measurement, in degrees, east or west of the reference meridian, which is called the prime, or Greenwich, meridian. Longitude has a minimum value of 0° at the prime meridian and a maximum value of 180° east and west (Fig. 17.4).

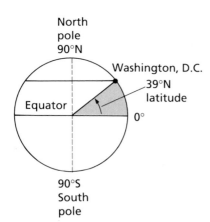

Figure 17.3 Diagram showing the latitude of Washington, D.C.

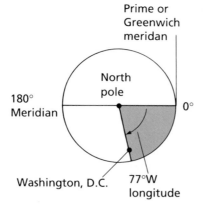

Figure 17.4 Diagram showing the longitude of Washington, D.C.

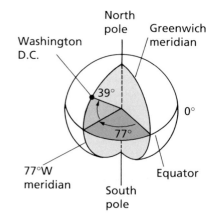

Figure 17.5 Diagram showing latitude and longitude of Washington, D.C.

The latitude and longitude of one point (Washington, D.C., 39°N, 77°W) are shown in Figs. 17.3 and 17.4. The latitude and longitude shown in Figs. 17.3 and 17.4 are combined in Fig. 17.5 and shown in a cutaway of the Earth. The **Greenwich,** or **prime, meridian** was chosen as the zero meridian because a large optical telescope was located at Greenwich, England, and because England ruled the seas at the time the coordinate system of latitude and longitude was originated. The primary purpose of the system at the time was to determine the location of ships at sea.

17.3 Time

The continuous measurement of time requires the periodic movement of some object as a reference. On October 8, 1964, the 12th General Conference of Weights and Measures, meeting in Paris, adopted an atomic definition of the *second* as the international unit of time. The definition is based on a change in energy levels of the cesium-133 atom. The exact wording of the definition for the atomic second is as follows: "The standard to be employed is the transition between the two hyperfine levels $F = 4$, $M_F = 0$ and $F = 3$, $M_F = 0$ of the fundamental state $2s_{1/2}$ of the atom of cesium-133 undisturbed by external fields, and the value 9,192,631,770 Hz is assigned." When a transition from one energy state to another occurs, the atom emits or absorbs radiation, whose frequency is proportional to the energy difference in the two states. The cesium-133 atom provides a highly accurate and stable reference frequency of 9,192,631,770 cycles per second, which can be referred to by electronic techniques. The National Bureau of Standards Frequency Standard, NBS-III, a cesium beam with a 3.66-m interaction region, is shown in Fig. 17.6.

For everyday purposes, we are interested in the Earth as a time reference, because our daily lives are influenced by the day and its subdivisions of hours, minutes, and seconds. The day has been defined in two basic ways. In the first definition, the solar day has been defined as the elapsed time between two successive crossings of the same meridian by the Sun. This period is also known as the **apparent solar day,** since this is what appears to happen. Because the Earth travels in an elliptical orbit, the orbital velocity of the Earth is not constant; therefore, the apparent days are not the same in duration. As a remedy for this situation, the mean solar day is computed from all the apparent days during a one-year period. The variation of the apparent solar day from the **mean solar day** is as much as 16 min.

In the second definition, the **sidereal day** has been defined as the elapsed time between two successive crossings of the same meridian by a star other than the Sun. Figure 17.7 illustrates the difference between the solar and sidereal days.

Because the Earth rotates 365.25 times during one revolution, the magnitude of the angle through which the Earth revolves in one day is

$$\frac{360°}{365.25 \text{ days}} = 0.985°$$

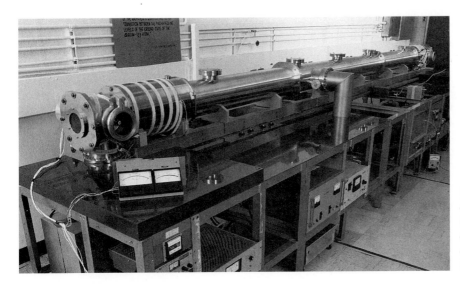

Figure 17.6 National Institute of Standards and Technology (formerly the National Bureau of Standards) cesium-beam frequency generator for establishing the time standard.

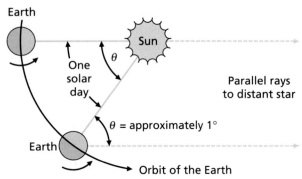

Figure 17.7 The difference between the solar day and the sidereal day.
One rotation of Earth on its axis with respect to the Sun is known as one solar day. One rotation of Earth on its axis with respect to any other star is known as one sidereal day. Note that Earth turns through an angle of 360° for one sidereal day and approximately 361° for one solar day.

or slightly less than 1° per day. The Earth must rotate through an angle of this same magnitude for the completion of one rotation in respect to the Sun. Therefore, the solar day is longer than the sidereal day by approximately 4 min, because the Earth rotates 360°/24 h, or 15°/h, or 1°/4 min.

The earliest measurement of solar time was accomplished with a simple device known as a gnomon (from the Greek *gnomon*, meaning "a way of knowing"), which is a vertical rod erected on level ground that casts a shadow when the Sun is shining (Fig. 17.8). The vertical pointer of the sundial is called a gnomon. By the third century B.C. the water clock had been invented. In hot and dry regions of the Earth sand was used to count the hours. The burning of candles was also used, because they burn at a fairly constant rate.

In the fourteenth century the first mechanical clock was invented, but it was not accurate. In the sixteenth century Galileo originated the plans for the construction of a pendulum clock. The story is written that while attending church, Galileo noticed a large hanging lamp swinging back and forth at regular intervals of time as measured by his pulse rate. This model provided him with the incentive to experiment with the pendulum and thus pave the way for the building of the first pendulum clock by Christian Huygens after Galileo's death.

Today, we have electric clocks, which are controlled by the frequency of the alternating current delivered by the power company. We also have wristwatches with quartz-crystal control, which are precise to one part in 10^9. Still greater precision is obtained by the atomic clocks mentioned earlier in this chapter.

The 24-h day, as we know it, begins at midnight and ends 24 h later at midnight. By definition, when the Sun is on the meridian, it is 12 noon *local solar time* at this meridian. The hours before noon are designated a.m. (**ante meridiem,** before midday and those after noon, p.m. (**post meridiem,** after midday). The time of 12 o'clock should be stated as 12 o'clock noon or 12 o'clock midnight, with the dates. For example, we should write 12 o'clock midnight, December 3–4, to distinguish that time from, say, 12 o'clock noon, December 4.

Our modern civilization runs efficiently because of our ability to keep accurate time. Since the late nineteenth century most of the countries of the world have adopted the system of **standard time zones.** This scheme theoretically divides the Earth into 24 time zones, each containing 15° of longitude or 1 hour, since the planet rotates 15° per hour. The first zone begins at the prime meridian, which runs through Greenwich, England, and extends $7\frac{1}{2}°$ each side of the prime meridian. The zones continue east and west from the Greenwich meridian, with the centers of the zones being multiples of 15°. The actual widths of the zones vary because of local conditions, but all places within a zone have the same time, which is the time of the central meridian of that zone. For example, Washington, D.C., is located at 77° W longitude, which is within $7\frac{1}{2}°$ of the 75° meridian. See Fig. 17.9.

Figure 17.10 shows the time and date on the Earth for any Tuesday at 7 A.M (PST), 8 A.M (MST), 9 A.M (CST),

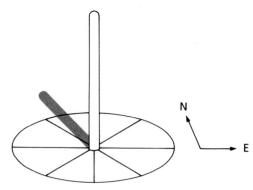

Figure 17.8 A gnomon.
This device is simply a vertical rod positioned so as to cast a shadow to indicate the time of day.

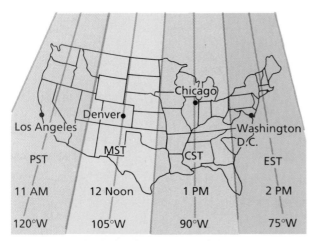

Figure 17.9 Time zones of the continental United States. Theoretical time zones are shown. Actual boundaries are irregular.

as you continue westward through additional time zones. A trip all the way around the Earth in a westward direction will mean the loss of 24 h, or one complete day. When you travel eastward, the opposite is true; that is, your watch will be 1 h slow for each zone, and the hour hand is set ahead 1 h.

A better understanding of why a day is lost in traveling around the Earth in a westward direction can be obtained if we take a make-believe trip. Suppose we leave Dulles International Airport in Washington, D.C., by jet plane at exactly 12 noon local mean solar time on Tuesday, December 5, and fly westward at a speed equal to the apparent westward speed of the Sun. Because Washington is located at 39°N latitude, the plane must travel the 39°N parallel westward at about 800 mi/h. As we leave the airport and fly westward, we observe the Sun out the left window of the plane about 32° above the southern horizon. One hour after leaving the airport, we notice the Sun can still be seen out the left window at the same altitude. Six hours later, with our watches indicating 7 P.M., the Sun still has the same apparent position as observed from the left window of the plane. Because the plane is flying at the same apparent speed as the Sun, the Sun will continue to be observed out the left window of the plane. Twenty-four hours later we arrive back in Washington with the Sun in the same apparent position. During the 24-h trip our time remained at 12 noon local mean solar time. If we had been observing the time with a sundial, it would have remained at 12 noon. We are aware of the passing of the 24 h because we kept track of the times with our watches, but we did not see the Sun set or rise, and we did not pass through 12 midnight; therefore, the time

and 10 A.M (EST). As the Earth turns eastward, the Sun appears to move westward, taking 12 noon with it. Twelve midnight is 180° or 12 h eastward of the Sun; and as 12 noon moves westward, 12 midnight follows, bringing the new day.

When you travel westward into a different time zone, the time kept by your watch will be 1 h fast or ahead of the standard time of the westward zone; therefore, you must move the hour hand back 1 h if the watch is to have the correct time. This process will be necessary

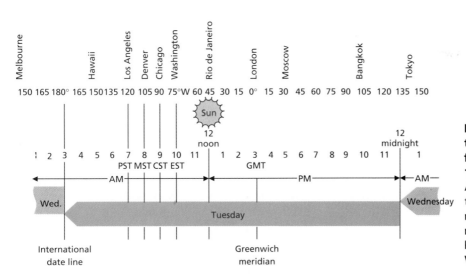

Figure 17.10 Diagram showing the times and dates on Earth for any Tuesday at 10 A.M. EST.
As time passes, the Sun appears to move westward; thus, 12 noon moves westward and midnight follows (180° or 12 hours) behind, bringing the new day, Wednesday, with it.

to us is still 12 noon Tuesday. To friends meeting us at the airport, it is 12 noon Wednesday.

To remedy situations like this one, the **International Date Line** (IDL) was established at the 180° meridian. When one crosses the IDL traveling westward, the date is advanced into the next day; and when one crosses the IDL traveling eastward, one day is subtracted from the present date.

If the local solar time is known at one longitude, the local time at another longitude can be determined by remembering that there are 15° for each hour of time, or 4 minutes for each degree. Should the calculation extend through midnight or should the International Date Line be crossed, the date would change.

One type of problem of practical importance is to find the time and date in a distant city when you know the time and date in a given city. This problem is encountered when you are trying to make a long-distance phone call and don't want to awaken someone in the middle of the night.

EXAMPLE 1 _____

What are the corresponding time and date in Tokyo (36°N, 140°E) when the time and date in Los Angeles (34°N, 118°W) are 6 A.M. PST, March 21?

Solution   The solution to a problem of this kind is easier to visualize when a diagram similar to Fig. 17.11 is used. This diagram has the north pole at the center, and the lines of longitude radiate outward from it as shown. Lines of longitude are drawn every 15° to correspond to the centers of the times zones.

In the figure the longitude of Los Angeles (118°W) is closest to the time zone centered on 120°W, and that of Tokyo (140°E) is closest to the one centered on 135°E. So the given time of 6 A.M. can be placed adjacent to the time zone center for Los Angeles, as shown in Fig. 17.11.

By definition, the meridian the Sun is on has the time of 12 noon. Because the Earth rotates eastward, the Sun appears to travel westward. Because it is 6 A.M. in Los Angeles, the Sun must be at 30°W because in 6 h the Sun will be at 120°W. Twelve midnight is always 180° from the Sun. Thus 12 midnight on March 21–22 is at 150°E, and Tokyo at 135°E has the time of 11 P.M., March 21. Midnight will arrive at Tokyo in one more hour. When you go east (counterclockwise), add an hour for each time zone; when you go west, subtract an hour. When you cross the International Date Line going west, add one day; when you cross the IDL going east, subtract one day. Also, change the date when you pass through midnight.

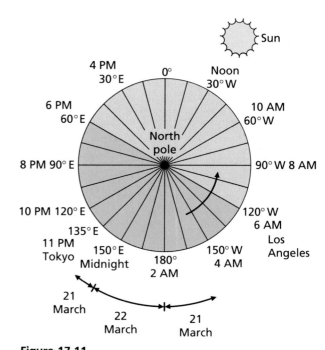

Figure 17.11
An example of finding the time and date in Tokyo, knowing that the time and date in Los Angeles is 6 A.M., Nov. 24. The North Pole is at the center of the circle.

During World War I the clocks of many countries were set ahead one hour during the summer months to give more daylight hours in the evening, thus conserving fuel used for generating electricity for lighting. This practice has now become standard for all but 3 or 4 of the 50 states of the United States. During the summer months in this country, time known as **Daylight Saving Time** (DST) begins at 2 A.M. on the first Sunday of April and ends at 2 A.M. on the last Sunday of October. The change to Daylight Saving helps conserve energy; it also reduced injuries and saves lives by preventing early-evening traffic accidents.

17.4 The Seasons

The spinning Earth is revolving around the Sun in an orbit that is elliptical yet nearly circular. When the Earth makes one complete orbit around the Sun, the elapsed time is known as one *year*. We are concerned with two different definitions of the year in this text. The **tropical year**, or the year of the seasons, is the time interval from one vernal equinox to the next vernal equinox. That is, the tropical year is the elapsed time between

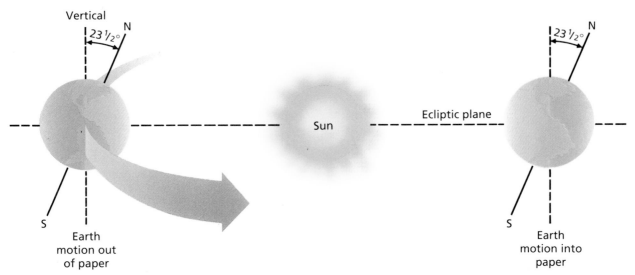

Figure 17.12
As Earth revolves around the Sun, Earth's axis remains tilted $23\frac{1}{2}°$ from the vertical. This inclination of the axis, in conjunction with Earth's motion, causes a change in seasons on Earth. When Earth is at the position shown with its motion out of the page, the northern hemisphere has summer and the southern hemisphere has winter. When Earth is at the position shown with its motion into the paper, the northern hemisphere has winter and the southern hemisphere has summer.

one northward crossing of the Sun above the equator and the next northward crossing of the Sun above the equator. In respect to the rotation period of the Earth, the tropical year is 365.2422 mean solar days.

The sidereal year is the time interval for the Earth to make one complete revolution around the Sun with respect to any particular star other than the Sun. The sidereal year has a period equal to 365.2536 mean solar days. This period is approximately 20 min longer than the tropical year. The reason for the difference will be explained later in Section 17.5.

The axis of the spinning Earth is not perpendicular to the plane swept out by the Earth as it revolves around the Sun, but it is tilted $23\frac{1°}{2}$ from the vertical, as illustrated in Fig. 17.12. This position of the axis in respect to the orbital plane produces a change in the Sun's overhead position throughout the year and causes our changing seasons. Figures 17.13 and 17.14 illustrate the apparent positions of the Sun over a period of one year.

In the summer in the Northern Hemisphere the Sun's rays are most direct on the Northern Hemisphere. Thus it is hotter in the summer when the Sun's rays are most direct, and it is colder in the winter when the Sun's rays

are the least direct. When it is summer in the Northern Hemisphere, it is winter in the Southern Hemisphere, and vice versa.

The noon Sun's overhead position is never greater than $23\frac{1°}{2}$ latitude, and the Sun always appears due south at 12 noon local solar time for an observer located in the continental United States. When the Sun is at $23\frac{1°}{2}$ north or south, it is at its farthest point from the equator. This farthest point of the Sun from the equator is known as the solstice (meaning the Sun stands still). The most northern point is called the **summer solstice,** and the most southern position is known as the **winter solstice.** This discussion applies to the Northern Hemisphere. In the Southern Hemisphere dates for summer and winter solstices are reversed from those shown in Fig. 17.13

As the Earth circles the Sun, the Sun's position overhead varies from $23\frac{1°}{2}$ north to $23\frac{1°}{2}$ south of the equator. When it is directly over the equator, the days and nights have 12 h each around the world. These dates are called the equinoxes. The **vernal equinox** occurs on or about March 21 and the **autumnal equinox** occurs on or about September 22 each year. These dates are labeled in Fig. 17.13.

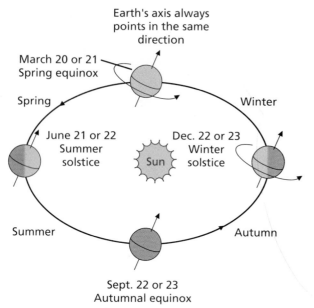

Figure 17.13 Earth's positions, relative to the Sun, and the four seasons.

As Earth revolves around the Sun, its north-south axis remains pointing in the same direction. On March 21 and September 21, the Sun is directly above the equator and everywhere on Earth has 12 hours of daylight and 12 hours of darkness. On June 21 the Sun's declination is $23\frac{1}{2}°$ N, and the northern hemisphere has more daylight hours than dark hours; it is then summer in the northern hemisphere and winter in the southern hemisphere. On December 23 the Sun's declination is $23\frac{1}{2}°$ S, and the southern hemisphere has more daylight hours than dark hours; it is then winter in the northern and summer in the southern hemisphere. The Sun is drawn slightly off center to indicate that Earth is slightly closer to the Sun during winter in the northern hemisphere than during summer.

When the Sun is observed at 12 noon local solar time, it is on the observer's meridian and appears at its maximum altitude above the southern horizon on that day for all observers north of the Sun. The angle measured from the horizon to the line of sight to the Sun at noon is called its **altitude,** and the angle from the **zenith** (position directly overhead) to the line of sight to the Sun at noon is called its **zenith angle.** See Fig. 17.15. The zenith is 90° from the **horizon** (the dividing line where the Earth and sky appear to meet); therefore, the sum of the zenith angle and the altitude is 90°. The altitude of the Sun can easily be determined by measurement with a sextant. If the Sun's position is known,

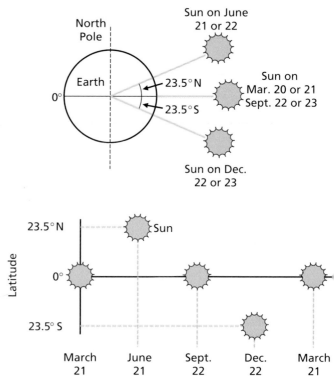

Figure 17.14 Diagrams of the Sun's position (degrees latitude) at different times of the year.

The upper drawing is an illustration of a greatly magnified Earth with respect to the Sun, showing the Sun's position on the dates indicated. The lower graph plots the Sun's position (degrees latitude) versus time (months).

the observer's latitude can be determined. The relation between the above terms is illustrated in Fig. 17.16.

The maximum and minimum altitudes of the Sun for an observer in Washington, D.C. (39°N), can be determined by using data from Fig. 17.16. The solutions are as shown in Figs. 17.17 and 17.18. The relationship between the two solutions is illustrated in Fig. 17.19. The altitude of the Sun for all other days of the year, as observed from Washington, would be between these two values. A similar solution will give the Sun's altitude from any latitude.

The seasons have a tremendous effect on the lives of everyone. Our yearly life cycles are ordered by the season's progressions. Many of our holidays were originally celebrated as commemorating a certain season of the year. The celebration that has evolved into our Easter holiday was originally a celebration of the coming of spring and a renewal of nature's life. Halloween originally commemorated the beginning of the winter season,

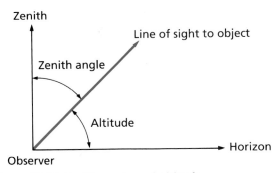

Figure 17.15 Zenith angle and altitude.
Because the zenith is perpendicular to the horizon, the zenith angle plus the altitude equals 90°.

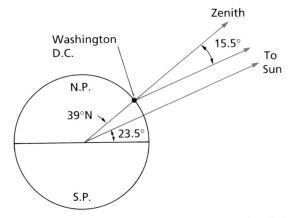

Figure 17.17 Finding the approximate altitude of the Sun as observed from Washington, D.C., on June 21.
Because the angle between the Sun and the observer is $39° - 23.5° = 15.5°$, the altitude of the Sun is $90° - 15.5° = 74.5°$.

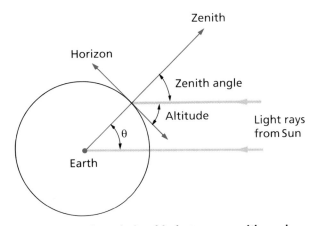

Figure 17.16 The relationship between zenith angle, altitude, horizon, and the angle θ.
Because the incoming rays of light from the Sun are parallel, the angle θ and the zenith angle are equal in magnitude.

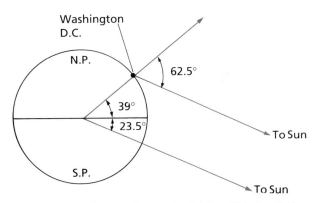

Figure 17.18 Finding the approximate altitude of the Sun as observed from Washington, D.C., on December 21.
Because the angle between Sun and observer is $39° + 23.5° = 62.5°$, the altitude of the Sun is $90° - 62.5° = 27.5°$.

while Thanksgiving commemorated the end of the harvest. The ancient festival of the winter solstice (on December 21 or December 22) and the beginning of the northward movement of the Sun has evolved into our Christmas holiday on December 25. The reasons for celebrating our various holidays have changed over the course of time, but the original dates were set by nature's annual timepiece—the movement of the Earth around the Sun.

17.5 Precession of the Earth's Axis

Most of us are acquainted with the action of a toy top that has been placed in rapid motion and allowed to spin about its axis. After spinning a few seconds, the top begins to wobble or do what physicists call "precess." See Fig. 17.20. The top, a symmetrical object, will continue to spin about an axis if the center of gravity remains above the point of support. When the center of gravity is not in a vertical line with the point of support, the

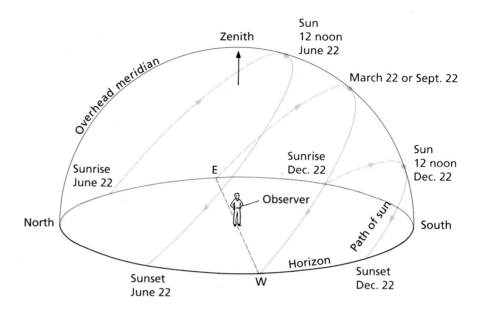

Figure 17.19
Diagram showing the apparent path of the Sun across the sky on June 21 and December 21, as observed from Washington, D.C. (39°N latitude).

axis slowly changes its direction. This slow rotation of the axis is called **precession.**

Because the Earth is spinning rapidly, it bulges at the equator and cannot be considered a perfect sphere. The moon and Sun apply a gravitational torque to the Earth; this torque tends to bring the Earth's equatorial plane into its orbital plane. Because of this torque, the axis of the Earth slowly rotates clockwise or westward about the vertical or the north ecliptic pole (Fig. 17.21). The period of the precession is 25,800 years; that is, it takes 25,800 years for the axis to precess through 360°.

Because the precession is clockwise and the Earth is revolving counterclockwise around the Sun, the tropical year is approximately 20 min shorter than the sidereal year. As the axis precesses, Polaris will no longer be the north star. The star Vega in the constellation

Lyra will be the north star some 12,000 years from now. The Southern Cross, a constellation of stars located within 27° of the present south celestial pole, will then be visible from Washington, D.C.

Our calendar is based on the seasons of the year. We want our summers to be warm and our winters to be cold. As the Earth precesses, we define June 21 to be the date when the Northern Hemisphere tilts toward the Sun at a maximum angle. Thus, as the Earth precesses, the stars seen at various seasons will change. This change is shown in Fig. 17.22. The stars we see on summer nights will slowly change within a period of 25,800 years. In the year A.D. 14,900 stars seen on summer nights will be our present winter night stars. Because of precession, the 12 constellations in the zodiac will slowly cycle through different months, with a one-month change occurring every 2150 years (25,800/12). The stars seen overhead on June 21 some 2150 years ago are seen overhead on July 21 today and will be seen overhead on August 21 in another 2150 years.

In the early 1970s a popular song called "The Age of Aquarius" took its title from the fact that the constellation Aquarius was moving from its January 21 to February 21 time slot of thousands of years ago toward the March 21 to April 21 time slot. Because older cultures started the year on March 21, the "age" or 2150-year period took on aspects of the March 21 to April 21 constellation. The song spoke of "the dawning of the age of Aquarius," which we see would slowly occur as the Earth precessed.

Figure 17.20 Precession of a top.

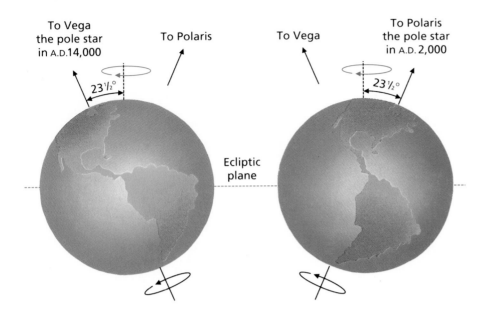

Figure 17.21 Precession of Earth's axis clockwise about a line perpendicular to Earth's orbit.
Earth's axis is presently pointing toward the star Polaris, which we call the pole, or north, star. In approximately 12,000 years the axis will be pointing toward the star Vega in the constellation Lyra.

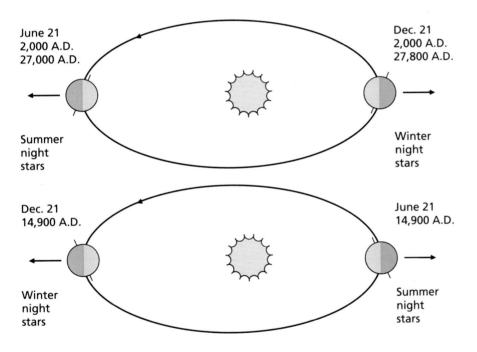

Figure 17.22
The precession of Earth's axis will cause future generations to see different stars in the summer than are seen today. After 12,900 years our winter night stars will be their summer night stars and vice versa. After another 12,900 years, the constellations will be similar to what we see now.

17.6 The Calendar

The continuous measurement of time requires the periodic movement of some object as a reference. Various lengths of time had a direct influence on the life of ancient men and women. Because of these influences, they had more than one reference for measuring the events of their lives. It is reasonable to believe that the first unit for the measurement of time was the day; because of the daily need for food, people spent the daylight hours hunting. During the dark hours they slept.

A longer period of reference was based on the periodic movement of the moon. Some societies probably used the moon as a basic division of time, because they were unable to count the days over a long period. The first appearance of the crescent moon and the time of the full moon were times of worship for many primitive tribes.

Our month of today originated from the periodic movement of the moon, which requires $29\frac{1}{2}$ solar days to orbit the Earth. The plans for the first calendar seem to have originated before 3000 B.C. with the Sumerians, who ruled Mesopotamia. Their calendar was based upon the motion of the moon, which divided the year into 12 lunar months of 30 days each. Because $30 \times 12 = 360$ days, and the year actually contains $365\frac{1}{4}$ days, corrections had to be made to keep the calendar adjusted to the seasons. How the Sumerians made the corrections remains a mystery, but their successors, the Babylonians, adjusted the length of the months and added an extra month when needed. This Babylonian calendar set the pattern for many of the calendars adopted by ancient civilizations.

The calendar we use today originated with the Romans. The early Roman calendar contained only 10 months, and the year began with the coming of spring. The months were named March, April, May, June, Quintilis, Sextilis, September, October, November, and December. The winter months of January and February did not exist. This was the period of waiting for spring to arrive. About 700 B.C. Numa, who reigned in Rome, added the month of January at the beginning of the year, before March; and February at the end of the year, after December. About 275 years later, in 425 B.C., the two months were changed to the present order.

The so-called Julian calendar was adopted in 45 B.C., during the reign of Julius Caesar. Augustus Caesar, who ruled the Roman Empire after Julius, renamed the month Quintilis as July in honor of Julius, and the month Sextilis as August in honor of himself. He also removed one day from February, which he added to August to make it as long as the other months.

The Julian calendar had 365 days in a year. In every year divisible by 4, an extra day was added to make up for the fact that it takes approximately $365\frac{1}{4}$ days for the Earth to orbit the Sun. Thus in 1985, 1986, and 1987 there are 365 days. In 1984 and 1988 there are 366 days. The Julian calendar was fairly accurate and was used for over 1600 years.

In 1582 Pope Gregory XIII (Fig. 17.23) realized that the Julian calendar was slightly inaccurate. The vernal equinox was not falling on March 21, and religious holidays were coming at the wrong time. A discrepancy was found, and the pope decreed that 10 days would be skipped to correct it. The discrepancy arose because

Figure 17.23 Pope Gregory XIII.
In 1582 he issued a proclamation to drop 10 days from the calendar and to have leap years 97 times (instead of 100) every 400 years. The Gregorian calendar is now used worldwide.

there are 365.2422 days in a year and not 365.25, as the Julian calendar used.

Pope Gregory set the calendar straight and devised a method to keep it correct. He decreed that every 400 years, 3 leap years would be skipped. The leap years to be skipped were the century years not evenly divisible by 400. The present corrections make the calendar accurate to one day in 3300 years. Our present-day calendar with these leap-year designations is called the **Gregorian calendar.**

Pope Gregory's reform of 1582 was not universally accepted. Most Protestant countries did not immediately go along. England and the American colonies finally changed their calendars in 1752, when there was an 11-day discrepancy. As a correction, September 2 was followed by September 14, and some people thought they were losing 11 days of their life. Other problems also arose when landlords asked for a full month's rent and banks sought to collect a full month's interest. In fact, riots were touched off in London over the change. When Russia and China accepted the Gregorian calendar in the early 1900s, it marked the first time that the whole world used the same calendar. There are other calendars still in use for religious purposes, but for civil events the Gregorian calendar is used around the world.

In our calendar we have seven days in each week. The origin of the seven-day week is not definitely known, and not all cultures have had a seven-day week. One possible origin of the seven-day week is that it takes approximately seven days for the moon to go from one phase to the next (e.g., new to first-quarter phase).

A more likely origin is the nighttime sky. As the ancients watched the sky night after night, there were exactly seven celestial bodies that moved relative to the fixed stars. These seven objects visible to the naked eye are the Sun, moon, and five visible planets: Mars, Mercury, Jupiter, Venus, and Saturn.

Our present days of the week can still be connected by their names to the Sun, moon, and five visible planets. Table 17.1 shows the heavenly object, English name, French name, and Saxon name for the seven days of the week. Note that either the English or French word is similar to the name of the heavenly object.

Table 17.1 The Days of the Week

Heavenly Object	English Name	French Name	Saxon Name
Sun	Sunday	dimanche	Sun's day
Moon	Monday	lundi	Moon's day
Mars	Tuesday	mardi	Tiw's day
Mercury	Wednesday	mercredi	Woden's day
Jupiter	Thursday	jeudi	Thor's day
Venus	Friday	vendredi	Fria's day
Saturn	Saturday	samedi	Saturn's day

Our English days, Tuesday, Wednesday, Thursday, and Friday, come from the words Tiw, Woden, Thor, and Fria, who were Nordic gods. Woden was the principal Nordic god; Tiw and Thor were the gods of law and war; and Fria was the goddess of love. Sunday, Monday, and Saturday retain their connection to the Sun, moon, and Saturn.

Learning Objectives

After reading and studying this chapter, you should be able to do the following without referring to the text:

1. State the latitude of the equator and North and South Poles.

2. State how the prime meridian and International Date Line are determined.

3. Calculate the date and time at any latitude and longitude, given the date and time at another latitude and longitude.

4. Explain, with the aid of a diagram, why we have different seasons of the year.

5. Tell when we have the vernal equinox, autumnal equinox, summer solstice, and winter solstice each year.

6. Compute the altitude of the Sun at noon 40°N latitude on March 21, June 21, September 22 and December 22.

7. Draw a diagram and use it to explain the precession of the Earth's axis.

8. Explain why we have 12 months in a year and 7 days in a week in our present calendar.

9. Explain when we have leap years in the Gregorian calendar.

10. Define and explain the important words and terms listed in the following section.

Important Words and Terms

Cartesian coordinate system
latitude
parallels
meridians
longitude
Greenwich meridian, prime meridian
apparent solar day
mean solar day
sidereal day

ante meridiem
post meridiem
standard time zones
International Date Line
Daylight Saving Time
tropical year
summer solstice
winter solstice
vernal equinox

autumnal equinox
altitude
zenith
zenith angle
horizon
precession
Gregorian calendar

Questions

Cartesian Coordinates

1. Name two fundamental features of every coordinate system.

2. Give three examples of a one-dimensional reference system.

3. (a) What two streets in your city or town divide the city into four quadrants?
   (b) Name the four quadrants.

4. Draw a three-dimensional reference system.

5. What is the angle between each of the three axes in Question 4?

Latitude and Longitude

6. What are the minimum and maximum values for latitude and longitude?

7. Can one travel continuously eastward and circle the Earth? Why or why not?

8. Can one travel continuously southward and circle the Earth? Why or why not?

9. How is 0° defined for latitude and longitude?

10. What is the name of a line of equal longitude?

Time

11. How are the boundaries of standard time zones determined?

12. How many time zones are there in the 48 contiguous (adjoining) states?

13. What do A.M. and P.M. mean?

14. Is it correct to state the time as 12 A.M.? Explain your answer.

15. What are some advantages of Daylight Saving Time?

16. What is the direction of rotation (clockwise or counterclockwise) of the shadow cast by a sundial located at (a) 30° N and (b) 30°S?

17. Explain how a gnomon can be used to determine true north.

18. When does Daylight Saving Time begin and end?

The Seasons

19. Determine when each of the following occur in the Northern Hemisphere: autumnal equinox, spring equinox, winter solstice, summer solstice.

20. Explain the difference between one sidereal year and one tropical year.

21. Why are there four different seasons each year?

22. How would the seasons be modified if the Earth's axis were tilted at 10° instead of 23.5°?

23. What is the altitude of Polaris (the north star) for an observer at the equator (0° latitude)? at the north pole (90°N)?

24. What is the altitude of Polaris (the north star) for an observer at Washington, D.C. (39°N)?

Precession of the Earth's Axis

25. Define and explain precession.

26. What evidence is there to support precession of the Earth's axis?

27. How long does it take for the Earth to precess one time?

The Calendar

28. What is the origin of the month?

29. What is the origin of the seven-day week?

30. How often was there a leap year in the Julian calendar?

31. How often is there a leap year in the Gregorian calendar?

32. What are the origins of the dates for Halloween (October 31) and Christmas (December 25)?

Exercises

Latitude and Longitude

1. How far away is the point at 90°N, 130°E from 90°N, 150°E?

2. Draw a diagram and explain why the points 60°N, 130°E and 60°N, 150°E are closer together than the points 30°N, 130°E and 30°N, 150°E.

3. What are the latitude and longitude of the point on the Earth that is opposite Washington, D.C. (39°N, 77°W)?
   Answer: 39°S, 103°E

4. What are the latitude and longitude of the point on the Earth that is opposite Tokyo (36°N, 140°E)?

5. One nautical mile is a length unit of one minute of arc of a great circle. Sixty minutes of arc equal one degree.
   (a) Determine the shortest distance in nautical miles between Washington, D.C. (39°N), and the equator.
   (b) The nautical mile is $\frac{69}{60}$ of a statute (land) mile. Determine the distance in part (a) in statute miles.

6. Suppose you start at Washington, D.C. (39°N, 77°W), and travel 300 nautical miles due north, then 300 nautical miles due west, then 300 nautical miles due south, then 300 nautical miles due east. Where will you arrive in respect to your starting point—at your starting point, or north, south, east, or west of your starting point?

Time

7. A professional basketball game is to be played in Portland, Oregon. It is televised live in New York beginning at 9 P.M. EST. What time must the game begin in Portland?

Answer: 6 P.M. PST

8. If the polls close during a presidential election at 7 P.M. EST in New York, what is the time in California?

9. If an Olympic event begins at 10 A.M. on July 28 in Los Angeles (34°N, 118°W), what time and date will it be in Moscow (56°N, 38°E)? *Answer:* 9 P.M., July 28

10. When it is 9 P.M. on November 26 in Moscow (56°N, 38°E), what time and date is it in Tokyo (36°N, 140°E)?

11. When it is 10 A.M. on February 22 in Los Angeles (34°N, 118°W), what time and date is it in Tokyo (36°N, 140°E)?

The Seasons

12. What is the altitude angle of the Sun on March 21 for someone at the north pole?

13. What is the altitude angle of the Sun for someone at 42°N latitude on (a) March 21 and (b) June 21?

Answer: (a) 48°

14. What is the altitude angle of the Sun for someone at 34°N latitude on (a) September 22 and (b) December 22?

15. What is the latitude of someone in the United States who sees the Sun at an altitude angle of 71.5° on June 21?

16. What is the latitude of someone in the United States who sees the Sun at an altitude angle of 31.5° on December 22?

17. How many days are in each of the following years: 1986, 1987, 1988, 1989, 1990, 1991, 1992, 2000, 2001, 2004, 2100, 2200, 2300, 2400?

18. Determine the month and day when the Sun is at maximum altitude for an observer at Washington, D.C. (39°N) What is the altitude of the Sun at this time?

Answer: on or about June 21

19. Determine the month and day when the Sun is at minimum altitude for an observer at Washington, D.C. (39°N)? What is the altitude of the Sun at this time? *Answer:* $27\frac{1}{2}°$

20. Is the difference between the maximum and minimum altitude of the Sun as determined in Exercises 18 and 19 equal to twice the angle the Earth's axis is tilted from the vertical?

The Moon

THE EXACT ORIGIN of the word *moon* seems to be unknown, but many writers believe the original meaning related to the measurement of time. We do know that the length of our present month is based upon the motion and the phases of the moon, that primitive people worshipped the moon, and that many societies today base their religious ceremonies on the new and full phases of the moon. We also know that the human reproductive cycle is synchronized to the lunar cycle, with the ovaries producing ova about every twenty-eight days.

On July 20, 1969, human beings first landed on the moon, and Apollo 11 astronauts placed a retroreflector (an optical reflector designed to return the reflected ray of a laser beam exactly parallel to the incident ray) array on the moon's surface. The retroreflector is part of a lunar-ranging experiment that measures the distance to the moon with an accuracy of 15 cm. Measurements with this high accuracy can be taken over long periods of time and will show the variation in the orbital distance of the moon in great detail. Such measurements can be used to (1) determine the rate of continental drift on Earth (latitude and longitude of a place can be determined with great accuracy); (2) detect any change in the location of the north pole; (3) determine the orbit of the moon with greater accuracy; and (4) determine whether the gravitational constant (*G*) is decreasing very slowly with time, because the universe is believed to be expanding.

The moon is a rather insignificant body if viewed from outside our solar system, but it is our largest natural satellite and appears as the second brightest object in the sky to the Earth observer because it is very close to us. The moon's average distance from Earth is about 240,000 mi. Because of the moon's nearness and its influence on our lives, this chapter is devoted to its study. Figure 18.1 shows an astronaut with the Lunar Rover collecting samples of the lunar surface.

◀ **The foothills of the Taurus mountains on the moon.**

Figure 18.1 An astronaut on the moon with the Lunar Rover collecting rock samples.

18.1 General Features

Earth's moon at its brightest is a wondrous sight as it reflects the Sun's light back to Earth. The moon appears quite large to Earth observers. In fact, our moon is the largest moon of any inner planet. Mercury and Venus have no moons and the moons of Mars are quite small. Our moon is the fifth largest in the solar system. A unique feature of Earth and the moon is that they are nearer in size than any other planet and its satellite.

The moon revolves around Earth in approximately $29\frac{1}{2}$ days, and it rotates at the same rate as it revolves. For this reason we see only one side of the moon. An observer on the side of the moon that faces Earth would always be able to see Earth, but the Sun would appear to rise and set and rise again once every $29\frac{1}{2}$ days. Thus all sides of the moon are heated by the Sun's rays.

The moon is spherical, with a diameter of 3476 km, a value slightly greater than one-fourth Earth's diameter. The slow rotation of the moon, coupled with the tidal bulge caused by Earth's gravitational pull on the solid material, produces an oblateness that is very small. The

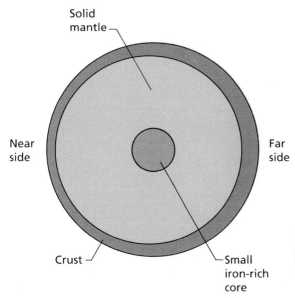

Figure 18.2
Geologists theorize that the moon consists of a solid mantle rich in silicate materials, a crust that varies in depth from about 40 miles on the near side to 80 miles on the far side from Earth, and a small, perhaps solid, iron-rich core.

Figure 18.3
This rugged terrain is typical of the far side of the moon. The large crater in the center of the picture has a diameter of about 81 km (50 mi).

best measurements indicate a difference of less than a mile between the polar and equatorial diameters.

The moon's mass is $\frac{1}{81}$ that of Earth, and its average density is 3.3 g/cm^3. Earth's average density is 5.5 g/cm^3. The surface gravity of the moon is only $\frac{1}{6}$ that of Earth. Therefore, your weight on the surface of the moon would be about $\frac{1}{6}$ of your weight at Earth's surface (see Fig. 1.4). The moon's interior is thought to be made up of a small, perhaps solid, iron-rich core, a solid mantle, and a crust that is about 40 mi thick on the near side and 80 mi thick on the far side (Fig. 18.2). The moon does not possess a magnetic field; at least, no magnetic field was detected by instruments carried by the Apollo astronauts. Surface rocks brought back from the moon show some magnetism, indicating that the moon had a slight magnetic field when the rocks solidified. The origin of this previous magnetic field is not known.

Except for the change in the phases of the moon, its most predominant feature is the appearance of the surface, which is marked with craters, plains, rays, rills, mountain ranges, and faults. These features vary in size, shape, and structure. The most outstanding are the craters that are clearly visible to an Earth observer with low-power binoculars or a telescope.

Craters

The word **crater** (Greek *krater*) means bowl shaped, and the lunar craters are great pits believed to be caused by the impact and explosion of small and large objects that have come from space. The craters are rather shallow (their depths are small in comparison with their diameters), and their floors are located below the lunar surface. See Fig. 18.3.

The slope of a rim with a width of about $\frac{1}{5}$ the width of the crater from crest to crest is greater on the inside than on the outside. Measuring the volume of the material in the rim and comparing it with the volume of the crater hole shows that they are approximately the same. This result supports the impact hypothesis.

Plains

There are thousands of craters on the lunar surface, ranging in diameter from a few feet to the 150-mi Clavius. The large flat areas called *maria* (an Italian word meaning "seas"), named by Galileo, are believed to be craters formed by the impact of huge objects from space and that later were filled with lava. These areas, which are now called **plains,** appear to be very dark because the moon's surface is a poor reflector. The surface reflects only 7% of the light received from the Sun. The plains, which are similar to black asphalt, reflect very little light. There are 14 major plains on the front side, which cover over 50% of the visible lunar surface. Most of the plains are located in the northern hemisphere and can be plainly seen with the unaided eye during the full phase of the moon.

The surface of the moon is a terrain of rolling rounded knolls composed of a layer of loose debris or soil called *regolith,* which has a depth less than 10 m on the flat lunar plains (*maria*). The lunar highlands, because they are older, have a thicker layer of regolith. The rock samples brought back by Apollo astronauts are similar to the volcanic rock found on the Earth.

Rays

Some craters are surrounded by streaks, or **rays,** that extend outward over the surface. They are believed to be pulverized rock that was thrown out when the crater was formed. The rays appears much brighter than the crater, and we know that powdered rock reflects light better than regular-sized rock. The rays also become darker with age. Photographs show that in cases where rays from one crater overlap the rays of another, the rays on top appear to be brighter.

The ray system of a crater has an average diameter of about 12 times the diameter of the crater. The lunar photographs also show that the ray systems are marked with small craters called secondary craters, which are believed to have been formed by debris thrown out from the primary crater during the explosion caused by an impinging object from space.

Rills

Another feature of the lunar surface is the existence of long narrow trenches, or valleys, called **rills.** They vary from a few feet to about 3 mi in width and extend hundreds of miles in length with little or no variation of width. Some rills are rather straight, while others follow a circular path. The rills have very steep walls and fairly flat bottoms that are as much as $\frac{1}{2}$ mi below the lunar surface. Moonquakes are thought to cause the formation of rills. A similar separation of the Earth's surface would be produced by an earthquake.

Mountain Ranges

The **mountain ranges** on the lunar surface have peaks as high as 20,000 ft, and all formations seem to be components of circular patterns bordering the great plains, or *maria.* This pattern indicates that they were not formed and shaped by the same processes as mountain ranges on the Earth, which were formed by internal forces.

The formation of craters and other surface features of the moon took place a long time ago when the solar system was filled with large amounts of matter, and these features are much the same now as they were when formed, because of the absence of water and a lunar atmosphere. There have been some changes resulting from the impact of projectiles from space.

A **fault** is a break or fracture in the surface of the moon along which movement has occurred. The motion along a fault can be vertical, horizontal, or parallel. Several faults are observed on the lunar surface. Figure 18.4 shows the moon at 18 and 22 days with a very large fault in Mare Nubium. This fault (called the Straight Wall) is about 60 mi long and 1200 ft high, and its side is inclined about 40° to the horizontal.

18.2 History of the Moon

Before the Apollo program to land an astronaut on the moon was begun in the early 1960s, very little was known about the origin and history of the moon. But our exploration of the moon has changed all that.

The first landing of the moon was July 20, 1969, when the landing craft of Apollo 11 descended to the moon's surface in Mare Tranquillitatis at 0.67° north, 23.49° east. After that, five other Apollo lunar landing missions (Apollo 12, 14, 15, 16, and 17) were completed. The Apollo astronauts collected and brought back to Earth 379 kg of lunar material and erected on the lunar surface 2104 kg of scientific instruments that will collect data for many years. The American Apollo program has now ceased because of budget considerations.

The rock samples brought back from the moon, such as those shown in Fig. 18.5, have enabled us to have a much better understanding of the moon's origin and history. Samples from the plains or lowlands have yielded ages considerably younger than samples from the highlands. Rocks from the highlands were formed between 3.9 and 4.4 billion years ago, whereas the rocks from plains or lowlands have ages between 3.8 and 3.1

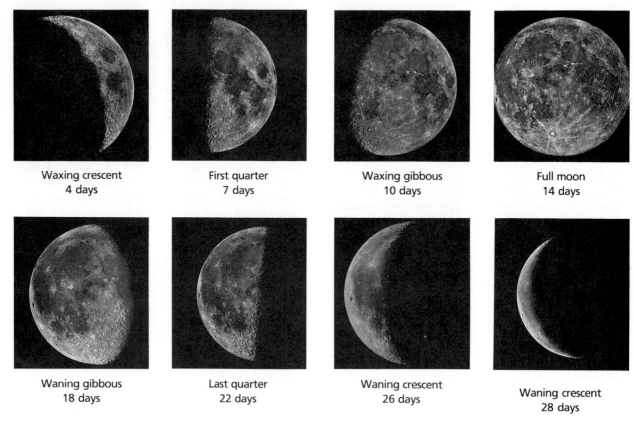

Waxing crescent 4 days	First quarter 7 days	Waxing gibbous 10 days	Full moon 14 days
Waning gibbous 18 days	Last quarter 22 days	Waning crescent 26 days	Waning crescent 28 days

Figure 18.4 Eight photographs showing the moon at different times of the lunar month.
The photos are arranged so that the view, from Earth, of the phases are the way they appear in a close-up view by the unaided eye. Compare these photos with Figure 17.9.

Figure 18.5 Scientists examine a lunar sample.
The samples are stored in an atmosphere of dry nitrogen, thus isolating the samples from oxygen and moisture (water) to prevent chemical reactions.

billion years. No rocks older than 4.4 billion years or younger than 3.1 billion years have been found.

Almost all of the craters on the moon are now known to have resulted from the bombardment of meteorites of various sizes. Because the moon has no atmosphere or water on its surface, very little erosion takes place on the moon. Once formed, a crater remains for billions of years or until a meteorite hits to form a crater on top of it. In contrast, Earth has only a few remaining meteorite craters. The craters left by meteorites striking Earth have been eroded away for the most part.

The plains and a few (about 1%) of the craters were produced by volcanic eruptions on the moon. The plains are composed of black volcanic lava that covered many craters. Most of the plains are on the near side of the moon. Many of them can be seen in Fig. 18.4. The fact that there were fewer volcanic eruptions on the moon's far side is probably correlated with the fact that the moon's crust is thicker there (see Fig. 18.2).

From the ages of moon rocks and other data we now have a fairly solid understanding of the moon's origin and history. However, as more samples are brought back from the moon and more studies are done, the theory may change.

The Earth and moon are both believed to have originated in the same part of the solar system at about the same time, 4.6 billion years ago. At this time the Earth and moon formed from the agglomeration of many smaller pieces of rock and other matter. The Earth was a much larger body, and it was formed from higher-density materials than the moon. The moon's surface was hot and molten until about 4.4 billion years ago.

The oldest rocks on the moon were formed about 4.4 billion years ago when the moon's crust was cool enough to solidify. From 4.4 to 3.9 billion years ago the moon was intensely bombarded by many meteorites that were still present near the Earth-moon system. This was the period when most of the moon's craters were formed.

During the period 3.9 to 3.1 billion years ago, the moon's interior had heated up enough from radioactive effects to cause volcanic eruptions to occur that formed the many plains. The lava flows from these eruptions covered much of the moon's lowlands. During this period meteorite bombardment became less intense, because fewer and fewer rock fragments were left near the Earth-moon system.

After 3.1 billion years ago the moon's mantle had become so thick that it could no longer be penetrated by molten rock, and the moon has been geologically quiet since that time. Meteorites have continued to bombard the surface and have formed a thin veneer of dust several feet thick on the moon's surface.

Much of our knowledge of the moon has been gained by analyzing rock samples. Another important avenue of investigation has been the study of moonquakes. The seismological data gathered have enabled us to construct diagrams like Fig. 18.2. Other studies have shown that there is practically no water at all in any form on the moon. In addition, no biological life has been found.

18.3 Lunar Motion

The Earth's moon revolves eastward around the Earth in an elliptical orbit once every $29\frac{1}{2}$ solar days or $27\frac{1}{3}$ sidereal days. The orbital plane of the moon does not

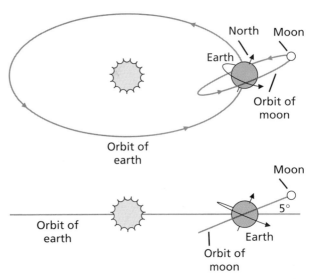

Figure 18.6 The relative motions of the moon and Earth. The top diagram is a view from above the ecliptic plane; the lower diagram is a view from on the ecliptic plane.

coincide with the orbital plane of the Earth but is tilted at an angle of approximately 5° with respect to the Earth's orbital plane. The moon rotates eastward as it revolves, making one rotation during one revolution. Figure 18.6 is an illustration of the eastward motion of the moon and the inclination of the orbital plane to the ecliptic.

Because the moon revolves in an elliptical orbit, the distance from the Earth to the moon varies as the moon revolves. At the closest point, called **perigee,** the moon is 221,463 mi from the Earth. At the farthest point, called **apogee,** the moon is 252,710 mi from the Earth. This change in distance of over 31,000 mi produces only a slight change in the apparent size of the moon as viewed from the Earth. The mean distance from Earth to the moon is 238,856 mi, but 240,000 mi will be used as the mean for solving problems in this text.

As we stated above, there are two different lunar months. The period of the moon in respect to a star other than the Sun is approximately $27\frac{1}{3}$ days; this is called the sidereal period, or **sidereal month.** This period is the actual time taken for the moon to revolve 360°. The period of the moon in respect to the Sun is approximately $29\frac{1}{2}$ days. This period is called the **synodic month,** or the month of the phases. The moon revolves more than 360° during the synodic period. See Fig. 18.7.

To an observer on Earth, the moon appears to rise in the east and set in the west each day. This apparent

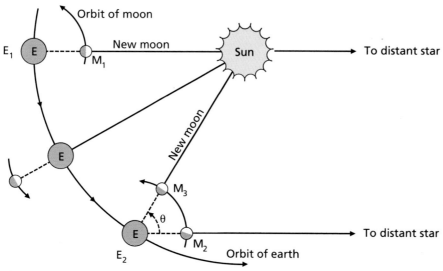

Figure 18.7
Diagram illustrating the difference between the sidereal and synodic months. With the Earth at position E_1 and the moon at position M_1, the Sun, moon, and Earth are all in the same plane. At this time the moon is in its new phase. As the Earth revolves eastward to position E_2, the moon has revolved through 360° in approximately 27 and $\frac{1}{3}$ days to position M_2, and one sidereal month—one revolution with respect to a distant star—has elapsed. The moon must revolve through 360° plus the angle θ before arriving at position M_3. At this time the Sun, moon, and Earth will be in the same plane, the moon will be in new phase, and one synodic month will have passed. The time for one synodic month is approximately 29 and $\frac{1}{2}$ days.

motion of the moon is due to the spinning of the Earth on its axis once each day. The times at which it rises and sets are discussed in the following section.

18.4 Phases

The most outstanding feature presented by the moon to an Earth observer is the periodic change in its appearance. One-half of the moon's surface is always reflecting light from the Sun, but only once during the lunar month does the observer on the Earth see all of the lighted half of the moon. Throughout most of the moon's period of revolution, only a portion of its lighted side is presented to us.

The starting point for the periodic, or cyclic, motion of the moon is arbitrarily taken at the new-phase position. The new phase of the moon occurs when the Earth, Sun, and moon are in the same plane, with the moon positioned between the Sun and Earth. They are not necessarily in a straight line. At this position the

dark side of the moon is toward the Earth and the moon cannot be seen from this planet. Because the Sun is on the observer's meridian with the moon, the new moon occurs at 12 noon local solar time.

The **new moon** occurs just for an instant—the instant it is on the same meridian as the Sun. We often speak, however, of a phase of the moon lasting for a full day of 24 h.

The moon revolves eastward from the new-phase position, and for the next $7\frac{3}{8}$ solar days (one-fourth of $29\frac{1}{2}$ days) it is seen as a waxing crescent moon. The term **waxing phase** means that the illuminated portion of the moon is getting larger; **waning phase** means that the illuminated portion is getting smaller as observed from Earth. A **crescent moon** is a moon in which less than one-quarter of the moon's surface is illuminated, as we observe it from the Earth. A **gibbous moon** occurs when more than one-quarter of the moon's surface appears illuminated, when seen by an Earth-based observer. Figures 18.8 and 18.9 illustrate how the phases occur and

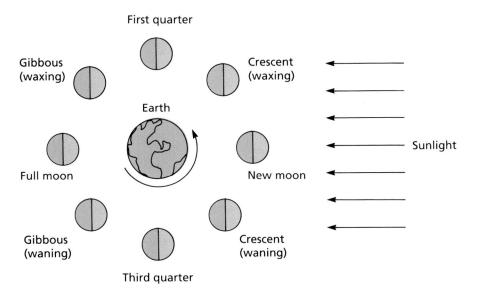

Figure 18.8 The phases of the moon as observed from a position in space above the north pole of Earth.

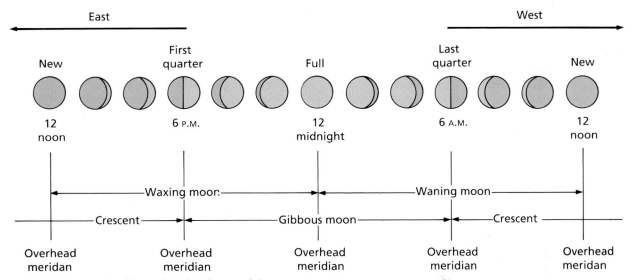

Figure 18.9 Drawing illustrating the phases of the moon as observed from the United States.
The observer is looking south; therefore, east is on the left. The Sun's position can be determined by noting the local solar time the moon is on the overhead meridian. The time period represented in the drawing is $29\frac{1}{2}$ days, or one synodic month.

appear to look to an observer on the Earth. Figure 18.4 shows eight views of the moon as the unaided eye would see it.

The moon is in the *waxing crescent phase* and appears as a crescent moon to an Earth-bound observer when the moon is less than 90° east of the Sun. The moon is in **first-quarter phase** when the moon is exactly 90° east of the Sun and appears as a quarter moon on the observer's meridian at 6 P.M. local solar time. The first-quarter phase occurs only for an instant, because the moon can only be 90° east of the Sun for an instant. See Fig. 18.8.

From the first-quarter position, the moon enters the *waxing gibbous phase* for $7\frac{3}{8}$ solar days. During this phase the moon appears larger than a quarter moon but less than a full moon. When the moon is exactly 180° east of the Sun, the moon will be in full phase and will appear as a **full moon** to the Earth-bound observer. The full moon appears on the observer's meridian at 12 midnight local solar time.

From the full-phase position the moon enters the *waning gibbous phase* and remains in the waning gibbous phase for $7\frac{3}{8}$ solar days. The appearance of the moon during this phase is the same as in the waxing gibbous phase, except that the illuminated side of the moon is toward the east and the moon is seen in the sky at a different time. When the moon is exactly 270° east of the Sun, the moon will be in the third-, or last-, quarter phase. The **last-quarter phase** appears on the observer's meridian at 6 A.M. local solar time, with the illuminated side of the moon toward the east.

From the last-quarter position the moon enters the *waning crescent phase* and remains in this phase for $7\frac{3}{8}$ solar days. During this phase the illuminated portion of the moon appears smaller than a quarter moon. Its appearance is the same as the waxing crescent moon, except that the illuminated side of the moon is toward the east, and the moon appears in the sky at a different time.

Figure 18.9 illustrates the moon's appearance and position above the southern horizon as observed from a northern latitude greater than $28\frac{1}{2}°$ north. The moon is shown in the first drawing on the left in the new phase position. The moon is shown on the observer's meridian at 12 noon local solar time and shaded black, illustrating that it cannot be seen at this time because it is on the same meridian as the Sun. The next two positions illustrate the waxing crescent phase. Note that the illuminated area of the moon is appearing larger in size for an Earth observer as the moon approaches the first-quarter phase, and that the illuminated side is toward the west where the Sun is located.

When the moon is in first-quarter phase, the moon is on the observer's meridian at 6 P.M. local solar time. The Sun will be at or near the western horizon at this time. The moon revolves eastward, entering the waxing gibbous phase, as shown by the next two positions. Note that the illuminated area of the moon is larger than a quarter moon and still increasing in size of face. The illuminated side is still toward the west.

The next position shows the moon at full phase and on the observer's meridian at 12 midnight. If the date is at the time of the spring or vernal equinox, the moon will rise on the eastern horizon at 6 P.M., when the Sun is setting in the west; and the moon will set at 6 A.M., as the Sun is seen rising on the eastern horizon.

The moon continues revolving eastward, entering the waning gibbous phase. Note that the size of the illuminated area is decreasing for an Earth observer and that the illuminated side of the moon is toward the east—just the opposite from the waxing gibbous moon. When the moon is 90° west of the Sun (same as 270° east of the Sun), it will be on the observer's meridian at 6 A.M. local solar time, as shown in the next position. The moon appears as a quarter moon, but note that the illuminated side is toward the east. The Sun will be rising at or near this time.

After the third-quarter phase the moon enters the waning crescent phase and the size of its face continues to decrease. Note that the illuminated side of the waning crescent moon is toward the east. The last position shows the moon back to the new-phase position.

Table 18.1 summarizes the times for the various phases of the moon to rise, be overhead, and set. An example of what an observer in the United States sees when looking at the first-quarter phase is shown in Fig. 18.10. Figures similar to 18.10 for the other phases can be drawn using the information in Table 18.1.

Because the moon revolves around the Earth every $29\frac{1}{2}$ solar days, it gains 360° in $29\frac{1}{2}$ days, or 12.2° per day on the Sun. Thus the moon is on the observer's meridian about 50 min later each day, because the Earth

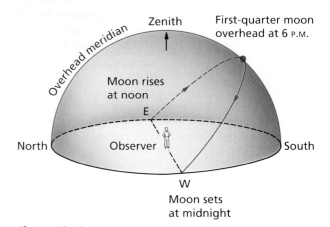

Figure 18.10
Diagram illustrating the first-quarter moon rising, on the overhead meridian, and setting during the time of the vernal or autumnal equinox for an observer in the United States. The side of the moon facing west is illuminated by sunlight.

Table 18.1 Times for the Various Phases of the Moon to Rise, Be Overhead, and Set

Phase	Approximate Rising Time	Approximate Time Overhead	Approximate Setting Time
New moon	6 A.M.	Noon	6 P.M.
First-quarter moon	Noon	6 P.M.	Midnight
Full moon	6 P.M.	Midnight	6 A.M.
Last-quarter moon	Midnight	6 A.M.	Noon

must rotate through 360° plus 12.2° before the moon appears on the overhead meridian. (See Fig. 18.11.) The average time of moonrise is thus delayed about 50 min each day. The actual time depends on the latitude of the observer, with greater variation noted by an observer in the higher latitudes. The variation depends upon the angle between the moon's path and the horizon.

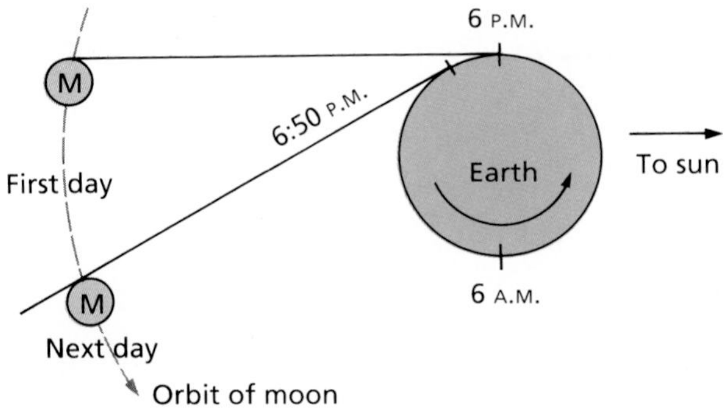

Figure 18.11
The moon rises 50 minutes later each day because as Earth rotates, the moon is revolving around Earth. For example, the full moon rises at about 6 P.M. on March 21 and at about 6:50 P.M. on March 22.

The approximate altitude of the full moon can be found by recognizing that the full moon will be on the opposite side of the Earth from the Sun (Fig. 18.12). Thus when the Sun is low in the sky in the winter, the full moon will be high in the sky. In the summer the Sun is high in the sky and the full moon is low in the sky.

18.5 Eclipses

The word **eclipse** means the darkening of the light of one celestial body by another. The Sun provides the light by which we see objects in our solar system. That is, nonluminous objects in our solar system are observed by reflected light from the Sun. Because the light from the Sun falls on objects in the solar system, objects cast shadows that extend away from the Sun. The size and shape of the shadow depend on the size and shape of the object and its distance from the Sun. The Earth and moon, being spherical bodies, cast conical shadows, as viewed from space.

If we examine the shadow cast by the Earth or moon, we discover two regions of different degrees of darkness. The darkest and smallest region is known as the **umbra.** See Figure 18.13. An observer located within this region is completely blocked from the Sun during a solar eclipse. The semidark region is called the

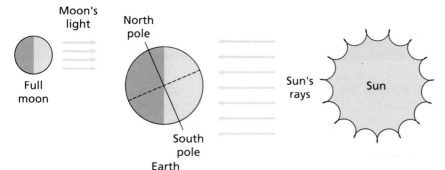

Figure 18.12 The full moon is on the opposite side of Earth from the Sun.
In this side view of Earth, full moon, and Sun in winter, the Sun's rays strike the southern hemisphere most directly. The moon's reflected light falls most directly on the northern hemisphere.

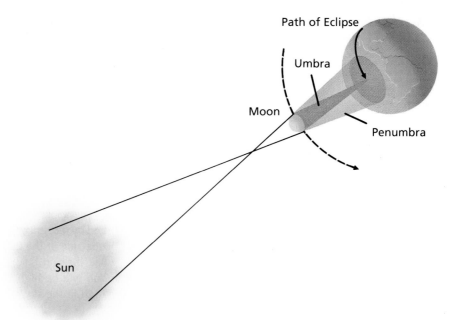

Figure 18.13 The positions of the Sun, moon, and Earth during a total solar eclipse. The umbra and penumbra are, respectively, the dark and semidark shadows cast by the moon on the surface of Earth.

Figure 18.14 An annular eclipse of the Sun.

penumbra. An observer positioned in this region can see only a portion of the Sun during a solar eclipse.

A **solar eclipse** occurs when the moon is at or near new phase and is in or near the ecliptic plane. When these two events occur together, the Sun, moon, and Earth are nearly in a straight line. The moon's shadow will then fall upon the Earth, and the Sun's rays will be hidden from those observers in the shadow zone. A **total eclipse** occurs in the umbra region and a **partial eclipse** in the penumbra region. See Figs. 18.13, 18.14, and 18.15.

The length of the moon's shadow varies as the

Figure 18.15 Solar corona photographed at total solar eclipse.

moon's distance to the Sun varies. The average length of the moon's umbra is 233,000 mi, which is slightly less than the mean distance between the Earth and moon. Because the umbra is shorter in length than the mean distance from the Earth to the moon, an eclipse of the Sun can occur in which the umbra fails to reach the Earth. An observer positioned on the Earth's surface directly in line with the moon and Sun sees the moon's disk projected against the Sun, and a bright ring, or annulus, appears outside the dark moon. This condition is called an **annular eclipse.** (See Fig. 18.14.)

Around the zone of the total (very dark) annular eclipse appears the larger semidark region of the pe-numbra. The penumbra region may be as large as 6000 mi in diameter at the surface of the Earth. The maximum diameter of the umbra at the Earth's surface is about 170 mi. This maximum value can exist only when the Sun is farthest from the Earth, which is in early July, and the moon is at perigee, or at its closest distance to the Earth.

The motion of the moon and Earth are such that the shadow of the moon moves generally eastward during the time of the eclipse with a speed of about 1000 mi/h. Thus the region of total eclipse does not remain long at any one place. The greatest possible value is about $7\frac{1}{2}$ min, and the average is about 3 or 4 min.

A **lunar eclipse** occurs when the moon is at or near full phase and is in or near the ecliptic plane. See Fig. 18.16. The Sun, Earth, and moon will be positioned in a nearly straight line, with the Earth between the Sun and moon. Thus the shadow formed by the Earth conceals the face of the moon. The average length of the Earth's shadow is about 860,000 mi, and the diameter of the shadow at the moon's position is great enough to place the moon in total eclipse for a time slightly greater than $1\frac{1}{2}$ h. A partial eclipse of the moon can last as long as 3 h 40 min.

The orbital plane of the moon is inclined to the ecliptic (the annual path of the Sun) at an angle slightly greater than 5°. Therefore, the path of the moon crosses the ecliptic at two points as it makes its monthly journey around the Earth. The points where the moon's path crosses the ecliptic are known as *nodes.* The point of crossing going northward is called the **ascending node,** and the point of crossing going southward is called the **descending node.** See Fig. 18.17. A solar or lunar eclipse can occur only at or near the nodal points, because the Earth, moon, and Sun must be in a nearly straight line. This positioning occurs only at or near the points where the moon crosses the ecliptic.

The orbital plane of the moon is precessing westward, or clockwise if viewed from above the orbital

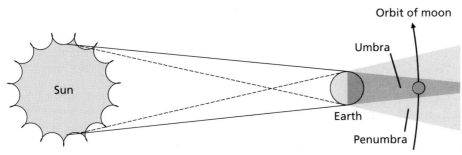

Figure 18.16 The positions of the Sun, Earth, and moon during a lunar eclipse.

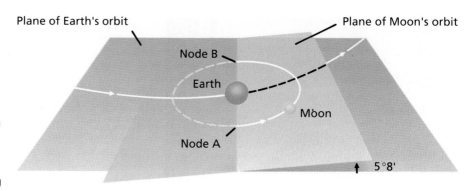

Figure 18.17 The 5-degree, 8-minute angle between the orbital planes of the moon and Earth.
Viewed from above the north pole, the moon and Earth both revolve counterclockwise around the Sun. Node A is called the ascending node, and Node B is called the descending node.

plane. The precession of the moon's orbit causes the nodal points to move westward along the ecliptic, making one complete cycle in 18.6 years.

Predicting total solar eclipses is complicated, because the line of nodes gradually moves westward and changes the nodes' direction in space. Data concerning the next five total solar eclipses is given in Table 18.2.

18.6 Tides

Anyone who has been to the seashore for a day's visit is aware of the rising and falling of the surface level of the ocean. The alternate rise and fall of the ocean's surface level is called the *tides.*

People related tides to the passage of the moon in the first century A.D., but all efforts to explain the phenomenon failed until the seventeenth century, when Newton applied his law of universal gravitation to the problem. He related, and explained, the alternate rise and fall of the ocean's surface level with the motions of the Earth, moon, and Sun. A few of the many factors contributing to the height that the ocean rises and falls

at a particular location are:

1. the rotation of the Earth on its axis
2. the position of the Earth, moon, and Sun in respect to one another
3. the varying distance between the Earth and moon
4. the inclination of the moon's orbit
5. the varying distance between the Earth and Sun
6. the variation in the shape of coastlines and relief of ocean basins.

There are two high and two low tides daily because of the moon's gravitational attraction and the motion of the moon and Earth.

An understanding of the two daily tides can be clarified by visualizing what shape the Earth and its surface of water would take if there were no external gravitational forces and the Earth had no motion (Fig. 18.18a). If the Earth did not rotate, there would be no centripetal force. On a nonrotating Earth, with no external gravitational forces, there would be no forces

Table 18.2 Total Solar Eclipses from 1990 Through 1995

Date	Time of Total Duration	Location Where Visible
July 22, 1990	2.6 min	Arctic regions, Finland
July 11, 1991	7.1 min	Brazil, Central America, Hawaii
June 30, 1992	5.4 min	South Atlantic
November 3, 1994	4.6 min	South America
October 24, 1995	2.4 min	South Asia

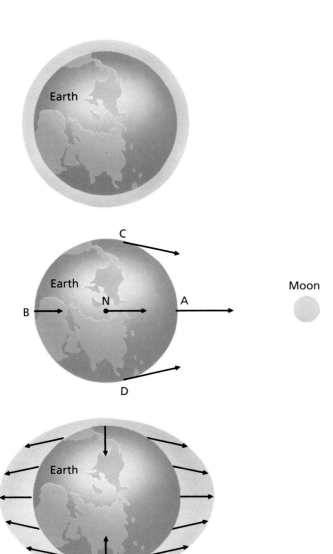

Figure 18.18
(a) *Top:* The shape of stationary Earth with all water surfaces unaffected by any external gravitational forces. (b) *Middle:* Gravitational forces of the moon (arrows) acting on Earth, which is considered as having a mass concentrated at a point. (c) *Bottom:* The resultant tidal forces (arrows) acting on Earth to produce the oblate (elongated at the center) contour shown in the diagram. This view of Earth is from above the North Pole. The tidal bulge is greatly exaggerated.

exerted on the Earth or the surface water and hence no tides. How, then, do the gravitational force of the moon and the motion of the Earth produce the tides?

The answer to this question can be found, and the reason for two daily tides at a given location on the Earth's surface can be explained, if the moon's mass is considered to be concentrated at a point (Fig. 18.18b). The Sun is also a factor in causing tides, but it is left out of this explanation to simplify the results. The magnitude and direction of the moon's gravitational forces acting on the Earth at points *A, B, C, D,* and *O* are shown. The magnitude is indicated by the length of the arrow drawn to represent the gravitational force. Note that the forces at *A, O,* and *B* are all in the same direction toward the moon, but the magnitudes are not the same. The force at *A* is the greatest because it is closest to the moon. Remember that Newton's law of gravitational attraction says the force between two masses is inversely proportional to the square of the distance between the two masses. Thus the force at *O* is less than the force at *A,* and the force at *B* is less than the force at *O.* Also, forces at *C* and *D* have approximately the same magnitude but are slightly less than the force at *O.* The direction of the forces at *C* and *D* are toward the moon, as shown. If all the forces shown are now added in reference to the force at *O,* at the center of the Earth, we obtain the results as shown in Fig. 18.18c. Thus the forces acting over a period of hours will produce movement of the water that results in tidal bulges in the oceans.

When the Sun, Earth, and moon are positioned in a nearly straight line, the gravitational force of the moon and Sun combine to produce higher high tides and lower low tides than usual. That is, the variations between high and low tides are greatest at this time. These tides of greatest variation are called **spring tides,** and they occur at the new and full phases of the moon. When the moon is at first- or third-quarter phase, the Sun and moon are 90° with respect to the Earth. At these times the tidal forces of the moon and Sun tend to cancel one another, and there is a minimum difference in the height of the surface of the ocean. In this case the tides are known as **neap tides.**

Note that two spring tides and two neap tides take place each lunar month, because the moon passes through each of its phases once each month. There is a spring tide at new moon, a neap tide at first quarter, another spring tide at full moon, and a second neap tide at last quarter. See Fig. 18.19.

The height of the tide also varies with latitude. See Fig. 18.20. The tide is highest at the moon's overhead position and on the other side of the Earth opposite the position of the moon. The time of high tide does not correspond to the time of the meridian crossing of the moon. The bulge is always a little ahead (eastward) of

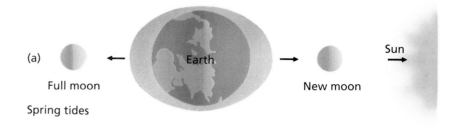

(a) Full moon Spring tides

First quarter moon

(b) Neap tides

Last quarter moon

Figure 18.19 The relative positions of earth, moon, and Sun at the times of spring and neap tides.
(a) During spring tides the Sun and moon are aligned, which causes more extreme tidal effects. (b) During neap tides, the Sun and moon act against each other, so the tides are more moderate.

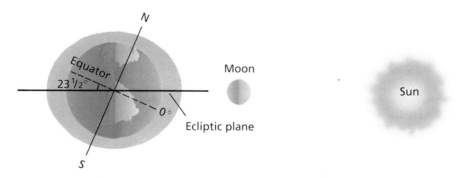

Figure 18.20 Spring tide occurring at time of summer solstice (Sun at greatest northern latitude) when the moon is in new phase.
Maximum height of bulge is at $23\frac{1}{2}$ degrees north and $23\frac{1}{2}$ degrees south, because the Sun is overhead at $23\frac{1}{2}$ degrees north. If a spring tide were to occur three months later, when the Sun is on the equator, the maximum height of the bulge would be in the equatorial region. The tidal bulge is exaggerated in this diagram.

the moon because of the Earth's rotation. Because the Earth rotates faster than the moon revolves, the Earth carries the tidal bulge forward in the direction it is rotating, which is eastward.

The action of the tides produces a retarding motion on the Earth's rotation, slowing it and lengthening the solar day about $\frac{1}{1000}$ second per century. Because the conservation of angular momentum applies, the decrease of the Earth's angular momentum must appear as an increase in the moon's angular momentum. A measurement of the moon's orbit shows that the semimajor axis is increasing about $\frac{1}{2}$ in per year. Thus 1 billion years ago the solar day was 2.8 h shorter, and the moon was 8000 mi closer to the Earth.

Learning Objectives

After reading and studying this chapter, you should be able to do the following without referring to the text:

1. State the approximate distance between the Earth and moon, the approximate diameter of the moon, the time for the moon to revolve around the Earth, and the time for the moon's rotation.

2. Explain why the moon has so many more craters than the Earth.

3. State what the plains are and what caused them.

4. Summarize the important periods in the moon's history.

5. Draw a diagram of the Earth, moon, and Sun when the moon is in the first-quarter phase, and predict when it will rise, be overhead, and set.

6. Draw a diagram to explain why the moon rises 50 min later each day.

7. Explain what causes a solar eclipse and why we do not usually have one when the moon is in the new phase.

8. Explain what causes a lunar eclipse and why they last longer than solar eclipses.

9. Explain why we have tides; in particular, explain why we have two high tides and two low tides each day.

10. Explain what causes spring tides and neap tides.

11. Define and explain the other important words and terms listed in the following section.

Important Words and Terms

crater	perigee	waning phase	eclipse	annular eclipse
plains	apogee	crescent moon	umbra	lunar eclipse
rays	sidereal month	gibbous moon	penumbra	ascending node
rills	synodic month	first-quarter phase	solar eclipse	descending node
mountain ranges	new moon	full moon	total eclipse	spring tides
fault	waxing phase	last-quarter phase	partial eclipse	neap tides

Questions

General Features

1. What is the numerical value of the moon's synodic period of revolution?   *Answer:*   29.530588 days

2. What is the diameter of the moon in (a) miles and (b) kilometers?

3. How does the surface gravity of the moon compare with the surface gravity of Earth?

4. What is the average distance between the moon and Earth in (a) miles and (b) kilometers?

5. Distinguish between the moon-surface features rays and rills.

6. What causes each of the following on the moon: (a) craters, (b) plains, (c) rays, and (d) rills?

7. Name the three major components of the moon's internal structure.

8. An examination of the planets and their moons reveals something exceptional about the Earth-moon system. What is this unique feature?

History of the Moon

9. What is the origin of the moon?

10. How long ago did the moon form?

11. Why can scientists learn more about the early history of our planet by studying rocks from the moon than by studying rocks from Earth?

Lunar Motion

12. What is the direction of rotation and revolution of the moon?

13. What is the period of rotation in respect to the period of revolution?

14. (a) What is the numerical value of the moon's sidereal period?
    (b) How does the sidereal period compare with the synodic period?
    (c) Explain the difference.

15. What is the numerical value of the angle between the moon's orbital plane and Earth's orbital plane?

16. Why does the moon appear to rise in the east and set in the west every day?

17. How often would an observer on the moon see (a) sunrise (b) earthrise?

Phases

18. How often do we have a full moon?

19. Determine when the following occur: (a) the new moon sets; (b) the full moon rises; (c) the last-quarter moon sets.

20. Why does the moon rise 50 min later each day?

21. Which phase of the moon (a) is overhead at 6 P.M., (b) sets at 6 A.M., and (c) rises at midnight?

22. What is the difference between a waxing and a waning moon?

23. Why is the full moon higher in the sky in winter than in summer?

Eclipses

24. State the positions of the Sun, moon, and Earth during a (a) solar eclipse and (b) lunar eclipse.

25. Distinguish between umbra and penumbra.

26. What is the period of precession of the moon's orbital plane? What is the direction of the precession?

27. Why do eclipses (solar and lunar) only take place at or near the nodal points of the moon's and Earth's orbital planes?

28. What is an annular eclipse?

29. Why doesn't an eclipse occur every time there is a new or full moon?

30. Why do lunar eclipses last much longer than solar eclipses?

Tides

31. What is the origin of tidal forces?

32. What are some factors that contribute to the height of ocean tides?

33. During which phases of the moon do spring and neap tides occur?

34. Why are there two high and two low tides each day?

Exercises

General Features

1. If a person weighs 600 N on the Earth, what is the person's weight on the moon?   *Answer:*   100 N

2. If a person weighs 120 lb on the Earth, what is the person's weight on the moon?

Lunar Motion

3. How many days are in 12 lunar months (synodic months)?

4. Are there more sidereal or synodic months in one year? State the number difference, and explain your answer.

Phases

5. At what longitude is a person who sees a full moon overhead when it is noon in Washington, D.C. (30°N, 77°W)?   *Answer:*   103°E

6. Referring to Exercise 5, at what longitude is a person who sees the full moon setting at the same time?

7. At what longitude is a person who has a new moon overhead when it is noon in Washington, D.C. (39°N, 77°W)?   *Answer:*   77°W

8. Referring to Exercise 7, at what longitude is a person who sees the new moon rising at the same time?

9. Consider a person in the United States who sees the first-quarter phase.
   (a) Which side of the moon is bright, east or west?
   (b) What phase does an observer in Australia see at the same time, and which side is bright?
       Answer: (b) first-quarter phase; west (left) side bright

10. Consider a person in the United States who sees the last-quarter phase.
    (a) Which side of the moon is bright?
    (b) What phase does an observer in Australia see at the same time, and which side is bright?

11. A book states that the moon rises about 50 min later each day. What is a more accurate figure?   *Answer:*   48.8 min

12. Draw a diagram similar to Fig. 18.10 showing the times the last-quarter moon rises and sets.

13. Determine the month and day when the full moon is at maximum altitude for an observer at Washington, D.C. (39°N). What is the altitude of the full moon at this time?

14. Determine the month and day when the full moon is at minimum altitude for an observer at Washington, D.C. (39°N). What is the altitude of the full moon at this time?

    Answer: June 21, 28.5° ± 5°

15. What is the phase of the moon that (a) rises at 4 P.M., (b) sets at 9 A.M., and (c) is on the meridian at 8 A.M.?

16. An observer at 28.5° South sees the moon on his zenith. If the moon appears on the southern horizon at this time for an observer in the Northern Hemisphere, what is the latitude of this observer?

17. An observer in Washington, D.C., sees a waxing crescent moon slightly north of the vernal equinox. Name the observer's season, and explain your answer.

Eclipses

18. Draw a diagram illustrating a total solar eclipse. Include the orbital paths of the Earth and moon, and indicate the approximate time of day the eclipse is taking place.

19. Draw a diagram illustrating a total lunar eclipse. Include the orbital paths of the Earth and moon, and indicate the approximate time of day the eclipse is taking place.

Tides

20. A high tide is occurring at Washington, D.C. (39°N, 77°W).
    (a) What other longitude is also experiencing a high tide?
    (b) What two longitudes are experiencing low tide?

    Answer: (a) 103°E (b) 13°E and 167°W

21. A low tide is occurring at Los Angeles (34°N, 118°W).
    (a) What other longitude is also experiencing a low tide?
    (b) What two longitudes are experiencing high tide?

The Universe

Do conservation laws dictate
A universe transformed
From spaceless, timeless, and nothing
To space, time, and something?

T HE STUDY OF the stars is the oldest science. Thousands of years ago under the clear desert skies of the Near East, people watched the stars in awesome wonder. The earliest scientists plotted the positions and brightnesses of the stars as Earth went through its calendar of seasons. The Sun and moon were worshipped as gods, and the days of the week were named after the Sun, moon, and visible planets, as discussed in Chapter 17. Yet it is only recently that we have understood the most basic nature of stars.

Sixty years ago not even Einstein himself knew what made the stars shine. Today, we know that all stars go through a cycle of stages. Stars are born, they radiate energy, and then they expand, contract, possibly explode, and eventually die out. Our knowledge of how all this happens has been made possible through our study of the atomic nucleus and by applying the laws of science in many diverse fields.

Our knowledge of the universe is also growing. We are learning more and more of its secrets, and we are beginning to understand how our present civilization fits into an overall scheme in the harmony of the universe. We now have an imperfect theory of evolution—a theory concerned not so much with the origin of human beings but with the origin of the elements of our solar system. We can now begin to understand how carbon, nitrogen, and oxygen atoms were produced in the stars. It is these atoms that were necessary to produce life itself. As these theories evolve, new mysteries appear; but the continued search for truth is invigorating.

The stars and galaxies of our universe give off many different kinds of electromagnetic radiation. This radiation was discussed in Chapter 6 and includes radio waves, microwaves, infrared, visible, ultraviolet, X-rays, and gamma rays. Up until about fifty years ago, we looked only at the visible light given off by the stars. With the advent of radio telescopes, quasars and pulsars were discovered in 1960 and 1968, respectively. Most of the other regions of the electromagnetic spectrum are absorbed by our atmosphere. Satellites and balloons going above our atmosphere have enabled us to study other forms of radiation emitted by stars, and new developments are occurring frequently. There is now the strong possibility that we have detected black holes using X-ray astronomy techniques.

Recent experiments in the microwave region of the spectrum confirm the finding that we live in an expanding universe that had its beginning 12 to 20 billion years ago. But will our universe continue to expand? What is the ultimate fate of the universe? We hope the answers to these questions will be forthcoming, but for now there is much to understand. This chapter attempts to summarize many of the basic ideas about the stars and galaxies that make up our universe.

19.1 The Sun

The Sun is a star, spherical in shape, with a diameter of 865,000 mi (over a hundred times greater than Earth's diameter). Viewed from Earth, at its mean distance, the Sun's angular diameter is approximately $\frac{1}{2}$ degree. This ordinary star of the Milky Way galaxy is the most important object in the solar system to us because it supplies heat, light, and other radiation for the processes of life on Earth. The Sun rotates about an axis every 25 Earth days and moves through space with its family of planets at a speed of approximately 250 km (150 mi)/s in the direction of the constellation Lyra. The Sun's equator is inclined about 7° from the orbital plane of Earth. The rotational period of the Sun given above is the period at its equator. The period of rotation is longer at higher latitudes.

A cross-sectional view of the Sun is shown in Fig. 19.1. The Sun's temperature is believed to be about 15 million K at its center, and the temperature decreases radially outward to the visible surface of the Sun, which is called the **photosphere.** The temperature of the photosphere has been measured at about 6000 K.

◄ **A sample of the universe.**
The Coma cluster, which is 70 to 140 parsecs away, contains thousands of galaxies.

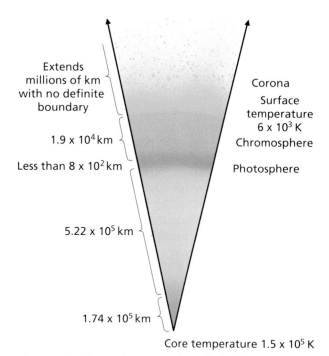

Figure 19.1 A radial cross-section of the Sun.

The interior of the Sun is so hot that individual atoms do not exist because high-speed collisions continually knock the electrons loose from the atomic nuclei. The interior is composed of high-speed nuclei and electrons moving about more or less independently, similar to a gas. A gas, you will recall, is composed of rapidly moving atoms or molecules. In the Sun we have very rapidly moving positively charged nuclei and negatively charged electrons. These high-speed charged nuclei and electrons form a fourth phase of matter called a **plasma.** The Sun is a plasma with an average density of 1.4 g/cm³. Note that this density is 1.4 times as great as water!

The Sun's photosphere is composed of about 94% hydrogen, 5.9% helium, and 0.1% of heavier elements, the most abundant being carbon, oxygen, nitrogen, and neon. We believe that the interior of the Sun has a similar composition, although we have no good experimental evidence for this belief. Note that when we discuss elements in stars, we are really speaking about the nuclei of the elements, because the electrons are stripped off the nuclei and are speeding around independently of the nuclei.

The photosphere, viewed through a telescope with appropriate filters, has a granular appearance. The granules are hot spots (about 100 K higher than the sur-

rounding surface) that are a few hundred miles in diameter and last only a few minutes. Extending more than 12,000 mi above the photosphere lies the **chromosphere** (*color* + *sphere*), which is composed mainly of hydrogen. The chromosphere can be seen as a thin red crescent for only the few seconds that a total eclipse shuts out the light from the Sun. At the time of a total solar eclipse, the chromosphere and photosphere are hidden by the moon, and the **corona** (outer solar atmosphere) can be seen as a white halo. See Fig. 18.15.

A very distinct feature of the Sun's surface is the appearance of sunspots. **Sunspots** are patches (some are thousands of miles in diameter) of cooler material on the surface of the Sun. Each has a central darker part, called the *umbra*, and a lighter border, called the *penumbra*. Figure 19.2 is a photograph of the whole solar

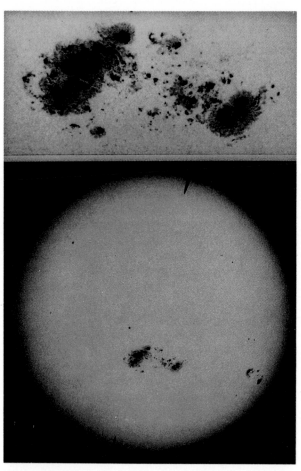

Figure 19.2 The very large sunspot group of March–April 1947.
The photo at the top is an enlargement of the large group just below the center of the bottom photo.

disk and an enlargement of a very large sunspot. These large sunspots last for several weeks before disappearing from view.

The number of sunspots appearing on the Sun varies over a 22-year period. A period begins with the appearance of a few spots or groups near 30° latitude in both hemispheres of the Sun. The number of spots increases, with a maximum generally between 100 and 200 occurring about four years later near an average latitude of 15°. As time passes, the number of spots decreases until, in about seven more years, only a few spots are observed near 8° latitude.

About this same time a few spots begin to appear at 30°, and the number of sunspots begins to increase again, indicating an 11-year cycle. But there is a notable difference. The sunspots have an associated magnetic field that is different in appearance from the previous ones. Studies indicate that if a sunspot has a north magnetic pole during the initial increase and decrease, the next 11-year cycle will show a south magnetic pole associated with the sunspot.

The 22-year or 11-year sunspot cycle has been observed since about 1715. Galileo saw sunspots through a telescope in 1610, and they were possibly observed even before that without telescopes. There were reports of their observation after Galileo, but during the period 1645 to 1715 hardly any sunspots were reported. During this 70-year period very few northern lights were seen in Northern Europe, and a "Little Ice Age" occurred during this time in Europe. The evidence seems to indicate that the Sun's activity is not always as regular as it often appears to be.

Another distinct feature of the Sun's surface is the appearance of **prominences** that seem to be connected with violent storms in the chromosphere. They are very evident to the astronomer during solar eclipses, at which time they appear as great eruptions at the edge of the Sun. They are red, have an associated magnetic field, and take many different shapes and forms. They may appear as streamers, loops, spiral or twisted columns, fountains, curtains, or haystacks. They extend outward for thousands of miles from the surface, occasionally reaching a height of 1 million miles. An extraordinarily large prominence is shown in Fig. 19.3.

The chief property of the Sun, of course, is the fact that it radiates, or gives off energy. However, it was not until 1938 that scientists came to understand the source of the radiation. We now know that the Sun radiates energy because of nuclear fusion reactions inside its core.

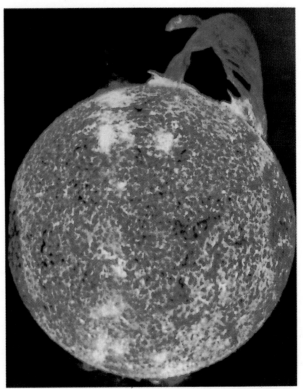

Figure 19.3 The Sun during a major solar eruption. This ultraviolet photograph showing several flares and a large prominence was taken from the Skylab orbiting space station.

The Sun's core is made up mostly of hydrogen nuclei or protons, or in nuclear notation, ^1_1H. These protons are moving at very high speeds and they occasionally fuse together, as shown in Fig. 19.4. The products of this nuclear reaction are a deuteron (proton and neutron together, or ^2_1H), a positive electron (or positron, designated $_{+1}e$), and a neutrino (designated by the symbol v). A **neutrino** is an elementary particle that has no charge, has no mass (or very little mass), travels at or near the speed of light, and hardly ever interacts with other particles such as electrons or protons. This first reaction is fairly rare and for this reason the Sun's hydrogen burns relatively slowly. In fact, we believe that the Sun has been radiating energy for about 5 billion years.

Once the deuteron is formed, it quickly reacts with a proton to form a helium-3 nucleus (^3_2He) and gamma rays, designated by the symbol γ. Next, two helium-3 nuclei fuse to form the more common helium-4 nucleus and two protons.

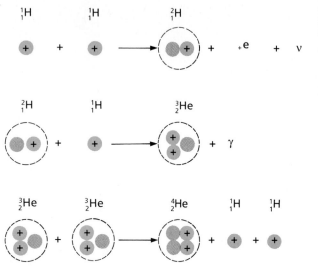

Figure 19.4

(a) *Above:* The reactions that make up the proton-proton chain. (b) *Right:* The Brookhoven solar-neutrino-experiment apparatus. (See Highlight, p. 380.)

In each of these three fusion reactions, mass is converted into energy and energy is liberated. These three reactions are called the **proton-proton chain** and can be written as

$${}_1^1H + {}_1^1H \longrightarrow {}_1^2H + {}_{+1}^0e + \nu + \text{energy} \qquad \text{(slow)}$$

$${}_1^2H + {}_1^1H \longrightarrow {}_2^3He + \gamma + \text{energy} \qquad \text{(fast)}$$

$${}_2^3He + {}_2^3He \longrightarrow {}_2^4He + {}_1^1H + {}_1^1H + \text{energy} \qquad \text{(fast)}$$

If we multiply the first two reactions by 2 and add both sides, we get the net reaction, which is

$${}_1^1H + {}_1^1H + {}_1^1H + {}_1^1H \longrightarrow$$
$${}_2^4He + 2({}_{+1}^0e) + 2\gamma + 2\nu + \text{energy}$$

In the net reaction four protons or hydrogen nuclei react to form a helium nucleus, two positive electrons (positrons), two high-energy gamma rays, two neutrinos, and a great deal of energy. The energy factor simply means that the particles on the right possess more kinetic and radiant energy than the particles on the left. In our reaction we have converted mass into energy via Einstein's formula $E = mc^2$.

Every second in the Sun's interior, 630 million tons of hydrogen are being converted into 625.4 million tons of helium with the release of the equivalent of 4.6 million tons of mass energy. Yet at this rate we expect the Sun to radiate energy for another 5 billion years or so.

19.2 The Celestial Sphere

A view of the stars on a clear night makes a deep impression on an observer. As we look into the night sky, the stars appear as bright points of light on a huge dome overhead. As the time of night passes, the dome seems to turn westward as part of a great sphere, with the observer at the center. The apparent motion of the stars is due to the eastward rotation of the Earth. The unaided eye is unable to detect any relative motion among the stars on the apparent sphere or to perceive their relative distances from the Earth. The stars all appear to be mounted on a very large sphere having a very large radius.

The huge moving (apparent motion) sphere has been named the **celestial sphere,** and the way it appears to the observer depends upon the observer's position on the Earth. An observer positioned at 90°N (the north pole) would see Polaris, the north star, overhead. From this latitude all stars on the celestial sphere appear to move in concentric circles about the north star, never going below the horizon (Fig. 19.5); they never set. If the observer is located at 40°N latitude, he or she would observe the north star 40° above the northern horizon, and all stars within 40° of the north star would appear to move in concentric circles, never going below the horizon (Fig. 19.6). Detailed observations reveal the celestial sphere to rotate (apparently) about an axis that

Figure 19.5
This long photo exposure shows star trails at the south celestial pole.

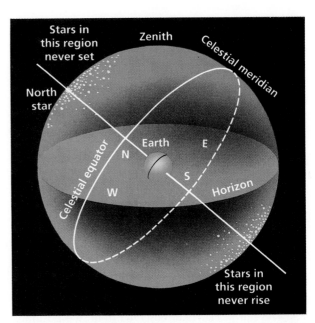

Figure 19.6 The celestial sphere, as seen by an observer on Earth at a latitude of 40°N.

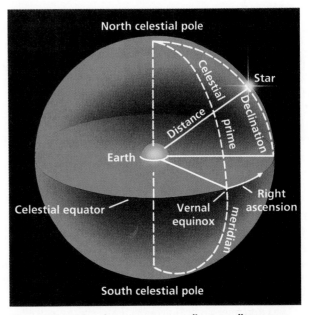

Figure 19.7 The three space coordinates: distance, declination, and right ascension.

is an extension of the Earth's polar axis, with the celestial equator lying in the same plane as the Earth's equator. See Fig. 19.6.

The position of a star or other object beyond our solar system is determined with the assignment of three space coordinates. The first and second are *declination* and *right ascension,* which are angular coordinates representing the direction of the star in respect to the Sun. The third coordinate is *distance,* which determines the star's linear distance from the Sun (Fig. 19.7).

Declination is the angular measure north or south of the celestial equator measured in degrees. It has a minimum value of zero at the celestial equator and increases to a maximum of 90° north and to a maximum of 90° south. The direction in respect to the equator is indicated by a plus (+) or a minus (−) sign. All angles measured north of the equator have (+) values, and all angles measured south of the equator have (−) values. For example, the star in Figure 19.7 has a declination angle of +37°.

Right ascension is the angular measure in degrees or hours, with the hours divided into minutes and sec-

onds. Right ascension begins with 0 h at the celestial prime meridian and continues eastward to a maximum value of 24 h or 360°, either of which coincides with the starting point. The **celestial prime meridian** is an imaginary half circle running from the north celestial pole to the south celestial pole and crossing perpendicular to the celestial equator at the intersection of the ecliptic at the point where the ecliptic passes northward.

The distance coordinate is usually measured in astronomical units, in parsecs, or in light-years. We have defined an **astronomical unit** as the mean distance of the Earth from the Sun, which is 92,955,700 mi as measured by radar. A **light-year** is the distance traveled by light in one year. One light-year equals approximately 6×10^{12} mi or 9.5×10^{12} km, calculated by multiplying the speed of light by the number of seconds in one year. One **parsec** is defined as the distance to a star when the star exhibits a parallax of one second of arc (Fig. 19.8). One parsec equals 3.26 light-years, or 206,265 astronomical units:

$$1 \text{ parsec} = 3.26 \text{ light-years}$$
$$= 206{,}265 \text{ astronomical units}$$

The star in Fig. 19.8, observed from two positions appears to move against the background of more distant stars. This apparent motion is, as you know, called parallax. The angle p measures the parallax in seconds of arc. The definition of a parsec provides an easy method for determining the distance to a celestial object, because merely taking the reciprocal of the angle p, measured in seconds, gives the distance in parsecs.

EXAMPLE

What is the distance to the star Proxima Centauri if the annual parallax is 0.762 second?

Solution

$$d = \frac{1}{0.762} = 1.31 \text{ parsecs}$$

There are prominent groups of stars in the celestial sky that appear to an Earth observer as distinct patterns. These groups, called **constellations,** have names that can be traced back to the early Babylonian and Greek civilizations. Although the constellations have no physical significance, today's astronomers find them useful

HIGHLIGHT

Solar Neutrinos

We now have a good understanding of the basic concept of how a star radiates energy. A star radiates energy because of nuclear fusion reactions inside its core. But do we have any direct evidence of this reaction? In an effort to get direct evidence of what goes on inside the Sun's core, scientists are conducting the solar neutrino experiment. The *solar neutrino experiment* is an experiment to detect neutrinos from the Sun's core.

The experiment is being conducted by Raymond Davis, Jr., of the Brookhaven National Laboratory. The data collected thus far by Davis do not agree with theoretical calculations. He is finding fewer neutrinos than calculated by a factor of 3 to 1.

Presently, Davis uses 100,000 gal of fluid C_2Cl_4 in a tank located deep within a South Dakota gold mine to detect the neutrinos. When a neutrino interacts with a chlorine atom (only high-energy neutrinos are able to do so), the atom is transformed into an atom of argon that can be removed from the tank by bubbling helium gas through the fluid. (See Fig. 19.4b.)

Davis plans to increase the sensitivity of his equipment to detect neutrinos by using the element gallium in place of chlorine. With gallium, low-energy as well as high-energy neutrinos will be detected. Perhaps the new experimental results will agree more closely with calculated values.

The major difference between photons and neutrinos is that photons react much more readily with matter than neutrinos. In fact, a photon produced at the center of the Sun will interact countless times and take possibly millions of years before it finally reaches the Sun's surface. A neutrino, on the other hand, will zip right through the Sun and Earth and most neutrino detection devices without interacting at all. Of course, occasionally a neutrino will interact with a detector, and we can say that a neutrino has been detected. In the solar neutrino experiment only a few neutrinos are detected in a month's time.

So far, the results of the solar neutrino experiment have been perplexing. They indicate that fewer neutrinos are being detected than should be according to current theory. Either our understanding of neutrinos needs to be improved or there are some unsolved problems with regard to our understanding of the structure and dynamics of the interior of a star.

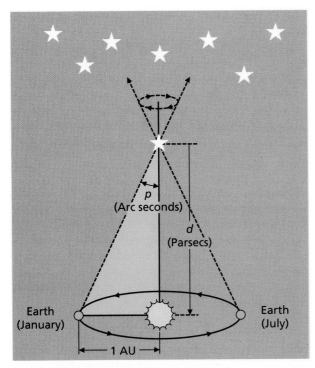

Figure 19.8 Annual parallax of a star.
The shaded angle represents the annual parallax of the star, measured in arc seconds. By definition, when angle *p* is equal to 1 arc second the distance *d* to the star is equal to 1 parsec. The basic relationship can be written *d = 1/p.*

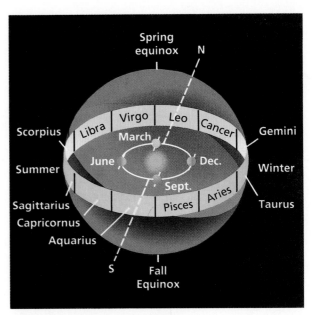

Figure 19.9 Signs of the zodiac.
The drawing illustrates the boundaries of the zodiacal constellations. Each of the 12 sections of the zodiac is 30° wide and 16° high, or 8° above and below the ecliptic plane.

in referring to certain areas of the sky, and in 1927 they set specified boundaries for the 88 constellations so as to encompass the complete celestial sphere.

We are all aware of the apparent daily motion of the Sun across the sky, and when the moon is visible, we are aware of its apparent motion. The constellations also appear to move across the sky from east to west, if one observes the stars for an hour or two. Their daily motion is due to the eastward rotation of Earth on its axis. The constellations also have an annual motion resulting from Earth's motion around the Sun. We observe the constellations Pisces, Aquarius, and Capricornus in the autumn night sky. In the winter months Orion, the Hunter, is seen, along with Gemini, Taurus, and Aries. Sagittarius, the Archer, is a summer constellation. For other summer constellations and the spring constellations, see Fig 19.9.

The **zodiac** is a section (actually a volume) of the sky extending around the ecliptic 8° above and 8° below the ecliptic plane (Fig. 19.9). The zodiac is divided into 12 equal sections, each 30° wide and 16° high. Each

section has its apex at the Sun and extends outward to infinity. The boundaries of the zodiac were specified such that the Sun, moon, visible planets, and most of the asteroids traveled within its limits. Occasionally, however, the planets Pluto and Venus are outside the boundaries of the zodiac.

19.3 Stars

Hipparchus of Nicaea, a Greek astronomer and mathematician, was antiquity's greatest known observer of the stars. He measured the celestial latitude and longitude of more than 800 stars and compiled the first star catalog, which was completed in 129 B.C. He assigned the stars, with respect to their brightness, to six **magnitudes.** The brightest ones were listed as stars of the first magnitude, those not quite as bright as second magnitude, the next less bright as third magnitude, and so on, down to the sixth magnitude.

A modified version of Hipparchus' scale is used today. When a comparison was made between a first-magnitude star and a sixth-magnitude star, the brighter first-magnitude star gave off about 100 times as much radiant energy. From this observation a definition of the magnitude scale was made, in which each magnitude difference is equal to the fifth root of 100. This definition can be written as

$$\text{magnitude difference} = \sqrt[5]{100} = 2.512$$

For example, a first-magnitude star is 2.512 times as bright as a second-magnitude star, and a second magnitude star is 2.512 times as bright as a third, and so on. On this scale the brightest object in the sky, our Sun, has a magnitude of -26.7; the full moon, a magnitude of -12.7; the planet Venus, a magnitude of -4.2; and the brightest star, Sirius, a magnitude of -1.43.

The energy output of a star is measured by its absolute magnitude. The **absolute magnitude** of a star is defined as the apparent magnitude a star would have if it were placed 10 parsecs (32.6 light-years) from Earth. The absolute magnitude of the Sun is about 4×10^{14} j/s, which is measured by determining the amount of energy falling on Earth's surface and knowing the distance from Earth to the Sun. If the annual parallax (from which distance is calculated) and the apparent brightness of a star can be measured, the absolute magnitude can be calculated.

The surface temperature of a star can be inferred from its color. Cool stars are red, hotter stars are yellow, then white, and the hottest stars are blue-white.

When the absolute magnitudes or brightnesses of stars are plotted against their surface temperatures or colors, we get an **H-R diagram,** named after Ejnar Hertzsprung, a Danish astronomer, and Henry Russell, an American astronomer. An H-R diagram for stars is shown in Fig. 19.10. Note that the temperature axis is reversed.

That is, the temperature increases to the left instead of to the right.

Most stars on an H-R diagram get brighter as they get hotter. These stars form the **main sequence,** a narrow band going from upper left to lower right. Stars above the main sequence that are cool and yet very bright must be unusually large to be so bright. So they are called **red giants.** Stars below the main sequence that are hot yet very dim must be very small, so they are called **white dwarfs.**

Stars are plasmas having a composition, by number of atoms, of about 94% hydrogen, 5.9% helium, and 0.1% other elements. Their surface temperatures range from about 2000 to 50,000 K. They vary in mass and size. Most have a mass of between 0.1 and 5 solar masses. Others range from 0.04 to 75 solar masses. Normal stars range in size from white dwarfs, which are about 8000 mi in diameter, to supergiants, which are over 400 million miles in diameter.

Most stars appear as multiple systems. A **binary star** system consists of two stars orbiting around each other. Systems with three or more stars also occur, but they are not nearly as common as binary stars.

There are many stars in the sky that are observed to vary in brightness over a period of time. The first such star noted was Delta Cephei in the constellation Cepheus. Delta Cephei varies in brightness with a period of approximately 5.4 days. From its least bright magnitude of 5.2, it gradually becomes brighter for about two days and attains a magnitude of 4.1, after which it decreases in brightness until it reaches its minimum in another 5.4 days. The cycle begins again and goes through the same sequence. The period of other cepheid variables varies from 1 to 50 days. Presently there are over five hundred known stars that vary in magnitude with a fixed period at between 1 and 50 days, and they are known as **cepheid variables** after Delta Cephei.

The importance of the cepheid variables is the fact that a definite relationship exists between the period and the average brightness. Generally, the longer the period, the brighter the cepheid variable is. By measuring the period, astronomers can calculate the absolute brightness. Then the distance can be computed from the absolute brightness. Thus astronomers can calculate the distance to any galactic system in which cepheid variables can be detected.

The period-luminosity relationship for cepheid-variable stars was discovered by an American astronomer, Henrietta Swan Leavitt (1868–1921), in 1912. This relationship provided Edwin P. Hubble (1889–1953), an American astronomer at the Mount Wilson Observatory

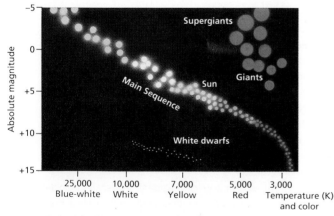

Figure 19.10 Hertzsprung-Russell diagram.
This diagram shows the absolute magnitude, spectral class, temperature, and color of the various classes of stars.

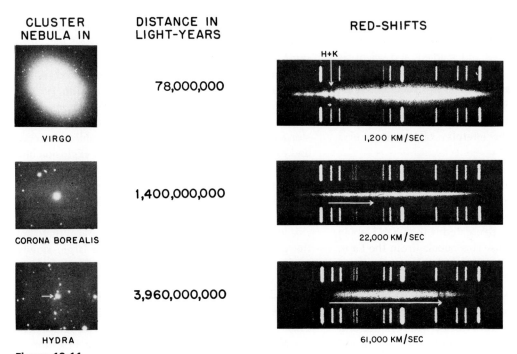

CLUSTER NEBULA IN	DISTANCE IN LIGHT-YEARS	RED-SHIFTS
VIRGO	78,000,000	1,200 KM/SEC
CORONA BOREALIS	1,400,000,000	22,000 KM/SEC
HYDRA	3,960,000,000	61,000 KM/SEC

Figure 19.11
On the left are photographs of three individual elliptical galaxies. From top to bottom the photographs show the galaxies at increasing distance from the observer. On the right, the spectrum (the broad white band) of each galaxy is shown between an upper and lower comparison spectrum. The H and K lines of ionized calcium are the two dark vertical lines in the galaxy's spectrum. The arrows indicate the shift in the calcium H and K lines. The red shifts are expressed as velocities.

in California, with the knowledge to show that observed white patches of light (called nebulae) were actually galaxies beyond the Milky Way.

The distance to more remote galaxies is calculated by means of the red shift in the galaxy's spectrum. When a photograph is taken of an excited element in the gaseous phase by a spectrograph located in an Earth-based laboratory, a normal spectrum is obtained. When a similar photograph is taken of a galaxy containing this same element, the lines of the spectrum may show a displacement of the normal lines toward the red end of the spectrum (longer wavelengths) or toward the blue end of the spectrum (shorter wavelengths). The displacement is due to the Doppler effect, and the direction of the displacement depends upon whether the galaxy is moving toward or away from the observer. Refer to Section 6.5 for a discussion of the Doppler effect.

The displacement for two lines on the calcium spectrum labeled H and K for three different galaxies is shown in Fig. 19.11. A small shift of the lines means a low velocity of recession, a greater shift means a greater velocity, and so on. The interesting point concerning the red shift is a correlation of velocity and the apparent brightness of the galaxy. The fainter the galaxy, the greater the velocity. Generally speaking, the farther away a galaxy is located, the less its apparent brightness. Thus, a velocity and magnitude correlation yields a velocity and distance correlation. This provides us with a means for measuring great distances that are far beyond the limits of cepheid variable observation.

When a galaxy is approaching Earth, the lines of the spectrum shift toward the blue end of the spectrum, indicating a shortening of wavelength. When the galaxy is receding from Earth, the lines shift to the red end of the spectrum, indicating a lengthening of the wavelength. See Fig. 19.11.

There are many stars that appear dim and insignificant but suddenly increase in brightness in a matter of hours by a factor of a hundred to a million. A star undergoing such a drastic change in brightness is called a **nova**, or new star. The star is unstable and throwing off an expanding shell of gas. The eruption may last for as many as 50 years before the star settles down to a stable state and becomes the original dim star. Several novae have been observed and studied, and scientists believe that the star is releasing internal energy faster than the surface can radiate. Thus in order to remain active, the star throws off the excess energy plus a small portion of its mass. One probable theory for novae is that they result from the interplay of two stars in a binary system.

Occasionally, a star explodes and throws off large amounts of material that may be so great that the star is destroyed. Such a gigantic explosion is known as a **supernova.** Only three supernovae have been in our galaxy. One of the most celebrated is the Crab Nebula in the constellation Taurus. This nebula (from the Latin word for cloud) is expanding at the rate of approximately 70 million miles per day. Because we know the average angular radius and the expansion rate, the original time of the explosion can be calculated. The result yields about 900 years, which agrees closely with Chinese and Japanese records that report the appearance of a bright new star in the constellation Taurus in A.D 1054.

On February 24, 1987, a supernova was observed in the Large Magellanic Cloud. It is the first supernova observed in this galaxy, which is the closest galaxy to our Milky Way. The Large Magellanic Cloud is 170,000 light-years away. The supernova was designated 1987A, because it was the first one to be observed in 1987.

The discovery of supernova 1987A is important to astronomers because they are observing a stellar explosion from the beginning, and they will be able to observe and examine the debris coming from the explosion and the remains of the star after the debris has cleared. Thus it will provide a history of a stellar explosion and supply valuable information concerning many branches of physics.

In addition to the novae and supernovae, there are the planetary nebulae, which possess a large, slowly expanding ringlike envelope, as shown in Fig. 19.12. These nebulae, as observed with a telescope, appear greenish. The nebula shown in Fig. 19.13, one of the brightest in the night sky, is known as a diffuse nebula. This type of nebula is more irregular and turbulent than the planetary nebulae. The Great Nebula in Orion is over 25 light-years in diameter and contains hundreds of stars that illuminate the dark material. Some cos-

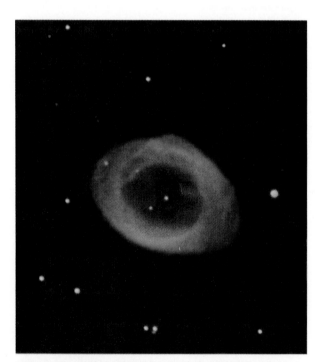

Figure 19.12 Ring nebula in the constellation Lyra.
Ring nebulae are spherical shells of gas ejected from, and expanding around, very hot stars. The shells expand about their parent stars at speeds of 20–30 km/s.

Figure 19.13 Great Nebula in Orion.
Considered the brightest of the diffuse nebulae, this one is located near the middle star in Orion's sword and is a gaseous and dusty nebula.

mologists (scientists who study the universe—i.e., its history, present structure, and future evolution) believe that the dust and gaseous material found in clouds such as these provide the material needed for the birth of new stars.

We now know that stars are born, radiate energy, expand, possibly explode, and then die. This is the general life cycle of a star. However, the exact details depend on a star's initial composition (the percentage amounts of hydrogen, helium, and heavier elements) and on its mass. The greater the mass of a star, the faster it moves through its life cycle.

The general evolution of a star with a mass typical of our Sun is shown in Fig. 19.14. The birth of a star, according to the best accepted theory, begins with the condensation of interstellar material, mostly hydrogen, because of the gravitational attraction between the interstellar material, radiation pressure from nearby stars, and supernova shock waves. The size of the star formed depends upon the total mass available, which in turn determines the rate of contraction. As the interstellar mass condenses and loses gravitational potential energy, the temperature rises and the material gains thermal energy. As the star continues to decrease in size, the temperature continues to increase until a thermonuclear reaction begins and hydrogen is converted to helium, as discussed in Section 19.1. The star's position in the main sequence of the H-R diagram, its temperature, and the radiation given off are determined by the mass of the star. On the main sequence of the H-R diagram, the star continues to convert hydrogen into helium. This process is called **hydrogen burning**, but it is a nuclear "burning," or fusion, not a chemical fire. The hydrogen-burning stage lasts for billions of years—possibly 10 to 15 billion years for a star like our Sun.

As the hydrogen in the core is converted into helium, the core begins to contract and heat up. This heats up the surrounding shell of hydrogen and causes the hydrogen burning in the shell to proceed at a more rapid rate. This rapid release of energy causes the star to expand and become a red giant.

Eventually, the core gets so hot that helium can fuse into carbon, and soon other nuclear reactions occur in which all elements are created. The creation of the nuclei of elements inside stars is called **nucleosynthesis.** The physics of the process becomes extremely complicated, but we believe that the cepheid-variable stage occurs next if conditions (mass and composition) are just right. We also believe that supernovae occur sometime after the red giant stage. The supernova stage occurs when reactions in the core proceed so rapidly that the

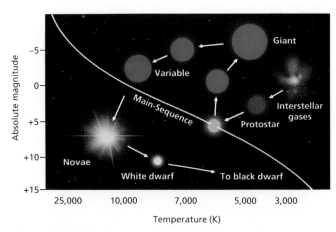

Figure 19.14 Hertzsprung-Russell diagram used to illustrate the evolution of a star with the same mass as our Sun.

star is blown almost completely apart, with only a remnant core remaining. The supernova stage probably does not occur for all stars.

Eventually, no more nuclear fusion reactions are possible and the star's nuclear energy has been spent. What is left of the star begins to gravitationally collapse and go to its end stage. The end stage of a star depends upon the star's mass at the end of its active life. Average- and small-mass stars become white dwarfs. When a star is a white dwarf, it is very small. It has gravitationally collapsed as far as it can and still obey the laws of physics. It is about the size of the Earth and is so dense that a single teaspoonful of matter weighs 5 tons. Because it is so small, it is a fairly dim star.

Large-mass stars (between 1.5 and 3.0 times the Sun's mass) have more gravitational attraction, and they collapse to a size of approximately 20 mi in diameter. The electrons and protons in this superdense star combine to form neutrons, and this **neutron star** is composed of about 99% neutrons. A teaspoonful of a neutron star would weigh 1 billion tons. Because the angular momentum of the star must be conserved, the small size of the neutron star dictates that it must be spinning rapidly. Rapidly rotating neutron stars give out radio radiation in pulses and are called **pulsars.**

Pulsars were first discovered in 1968 using radio telescopes similar to the one shown in Fig. 19.15. They pulse with a constant period that may be between $\frac{1}{30}$ and 4 seconds. The fastest-spinning pulsar is located at the center of the Crab Nebula. This pulsar is identified with the remains of the supernova that took place in A.D. 1054. This supernova was recorded by the Chinese

Figure 19.15 A 140-foot radio telescope.

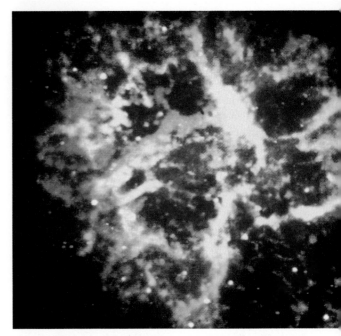

Figure 19.16 Crab nebula in the constellation Taurus.

19.4 Gravitational Collapse and Black Holes

The possibility of a rather remarkable phenomenon has been proposed to explain some strange astronomical findings. This phenomenon may or may not explain these discoveries, but it is interesting in its own right. The phenomenon is called **gravitational collapse** and is the collapse of a very massive body because of its attraction for itself. Recall that Newton's law of universal gravitation was given as

$$F = G\frac{m_1 m_2}{r^2} \qquad (19.1)$$

where F = force of gravity between m_1 and m_2,
  m_1 = a mass,
  m_2 = a second mass,
  G = a universal constant,
  r = distance between the centers of m_1 and m_2.

From Eq. 19.1 it is evident that if the distance r is made smaller, the force of gravitation increases.

In a large mass such as Earth or the Sun, parts are always attracting each other. Normally in large masses, electromagnetic forces tend to keep the various particles (e.g., atoms) apart. What would happen if the body were so massive that the gravitational forces of attraction were stronger than the electromagnetic forces of repulsion? All parts would be drawn closer together. Ac-

and must have been quite bright. The period of this pulsar is slowly getting larger, indicating that the rotating neutron star is gradually slowing down.

If the remaining core is greater than about three times the mass of the Sun, the star is believed to end up as a gravitationally collapsed object, even smaller and more dense than a neutron star. Such a star is so dense that light cannot escape from its surface because of its intense gravitational field. Thus it would appear black and is called a black hole. See Section 19.4.

After a star explodes in a supernova and the core goes into its end stage, what becomes of the ejected material? We believe that this material is thrown into space (see Fig. 19.16) and eventually becomes seed material for future stars. In fact, because of its surface composition and age, we believe our own Sun is a second-generation star. That is, one or more stars went through their life cycles and exploded as supernovae. The ejected material later coalesced again into our Sun and solar system. Because the universe is believed to be 15 to 20 billion years old and our Sun is only about 5 billion years old, this idea of the Sun being a second-generation star is generally accepted.

cording to Eq. 19.1, the forces increase as the distance *r* decreases. The closer the particles move together, the greater the gravitational force becomes. The whole mass would continue to contract until the original volume of mass becomes a fantastically massive point called a **singularity.**

The singularity is surrounded by an invisible spherical boundary known as the **event horizon.** Any matter or radiation within the event horizon cannot escape the influence of the singularity. See Fig. 19.17 for a sketch of a singularity and its event horizon. The value *R*, the radial distance the event horizon is located from the singularity, can be determined by equating the escape velocity formula to the velocity of light. We obtain

$$R = \frac{2GM}{c^2}$$

where G = the universal gravitational constant,
   M = the mass,
   c = the velocity of light.

Anything located within the event horizon cannot escape. The event horizon is a one-way boundary because matter and radiation can enter but cannot leave. Thus space inward from the event horizon is a **black hole.**

Because most stars are rotating, a black hole will probably also be rotating. And because angular momentum is conserved, the rotating rate will increase as the star gets smaller, so the rotation of the black hole will be extremely rapid. Thus rather than being spherical, a black hole is likely to be disk shaped.

Because nothing can escape the influence of a black hole, how are they detected? One method is the detection of X-rays coming from the vicinity of a black hole. The detection of X-rays by Earth satellites from a double-star system in which one companion may be a black hole has provided astronomers with data that black holes probably do exist. For example, the supergiant star NDE226868, which is located some 8000 light-years from Earth in the constellation Cygnus, has a companion called Cygnus X-1 that cannot be detected directly, indicating that it may be a black hole. The detected X-rays coming from the double-star system are generated by gases captured by Cygnus X-1 from the supergiant star. The captured gases, accelerating to extremely high speeds, go into orbit around Cygnus X-1, forming a spiraling flat disk of matter called an *accretion disk.* These gases become extremely hot because of internal friction and emit high-energy X rays. See Fig. 19.18.

Calculations from experimental data indicate that Cygnus X-1 is smaller than Earth and more than six times as massive as the Sun. This object is too massive

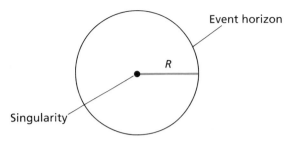

Figure 19.17 Physics of a black hole.
The black dot represents a singularity surrounded by the event horizon at a distance *R*. Anything located within the event horizon cannot escape. Thus, space inward from the event horizon is known as a black hole.

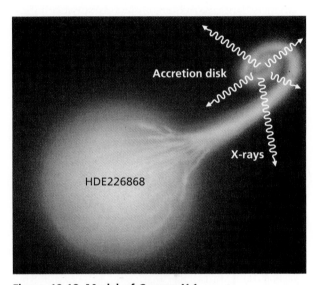

Figure 19.18 Model of Cygnus X-1.
The blue supergiant star HDE226868 ejects material, some of which is captured by its unseen companion (assumed to be a black hole), forming an accretion disk. X-rays generated at the inner edge of the disk have been detected by astronomers.

to be a white dwarf or a neutron star, leaving the possibility that Cygnus X-1 is a black hole. Many astronomers believe that Cygnus X-1 is probably a black hole. A binary system similar to Cygnus X-1 has been identified in the Large Magellanic Cloud, and other systems probably exist. Future observations may reveal them.

19.5 Galaxies

The Sun and its satellites occupy a very small volume of space in a very large system of stars known as a galaxy (Greek *galaxias*, Milky Way). A **galaxy** is an

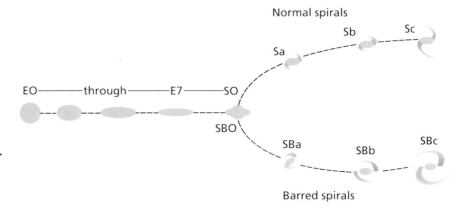

Figure 19.19 Galactic systems.
Sequence of elliptical, normal
spiral, and barred spiral
galaxies.

extremely large collection of stars occupying an extremely large volume of space. A galactic system is classified as irregular, spiral, or elliptical, depending upon how it appears when photographed.

The astronomer Edwin P. Hubble established a system of classifying galaxies according to their appearance in photographs. The system starts with spherical ellipticals, spreads out with increasing flatness, then branches off into a normal spiral sequence and a barred spiral sequence, with a scattering of irregulars outside the two branches (see Fig. 19.19).

The type Sa spirals have their spiral arms closely wound to the central region (see Fig. 19.20); whereas the Sb spirals have arms that spread out more from the center (see Fig. 19.21); and the Sc spirals have very loose spiral arms. The S0 type links the normal spirals to the smooth ellipticals. The barred spirals, which are distinguished by a broad bar that extends outward from opposite sides of the central region, follow similar unwinding of the spiral arms (see Fig. 19.22).

The irregular-type galaxies have no definite form but take on the appearance of shapeless clouds. Examples of this type are the two Magellanic Clouds that can be easily seen with the unaided eye from the Southern Hemisphere. They are named in honor of Magellan, who reported seeing them on his famous voyage around the world. See Figs. 19.23 and 19.26.

In a given volume of space there are more elliptical galaxies than spiral galaxies, while the irregular types account for only about 3% of all galaxies. The elliptical galaxies are made up of older stars and are usually dimmer than the spirals. The spiral galaxies account for over 75% of the brighter galaxies observed.

Carl Seyfert, an American astronomer at the Mount Wilson Observatory in California, reported in 1944 that a few spiral galaxies exhibit very bright centers and their spectra show broad emission lines that indicate that hot

Figure 19.20 The Sombrero galaxy, Type Sab spiral.
The dark band across the galaxy's center is composed of dust and gas.

Figure 19.21 Spiral galaxy NGC 2997, Type Sc.

gas is present and expanding at very rapid rates. This evidence indicates that violent activity is taking place in the central core of the galaxies. Large amounts of energy are being released from these central cores, and the best theory of the source of this energy is the gravitational collapse of an enormous amount of matter. Over a hundred of these galaxies have now been reported and are known as **Seyfert galaxies.** See Figure 19.24.

Our local galaxy is known as the **Milky Way,** and it is of the spiral Sb type. The Milky Way contains some 100 billion (10^{11}) stars and has an appearance similar to that of the Great Nebula in Andromeda (Fig. 19.25). Our galaxy (Fig. 19.26) is about 10^5 light-years in diameter and has a maximum thickness in the central region of approximately 10^4 light-years. It is rotating eastward or counterclockwise as viewed from above. The period of rotation, which is not the same for all regions of the galaxy, is more than 2×10^8 years for the region that contains our solar system. Our solar system is located near the plane of rotation about 3.0×10^4 light-years from the galactic center and is moving at a rate of approximately 150 mi/s in the direction of the constellation Lyra.

Figure 19.22 Barred galaxy NGC 4548, Type SBb.

Figure 19.24 The Seyfert galaxy NGC 4151.
Notice the very bright nucleus. If the galaxy were at a very extreme distance, only the bright nucleus would be visible. Perhaps quasars are the tremendous energetic sources in the nuclei of Seyfert galaxies.

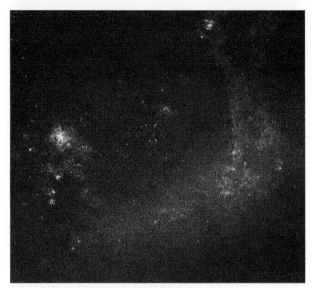

Figure 19.23 The Large Magellanic Cloud.
This irregular galaxy (the closest galaxy to our Milky Way) is only 150,000 light-years away. The huge bright region, which is 800 light-years in diameter, is the Tarantula Nebula.

Figure 19.25 The spiral galaxy M31 in Andromeda.
M31 is a Type Sb galaxy 2.25 million light-years from Earth. Two elliptical galaxies, NGC 205 (lower right) and NGC 221-M32, are also shown.

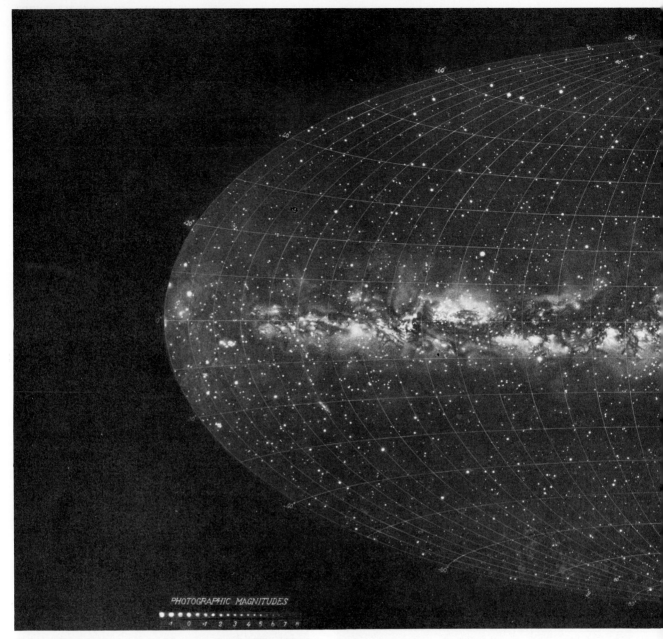

PHOTOGRAPHIC MAGNITUDES

Figure 19.26 Knut Lundmark's map of the Milky Way.
This panorama of the Milky Way is a composite made from many photographs. It illustrates the Milky Way on an Aitoff projection, and accurately shows 7000 stars with known coordinates.

The galactic equator, a great circle positioned half-way between the galactic poles, is inclined about 62° from the celestial equator (Fig. 19.27). The north galactic pole has a right ascension of 12 h 40 min and a declination of +28°. The south galactic pole has a right ascension of 0 h 40 min and a declination of −28°.

A galaxy is composed of individual atoms and molecules, dust, planets, stars, multiple stars, and star clusters.

The closest star to our solar system is Alpha Centauri, a triple star, located about 4.3 light-years away, with right ascension 14 h 36 min and −60°38′ declination. The next closest star is Barnard's star, named after Edward M. Barnard (1857−1923), an American astronomer who first observed the fast motion of the star in 1916. This star has a faster apparent motion than any other known star, moving at a rate of approximately 10 sec-

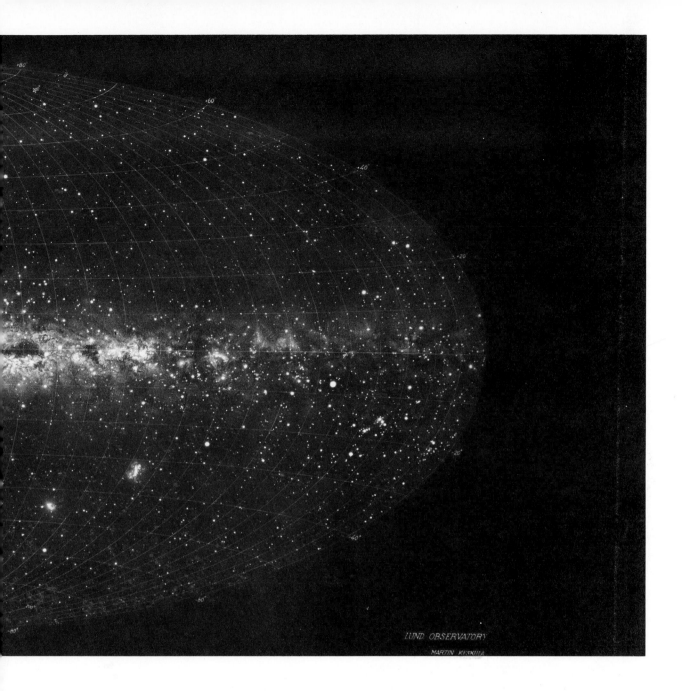

onds of arc per year. The average rate for visual stars is about $\frac{1}{10}$ second per year. Barnard's star is approximately 5.9 light-years from our Sun and is located near 18 h right ascension and $+4°$ declination.

In the immediate neighborhood of the Milky Way and confined to an ellipsoidal volume of space—some 9×10^5 parsecs for the major axis and about 8×10^5 parsecs for the minor axis—is located a small group of galaxies known as the **Local Group.** There are at least 21 known members of the group, and others are believed to exist that have not been detected because of their low magnitude. Our Milky Way is a member located near one end of the major axis. Messier 31 (Great Nebula in Andromeda, Fig. 19.25) is also a member and is positioned near the opposite end of the major axis.

The galaxies of the Local Group seem to be moving

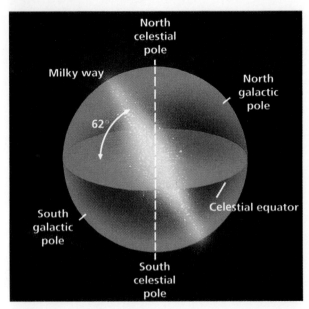

Figure 19.27 Diagram illustrating the inclination of the Milky Way to the celestial equator.

with random motions. The two Magellanic Clouds are moving away from our galaxy, and several others, including M31, are moving toward our galaxy.

The galaxies astronomers photograph throughout the vast volume of the universe are lumped together in **clusters** that vary in size and number. See the chapter-opening photograph of a coma cluster. The clusters, classified as regular or irregular, range in size from 3 to 15 million light-years in diameter. Some clusters contain a few galaxies such as our Local Group, and others contain thousands. As Hubble found regarding the distribution of galaxies, astronomers have discovered that the distribution of clusters is isotropic and homogeneous.

The galactic clusters lump together into what are called superclusters. That is, the **superclusters** are clusters of clusters. These superclusters have diameters as large as 300 million light-years and masses equal to or greater than 10^{15} solar masses. Our Local Group and adjoining groups and clusters such as the Virgo cluster form what is called the Local Supercluster.

When Edwin Hubble began looking at Doppler shifts of galaxies, he found nothing surprising at first. In the Local Group the Doppler shifts were small: some were blue and some were red. But as he looked at galaxies farther and farther away, he found only red shifts. In fact, the farther away the galaxy, the larger the red shift. From these measurements the distances to remote galaxies can be determined by converting the observed red shift of the galaxy to radial velocity and then plotting

the logarithm of the velocity against the apparent magnitude. See Fig. 19.28. Hubble's discovery, now known as **Hubble's law,** can be written as

$$v = Hd$$

where v = recessional velocity of the galaxy,
           d = distance away from the galaxy,

$$H = \text{Hubble's constant} = \frac{50 \text{ km/s}}{10^6 \text{ parsec}}$$

The Hubble constant is believed to have a value of 50 to 100 km/s per million parsecs. For example, if H is 50 km/s per million parsecs, the observed galaxy is moving away from our Sun at a speed of 50 km/s for every 1 million parsecs the galaxy is from our Sun. Hubble's law is extremely important because it gives astronomers vital information about the structure of the universe.

Astronomers and other scientists determine the structure of the universe by detecting and analyzing electromagnetic waves that come from the stars and galaxies that occupy the universe. From the collected data scientists contemplate the way in which these objects are distributed throughout the vast volume of the universe. Scientists estimate that there are at least 10^9 (1 billion) galaxies within range of astronomers' optical telescopes that can be photographed.

Even if they were inclined to do so, astronomers would not have the time to photograph the tremendous volume of the universe to obtain photographs of all galaxies. Instead, the astronomers' model of the structure of the universe is based upon a sampling of different

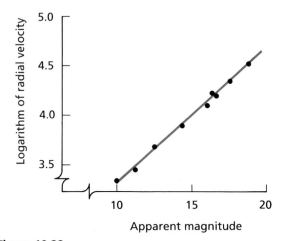

Figure 19.28
This graph shows the relationship (Hubble's law) between the logarithm of the radial velocity of a galaxy and its distance from the Milky Way. Recall that apparent magnitude and distance are related.

regions of the universe. Using the 60-in and 100-in reflecting telescopes on Mount Wilson, Hubble obtained photographs of more than 1200 sample regions of space, counted some 44,000 galaxies, and estimated that it would be possible to photograph 100 million galaxies. After correcting his data for such things as the obstruction presented by the Milky Way and interstellar dust, he concluded that when observation of the universe is made over a large volume, the distribution of galaxies is isotropic and homogeneous. That is, when we observe a large volume of space, we observe as many galaxies in one direction as in any other, and the observations are the same at all distances. This concept of the uniformity of the universe is known as the **cosmological principle** and is the basic assumption for most theories of cosmology.

19.6 Quasars

In about 1960 radio astronomers, using their newly developed high-resolution (ability to distinguish the separation of two points) radio telescopes, began detecting extremely strong radio signals from sources having small angular dimensions. The sources were named **quasars**— a shortened term for "quasistellar radio sources."

Hundreds of quasars have now been detected (see Fig. 19.29), and they have two important characteristics. First, quasars have extremely large red shifts. This fact indicates (based on Hubble's law) that they must be very far away. When their brightness was considered, however, scientists showed that if they were indeed extremely distant, quasars must be extremely powerful sources of energy. The mystery of quasars then becomes a question of energy. Quasars have an energy output equivalent to about 10,000 times that of a typical spiral galaxy.

Quasars are considered to be the most distant objects in the universe. The quasar OH 471 has an enormous red shift that, according to Hubble's law, places the quasar 18 billion light-years from our solar system. In addition to having enormous red shifts, quasars emit tremendous amounts of electromagnetic radiation (energy) at all wavelengths, and most of the radiation varies in intensity (some regular, some random) over a range of from a few days to years. Quasars are very small, appear to be blue, and have absolute magnitudes down to −25. Although not conclusive, the best theoretical evidence points to the theory that the "powerhouse" of quasar energy is a supermassive black hole at their center.

Recent evidence indicates that quasars are somehow related to the centers of galaxies. As mentioned in Sec-

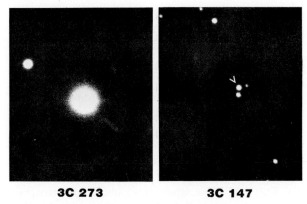

Figure 19.29 Quasistellar radio sources.
Note the jetlike prong of 3C273. This quasar has an apparent visual magnitude of 12.8 and can be seen with a good 8-inch telescope.

tion 19.5, Seyfert galaxies have very bright centers, and the broad emission lines present in their spectra indicate that they contain gas that is extremely hot. Because quasars are very distant objects, perhaps they are the bright nuclei of spiral galaxies. The spiral arms of a galaxy would not be visible at a great distance from us.

19.7 Cosmology

Cosmology is the study of the entire universe. The human mind first observes phenomena and then develops concepts, based on those observations, that deal with the origin, evolution, and structure of the universe. When we construct such concepts, we must remember that any concept of the universe conceived by the human mind is the way the universe is *conceived*, which is not necessarily the way it actually *is*.

The concept presently accepted by most astronomers is called the Big Bang. This concept has received broad acceptance because experimental evidence has supported the concept in three major area:

1. Astronomers observe galaxies that show a shift in their spectrum lines toward the low-frequency (red) end of the electromagnetic spectrum. This red shift is known as the **cosmological red shift.**

2. They detect a cosmic background radiation coming from space in all directions, commonly referred to as the 3-K cosmic microwave background.

3. They observe a mass ratio of hydrogen to helium of 3 to 1 in stars and interstellar matter.

The red shift indicates an expanding universe. To explain this observation, astronomers conceived the idea that the universe began with a violent event called the

Big Bang and we are now seeing the universe expanding from that event.

Because the event came before the expansion, we can calculate the time that has elapsed since the event. Astronomers observe galaxies receding from each other and assume they have been receding from each other since the event. To determine the maximum time (t_m), let d represent the distance to the most remote galaxy that we observe, and assume that the rate of receding has always been congruent with what we presently observe. The receding velocity (v_r) is given by Hubble's law. That is, $v_r = Hd$, where H is Hubble's constant. We know, by definition, that velocity is distance divided by time ($v = d/t$).

Thus $v_r = Hd = d/t_m$. Canceling and rearranging, we have

$$t_m = \frac{1}{H}$$

Substituting the minimum value for Hubble's constant,

$$H = \frac{50 \text{ km/s}}{10^6 \text{ parsecs}}$$

we have

$$t_m = \frac{1}{\dfrac{50 \text{ km/s}}{10^6 \text{ parsecs}}}$$

Because 3.086×10^{19} km $= 10^6$ parsecs,

$$t_m = \frac{1}{\dfrac{50 \text{ km/s}}{3.086 \times 10^{19} \text{ km}}}$$

$$= 6.173 \times 10^{17} \text{ s}$$

$$= 1.957 \times 10^{10} \text{ years}$$

$$= 19.57 \text{ billion years.}$$

Because we used the minimum value for Hubble's constant, this answer represents the maximum age of the universe.

What is the origin of all matter and radiation in the universe? One possible answer is that it came from nothing. When astronomers take a relativistic view of the universe, space and time change. At the beginning of the Big Bang, which is considered the origin of the universe, space and time did not exist. Thus, no space, no time, no anything (maybe energy) expanded violently creating space, time, and something. It seems meaningless to conceive of nothing, but out of nothing (a vacuum) comes something when we theoretically allow space to expand and time to flow. Thus the possibility exists for the transformation from timeless, spaceless, and nothing to space, time, and something.

Presently, what the something might have been at the beginning of the Big Bang is only speculation. General relativity fails to provide answers for a universe of infinite density and curvature. Scientists theorize that all kinds of particles existed in equilibrium with radiation during the first fractional part of a second that the Big Bang took place. Particles and antiparticles were being produced in pairs from photons, and annihilating and reconverting to photons. But by the time the explosion was 1 s old and the temperature about 10^{10} K, matter as we know it today (electrons, protons, neutrons, and neutrinos) existed. About 1 million years after the Big Bang the temperature had dropped below 3000 K, and the particles could combine to form hydrogen, helium, and a trace of deuterium. The ratio of hydrogen to helium was then 3 to 1 by mass, which is the value presently observed in stars and interstellar matter.

People often ask, "Where did the Big Bang take place?" This question has no meaning, because at the moment of the explosion the makeup of the Big Bang was the entire universe. The explosion did not throw matter off into space. Space expands with time and forms the universe. Thus time and space do not exist outside the universe.

To comprehend an expanding three-dimensional universe of galaxies, observe the expanding two-dimensional surface of a toy rubber balloon speckled with small dots (representing galaxies) as the balloon is being inflated. Think of yourself as standing where a dot is located on the surface of the balloon. As the balloon expands, all dots are observed to recede from you. It makes no difference which dot you choose to observe from; the result is the same. There is no central point on the surface for the dots. Likewise, there is no central point for the galaxies in our three-dimensional universe. Note also that the surface of the balloon has no edge: You could examine every point on the surface of the balloon and find no edge. Similarly, there is no edge to our three-dimensional universe.

As early as 1948, scientists predicted that residual radiation from the early universe should be present here and now, and the radiation should be at radio wavelengths. The radiation should also resemble the radiation from a black body at a temperature of a few degrees above absolute zero. Because the radiation from the early universe was everywhere at once, it should fill the entire universe and be isotropic. That is, the radiation is the same in any direction that we observe.

This radiation was first detected in 1965 by Arno Penzias and Robert Wilson at Bell Telephone Laboratories in New Jersey. They were testing a new microwave horn antenna used to make measurements of the absolute intensity of microwave radiation coming from certain regions of the Milky Way. After debugging and accounting for known static sources in their equipment, there remained a source of static, coming from every direction of space, that they were unable to locate. The static was received equally from space at all times from any direction. After consulting with scientists at Princeton University who were designing and building equipment to detect the dying glow of the Big Bang, they concluded that the static (radiation) they were annoyed by was the residual radiation from the early universe.

Today, the dying glow is called the **3-K cosmic background radiation,** and its presence is considered evidence of an ancient, extremely hot universe. This residual radiation is the greatly red-shifted radiation of the extremely hot universe that existed about 1 million years after the Big Bang.

Will the universe continue to expand forever? Presently, astronomers do not know the answer to this question, but the key to our understanding lies in determining the average density of matter in the universe. If the average density of matter is great enough, then space has a positive curvature and we live in a closed universe. That is, gravity in the universe is great enough to stop the expansion and eventually all matter will collapse in what is called the **Big Crunch.** If the average density of matter is too small, then space has a negative curvature and we live in an open universe that will continue to expand forever.

Present data for the value of the average density of matter are not precise enough to determine whether space has a positive or a negative curvature. Thus present knowledge of the average density seems to indicate that we are living in a nearly flat universe.

Today, astronomers are applying the laws of physics and using the techniques of modern computer technology to probe the near and distant volumes of space, seeking knowledge of its matter, radiation, composition,

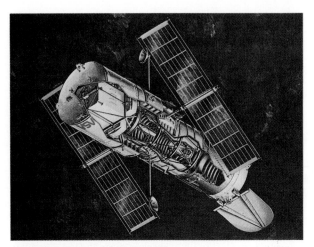

Figure 19.30 An artist's depiction of NASA'S Hubble Space Telescope.
This instrument will allow scientists to look seven times farther into space than ever before. The telescope is to be launched into orbit around Earth by the Space Shuttle sometime in 1990. The telescope's mirror is 2.4 m (7.8 ft) in diameter.

motion, and structure. They are using data collected by sophisticated land-based optical telescopes and others that operate from space (see Figure 19.30), plus an array of new devices operating in the X-ray, gamma ray, microwave, radio, ultraviolet, and infrared regions of the electromagnetic spectrum. Space probes have given astronomers a great deal of information about the solar system: 12 astronauts have gone to the moon; robots to Mars have chemically analyzed the Martian surface; flyby missions have examined the clouds and surface of Venus; spacecraft on flyby missions have photographed and transmitted data on Jupiter, Saturn, and Uranus; and ongoing space probes will continue on to Neptune and interstellar space.

The search for knowledge about the universe will continue. By the year 2000 we may know the course our universe will follow and the distant future of galaxies.

Learning Objectives

After reading and studying this chapter, you should be able to do the following without referring to the text:

1. Define the term *plasma*, and describe the composition of the Sun's interior.

2. Describe the basic process by which a star, such as the Sun, radiates energy.

3. Define the term *neutrino*, and explain why the solar neutrino experiment is so important.

4. Describe the gravitational collapse of a star.

5. Draw an H-R diagram and label the axes and the major features.

6. State the major events in the birth, life, and death of a star.

7. List the three possible end stages of a star and the mass a star would have to have to fall into each category.

8. State the Sun's age and the age of the universe.

9. State Hubble's law in words.

10. State and give experimental evidence to support the Big Bang theory.

11. Define and explain the other important words and terms listed in the following section.

Important Words and Terms

photosphere	light-year	supernova	clusters
plasma	parsec	hydrogen burning	superclusters
chromosphere	constellations	nucleosynthesis	Hubble's law
corona	zodiac	neutron star	cosmological principle
sunspots	magnitudes	pulsars	quasars
prominences	absolute magnitude	gravitational collapse	cosmology
neutrino	H-R diagram	singularity	cosmological red shift
proton-proton chain	main sequence	event horizon	Big Bang
celestial sphere	red giants	black hole	3-K cosmic background radiation
declination	white dwarfs	galaxy	Big Crunch
right ascension	binary star	Seyfert galaxies	
celestial prime meridian	cepheid variables	Milky Way	
astronomical unit	nova	Local Group	

Questions

The Sun

1. What is a plasma?

2. What is the temperature of the Sun (a) at the center and (b) at the surface?

3. Are sunspots cyclic?

4. What makes the Sun radiate energy?

5. What are neutrinos and photons, and how are they different?

6. Why are the results of the solar neutrino experiment perplexing?

The Celestial Sphere

7. What is the celestial sphere?

8. Define *declination* and give an example of it.

9. Define *right ascension* and give an example of it.

10. What is the celestial prime meridian?

11. Define each term: (a) light-year, (b) parsec, and (c) astronomical unit.

12. Describe the zodiac.

Stars

13. On an H-R diagram, give the position of each of the following: (a) the Sun, (b) the main sequence, (c) red giants, (d) cepheid variables, and (e) white dwarfs.

14. List the following possible stages of a star in the correct order: hydrogen burning, cepheid variable, white dwarf, gravitational accretion, red giant.

15. What is a pulsar?

16. What are cepheid variables? Why are they important to astronomers?

17. Distinguish between the apparent and absolute magnitude of a star.

18. What is a binary star system? Do most stars appear as a binary system?

Gravitational Collapse and Black Holes

19. Describe the gravitational collapse of a star.

20. What is a singularity?

21. What factors determine the radial distance of the event horizon?

22. What is a black hole?

23. Describe a method for detecting black holes.

24. How does the radial distance of the event horizon of a black hole vary in respect to the (a) mass of a collapsing star and (b) velocity of light?

Galaxies

25. What is a galaxy? Name three.

26. How are galaxies classified?

27. Name the three major classifications of galaxies.

28. State the structure and dimensions of the Milky Way.

29. Distinguish between isotropic and homogeneous distribution of galaxies.

Quasars

30. What is a quasar?

31. What information is there to support the belief that quasars are the most distant objects observed by astronomers?

32. In what respect do quasars resemble stars?

33. In what respect do quasars differ from stars?

Cosmology

34. State the cosmological principle.

35. Explain the Big Bang model of the universe.

36. State three experimental facts that support the Big Bang model.

37. What property of the universe can be determined by taking the reciprocal of Hubble's constant?

38. How old is (a) the Sun and (b) the universe?

39. Why is the universe thought to be expanding?

Exercises

The Sun

1. The mass of the Sun is 2×10^{30} kg, or in terms of Earth weight, 2.2×10^{27} tons. Compute how many years it would take to convert one-tenth of the Sun's hydrogen if 630 million tons are converted into helium and energy every second. *Answer:* 11 billion years

2. Perform the calculations mentioned in the book to derive the net reaction

$${}^1_1H + {}^1_1H + {}^1_1H + {}^1_1H \longrightarrow$$
$${}^4_2He + 2(_{+1}e) + 2\gamma + 2\nu + \text{energy}$$

from the reactions for the proton-proton chain.

The Celestial Sphere

3. Calculate the number of miles in a light-year, using 186,000 mi/s as the speed of light. *Answer:* 5.87×10^{12} mi

4. Calculate the number of meters in a light-year, using 3×10^8 m/s as the speed of light or referring to the answer to Exercise 3.

5. How many miles away is the closest star, Alpha Centauri, which is 4.3 light-years away? *Answer:* 2.47×10^{13} mi

6. How many parsecs away is Alpha Centauri, which is 4.3 light-years away?

7. How long does it take light to reach us from Alpha Centauri, which is 4.3 light-years away?

8. Find the distance in parsecs to a star with a parallax of 0.2 second.

9. Find the distance in light-years to a star with a parallax of 0.2 second. *Answer:* 16.3 light-years

Stars

10. How much brighter is a star of absolute magnitude $+1$ than a star of absolute magnitude (a) $+2$ and (b) $+6$?

11. How many times brighter is a cepheid variable of absolute magnitude -3 than a white dwarf of absolute magnitude $+7$? *Answer:* 10,000 times

12. (a) What is the color index of a star with a photographic magnitude of 3 and a visual magnitude of 3.3?
    (b) What is the color of this star? *Answer:* (b) blue-white

13. The Crab Nebula is expanding at a rate of 70 million miles per day. If it is the remnants of a supernova of A.D. 1054, how many miles in diameter is it today? *Answer:* 47×10^{12} mi, or 8 light-years

Gravitational Collapse and Black Holes

14. Determine the radial distance of the event horizon of a black hole formed with the gravitational collapse of a star having a mass of 15×10^{30} kg.

Galaxies

15. Suppose our universe contained 100 billion galaxies with 100 billion stars in each.
    (a) How many stars would there be?
    (b) Is this number greater or less than Avogadro's number (6×10^{23})? *Answer:* (b) less

16. It takes the Sun 2×10^8 years to go around the center of the galaxy. How many times has it (and our solar system) gone around the galaxy during its life of 5 billion years? *Answer:* 25 times

Cosmology

17. Determine the age of the universe, using a value for Hubble's constant of 100 km/s per 10^6 parsecs.

18. Hubble's constant is estimated to be between 50 and 100 km/s per 10^6 parsecs. Use the average of these two values and determine the age of the universe.

CHAPTER 20

The Atmosphere

> . . . this most excellent canopy, the air, look you, this brave o'erhanging firmament, this majestical roof fretted with golden fire. . . .
>
> —Shakespeare, *Hamlet*

OUR ATMOSPHERE (from the Greek *atmos*, "vapor", and *sphaira*, "sphere") is the gaseous shell, or envelope, of air that surrounds the Earth. Just as certain sea creatures live at the bottom of the ocean, humans live at the bottom of this vast atmospheric sea of gases.

In recent years the study of the atmosphere has expanded because of advances in technology. Every aspect of the atmosphere, from the ground to outer space, is now investigated in what is called **atmospheric science.** The older term of **meteorology** (from the Greek *meteora,* "the air") is now more commonly applied to the study of the lower atmosphere. The continuously changing conditions of the lower atmosphere are what we call weather. Lower atmospheric variations are monitored daily, and changing patterns are studied to help predict future conditions.

The phenomena of the atmosphere are important to many branches of science. They produce changes on the surface of the Earth that concern the geologist, as will be discussed in later chapters. The astronomer longs for a view of the stars that is not obscured by the atmosphere and must consider the effect of atmospheric phenomena on observations. Yet while we attempt to learn the workings of the atmosphere and to even effect local changes for our own benefit, we indiscriminately pollute the atmospheric blanket that protects and sustains life.

The condition of our atmosphere has become a topic of major interest with respect to ecology, the study of living organisms in relationship to their environment. Because the air we breathe is such an integral part of our environment, a knowledge of the atmosphere's constituents, properties, and workings is needed to understand and appreciate current environmental problems. In other words, what are some of the normal conditions and other phenomena of the atmosphere?

20.1 Composition

The air of the atmosphere is a mixture of many gases. In addition, the air holds many suspended liquid droplets and solid particles. However, only two gases comprise about 99% of the volume of air near the Earth. From Fig. 20.1 and Table 20.1, we see that air is primarily composed of nitrogen (78%) and oxygen (21%), with nitrogen being about four times as abundant as oxygen. The other main constituents are argon and carbon dioxide.

Minute quantities of many other gases are found in the atmosphere, along with dust, pollen, salt particles, and so on. Some of these gases, especially water vapor and carbon monoxide, vary in concentration, depending on conditions and locality. The amount of water vapor in the air depends to a great extent on the temperature, as will be discussed later. Carbon monoxide is a by-product of incomplete combustion. A sample of air taken near a freeway, for example, will contain a concentration of carbon monoxide considerably higher than that normally found in the atmosphere. Carbon dioxide, another by-product of combustion, is also present in abnormally high concentrations in such populated areas as the Los Angeles basin.

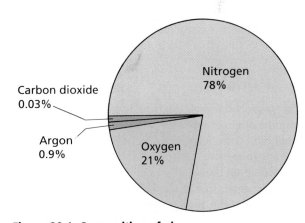

Figure 20.1 Composition of air.
A graphic representation of the volume composition of the major constituents of air.

◄ Hot-air balloons floating in the atmosphere.

Table 20.1 Composition of Air

Nitrogen	N_2	78%	(by volume)
Oxygen	O_2	21%	
Argon	Ar	0.9%	
Carbon dioxide	CO_2	0.03%	

Others (traces)		*Others (variable)*
Neon	Ne	Water vapor (H_2O) 0–4%
Helium	He	Carbon monoxide (CO)
Methane	CH_4	Ammonia (NH_3)
Nitrous oxide	N_2O	Solid particles—dust,
Hydrogen	H_2	pollen, etc.

In general, however, the relative amounts of the major constituents of the atmosphere remain fairly constant. Nitrogen, oxygen, and carbon dioxide are involved in the life processes of plants and animals and are continually taken from the atmosphere and replenished as by-products of the various processes. Animals breathe in oxygen and expel carbon dioxide, while plants convert carbon dioxide to oxygen. Nitrogen is also assimilated by some plants and released in organic decay. Thus nature operates in such a way as to maintain the balance of the various gases in the atmosphere.

On the other hand, human beings may contribute to changes in the atmosphere and perhaps eventually create imbalance. Recent concern has arisen in this regard because of increased amounts of CO_2 being poured into the atmosphere. Consider the amount of burning required to heat the houses in the United States alone in the winter. Add to this the industrial burning for heat and energy. The resultant volume of gases released into the atmosphere is considerable and gives cause for concern, as will be discussed in Chapter 24.

20.2 Origin

The origin of the atmosphere is generally associated with the origin of the Earth, which is some 4 to 5 billion years old. There are several theories of *cosmogony* that attempt to explain the origin of the universe. Without speculating on the validity of any one theory, scientists generally agree that at the time of the creation of our solar system, the planets were extremely hot masses, because their creation by any means would require enormous amounts of energy.

In the beginning, then, the Earth was a molten mass surrounded by hot gases, probably methane, ammonia, and hydrogen compounds. Because of their high temperatures and inherent kinetic energy, large amounts of these gases escaped into space. As the Earth cooled, gases that were dissolved in the molten mass were released, giving rise to H_2O, N_2, and CO_2 in our atmosphere. The water vapor condensed and reevaporated, causing cooling and solidification of the Earth's crust, with subsequent erosion and formation of bodies of water. The gases cooled along with the planet and, being then less energetic, were retained by gravitational attraction. At this time life began to appear.

How life occurred is still a mystery. However, scientists have shown that the fundamental building blocks of life, the amino acids, can be constructed from combinations of methane, hydrogen, ammonia, and water, all of which were available on the cooling Earth. (See Chapter 15, the Miller-Urey experiment.) In any event, the appearance of plant life allowed the formation of oxygen and the evolution of animal life. Plants produce oxygen by **photosynthesis**, the process by which CO_2 and H_2O are converted into sugars (carbohydrates) and O_2, using the energy from the Sun (Fig. 20.2).

Figure 20.2 Photosynthesis.
Energy from the Sun is necessary in the photosynthesis process whereby plants produce sugars and oxygen from water and carbon dioxide.

The key to photosynthesis is the ability of chlorophyll, the green pigment in plants, to convert sunlight into chemical energy. In chemical notation the photosynthesis process may be expressed as

$$(n)\ CO_2 + (n)\ H_2O \xrightarrow[\text{Chlorophyll}]{\text{Sunlight}} (CH_2O)_n + (n)\ O_2$$

where the n's indicate the numbers that balance the chemical equation. Photosynthesis accounts for the liberation of approximately 130 billion tons of oxygen into the air annually, with some 2000 billion tons of CO_2 being involved in the process. Over half of the photosynthesis takes place in the oceans, which contain many forms of green plants. Originally, oxygen may have come from the dissociation of water.

Over millions of years the Earth's atmosphere evolved into its present composition. The atmosphere of other planets evolved in different fashions. For example, the atmospheres of Jupiter and Saturn contain mainly ammonia and hydrogen, while Venus has a considerable amount of CO_2. Mercury, however, lost virtually all of its atmosphere, as did our moon.

The retention of an atmosphere depends on a balance between gravity and the thermal kinetic energy of the atmospheric gases. If the energy input (from the Sun) is sufficient to give the gas molecules enough kinetic energy so that some continually escape from the gravitational attraction of the body, then any original atmosphere will eventually be lost.

20.3 Vertical Structure

To distinguish different regions of the atmosphere, we look for variations that occur with height or altitude. The changes in physical properties can be used to define vertical divisions of the atmosphere. Some of the atmospheric properties that show vertical variations that we will examine are: (1) density and pressure, (2) temperature, (3) homogeneity and heterogeneity of gases, (4) ozone and ion concentrations.

Density and Pressure

The atmosphere extends upward with continuously decreasing density (mass per volume). The greater density near the Earth's surface is due to gravitational attraction and compression of the air. As a result, over half of the mass of the atmosphere lies below an altitude of 11 km (7 mi) and almost 99% lies below an altitude of 30 km (19 mi). The air becomes quite "thin" at higher altitudes.

At a height of 320 km (200 mi) above the Earth, the density of the atmosphere is such that a gas molecule may travel a distance of 1.6 km (1 mi) before encountering another gas molecule.

There is no clearly defined upper limit of the Earth's atmosphere. It simply becomes more and more tenuous and merges into the interplanetary gases, which may be thought of as part of the extensive "atmosphere" of the Sun. Because of the continuous decrease in density, an outermost atmospheric limit can be placed anywhere from 480 to 960 km (300 to 600 mi) from the surface of the Earth.

Closely related to density is atmospheric pressure (force per area). The pressure at a particular altitude is effectively a measure of the weight (or amount) of gas above that location. Just as a sea diver experiences increased pressure when descending, because of the weight of the water above him, the atmospheric pressure is greater near the Earth because of the weight of the air above. Like density, the pressure varies vertically, decreasing with increasing altitude. This decrease occurs continuously (Fig. 20.3) and does not have distinct boundaries. Consequently, vertical divisions of the atmosphere are not made on the basis of pressure.

Temperature

In measuring the temperature of the atmosphere versus altitude, we find that distinct changes do occur. These distinctions lead to major divisions of the atmosphere based on temperature variations. Near the Earth's surface the temperature of the atmosphere decreases with increasing altitude at an average rate of about $6\frac{1}{2}°C/km$ ($3\frac{1}{2}°F/1000$ ft) up to about 16 km (10 mi), as illustrated in Fig. 20.3. This region is called the **troposphere** (from the Greek *tropism*, "to change"). The troposphere contains over 80 percent of the atmospheric mass and virtually all the clouds and water vapor. There is continual mixing and a great deal of change. The atmospheric conditions of the lower troposphere are referred to collectively as **weather.** Changes in the weather reflect the local variations of the atmosphere near the Earth's surface.

The actual depth of the troposphere varies from about 16 km (10 mi) at the equator to about 8 km (5 mi) at the poles, with an average thickness of about 11 km (7 mi). The lower troposphere was the only region of the atmosphere investigated until the twentieth century. Balloonists ascended to the upper regions of the troposphere only in the 1930s. Many unexpected problems

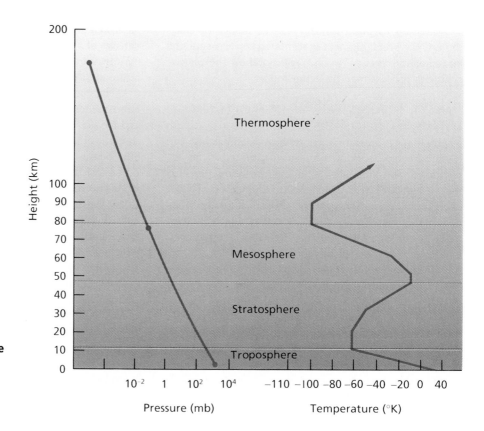

Figure 20.3 Vertical structure of the atmosphere.
Divisions of the atmosphere are based on variations in physical properties.

arose. The balloonists encountered difficulty in breathing, the freezing of valves necessary for descent, and loss of the use of limbs due to frostbite. At the top of the troposphere the temperature falls to −45° to −50°C (about −50° to −60°F). Beyond these altitudes, unmanned balloons and rockets were used for exploration. Instruments on board recorded the atmospheric properties. More recently, satellites have been used to make atmospheric observations (see Chapter 23).

Above the troposphere the temperature of the atmosphere increases nonuniformly up to an altitude of about 50 km (30 mi). See Fig. 20.3. This region of the atmosphere, from approximately 16 to 50 km (10 to 30 mi) in altitude, is called the **stratosphere** (from the Greek *stratum*, "covering layer"). Together, the stratosphere and the troposphere account for about 99.9% of the mass of the atmosphere.

The temperature of the atmosphere then decreases rather uniformly to a temperature of about −95°C (−140°F) at an altitude of 80 km (50 mi). This region, between 50 to 80 km (30 to 50 mi) in altitude, is called the **mesosphere** (from the Greek *meso*, "middle"). The average rate of temperature decrease with increasing altitude in this region is on the order of about $3\frac{1}{2}$°C/km (1.9°F/1000 ft).

Above the mesosphere the thin atmosphere is heated intensely by the Sun's rays and the temperature climbs to over 1000°C (about 1800°F). This region extending to the outer reaches of the atmosphere is called the **thermosphere** (from the Greek *therme*, "heat"). The temperature of the thermosphere varies considerably with solar activity.

Homogeneity and Heterogeneity of Gases

The mixture of atmospheric gases provides another means of atmospheric division. In the lower portion of the atmosphere, turbulence continually mixes the gases so that the air is a homogeneous mixture. Thus this region is called the **homosphere** (*homo*, meaning "same").

Above the homosphere, beginning at about 88 to 96 km (55 to 60 mi) from the Earth's surface, is the **heterosphere** (*hetero*, meaning "different"). Because no turbulence exists there to keep the gases mixed, they settle out in a layered structure, as illustrated in Fig.

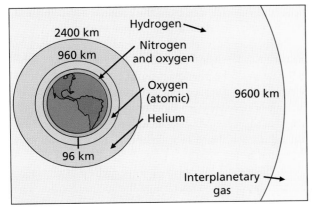

Figure 20.4 Heterosphere.
An illustration of the heterogeneous layering of the gases in the upper atmosphere based on satellite data.

20.4. In the layer above 96 km (60 mi) the oxygen is in atomic form (O) instead of the normal diatomic form (O_2) found in the lower atmosphere.

This settling is analogous to sand suspended in water. As long as the mixture is stirred, the sand remains in a uniform suspension. However, if the mixing ceases, the sand settles out according to particle size and weight, much the same as the gases of the heterosphere.

Ozone and Ion Concentrations

The atmosphere is also divided into the **ozonosphere** and **ionosphere.** Their boundary is approximately the same as the homosphere and heterosphere, respectively. As the names imply, there are concentrations of ozone and ions in these regions.

Ozone (O_3) is formed by the dissociation of molecular oxygen and the combining of the atomic oxygen with the molecular oxygen:

$$O_2 + energy \longrightarrow O + O$$
$$O + O_2 \longrightarrow O_3$$

At high altitudes energetic ultraviolet radiation from the Sun provides the energy necessary to dissociate the molecular oxygen. Oxygen, however, becomes less abundant at higher altitudes, so the production and concentration of ozone depends on the appropriate balance of ultraviolet radiation and oxygen molecules. The optimum condition occurs at an altitude of about 30 km (20 mi), and in this region the central concentration of an ozone layer is found as illustrated in Fig. 20.5.

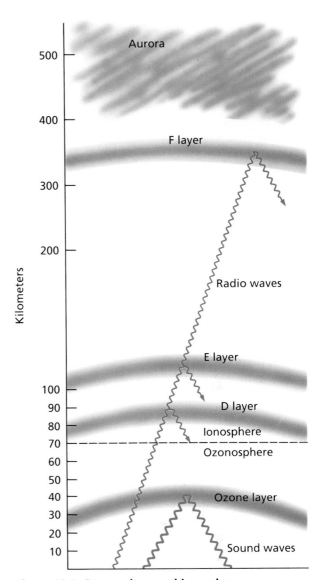

Figure 20.5 Ozonosphere and ionosphere.
These atmospheric regions are based on ozone and ion concentrations. The warm-air layer, because the ozone in it absorbs ultraviolet radiation, reflects sound waves, while the ion layers reflect radio waves.

The ozone layer is a broad band of gas that extends through nearly all of the stratosphere. Ozone is very unstable in the presence of sunlight, and when it absorbs ultraviolet radiation, it is again dissociated into molecular and atomic oxygen. A free oxygen atom may recombine with another oxygen molecule to again form ozone. However, an oxygen atom may meet an ozone molecule and form two ordinary oxygen molecules ($O + O_3 \rightarrow O_2 + O_2$), thus destroying the ozone.

These processes go on simultaneously in the ozone layer so that a balance between ozone production and destruction is maintained.

There is little ozone present near the Earth's surface. It is sometimes formed by electrical discharges and is easily detected by a distinct pungent smell from which it derives its name (Greek *ozein,* "to smell"). In some areas, e.g., Los Angeles, ozone is classified as an air pollutant. It is found in relatively high concentrations resulting from photochemical reactions of air pollutants. Such reactions give rise to photochemical smog, as will be discussed in Chapter 24.

The ozone layer in the stratosphere acts as an umbrella that shields life against harmful ultraviolet radiation by absorbing most of the short wavelengths of this radiation. The portion of the ultraviolet radiation that gets through the ozone layer burns and tans our skins in the summer. Were it not for the ozone absorption, we would be badly burned and find the sunlight intolerable.

Recently, concern was expressed over the effect on the ozone layer by gaseous chlorofluorocarbon compounds used as propellants in spray aerosol cans. Through certain chemical reactions these chlorofluorocarbons react to change ozone to O_2. Appreciable concentrations of chlorofluorocarbons in the ozonosphere could conceivably reduce the ozone concentration and permit large amounts of harmful ultraviolet radiation to reach the Earth. This potential pollution hazard will be discussed further in Chapter 24.

Because the ozone layer absorbs the energetic ultraviolet radiation, one can expect an increased temperature in the ozonosphere. A comparison of Figs. 20.3 and 20.5 shows that the ozone layer lies in the stratosphere. Hence the ozone absorption of ultraviolet radiation provides an explanation for the temperature increase in the stratosphere, as opposed to the continually decreasing temperatures versus altitude experienced in the neighboring troposphere and mesosphere.

This warm region of ozone concentration was first investigated by means of sound. Samuel Pepys, the noted English writer, recorded the phenomenon of being able to hear, in London, the cannons of the fleet in the English Channel during a battle of the Dutch Trade War of 1666, while people on the coast heard nothing. Similar events were observed during the Civil War and World War I. People reasoned that the sound of the cannon must have been reflected back to Earth by something in the atmosphere. Just as light is reflected from water, perhaps the sound was reflected from the boundary of warm and cold air when incident at the proper angle.

Subsequent investigations revealed the existence of the warm air layer and established the presence of ozone as the cause of the phenomenon. In cases in which sound is heard at a distance but not nearby, sound waves moving horizontally are blocked, for example, by a hill; whereas those moving vertically are reflected by the layered air and heard at a distance. This explanation is illustrated in Fig. 20.6.

In the upper atmosphere the energetic particles from the Sun cause the ionization of the gas molecules. For example,

$$N_2 + \text{energy} \longrightarrow N_2^+ + e^-$$

The electrically charged ions and electrons are trapped in the Earth's magnetic field and form ionic layers in the region of the atmosphere referred to as the *ionosphere.* This region is sometimes also referred to as the *Heaviside layer,* after Oliver Heaviside, a self-taught British physicist who in 1902 predicted the existence of an ionized layer of air and its effects on radio transmission.

Variations in the ion density with altitude give rise to the labeling of three regions, or layers—D, E, and F (Fig. 20.5). The D layer strongly absorbs radio waves below a certain frequency. Radio waves with frequencies above this value pass through the D layer but are reflected by the E and F layers, up to a limiting frequency.

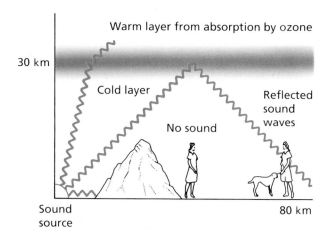

Figure 20.6 Atmospheric sound reflection.
An illustration of how sound waves may be reflected at the boundary of warm air associated with the ozone layer and be heard many miles away from the sound source.

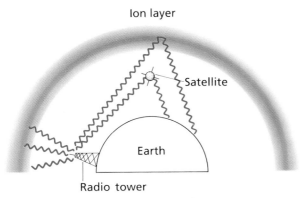

Figure 20.7 Global radio transmission.
Radio waves, which normally travel in a straight line, are reflected around the curvature of the Earth by ion layers. Ionic disturbances from solar activity may affect the ion density and disrupt communications. Transmission via satellite is not as severely affected by ionic disturbances.

Figure 20.8 Aurora.
The Aurora Borealis, or Northern Lights, as seen in Alberta, Canada.

Hence the ionosphere controls global radio communications by the reflecting of waves from ionic layers, as illustrated in Fig. 20.7.

At night, without the influx of solar radiation, most of the electrons and ions in the D and E layers recombine, and these layers virtually disappear. The F layer exhibits a smaller variation.

The reflection of radio waves from the ionic layers depends on uniformity in the density of the layer. Should a solar disturbance produce a shower of energetic particles that upsets this uniformity, a communications "blackout" may occur. Radio transmission is then reduced to line-of-sight transmission as reflections from ionic layers are impaired. These disturbances caused considerable inconvenience in early worldwide communications. Cables were laid across the oceans to keep communications open, and now satellites relay radio and television communications.

Solar disturbances are also associated with beautiful displays of light in the upper atmosphere of the polar regions called northern lights, or **aurora borealis.** See Fig. 20.8. The Southern Hemisphere tends to be forgotten by people living north of the equator. However, light displays of equal beauty occur in the southern polar atmosphere and are called **aurora australis.**

The location and height of the aurora depend on the interaction of the incoming charged particles and the Earth's magnetic field. In general, the particles are deflected toward the Earth's magnetic poles (our polar regions) over which the majority of the auroras occur. The Norsemen thought the auroras were the reflections from shields in Valhalla. Today, however, we know that they are associated with the excitation of oxygen and nitrogen molecules in the atmosphere by charged particles (electrons and protons) and the recombination of electrons and ions. The occurrence of auroras has been correlated with solar activity, which gives rise to an influx of ionizing radiation to the Earth's upper atmosphere.

20.4 **Energy Content**

The Sun is by far the most important source of energy for the Earth and its atmosphere. At an average distance of 93 million miles from the Sun, Earth intercepts only a small portion of the vast amount of the solar energy emitted. This energy traverses space in the form of radiation, and the portion incident on the Earth's atmosphere is called **insolation** (*in*coming *sol*ar radi*ation*).

Because the Earth is tilted on its axis $23\frac{1}{2}°$ with respect to the plane of its orbit about the Sun, the insolation is not evenly distributed over the Earth's surface (Fig. 20.9). This tilt, coupled with the Earth's revolution around the Sun, gives rise to the seasons.

Figure 20.9 Seasonal variation of insolation. The tilt of the Earth's axis relative to the plane of its orbit causes the incoming solar radiation to be distributed differently on the Earth's surface at different times of the year. In December, the southern hemisphere receives more direct rays from the Sun than does the northern hemisphere. In June, the situation is reversed.

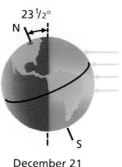

December 21

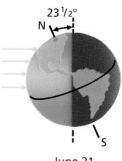

June 21

Although the Earth is closest to the Sun in January and farthest away in July, the Northern Hemisphere experiences colder weather in January because this part of the Earth is tilted away from the Sun. As a result, the days are shorter and the radiation received north of the equator is distributed over a larger area, accounting for less warming. The Southern Hemisphere at this time is enjoying warm weather. Six months later, in July, the situation is reversed for the hemispheres. Thus the tilting and resultant distribution of insolation gives rise to seasonal variations and their reversal in the Northern and Southern hemispheres.

The average summer temperature of the Southern Hemisphere is slightly higher than that of the Northern Hemisphere. This difference is, in part, due to the fact that the Earth is 5 million kilometers (about 3 million miles) closer to the Sun in January when it is summer in the Southern Hemisphere than in July because of its elliptical orbit. However, more water and less land mass in the Southern Hemisphere mean less summer-winter temperature variation.

The average intensity of solar radiation outside the Earth's atmosphere at the mean distance of the Earth from the Sun is 20 kcal/m^2·min, or 3.1×10^9 kcal/mi^2·h. This quantity is referred to as the **solar constant** and expresses the amount of energy received at normal incidence on an area per unit time. For example, let's assume that this energy intensity is received at the Earth's surface and that the area of a sunbather's back is 0.30 m^2. Every minute, he or she will receive, on the average, 6.0 kcal of energy, because

$$20 \text{ kcal/m}^2 \cdot \text{min} \times 0.30 \text{ m}^2 = 6.0 \text{ kcal/min}$$

If the bather remains in the Sun for 30 min, he or she will receive a total amount of energy of

$$6.0 \text{ kcal/min} \times 30 \text{ min} = 180 \text{ kcal}$$

This amount of energy can raise the temperature of over 9 cups of water from room temperature (20°C) to its boiling point. (See Section 5.3.) However, only about 50% or less of the insolation reaches the Earth's surface, depending on atmospheric conditions.

In considering the energy content of the atmosphere, one might think that it comes directly from insolation. However, surprisingly enough, most of the direct heating of the atmosphere comes not from the Sun but from the Earth. To understand this result, we need to examine the distribution and disposal of the incoming solar radiation (Fig. 20.10).

About 33% of the insolation received is returned to space with no appreciable effect on the atmosphere as a result of reflection by clouds, scattering by particles in the atmosphere, and reflection from terrestrial surfaces, such as water, ice, and variable ground surfaces. The reflectivity, or the average fraction of light a body reflects, is known as its **albedo** (from the Latin *albus*, "white").

Thus the Earth has an albedo of 0.33, or reflects about one-third of the incident sunlight. The brightness of the Earth as viewed from space depends on the amount of sunlight that is reflected, and clouds play an important role (Fig. 20.11). In comparison, the moon with its dark surface and no atmosphere has an albedo of only 0.07, i.e., reflects only 7% of the insolation. Even if the Earth were as small as the moon, it would appear about five times brighter than the moon to an astronaut in space.

Scattering of insolation occurs in the atmosphere from gas molecules of the air, dust particles, water drop-

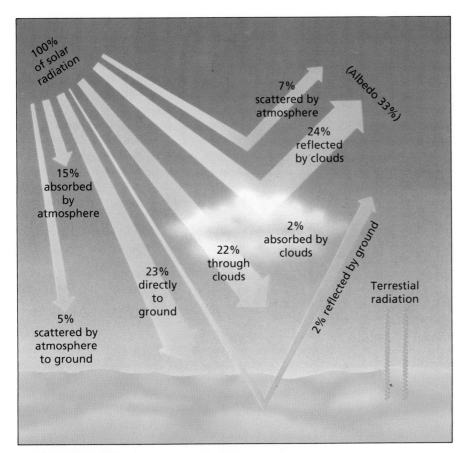

Figure 20.10 Insolation distribution.
An illustration of how the incoming solar radiation is distributed. The percentages vary somewhat, depending on atmospheric conditions.

Figure 20.11 Planet Earth.
The brightness of the Earth, as seen from space, depends on the amount of sunlight that is reflected, and clouds play an important role.

lets, etc. By scattering, we mean the absorption of incident radiation and its reradiation in all directions. As shown in Fig. 20.10, some of the scattered radiation is scattered back into space, and some is scattered toward the Earth.

The important type of scattering for our discussion is the so-called **Rayleigh scattering,** named after Lord Rayleigh (1842–1919), the British physicist who developed the theory. Lord Rayleigh showed that the scattering for particles of molecular size was proportional to $1/\lambda^4$, where λ is the wavelength of the incident light. That is, the longer the wavelength of the radiation, the less the scattering. It is this scattering that gives rise to the color of the sky. See the chapter Highlight.

About 15% of the insolation is absorbed directly by the atmosphere. Most of this absorption is accomplished by ozone, which removes the ultraviolet radiation, and by water vapor, which absorbs strongly in the infrared region of the spectrum. The major portion of the solar spectrum lies in the narrow visible region. Nitrogen and oxygen, which compose the bulk of the atomsphere, are practically transparent to visible radiation, so the major portion of the solar radiation passes through the atmosphere without appreciable absorption.

After reflection, scattering, and direct absorption of insolation by the atmosphere, about 50% of the total incoming solar radiation reaches the Earth. This radiation goes into terrestrial surface heating, primarily through the absorption of visible radiation. As we noted previously, the atmosphere—in particular, the troposphere—derives most of its energy content from the Earth. This energy absorption is accomplished in three main ways. In order of decreasing contribution, they are (1) absorption of terrestrial radiation, (2) latent heat of condensation, and (3) conduction from the Earth's surface.

HIGHLIGHT

Why the Sky Is Blue and Sunsets Are Red

The gas molecules of the air account for most of the scattering in the visible region of the spectrum. In the visible spectrum the wavelength increases from violet to red. The blue end of the spectrum is therefore scattered more than the red end. (The colors of the visible spectrum—and the rainbow—may be remembered with the help of the name of ROY G. BIV—red, orange, yellow, green, blue, indigo, and violet.)

As the sunlight passes through the atmosphere, the blue end of the spectrum is preferentially scattered. Some of this scattered light reaches the Earth, where we see it as blue skylight (Fig. 20.12). Keep in mind that all colors are present in skylight, but the dominant wavelength or color lies in the blue. You may have noticed that the skylight is more blue directly overhead or high in the sky and less blue toward the horizon, becoming white just above the horizon. You see these effects because there are fewer scatterers along a path through the atmosphere overhead than toward the horizon, and multiple scattering along the horizon path mixes the colors to give the white appearance. If the Earth had no atmosphere, the sky would appear black, other than in the vicinity of the Sun.

Because Rayleigh scattering is greater the shorter the wavelength, you might be wondering why the sky isn't violet, since this color has the shortest wavelength in the visible spectrum. Violet light is scattered, but the eye is more sensitive to blue light than to violet light; also, sunlight contains more blue light than violet light. The greatest color component is yellow-green, and the distribution generally decreases toward the ends of the spectrum.

The scattering of sunlight by the atmospheric gases *and* small particles give rise to red sunsets. One might think that because the sunlight travels a greater distance through the atmosphere to an observer at sunset, most of the shorter wavelengths would be scattered from the sunlight and only light in the red end of the spectrum would reach the observer. However, the dominant color of this light, were it due solely to molecular scattering, would be orange. Hence there must be additional scattering by small particles in the atmosphere that shifts the light from the setting (or rising) Sun toward the red. Foreign particles (natural or pollutants) in the atmosphere are not necessary to give a blue sky and even detract from it. Yet such particles are necessary for deep red sunsets and sunrises.

The beauty of red sunrises and sunsets is often made more spectacular by layers of pink-colored clouds. The cloud color is due to the reflection of red light.

Larger particles of dust, smoke, haze, and those from air pollution in the atmosphere may preferentially scatter long wavelengths. These scattered wavelengths, along with the scattered blue light due to Rayleigh scattering, can cause the sky to have a milky blue appearance—white being the presence of all colors. Hence the blueness of the sky gives an indication of atmospheric purity. Cloud droplets and raindrops scatter even longer wavelengths. This fact is used in the principle of weather radar, which is an important means of weather monitoring, as we shall learn when we examine weather forecasting in Chapter 23.

of this absorption, the lower atmosphere would become warmer and effectively hold in the Earth's heat, thus insulating the Earth. With additional insolation, the Earth would become warmer and its temperature would rise. However, according to the previous temperature-wavelength relationship, the greater the temperature, the shorter the wavelength of the emitted radiation; and so the wavelength of the terrestrial radiation would be shifted to shorter wavelengths.

The wavelength would eventually be shifted to a "window" in the absorption spectrum where little or no absorption would take place and the terrestrial radiation could pass through the atmosphere. Thus the Earth would lose energy; the temperature of the Earth would decrease and the terrestrial radiation would be shifted to a longer wavelength, which would be absorbed by the atmosphere. Averaged over the total spectrum, the selective absorption of the atmospheric gases plays an important role in maintaining the Earth's average temperature.

Humans radiate energy in the infrared region, as does the Earth. This radiation is not visible to the human eye and requires special equipment to be detected. Warm, infrared-emitting objects may be detected in the dark; hence infrared detection has come to be associated with "seeing in the dark." Because the wavelength of the radiation is dependent on the temperature of the source, infrared photographs taken from satellites are used to study temperature variations in the Earth's crust arising from such things as volcanic activity and geothermal regions (Fig. 20.14).

Latent Heat of Condensation

Approximately 70% of the Earth's surface is covered by water. Consequently, a great deal of evaporation occurs because of the isolation reaching the Earth's surface. You will recall from Chapter 5 that the latent heat of vaporization for water is 540 kcal/kg. That is, 540 kcal of heat energy are required to change 1 kg of water from a liquid to a gaseous state. Thus a large amount of energy is transferred to the atmosphere in the form of latent heat. This energy is released in the atmosphere

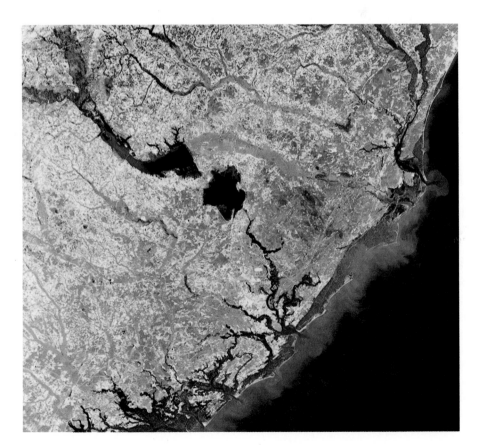

Figure 20.14 Infrared analysis.
An infrared photo, taken from a satellite, that shows the variation in temperature in the Earth's surface. Shown here is most of the coast of South Carolina, from Myrtle Beach (near upper right corner) south almost to Parris Island. Inland lakes and rivers are clearly visible.

with the condensation of water vapor in the formation of clouds, fog, dew, etc.

Conduction from the Earth's Surface

A comparatively smaller but significant amount of heat energy is transferred to the atmosphere by conduction from the Earth's surface. Because the air is a relatively poor heat conductor, this process is restricted to the layer of air in direct contact with the Earth's surface. The heated air is then transferred aloft by convection. Thus the temperature of the air tends to be greater near the surface of the Earth and decreases gradually with altitude.

The transfer of heat by conduction depends on the temperature difference between the air and the Earth's surface. (Heat energy may be transferred from the air to the ground if the latter is colder.) The heat transfer to the atmosphere varies for land and water surfaces because, in general, they have different temperatures. The specific heat of water (Section 5.3), is on the order of four times greater for water than for land. Therefore, water heats more slowly than does land given equal amounts of energy. This result is augmented by water being transparent, so the Sun's rays can penetrate, and by currents that affect mixing. Conversely, water cools more slowly than does land. Hence, neglecting seasonal influences, the land surface is warmer during the day, and the water is warmer at night. On a large scale, this effect also occurs seasonally, with summer and winter corresponding to day and night.

Because of the large surface area of the Earth, significant amounts of heat are transferred from land and sea to the atmosphere by conduction. The redistribution of this heat energy by atmospheric convection plays an important role in weather and climate conditions.

20.5 Atmospheric Measurements

Measurements of the atmosphere's properties and characteristics are very important in its study. Its properties are measured daily, and records have been compiled over many years. Meteorologists study these records with the hope of observing cycles and trends in the atmospheric behavior so as to better understand and predict its changes. We listen to the daily atmospheric readings to obtain a qualitative picture of the conditions we may experience that day. Fundamental measurements made in the atmosphere include those of (1) temperature, (2) pressure, (3) humidity, (4) wind speed and direction, and (5) precipitation.

Temperature

Temperature is related to heat energy. In matter it is a measure of the molecular activity or the kinetic energy of the molecules. The kinetic theory of gases shows the energy of gases to be directly proportional to the temperature (Section 5.7). The higher the temperature, the more energetic the gas molecules are. When we measure the temperature of a gas, some of the molecules collide with the temperature-measuring device and impart to it some of their energy. The more the molecular activity, the more collisions there are and the greater the energy transfer is, which results in a higher temperature reading.

The liquid-in-glass **thermometer** is the most common device used to measure temperature. As we discussed in Section 5.1, this thermometer consists of alcohol or mercury enclosed in a glass bulb that is attached to a capillary called the bore. The liquid expands linearly up the bore when heated, and the temperature is read from a calibrated scale on the glass of the capillary. The common temperature scales are Fahrenheit and Celsius. A thermometer in contact with the gases of the atmosphere records the air temperature, which is a measure of the energy transfer resulting from collisions of the air molecules with the glass bulb of the thermometer.

Heat transfer by radiation to a thermometer may give rise to a higher temperature reading. As a result, a thermometer in sunlight may read a comfortable 70°F (21°C), but the air may feel quite cool. A truer air temperature is often expressed in the summer by saying it is so many degrees "in the shade," implying the thermometer is not exposed to the direct rays of the Sun.

Pressure

Pressure is defined as the force per unit area ($p = F/A$). At the bottom of the atmosphere we experience the resultant weight of the gases above us. Because we experience this weight before and after birth as a natural part of our environment, little thought is given to the fact that every square inch of our bodies sustains an average weight of 14.7 lb at sea level, or a pressure of 14.7 lb/in². We refer to 14.7 lb/in² as one atmosphere of pressure.

One of the first investigations of atmospheric pressure was initiated by Galileo. In attempting to pipe water

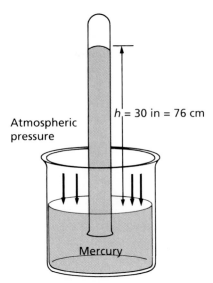

Figure 20.15 Principle of the mercury barometer.
The external air pressure on the surface of the pool of mercury supports the mercury column in the inverted tube. The height of the column thus depends on the atmospheric pressure and provides a means of measuring it.

Figure 20.16 Mercury barometers.
Two mercury barometers mounted on a wall in a laboratory.

to elevated heights by evacuating the air from a tube, he found that it was impossible to sustain a column of water over 10.4 m (34 ft) in height. Evangelista Torricelli (1608–1647), who was Galileo's successor as professor of mathematics in Florence, pointed out the difficulty through the invention of a device that exhibited the height of a liquid column to be dependent on atmospheric pressure. This device, a **mercury barometer** (from the Greek *baros*, "weight"), is still used to measure the atmospheric pressure. A filled tube of mercury was inverted into a pool of mercury. Although some ran out, a column of mercury 76 cm (30 in) high was left in the tube, as illustrated in Fig. 20.15.

A modern version of a mercury barometer is shown in Fig. 20.16. Because the column of mercury has weight, some force must hold the column up. The only available force is that of the atmospheric pressure on the surface of the mercury pool.

The pressure is related to the density ρ of the mercury and the height h of the column by the equation

$$p = \rho g h \qquad (20.1)$$

where g is the acceleration due to gravity. Hence knowing the height of the column, the density, and the acceleration due to gravity, we may calculate the pressure.

Note that on radio and TV weather reports the barometric reading is expressed in length units, so many "inches" (of mercury), because the height of the barometer column is directly proportional to the pressure.

The density of mercury is 13.6 g/cm³, and $g = 980$ cm/s²; hence

$$p = \rho g h = (13.6)(980) \times h \text{ (cm)}$$

One atmosphere of pressure supports a column of mercury approximately 30 in in height (actually 29.92 in). Thirty inches is approximately 76 cm; thus

$$p = 13.6 \text{ g/cm}^3 \times 980 \text{ cm/s}^2 \times 76 \text{ cm}$$
$$= 1.0 \times 10^6 \text{ dynes/cm}^2$$

This answer is an approximate value because of the approximate values used in the equation.

In the SI system pressure is measured in newtons per square meter (N/m²), which is given the special name of *pascal* (Pa) in honor of the French scientist Blaise Pascal (1623–1662). For example, we can inflate automobile tires to a pressure of 30 lb/in² (30 "pounds" pressure), or about 200 kPa (kilopascals). The pressure of one atmosphere is about 10^5 Pa. However, this unit is not commonly used in reporting atmospheric pressure. Also, a special name has been given to a mercury length unit. One millimeter of mercury (Hg) is called a *torr*, in honor of Torricelli, so 760 mm Hg = 760 torr.

The following list summarizes these various pressure and length units.

$$1 \text{ atmosphere (atm)} = 1,013,250 \text{ dyn/cm}^2$$
$$= 1.013 \times 10^5 \text{ Pa}$$
$$= 14.7 \text{ lb/in}^2$$
$$= 30 \text{ in Hg}$$
$$= 760 \text{ mm Hg} = 760 \text{ torr}$$

We are now in a position to understand Galileo's difficulty in not being able to sustain a column of water over 10.4 m in height. Any liquid may be used in a barometer; however, the column heights will vary because of the different densities of the liquids. Water has a density of 1 g/cm³, which is 13.6 times lighter than mercury. The atmospheric pressure then should support a column of water 13.6 times higher than the 76 cm of mercury. Thus

$$13.6 \times 76 \text{ cm} = 1033.6 \text{ cm} \ (1 \text{ m}/100 \text{ cm})$$
$$= 10.336 \text{ m, or, rounded off, } 10.4 \text{ m}$$

Hence the atmosphere does not have sufficient pressure to support a column of water over 10.4 m, just as Galileo and Torricelli found.

Meteorologists use another unit, the **millibar** (mb), to record pressure. A *bar* is defined as 10^6 dyn/cm² or 10^5 Pa, and 1 bar = 1000 mb. Because 1 atm is approximately 10^6 dyn/cm², we have

$$1 \text{ atm} \approx 1 \text{ bar or } 1000 \text{ mb}$$

and normal atmospheric pressures are of the order of 1000 mb. Actually,

$$1 \text{ atm} = 1013.25 \text{ mb}$$

Daily atmospheric disturbances also cause the barometric pressure to vary. These variations are continually monitored by weather stations and are important in predicting weather changes, as we shall learn.

Because mercury vapor is toxic and a column of 30

Figure 20.17 Aneroid barometer.
Atmospheric-pressure changes on a sensitive diaphragm are reflected on the dial face of the barometer. Fair weather is generally associated with high barometric pressure and rainy weather with low barometric pressure. (The reason will be discussed in a later chapter.)

in of mercury is awkward to handle, another type of barometer in common use is the **aneroid** (without fluid) **barometer.** It is a mechanical device having a metal diaphragm that is sensitive to pressure (Fig. 20.17). A pointer is used to indicate the pressure changes on the diaphragm. Aneroid barometers with dial faces are common around the home and are usually inlaid in wood with a dial thermometer for decorative display on the wall. Also, the altimeters used by airplane pilots and sky divers are really aneroid barometers. The pressure reading decreases rather uniformly with height in the troposphere. When the barometer dial face is replaced with one calibrated inversely in height, the barometer becomes an altimeter.

Atmospheric pressure quickly becomes evident to us when sudden changes are effected. A relatively small change in altitude will cause our ears to "pop," because the pressure in the inner ear does not equalize as quickly, which puts force on the eardrum. When the pressure equalizes (swallowing helps), the ears pop. Airplanes are equipped with pressurized cabins that maintain the normal atmospheric pressure on our bodies. The internal pressure of the body is accustomed to the external pressure of 14.7 lb/in². Should this pressure be reduced, the excess internal pressure may be evidenced in the form of a nosebleed.

Another effect of high altitude, or lower atmospheric pressure, is the reduction of the boiling points of liquids. For example, at the top of Pike's Peak (elevation 4300 m, or 14,110 ft) the atmospheric pressure is about 600 torr (mm Hg), and water boils at about 94°C rather than at 100°C. This result has an effect on cooking times. A pressure cooker will help. Why?

A frequent use of atmospheric pressure is in drinking through a straw. Most people think that sucking on the straw draws the liquid up the straw. Actually, the sucking action reduces the air pressure in the straw by removing the air molecules. The atmospheric pressure on the liquid's surface, external to the straw, then pushes the liquid up the straw. It should be evident from Galileo's experiment that water would not rise in a soda straw higher than 34 ft—even if you had a straw that long!

Humidity

Humidity is a measure of moisture, or water vapor, in the air. It affects our comfort and indirectly our ambition and state of mind. In the summer many homes have the hum of a dehumidifier that removes moisture from the air, while in the winter exposed pans of water may be strategically placed to allow the water to evaporate into the air. There are several ways to express humidity.

Absolute and Relative Humidity

Absolute humidity is simply the amount of water vapor in a given volume of air. In the United States, it is normally measured in grains per cubic foot using the British system of units.* An average value is on the order of 4.5 gr/ft³.

The most common method of expressing the water vapor content of the air is in terms of relative humidity. **Relative humidity** is the ratio of the actual moisture content of a volume of air to its maximum moisture capacity at a given temperature. Expressed in terms of a percentage, it is

$$(\%) \, RH = \frac{AC}{MC} \quad (\times \, 100\%) \quad (20.2)$$

where RH is the relative humidity,
        AC is the actual moisture content of the air,
        MC is the maximum moisture capacity of the air.

The actual moisture content is just the absolute humidity, or the amount of water vapor in a given volume of air. The maximum moisture capacity is the maximum amount of water vapor that the volume of air can hold *at a given temperature.*

Relative humidity is essentially a measure of how "full" of moisture a volume of air is at a given temperature. For example, if the relative humidity is 0.50, or 50%, then a volume of air is "half full," or contains half as much water vapor as it is capable of holding at that temperature.

To better understand how the water vapor content of air varies with temperature, consider an analogy with a saltwater solution. Just as a given amount of water at a certain temperature can dissolve so much table salt, a volume of air at a given temperature can hold only so much water vapor. When the maximum amount of salt is dissolved in solution, we say the solution is saturated. This condition is analogous to a volume of air having its maximum moisture capacity.

The addition of more salt to a saturated solution results in salt on the bottom of the container. However, more salt may be put into solution if the water is heated. Similarly, when air is heated, it can hold more water vapor. That is, warm air has a greater capacity for holding water vapor than does colder air.

Conversely, if the temperature of a nearly saturated salt solution is lowered, at a certain temperature the solution will become saturated. Additional lowering of the temperature will cause the salt to come out of solution, because it will be oversaturated. In an analogous fashion, if the temperature of a sample of air is lowered, it will become saturated at some temperature.

The temperature to which a sample of air must be cooled to become saturated is called the **dew point.** Hence at the dew point (temperature) the relative humidity is 100%. Why? Cooling below this point causes oversaturation and may result in condensation and loss of moisture in the form of precipitation.

Measurement of Humidity

Humidity may be measured by several means. For example, *absorption hygrometers* employ chemicals such as $CaCl_2$ or $Ca(NO_3)_2$ that absorb water vapor from the air and may be analyzed quantitatively. Other chemicals give a qualitative indication of the humidity by a change in some physical characteristic such as color.

---

* The small amount of moisture is expressed in grains (gr), which is a small unit (1 lb = 7000 gr). Medicines are sometimes measured in grains—e.g., a 5-gr aspirin tablet.

The most common method of measuring humidity uses the **psychrometer.** This instrument consists of two thermometers, one of which measures the air temperature while the other has its bulb surrounded by a wick that is kept wet. These thermometers are referred to as the dry bulb and wet bulb, respectively. They may be simply mounted, as in Fig. 20.18.

For accurate measurements air must move across the thermometer bulbs to give an average condition for both bulbs. The bulbs will be near equilibrium after standing for some time. Fanning may be used for air circulation, which should be continued until the wet-bulb reading reaches its lowest point. The dry bulb measures the air temperature, while the wet bulb has a lower reading that is a function of the amount of water vapor in the air.

Consider the human body's cooling mechanism. The body perspires and the evaporation of the perspiration causes cooling because of the removal of latent heat (Section 5.3). If the humidity is high and the air contains a lot of water vapor, it is difficult for evaporation from the body to take place. As a result, we say it is a hot and muggy day. However, if the humidity is low, evaporation can occur easily. We then feel cool and comfortable.

The wet bulb of the psychrometer works in a similar fashion. If the humidity is high, little water is evaporated and the wet bulb is only slightly cooled. Consequently, the temperature of the wet bulb is only slightly lower than the temperature of the dry bulb, or its reading is slightly depressed. If, however, the humidity is low, a great deal of evaporation will occur, accompanied by considerable cooling, and the wet bulb reading will be considerably depressed. Hence the temperature difference of the thermometers, or the depression of the wet bulb reading, is a measure of the humidity.

The relative humidity, the maximum moisture capacity, and the dew point (which depends on the moisture content and hence the relative humidity) may be read directly from Tables 1 and 2 in Appendix VII when the psychrometer readings are known. Consider an example in which the dry bulb has a reading of 80°F and the wet bulb a reading of 73°F. The depression of the wet bulb is then

$$80° - 73° = 7°$$

We find the dry-bulb temperature in the first column of Table 1, and the maximum moisture capacity for that temperature is 10.9 gr/ft³, as given in the adjacent column. (Recall that 1 lb = 7000 gr.)

The relative humidity is obtained from Table 1 by finding the depression of the wet bulb in the top row of the table and moving down the column to the row that corresponds to the dry-bulb reading. The relative humidity is found to be 72% for the above depression. The dew point temperature is obtained by the same procedure from Table 2 in Appendix VII, and is 70°F.

With this information, the actual moisture content may be calculated, using Eq. 20.2,

$$RH = \frac{AC}{MC}$$

where the relative humidity is expressed in decimal form. We obtain

$$AC = RH \times MC = 0.72 \times 10.9 \text{ gr/ft}^3 = 7.8 \text{ gr/ft}^3$$

This value is the actual moisture content at 80°F and should correspond to the maximum moisture capacity

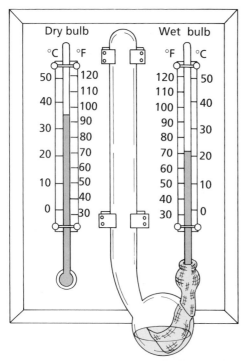

Figure 20.18 Psychrometer and relative humidity. The dry bulb of a psychrometer records the air temperature, which is greater than that shown by the wet bulb because of evaporation. The lower the humidity, the greater the evaporation and the greater the difference in the temperature readings. The temperature difference between the two thermometers is thus inversely proportional to the humidity and provides a means of measuring it.

of the air at 70°F (dew point temperature), as can be seen from the first column in Table 1.

At the dew point the air is saturated and the relative humidity is 100%. The actual moisture content is then equal to the maximum moisture capacity ($AC = MC$).

To visualize these concepts, consider a container of water that is partially full. The sides of this particular container are quite marvelous, inasmuch as they shrink when the temperature decreases. That is, the container's capacity, or the amount of water it is capable of holding, decreases. At 80°F the container has 7.8 gr of water but is capable of holding 10.9 gr. Hence the container is

$$\frac{7.8}{10.9} \times 100 = 72\% \text{ full}$$

as shown in Fig. 20.19.

But if the temperature is lowered to 70°F, the sides of the container shrink so that it can hold only 7.8 gr.

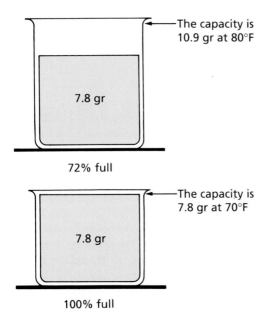

72% full

100% full

Figure 20.19 Container analogy of relative humidity. At 80°F, the beaker can hold 10.9 grains of water, but with only 7.8 grains is 72 percent full. A cubic foot of air at 80°F can hold 10.9 gr, but with only 7.8 gr, the relative humidity is 72 percent. As the temperature decreases, the sides of the beaker shrink and at 70°F it has a capacity of 7.8 gr and is 100 percent full. If the temperature of the cubic foot of air is reduced to 70° (the dew point), it can hold only 7.8 gr and the relative humidity is 100 percent.

At this special temperature (dew point), the container is 100% full (or saturated). Should the temperature be lowered still further and the sides shrink further, the water would overflow. In the analogous situation, the air generally loses its moisture in the form of precipitation when cooled below the dew point.

Temperature and humidity are important factors in influencing body comfort. As we mentioned earlier, the efficiency of the body's evaporation-cooling mechanism depends on humidity. Hence the apparent temperature, or how hot or cool one feels, depends on both temperature and humidity. Table 20.2 shows this relationship in terms of the apparent temperatures most people feel. For example, if the air temperature is 90°F and the relative humidity is 80%, the apparent temperature is 113°F. Note that the apparent temperature in the table is lower than the air temperature for low humidities. Why is that?

Keep in mind that the relative humidity reported on the weather report is the *outdoor* relative humidity. This value may bear little relationship to the value of the indoor relative humidity, particularly during cold months when the house is heated.

Wind Speed and Direction

Wind speed is measured by an **anemometer.** This instrument consists of three or four cups each attached to a rod that is free to rotate. The cups catch the wind. The greater the wind speed, the faster the anemometer rotates.

A **wind vane** indicates the direction from which the wind is blowing. This instrument is simply a freely rotating indicator that, because of its shape, lines up with the wind and points the wind direction (the direction the wind is coming *from*). Wind vanes designed around a metal rooster were once quite popular in rural areas and are usually found atop the barn. They are commonly called "weather vanes." An anemometer-and-wind-vane fixture is shown in Fig. 20.20. These instruments are common at airports where wind speed and direction are important.

Both the direction and a rough estimate of the wind s speed are given by the "wind sock," commonly seen at small airports. A rotating canvas "sock" that catches the wind can usually be seen from small aircraft or from the ground; the way it points indicates the direction in which the wind is blowing. The sock's fullness provides an estimate of wind speed.

Table 20.2 Determining the Apparent Temperature

Relative Humidity	Air Temperature (°F)										
	70	**75**	**80**	**85**	**90**	**95**	**100**	**105**	**110**	**115**	**120**
	Apparent Temperature (°F)										
0%	64	69	73	78	83	87	91	95	99	103	107
10%	65	70	75	80	85	90	95	100	105	111	116
20%	66	72	77	82	87	93	99	105	112	120	130
30%	67	73	78	84	90	96	104	113	123	135	148
40%	68	74	79	86	93	101	110	123	137	151	
50%	69	75	81	88	96	107	120	135	150		
60%	70	76	82	90	100	114	132	149			
70%	70	77	85	93	106	124	144				
80%	71	78	86	97	113	136					
90%	71	79	88	102	122						
100%	72	80	91	108							

Note: Find the temperature in the top row and the relative humidity in the left column. Read the apparent temperature where the corresponding column and row intersect. For example, if the air temperature is 90°F and the relative humidity is 80%, the apparent temperature that one feels is 113°F.

Figure 20.20 Anemometer and wind vane.
An array of two anemometers and two wind vanes are shown here. The shapes of the wind vanes cause them to point in the direction from which the wind is blowing.

Precipitation

The major forms of precipitation are rain and snow. Rainfall is measured by a **rain gauge.** This device may simply be a container with vertical sides marked in inches that is placed in an open area. After a rainfall the rain gauge is read, and the amount of precipitation is reported as so many inches of rain.* The assumption is that this much rainfall is distributed relatively evenly over the surrounding area.

If the precipitation is in the form of snow, the depth of snow (where not drifted) is reported in inches. The actual amount of water received depends on the density of the snow. If one wishes to record the water amount, a rain gauge is sprayed with a chemical that melts the snow, and the actual amount of water is reported in inches. More elaborate rain-measuring instruments automatically measure and record the rainfall or snowfall.

---

* In the United States, meteorologists commonly report precipitation in inches. Only two other countries, Brunei and the Union of Myanma (formerly known as Burma) use this system; all others report precipitation in centimeters.

Learning Objectives

After reading and studying this chapter, you should be able to do the following without referring to the text:

1. State the nature and general composition of air.

2. Explain why some planets have atmospheres and others do not.

3. Explain how the oxygen–carbon dioxide balance of the atmosphere is maintained.

4. Distinguish the properties that produce the vertical divisions of the atmosphere.

5. State the divisions of the atmosphere based on temperature.

6. Explain the distribution of insolation.

7. State and explain the processes by which the lower atmosphere obtains the major portion of its energy content.

8. Explain the greenhouse effect.

9. Name the atmospheric properties that are commonly measured, and describe how they are measured.

10. Define or explain the important words and terms listed below.

Important Words and Terms

atmosphere	mesosphere	aurora borealis	thermometer	psychrometer
atmospheric science	thermosphere	aurora australis	mercury barometer	anemometer
meteorology	homosphere	insolation	millibar	wind vane
photosynthesis	heterosphere	solar constant	aneroid barometer	rain gauge
troposphere	ozonosphere	albedo	absolute humidity	
weather	ionosphere	Rayleigh scattering	relative humidity	
stratosphere	ozone	greenhouse effect	dew point	

Questions

Composition

1. What is the difference between atmospheric science and meteorology?

2. What is the composition of the air you breathe?

3. Humans inhale oxygen and expel carbon dioxide. How is the relatively constant oxygen content of the atmosphere maintained?

Origin

4. Oxygen was probably not present in the primordial atmosphere. What was the source(s) of our atmospheric oxygen content?

5. Why is the plant pigment chlorophyll so important?

6. Why do the Earth and other planets have atmospheres, while Mercury and our moon do not?

Vertical Structure

7. Describe how the temperature of the atmosphere varies in each of the following regions: (a) the troposphere, (b) the stratosphere, (c) the mesophere, and (d) the thermosphere.

8. What is the basic distinction between the homosphere and the heterosphere?

9. How were the ozone and ion layers detected and investigated?

10. Of what importance is the atmospheric ozone layer?

11. What is believed to be the cause of the displays of light called auroras?

Energy Content

12. What is meant by insolation, and what is the solar constant?

13. From what source does the atmosphere receive most of its *direct* heating, and how is this heating accomplished (three methods)?

14. The maximum insolation is received daily around noon and yearly near June 21 (in the Northern Hemisphere). Why, then, is the maximum daily temperature around 2 or 3 P.M. and why is August the hottest month?

15. (a) Why is the sky blue?
    (b) In terms of Rayleigh scattering, explain why it is advantageous to have amber fog lights and red taillights on automobiles.

16. Explain what is meant by the "greenhouse effect".

17. How does the selective absorption of atmospheric gases provide a thermostatic action for the Earth?

18. Explain why the temperature of the troposphere decreases with altitude and why the temperature of the stratosphere increases.

Atmospheric Measurements

19. What is the principle of the liquid barometer, and what is the height of the barometer column for one atmosphere of pressure?

20. At high altitudes, would boiled food have to be cooked a greater or shorter amount of time than at sea level? Why? Explain the principle of a pressure cooker.

21. Why does water condense on the outside of a glass containing an iced drink?

22. How is it possible that it may be raining yet the relative humidity reported to be less than 100%? Explain the principle of the psychrometer.

23. Why is the apparent body temperature on a hot, humid day greater than the air temperature even when a person is in the shade?

24. Why is the apparent body temperature below that of the air temperature when the humidity is low? (See Table 20.2.)

25. Which way, relative to the wind direction, does a finned wind vane point? A wind sock?

Exercises

Vertical Structure

1. On a vertical scale of altitude in kilometers (km) above the Earth's surface at sea level, locate the heights of the following (the heights not listed may be found in the chapter).
   (a) the top of Pike's Peak
   (b) the top of Mt. Everest—29,000 ft
   (c) commercial airline flight—35,000 ft
   (d) supersonic transport (SST)—65,000 ft
   (e) communications satellite—40 mi
   (f) the E and F ion layers
   (g) auroral displays
   (h) syncom satellite—22,000 mi (satellite with synchronous period to the Earth's rotation, so it stays over one location)
   Note: Conversion factors are found on the inside back cover.

2. Express the thicknesses of the stratosphere, mesosphere, and thermosphere in terms of the thickness of the troposphere. What do these comparisons tell you?

3. If the air temperature is 70°F at sea level, what is the temperature at the top of Pike's Peak (14,000-ft elevation)? (*Hint:* The temperature decreases uniformly in the troposphere.) *Answer:* 21°F

4. A radio wave is directed vertically upward. The time lapse for its return is 7.0×10^{-4} s. From what ionic layer was the wave reflected? (*Hint:* Electromagnetic waves travel at the speed of light, 3.0×10^5 km/s, or 186,000 mi/s).

Energy Content

5. How much solar energy does the Earth's atmosphere intercept in 1 h? Assume the thickness of the atmosphere to be 600 mi. (*Hint:* Use the solar constant; consider the cross-sectional or exposed area of a sphere to be that of a circle, $A = \pi r^2$. The radius of the solid Earth is about 4000 mi.) *Answer:* 2.0×10^{20} cal

6. A satellite in orbit around the Earth near the top of the atmosphere has a solar panel array with an area of 10^4 cm² directed toward the Sun. Assuming that all of the incident insolation is absorbed, how much energy will the solar panels receive in 1 h? Express your answer in joules. *Answer:* 5×10^6 J

7. Using the facts that the energy of an emitted photon of light is $E = hf$ and that this energy is also proportional to the temperature (T) of the radiating source, show that the wavelength of the light is inversely proportional to the temperature—i.e., $\lambda \propto 1/T$.

Atmospheric Measurements

Pressure

8. A mercury barometer has a column height of 75 cm. What is the pressure in (a) dyn/cm², (b) mb, and (c) torr? *Answer:* (a) 999,600 dyn/cm² (b) 999.6 mbar (c) 750 torr

9. What would be the height of the barometer in Exercise 8 if a liquid with a density of 6.8 g/cm³ were used instead of mercury? *Answer:* 150 cm

10. The palm of a hand is about 3.0 in by 4.0 in. How many pounds of force is exerted on the palm by the atmosphere at sea level? *Answer:* 176 lb

11. The density of air at sea level is 1.2×10^{-3} g/cm³. If the atmosphere had this value as a constant density with altitude, what would be the approximate total height of the atmosphere in kilometers? (*Hint:* Use the pressure-height relationship for barometer.) *Answer:* 8.5 km

Humidity

12. On a day when the air temperature is 75°F, the wet-bulb reading of a psychrometer is 68°. Find each of the following:
    (a) relative humidity
    (b) dew point
    (c) maximum moisture capacity of the air
    (d) actual moisture content of the air
       Answer: (a) 70% (b) 64°F (c) 9.4 gr/ft^3 (d) 6.6 gr/ft^3

13. A psychrometer has a dry-bulb reading of 95°F and a wet bulb reading of 90°F. Find each of the quantities asked for in Exercise 12.

14. On a winter day a psychrometer has a dry-bulb reading of 35°F and a wet-bulb reading of 29°F.
    (a) What is the actual moisture content of the air?
    (b) Would the water in the wick of the wet bulb freeze? Explain.   *Answer:* (a) 1.1 gr/ft^3

15. On a very hot day with an air temperature of 105°F, the wet-bulb thermometer of a psychrometer records 102°F.

    (a) What is the actual moisture content of the air?
    (b) How many degrees would the air temperature have to be lowered for the actual moisture content to be equal to the maximum moisture capacity (i.e., 100% relative humidity)?   *Answer:* (a) 21.1 gr/ft^3 (b) 4°F

16. On a day when the air temperature is 85°F and the relative humidity is 80%, what are (a) the actual moisture content of the air and (b) the apparent temperature most people would feel?   *Answer:* (a) 10.2 gr/ft^3 (b) 97°F

17. What would be the apparent temperatures one would feel on days when a psychrometer has (a) a dry-bulb reading of 100°F and a wet-bulb reading of 91°F, and (b) a dry-bulb reading of 75°F and a wet-bulb reading of 60°F?
          Answer: (a) 144°F (b) 74°F

18. The dry-bulb and wet-bulb thermometers of a psychrometer both read 75°F. What are (a) the relative humidity, (b) the actual moisture content of the air, and (c) the apparent temperature?   *Answer:* (b) 9.4 gr/ft^3 (c) 80°F

CHAPTER 21

Winds and Clouds

> Who has seen the wind?
> Neither you nor I:
> But when the trees bow
> down their heads,
> The wind is passing by.
> —C. Rossetti

WINDS AND CLOUDS are common atmospheric observances. As we all know, wind is air in motion and a cloud is water droplets suspended in the air. However, there is a great deal more to these atmospheric phenomena that play important roles in our weather and environment.

Wind is the horizontal motion of air, or air motion along the Earth's surface. Vertical air motions are referred to as updrafts and downdrafts, or collectively as **air currents.** Air movement is the agent of transportation that redistributes the energy of the lower atmosphere and brings the warming influences of spring and summer and the cold chill of winter.

The effects of winds and air currents on our environment are often overlooked. Pollen carried by winds is fundamental to nature. Animals sniff the wind for scents of danger, while the transportation of radioactive particles and other contaminants may result in environmental problems. Before the invention of the combustion engine, people commonly harnessed the wind for use in transportation and developed devices that put the wind to work—for example, windmills. Occasionally, the wind causes death and destruction. Also, winds cause erosion of the Earth's surface and influence ocean currents (Chapter 29).

Winds and air currents are an integral part of cloud formation. **Clouds** are buoyant masses of visible water droplets or ice crystals. Have you ever wondered what keeps clouds afloat or how they are formed? These processes are important weather phenomena that indicate atmospheric conditions. The size, shape, and behavior of clouds are useful keys to the weather. With a little practice one may "read" the sky and predict with surprising accuracy the forthcoming weather. Cloud signs were well known to observers in ancient times, and cloud observations were probably the basis for the first attempts at weather forecasting.

Rain and other forms of precipitation originate in clouds, and this process is fundamental to life itself, because all plants and animals need water to survive. Clouds and air movement form key parts of the hydrologic cycle (Chapter 29) by which moisture is distributed over the Earth.

So as you can see, an understanding of winds and clouds is essential in the study of meteorology.

21.1 Causes of Air Motion

Winds and air currents require the air to be in motion. But what causes the air to move? As in all dynamic situations, forces are necessary to produce motion and changes in motion. The gases of the atmosphere are subject to two primary forces: (1) gravity and (2) pressure differences due to temperature variations.

The force of gravity is vertically downward and acts on each gas molecule. Although this force is often overruled by forces in other directions, the downward gravity component is ever-present and accounts for the greater density of air near the Earth.

Because the air is a mixture of gases, its behavior is governed by the gas laws of Chapter 5 and other physical principles. The pressure of a gas is directly proportional to its temperature. As defined in Section 20.5, pressure is force per unit area ($p = F/A$). Thus a temperature variation in air generally gives rise to a difference in pressure or force. It is the differences in pressures resulting from temperature differences in the atmosphere that give rise to primary air movement, both locally and on a large scale. A pressure difference corresponds to an unbalanced force; and when there is a pressure difference, the air moves from a high- to a low-pressure region.

The pressures over a region may be mapped by taking barometric readings at different locations. A line drawn through the locations (points) of equal pressure is called an **isobar.** Because all points on an isobar are of equal pressure, there will be no air movement along an isobar. The wind direction will be at right angles to

◄ The heavy winds of a tropical storm bend palm trees on an island coastline.

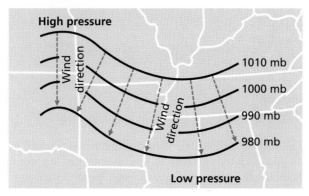

Figure 21.1 Isobars.
Isobars are lines drawn through locations having equal atmospheric pressures. The air motion, or wind direction, is perpendicular to the isobars from a region of high pressure (greater isobars) to a region of low pressure (lower isobars).

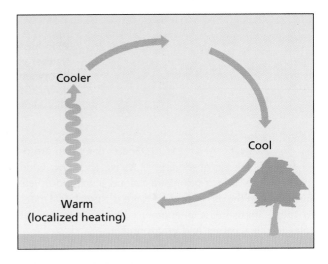

Figure 21.2 Thermal circulation.
Localized heating, which cause the air in the region to rise, initiates the circulation. As the warm air rises and cools, cooler air near the ground moves horizontally into the region vacated by the rising air. The upper, still-cooler air then descends to occupy the region vacated by the cool air. These motions set up a convection cycle.

the isobar in the direction of the lower pressure, as illustrated in Fig. 21.1.

Topographically, an isobar is analogous to a line of constant elevation. The isobars of greater value correspond to higher elevations. Just as a ball moves down a hill to a lower elevation, the air moves down a pressure hill, or pressure gradient, toward a lower isobar. An isobar of minimum pressure represents the bottom of a pressure "valley." This region is referred to as a low-pressure trough. Remember that the isobars in Fig. 21.1 represent the pressure readings at different locations. The vertical analogy is meant only as an aid in understanding the air motion along a surface of reasonably constant elevation.

Recall from Chapter 5 that the pressure *and* volume of a gas are related to its temperature ($pV \propto T$). A change in temperature, then, causes a change in the pressure and/or volume of a gas. With a change in volume, there is also a change in density, because $\rho = m/V$. Hence regions of the atmosphere with different temperatures may have different air pressures and densities. As a result, localized heating sets up air motion in a **convection cycle** or gives rise to **thermal circulation.**

To understand this concept, consider the lower left part of the cycle in Fig. 21.2. A warm land area may heat the immediate air by conduction. The warmed air expands, becomes less dense and buoyant, and rises, creating an updraft. For example, we can "see" warm air rising from a blacktop road in the summer. We see the rising "heat waves" as a result of the refraction of

light. Regions with different air temperatures and densities have different indices of refraction, and the refraction or "bending" of the light allows the rising air to be "seen."

The buoyant warm air rises and cools as it expands. Because the air rises from the heated region, the pressure is lowered here and cooler air moves horizontally toward this region of lower pressure. The region vacated by the horizontally moving air is then filled by descending cool air. These motions set up a convection cycle and the thermal circulation of air. Thermal circulation and convection cycles are important atmospheric mechanisms, as we shall see.

Once the air has been set into motion, velocity-dependent forces act. These secondary forces are (1) the Coriolis force and (2) friction.

The **Coriolis force,** named after the French engineer and mathematician who first described it, results because on the Earth an observer is in a rotating frame of reference. This force is sometimes referred to as a pseudoforce or false force, because it is introduced to account for the effect of the Earth's rotation.

Humans tend to be egocentric. We commonly consider ourselves to be motionless, although we are on a rotating Earth that has a surface velocity of about 1600 km/h (1000 mi/h) near the equator. Newton's laws of motion apply to nonaccelerating or inertial reference frames and may be used for ordinary motions on the Earth without correction. However, for high velocities or huge masses such as the atmospheric gases, the correction for the Earth's rotation becomes important.

To help you understand this effect, imagine a high-speed projectile being fired from the north pole southward along a meridian (Fig. 21.3). While the projectile travels southward, the Earth rotates beneath it, and hence it lands to the west of the original meridian. But to an observer at the north pole looking southward along the meridian, it appears that the projectile is deflected to the right. By Newton's laws, this deflection requires a force, and we call it the Coriolis force, even though no such "force" exists. Hence the Coriolis force is a pseudoforce that was invented so that the effect would be consistent with the laws of motion.

Taking into account surface velocities, projectiles at locations other than the north pole can also be shown to apparently be deflected to the right in the Northern Hemisphere. By similar reasoning, moving objects in the Southern Hemisphere appear to be deflected to the left.

Hence we say, because of the Coriolis force, that moving objects are deflected to the right in the Northern Hemisphere and to the left in the Southern Hemisphere, as observed in the direction of motion. The Coriolis force is at a right angle to the motion of the object, and its magnitude varies with latitude; it is zero at the equator and becomes greater toward the poles.

Consider this effect on wind motion. Initially, air moves toward a low-pressure area (a "low") and away from a high-pressure area (a "high"). Because of the Coriolis force, the wind is deflected, and in the Northern Hemisphere the wind rotates counterclockwise around a low and clockwise around a high as viewed from above (Fig. 21.4). These disturbances are referred to as cyclones and anticyclones, respectively. The rotations around cyclone lows and anticyclone highs are reversed in the Southern Hemisphere. Why? Water motion or currents in the oceans are also affected by the Coriolis force.

Friction, or drag, can also cause the retardation or deflection of air movements. Just as a liquid has a flow resistance caused by the internal friction of its molecules, moving air molecules experience frictional interactions among themselves or with terrestrial surfaces. The

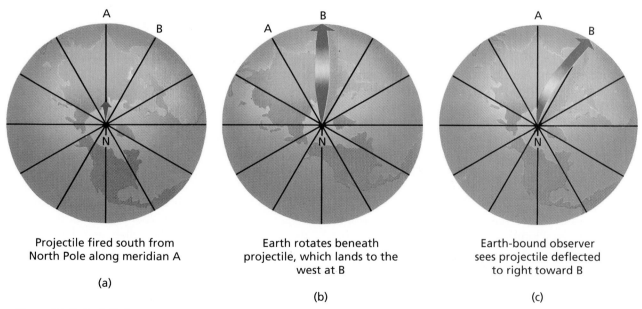

Projectile fired south from North Pole along meridian A

(a)

Earth rotates beneath projectile, which lands to the west at B

(b)

Earth-bound observer sees projectile deflected to right toward B

(c)

Figure 21.3 Coriolis force.
(a) Imagine someone firing a projectile from the North Pole at A. (b) Earth turns while the projectile is in flight and it lands to the right at B. This is the situation as you would see it viewing Earth from space over the North Pole. (c) For an observer on Earth, if the projectile lands at B it must have been deflected by some force, as required by Newton's laws. The Coriolis force was invented to account for this deflection.

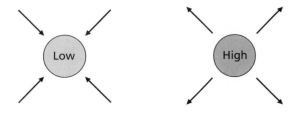

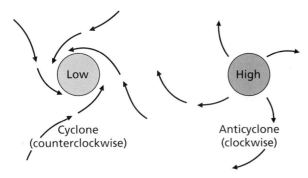

Figure 21.4 Effects of the Coriolis force on air motion. In the Northern Hemisphere, the Coriolis deflection to the right produces counterclockwise air motion around a low and clockwise rotation about a high (as viewed from above). The rotations are reversed in the Southern Hemisphere.

opposing frictional force along a surface is in the opposite direction of the air motion, and its magnitude depends on the air speed. As a result, winds moving into a cyclonic disturbance may be deflected in the direction opposite to that of the cyclonic rotation.

21.2 Local Winds and World Circulation

Air moves in all directions. Vertical air motion is important in cloud formation and precipitation, as will be discussed shortly. As you know, horizontal air movements near the surface of the Earth produce the winds that "blow," causing their presence to be felt and noticed. A wind is named after the direction or region *from* which it comes. For example, a wind blowing from north to south is referred to as a north wind. A wind blowing out to sea from the land is called a land breeze.

The strength of the wind is determined by its speed. As previously mentioned, wind speed is measured by

an anemometer; however, an estimate of the wind speed may be obtained from Table 21.1.

Wind is an important factor in how cold we feel. At moderately low temperatures in a brisk wind, we may feel extremely cold. The wind promotes the loss of body heat, which adds to the chilling effect. For example, more body heat may be lost when the temperature is 20°F and the wind speed is 18 mi/h than when the temperature is 0°F and there is no wind. Thus the temperature alone does not tell us how we should dress when going out of doors.

The effect of wind on how cold we feel may be expressed by a wind-chill index, commonly called the **wind-chill factor.** This is not to say that the wind-chill factor indicates how "chilly" one would feel, because our feeling depends on other things besides wind and temperature, such as state of nourishment, individual metabolism, and protective clothing. However, the wind-chill index in Fig. 21.5, which describes the cooling power of wind on exposed flesh at various temperatures, is a good guide about what clothing would be needed for protection on a cold, windy day.

Atmospheric pressure differences involved in thermal circulations due to geographical features give rise to local winds. As we discussed in Chapter 20, land areas heat up more quickly than water areas do. The heating of the land area gives rise to a convection cycle. As a result, during the day when the land is warmer than the water, a lake or **sea breeze** is experienced, as shown in Fig. 21.6. You may have noticed these daytime sea breezes at an ocean beach.

At night the land loses its heat more quickly than the water, and the air over the water is warmer. The convection cycle is then reversed and at night a **land breeze** blows. Sea and land breezes are sometimes referred to respectively as onshore and offshore winds.

This local heating effect is also experienced in mountain valleys. During the day the air in contact with the mountain slopes heats up and rises, producing a valley breeze. At night the slopes cool faster than the air at the same elevation over the valley, and the cool air descends the slopes, giving rise to a mountain breeze.

The heating effect that produces thermal circulation also applies on a large scale. Seasonal heating and cooling of large continental land masses initiate convection cycles. In the summer the air over the heated continents rises, creating low-pressure areas into which air flows. In winter, high-pressure areas exist over the continents, and air flows outward toward the warmer low-pressure areas over the oceans.

Table 21.1 Wind Speeds and Descriptions

Average Speed (mi/h)	Signs	Terminology
0	Smoke rises vertically	Calm
1–3	Smoke deflected	Light air
4–7	Leaves rustle	Slight breeze
8–12	Leaves and twigs move	Gentle breeze
13–18	Small branches move, dust rises	Moderate breeze
19–24	Small trees sway	Fresh breeze
25–31	Large branches move, telephone lines hum	Strong breeze
32–38	Trees in motion	Near gale
39–46	Twigs broken, walking impaired	Gale
47–54	Slight damage; e.g., to roofs and TV antennas	Strong gale
55–63	Trees uprooted, much damage	Storm
64–74	Widespread damage	Violent storm
> 74	Tropical storm or blizzard	Hurricane

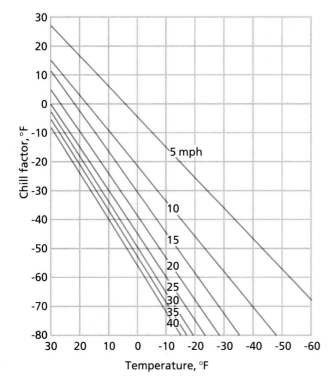

Figure 21.5 Wind-chill index.
To use this graph, find the air temperature on the bottom scale and move vertically upward to the appropriate wind-speed line. Then move horizontally to the left scale to find the chill factor.

The winds of these cycles are most pronounced on the Asian continent and are called **monsoons** (from the Arabic *mausin*, "season"). During the summer, sea air moves toward the heated continent and is called the summer monsoon. As the moist sea air travels inland toward the mountains, precipitation occurs. The sea wind or summer monsoon is associated with the wet season in southern and eastern Asia. In the winter the cycle is reversed, and then a cold dry wind blows from the mountains toward the sea. This prevailing land wind is called the winter monsoon.

Other local winds depend on the geography and acquire names particular to that area. An example is the **chinook,** or "snow eater," experienced on the eastward slopes of the Rocky Mountains. Similar winds are called *föhns* in the Alps. The winds from the Pacific Ocean blow toward and up the westward slopes of the Rockies (Fig. 21.7). Because the temperature decreases with altitude, much of the moisture is lost as rain or snow on the westward slopes and near the top of the mountains. As a result, dry air descends the eastern slopes, becoming warm with descent. This chinook wind of warm, dry air may cause a rapid rise in temperature and a quick melting of the winter snow—hence the name "snow eater."

Air motion changes locally with the altitude, the geographical features, and the seasons. However, the Earth does possess a general circulation pattern. If the Earth were completely covered by water or land and did not rotate, convection cycles would circulate

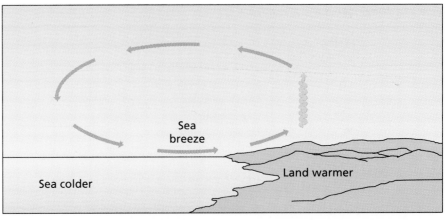

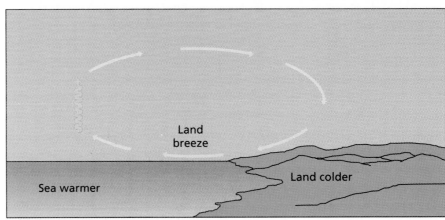

Figure 21.6 Daily convection cycles over land and water.
During the day, the land heats up more quickly than a body of water, which sets up a convection cycle, as illustrated in the upper drawing. At night the land cools more quickly than the water, which reverses the cycle, as shown by the lower drawing.

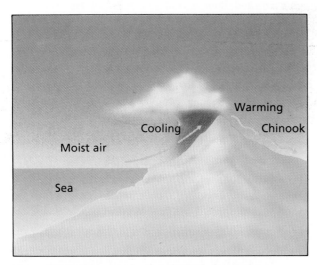

Figure 21.7 Topography for chinook formation.
After the ascending air loses moisture on the windward side of the mountains, the descending dry air is warmed on the leeward side of the mountains.

the surface air from the cold polar regions toward the equator. In this situation the prevailing winds would always assume a general north-south direction.

However, because of the Coriolis force, land and sea variations over the Earth's surface, and other complicated reasons, the hemispheric circulation is broken up into six general convectional cycles, or pressure cells. The Earth's general circulation structure is shown in Fig. 21.8. Many local variations occur within the cells, which shift seasonally in latitude because of variations in insolation. However, the prevailing winds of this semipermanent circulation structure are important in influencing general weather movement around the world.

A low-pressure belt exists in the equatorial region because of rising warm air. Direct insolation heats this region, and the air motion for the most part is rising air currents of hot, humid air. As a result, there is a vast equatorial low-pressure zone. The surface winds in this zone are light and variable, and often calm. These latitudes are referred to as the **doldrums.** Early sailing

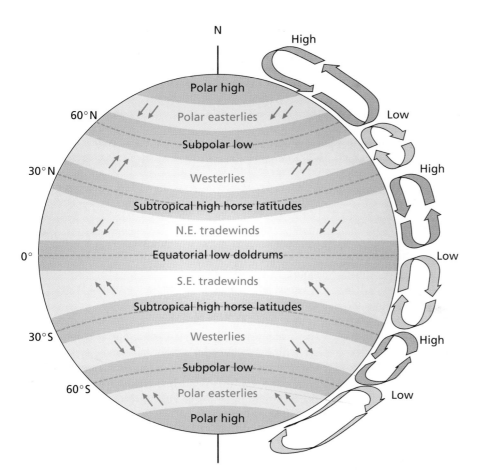

Figure 21.8 The Earth's general circulation structure.
For rather complicated reasons, the Earth's general circulation pattern has six convectional cycles. See text for description.

ships were often becalmed in the doldrums and floated listlessly for lack of wind. We use the expression "being in the doldrums" to describe a feeling or mood of bored listlessness.

The air rising from the equatorial low divides and travels horizontally north and south at high altitudes. At about 30°N and 30°S latitudes, the air cools and descends toward the surface. Hence the subtropical regions near 30°N and 30°S are characterized by descending air, which makes them high-pressure zones. There is little horizontal surface air movement. These regions became known as the **horse latitudes,** because early sailing ships were becalmed in these latitudes and their cargoes of horses had to be eaten for food or thrown overboard to conserve fresh water. These high-pressure regions of descending, warming air are the locations of some of the world's great deserts, e.g., the Sahara in North Africa and the Kalahari in southwest Africa.

The air descending in the horse latitudes moves toward the equator, completing the thermal convection cycle (see Fig. 21.8). The surface winds between the horse latitudes and the doldrums are known as the **trade winds.** These winds are generally regular and steady, and they were of considerable importance in establishing the early trade routes of sailing ships. Because of the Coriolis force, the trade winds blow, in general, from the northeast in the Northern Hemisphere and from the southeast in the Southern Hemisphere.

We might expect another single thermal circulation cell between the horse latitudes and the poles. However, for rather complicated reasons, there are two pressure cells. One is between the regions of subtropical highs of the horse latitudes (30°N and 30°S) and subpolar lows near 60°N and 60°S. In the latitudes between 30° and 60° the **westerlies** prevail. The general direction of these surface winds is again due to the Coriolis force. Because the 48 conterminous states of the United States lie in a westerly wind zone, the general movement of weather conditions across the country is from west to east.

These west winds blew the early sailing ships back across the Atlantic to Europe. The homeward wind was christened the "brave West wind." Voyages to the New World were made with the help of the gentle northeast trade winds. Care was taken to avoid the horse latitude region near 30°N, which is the latitude of the southeastern United States, by keeping within the northeast trade wind zone. We can now understand why Columbus, in sailing from Spain, arrived in the West Indies and why Spanish and Portuguese explorations were done in Central and South America.

The other sailing route to America was the northern route with the aid of the **polar easterlies.** These winds are the surface winds of the pressure cell between the subpolar lows at 60° latitude and the polar highs. By use of this route the Norse, French, and English explorations were made along the northeastern coast of Canada and the United States.

The general wind circulation has a major influence on climate. Prevailing winds affect the ocean currents, which redistribute the ocean's heat energy in much the same way as do winds in the atmosphere. There is still much more to be learned about the general circulation of the air, because its movement is sparsely monitored over many regions of the Earth.

The extent of the circulation patterns was widely noticed during early nuclear test periods when the prevailing winds carried radioactive fallout around the world. Food supplies began to show unnaturally high concentrations of radioactive materials. Of chief concern was milk and its strontium 90 concentration, which could potentially affect the consumer, especially infants. Indirectly, therefore, the Earth's general air circulation was responsible for the 1963 Nuclear Test-Ban Treaty that limited nuclear testing to underground explosions (for those countries that signed it).

21.3 Jet Streams

In the upper troposphere are fast-moving "rivers" of air called **jet streams.** The boundaries of the jet streams appear to be reasonably well defined as they move in the atmosphere in a complex fashion. They were first noted in the 1930s but did not receive much attention until a decade later, during World War II. Pilots reported that their aircraft had been held motionless by an "invisible hand" while at full throttle. The invisible hand

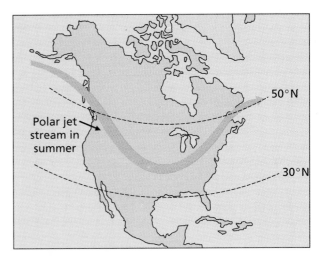

Figure 21.9 Jet stream.
A typical jet-stream pattern across the United States.

was really a jet stream head wind. Speed records were set with the help of jet stream tail winds.

The Japanese also knew of the jet streams. Near the end of the war they launched balloons carrying incendiary bombs into the jet stream blowing toward the United States across the Pacific Ocean. Special devices were automatically adjusted so that the balloons maintained the proper altitude until the mainland was reached. However, the balloons were widely scattered, and only a small percentage of them reached the West Coast, doing little damage.

Several jet streams meander like rivers around each hemisphere. The behavior of jet streams is variable and not well understood. The so-called polar jet stream moves from west to east across the United States (Fig. 21.9). It varies in altitude and latitude with the seasons. In the summer the jet stream is found in the region of 50° latitude and at an altitude of about 10 km (6 mi). In the winter the altitude decreases to below 10 km and the general location is in the region of 30° latitude. It is this jet stream that breeds winter storms and blizzards over the United States.

Mappings of the jet stream give its general dimensions and show that the wind speed increases toward the center of the stream. The air stream is 40–160 km (25–100 mi) wide and 3 km (about 2 mi) deep. As the center of the stream is approached, the winds may reach speeds of 400–500 km/h (250–300 mi/h). The general direction is mainly from west to east but shows considerable variation owing to deflection by the Coriolis force as well as by frictional forces.

Jet streams are thought to result from the general circulation structure in the regions where great high- and low-pressure areas meet. Under the proper conditions the additive circulation of a high and low could produce winds of exceedingly high speeds. The jet streams are relatively recent to meteorological observation, and our understanding of their effects on the weather is currently somewhat vague. Variations in their seasonal migrations have been associated with the severity of winters.

Some meteorologists also have speculated that intersecting jet streams may contribute to the formation of the destructive vortices of winds we call tornadoes (Section 22.3). As more is learned of their behavior, these rivers of air in our atmosphere may become routes of air transportation much as the rivers of the Earth are—especially downstream.

21.4 Cloud Classification

Clouds are classified according to their *shape, appearance, and altitude.* There are four basic root names: **cirrus** (Ci), meaning "curl" and referring to wispy, fibrous forms; **cumulus** (Cu), meaning "heap" and referring to billowy, round forms; **stratus** (St), meaning "layer" and referring to stratified or layered forms; and **nimbus** (Nb), referring to a cloud from which precipitation is occurring or threatens to occur. These root forms are then combined to describe the types of clouds and precipitation potential.

When classified according to height, clouds are sep-arated into four families: (1) **high clouds,** (2) **middle clouds,** (3) **low clouds,** and (4) **clouds with vertical development.** These families are listed in Table 21.2 along with their approximate heights and the cloud types belonging to each family. An illustration is provided for the cloud types in subsequent photographs.

High Clouds

The high clouds are made up of ice crystals because of the temperature at their altitudes. The **cirrus** member of this family is the well-known wispy mare's tail or artist's brush. They appear as though an artist had made short curling strokes with white paint on a blue background. The **cirrocumulus** clouds occur in layered patches and are referred to as mackerel scales because of their fish-scale–like structure. The cirrocumulus cloud pattern gives rise to the term *mackerel sky.* High winds at these altitudes often give the cirrocumulus clouds a wavy or ripple appearance. Cirrus and cirrocumulus clouds are shown in Fig. 21.10.

The **cirrostratus** clouds are in the form of a thin veil of ice crystals that may partially or completely cover the sky. The ice veil often gives rise to the effect of solar and lunar halos caused by the scattering of light by the ice crystals (Fig. 21.11).

This effect is analogous to a light-bulb placed behind a frosted pane of glass. Occasionally, there are enough brightly reflecting faces of ice crystals to produce a very large halo around the Sun. In some instances two bright-

Table 21.2 Cloud Families and Types

Family	Types	Illustration	Meteorological Symbols*
High clouds (above 6 km)	Cirrus (Ci)	Fig. 21.10	
	Cirrocumulus (Cc)	Fig. 21.10	
	Cirrostratus (Cs)	Fig. 21.11	
Middle clouds (1.8–6 km)	Altostratus (As)	Fig. 21.12	
	Altocumulus (Ac)	Fig. 21.13	
Low clouds (ground level–1.8 km)	Stratus (St)	Fig. 21.14	
	Stratocumulus (Sc)	Fig. 21.17	
	Nimbostratus (Ns)	Fig. 21.18	
Clouds with vertical development (5–18 km; see Section 21.5)	Cumulus (Cu)	Fig. 21.19	
	Cumulonimbus (Cb)	Fig. 21.20	

* Different symbols describe variations in cloud types. Consult meteorology texts for complete listings.

Figure 21.10 Cirrus and cirrocumulus clouds.
Artist's-brush cirrus are to the left and the mackerel-scale cirrocumulus are to the right.

Figure 21.12 Altostratus clouds.
Thick, gray altostratus clouds hide the Sun.

Figure 21.11 Cirrostratus clouds.
The clouds cover the sky and are evidenced by a lunar halo.

Figure 21.13 Altocumulus clouds.
These clouds are often rolled and arranged in flattened layers by moving air.

colored patches appear on each side of the Sun just outside the regular halo. These phenomena are known as *parhelia* (side suns); they are more commonly referred to as sun dogs.

Middle Clouds

The middle clouds are distinguished by names with the prefix *alto.* They vary considerably in shape and thickness. The **altostratus** member of this cloud family consists of layered forms of varying thickness (Fig. 21.12).

A thick layer may appear gray and dark. They often hide the Sun or moon and cast a shadow. When their thickness is such that they are translucent to the Sun or moon, an indistinctly defined halo, referred to as a corona, may be observed around the Sun or moon.

Altocumulus clouds have several varieties. They appear commonly as woolly patches or as rolled, flattened layers, as shown in Fig. 21.13. The wavy air motion resulting from air passing up and over mountains may form altocumulus clouds in a lenticular, or lens, shape on the leeward side of the mountain (see Figs. 21.23

Figure 21.14 Stratus clouds.
Low-lying stratus clouds are sometimes called high fogs.

Figure 21.15 Advection fog.
Formed when moist air moves over a cool surface, an advection fog rolls in around San Francisco Bay's Golden Gate bridge.

and 21.24). These clouds have a distinct "flying saucer" appearance.

Low Clouds

The **stratus** clouds of the low-cloud family are thin layers of water vapor. They may appear dark and are common in the winter, giving rise to the sky's "hazy shade of winter." Stratus clouds are sometimes referred to as high fogs (see Fig. 21.14).

Because fog consists of visible water droplets, it falls under the definition of a cloud. In some respects ground fog may be thought of as a low-lying stratus cloud. Most fogs are either advection or radiation fogs. **Advection fog** forms when moist air moving over a colder surface is cooled below the dew point and condensation occurs. Advection fogs "roll in," as shown in Fig. 21.15. **Radiation fog** results from stationary air overlying a surface that cools. It occurs typically in valleys, as in Fig. 21.16.

The **stratocumulus** clouds are long layers of cottonlike masses (Fig. 21.17). The rounded masses sometimes blend together to give the stratocumulus clouds a wavy appearance. When the low clouds become dark and given to precipitation, they take on the generic name of **nimbostratus.** See Fig. 21.18.

Vertical Development

The clouds of massive grandeur are those with vertical development. Formed by rising air currents, the billowy

Figure 21.16 Radiation fog.
Such fogs are commonly formed overnight in valleys when radiative heat loss cools the ground. The nearby air is also cooled, giving rise to condensation.

cumulus clouds are a common sight on a typical summer day (Fig 21.19). When the cumulus clouds turn dark and forecast an impending storm, they are referred to as **cumulonimbus** clouds. These clouds are often called thunderheads (Fig. 21.20).

Winds and air currents cause many variations in cloud shapes. The following words are used to describe these shapes: *fracto*, meaning "broken," applied to stratus and cumulus clouds; *congestus*, meaning "crowded" or "heaped," applied to cumulus clouds; *humilis*, meaning "lowly" or "poorly formed," also applied to cumulus

Figure 21.17 Stratocumulus clouds.
These clouds appear as long layers of cotton-like masses.

Figure 21.19 Cumulus clouds.
These billowy, white clouds are commonly seen on a clear day.

Figure 21.18 Nimbostratus clouds.
Dark nimbostratus clouds are given to precipitation.

Figure 21.20 Cumulonimbus cloud.
This huge cloud of vertical development has a dark nimbus lower portion, from which precipitation is occurring or threatening to occur. Such dark clouds are sometimes called thunderheads.

clouds; *mammatus,* referring to mammary-shaped clouds, as shown in Fig. 21.21.

Cloud height, or the cloud "ceiling," is important in meteorology and particularly in aviation. The height of the cloud coverage is determined by a ceilometer that operates on a reflection principle. Clouds are good reflectors of radiation. By observing the time elapsed between the emission and detection of reflected light, and knowing the speed of light, one can calculate the height of a cloud.

Figure 21.21 Cumulonimbus mammatus clouds.
This mammatus-type cloud forms on the lower surface of a cumulonimbus cloud.

21.5 Cloud Formation

To be visible as droplets, the water vapor in the air must condense. Condensation requires a certain temperature—namely, the dew point. Hence if moist air is cooled below the dew point, the water vapor contained therein will generally condense into fine droplets and form a cloud.

The air is continually in motion, and when an air mass moves into a cooler region, cloud formation may take place. Because the temperature of the troposphere decreases with height, cloud formation is associated with the vertical movement of air. In general, clouds are formed by vertical air motion and are shaped and moved about by horizontal air motion, or winds.

Vertical air motion may occur by different means, such as the vertical displacement of light warm air by heavier cold air, or by the localized heating of air, as in the case of a convection cycle. When a mass of air is heated locally, it rises. Because the heated air is less dense than the surrounding cooler air, it is therefore buoyant.

As the warm air mass ascends, it becomes cooler as a result of further expansion. It loses heat because internal heat energy is used to do work in expanding the rising air mass against the surrounding stationary air. The rate at which the air temperature decreases with height is called the **lapse rate.** The normal lapse rate in stationary air in the troposphere is on the order of $6\frac{1}{2}$°C/km ($3\frac{1}{2}$°F/1000 ft). This value may vary with latitude and changing atmospheric conditions—e.g., seasonal changes.

Because energy is used in the expansion of a warm air mass, a rising air column has a greater lapse rate than does the surrounding stationary air. Thus rising air cools more quickly. When a rising air mass cools to the same temperature as the surrounding stationary air, their densities become equal. The rising air mass then loses its buoyancy and is said to be in a *stable condition.* A heated air mass rises until stability is reached, and this portion of the atmosphere is referred to as a **stable layer,** that is, a layer of air of uniform temperature and density.

Clouds are formed when water vapor in rising air condenses into droplets that can be seen. If the rising air reaches its dew point before becoming stable, condensation occurs at that height; and the rising air carries the condensed droplets upward, forming a cloud. Hence when localized heating of an air mass occurs, one of the following happens:

1. The rising air reaches stability without condensation; that is, the rising air cools to the same temperature as the surrounding stationary air and becomes stable at some height without reaching the dew point. No condensation takes place and no cloud forms.

2. The temperature of the rising air cools to the dew point, condensation occurs, and a cloud begins to form. See Fig. 21.22. The air, still buoyant, continues to rise but cools more slowly because of heating from the latent heat of condensation. It rises until its temperature becomes equal to the surrounding stationary air.

The height at which condensation occurs is the height of the base of the cloud. The vertical distance between the level where condensation began and the level where stability is reached is the height, or thickness, of the cloud mass. We assume, for purposes of this discussion, that no precipitation occurs.

Figure 21.22 Vertical cloud development.
The cloud begins to form at the elevation at which the rising air reaches its dew point and condensation occurs. The vertical development continues until the rising air stabilizes.

By inverse reasoning of the above cloud-forming process, we see that if a cloud were carried downward by air currents, it would evaporate into the unsaturated air below the height where the cloud's temperature is equal to the dew point. Hence cloud formation indicates updrafts and dissipating clouds indicate downdrafts. Once formed, the wind shapes the clouds. Some examples of this wind action are shown in Figures 21.23 and 21.24.

Stable layers are the limits to which air may rise. Should stability occur at a low altitude, the air is held near the surface of the Earth. Under special atmospheric conditions, such as rapid radiative cooling near the Earth, the temperature may locally increase with altitude. The lapse rate is then said to be *inverted*, giving rise to a **temperature inversion** (Fig. 21.25). Local temperature inversions occur in several ways. Most common are radiation and subsidence inversions.

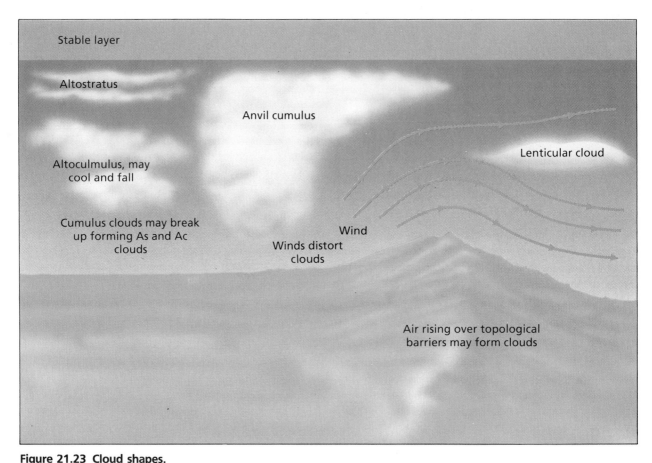

Figure 21.23 Cloud shapes.
An illustration of the formation of some different cloud shapes. Clouds are formed and shaped by vertical ascent and air motion.

Figure 21.24 Lenticular altocumulus clouds.
The wavy air motion resulting from air rising over a topo-
graphical barrier sometimes forms lens-shaped clouds.
Note the flying-saucer appearance of these clouds.

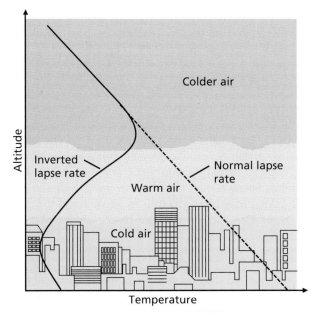

Figure 21.25 Temperature inversion.
Normally, the lapse rate near the Earth decreases uniformly with increasing altitude. However, radiative cooling
of the Earth can cause the lapse rate to become inverted,
and the temperature then *increases* with altitude (usually
below one mile). A similar condition may occur from the
subsidence of a high-pressure air mass.

Radiation temperature inversions occur daily. They
are associated with the Earth's radiative heat loss. The
ground is heated by insolation during the day, and at
night it cools by radiating heat back into the atmosphere
(see Section 20.4). If it is a clear night, the land surface
and the air near it cool quickly. The air some distance
above the surface, however, remains relatively warm,
thus giving rise to a temperature inversion. Radiation
fogs provide common evidence of this cooling effect in
valleys.

Subsidence temperature inversions occur when a
high-pressure air mass moves over a region and becomes
stationary. As the dense air settles, it becomes compressed and heated. If the temperature of the compressed
air layer exceeds the temperature of the air below it,
then the lapse rate is inverted, similar to what is shown
in Fig. 21.25.

21.6 Condensation and Precipitation

In the previous section on cloud formation, we stated
that condensation occurs when the dew point is reached.
We assumed that all the essentials for condensation were
present. However, it is quite possible for an air mass
containing water vapor to be cooled below the dew
point without condensation occurring. In this state the
air mass is said to be **supersaturated,** or **supercooled.**

How, then, are the visible droplets of water formed?
You might think that the collision and coalescing of the
water molecules would form a droplet, but this event
would require the collision of millions of molecules.
Moreover, only after a small droplet has reached a critical
size will it have sufficient binding force to retain additional molecules. The probability of a droplet forming
by this process is quite remote.

Instead, the water droplets form around microscopic
foreign particles already present in the air. These particles
on which the droplets form are called **hygroscopic nuclei.** They are present in the air in the form of dust,
combustion residue, salt from seawater evaporation, and
so forth. Because foreign particles initiate the formation
of droplets that eventually fall as precipitation, the previously mentioned mechanism for cleansing the atmosphere is readily understood.

Liquid water may be cooled below the freezing point
without the formation of ice if it does not contain the
proper type of foreign particles to act as ice nuclei. For
many years scientists believed that ice nuclei could be
just about anything, such as dust. However, research has

shown that "clean" dust without biological materials from plants or bacteria would not act as ice nuclei. This discovery is important, because precipitation involves ice crystals, as will be discussed shortly.

The droplets formed by the nucleation process are very minute, with diameters on the order of 5–200 μm. (One micrometer, μm, or "micron", is one-millionth of a meter.) Droplets of this size fall very slowly with an approximate speed of about 5 m/min, and form a fine drizzle. Because the condensation is formed in updrafts, the droplets are readily suspended in the air as a cloud. For precipitation, larger droplets or drops must form. This condition may be brought about by two processes: (1) coalescence or (2) the Bergeron process.

Coalescence

Coalescence is the formation of drops by the collision of the droplets, with the result that the larger droplets grow at the expense of the smaller ones. The efficiency of this process depends on the variation in the size of the droplets. Raindrops vary in size, reaching a diameter of approximately 7 mm. A drop 1 mm in diameter results from coalescing a million droplets of 10-μm diameter, but only 1000 droplets of 100-μm diameter would be needed. We thus see that having larger droplets greatly enhances the coalescence process.

Bergeron Process

The **Bergeron process,** named after the Swedish meteorologist who suggested it, is probably the more important process for the initiation of precipitation. This process involves clouds that contain ice crystals in their upper portions and have become supercooled in their lower portions (Fig. 21.26). Mixing or agitation of the clouds allows the ice crystals to come into contact with the supercooled vapor. Acting as nuclei, the ice crystals grow larger from the condensing vapor. The ice crystals melt into large droplets in the lower portion of the clouds and coalesce to fall as precipitation. Air currents are the normal mixing agents.

Note that there are three essentials in the Bergeron process: (1) ice crystals, (2) supercooled vapor, and (3) mixing. **Rainmaking** is based on the essentials of the Bergeron process.

The early rainmakers were mostly charlatans. With much ceremony they would beat on drums or, more sophisticatedly, fire cannons and rockets into the at-

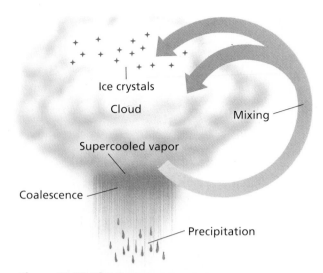

Figure 21.26 The Bergeron process.
The essence of the Bergeron process is the mixing of ice crystals and supercooled vapor, which produces water droplets and initiates precipitation.

mosphere. Explosives may supply the agitation or mixing for the Bergeron process, providing the other two essentials are present.

However, modern rainmakers approach the problem differently. There are usually enough air currents present for mixing, but the ice-crystal nuclei may be lacking. To correct this, they "seed" the cloud with silver iodide crystals or dry-ice pellets (solid CO_2). The silver iodide crystals have a crystal structure similar to that of ice and provide a substitute for the ice crystals. The silver iodide crystals are produced by a burning process. Burning may be done on the ground, with the silver iodide crystals being carried aloft by the rising warm air, or the burner may be attached to an airplane and the process carried out in the cloud to be seeded.

Dry-ice pellets are dropped into the cloud from an airplane. The temperature of solid dry ice is $-79°C$, and it quickly sublimates—that is, it goes directly from the solid to the gaseous phase. Rapid cooling associated with the sublimation triggers the conversion of supercooled cloud droplets into ice crystals. Precipitation then may occur if this part of the Bergeron process has been absent. Also, the released latent heat from the ice-crystal formation is available to set up convection cycles, which facilitate mixing. Seeding is receiving increasing attention in initiating the precipitation of fog, because fog frequently hinders airport operations.

Types of Precipitation

Precipitation requires that water vapor condense, a process that results when the vapor in ascending air cools. Three mechanisms by which air rises, thus allowing for precipitation, are convectional, orographic, and frontal. These mechanisms are illustrated in Fig. 21.27.

Convectional precipitation is a result of convection cycles. This type of precipitation predominantly occurs in the summer, because localized heating is required to initiate the convection cycle. Condensation may occur quickly because of the convectional updrafts, and the precipitation is usually confined to the local area. The sudden summer shower is an example.

Orographic precipitation arises when air is forced to rise because of land forms such as mountain ranges. The wind blows along the surface of the Earth and ascends along geographical variations. The ascending wind may give rise to orographic precipitation on the windward side of the mountain. (See Fig. 21.7.)

Frontal precipitation results from the meeting of air masses of different temperatures. Moving warm air flows up and over the cooler air because the warm air is lighter. This ascent is associated with horizontal motion, and the warm air travels upward at an angle. As a result, the cooling is usually less rapid than in the vertical convection motion.

The boundary between two air masses is called a *front* and is characterized by varying degrees of precipitation and storms. Fronts are discussed in Chapter 22.

The type of precipitation depends on atmospheric conditions, and it can occur in the form of rain, snow, sleet, hail, dew, or frost. **Rain** is the most common form of precipitation in the lower and middle latitudes. The formation of large water droplets that fall as rain has been described previously. A beautiful aftereffect sometimes observed, the **rainbow,** is discussed in the chapter Highlight.

If the dew point is below 0°C, the water vapor freezes on condensing, and the ice crystals that result fall as **snow.** In cold regions these ice crystals may fall individually, but in warmer regions the crystals become stuck together, forming a snowflake that may be as large as 1 in across. Because ice crystallizes in a hexagonal (six-sided) pattern, most snowflakes are hexagonal.

Frozen rain, or pellets of ice in the form of **sleet,**

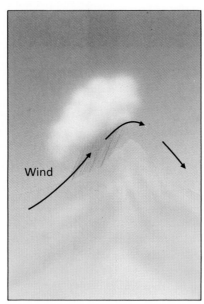
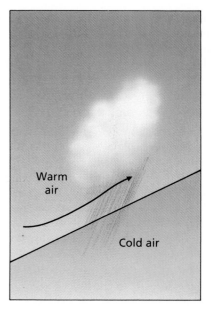

Convectional   Orographic   Frontal

Figure 21.27 Types of air ascension and precipitation.
Air may rise because of convectional heating, motion over some object, or the meeting of air masses having different temperatures, with the warmer, less dense air being displaced up and over the colder air. Precipitation resulting from these processes is referred to as convectional, orographic, and frontal, respectively.

The Rainbow

A beautiful atmospheric phenomenon commonly seen after rain is the rainbow. The colorful arc of a rainbow across the sky is the result of several optical effects: refraction, internal reflection, and dispersion (see Chapter 7). But the conditions must be just right. As we all know, a rainbow is seen after a rain but not after *every* rain.

Following a rain, there are many tiny water droplets in the air. Sunlight incident on the droplets produces a rainbow. But whether a rainbow is seen depends on the relative positions of the Sun and the observer. As you may have noticed, the Sun is generally behind you when you see a rainbow.

To understand the formation and observation of a rainbow, consider what happens when sunlight is incident on a water droplet. On entering the droplet, the light is refracted and then dispersed into component colors as it travels in the water (Fig. 21.28a). When the light enters the droplet above the critical angle, it is internally reflected and the color components emerge from the droplet at slightly different angles. Because of the conditions for refraction and internal reflection, the component colors lie in a narrow range of 40° to 42° for an observer on the ground.

Thus you see the display of colors only when the Sun is positioned so that the dispersed light is reflected to you through these angles. With this condition satisfied and an abundance of water droplets in the air, you see the colorful arc of a *primary rainbow* with colors running vertically from blue to red (Fig. 21.28a and c).

Occasionally, conditions are such that you see sunlight that has undergone two internal reflections in water droplets. The result is a vertical inver-

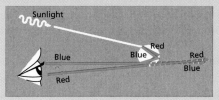

(a) Primary reflection

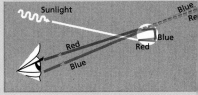

(b) Secondary reflection

Figure 21.28 Rainbow formation.
Sunlight may be internally reflected once (a) or twice (b) in a water droplet. The dispersion of the sunlight in the droplet produces the separation of colors and an observer may see an arc or bow of colors in a particular angular region (c). Both the Sun and the observer must be properly positioned in order for the observer to be able to see a rainbow, as in the photo below.

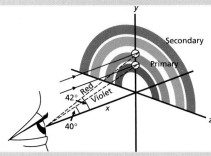

(c) Primary and secondary rainbows

sion of colors in a higher, fainter, and less frequently seen *secondary rainbow* (Fig. 21.28 b and c). Note the bright region below the primary rainbow. Light from the rainbows combines to form this illuminated region. (See the accompanying photo of primary and secondary rainbows.)

The arc length of a rainbow that you see depends on the altitude (angle above the horizon) of the Sun. As the altitude of the Sun increases, you see

less of the rainbow. On the ground you cannot see a (primary) rainbow if the altitude of the Sun is greater than 42°. The rainbow is below the horizon in this case. However, if your elevation for viewing a rainbow is increased, you see more of the rainbow arc. For instance, airplane passengers commonly view a completely circular rainbow, similar to the miniature "rainbow" that can be seen in the mist produced by a lawn sprayer.

occurs when rain falls through a cold surface layer of air and freezes or, more likely, when the ice pellets fall directly from the clouds without melting before striking the Earth. Large pellets of ice, or **hail,** result from successive vertical descents and ascents in vigorous convection cycles associated with thunderstorms. Additional condensation on successive cycles into supercooled regions that are below freezing may produce layered, structured hailstones the size of baseballs. See Fig. 21.29.

Dew is formed by atmospheric water vapor condensing on the Earth's surface. The land surfaces cool quickly at night, and the temperature near the surface may fall below the dew point. Water vapor condenses on the available surfaces such as blades of grass, giving rise to the "early morning dew."

If the dew point is below freezing, the water vapor condenses on objects in the form of frost. The white frost, sometimes referred to as hoar frost, is composed of ice crystals. **Frost** is not frozen dew but results from the direct change of water vapor into ice.

Interestingly, research has shown that frost is a result of bacteria-seeded ice formation. Without two common types of bacteria on leaf surfaces, water will not freeze at 0°C but can be supercooled to −6° to −8°C. These bacteria exist on plants, fruit trees, and so on, throughout the United States and serve as nuclei for frost formation.

With frost damage to crops and fruits exceeding $1 billion a year, scientists are exploring techniques to

Figure 21.30 Antifrost bacteria.
Genetically altered, frost-fighting bacteria being sprayed on a newly planted potato field. Such initial tests were successful in the reduction of frost formation.

Figure 21.29 Hailstones.
The successive vertical ascents of ice pellets into supercooled air and regions of condensation produce large, layered "stones" of ice.

prevent the formation of bacteria-seeded frost. One method involves the development of genetically engineered bacteria altered so that they no longer trigger ice formation. Researchers believe that a protein on the surface of the bacterium acts as the seed for the formation of frost ice crystals. By genetically removing the gene that serves as the blueprint for that protein, they hope to make "frost-free" bacteria.

Field trials for genetically altered bacteria were blocked for some time by legal actions. Some concerned citizens believe that the regular ice-seeding bacteria blown into the atmosphere may be important in precipitating rain and snow. Should the genetically engineered bacteria get into the atmosphere, people fear that the bacteria might alter the climate.

However, in the spring of 1987 antifrost bacteria were sprayed on test fields of strawberry and potato plants (Fig. 21.30). The bacteria did reduce the frost damage, and there was no evidence that the microbes had spread outside the test plots.

Learning Objectives

After reading and studying this chapter, you should be able to do the following without referring to the text:

1. Distinguish between winds and air currents.
2. State the forces that produce air motion.
3. Explain the Coriolis force and its effect on air motion around highs and lows.
4. Describe several local winds.
5. Describe the Earth's general air circulation structure.
6. Discuss jet streams.

7. Tell how clouds are classified.
8. Name and describe the various types of common clouds.
9. Explain the mechanisms by which clouds are formed.
10. State the three classifications of precipitation based on air ascension.
11. Describe the different common forms of precipitation.
12. Explain how rainbows are formed.
13. Define and explain the other important words and terms listed in the next section.

Important Words and Terms

wind	horse latitudes	cirrostratus	coalescence
air currents	trade winds	altostratus	Bergeron process
clouds	westerlies	altocumulus	rainmaking
isobar	polar easterlies	advection fog	convectional precipitation
convection cycle	jet streams	radiation fog	orographic precipitation
thermal circulation	cirrus	stratocumulus	frontal precipitation
Coriolis force	cumulus	nimbostratus	rain
friction	stratus	cumulonimbus	rainbow
wind-chill factor	nimbus	lapse rate	snow
sea breeze	high clouds	stable layer	sleet
land breeze	middle clouds	temperature inversion	hail
monsoons	low clouds	supersaturated	dew
chinook	clouds with vertical development	supercooled	frost
doldrums	cirrocumulus	hygroscopic nuclei	

Questions

Causes of Air Motion

1. What are the primary and secondary forces of air motion, and what is the basic distinction between the primary and secondary forces?
2. Explain how convection cycles are set up near the Earth.
3. (a) Explain why an object moving along a meridian in the Southern Hemisphere is deflected to the left.
   (b) Why is the Coriolis force zero at the equator?
4. Distinguish any differences between cyclones and anticyclones in the Northern and Southern hemispheres and explain.

Local Winds and World Circulation

5. In what direction does (a) a north wind and (b) a west wind blow?
6. Compare land and sea breezes with valley and mountain breezes. Explain seasonal monsoon effects.

7. What are the doldrums and the horse latitudes?
8. (a) What is the general wind direction for the conterminous United States and why?
   (b) Generally speaking, on which side of town would it be best to build a house so as to avoid smoke and other air pollutants generated in the town?
   (c) Should the prevailing wind direction be of any consideration in the heating plan and insulation of a house?
9. Explain how the wind zones of the Earth's general circulation pattern influenced the exploration and settlement of the Americas.

Jet Streams

10. What are jet streams, and how do they vary seasonally?
11. How are jet streams thought to be related to the pressure cells of the Earth's general circulation structure?

Cloud Classification

12. Name the cloud family for each of the following:
    (a) nimbostratus   (b) cirrostratus
    (c) altocumulus   (d) cumulonimbus

13. Name the cloud type associated with each of the following:
    (a) mackerel sky   (b) parhelia
    (c) solar or lunar corona   (d) the hazy shade of winter
    (e) thunderhead

Cloud Formation

14. What condition(s) is (are) necessary for cloud formation?

15. What determines the thickness of a cloud of vertical development?

16. Which is heavier, a cubic foot of dry air or a cubic foot of moist air at standard temperature and pressure? (*Hint:* Consider the fact that some of the molecules of the gases in dry air are replaced with molecules of water.)

17. How is an inverted lapse rate associated with air pollution conditions?

Condensation and Precipitation

18. What are the principles and methods of modern rain-making?

19. What are three types of precipitation based on mechanisms by which air rises?

20. (a) Is frost frozen dew?
    (b) How are large hailstones formed?

21. (a) What are the colors of the rainbow and how are rainbows formed?
    (b) Why aren't rainbows seen after every rain?

Exercises

Local Winds

1. Determine the wind-chill factor for each of the following:
   (a) a temperature of 20°F and a wind speed of 20 mi/h
   (b) a temperature of 30°F and a wind speed of 25 mi/h
   Answer: (a) $-10°$F

Cloud Classification

2. Sketch typical clouds from each of the four cloud families as a function of altitude.

Cloud Formation

Note: In finding the air temperature T at a given altitude or height h for a lapse rate R, use the simple formula $T = T_o - Rh$, where T_o is the temperature of the air at the level from which the height is measured—i.e., $h = 0$. For example, if the stationary ground temperature is 90°F, then at an altitude of 4000 ft and assuming a normal lapse rate, $T = T_o - Rh = 90°F - (3\frac{1}{2}°F/1000 \text{ ft})(4000 \text{ ft}) = 76°F$.

3. If the stationary air temperature near the ground is 65°F, what is the temperature of the air at 4000 ft? Assume a normal lapse rate.
   Answer: 51°F

4. If the stationary ground air temperature is 20°C (68°F), what is the temperature of the air at an altitude of (a) 1 km and (b) 1 mi?
   Answer: (a) 13.5°C (b) 49.5°F

5. The stationary ground air temperature is 70°F. What is the air temperature at the top of (a) Pike's Peak ($h \approx 14{,}000$ ft) and (b) Mt. Everest ($h \approx 29{,}000$ ft). Assume a normal lapse rate.
   Answer: (a) 21°F (b) $-31.5°$F

6. If the stationary air temperature near the ground is 65°F, what would be the air temperature outside (a) a commercial aircraft at 30,000 ft and (b) an SST (supersonic transport) at 50,000 ft? Assume a normal lapse rate.
   Answer: (a) $-40°$F (b) $-110°$F

7. What is the temperature at the top of the troposphere ($h = 16$ km), assuming a normal lapse rate and a stationary ground air temperature of 25°C? *Answer:* $-79°$C

8. What is the normal lapse rate in °F per mile?
   Answer: 18.8°F/mi

9. If the stationary ground temperature is 67°F, above what altitude will the clouds be in the form of ice crystals?
   Answer: 10,000 ft

10. Clouds are observed at an altitude of 2500 m. If the stationary ground air temperature is 20°C, would you expect the clouds to be water droplets or ice crystals? Justify your answer.

11. At what altitude would the air temperature be $-32°$C if on a particular day the stationary ground air temperature is 20°C. Assume a normal lapse rate. *Answer:* 8 km

12. Recall that the Celsius and Fahrenheit temperatures are equal at $-40°$. Assuming a normal lapse rate, at what altitude would this temperature occur if the stationary ground air temperature were 20°C (68°F)? Give answer in both kilometers and miles. *Answer:* 9.2 km, 5.8 mi

Air Masses and Storms

IF THE AIR of the troposphere were static, there would be little change in the local atmospheric conditions that constitute the weather. The dynamics of air movement in the form of winds were discussed in Chapter 21. In this chapter we shall consider the movements of large masses of air that have distinguishing physical characteristics.

The air we now breathe may have been far out over the Pacific Ocean a week ago. As air moves into a region, it brings with it the temperature and humidity that are mementos of its origin and travel. Cold, dry arctic air may cause sudden drops in the temperatures of the regions in its path. Warm, moist air from the Gulf of Mexico may bring heat and humidity to some regions and make the summer seem unbearable. Thus moving air transports the physical characteristics that produce changes and influence the weather. A large mass of air can influence the weather of a region for a considerable period of time or have only a brief effect. The movement of air masses depends a great deal on the Earth's air-circulation structure and its seasonal variations.

When air masses meet, the variations of their properties may trigger storms along their boundary. The types of storms depend on the properties of the air masses. Also, local variations within an air mass can give rise to storms. Storms can be violent and sometimes destructive. They remind us of the vast amount of energy contained in the atmosphere and also of its capability. The variations of our weather are therefore closely associated with air masses and their movements and interactions.

22.1 Air Masses

As we know, the weather varies and changes a great deal. However, we often experience several days of relatively uniform weather conditions. Our general weather conditions depend in large part on large air masses that move across the country.

When a body of air takes on physical characteristics that distinguish it from the surrounding air, it is referred to as an **air mass.** The main distinguishing characteristics are temperature and moisture content. A mass of air remaining for some time over a particular region takes on the physical characteristics of the surface of the region. So an air mass forms, usually near large bodies of land or water, with rather uniform properties that are primarily dependent on the uniform surface over which it is found.

The region in which an air mass derives its characteristics is called its **source region.** The time required for an air mass to take on its source region's characteristics depends on the surface of the region. If the surface area is a warm body of water, convection currents mix the air and general uniformity is attained relatively quickly. In this case the resulting air mass is warm and moist. However, if the surface area is a cold land mass, the air is more stable. In this situation the air mass takes a longer time to become cold and dry.

An air mass eventually moves from its source region, bringing its characteristics and a change in weather to the regions in its path. As an air mass travels, its properties may become modified because of local variations. For example, if Canadian polar air masses did not become warmer as they traveled southward, Florida might experience extremely cold winters.

Whether an air mass is termed cold or warm is relative to the surface over which it moves. Quite logically, if an air mass is warmer than the surface, it is referred to as a **warm air mass.** If colder, it is called a **cold air mass.** Remember, though, that these terms are relative. The cold and warm prefixes do not always imply cold and warm weather. For example, a cold air mass in winter usually brings cold, frigid weather. However, the weather associated with a cold air mass of 21°C (70°F) traveling over a 27°C (80°F) surface may be quite pleasant. Such a cold air mass may bring welcome relief in the hot summer months.

◄ **A severe storm, including tornado and lightning.**

Table 22.1 Air Masses That Affect the Weather of the United States

Classification	Symbol	Source Region
Maritime arctic	mA	Arctic regions
Continental arctic	cA	Greenland
Maritime polar	mP	Northern Atlantic and Pacific oceans
Continental polar	cP	Alaska and Canada
Maritime tropical	mT	Caribbean Sea, Gulf of Mexico, and Pacific Ocean
Continental tropical	cT	Northern Mexico, southwestern United States

Air masses are classified according to the surface and general latitude of their source regions:

Surface	*Latitude*
Maritime (m)	Arctic (A)
Continental (c)	Polar (P)
	Tropical (T)
	Equatorial (E)

The surface of the source region is abbreviated by a small letter and gives an indication of the moisture content of an air mass. An air mass forming over a body of water (maritime) would naturally be expected to have a greater moisture content than one forming over land (continental).

The general latitude of the source region is abbreviated by a capital letter and gives an indication of the temperature of an air mass. For example, mT designates a maritime tropical air mass. A list of the air masses that affect the weather of the United States is given in Table 22.1, along with their source regions, which are illustrated in Fig. 22.1.

The movement of air masses is influenced to a great extent by the Earth's general circulation patterns discussed in Chapter 21. Because the United States lies predominantly in the westerlies zone, the general movement of air masses, and hence the weather, is from west

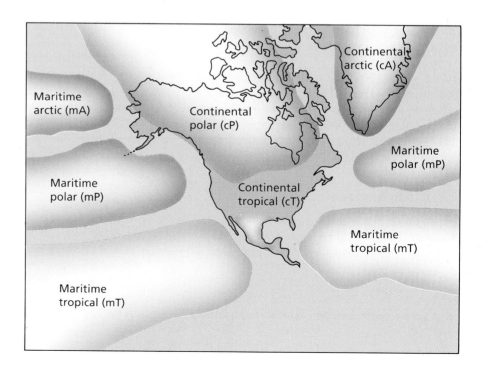

Figure 22.1 Air-mass source regions.
This map shows the source regions for the air masses of North America.

to east across the continent. The circulation zones vary to some degree in latitude with the seasons, and the polar easterlies may also move air masses into the United States during the winter.

Air masses are huge and cover thousands of square miles. Their movement is the prime factor in the redistribution of moisture and heat energy. A maritime tropical air mass formed in the Gulf of Mexico, for example, may bring warm, wet weather along the eastern coast of the United States into New England.

22.2 Fronts and Cyclonic Disturbances

As defined previously, the boundary between two air masses is called a **front**. A **warm front** is the boundary of an advancing warm air mass over a colder surface, and a **cold front** is the boundary of a cold air mass moving over a warmer surface. These boundaries, called **frontal zones,** may vary in width from a few miles to a wide zone of over 160 km (100 mi).

It is along these fronts, which divide air masses of different physical characteristics, that drastic changes of weather occur. Turbulent weather and storms usually define a front. When this turbulent weather extends for some distance horizontally, it is referred to as a **squall line.** A squall line is shown in Fig. 22.2. The series of storms that form along a front are referred to as line squalls.

The degree and rate of weather change depend on the difference in temperatures of the air masses and the

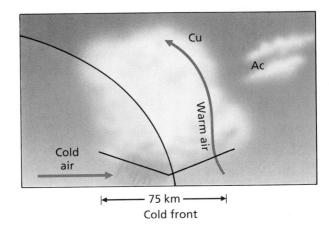

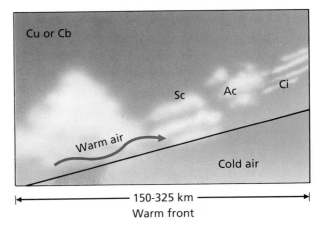

Figure 22.3 Side views of cold and warm fronts.
Notice in the upper diagram the sharp, steep boundary that is characteristic of a cold front. The boundary of a warm front, as shown in the lower diagram, is less steep. As a result, different cloud types are associated with the approach of the two types of fronts.

Figure 22.2 Squall line.
Clouds outline a squall line between fronts.

vertical slope of the front. A cold front moving into a warmer region causes the lighter, warm air to be displaced upward over the front. The lighter air of an advancing warm front cannot displace the heavier, colder air as readily and may move slowly over the colder air.

Cold air is associated with high pressure, and the downward divergent air flow in a high-pressure region in general gives a cold front a greater speed than a warm front. A cold front may have an average speed of 30 to 40 km/h (20 to 25 mi/h), while a warm front moves at 15 to 25 km/h (10 to 15 mi/h). The features of cold and warm fronts are illustrated in Fig. 22.3.

Cold fronts have sharper vertical boundaries than warm fronts, and warm air is displaced faster by an advancing cold front. As a result, cold fronts are accompanied by more violent or sudden changes in the weather. The sudden decrease in temperature is often describe as a "cold snap." Dark altocumulus clouds mark the cold front's approach. The sudden cooling and rising warm air may set off rainstorm or snowstorm activity along the front. The effects of an advancing cold front are described in the newspaper report below.

Cool Air Triggers Storms In Mid-U.S.

By THE ASSOCIATED PRESS

Cool air rolling across the nation's midsection dropped temperatures to record mid-June lows in the Plains today and triggered severe storms from Texas to Michigan.

Six young people were missing off lower Michigan's eastern shore after winds up to 65 miles an hour, and heavy rains, hit the Saginaw Bay area of Lake Huron. The group went out in a 13-foot boat Thursday afternoon, apparently to water ski.

Wind, hail and drenching rains were widespread along the storm belt.

Temperatures skidded 20 to 25 degrees in an hour in some areas as the cool front passed. Nearly 4 inches of rain soaked the Talpa, Tex., area, 40 miles south of Abilene. Almost $2\frac{1}{2}$ inches poured into Nashville, Ill., about 50 miles southeast of St. Louis.

Small hailstones piled to a depth of 2 to 3 inches during a thunderstorm at Ardmore, Okla., late Thursday. Hailstones the size of golfballs smashed out windows in the downtown section of Winters, Tex., 45 miles northeast of Abilene.

A tornado swooped into the community of Oak Creek Lake, north of San Angelo, Tex., and overturned four house trailers. No injuries were reported.

West of the storm belt, 3 inches of wet snow fell on Helena, Mont., Thursday evening. Flurries dusted scattered areas of the northern Rockies before daybreak.

The temperature fell to 42 at North Platte, Neb., before midnight, a record low for June 12 there.

Frost or freeze warnings were in effect, from Montana to northern Minnesota. Lewistown, Mont., chilled down to 29 well before dawn.

A warm front may also be characterized by precipitation and storms. Because its approach is more gradual, it is usually heralded by a period of lowering clouds. Cirrus and mackerel scales drift ahead of the front, followed by the alto clouds. As the front approaches, cumulus or cumulonimbus clouds resulting from rising air produce precipitation and storms. Most precipitation occurs before the front passes.

Storms are also associated with cyclonic, or rotational, disturbances along active fronts. **Wave cyclones** are formed by air moving in opposite directions along a front. The development of a wave cyclone is illustrated in Fig. 22.4. The graphic symbol for a cold front is ⎓⎓⎓⎓⎓. and for a warm front ⌒⌒⌒⌒⌒. The side of the line with the symbols indicates the direction of advance.

Friction and shear forces between the moving air masses cause a variation from their parallel motions and start a circulation between them. This action may be demonstrated by holding a pencil between the palms of the hands. When the palms are moved in opposite directions, a rotational motion of the pencil is produced. The analogous frontal motion is shown in Figs. 22.4a, 22.4b, and 22.4c. Accompanying this frontal motion is the rising of warm air, and a low-pressure area develops at the crest of the "wave," as illustrated in Fig. 22.4d. Air moving into this low-pressure region is deflected by the Coriolis force, producing cyclonic motion (see Chapter 21).

As a faster-moving cold front advances, it may overtake a warm air mass and push it upward. The boundary between these two masses is referred to as an **occluded front** (⎓⎓⎓⎓⎓). The cold front occludes, or closes off, the warm air along the occluded front. Horizontal views of Figs. 22.4d and 22.4e, are shown in Fig. 22.5. There are two types of occluded fronts. When the advancing cold air is colder than the air ahead of it, as in Fig. 22.5, the occluded front is called a **cold front occlusion.** That is, a cold front advances under a warm front. If the situation is reversed, and the air ahead is colder than the advancing air, the occluded front is referred to as a **warm front occlusion.** In this case a warm front advances up and over a cold front.

As the cyclonic wave fully develops, a low-pressure area or storm center spins off. See Fig. 22.4f. The low moves away, carrying with it rising air currents, clouds, precipitation, and generally bad weather. Lows, or cyclones, are therefore generally associated with poor weather. Lows may be of a local nature or cover hundreds of square miles, depending on the extent and conditions of the frontal activity.

Highs, or anticyclones, on the other hand, are generally associated with good weather. Developing in regions of divergent air flow, highs are relatively stable because of their descending air motion. As a result, dry air and a lack of precipitation usually characterize a high. There are exceptions, however, because the characteristics of a high depend on the conditions under which it develops.

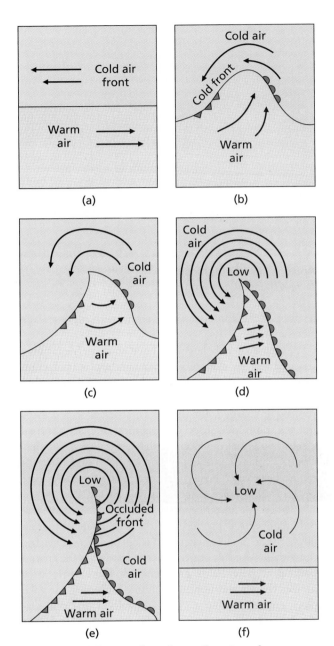

Figure 22.4 Wave-cyclone formation, top view.
Antiparallel wind motions produce the rotational motion for the development of wave cyclones (a through d). An occluded front occurs when one front overtakes the other (e). As a result of the rotational action, a low-pressure wave cyclone spins off (f).

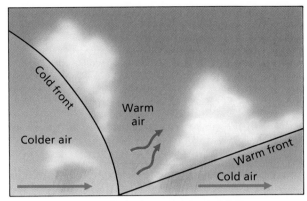

Horizontal view of Fig. 22.4(d)

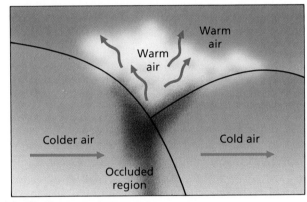

Horizontal view of Fig. 22.4(e)

Figure 22.5 Occluded-front development.
Side views of parts (d) and (e) of Fig. 22.4. The advancing colder air overtakes the retreating cold air and forces the intervening warm air upward, occluding it from the surface.

22.3 Local Storms

Storms are atmospheric disturbances associated with vertical air motion that may develop locally within a single air mass or may be due to frontal activity between two air masses. Several types of storms, distinguished by their intensity and violence, will be considered.

Rainstorms

The rate of rainfall depends upon the rate of the condensation and coalescence process, as discussed in Section 21.6. The duration of a rainfall depends on the amount of water vapor available. Should air rise rapidly, condensation and coalescence may occur quickly. The

Because of their influence on the weather, the movements of highs and lows are closely observed. Opposing fronts may balance each other so that no movement occurs. This case is referred to as a **stationary front** (△▽△▽△▽).

strong updraft will suspend the water droplets until they have coalesced into large drops. If there is a considerable amount of water vapor present, a heavy downpour of rain, called a **rainstorm,** results. Otherwise, the rainfall may be just a sudden shower that begins and ends quickly. Storms with rainfalls of 1 to 2 in/h are not uncommon.

When the convection currents cease or the cloud has an insufficient moisture content, the rainfall stops. However, if water vapor is continuously fed into the precipitating cloud, a prolonged, heavy rainfall occurs. This condition is referred to as a **cloudburst,** and the water runoff may produce a **flash flood.**

When wind accompanies a heavy rain, the storm may become violent with blowing "sheets" of rain. It beats against windows and quickly fills storm sewers. Wind damage in the form of broken tree limbs and downed power lines is common.

Thunderstorms

The **thunderstorm** is a rainstorm distinguished by thunder and lightning and sometimes hail. Thunderstorms may result from frontal cyclonic disturbances or strong local heating. Strong updrafts of air that produce the rain cloud are essential for thunderstorm formation.

In the upper portions of the resulting cumulonimbus cloud, or thunderhead, the water vapor becomes supercooled. With proper nucleation, raindrops form and fall; however, some may evaporate before reaching the ground. This evaporation results in a cooling of the surrounding air, and the cool air descends, producing downdrafts. These spread out along the ground and account for the cooling and the winds associated with thunderstorms (Fig. 22.6). Hail may result from successive ascents and descents of raindrops if the supercooled region is below freezing.

The **lightning** associated with a thunderstorm is a discharge of electrical energy. The electrical nature of lightning was demonstrated by Ben Franklin in his famous kite-flying key experiment. As he flew a kite in a thunderstorm, he drew sparks with his knuckles from a key hanging at the end of the kite cord.

Franklin was extremely lucky in the performance of this experiment in that it was a miracle that he was not electrocuted. Under no circumstances should duplication of this experiment be attempted. People have been killed trying to duplicate it—not only from lightning but from a modern hazard that did not exist in Franklin's day—

Figure 22.6 Thunderhead.
A dark thundercloud with rain.

contact of the conducting kite string with high-voltage electric lines.

A raindrop is believed to have an outer, negatively charged electric sheath. When a raindrop is broken up or partially evaporates, negatively charged droplets become separated from the drop, leaving the remainder positively charged. By this process or other processes not yet fully understood, a separation of charge and an electric potential develops in the turmoil of a thundercloud. Typically, the upper part of the cloud carries a preponderance of positive charge, while the lower part of the cloud carries a net negative charge.

When the electric potential is of sufficient magnitude, lightning is produced in the same manner that an electric spark arcs between two wires of different electric potentials. Air is a poor conductor, but when the charge buildup or potential is great enough, the electric force ionizes the air and lightning occurs. Lightning can take place entirely within a cloud (intracloud or cloud discharges), between two clouds (cloud-to-cloud discharges), between a cloud and the Earth (cloud-to-ground or ground discharges), or between a cloud and the surrounding air (air discharges) (Fig. 22.7). Lightning has even been reported to occur in clear air, apparently giving rise to the expression "a bolt from the blue." When it occurs below the horizon or behind clouds, lightning often illuminates the clouds with flickering flashes. This type commonly occurs on a still summer night and is known as **heat lightning.**

Although the most frequently occurring form of lightning is the intracloud discharge, of major concern is the lightning between a cloud and the Earth, which

Figure 22.7 Lightning.
Lightning discharges can occur between a cloud and
Earth, between clouds, and within a cloud.

is a huge reservoir of charges. Because the bottom of a
thundercloud generally is negatively charged, an opposite charge is induced on the Earth's surface and an
electric potential causes the electrical breakdown of
the air.

Each lightning stroke begins with a weakly luminous
(nearly invisible) downward discharge called a *leader*.
When the leader makes contact with the positive charges
near the Earth, large numbers of electrons flow to the
ground. This event is followed immediately by the bright
lightning flash, or *return stroke*, propagating upward as
a result of electron flow from the Earth to the cloud
along the original ionized path. We visually perceive
the stroke to be heading downward.

The shorter the distance from the cloud to the
ground, the more easily the electric discharge takes place.
For this reason lightning often strikes trees and high
buildings. It is inadvisable, therefore, to take shelter from
a thunderstorm under a tree. A person in the vicinity
of a lightning strike may experience an electric shock
that causes breathing to fail. In this case mouth-to-mouth
resuscitation or some other form of artificial respiration
should be given immediately and the person kept warm,
as treatment for shock. (See the boxed feature on lightning safety.)

Because barns and other tall buildings are the occasional targets of lightning, a common practice to avoid
lightning damage is the installation of lightning rods. A
lightning rod consists of a metal rod that extends above

Lightning Safety

If you are outside during a thunderstorm and feel
an electrical charge, as evidenced by hair standing
on end or skin tingling, what should you do? Fall
to the ground fast! Lightning may be about to
strike. Statistics show that lightning kills, on the
average, 200 people a year in the United States
and injures another 550. Most deaths and injuries
occur at home. Indoor casualties occur most frequently when people are talking on the telephone,
working in the kitchen, doing laundry, or watching
TV. During severe lightning activity, the following
safety rules are recommended.

Stay indoors away from open windows, fire-
places, and electrical conductors such as sinks and
stoves. Do not use electrical plug-in equipment
such as radios, TVs, and lamps. Also, avoid using
the telephone. Lightning may strike the telephone
lines outside.

Should you be caught outside, seek shelter in
a building. If no buildings are available, seek protection in a ditch or ravine.

the structure and is connected by a conductor to the
Earth. It is thus likely that the higher lightning rod,
rather than the building will make contact with the
discharge leader and conduct it harmlessly to Earth.

The old cliché that lightning never strikes twice in
the same place is practically true, because there is usually
nothing left of that spot for the lightning to strike again.
Actually, lightning often strikes more than once in the
same spot. The Empire State Building is struck by lightning, on the average, more than twenty times a year,
but it is protected by an elaborate lightning rod system.
(See Fig. 22.8.)

Bolts of lightning have tremendous energies. The
temperature in the vicinity of a lightning flash is estimated to be of the order of 15,000–30,000°C. This value
is quite large compared with the Sun's surface temperature of 6000°C. The awesome electrical properties of
lightning become clear when you compare it with ordinary household electricity, which has up to 240 V and
100–200 A of current available at the main service panel.
A typical lightning discharge has from 10–100 million
volts and up to 300,000 A of current. It can leap up to
a mile or more.

A lightning flash's sudden release of energy explo-
sively heats the air, producing the compressions we hear

Figure 22.8 Lightning strikes twice (and more).
The Empire State Building is struck by lightning, or the average, more than 20 times a year.

as **thunder.** When heard at a distance of about 100 m (330 ft) or less from the discharge channel, thunder consists of one loud bang, or "clap." When heard at a distance of 1 km (0.62 mi) from the discharge channel, thunder generally consists of a rumbling sound punctuated by several large claps. In general, thunder cannot be heard at distances of more than 25 km (16 mi) from the discharge channel.

Presumably, the loud bang of thunder heard when one sees a nearby lightning flash is due to the strong sound wave from the channel base. For an observer at a distance of 1 km from the lightning channel, the initial loud bang is refracted overhead because of temperature variations of the air, and the discharge begins with a rumble. At distances greater than 25 km, the sound is refracted above the observer.

Because lightning flashes generally occur near the storm center, the resultant thunder provides a method of easily approximating the distance to the storm. Light travels at approximately 297,600 km (186,000 mi)/s, and the lightning flash is seen without any appreciable time lapse. Sound, however, travels at approximately $\frac{1}{3}$ km ($\frac{1}{5}$ mi)/s, and a time lapse occurs between seeing the lightning flash and hearing the thunder. This phenomenon is also observed when watching someone at a distance fire a gun at night or hit a baseball. The report of the gun or the "crack" of the bat is always heard after the flash of the gun is observed or the baseball is well on its way.

By counting the seconds between seeing the lightning flash and hearing the thunder, you can estimate the distance of the storm center. For example, if 5 s elapsed, then the storm center would be approximately 1.6 km away, taking the velocity of sound as $\frac{1}{3}$ km/s; if 8 s elapsed, the storm would be $2\frac{1}{2}$ km away.

Ice Storms and Snowstorms

When a warm air mass overrides a cold air mass, as in Fig. 22.3, rain may form. The rain falls to Earth through the underlying cold air. If the temperature of the Earth's surface is below 0°C and the raindrops do not freeze before striking the Earth, the rain will freeze on the cold objects on which it falls. A layer of ice builds up on the objects exposed to the freezing rain. The resultant glaze is referred to as an **ice storm.** The ice layer may build up to over half an inch in thickness, depending on the magnitude of the rainfall (Fig. 22.9). Viewed in sunlight, the ice glaze produces a beautiful winter scene with the ice-coated landscape glistening in the Sun.

Other aspects of an ice storm are not so beautiful. Damage can be quite severe. Tree branches and power lines snap and fall from the weight of their ice coatings; animals may become frozen statues; transportation by foot or vehicle becomes extremely hazardous; and injuries usually result.

Snow is made up of ice crystals that fall from ice clouds. A **snowstorm** is an appreciable accumulation of snow. What may be considered a severe snowstorm in some regions may be thought of as a light snow in others where snow is more prevalent. For example, 4–6 in of snow can paralyze areas like New York City and Washington, D.C.; however, in areas like upstate New York, Michigan, and Wisconsin, this accumulation is just an incidental snowfall.

When a snowstorm is accompanied by high winds and low temperatures, the storm is referred to as a

Figure 22.9 Ice storm.
Rain freezes on cold surfaces, giving rise to layers of ice. With enough accumulation, tree limbs and utility lines sag under the weight of the ice and can snap or break.

Figure 22.10 Snow drifts.
Wind can pile snow up into deep drifts.

blizzard. The winds whip the fallen snow into blinding swirls. Visibility may be reduced to a few inches. For this reason a blizzard is often called a *blinding* snowstorm. The swirling snow causes a loss of one's sense of direction, and people have become lost within a few feet of their homes. The wind may blow the snow across level terrain, forming huge drifts against some obstructing object, as shown in Fig. 22.10. Drifting is common on the flat prairies of the western United States.

Snowstorms without appreciable winds may be equally severe because of huge quantities of snow. A single inundation of snow in western New York in 1966 measured nearly 100 in. The region to the south and east of the Great Lakes is often termed a snow belt because of the large quantities of snow it receives, especially in the early part of winter. The lakes are still comparatively warm at this time of the year, and dry continental polar air masses moving in from the northwest pick up large quantities of water from evaporation.

Orographic ascension occurs as the air masses move eastward over the elevated land area around the lakes. Hence the moist air masses are cooled and the conditions are favorable for snow. During the particularly severe winter of 1977, Buffalo received a total of nearly 200 in of snow. As the winter progresses, the water of the lakes cools and evaporation is less, thus reducing the probability of snowstorms.

For some extreme weather conditions, see the boxed feature on weather records on page 454.

Tornadoes

The tornado is the most violent of storms.

Although it may have less *total* energy than other storms (see Table 22.2), the concentration of its energy in a relatively small region gives the tornado its violent distinction. Characterized by a whirling, funnel-shaped cloud that hangs from a dark cloud mass, the tornado is commonly referred to as a **twister.** See Fig. 22.11.

Tornadoes occur around the world but are most prevalent in the United States and Australia. In the United States most tornadoes occur in the Deep South and in the broad, relatively flat basin between the Rockies and the Appalachians. But no state is immune (Fig. 22.12). The peak months of activity are April, May, and June, with southern states usually hit hardest in winter and spring, northern states in spring and summer. However,

---

Weather Records

To give you an idea of weather extremes, here are a few recorded values from around the world and in the United States.

Temperature
High: 58°C (136°F), El Aziza, Libya, September 13, 1922
      57°C (134°F), Death Valley, California, July 10, 1913
Low: −89°C (−129°F), Vostok, Antarctica, July 21, 1983
      −62°C (−80°F), Prospect Creek, Alaska, January 23, 1971
      −57°C (−70°F), Rogers Pass, Montana, January 20, 1954
Drop: 55°C in one day (7° to −49°C), Browning, Montana, January 23−24, 1916
Rise: 27°C in 2 min (−20°C at 7:30 A.M. to 43°C at 7:32 A.M.), Spearfish, South Dakota, January 22, 1943

Precipitation
Rain
1 min: 3.1 cm (1.23 in), Unionville, Maryland, July 4, 1956
24 h: 188 cm (74 in), Cilaos La Reunion, Indian Ocean, March 15−16, 1952
      109 cm (43 in), Alvin, Texas, July 26, 1979
1 month: 930 cm (366 in) Cherrapunji, India, July 1981
1 year: 2647 cm (1042 in), Cherrapunji, India, August 1860−July 1861
171 months (> 14 years): 0.08 cm (0.03 in), Arica, Chile, October 1903−January 1918

Snow
24 h: 193 cm (76 in), Silver Lake, Colorado, April 14−15, 1921
1 week: 480 cm (189 in), Mt. Shasta Ski Bowl, California, February 13−19, 1959
1 month: 991 cm (390 in), Tamarack, California, January 1911
1 season: 2850 cm (1122 in), Paradise Ranger Station, Mt. Rainier, Washington, 1971−1972
Greatest depth on ground: 1146 cm (451 in), Tamarack, California, March 11, 1911

---

tornadoes have occurred in every month at all times of the day or night. A typical time of occurrence is on an unseasonably warm, sultry spring afternoon between 3 and 7 P.M.

Most tornadoes travel from southwest to northeast, but their direction of travel can be erratic and may change suddenly. They travel at an average speed of 48 km (30 mi)/h, but speeds up to 112 km (70 mi)/h have been reported. Tornadoes are classified into three general categories: *weak, strong,* and *violent.* Most (over 60%)

Table 22.2 Comparison of Average Storm Characteristics

	Cyclones	Hurricanes	Tornadoes	Thunderstorms
Wind speed	0−80 km/h	118−320 km/h	160−480 km/h	32−50 km/h
Width	800−1600 km	480−960 km	200 m	1.6−32 km
Duration	Week or more	Week	Few minutes	An hour or less
Approx. kinetic energy, based on wind velocity (joules)*	10^{14}	10^{15}	10^{10}	10^{9}

* Energy of atomic bomb ≈ 10^{14} J.

Figure 22.11 Tornado.
This tornado funnel has touched down and is moving toward the shopping center.

Figure 22.12 Peak month of tornado activity by state in the conterminous United States.

of the tornadoes that occur each year fall into the weak category, with wind speeds of 160 km (100 mi)/h or less. About one out of every three (30%) tornadoes is classified as strong. Wind speeds reach about 320 km (200 mi)/h and strong tornadoes have an average path length of 15 km (9 mi) and a width of 183 m (200 yd). Only about 2% are violent tornadoes. These extreme tornadoes can last for hours and have an average path length of 42 km (26 mi) and a width of 390 m (425 yd). The largest of these may exceed a mile or more in width, with wind speeds approaching 480 km (300 mi)/h.

The complete mechanism of tornado formation is not known, because of its many variables. One essential component, however, is rising air, which occurs in thunderstorm formation and in the collision of cold and warm air masses. This condition is frequently fulfilled during the spring and summer in the midwestern United States, where tornadoes are common. In the case of air masses, warm air is forced upward when it collides with a cold air mass. The rapid rising of the warm air, coupled with the frictional and Coriolis forces acting on moving cooler air, produces a circulating spiral. Intersecting jet streams are also thought to contribute to the vortex formation, as shown in Fig. 22.13.

As the ascending air cools, clouds are formed that are swept to the outer portions of the cyclonic motion and outline its funnel form. Because clouds form at certain heights (Section 21.5), the funnel may appear well above the ground. Under the proper conditions a full-fledged tornado develops. The winds increase and the air pressure near the center of the vortex is reduced as the air swirls upward. When the funnel is well developed, it may "touch down" or be seen extending up from the Earth as a result of dust and debris picked up by the whirling winds.

The alerting system for tornadoes has two phases. A **tornado watch** is issued when atmospheric conditions indicate that tornadoes may form, and a **tornado warning** is issued when a tornado has actually been sighted or indicated on radar. The similarity between the terms *watch* and *warning* is sometimes confusing. Remember that you should *watch* for a tornado when the conditions are right; and when you are given a *warning*, the situation is dangerous and critical—no more watching.

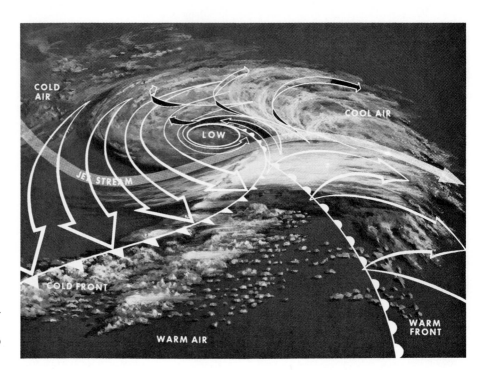

Figure 22.13 Conditions favorable for tornado formation.
Jet-stream activity and the cyclonic motion associated with frontal movement combine to produce tornadoes.

Despite radar, satellites, and other sophisticated instruments, most tornado reports are made by volunteer spotters who relay the word of sightings to official warning centers. More advanced warning of these dangerous storms may be available with the installation of new Doppler radar systems (see Chapter 23).

Most of the injuries and some of the damage done by a tornado come from flying debris in the high-speed winds. For many years people incorrectly believed that buildings literally exploded when hit by a tornado. They reasoned that the low-pressure tornado would cause a sudden outside pressure drop, and the resulting pressure difference with the normal pressure inside a structure would push the walls outward. The recommended course of action on the approach of a tornado was to open windows to equalize the pressure and prevent damage.

However, research has shown that buildings are far from airtight, even with the windows and doors closed. There are enough cracks and holes for pressure equalization. What happens in the simplest case is that the tornado's winds push the windward wall inward. The high-speed air flow over the roof lifts the roof off (much like the lift when air flows over an airplane wing), and the other walls fall outward as a result of the winds (Fig. 22.14). The action now recommended is to forget the windows and seek shelter before the tornado strikes. Flying glass from shattered windows can cause serious

Figure 22.14 The "lifting" effect of tornadoes.
The roofs of buildings are lifted up and blown away by the high-speed winds of a tornado.

injury. Moreover, at night or during heavy rain, the only clue to a tornado's presence may be its roar, and the time spent in opening windows may be critical in terms of seeking shelter. (See the boxed feature on tornado safety.)

The wind damage accompanying a tornado is enormous. A wind speed of 400 km (250 mi) or more per hour is frightening and almost inconceivable. Along with the havoc and destruction caused by the wind come stories of remarkable occurrences of a less serious nature. Chickens have been picked up by tornadoes, plucked clean, and redeposited otherwise unharmed. Livestock and barns have been lifted over fences into a neighbor's field. Newspapers have been driven deep into the wood of trees. The documented feats of tornadoes are endless.

A funnel cloud occurring over water is known as a **waterspout.** Although the funnel appears to be a spout of water rising up from the water surface, the upper portions of the waterspout are chiefly composed of whirling clouds, like that of a land tornado. Dust devils, or whirling columns of dust that are common to dry areas, are sometimes looked upon as minitornadoes, but they are not.

22.4 Tropical Storms

Tropical storms, or tropical cyclones, are more violent than the cyclones discussed previously (see again Section 22.2). A tropical storm becomes a **hurricane** when its wind speed exceeds 118 km (74 mi)/h. The hurricane is known by different names in different parts of the world. In southeast Asia it is called a **typhoon.** In the region of the Indian Ocean **cyclone** is the term used. A particularly destructive cyclone hit Bangladesh in 1985, with the loss of thousands of lives. Off the Australian coast the tropical cyclone is a **willy-willy** (Fig. 22.15). Re-

gardless of the name, the hurricane is characterized by high-speed rotating winds, whose energy is spread over a large area. A hurricane may be 480–960 km (300–600 mi) in diameter and have wind speeds of 118–320 km (74–200 mi)/h.

Hurricanes form over the tropical oceanic regions where the Sun heats huge masses of moist air. An ascending spiral motion results, in the same manner as described for tornado formation. When the moisture of the rising air condenses, the latent heat provides additional energy and more air rises up the column. This latent heat is a chief source of the hurricane's energy and is readily available from the condensation of the moist air of its source region. The growth and development of a hurricane is shown in the satellite photograph of Fig. 22.16.

Unlike the tornado, a hurricane gains energy from its source region. As more and more air rises, the hurricane grows, accompanied by clouds and increasing winds that blow in a large spiral around a relatively calm, low-pressure center—the eye of the hurricane (Fig. 22.17). The eye may be 32–48 km (20–30 mi) wide, and ships sailing into this area have found that it is usually calm and clear, with no indication of the surrounding storm. The air pressure is reduced 6–8% (to about 71 cm (28 in) of Hg) near the eye. Hurricanes move rather slowly at a few miles per hour.

Covering broad areas, hurricanes can be particularly destructive. There are winds of at least 118 km/h, but they can be much greater, up to 190–210 km (120–130 mi)/h, which are *very* dangerous. Mobile homes are particularly vulnerable to hurricane winds. The greatest threat from a hurricane's winds comes from their cargo of debris—a deadly barrage of flying missiles such as lawn furniture, signs, roofing, and metal siding (Fig. 22.18).

Hurricane winds do much damage, but drowning is the greatest cause of hurricane deaths. As the eye of the hurricane comes ashore, or "makes landfall," a great dome of water called a **storm surge,** often over 80 km (50 mi) wide, comes sweeping across the coast line. It brings huge waves and storm tides that may reach 8 m (25 ft) or more above normal. The rise may come rapidly, flooding coastal lowlands. Nine out of ten hurricane casualties are caused by the storm surge. The torrential rains that accompany the hurricane produce sudden flooding as the storm moves inland. As its winds diminish, rainfall floods constitute the hurricane's greatest threat.

Once cut off from the warm ocean, the storm begins to die, starved for water and heat energy and dragged

Figure 22.15 Tropical-storm regions of the world.
Tropical storms are known by different names in different parts of the world.

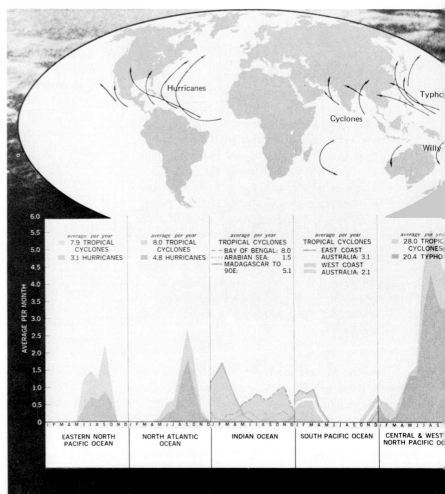

Figure 22.16 Hurricane development.
The development of a hurricane, as photographed by satellite.

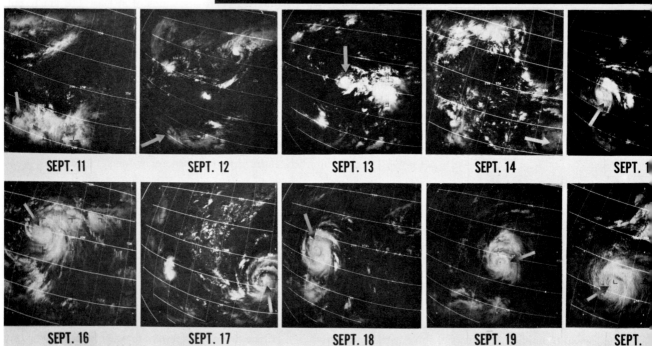

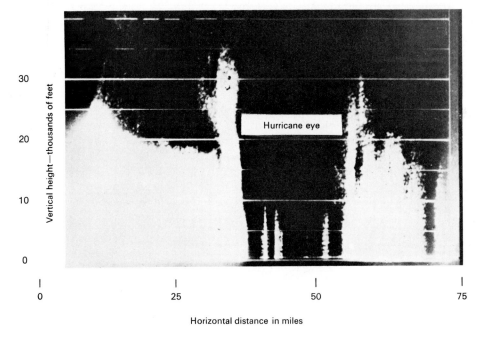

Figure 22.17 A hurricane's eye, as seen by satellite and radar.
The lower photo is a radar profile of a hurricane. Note the generally clear eye about 35 to 55 miles from the radar station and the vertical buildup of clouds (over 35,000 ft) at the near side of the eye.

Figure 22.18 Hurricane damage.
An aerial view of housing destruction after a hurricane.

apart by friction as it moves over the land. Even though a hurricane weakens rapidly as it moves inland, the remnants of the storm can bring 6–12 in of rain or more to the areas they cross.

Tropical storms form in the regions between 5° and 20° latitude in the late summer (Fig. 22.19). At latitudes less than 20° the stabilizing influence of the high-pressure horse latitudes near 30° is small, and large quantities of moisture are evaporated from the warm tropical oceans. Thus in these regions the conditions for hurricane formation are optimum.

The breeding grounds of the hurricanes that affect the United States are in the Atlantic Ocean southeast of the Caribbean Sea. As the hurricane forms, it moves westward with the trade winds, traveling about 160–320 km (100–200 mi) a day. The Coriolis force acts on this air movement, deflecting it toward the northwest. As a result, the hurricanes move away from the equator and into the Gulf of Mexico as they travel westward. Monsoon-type effects may also have some influence on the hurricane movement toward land.

Those hurricanes that do strike the United States usually hit along the Gulf and south Atlantic coasts. These areas may be lashed by several hurricanes during the late summer hurricane season. Once over a land mass and deprived of its moist-air source of energy, the hurricane dissipates. However, if the hurricane proceeds far enough northward, it may be blown back into the Atlantic by the prevailing westerlies that lie above 30°N. As a result, the coastal area may be hit by a hurricane both coming and going. (See Fig. 22.19.)

During the hurricane season the area of their formation is constantly monitored by satellite. Because of the long time required for their development and migration, the hurricane requires a great deal of observation. Radar-equipped airplanes, or "hurricane hunters,"

Figure 22.19 Hurricane paths.
Hurricanes formed in the Atlantic Ocean generally travel northwest until they are blown eastward by the prevailing westerlies. As a result, hurricanes often strike the coastal areas of the southeastern United States both coming and going.

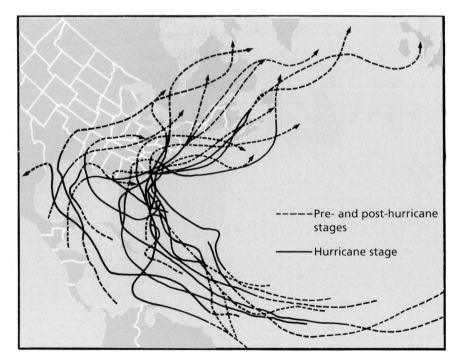

- - - - Pre- and post-hurricane stages

——— Hurricane stage

Table 22.3 The Five-Year List of Names for Atlantic Storms

1990		1991		1992		1993		1994	
Ana	Larry	Allen	Lisa	Arlene	Lenny	Alberto	Leslie	Alicia	Luis
Bob	Mindy	Bonnie	Mitch	Bret	Maria	Beryl	Michael	Barry	Marilyn
Claudette	Nicolas	Charley	Nicole	Cindy	Nate	Chris	Nadine	Chantal	Noel
David	Odette	Danielle	Otto	Dennis	Ophelia	Debby	Oscar	Dean	Opal
Elena	Peter	Earl	Paula	Emily	Philippe	Ernesto	Patty	Erin	Pablo
Frederic	Rose	Frances	Richard	Floyd	Rita	Florence	Rafael	Felix	Roxanne
Gloria	Sam	Georges	Shary	Gert	Stan	Gilbert	Sandy	Gabrielle	Sebastien
Henri	Teresa	Hermine	Tomas	Harvey	Tammy	Helene	Tony	Hugo*	Tanya
Isabel	Victor	Ivan	Virginia	Irene	Vince	Isaac	Valarie	Iris	Van
Juan	Wanda	Jeanne	Walter	Jose	Wilma	Joan	William	Jerry	Wendy
Kate		Karl		Katrina		Keith		Karen	

* Because Hurricane Hugo, of September 1989, was a particularly devastating storm, this name will probably be retired and replaced by another name beginning with *H*.

seek out and track tropical storms and hurricanes. Prior to the advent of satellites, hurricanes were almost exclusively monitored by aircraft and surface ships. This is a dangerous business. However, in order to predict a hurricane's behavior, on-the-spot weather data not furnished by satellites, such as air pressure and wind velocity, are essential. These observations permit advanced hurricane warnings along coastal areas, which save lives and reduce property damage. The course of a hurricane is constantly observed, because it may suddenly change its path and strike elsewhere with seemingly malicious intent.

Like that for the tornado, the hurricane alerting system has two phases. A **hurricane watch** is issued for a coastal area when there is a threat of hurricane conditions within 24–36 h. This alert does not mean that hurricane conditions are a certainty, but a watch does caution those people who would be affected so that they can be prepared to act quickly. A **hurricane warning** indicates that hurricane conditions are expected within 24 h and appropriate precautionary actions should be taken.

For several hundred years many hurricanes in the West Indies were named after the particular saint's day on which the hurricane occurred. In 1953 the National Weather Service began to use female names for tropical storms and hurricanes. This practice began in World War II when military personnel named typhoons in the western Pacific after their wives and girl friends. Under this method the storms were named in alphabetical order. The first hurricane of a season received a name beginning with A, such as Alma; the second one was given a name beginning with B, such as Beulah; and so on.

The practice of naming hurricanes solely after women was changed in 1979 when men's names were included in the lists. A five-year list of names for Atlantic storms is given in Table 22.3. A similar list is available for eastern Pacific storms. Names beginning with the letters Q, U, X, Y, and Z are excluded because of their scarcity. The name lists have an international flavor because hurricanes affect other nations and are tracked by the public and the weather services of countries other than the United States. The lists are recycled. For example, the 1990 list will be used again in 1995.

Learning Objectives

After reading and studying this chapter, you should be able to do the following without referring to the text:

1. Distinguish the different air masses, particularly those that affect the weather in the United States.

2. Describe the different kinds of fronts and their associated characteristics.

3. Explain the formation of a wave cyclone.

4. Describe the formation and characteristics of a thunderstorm.

5. Explain the cause of ice storms.

6. Discuss the formation and effects of tornadoes.

7. Distinguish between tropical storms and hurricanes, and describe their formation and effects.

8. Define and explain the other important words and terms listed in the next section.

Important Words and Terms

air mass	warm front occlusion	tornado
source region	stationary front	twister
warm air mass	rainstorm	tornado watch
cold air mass	cloudburst	tornado warning
front	flash flood	waterspout
warm front	thunderstorm	hurricane
cold front	lightning	typhoon
frontal zones	heat lightning	cyclone
squall line	thunder	willy-willy
wave cyclones	ice storm	storm surge
occluded front	snowstorm	hurricane watch
cold front occlusion	blizzard	hurricane warning

Questions

Air Masses

1. How are air masses classified? Explain the relationship between air-mass characteristics and source regions.

2. What air masses affect the weather in the United States?

3. What would be the abbreviated designation of an air mass that developed over Siberia? over northern Africa? over the Arabian Sea?

Fronts and Cyclonic Disturbances

4. What is a front? List the meteorological symbols for four types of fronts.

5. Describe the characteristics and weather associated with cold and warm fronts. What is the significance of the sharpness of their vertical boundaries?

6. Assume a low-pressure air mass approaches and advances over your location. Give the variations in barometric readings and winds you would expect to occur.

7. Explain why the household barometer often lists descriptives adjectives, such as stormy, unsettled, and fair, on the barometer's face instead of direct pressure readings.

Local Storms

8. How do conditions for lightning discharges occur?

9. What type of first aid should be given to someone suffering from the shock of a lightning stroke?

10. What is thunder, and why is it sometimes heard as a crash and other times as a rumble?

11. An ice storm is likely to result along what type of front? Explain.

12. What is the most violent of storms?

13. Describe the formation and characteristics of a tornado.

14. Distinguish between a tornado watch and a tornado warning.

15. For some years we have been told to open the windows of a house on the approach of a tornado. Now we are told to forget the windows and seek shelter. Why did the recommendation change?

16. When a tornado approaches, what should the residents do in (a) a house with a basement, (b) a house without a basement, and (c) a mobile home?

Tropical Storms

17. What is the major source of energy for a tropical storm?

18. When does a tropical storm become a hurricane?

19. What is the major cause of damage to life and property when a hurricane "makes landfall"?

20. Why do hurricanes often enter the Gulf of Mexico and then move back toward the northeast?

21. How are tropical storms and hurricanes named?

Exercises

Air Masses

1. Locate the source regions for the following air masses that affect the United States: (a) cA (b) mP (c) cT (d) mT (e) cP

2. What would be the classifications of air masses forming over the following areas?
   (a) Sahara Desert   (b) Antarctic Ocean
   (c) Greenland   (d) Mid-Pacific Ocean
   (e) Siberia

Fronts

3. About how far does the average cold front travel in one day?   *Answer:*   800 km (500 mi)

4. About how far does the average warm front travel in one day?   *Answer:*   500 km (300 mi)

5. Sketch top and horizontal views of a warm front occlusion.

Local Storms

6. Using the ideal gas law (Chapter 5), approximate the increase in pressure in the vicinity of a lightning flash. (*Hint:* Assume a constant volume.)
   Answer: A factor of 50

7. If thunder is heard 4 s after the observation of a lightning flash, approximately how far away is the storm center?
   Answer: 1.3 km

8. While picnicking on a summer day, you hear thunder 11 s after you see a lightning flash from an approaching storm. Approximately how far away is the storm?
   Answer: 3.7 km

9. How far does an average tornado travel in 30 min?
   Answer: 24 km

10. Using the average speed of tornadoes and the average path length of a strong tornado, compute the touchdown time.   *Answer:*   19 min

11. If a violent tornado moved with an average speed of 80 km (50 mi)/h, what would be its touchdown time with an average path length?   *Answer:*   32 min

Tropical Storms

12. Estimate how far a hurricane moves in one hour.
   Answer: Approximately 6.7–13.3 km

CHAPTER 23

Weather Forecasting

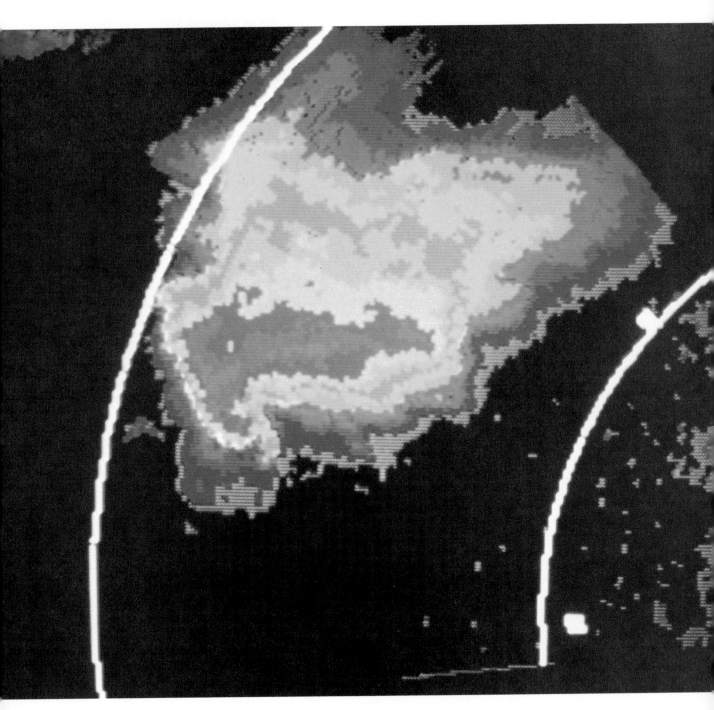

When ye see a cloud rise out of the west, straight way ye say
There cometh a shower, and so it is
And when ye see the southwind blow, ye say
There will be heat; and it cometh to pass.

—Luke 12:54–55

HUMAN BEINGS have long desired to control the weather. However, little progress has been made toward achieving this desire; and as the next best thing, the preoccupation of observing and attempting to predict the weather developed. The influence of the weather on everyday activity has grown to the extent that this preoccupation has become the sole occupation of many modern meteorologists, who are known as weather forecasters.

Weather forecasters are much maligned. When the weather fails to follow the forecast, they are the target of many unflattering remarks. Bad weather that spoils a planned outing is blamed on the forecaster's shortsightedness, if not on the forecaster. Although meteorologists are scapegoats, they do a commendable job and on the average are correct about 80% of the time.

The accurate prediction of atmospheric behavior is a difficult task. Changes in the weather are governed by scientific principles. However, the atmosphere is complex and contains so many variables that meteorologists must combine empirical and scientific knowledge in making reasonable forecasts. Weather data are collected and processed by the most modern means. From the current weather conditions, future conditions are projected, according to established behavior patterns. This chapter will deal with a few of the many aspects of weather forecasting.

23.1 The National Weather Service

The United States **National Weather Service** * (formerly the Weather Bureau) is the federal organization that provides national weather information. It forms part of the National Oceanic and Atmospheric Administration (**NOAA**—pronounced "Noah"), which was created within the U.S. Department of Commerce in 1970.

The original U.S. Weather Bureau grew out of the Army Signal Corps, which maintained an early telegraph system that was used for weather reporting. It became a part of the Department of Agriculture in 1891 and then part of the Department of Commerce in 1940. The Weather Bureau's early activities were primarily directed toward weather forecasting as an aid to agriculture. However, the greatest impetus for the bureau's growth came with the development of aviation, for which up-to-date weather reports are vital. Today, the Weather Bureau is known as the National Weather Service, and its activities are concerned with every phase of the weather.

The nerve center of the National Weather Service is the **National Meteorological Center** (NMC) located just outside Washington, D.C., in Suitland, Maryland. It is the NMC that receives and processes the raw weather data taken at numerous weather stations. Currently, complete weather observations are collected at 260 National Weather Service facilities by some 1200 people. There are also aircraft and radar reports, upper-air monitoring, and data from weather satellites.

The NMC analyzes and makes forecasts from the received data. Other central National Weather Service organizations deal with specialized weather conditions, such as hurricanes, tornadoes, and severe storms. From these central organizations, data go to Forecast Offices, which have the responsibility for warnings and forecasts for states, or large portions of states, and assigned zones. State forecasts are issued twice daily for a time period up to 48 h. An extended outlook, up to five days, is issued daily for these same areas. The forecasts organizational structure is shown in Fig. 23.1.

Weather Service Offices represent the third echelon of the forecast system. Local forecasts are adaptions of state forecasts. They are issued to meet local requirements and to provide general weather information to the public. The local forecasts are distributed by telephone services, and by radio and television broadcasts,

---

* Because the National Weather Service issues its surface weather reports using British units, this chapter will present weather statistics in British units first, followed by SI units, where appropriate, in parentheses.

◀ **Doppler radar screen for forecasting weather.**

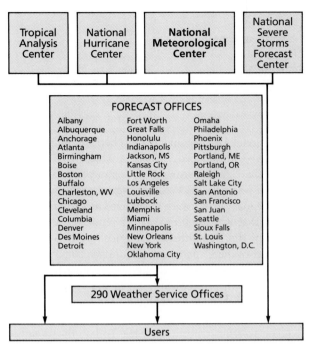

Figure 23.1 National Weather Service forecast organization.

including VHF–FM (very high frequency–frequency-modulated) radio stations near major population centers that transmit weather information continuously 24 h a day 365 days a year.

Let's take a closer look at how forecasts are made and distributed. The operations of the NMC are almost completely automated, and data are handled by high-speed computers, printers, and communication systems. Analyses are made twice daily at 0000 hours and 1200 hours Greenwich Mean Time (7 A.M. and 7 P.M. EST). A preliminary analysis is made after $1\frac{1}{2}$ h of data reception. An operational analysis is made after $3\frac{1}{2}$ h of data collection, when 80% of all incoming data have been received. The computer takes a half hour to analyze the information, and the analysis is completed before the data are 4 h old. NMC estimates the same analysis would take five people working 8 h to complete. Manual analyses are made on certain portions of the data as a check on the computer, as shown in Fig. 23.2.

Once the analyses are completed, three somewhat different simulations of the weather begin. The simulations use the basic laws of physics and some statistics to calculate values of temperature, wind, humidity, and rain at equally spaced locations over the entire globe and at a number of different altitudes. The results are

sent out, almost untouched by human hands, to National Weather Service field offices and other users for guidance in making forecasts. Quality control is done in the field. The forecaster must estimate possible errors and use the results that appear to be best for his or her purpose, with other knowledge, to make forecasts.

The distribution phase was once done by the teletype circuits that brought in the data. Facsimile circuits transmit charts and 2500 weather maps. However, the National Weather Service now uses a data-handling system known as AFOS (Automation of Field Operations and Services). The AFOS system uses minicomputers and TV-type displays in a network of more than 200 automated weather offices. The network has four loop circuits in different regions of the country and is centrally linked to the NMC.

AFOS has done away with the older system of teletypewriters and facsimile machines and the enormous quantities of paper they generated. Instead, weather information from the minicomputer systems is displayed on video screens (Fig. 23.3). However, the AFOS system is outdated and will be obsolete in a short time. The present system was not designed to take advantage of

Figure 23.2 Manual analysis.
Manual analyses are done on portions of the weather data as a check on the computer.

data from satellites and the relatively new Doppler radar, which will be discussed in the next section. The National Weather Service will replace the AFOS system with a new system in the 1990s. It will be designed to supply forecasters with all essential weather data.

The National Weather Service is dedicated to getting the weather information to the public by the fastest means available. However, there is an exception to the prompt distribution of weather information. In wartime weather data are of great military value and become classified information. During World War II weather reports were not made public except in cases of severe weather, in which warnings were necessary to prevent property damage and loss of life. The news media were prohibited from mentioning anything that pertained to the weather. Farmers were simply told it would be a good day to make hay or a rained-out ball game was called off due to conditions beyond control. As one sportscaster put it, "Well, folks, I can't say anything about the weather, but that isn't perspiration on the pitcher's face."

As technology increases and facilities expand, better weather forecasts become available. The predictions of

the National Weather Service meteorologists are estimated to be about 80% accurate. This accuracy, of course, depends on the weather element and location. For example, predictions of no rain in parts of Arizona during the summer are almost 100% correct.

The method of reporting the weather has also changed over the years. In the past, forecasts of precipitation used terms such as *probable, likely* or *occasional.* Precipitation forecasts are now expressed in "percent probabilities," such as a 70% chance of rain. The percent probability given in such a forecast results from a consideration of two quantities: (1) the probability that a precipitation-producing storm will develop in, or move into, the forecast area; and (2) the percentage of the area that the storm is expected to cover. Thus, in summer when storms (because they are convectional) tend to be more isolated, or scattered, the probability that your immediate area will get rain tends to be less than in the winter when frontal storms are more prevalent. The percentage probabilities give the public a better indication of what the weather might be.

Although many hours of work and endeavor go into making the daily weather forecasts as accurate as possible, they sometimes fail to hold true. But after all, if the forecasts were always correct, wouldn't it take a bit of the excitement out of life?

23.2 Data Collection and Weather Observation

In 1870 weather data were taken by the Army Signal Corps at 24 stations. Today, weather observations are collected from approximately 1000 land stations, 6 fixed ocean stations, and several hundred merchant vessels of all nationalities. Many of the land stations are associated with airport operations. Volunteer observers also supply climatology data from about 12,000 substations around the country.

Weather stations are concerned with the basic meteorological measurements of temperature, pressure, humidity, precipitation, and wind direction and speed, as discussed in Chapter 20. The instruments used are also basically the same but have official specifications and in some cases may be automated to supply continuous readings.

A typical installation for data collection is shown in Fig. 23.4. The hut in the background is an instrument shelter. It is constructed with louvered sides, a ventilated floor, and a double roof with an air space between. This construction permits air measurements to be taken free

Figure 23.3 AFOS monitor.
The AFOS system uses minicomputers and TV-type displays in a network of more than 200 automated weather offices.

Figure 23.4 Remote weather-data-collection installation.
An instrument shelter, which contains temperature- and humidity-measuring devices (lower left), and a rain-and-snow gauge (right). A technician is making adjustments on the anemometers below the wind vane on the pole.

from insolation influences. The **instrument shelter** may contain a thermograph, a maximum-minimum thermometer, and a psychrometer. A thermograph is an automatic temperature-recording device. Quite often these automated instruments are connected electronically to nearby offices.

A hydrothermograph is sometimes available and records the temperature as well as the relative humidity by some hygrometric means, thus eliminating the need for the psychrometer. The relative humidity and temperature measurements are taken at a specified distance of 6 ft above ground level. The maximum and minimum daily temperatures are important weather observations. They may be measured with a set of maximum-minimum thermometers.

The **maximum thermometer** is a mercury thermometer with a constriction in the lower part of its capillary bore (Fig. 23.5, left scale). The pressure of the expanding mercury in the bulb causes the mercury to pass through the constriction as the temperature increases. When the temperature decreases, the mercury is unable to pass back through the narrow constriction. Thus the column of mercury above the constriction indicates the highest temperature reached.

The maximum thermometer is reset by shaking. Shaking forces the mercury back into the bulb and readies the thermometer for a new reading. Clinical thermometers are of similar construction and are maximum thermometers. A common error in taking one's temperature at home is forgetting to reset the thermometer.

The **minimum thermometer** (see Figure 23.5, right scale) is an alcohol thermometer that contains a thin, colored-glass, dumbbell-shaped rod in its bore called the *index*. The surface tension of the alcohol surface draws the index with it to the lowest point of descent of the liquid column surface. When the temperature increases

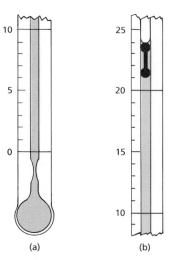

Figure 23.5 A combination maximum-and-minimum thermometer.
Left scale: The constriction in the capillary bore prevents the mercury from returning to the bulb, so the thermometer retains its highest reading. *Right scale:* Surface tension draws the index rod to the lowest descent of the liquid column surface. When the temperature increases and the alcohol expands, the liquid flows around the index, leaving it at the lowest temperature reading. See text for description.

and the alcohol expands, the liquid flows around the index, leaving it at the lowest position. Thus the upper end of the index marks the lowest or minimum temperature reached. In operation, the minimum thermometer is positioned horizontally. Positioned vertically, the glass index would fall through the liquid to the bottom of the tube. The thermometer is reset by tilting it downward.

Another type of maximum-minimum thermometer that gives both readings uses a U tube with two liquids. The U tube contains mercury in the bottom and a clear, expanding liquid above. There is a reservoir of this liquid at the top of the left side of the U tube. As the temperature increases, the liquid inside expands and forces the mercury column *down* on the left side, and *up* on the right side, carrying with it the maximum index.

As the temperature decreases, the liquid contracts, causing the mercury column of the left side of the U tube to rise and carrying with it the minimum index. The maximum index remains at its highest graduation or maximum temperature, and the minimum index is carried to its lowest graduation or minimum temperature. (Note the inverse scale on the left side.) The indices, which are magnetic, are reset with a small ceramic magnet shown in the holder at the top.

The tall, cylindrical object on the right in Fig. 23.4 is an 8-in **rain gauge.** The gauge automatically weighs and records the accumulated precipitation. The collection bucket of the rain gauge is mounted on a weighing mechanism having a scale that converts the weight of the rain to equivalent inches. The rain gauge is also used for snow measurements. The collection bucket is sprayed with special chemicals that melt the snow so its water equivalent can be measured.

Open exposure and immediate locale are not critical for barometric readings, as they are for the rain gauge, and barometers are usually kept indoors because of their intricate construction.

The National Weather Service has begun work on a new system of data collection. The Automatic Surface Observations Systems (ASOS) program is developing modular units that will be used to monitor weather conditions automatically. Modern technology will be used to acquire, process, and distribute the surface observations to various forecasting systems. Current plans call for over 200 units to be deployed across the country in the 1990s.

Radar (*ra*dio *d*etecting *a*nd *r*anging) is used to detect and monitor precipitation, especially that of severe storms. Radar operates by sending out electromagnetic waves and monitoring the returning waves that have been reflected back from some object. In this manner the location of the object may be determined. The objects of interest in weather observations are storms and precipitation. Continuous radar scans are now commonly seen on TV weather reports.

There are 230 conventional weather radars deployed across the United States. Radar installations are located mainly in the tornado belt of the midwestern United States and along the Atlantic and Gulf coasts where hurricanes are probable. Additional radar information is obtained from air traffic control systems at various airports. The NMC is linked to radar stations by telephone circuits, and its meteorologists can view the current radar scans across the country on television screens.

A more advanced radar system has been developed and a network of about 160 new weather radars is being planned for the 1990s. The NEXRAD (Next Generation Radar) program employs **Doppler radar.** Like conventional radar, Doppler radar measures the distribution and intensity of precipitation over a broad area. However, Doppler radar has the additional ability of measuring wind speeds. It is based on the Doppler effect (Chapter 6), the same principle used in police and highway patrol radar to measure the speeds of automobiles.

Radar waves are reflected from raindrops in storms. The direction of a storm's wind-driven rain and hence a wind "field" of the storm region can be mapped. This map provides strong clues, or signatures, of developing tornadoes. Conventional radar can detect the hooked signature of a tornado only after the storm is well developed (Fig. 23.6).

Doppler radar can penetrate a storm and monitor its wind speeds. With a wind-field map, a developing tornado signature can be detected much earlier. In field tests forecasters using Doppler radar were typically able to predict tornadoes 20 min before they touched down, as compared to just over 2 min for regular public warnings. The new system will no doubt save many lives with this increased advance warning time.

Upper-air observations are important in the collection of meteorological data. These observations are made chiefly with radiosondes (radio sounding equipment). A **radiosonde** is a small package of meteorological instruments combined with a radio transmitter. It is carried aloft by balloons, and data are transmitted to ground receiving stations (Fig. 23.7). The wind direction and speed may also be obtained by tracking the flight of the radiosonde.

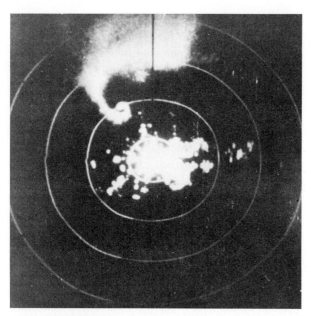

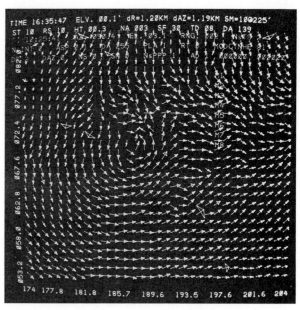

Figure 23.6 Radar.
Left: Conventional radar scan showing the characteristic hooked signature of a tornado.
Right: Doppler radar display showing wind-field patterns.

Figure 23.7 Radiosonde.
The radiosonde instrument package held by the technician is about to be carried aloft by a weather balloon.

The carrier balloon will eventually burst, and the radiosonde descends slowly by means of a small parachute. Parachute descent prevents damage to the instruments and anything in the radiosonde's path on landing. Directions on the radiosondes request the finder to return the radiosonde to the National Weather Service, where they are reconditioned and flown again. Approximately 150 National Weather Service offices engage in radiosonde observations.

Rockets are also used in upper-atmosphere data collection, but probably the greatest progress in general weather observation has come with the advent of the weather satellite. Before satellites, weather observations were unavailable for more than 80% of the globe. The first weather picture was sent back from space on April 1, 1960, from the 118-kg (260-lb) TIROS-1 (*Television Infrared Observation Satellite*). The first fully operational weather satellite system was in place by 1966. These early pole-orbiting (traveling from pole to pole) satellites at altitudes of several hundred miles monitored only a limited area below the orbital path. It took almost three orbits to photograph the entire conterminous United States (Fig. 23.8).

Today, a fleet of GOES's (*Geostationary Orbiting Environmental Satellite*), which orbit at fixed points above the equator, and polar-orbiting satellites—including a more recent 1040-kg TIROS—provide an almost continuous picture of weather patterns all over the globe. At an altitude of about 36,800 km, the GOES orbiters have the same orbital period as that of the Earth's rotation and hence are "stationary" over a particular location. At this altitude the GOES's can send back pictures of large portions of the Earth's surface, as shown in Figure 23.9.

Geographic boundaries and longitude and latitude grids are prepared by a computer and electronically combined with the picture signal so the areas of particular weather disturbances can be easily identified.

With satellite photographs meteorologists have a panoramic view of the weather conditions. The dominant feature of the photograph is, of course, the cloud cov-erage. However, with the aid of radar, which uses a wavelength that picks up only precipitation, the storm areas are easily differentiated from regular cloud cov-erage. Successive photographs of these storm centers give an indication of their movement, growth, and be-havior, thus aiding weather forecasting.

Another important feature of the satellite is **infrared measurements.** As discussed in Chapter 20, the energy and wavelength of emitted radiation depend on the temperature of the radiator. Objects with normal tem-peratures emit radiation energy in the infrared region of the spectrum. Hence using a spectrometer to deter-mine the wavelength in this region provides a method of determining the temperature. See Fig. 23.10.

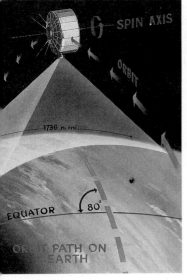

Figure 23.8 A wheel satellite.
Left: The wheel spacecraft for TOS (Trans-Orbiting Satellite) rotated as its cameras scanned a portion of the Earth and adjacent atmosphere. *Above:* Combined scans gave a wide view of the Earth's weather picture. Note the opposite cyclonic circulations in the northern and southern hemispheres.

Figure 23.9 Satellite image.
A GOES weather picture for the United States. A hurricane is clearly visible off the North Carolina coast, producing heavy rains to the south. A band of frontal clouds stretches to the north of the hurricane, and a few clouds appear over the Rockies. Most of the country is clear, with the exception of a band of frontal clouds extend-ing from North Dakota to Michigan, producing cold rain.

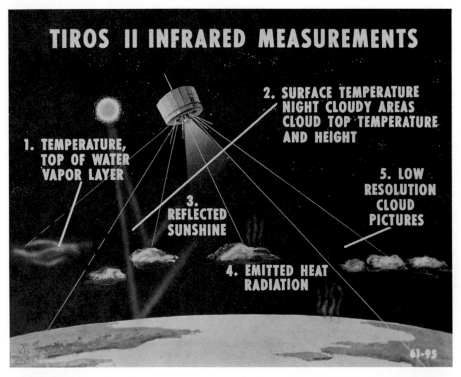

Figure 23.10 Infrared measurements.
Some satellites monitor the infrared radiation of objects, thereby measuring their temperatures, which is proportional to the wavelength of the radiation.

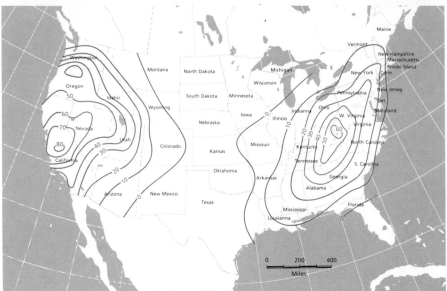

Figure 23.11 Air-pollution potential.
The potential lines show the number of advisory forecasts issued for stagnation condition.

Normally, temperature measurements versus altitude are taken by conventional methods using radiosondes. However, infrared spectrometers now permit these measurements to be taken from lower-altitude, pole-orbiting satellites. This method offers a great advantage, because satellites can cover all areas of the Earth, some of which have never been monitored by radiosonde because of their inaccessibility. This method represents a major step in making the meteorologists' global weather picture more complete.

One of the more recent down-to-earth atmospheric observations is the **air pollution potential** (Fig. 23.11). As we discussed in Chapter 21, the pollution problem becomes severe when certain atmospheric conditions prohibit the dispersal of pollutants. Using upper-air observations, the NMC prepares wind and air-current data

Figure 23.12 Air pollution.
Despite massive efforts—and some success—it is still a problem.

that are analyzed for potential pollution conditions. Advisory forecasts are issued to areas in which stagnation conditions are expected to persist for at least 36 h (Fig. 23.12). The air pollution potential provides an opportunity for air pollution control and research, particularly in studying the influence of meteorological factors on the pollution conditions observed (see Chapter 24).

Another increasingly common atmospheric observation not related to weather forecasting is the pollen count. These counts are made by health officials and are issued to give hay fever sufferers an indication of the amount of airborne pollen that may aggravate their condition.

23.3 Weather Maps

To help analyze the data taken at the various weather stations, meteorologists prepare maps and charts that display the weather picture. Because these charts present a synopsis of the weather data, they are referred to as **synoptic weather charts.** There are a great variety of weather maps with various presentations. However, the most common are the daily weather maps issued by the U.S. National Weather Service. A set consists of a surface weather map, a 500-mb height contour map, a map of the highest and lowest temperatures, and one showing

the precipitation areas and amounts. Examine the charts in Fig. 23.13 carefully.

The Surface Weather map (Fig. 23.13a) presents station data and analysis for 7 A.M. EST for a particular day. This map shows the major frontal systems and high- and low-pressure areas at the constant surface level. Isobars are represented by solid lines, and prevalent isotherms (the lines of constant temperature) are indicated by dashed lines (in degrees Fahrenheit).

The Height Contour map gives a representation of a surface of constant pressure (Fig. 23.13b) of 500 mb. The Temperature map (Fig. 23.13c) gives the highest and lowest temperatures for weather stations throughout the country. The Precipitation Areas and Amounts map (Fig. 23.13d) gives the complete precipitation picture. Major precipitation areas are shaded.

On the Surface Weather map a station model, as shown in Fig. 23.14, gives the pertinent weather data at the station's location. The stations shown in Fig. 23.13 are only a fraction of those included in the National Weather Service's operational weather maps. Also, the station model information has been abbreviated.

A surface map shows isobars at surface level, whereas a height-contour map shows height contours at a constant pressure (usually 500 mb). The contours indicate the heights of the designated pressure in feet above sea level. This map is analogous to a topographical map that indicates the contour of the Earth's surface. If constant-pressure maps were three-dimensional and viewed from the side, the hills and valleys represented by the contours would be the surfaces of equal barometric readings.

On the height-contour map the wind symbols show the wind directions and speeds at the 500-mb level. Isotherms also appear on the height-contour maps. Because many weather conditions and changes depend on the dynamics of the upper atmosphere, these maps are important in forecasting future conditions. Note the correspondence of the low-pressure regions on the height-contour map and areas of precipitation on the precipitation map in Fig. 23.13.

From a knowledge of the general behavior of fronts, cyclonic motion, and air masses studied in the previous chapters, reasonable weather forecasts may be made from surface maps. Forecasting is not always easy, because the weather conditions may be quite complex. For example, note the complexity of the frontal systems over the northwestern United States and Canada shown in Fig. 23.15. The student may find it difficult to predict the weather changes for that region.

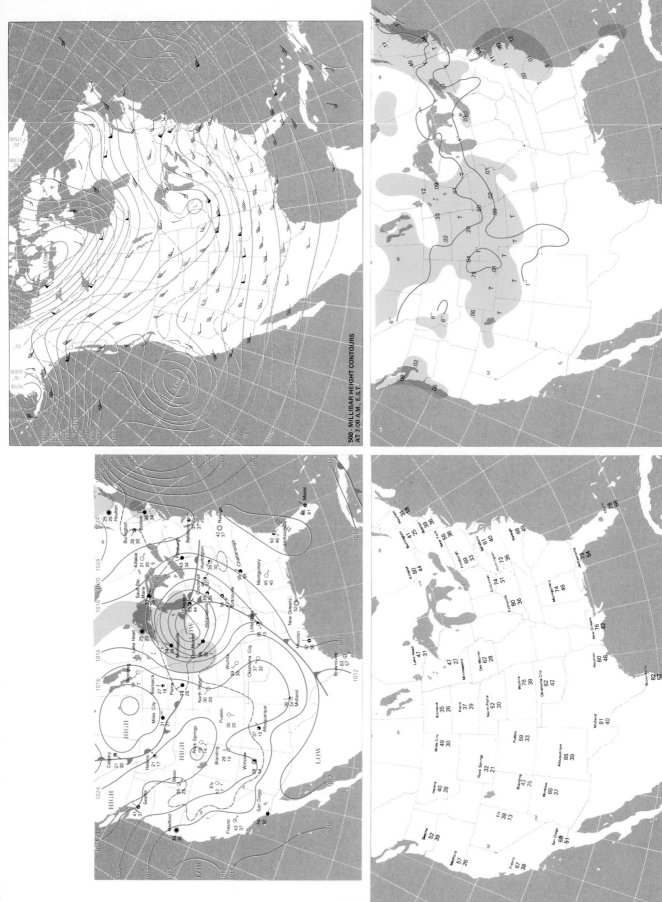

500 - MILLIBAR HEIGHT CONTOURS
AT 7:00 A.M., E.S.T.

Figure 23.13 Daily weather maps.
Weather maps are issued daily by the National Weather Service. The maps from which these

474

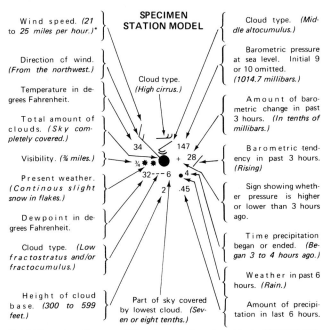

Wind speed. (21 to 25 miles per hour.)*

Direction of wind. (From the northwest.)

Temperature in degrees Fahrenheit.

Total amount of clouds. (Sky completely covered.)

Visibility. (¾ miles.)

Present weather. (Continuous slight snow in flakes.)

Dewpoint in degrees Fahrenheit.

Cloud type. (Low fractostratus and/or fractocumulus.)

Height of cloud base. (300 to 599 feet.)

SPECIMEN STATION MODEL

Cloud type. (High cirrus.)

Part of sky covered by lowest cloud. (Seven or eight tenths.)

Cloud type. (Middle altocumulus.)

Barometric pressure at sea level. Initial 9 or 10 omitted. (1014.7 millibars.)

Amount of barometric change in past 3 hours. (In tenths of millibars.)

Barometric tendency in past 3 hours. (Rising)

Sign showing whether pressure is higher or lower than 3 hours ago.

Time precipitation began or ended. (Began 3 to 4 hours ago.)

Weather in past 6 hours. (Rain.)

Amount of precipitation in last 6 hours.

34   147
¾ ❄ ☀
32 – – 6   + 28
2   .45
4

*Each full flag on the staff represents 10 knots (nautical miles per hour, 1 knot = 1.15 statute mile).

Figure 23.14 Station model.
The weather data from a reporting station are arranged in this model format.

At the National Meteorological Center a computer projects the weather conditions based on stored empirical knowledge, statistical methods, and continuous data. A **prognosis chart** (prog chart) is then printed out, based on the computer's analysis of what the weather conditions will be at some time in the future. However, some unpredictable atmospheric occurrence may take place at any time and make the projection incorrect. One should now begin to realize how difficult it is to accurately forecast the weather.

Figure 23.13 contains a classical cyclonic formation, with a stationary front extending toward the northwest into Canada. In the low-pressure occluded front region the conditions are favorable for convective and frontal precipitation, and its occurrence is indicated by the colored region over Nebraska, Iowa, and Minnesota. Moreover, the counterclockwise circulation around the low may draw in moist air from the Great Lakes region. The clockwise rotation of the adjacent highs to the west carries the precipitation farther west into Wyoming, as shown on the precipitation-areas map. The cyclonic disturbances off the east coast contribute to the precipitation in that area.

To forecast the next day's weather, one must consider the probable movement of the fronts and pressure areas. Localized forecasts are more concerned with the internal dynamics of these phenomena. Because of the general west-to-east movement of weather across the United States and the circulation around a low-pressure area, the low-pressure system over the Midwest in Fig. 23.13 should move northeast, carrying its precipitation with it. The high-pressure areas would be expected to move over the middle United States, and the low-pressure region off the California coast would move inland. The speeds of the high- and low-pressure areas vary. Generally, highs and lows cross the United States in three to five days.

The northeastern United States would experience a warm front, which usually indicates an advancing low. The center of a low-pressure region is normally 0.5 in (1.27 cm) of mercury or more below normal atmospheric pressure and increases radially outward. Hence, aside from the small deflection of the Coriolis force, the winds are generally toward the center of the low.

As a low from the west approaches an observer, the barometer falls and the winds are generally from the east. As the low passes, the barometer begins to rise and the winds shift, coming from the west. In Fig. 23.13 these conditions should occur in the Great Lakes region, over which the low should pass.

The wind conditions are reversed for the passage of a high and would be expected in the northern plains states. The southeastern United States should feel the effect of the advancing cold front. The upward displacement of the warm air by the advancing cold front will produce a slight fall in the barometric pressure, followed by a sharp rise as the front passes, because of the incoming high. Little change would be expected for the northwest states, which will remain under the influence of high pressure. California may expect precipitation from the advancing low.

Figure 23.15 shows the next day's weather maps, which prove these predictions to be fairly accurate. The entire central United States is now dominated by high pressure. Highs are characterized by descending air. The descending air becomes warm and can hold more moisture, so precipitation will not occur. In the summer these conditions are responsible for the high humidity and frequent droughts in the central United States.

Note also that high-pressure areas are usually larger than low-pressure areas, with widely spaced and less streamlined isobars. In the winter a high may enter and move across the United States in the form of a cold, cP air mass. This occurrence is evident in the maps of Fig. 23.16. Hence, you can see that weather forecasting is,

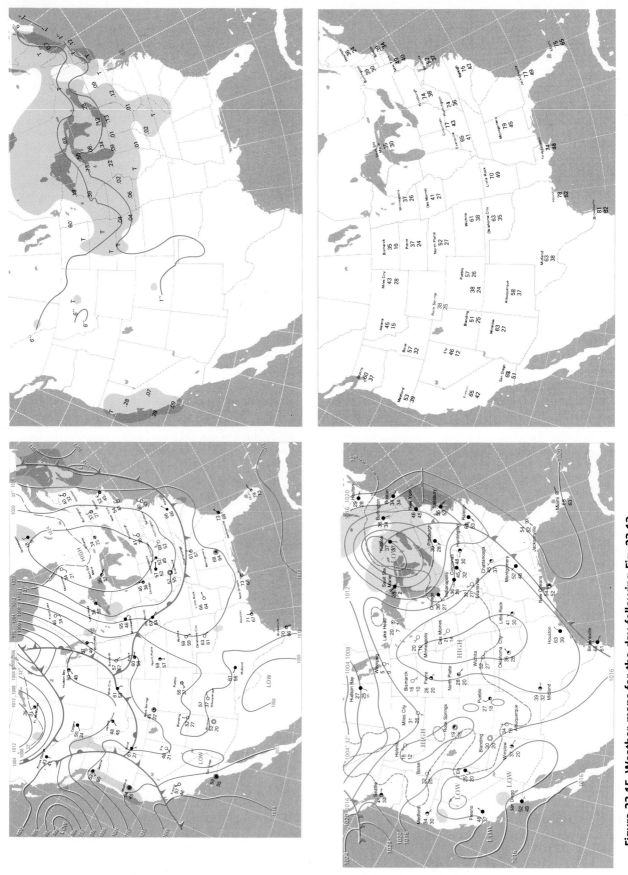

Figure 23.15 Weather maps for the day following Fig. 23.13.
Note the complex array of frontal systems on the surface map. Complicated atmospheric conditions make weather forecasting a difficult task.

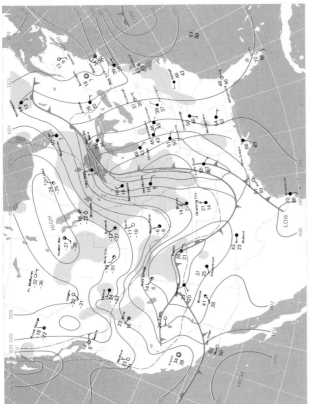

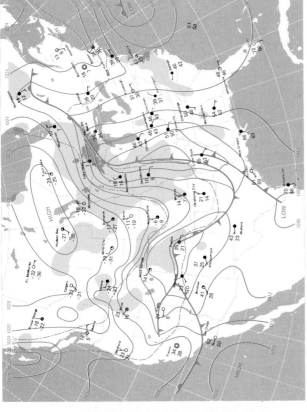

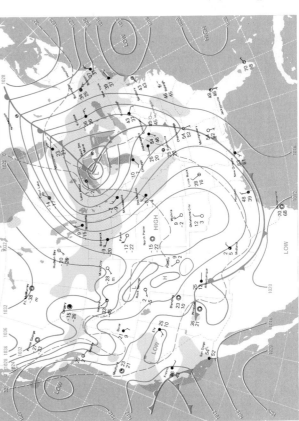

Figure 23.16 Cold-front advance.
A series of three surface maps show the advance of a cold, high-pressure cP air mass across Canada and the United States. Notice the temperature and precipitation variations as the air mass moves across the country. (*Note:* Temperature and dewpoint numbers normally appear to the left of the cloud-cover dot. However, in this map, some of them have been placed to the right of the dot, so as not to interfere with isobars.)

to a large part, the projection of the movement of fronts, pressure areas, and air masses. This process is complicated by a multitude of local variations and influences.

23.4 Folklore and the Weather

Folklore and legends have been associated with the weather from the beginning of humanity. People in the early civilizations worshipped deities whose actions were manifested by weather phenomena. Myths were created to account for those occurrences that could not be otherwise explained. Thunder and lightning, for example, indicated the presence of the violent god Thor. Clouds were the cattle of the sun god Apollo, grazing in the meadows of heaven.

> Men judge by the complexion of the sky
> The state and inclination of the day.
> Shakespeare (*Richard II*)

As human beings progressed and gave up the mythological interpretations of the weather, they began associating its behavior with phenomena they observed. In this manner a similar observation might allow a prediction of future weather conditions. It is these observations that have come down to us as folklore. Usually in the forms of sayings or verse, they have been handed down by word of mouth from generation to generation and often appear in literature. Some of these sayings are well founded and can be explained scientifically. Others, seemingly without any valid explanation, live on, much the same as superstitions. A few of these well-known sayings will be examined in this section. Those related to the meteorological explanations of the previous chapters will be pointed out. The others will be left to the imagination of the reader.

Many amateur weather forecasters closely observe the activities of birds and animals. The antics and instincts of these creatures are often without explanation. However, they form an essential part of folklore. Ducks flapping their wings, the braying of donkeys, or the bolting of horses, for example, are thought by some to be indicative of stormy weather. Loggers in the Pacific Northwest are said to predict snow two or three days before a blizzard by watching the gathering of elks in the shelter of trees. Fiddler crabs have been observed to retreat to inland burrows days before a hurricane's arrival.

Other observations are of a seasonal nature. Marks on a caterpillar or the amount of food stored by squirrels are used to predict the severity of the coming winter. The southward flight of birds may indicate an early winter, while wild geese flying north is taken as a sure sign that warm weather is coming soon.

Probably the most celebrated weather-predicting animal is the groundhog. Legend has it that the groundhog emerges from his winter hibernation each year on February 2 to check on the approach of spring. If he sees his shadow, he returns to his burrow, indicating that six more weeks of winter can be expected. This bit of folklore is greatly publicized each year by the town of Punxatawney, Pennsylvania, where the local groundhog, "Punxatawney Phil," is said to be the superior forecaster. It is somewhat surprising that his shadow, which occurs when the Sun is shining, should warn the groundhog of continuing winter, but that's the legend.

More dependable observations result from the Sun and its effects.

> Above the rest, the sun who never lies
> Foretells the change of weather in the skies.
> Vergil

There are several sayings about the Sun and the weather indications it gives. Among them are the following:

> The weary sun hath made a golden set
> And by the bright tracks of his fiery car
> Gives token of a goodly tomorrow.
> Shakespeare (*Richard II*)

> If the red sun begins his race
> Be sure the rain will fall apace.

> The Pharisees also with the Sadducees came and tempting desired him that he would show them a sign from heaven. He answered and said unto them, When it is evening ye say, It will be fair weather for the sky is red. And in the morning, It will be foul weather today for the sky is red and lowering. Oh ye hypocrites, ye can discern the face of the sky, but can ye not discern the signs of the times.
> Matthew 16:1—4

The weather predictions alluded to in the Bible are more commonly stated:

> Red sky at night, sailors delight
> Red sky in the morning, sailors take warning.

"Rainbow" is often substituted for "red sky" in the above saying. All these sayings involve the red sky, which we often see at sunrise and sunset. What then causes the Sun and sky to appear red? When the Sun is on the horizon, the sunlight travels farther through the atmosphere to reach us than when the Sun is overhead. The blue portion of the sunlight is normally scattered (Section 20.4). However, if the air contains impurities such as dust, more of the longer wavelengths are scattered as the sunlight travels near the surface of the Earth, with the result that only the red portion of the spectrum may reach an observer. This, of course, makes the sky and Sun appear red. (See Chapter 20.)

The condition is enhanced if there is a stable high-pressure region between the observer and the Sun, as the high pressure holds air contaminants near the Earth and scattering is increased. As discussed in Section 22.2, highs are generally associated with good weather, so if one sees a red sky in the evening, it is quite probable that there is a high-pressure area to the west, and good weather will accompany it as it moves eastward, delighting sailors in the westerly wind zone where this weather movement applies. Should the red sky occur at sunrise in the east, the high has probably passed and will usually be followed by a low-pressure system, which is generally associated with poor weather.

Another saying describing an observed phenomenon of sunlight is as follows:

When the sun draws water, rain will follow.

The Sun's rays are often blocked by dense clouds. Occasionally, a cloud may be thin enough in a small area to show sunlight, or a small break in the clouds may allow the sunlight to shine through. In the humid air associated with the clouds, the sunlight may be scattered by fine water droplets or other particles, giving rise to diffuse reflection and creating the effect of a fanlike ray of sunlight extending from the Earth to the cloud. This effect may also be observed when sunlight shines through dense leaf coverage in a thickly wooded area or in a flashlight or searchlight beam at night. There is no drawing up of water, but clouds and particles *are* available for the production of rain.

Mackerel sky and mare's tails
Make lofty ships carry low sails.

Trace in the sky the painter's brush
The winds around you soon will rush.

The cirrus and cirrocumulus clouds referred to in these sayings usually precede the approach of a warm front (Fig. 22.3). As the front approaches, the warm air rising over the cold air mass is likely to produce the winds predicted by these verses.

The wind in the West
Suits everyone best.

In general, our good weather comes from the west, as opposed to north and east, so the saying is to some degree valid. Remember, however, that some locations on the Earth receive very poor weather from the west. Hence the saying is not universal.

When the morn is dry
The rain is nigh
When the morn is wet
No rain you get.

Or in different form and meter:

When the grass is dry at morning light
Look for rain before the night;
When the dew is on the grass
Rain will never come to pass.

The reasoning behind these rhymes is obvious. If condensation in the form of dew has occurred, the air will have a lower relative humidity and rain will be unlikely. However, a better last line to the second poem might read "Rain will *seldom* come to pass," as the air at higher elevations may have sufficient moisture to produce rain. Also, the lack of dew in the morning is not a positive indication of rain. The general validity of these sayings is therefore questionable.

The moon also shares considerable prominence in weather folklore. Many of its related sayings are unfounded. Among these are the indication of rain when "the new moon holds the old moon in its arms." This saying refers to the time when the crescent new moon is near the lower portion of the moon and appears to hold the upper darkened portion. The rain supposedly results from water being spilled from the saucer-shaped new moon. Such sayings are best answered by this countersaying:

Moon and weather may change together
But a change of the moon does not change the weather.

More reasonable predictions are given by the following sayings:

Clear moon, frost soon.

Or:

A ring around the moon is a sure sign of rain.

Or:

When the stars begin to huddle
The Earth will soon become a puddle.

The first of the above sayings refers to the lack of cloud coverage that acts like an insulation to keep the Earth warm. In the absence of clouds, the moon is clearly seen and the land masses cool quickly, making frost likely. The latter two sayings refer to the appearance of the moon and stars as viewed through cirrostratus clouds. The moon appears with a diffuse halo (Fig. 21.11), while the indistinct stars appear to be closer together. Cirrostratus clouds normally precede an approaching warm front, which is accompanied by turbulent weather, usually in the form of rain.

Although not directly related to meteorology, the moon and its phases are often referred to for planting crops. Many gardeners follow the *Farmer's Almanac*, which gives the periods of the proper phases of the moon for planting the proper crops. In general, crops that produce above ground are to be planted in the "light of the moon," or in the waxing phase. Crops that produce below the surface are to be planted in the "dark of the moon," or in the waning phase. For example, it has been reported that a certain gardener had difficulty in keeping dirt on potatoes, which were planted in the light of the moon.

Wood is also said to be affected by the phases of the moon. A board lying on the ground will supposedly curl up at the ends in the light of the moon and stick firmly to the ground in the dark of the moon. Whether this saying is true is subject to doubt, but some builders will put on wooden shingles only during the waning phase or dark of the moon.

Other weather sayings include the following:

Sound travelling far and wide
A stormy day will betide.

Sound waves may be reflected by air layers of different temperatures in the atmosphere. As a result, the sound is heard at a considerable distance, where it is reflected back to Earth. A reflecting cold air layer may also be the source of precipitation, accounting for the stormy day.

Rain before seven, stops before eleven.

This is a reasonably safe prediction for convectional precipitation. If it were raining prior to 7 A.M., there would probably be little moisture left in the precipitating cloud by 11 A.M. Moreover, the Sun would be rising high in the sky and cause the temperature to rise above the dew-point temperature. The clouds would then dissipate and the rain cease, which would bear out the prediction. However, for frontal precipitation, which is controlled by huge air masses, the prediction may not prove to be so accurate.

Leaves turn silver before a rain.

When the wind blows the leaves so that the shiny underside is exposed, they take on a silvery appearance. This turning up of the leaves may result from vertical air motion, which may cause cloud formation. If a nimbus cloud develops, the leaves' prediction of rain will be fulfilled.

It smells like rain.

Before a rain there may be a musty, or earthy, smell in the air. Because convectional precipitation is associated with low pressure and rising air, air and gases in the ground may diffuse out, giving rise to the "smell of rain."

Snow on the ground for three days is waiting for another.

If snow remains on the ground for three days, it is obviously quite cold and additional precipitation is likely to be in the form of snow.

Several common sayings that are without scientific foundation are:

Rain on the first Sunday,
rain every Sunday [of the month].

Rain on Easter Sunday,
rain for seven straight Sundays.

Rain on Good Friday,
the Saint is pouring water on a flat rock.

The latter demonstrates how the meanings of folklore sayings may be disguised. Its interpretation is that if it rains on Good Friday, the following summer will be dry, such that the summer rains will run off the hard, dry ground as though it were a rock.

Another special day for rain is St. Swithin's Day, July 15. If it rains on St. Swithin's Day, the ancient legend has it, 40 days of rain will follow. St. Swithin was an English bishop whose wish it was to be buried in the open churchyard. When sainted and moved into the church, his spirit protested with 40 days of rain. A rain occurring on the anniversary of this event will supposedly arouse St. Swithin's spirit for a repeat performance.

Certain weather occurrences are given special names. A period of warm weather in October or November is sometimes referred to as **Indian Summer.** The seasonal cooling at this time of year gives rise to low-lying fogs. It is said that thin fogs lying near the tops of corn shocks reminded the early settlers of smoke coming from Indian tepees, hence the name Indian Summer. There is nothing uncommon about a brief warm "spell" this late in the year, as the season changes from autumn to winter.

The special significance of this period of warm weather is no doubt largely psychological. People are aware of the coming harsh winter and are sentimental toward this last warm reminder of the pleasant summer. The fondness for summer prompts a special name for this last trace. A wintry period in April is not so honored as Indian Winter but is deplored and quickly forgotten. More significance is given to the break in winter called the January thaw, which often occurs in February. A cold snap during the first few days of spring, when blackberries usually bloom, is sometimes called Blackberry Winter.

Some people are able to forecast a change in the weather or rain by an ache in the knee or some other joint. This ability was thought to be a joke for some time; however, it is now believed that changes in the pressure and humidity affect the aches and pains of rheumatic joints. Others who are not afflicted may rely on folklore to help predict weather behavior. The date of the first snow is sometimes taken as the number of snows that will occur during the winter. Those who believe in this prediction become very discriminating between an actual snow and a noncountable flurry, as the number of snows approaches the predicted number.

Folklore, whether true or unfounded, will exist as long as people talk about the weather, and weather is one of our most talked-about subjects. When we are without anything to say or at a loss for words, it is a favorite topic that comes to our aid. Everyone comments on the weather, while an old English proverb surmises:

Weather is the discourse of fools.

Learning Objectives

After reading and studying this chapter, you should be able to do the following without referring to the text:

1. Describe the organization and functions of the National Weather Service.

2. Explain how weather data are processed and how forecasts are distributed.

3. State the instruments, their operation, and the data collected in a typical weather data station.

4. Describe the operation and data collected by radar, radiosondes, and weather satellites.

5. Explain the different types of synoptic weather charts.

6. Describe how forecasts are made from weather maps.

7. Distinguish between folklore sayings with scientific merit and those without.

8. Define and explain the other important words and terms listed in the next section.

Important Words and Terms

National Weather Service	minimum thermometer	infrared measurements
NOAA	rain gauge	air pollution potential
National Meteorological Center	radar	synoptic weather charts
instrument shelter	Doppler radar	prognosis chart
maximum thermometer	radiosonde	Indian Summer

Questions

The National Weather Service

1. What is NOAA?

2. Has the United States always had a National Weather Service or a similar organization?

3. How are weather data processed and forecasts made? How often are forecasts made?

4. What are the locations of the Forecast Office and the Office of the National Weather Service nearest your hometown? your college or university?

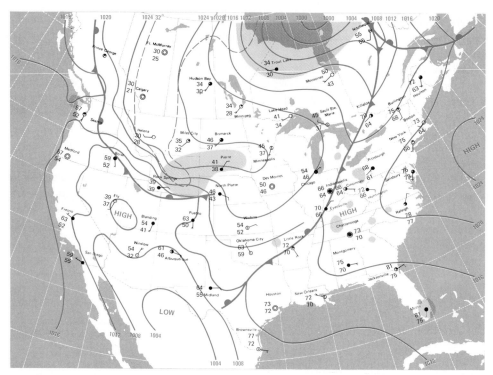

Figure 23.17 Weather forecasting.
See Exercises 2 and 5.

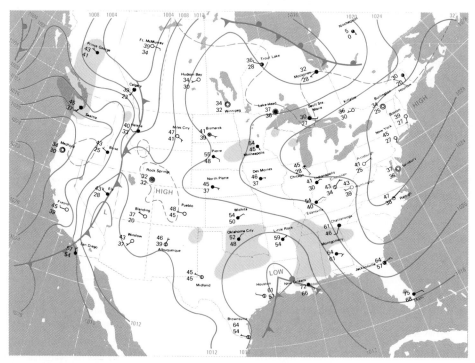

Figure 23.18 Weather forecasting.
See Exercise 7.

Data Collection and Weather Observation

5. What type of data are taken at a typical weather data station?

6. Explain the operation of a maximum-minimum thermometer.

7. What does the word *radar* mean?

8. Distinguish between conventional and Doppler radar.

9. What is a radiosonde?

10. How are temperature profiles obtained via satellite?

11. What is a GOES, and why is it "stationary"?

12. What regions of the United States have high air-pollution potential?

Weather Maps

13. What is a synoptic weather map and what information is given in the station models?

14. Explain a 500-mb height-contour map.

15. How can weather forecasts be made from weather maps?

Folklore and the Weather

16. Are folklore sayings concerning the weather reliable? Do they have any scientific merit?

17. How do folklore sayings originate?

18. What are some of the major elements of observation in folklore sayings?

Exercises

Weather Maps

1. Given the following data, plot a weather station model.
   Temperature: 68°F
   Sky: One-half covered with cirrus, cirrocumulus, and a few stratocumulus clouds
   Winds: 15 mi/h out of the northeast
   Visibility: 3 mi
   Psychrometer wet bulb: 66°F
   Barometer: 1012.7 mb, falling from 1013.4 in the last 3 h
   Started raining 2 h ago, with current accumulation of 0.15 in.

2. From Fig. 23.17 calculate the approximate average speed of the high-pressure cP air mass moving from western Canada into the central United States.
   Answer: 500 to 1000 mi/day

3. Describe the weather conditions and formations as best you can from the satellite photograph of the United States in Fig. 23.9.

4. Prepare a local weather forecast for your area from Fig. 23.13. Use Fig. 23.15 to check your predictions.

5. From the surface map in Fig. 23.17, what are the weather conditions across the country? Give particular attention to weather changes and variations. Check your answer with the newspaper report given in Section 22.2.

6. Prepare a general forecast for the United States from the last surface map in Fig. 23.13.

7. Prepare a 24-h forecast for the U.S. east coast from the surface map in Fig. 23.18. Give the reasons for your predictions.

Folklore and the Weather

8. Examine the following weather sayings and explain their meteorological merit and meaning, if any:
   (a) A red sun has water in his eye.
   (b) Two full moons in a calender month bring on a flood.
   (c) Mackerel clouds in the sky, expect more wet than dry.
   (d) February rain is only good to fill ditches.
   (e) Candles burn dim before rain.
   (f) March comes in like a lion and goes out like a lamb.
   (g) A year of snow, a year of plenty.
   (h) Dew long on the ground, Jack Frost will be around.
   (i) It's too cold to snow.

Pollution and Climate

TO MAKE A COMPLETE survey of the atmosphere, one cannot neglect the problem of air pollution. The various pollutants being expelled into the atmosphere may have direct effects on our health and living conditions and may influence our local weather conditions and perhaps the global climate.

Climate may also be affected by land and water pollution. Pollution effects are often coupled. For example, we hear a great deal about *acid rain.* This type of pollution involves both the atmosphere and lakes and streams. We indiscriminately and directly pollute the landscape and our rivers and streams with solid and liquid wastes. Removing foliage causes erosion and silt flows into lakes, upsetting ecological balances.

Of course, there is natural erosion, and gases and particulate matter are spewn into the air from volcanic eruptions and lightning-initiated forest fires. These events are natural phenomena over which we have little or no control. By **pollution** we mean any atypical contributions to the environment resulting from the activities of human beings.

In this chapter we shall consider some of the sources of pollution and the environmental effects. We will also investigate concerns that the pollution of today may affect the climate of the future.

24.1 Pollutants

Air Pollutants

Air pollution results primarily from the products of combustion and industrial processes that are released into the atmosphere. It has long been a common practice to vent these wastes into the atmosphere, and the resulting problems are not new, particularly in areas of population concentrations.

Smoke and soot from the burning of coal plagued England over seven hundred years ago. London recorded air pollution problems in the late 1200s, and particularly smoky types of coal were taxed and even banned. The problem was not alleviated, and in the middle 1600s Charles II was prompted to commission one of the outstanding scholars of the day, Sir John Evelyn, to make a study of the situation. The degree of London's air pollution at that time is described in his report, *Fumifugium* (a Latin term, generally meaning *On Dispelling of Smoke;* see Fig. 24.1).

. . . the inhabitants breathe nothing but impure thick mist, accompanied with a fuliginous and filthy vapor, corrupting the lungs. Coughs and consumption rage more in this one city (London) than in the whole world. When in all other places the aer is most serene and pure, it is here eclipsed with such a cloud . . . as the sun itself is hardly able to penetrate. The traveler, at miles distance, sooner smells than sees the City.

However, the Industrial Revolution was about to begin, and Sir John's report was ignored and so gathered dust (and soot). Another indication of the polluted air in parts of seventeenth century England is given by the quotation from Shakespeare at the beginning of this chapter, who was drawing on his own experience.

As a result of this air pollution, London has experienced several disasters involving the loss of life. Thick fogs are quite common in this island nation, and the combination of smoke and fog forms a particularly noxious mixture known by the contraction of *smoke-fog,* or **smog.**

The presence of fog indicates that the temperature of the air near the ground is below the dew point, and with the release of latent heat, there is possibly a temperature inversion (see Section 21.5). The gases and smoke are then held near the ground. Continued combustion causes the air to become polluted with smog. Smog conditions are particularly hazardous for persons with heart and lung ailments. Smog episodes in various parts of the world have contributed to numerous deaths (Fig. 24.2).

◄ **One of the many sources of industrial pollution.**

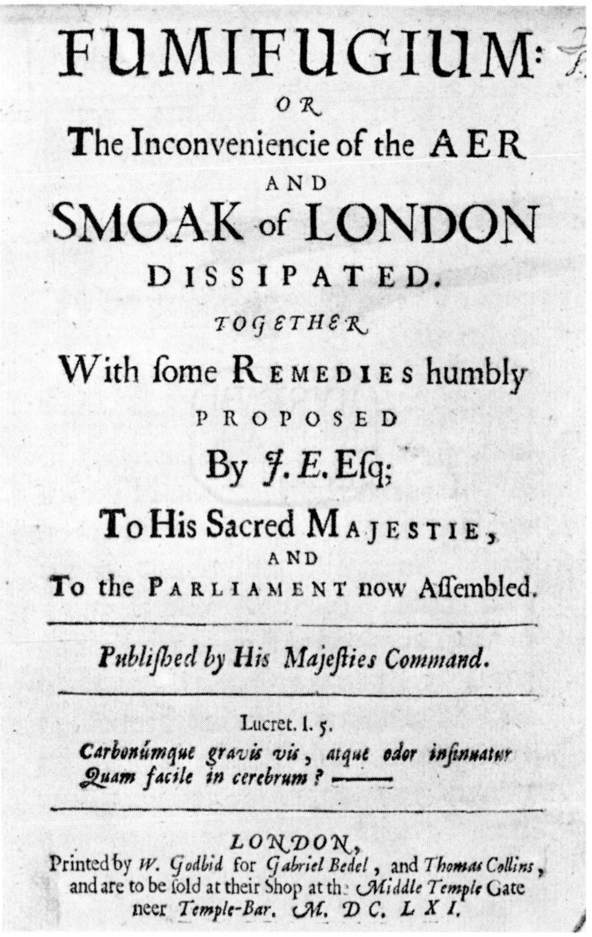

FUMIFUGIUM:

OR

The Inconveniencie of the AER

AND

SMOAK of LONDON

DISSIPATED.

TOGETHER

With some REMEDIES humbly

PROPOSED

By *J. E.* Esq;

To His Sacred MAJESTIE,

AND

To the PARLIAMENT now Assembled.

Published by His Majesties Command.

Lucret. l. 5.

Carbonúmque gravis vis, atque odor insinuatur
Quam facile in cerebrum? ———

LONDON,
Printed by *W. Godbid* for *Gabriel Bedel*, and *Thomas Collins*,
and are to be sold at their Shop at the *Middle Temple* Gate
neer *Temple-Bar. M. D C. L X I.*

Figure 24.1 The title page of Sir John Evelyn's 1661 book on air pollution.
The Latin quotation from the Roman poet Lucretius (97–53 B.C.) near the bottom of the page may be translated "How easily the heavy potency and odor of the carbons sneak into the brain."

The major source of air pollution is the combustion of fuels, which are primarily **fossil fuels**—coal, gas, and oil. More accurately, air pollution results from the incomplete combustion of impure fuels. Technically, combustion (burning) is the chemical combination of certain substances with oxygen. Fossil fuels are the remains of plant and animal life and are composed chiefly of carbon and compounds of carbon and hydrogen, called hydrocarbons.

If a fuel is pure and combustion is complete, the products, CO_2 and H_2O, are not usually considered pollutants, because they are part of the natural atmospheric cycles. For example, if carbon (coal) or methane (natural gas), CH_4, are burned completely, the reactions are

$$C + O_2 \longrightarrow CO_2$$

$$CH_4 + 2\,O_2 \longrightarrow CO_2 + 2\,H_2O$$

However, if fuel combustion is incomplete, the products may include carbon (soot), various hydrocarbons, and carbon monoxide (CO), all of which may contribute to air pollution. **Carbon monoxide** results from the incomplete combustion (oxidation) of carbon. That is,

$$2\,C + O_2 \longrightarrow 2\,CO$$

But even increased concentrations of CO_2 can affect our environment. Carbon dioxide combines with water to form carbonic acid, a mild acid we all drink in the form of carbonated beverages (carbonated water):

$$CO_2 + H_2O \longrightarrow H_2CO_3$$

Carbonic acid is a natural agent of chemical weathering in geologic processes (see Chapter 29). But as an indirect product of air pollution, increased concentrations may also aid in the corrosion of metals and react with certain materials, causing decomposition (Fig. 24.3). Also, there is some concern that the increased CO_2 content of the atmosphere may cause a change in global climate through the greenhouse effect. This concern will be considered later in the chapter.

Oddly enough, some by-products of complete combustion contribute to air pollution, viz., nitrogen oxides (NO_x). **Nitrogen oxides** are formed when combustion temperatures are high enough to cause a reaction between nitrogen and the oxygen of the air. This reaction typically occurs when combustion is nearly complete, a condition that produces the high temperature, or when combustion takes place at high pressure—e.g., in the cylinders of automobile engines. These oxides, normally NO (nitric oxide) and NO_2 (nitrogen dioxide), can combine with water vapor in the air to form nitric acid and contribute to acid rain, which will be discussed shortly.

Nitrogen oxides can also cause lung irritation and are a key substance in the chemical reactions producing "Los Angeles" smog. This is not the classical London smoke-fog variety but a smog resulting from the chemical reactions of hydrocarbons with oxygen in the air and other pollutants in the presence of sunlight; it is called **photochemical smog.** Since it was first identified in Los Angeles, it is often referred to as Los Angeles smog. Over 13 million people live in the Los Angeles area, which is in the form of a basin with the Pacific Ocean on one side and mountains on the other (east).

Figure 24.2 Smog.
A scene in Donora, Pennsylvania, taken in 1949. During a five-day smog episode in 1948 in this Monongahela River Valley town, located 20 mi southeast of Pittsburgh, hundreds of people became ill and at least 20 died.

Figure 24.3 An effect of air pollution.
Acid-rain damage to a statue at a museum in Chicago.

This topography makes air pollution and temperature inversions a particularly hazardous combination. A temperature inversion can essentially put a "lid" on the city, which then becomes engulfed in its own fumes and exhaustive wastes.

Los Angeles has its share of temperature inversions, which may occur as frequently as 320 days per year. These inversions, a generous amount of air pollution, and an abundance of sunshine set the stage for the production of photochemical smog (see Fig. 24.4).

In contrast to the smoke-fogs of London, photochemical smog produces eye irritation and contains many more dangerous contaminants. They include organic compounds, some of which may be **carcinogens,** or substances capable of producing cancer. One of the best indicators of photochemical reactions, and a pollutant itself, is **ozone** (O_3), which is found in relatively large quantities in polluted air. In Los Angeles, air pollution warnings of various degrees are given on the basis of ozone concentrations in the air.

For the generation of the amount of ozone in some polluted areas, a process other than dissociation of O_2 and the combination of O_2 and O, which occurs in the ozonosphere, must be involved (see Section 20.3). The energetic particles responsible for ozone production in

Figure 24.4 Photochemical smog scene in Los Angeles.

the stratosphere do not reach the Earth's surface, so they are not responsible. Scientists think part of the answer involves nitrogen dioxide, which in the presence of certain pollutants is involved in a series of chemical reactions that produce ozone.

Both O_3 and NO_2 are potentially dangerous to plants and animals. NO_2 has a pungent, sweet odor and is yellow-brown. During the peak rush hour traffic, it is often evident in a whiskey-brown haze over large cities. NO_2 reacts with water vapor in the air to form nitric acid (HNO_3), which is very corrosive.

Fuel Impurities

Fuel impurities occur in a variety of forms. Probably the most common impurity in fossil fuels and the most critical to air pollution is **sulfur.** Sulfur is present in fossil fuels in different concentrations. A low-sulfur fuel has less than 1% sulfur content; and a high-sulfur fuel, greater than 2%. When fuels containing sulfur are burned, the sulfur combines with oxygen to form sulfur oxides (SO_x), the most common being **sulfur dioxide** (SO_2):

$$S + O_2 \longrightarrow SO_2$$

A majority of SO_2 emissions comes from the burning of coal and an appreciable amount from the burning of fuel oils. Coal and oil are the major fuels used in the generation of electricity. Almost half of the SO_2 pollution in the United States occurs in seven northeast industrial states.

Sulfur dioxide in the presence of oxygen and water can react chemically to produce sulfurous and sulfuric acids. Sulfurous acid (H_2SO_3) is mildly corrosive and is

used as an industrial bleaching agent. Sulfuric acid (H_2SO_4), a very corrosive acid, is a widely used industrial chemical. In the atmosphere these sulfur compounds can cause considerable damage to practically all forms of life and property. Anyone familiar with sulfuric acid will be able to appreciate its undesirability as an air pollutant. Sulfuric acid is the electrolyte used in car batteries.

The sulfur pollution problem is receiving considerable attention because of the occurrence of **acid rain.** Rain is normally acidic as a result of carbon dioxide combining with water vapor in the air to form carbonic acid. However, sulfur oxide and nitrogen oxide pollutants cause precipitation from contaminated clouds to be even more acidic, giving rise to acid rain (and also acid snow, sleet, fog, and hail). The problem is most serious in New York, New England, and Canada, where pollution emissions from the industrialized areas in the midwestern United States are carried by the general weather patterns. Other areas are not immune. Acid rain now occurs in the Southeast, and acid fogs are observed on the west coast.

Rainfall with a pH of 1.4 has been recorded in the northeastern United States. This value surpasses the pH of lemon juice (pH 2.2). In Canada a monthly rainfall had an average pH of 3.5, which is as acidic as tomato juice (pH 3.5). The yearly average pH of the rain in these affected regions is of the order of 4.2 to 4.4. (See Chapter 14 for a discussion of pH.) In addition to acid rain, there are acid snows. Over the course of a winter acid precipitations build up in snowpacks. During the spring thaw and runoff, the sudden release of these acids gives an "acid shock" to streams and lakes.

Acid precipitations lower the pH of lakes, which threatens aquatic plant and animal life. Most fish species die at a pH of 4.5 to 5.0. As a result, many lakes in the northeastern United States and Canada are "dead" or are in jeopardy of dying. Natural buffers in area soils tend to neutralize the acidity, so waterways and lakes in an area don't necessarily match the pH of the rain. However, the neutralizing capability in some regions is being taxed, and the effects of acid rain are increasing.

Thus air pollution can be quite insidious and can consist of a great deal more than the common particulate matter in the form of smoke, soot, and fly ash, which blackens buildings and the weekly wash if it is hung outside. Pollutants may be in the form of mists and aerosols. Some pollutants are metals, such as lead and arsenic. Approximately 100 atmospheric pollutants have been identified, and 20 of them are metals that come primarily from industrial processes.

Land and Water Pollutants

Land and water pollution are difficult to separate because of their interactions. Often we do not consider land pollution, but it can take a variety of forms.

One of the chief sources of pollution of our countryside is improper solid waste disposal. **Solid waste** is a general term that is applied to any normally solid material resulting from human or animal activities that is useless or unwanted. More than 12 million tons of solid waste are produced each day in the United States.

Modern packaging technology and disposable items have contributed greatly to urban wastes. "Waste not" is no longer a popular adage. The domestic solid waste per capita in this country has more than doubled from its 1920 value of 1.3 kg (2.9 lb) per person per day. If you have had to empty the garbage and trash for an average family, you are aware of the magnitude of the solid waste generated.

Land pollution is evident in the form of litter. **Littering** is the indiscriminate disposal of solid waste on public and private property. Tons of litter are strewn along streets and highways and in parks and along beaches. Some wastes, such as food scraps and paper items, are **biodegradable** (capable of being decomposed by bacteria) and will relatively quickly decompose and return to the soil. However, other items, such as plastics, are nonbiodegradable and are semipermanent. Some show little or no sign of weathering, as in the case of chemical weathering of metal wastes and plastics.

Solid-waste disposal is a serious problem, particularly in metropolitan areas where the concentrations of solid waste are enormous. One method of solid-waste disposal is in **sanitary landfills,** where refuse is covered with a layer of soil within a day after being deposited. However, landfills still have potential pollution danger. The seepage of water through the wastes can leach out undesirable components that pollute the groundwater.

Some cities along the ocean use the convenience of this body of water to dispose of their solid waste. For example, solid wastes from New York City are hauled offshore by barge and dumped into the ocean. The use of the oceans as a dumping ground was brought to light recently with the washing up of hospital wastes on beaches. Incineration is also a common method of combustible solid waste disposal, but air pollution is associated with this method.

Activities in construction and mining often lead to the removal of trees and vegetation that once prevented erosion. The resulting erosion can lead to silting in lakes and streams. Although a natural process, increased silting

Figure 24.5 Silting.
Increased erosion resulting from human activities can give rise to silting that fouls streams and lakes, causing severe damage to animal and plant life.

can have a profound effect on the environments of lakes and ponds into which streams drain (Fig. 24.5). The suspended sediment reduces the extent to which sunlight penetrates into the water. With less sunlight, the level of photosynthesis by submerged vegetation is lower, thus reducing the production of oxygen and the food supply for fish.

Mining operations can give rise to land and water pollution. The minerals and fuels needed by society are mined from the Earth. Many such deposits lie well below the surface and are reached by deep mines. Other deposits lying near the surface are surface mined. Large excavating equipment of the type shown in Fig. 24.6 removes the overlying earth (overburden) to expose the commercially valuable materials. Surface mining is faster and cheaper than deep mining. Properly treated and managed, the land after being surface mined can be returned to its natural condition. Unclaimed, it remains an ugly scar on the Earth's surface and a source of stream-fouling sediment and leached pollutants.

The mining of coal accounts for most of the disturbed acreage. Sulfur compounds are usually associated with coal deposits and may give rise to acid drainage in the water coming from mines. Sand and gravel are the second-largest mining activity. Stone, gold, phosphates, iron ore, and clay follow. The relative proportion of mining done in each state is shown in Fig. 24.7.

There is a great variety of water pollutants. Many substances are discharged into rivers and streams, and

Figure 24.6 Surface-mine operation.
The large dragline shovel in the background removes the overburden and the smaller shovel in the foreground loads the coal into transports.

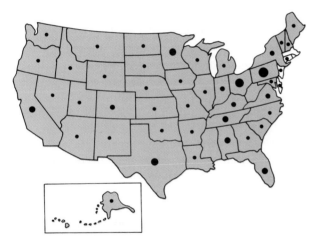

Figure 24.7 Acreage disturbed by surface mining in the United States.
The dots on the map show the proportionate surface-mined acreage in each state. Only about one-third of the total mined area has been reclaimed.

others deposited on land are leached into water systems. However, unlike air pollution discharges that are dispersed and mixed over large regions, water pollutants are relatively localized and easily identified and studied. Rather than discuss the major water pollutants here, we will consider them in the following section, along with their sources.

24.2 Sources of Pollution

Air Pollution

Figure 24.8 shows the sources and relative magnitudes of atmospheric pollutants. As the figure shows, transportation is the major source of total air pollutants. The United States has a mobile society, with over 100 million registered vehicles powered by the internal-combustion engine.

There are various types of internal combustion engines, but most common is the gasoline engine used in the automobile. The completeness of combustion depends primarily on the air-fuel mixture. The proper **air-fuel ratio** for near complete combustion is 15:1. When the car is accelerating, the air-fuel ratio is on the order of 12:1; while cruising, 13:1; and while idling and decelerating, as in stop-and-go city traffic, 11:1. When the ratio is less than 15:1, substantial amounts of CO and hydrocarbons are formed. However, the emission of nitrogen oxides is greater when the air-fuel ratio is 15:1.

Considerable efforts are being made by automobile manufacturers to reduce these emissions through engine modifications and antipollution devices. You should now readily understand why you are urged to keep your automobile engine properly tuned. A properly tuned engine operates nearer to the correct air-fuel ratio, which

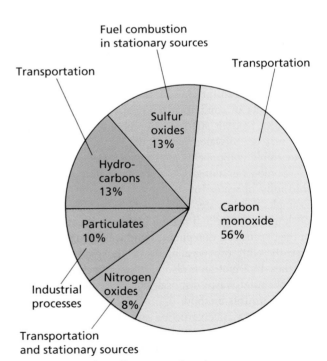

Figure 24.8 Air pollutants and major sources.

Figure 24.9 Transportation air-pollution sources.
Vast amounts of air pollution come from cars and trucks (*top*) and aircraft (*bottom*).

have been reduced and "regular" gasoline will eventually be phased out.

Figure 24.8 shows that the other major sources of air pollution are stationary sources and industrial processes. Stationary sources refer mainly to electrical generating facilities. These sources account for the majority of the sulfur oxide (SO_x) pollution. This pollution results primarily from the burning of coal, which always has some sulfur content.

Water Pollution

> I counted two-and-seventy stenches,
> All well defined, and several stinks.
> The river Rhine, it is well known,
> Doth wash your city of Cologne;
> But tell me, nymphs! what power divine
> Shall henceforth wash the river Rhine?
> (From the poem *Cologne* by
> Samuel Taylor Coleridge, 1828)

The direct pollution of land should be evident, so we will concentrate on water pollution. Water pollution can occur in many different forms and originate from many different sources, such as the acidic mine drainage discussed previously. The major sources and types of pollution are summarized in Table 24.1. Let's consider each source briefly.

Agricultural Water Pollution

Water pollution from agricultural sources is usually not direct but rather enters the water supply via runoff and groundwater absorption in areas of agricultural activity. In modern agriculture, pesticides and fertilizers are essential in producing the yield of food and other crop products required for the world population. Without **pesticides**, tons of agricultural products would be lost each year to insects.

Prior to 1972 the most widely used pesticide was DDT (dichloro-diphenyl-trichloro-ethane). DDT is not biodegradable, and as result of extensive worldwide use, it accumulated in the environment. DDT residue was

minimizes emissions (see Fig. 24.9). However, even with the current efforts to reduce pollution emissions from automobiles, greater pollution is predicted in the future because of the increasing number of motor vehicles.

Another source of pollution comes from the lead additive used in the gasoline as an "antiknock" agent. The engines of new automobiles are designed to use lead-free gasoline. The levels of lead in leaded gasoline

Table 24.1 Sources and Types of Water Pollution

Agricultural	Industrial	Domestic
Pesticides	Chemical	Detergents
Fertilizers	Thermal	Organic wastes
Animal wastes	Radioactive	

found in animals, entering through the biological food chain. Concentrations were even found in penguins in Antarctica. The government banned the use of DDT in 1972. Because the United States was the major supplier, this ban was in effect a worldwide ban.

Other pesticides include arsenic compounds of calcium, lead, and mercury. There are some 34,500 registered pesticide products composed of one or more of 900 chemical compounds. Although pesticides are usually sold for a specific pest, they often kill nonpest species and have side effects on the growth and reproduction of birds and fish.

Farmers in the United States apply millions of tons of chemical **fertilizers** to their fields each year. Water pollution results from the **phosphates** (PO_4^{4-}) and **nitrates** (NO_3^-) present in the fertilizers. These substances enter the water supply primarily through water runoff and through the erosion of topsoil. In sufficient concentration nitrates are toxic to humans and animals. Normal water-purifying procedures do not remove nitrate contaminants.

Nitrates and phosphates are both believed to contribute to the excessive growth of microscopic plant algae in lakes. This effect will be considered in detail in the discussion of domestic detergents, a major source of phosphates in water pollution.

Animal wastes are a source of water pollution with potential health hazards. Because of the trend toward raising animals in feed lots, disposal of these wastes has become a problem (Fig. 24.10). Table 24.2 gives an

Figure 24.10 Cattle feedlot.
Commercial beef cattle no longer roam the open range. Instead, they are raised in feed lots such as this one near Imperial Valley, California. Such concentrations of livestock create enormous amounts of animal waste, which leaches into the soil and produces major water-pollution problems.

Table 24.2 Animal Waste Production

Source	Kilograms/Day
Human being	0.15
Cow	24*
Pig	2.7
Sheep	1.1
Chicken	0.18

Source: Soil Conservation Service, U.S. Dept. of Agriculture.
* Perhaps you, like the author, thought this was an inordinate amount, but a county agricultural agent said it is an accurate figure.

indication of the magnitude of the problem. The data in the table show that a feed lot handling 1000 head of cattle would have the same weight-equivalent waste-disposal problem as that of a city with a population of 157,000. The city would probably have a sewage treatment plant; the feed lot would probably not.

Traditionally, the disposal of animal waste has been as a fertilizer. However, studies show that compared with the benefits of modern chemical fertilizers, the benefits of soil fertilization from animal wastes do not justify the costs of hauling and applying it. Thus the livestock owner must contend with a daily supply of a product that cannot be sold, given away, or burned, and that may cause water pollution in water runoff. The effects of this pollutant will be discussed with those of domestic waste.

Industrial Water Pollution

Many industries by their very nature generate a lot of wastes. For many years industry commonly discharged wastes into rivers and streams (Fig. 24.11). The streamflow then may have been adequate to dilute and carry away the wastes without environmental damage. But with the increase in industry and population in recent decades, industrial wastes have become a serious problem.

Many responsible companies have instituted programs to reduce or eliminate pollution. Also, governmental agencies monitor the disposal of wastes and seek new methods of waste disposal. While solutions are being sought, industry continues to be the largest user of water, with 240,000 industries in the United States using this resource. The amount of water use, of course, depends on the nature of the industry. The largest use by far is for cooling in electrical generation. A com-

Figure 24.11 Industrial pollution.
Chemical-plant effluent in a stream.

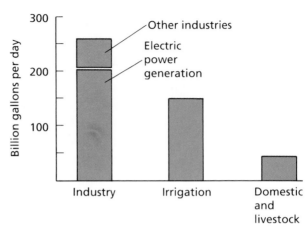

Figure 24.12 The relative use of water by economic sector in the United States.

parison is shown in Fig. 24.12, along with the relative uses in other sectors.

The number of industrial chemical wastes is legion. Let's consider only one major potential pollutant, **mercury.** Mercury has many industrial uses (as many as 3000), and compounds of mercury are used extensively as pesticides. Mercury in liquid form is not overly dangerous unless ingested. However, mercury vapor is highly toxic and is readily absorbed through the lungs. Metals such as mercury tend to be retained in the body for long periods of time, so there is a cumulative effect from otherwise small, harmless doses.

Mercury compounds were employed in the treatment of beaver pelts used in the hat industry, and the early symptoms of mercury poisoning (giddiness, silliness, and strange behavior) resulting from action on the nervous system were common among the workers— hence the phrase, "mad as a hatter." The symptoms of mercury poisoning are typified by the Mad Hatter in Lewis Carroll's *Alice in Wonderland.*

Mercury pollution in water poses a severe health threat to the inhabitants of areas in which fish is a main part of the diet. Such a case occurred in the coastal town of Minamata, Japan, where a plastics manufacturing plant

discharged quantities of wastes containing mercury into Minamata Bay. Between 1953 and 1960 some 110 people, mostly from the families of fishermen, died or were severely disabled after eating fish caught in the mercury-polluted waters.

Another source of chemical pollution that has come to light in recent years is chemical waste burial. At one time chemical companies buried their wastes without standards or specifications. Chemical wastes seeping from the containers have contaminated groundwater and have been directly toxic to nearby inhabitants. In some cases homes have been built unknowingly on the sites of old waste dumps. One such incident, at Love Canal in western New York, was publicized in 1980. Two years later Times Beach, Missouri, was in the public spotlight when the federal government had to evacuate and resettle its entire population. The extent of these hazardous waste sites in illustrated in Fig. 24.13.

A great deal of the water required by industry is used in cooling processes. In fact, 94 percent of the water used by industry is for this purpose. The major cooling application of water is in thermoelectric power generation, which explains the disproportionate share of this user shown in Fig. 24.12. Other industries, such as oil refining and steelmaking, also use an appreciable amount of water as a coolant.

In thermoelectric power generation, steam is used to drive turbines in both conventional fuel and nuclear methods. Water is used to cool the steam from the turbine. The cooled steam condenses, and so releases the back pressure on the turbine and increases its efficiency. The water for cooling does not have to be of

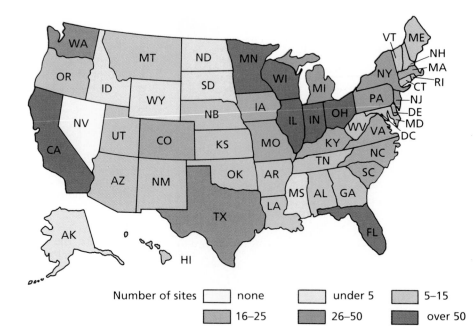

Figure 24.13 Hazardous-waste sites.
The number of targeted waste sites identified for cleanup under the national-trust "Superfund."

Number of sites: none | under 5 | 5–15 | 16–25 | 26–50 | over 50

high quality. In some locations salt water is used as the coolant, but this process requires that parts of the equipment be made of special materials because of the salt water's corrosive properties.

The temperature rise of water used in electric generation cooling is between 10° and 30°C, depending on plant design and operation. Nuclear power plants require more cooling than conventional (fossil fuel) plants of equivalent size. Either the temperature of the effluent water from nuclear plants is greater or a greater volume of water must be used—as much as 50% more.

The **thermal pollution** problem associated with power generation results from the return of the heated water to rivers and lakes. The heated water causes the normal temperature of the lake or a portion of the river to rise, which is a type of pollution. One might not think that a few degrees rise in the temperature of a body of water would create an environmental problem. However, the thermal aspect of an ecologic balance is very delicate. A temperature rise in a natural water formation causes adverse conditions for some aquatic life, reduces the dissolved oxygen content of the water, and accelerates plant growth and other biologic processes.

The most observable effect of thermal pollution is that on aquatic life, particularly fish. The amount of dissolved oxygen in water decreases with increasing temperature. Fish absorb oxygen from the water passing through their gills and may die in water with a reduced oxygen content. This possibility is particularly likely in streams with thermal pollution in the summertime when the water temperature rises naturally (Fig. 24.14). The environmental effects of thermal pollution can be quite severe and far reaching; however, thermal pollution can be readily prevented by cooling processes.

Figure 24.14 Fish kill.
Fish kills can be caused by chemical or thermal pollution.

Radiation pollution arises from the use of radio-active materials. This type of pollution immediately brings to mind nuclear power plants. However, they are not the greatest source of radiation pollution. The three main sources of radiation pollution, in order, are coal-fired power plants, nuclear power plants, and nuclear explosions. Radiation from coal-fired power plants results from natural radioactive materials in coal. The radiation escaping from nuclear power plants is less than that from coal-fired plants and is continuously monitored to be sure it is under the maximum permissible levels.

However, permissible levels are the subject of much debate and disagreement. Some people feel that any radiation is too much. The possibility of radiation pollution resulting from accidents in nuclear power plants is extremely low, but it is always a possibility. Probably the best-known examples of nuclear reactor accidents occurred at Three Mile Island in the United States (Pennsylvania) and, more recently, at Chernobyl in the Soviet Union.

Direct sources of radiation pollution are nuclear explosions. In the past, test explosions were usually done in the atmosphere. These tests are now banned by the Nuclear Test Ban Treaty, signed by the major nuclear powers in 1963. Most test explosions are presently carried out underground, which reduces the risk of radiation pollution. But not all nations signed the Test Ban Treaty. Several exceptions are France, India, and China.

Another industry-related source of pollution is oil spills. These may arise from leaks in offshore oil wells or from leaks and discharges from ships used in transporting oil. Major incidents of both types have occurred. The most recent spill—in fact, the worst one in history—was the 1989 Alaskan oil spill, which resulted from a supertanker running aground on a reef and leaking some 11 million gallons of crude oil into Prince William Sound, southeast of Anchorage. The environmental effects of such oil spills are devastating and far-reaching.

Domestic Water Pollution

Pollution from domestic sources increases as our population increases. No one wants to be labeled a polluter, particularly at home. But we all contribute, inadvertently and unknowingly. We are now finding that even with accepted disposal methods there are long-term effects of water pollution that are evident only after several years. Of course, immediate health hazards can result from improper and unsanitary disposal practices. In the case of domestic pollution it may be said, quite truthfully, that pollution prevention begins at home. The two major

sources of domestic pollution are detergents and human organic wastes.

A **detergent** in the general sense is a cleansing agent. Ordinary soap, which is a cleansing agent and detergent, is made by combining either fats or oils with alkalies such as sodium or potassium hydroxide. The complex soap molecules in water can effectively interact with grease and oil so that both are carried away with dirt in this solution. This is the detergent action, or detergency, of soap.

Few problems occur with ordinary soap in wastes, because it is normally degraded into harmless substances by bacterial action in sewage. However, the detergent action of soap is impaired by dissolved minerals in water. To accommodate the consumer, industry invented synthetic detergents that react with the dissolved minerals without greatly impairing the cleansing action. But society quickly discovered that the synthetic detergents were nonbiodegradable and not removed by accepted waste-treatment methods.

As a result, the synthetic detergents persisted when discharged into streams and sometimes reappeared as shown in Fig. 24.15. The manufacturing of these "hard" detergents was voluntarily stopped in the mid-1960s. Through chemical modification of the molecular structures, other synthetic detergents were made that were biodegradable. After sufficient time these "soft" detergents are completely broken down by bacterial action. There is still some concern, however, about possible,

Figure 24.15 Suds from detergents.
The use of "hard" detergents once caused such scenes. We have now stopped using hard detergents, replacing them with "soft," biodegradable equivalent products.

yet-unknown effects of the decomposition products on the environment. Detergent sales in the United States are excess of $1–2 billion annually.

Another more obvious pollution problem arises from **detergent builders.** These substances are added to the detergent to make its cleaning action more efficient. In some cases detergent builders account for as much as 40% of the detergent weight. Phosphate compounds are commonly used as detergent builders. The phosphates form compounds with the Ca, Mg, and Fe ions in hard water. This process prevents the ions from reacting with the soap and reducing its detergency. As expected, the waste water has a high phosphate content. Unfortunately, the phosphates are not removed by ordinary waste-treatment processes.

Environmental problems arise from phosphate waste water in lakes, where the phosphates act as a nutrient for microscopic plant algae. A normal amount of algae is good for a lake, because oxygen is added to the water by algae through photosynthesis. Algae also serve as food for fish, which means that the alga population is kept in a natural balance. However, the phosphates from detergents cause the algae to multiply rapidly, or "bloom." The excess algae bloom dies and in time decays as the result of bacterial action. Decay reduces the oxygen content of the water, and so the aquatic animal life of the lake is affected. Thus with the natural balance upset, the lake begins to "die." It becomes covered with algae mats and is so deficient in oxygen that only primitive animal life forms can survive.

The process by which excess algae growth occurs is called **eutrophication** (from the Greek *eutrophos*, "well nourished"). Eutrophication is also a natural process and takes place over thousands of years, gradually turning lakes into marshes. However, human beings hasten the natural aging process of lakes by introducing phosphates, nitrates (fertilizers), and other pollutants. The classic example is Lake Erie, which has aged as much in the past 50 years as it might have normally aged in 15,000 years.

The second major cause of domestic pollution is human organic waste. Large quantities of human wastes, along with the other unwanted liquids of human living, are generally referred to as sewage. Sewage is composed of over 99% water and only 0.02 to 0.04% organic solids. Any concentration of people creates a sewage-disposal problem, and rivers and streams are usually relied upon to carry away waste products (Fig. 24.16).

Although generally banned, the flushing of raw sewage into rivers and streams can cause severe problems. The most critical effect for people is the possibility of

Figure 24.16 Sewage pollution.
Raw-sewage discharge from a sewer overflow.

waterborne diseases such as cholera, typhoid fever, and dysentery. In public systems, water is treated and its quality checked prior to consumption. Private wells, on the other hand, are not frequently checked, and can become contaminated through groundwater seepage.

24.3 The Costs of Pollution

The costs of air pollution are difficult to assess. Visible evidence of the costs resulting from air pollution comes from reduced real estate values, repairs and replacements from corrosion, cleaning costs, animal and agricultural damages, and many more sources, not excluding the effects on human health. Some examples of air-pollution damage are shown in Fig. 24.17.

The corrosion of metals and other building materials is 3 to 30 times greater in cities than in rural areas, depending on the type of pollutant. Corrosive action also depends on the amount of moisture in the air. The rate of corrosion increases sharply when the relative humidity is over 50%.

Pollution effects range from damage to seemingly permanent stone structures to damage to electrical insulators. Concentrations of ozone in photochemical smog can cause the deterioration and cracking of rubber, a common electrical insulator. This deterioration can cause electrical problems and damage. In some cities where photochemical smog is severe, car owners can purchase special anti-ozone tires for their automobiles to help alleviate the problem. Also, dust and particulate matter soil our clothes and home furnishings and can

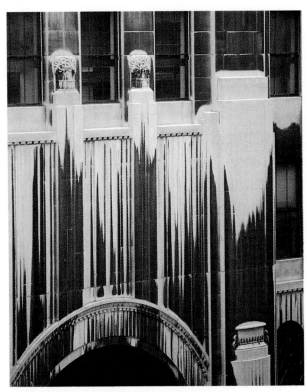

Figure 24.17 Air-pollution damage.
The whitish streaks on this building are not icicles but the effects of air pollution.

Figure 24.18 Copper Basin, Tennessee.
SO$_2$ emissions from the copper-smelting process killed virtually all plant life in the surrounding area, leading to erosion. Emissions controls and land reclamation have helped reverse the situation.

cause damage to delicate instruments. To alleviate some of this problem, we pay for the cost of filtering the air.

Crops, shrubs, trees, and other plants in general suffer from air pollution. There are many cases of industrial pollution having effects on the flora of an area. The classic example is Copper Basin, Tennessee, where a copper smelting plant was built shortly after the Civil War. Its production increased in the late 1800s and consequently, so did its SO$_2$ emissions. In time these deadly fumes killed the plant life of 30,000 acres of the surrounding timberland (Fig. 24.18). Plants are particularly vulnerable to gaseous pollutants. Gases enter the microscopic openings, called stomata, on the underside of the plant leaves and attack the plant cells.

The preceding examples are but a few of the effects of air pollution that result in government estimates for the average yearly cost of air pollution in the United States of about $75 per person. With a population of over 240 million, simple arithmetic shows that the total yearly cost exceeds $18 billion. This figure includes estimates of $100 million for paint damage, $800 million for laundering, $240 million for car washing, and $500 million in damage to livestock and agriculture.

Air pollution has severe effects not only on the objects of our physical environment but also on ourselves. The cost of air pollution on human health is difficult to estimate. How much is your health worth?

As in determining the cost of clean air, determining the cost of clean water is also an insurmountable task. Millions of dollars are spent each year in the treatment of municipal water supplies to make the water safe for human consumption. However, this amount is minor compared with the monetary expenditures needed to alleviate the water-pollution problems discussed in the previous section. These problems are interwoven with land management and waste disposal. The sheer magnitude of these factors makes the analysis of even the indirect costs of clean water enormously complex. Monetary valuations cannot be placed on the effects of water pollution on the environment and on human health.

Because of the large variety of water pollutants, many measures are required to ensure clean water. Specific treatment may be required for some pollutants, particularly toxic industrial wastes. Space limitations in this text do not permit a detailed consideration of the various methods of pollution prevention and water treatment, so we will consider only the general aspects of one major problem: sewage treatment.

Sewage constitutes a large part of water pollution and is an important consideration because of potential health hazards from organic wastes. How is sewage handled, and how is it treated to avoid pollution problems? In rural areas vaults and septic tanks are the normal means of waste disposal and treatment for private homes. Septic tanks are storage containers placed in the ground into which water and human wastes flow or are deposited. Bacterial action breaks down the organic matter, and the water seeps into the ground directly or through subsurface leach beds. However, the lifetimes of these systems are finite, because the soil eventually becomes clogged with nonbiodegradable particulate matter. Also, the septic tank–leach bed is not a good system in areas with heavy clay soils.

As the water percolates through the ground toward the water table, it is further purified by natural processes. However, proper planning is required so that the effluent does not seep into and contaminate nearby sources of drinking water.

In cities and muncipalities, sewers are used to collect waste water (and organic contents) and deliver it to treatment plants for processing before it is discharged into oceans, lakes, or rivers. Some city dwellers may be unaware of how sewage is disposed of, but all householders are intimately aware of the sewage assessments in terms of the bills they must pay. There are two general types of sewer systems—combined and separated systems.

A **combined sewer system** uses the same network as both a sanitary and storm sewer. At one time cities and towns generally used combined systems because the original sewer installation was for storm and rain-water runoff. Human-waste disposal was usually an individual home problem and handled as described previously. With small populations, the receiving streams sufficiently diluted the sewage, and microorganisms consumed or converted the sewage into less noxious products. Large populations, however, produced excessive sewage loads and upset and impaired this process. When the population became sufficiently large that water supplies were threatened with contamination, the original storm sewer system was pressed into combined service. Today, sewage-treatment plants have been installed in most cities, but some raw sewage still seeps into our waterways.

Combined sewers usually had overflow bypasses that diverted part of the load during a rainstorm when the water volume was much greater than usual. This feature safeguarded treatment plants from overload, but the part of the water bypassed usually flowed directly into the receiving streams and, of course, included some amount of human wastes.

Separated sewer systems provide separate networks for sanitation and storm water. The storm-sewer water does not require treatment and may be run directly into the receiving system. The sanitary sewer runs to the treatment plant.

Through applied technology waste-treatment plants essentially speed up the natural processes by which water purifies itself. The two basic methods of treating ordinary municipal wastes are called primary and secondary treatments. In the **primary sewage treatment** there are four main steps, as illustrated in Fig. 24.19. First, the large objects are removed by passing the effluent through screens with openings of a fraction of an inch. This debris is usually removed from the screens and discarded, but some plants have a disintegrator that grinds or breaks the coarse material into smaller pieces. Beyond the first-stage screening process is a grit chamber in which sand,

Figure 24.19 The four steps of primary sewage treatment.
(1) Large objects are removed by screens. (2) Heavier materials settle out in a grit chamber. (3) Suspended solids are removed in a sedimentation tank. (4) Bacteria and odors are destroyed by chlorination.

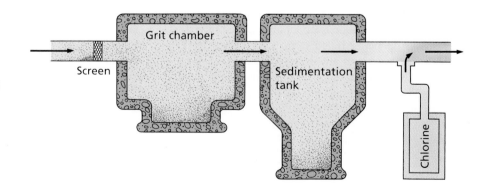

gravel, and heavy, coarse materials settle out. The sediment of the grit chamber may be used for landfill or construction purposes. The grit-chamber stage is particularly important in combined sewer systems with a heavy grit load.

The third step of primary treatment is a sedimentation tank in which suspended solids are removed. This sediment is known as raw sludge and has a high organic content. In some plants the raw sludge is dried and sold or given away as fertilizer. In more sophisticated plants the sludge is put into a heated digestion tank where bacteria digest the solids, producing gaseous by-products. About two-thirds of the gas is methane that may be sold or used as fuel in generating electrical power for the treatment plant.

The final step is the chlorination of the remaining liquid. Chlorination kills disease-causing bacteria and also reduces odors. The product of this primary treatment is then placed into the receiving stream, which dilutes and carries it away.

Some cities in the United States give only primary treatment to their sewage. However, if a community must deal with sewage in large amounts and with a variety of wastes, secondary treatment is recommended. **Secondary sewage treatment** removes up to 90% of the suspended organic matter. This process is accomplished by adding an additional step to the primary treatment between the sedimentation tank and the chlorination. This secondary step is done by either of two processes—trickle-bed filters or activated sludge.

In the **trickle-bed filter process** the sewage effluent is trickled through a bed of stones several feet deep. A

Figure 24.20 Sewage-treatment-plant trickle-filter bed. Bacteria gather and multiply on the stones until they have the capacity to consume most of the organic matter in the sewage effluent.

trickle filter bed is shown in Fig. 24.20. Bacteria gather and multiply on the stones until they have the capacity to consume most of the organic matter in the sewage effluent. A relatively new type of trickle-bed filter is built above ground and uses a plastic material instead of rocks.

The current trend favors the use of an **activated sludge process** for the secondary treatment step. In this process the sewage effluent passes into an aeration tank and then into a large container where it mixes with sludge heavily laden with bacteria (Fig. 24.21). The aeration oxygenates the effluent and speeds up the bacterial breakdown of the organic matter. Though more costly,

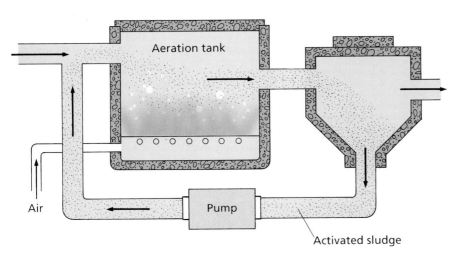

Figure 24.21 Activated-sludge process in secondary sewage treatment. Bacteria in the sludge break down the organic matter in the sewage effluent that was not removed by the primary treatment. The aeration oxygenates the effluent-sludge mixture to replace the oxygen removed in the bacterial action and speeds up the process.

Figure 24.22 The ultimate goal—pure water.

the activated sludge process is more efficient and is a closed system that reduces odors.

The sewage treatment methods just described are generally adequate for normal sewage. However, other treatments are necessary for sewage with complex industrial wastes. Some treatments may be specific for a particular type of chemical waste that cannot be broken down by the bacterial action (nonbiodegradable) of conventional methods. Precipitation and coagulation methods have been used for years to treat industrial wastes. Chemicals are added to precipitate dissolved materials or to coagulate fine, suspended particles. The precipitate or coagulate is then allowed to settle out in a sedimentation tank.

Another special treatment for organic matter that resists normal bacterial action is adsorption. The waste effluent is passed through beds of activated charcoal, which adsorbs as much as 98% of the organic material. Obviously, all of these treatment methods add to the cost of clean water (Fig. 24.22).

24.4 Pollution, Weather, and Climate

Pollution and Weather

Although we sometimes purposely attempt to modify the weather (e.g., rainmaking or fog dissipation), the effects resulting from air pollution are inadvertent modifications of weather and climate. The most immediate modifications, of course, occur near the sources of pollution. Because air pollution results from the activities of human beings, the major concentrations of pollutants are in and around population centers or industrial areas.

Cities have always had somewhat different weather conditions than those of the surrounding countryside. The buildings of a city obstruct the wind and absorb much more insolation, because their surface area is greater than that of the ground they cover. Also, the heat energy is radiated between buildings rather than back into the atmosphere. As a result of such effects, the temperature and heat content in cities are higher than in outlying rural areas. This situation gives rise to what is known as the urban **heat-island effect.**

Because of the concentrated heat of the city, warm air rises from it, carrying combustive wastes and other air pollution. As the rising warm air expands and cools, it flows out over the edges of the city, where it cools further and sinks. This action sets up a thermal circulation cell, as shown in Fig. 24.23. The cooler air on the outskirts of the city flows into the center in the convection cycle, but this is in part cooled polluted city air. Thus the heat-island effect sets up a self-contained thermal circulation system in which the air is continually polluted.

In the absence of winds—e.g., frontal air movement—to break up the circulation system and sweep the pollution away, serious pollution conditions can arise in cities. Pollution is particularly serious if a high-pressure air mass moves over the city and becomes stationary. In this windless condition, along with possible topographical effects, such as a city being in a river valley, the air of a city becomes continually polluted by its own activities, which may eventually have to be curtailed. This condition is not uncommon in several U.S. cities, typically in the summer as a result of a "stagnant" air mass.

You might think that additional insolation on an urban area would supply sufficient heating and convection to break up the self-contained circulatory system of the heat-island effect. This result is a possibility, but it is counterbalanced by the formation of a **dust dome** over the city (Fig. 24.23). City pollutants contain large amounts of aerosols and particulate matter. As the polluted air moves up and spreads out over the city, it forms an often-discernible dust dome. The dust concentration may be several hundred to a thousand times that of rural air. This polluted ceiling decreases the amount of insolation that penetrates the city atmosphere and prevents the Sun from efficiently heating the metropolitan area to break up the circulation cell of the urban heat island. The city must then wait for outside air movements to sweep away its pollution.

Thus air pollution causes cities to have modified weather conditions. They are manifested not only in a

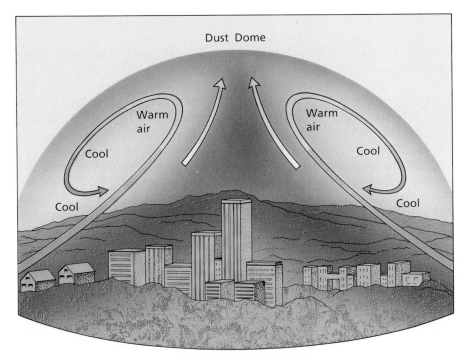

Dust Dome

Warm air

Cool

Cool

Warm air

Cool

Cool

Figure 24.23 Heat-island effect.
Certain conditions produce circulation patterns that create a dust dome over a city.

temperature increase and attenuated insolation, as just mentioned, but also in increased cloudiness and precipitation. The updrafts and concentration of particles give rise to cloud formation and precipitation (Chapter 21). Records show that cities receive several percent more precipitation than nearby rural areas. Also, with drains and sewer systems, the water runoff is more rapid in cities, which results in a lower relative humidity near the surface.

Climate

Although not as immediately obvious, there is a general belief that there are (and will be) changes in the global climate brought about by atmospheric pollution. **Climate** is the long-term average weather conditions of a region. Some regions are identified by their climates. For example, when someone mentions Florida or California, you usually think of a warm climate, and Arizona is known for its dryness and low humidity. Because of such favorable conditions, the climate of a region often attracts people to live there, and thus the distribution of population (and pollution) is affected.

Dramatic climate changes have occurred throughout the Earth's history. Probably the most familiar is that of the Ice Age, when glacial ice sheets advanced southward over the North American continent. The most recent ice age occurred some 10,000 years ago and came as far south as the northern coterminous United States. Evidence from ocean sediment cores supports the theory that dramatic changes in global climate result from subtle, regular variations in the Earth's orbit around the Sun. For example, one cyclic variation is the advance of the Earth in its elliptical orbit. That is, the Earth's closest approach to the Sun occurs at different times of the year in a cycle of 23,000 years. The Earth and the Sun are now closest in January; in 10,000 years they will be closest in July. The result will be cooler summer temperatures, less snow melting, and a growth of the polar ice caps. Such a change could slowly lead into a new ice age.

However, climatic fluctuations continually occur on a smaller scale than that of the Ice Age. For example, in the past several decades there has been a noticeable southward shift of world climate. The Sahara Desert is reportedly advancing southward as much as 48 km (30 mi) a year in some places, and rainfall in some regions is barely half of what it was twenty years ago, giving rise to great bands of land with drought conditions. The ice borders of the northern polar cap appear to be paralleling the shifts of wind patterns and climatic zones, producing record-low temperatures. And apparently, while the Earth's average total rainfall shows little variation, more rain is now falling in the Southern Hemisphere.

The question being asked today is whether or not air pollution may be responsible for some of the observed climate variations. It is reasonably well established that the climate of cities can differ from that of the surrounding countryside, as previously noted; however, the extent of the effects of human activities on global climate is not clear. For example, from 1880 to 1940 the average annual temperature of the Earth's surface increased by about 0.6°C (1.1°F). Since 1940 the average temperature has decreased by about 0.3°C (0.55°F). Associated with this lowering temperature has been a shift in the frost and ice boundaries, a weakening of zonal wind circulation, and marked variations in the world's rainfall pattern. However, this pattern appears to be changing, possibly as a result of the depletion of the ozone layer, as will be discussed shortly.

Global climate is sensitive to atmospheric contributions that affect the radiation balance of the atmosphere. These contributions include the concentrations of CO_2 and other "greenhouse" gases, the particulate concentration, and the extent of cloud cover, which affect the Earth's albedo. Air pollution and other human activities do contribute to changes in these conditions. Scientists are now trying to understand climate changes by using various models. These models of the workings of the Earth's atmosphere and oceans allow them to compare theories, using historical data on climate changes; however, specific data are scant.

More recent natural occurrences have received a great deal of study. In 1982–83 the "Big El's"—El Niño and El Chichón—provided data. **El Niño** is a Pacific current that sporadically rises off the coast of Peru and Ecuador (the western coast of South America). El Niño is the Spanish name for "The Child," so named because the current usually occurs around Christmas. Every 8 to 10 years El Niño grows more intense, and its nutrient-poor waters destroy rich coastal fishing areas.

The 1982–83 El Niño was a record breaker. The ocean waters were 7–8°C warmer than usual for more than a year. This El Niño was connected to a failure of the equatorial trade winds that normally push the warm waters away from the South American coastline into the Pacific. Without these winds, the warm currents pushed against the coast, bringing heavy rains to normally dry areas. The arid regions of the coastal countries received more than a hundred times the normal precipitation. On the other side of the Pacific, Australia and the Philippines had droughts.

In March 1982 the volcano **El Chichón** in Mexico erupted in one of the century's most significant outbursts (Fig. 24.24). The eruption sent debris 42 km (26 mi) into the atmosphere. (In contrast, the eruptions of Mount St. Helens sent debris about 19 km (12 mi) high. See Chapter 25.) Within a month the volcanic veil had crept over North America. The heavy ash fell back to Earth, but a layer of fine particles, sulfuric acid, and salt remained. El Chichón was situated over a natural salt dome.

Volcanic eruptions can affect the climate through changes in the albedo. For example, the 1815 eruption of the volcano Tambora, located on an island just east of Java, vented an estimated 146 km³ (35 mi³) of particle debris into the air. Fine volcanic dust was circulated

Figure 24.24 El Chichón eruption.
The 1982 eruption of this Mexican volcano sent debris 42 km (26 mi) into the atmosphere.

around the Earth by global wind patterns, and the winter of 1816 was unseasonably cold because of the change in albedo. New England farmers called 1816 "the year without summer," with frosts in June and July.

A similar albedo concern has been dubbed **nuclear winter.** Nuclear war could plunge the Earth into severe cold. Extensive fires and massive quantities of smoke could shut off sunlight and cool the Earth.

Of course, we can't do as much about volcanic eruptions as we can about the potential of nuclear winter; however, the possible climatic effects are a concern. Such eruptions could perhaps affect the jet streams, which could lead to climatic changes. Also, with volcanic gases being projected into the stratosphere, scientists are concerned about possible chemical reactions with and depletion of the ozone layer. Other environmental concerns about the ozone layer are discussed in the chapter Highlight.

Thus, **particulate pollution** could contribute to changes in the Earth's thermal balance by decreasing the transparency of the atmosphere to insolation. An increase in the albedo would result in lower surface temperatures. The effect of particulate pollution depends on the number and size of the particles. Small particles can also cause increased absorption of the outgoing infrared terrestrial radiation.

Contributions to an albedo change could also come from an increase in cloud coverage that might result from an abundance of particle nuclei. Recall from Section 20.4 that clouds are good reflectors of solar radiation, and thus they play a major role in the Earth's albedo. Increased aircraft activity in the upper troposphere in recent decades may have resulted in an increase in cirrus clouds.

Supersonic transport **(SST)** aircraft operating in the lower stratosphere have also caused some concern because of particulate and gaseous emissions from their jet engines (Fig. 24.25). The hourly combustion of fuel by an SST releases about 73,000 kg of water vapor and 63,000 kg of carbon dioxide into the lower stratosphere. In the troposphere, precipitation processes act to "wash" out particle and gaseous pollutants, but there is no snow or rain washout mechanism in the stratosphere. Also, the stratosphere is a region of high chemical activity, and chemical pollutants (e.g., NO_x and hydrocarbons) could possibly give rise to climate-changing reactions.

Albedo considerations also arise from activities that affect the Earth's surface. Thus urbanization and agriculture affect the surface albedo.

Figure 24.25 Stratospheric pollution.
The supersonic transport (SST), produces a great deal of noise pollution during take off and landing, and it also releases pollutants into the lower stratosphere, where there is no weather to wash them back to Earth.

There is also a temperature-increase pollution aspect. Vast amounts of CO_2 are being expelled into the atmosphere as a result of the combustion of fossil fuels. As we discussed in Section 20.4, CO_2 and water vapor play important roles in the Earth's energy balance, because of the greenhouse effect. An increase in the atmospheric concentration of CO_2 could alter the amount of radiation absorbed from the Earth's surface and produce an increase in the Earth's temperature. An early calculation of the effect of such an increase was made in 1896 by Svante Arrhenius. Arrhenius's calculations showed that a doubling of the atmospheric **CO_2 concentration** would increase the Earth's surface temperature by 5°–6°C. Arrhenius's concern arose from the CO_2 escaping from volcanoes, not from automobile exhaust emissions.

During the nineteenth century the atmospheric CO_2 content increased by about 10 percent, as determined from old records. During the twentieth century, because of larger populations and more combustion, the increase has been even greater. It appears that the Earth will be a warmer place in the twenty-first century, with climate changes that will affect agriculture, water resources, and sea level. Recent studies predict that the atmospheric CO_2 content will double by the year 2065, with an accompanying temperature increase of 1.5°–4.5°C. Note that Venus has a rich CO_2 atmosphere—200,000 times more than that of Earth—and a surface temperature of

HIGHLIGHT

Atmospheric Ozone

In 1974 scientists in California warned that CFCs (chlorofluorocarbons) might seriously damage the ozone layer through depletion. Observations generally supported this prediction, and in 1978 the United States put a ban on the use of these gases in aerosol spray cans. Even so, more than 635,000,000 kg of CFCs continue to leach into the atmosphere each year, primarily from refrigerants, plastic foam containers (such as those used to keep fast foods warm), and spray propellants manufactured in other countries.

The major CFCs are called CFC-11 (CFCl$_3$) and CFC-12 (CF$_2$Cl$_2$). When released, they slowly rise into the stratosphere (a process that takes 20 to 30 years). In the stratosphere the CFC atoms are broken apart by ultraviolet radiation, releasing reactive chlorine atoms. These atoms in turn react with and destroy ozone molecules in a repeating cycle:

$$CFC \xrightarrow{uv} Cl$$

$$Cl + O_3 \longrightarrow ClO + O_2$$
(Ozone)

$$ClO + O \longrightarrow Cl + O_2$$

Notice that the chlorine atom is again available for reaction. These atoms may remain in the atmosphere for a year or two. In this time a single Cl atom may destroy as many as 100,000 ozone molecules.

Measurements indicate that the concentrations of CFC-11 and CFC-12 have more than doubled in the past 10 years. Worldwide ozone levels have declined an estimated 3–7 percent over the past few decades. Part of this decline is the result of normal fluctuations. But in 1985 scientists announced that they

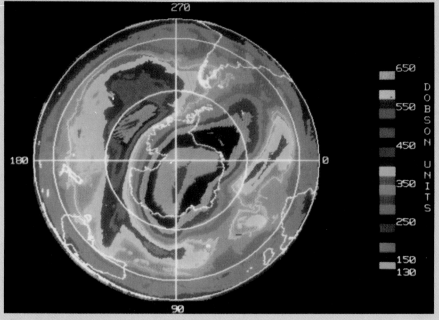

Figure 24.26 Ozone hole over the South Pole.
This satellite spectrometer picture shows the ozone "hole" as a gray, blue, and black oval generally covering the Antarctic. The hole is surrounded by regions of high total ozone (yellow, brown, and green in color).

measured a 40–50 percent reduction of the ozone over the south pole. (see Fig. 24.26).

This "hole" in the ozone layer opens up annually during the southern springtime months of September or October. Scientists speculate that the Antarctic hole is the result of chemical processes, probably involving CFCs. Ice particles in high-altitude clouds in the stratosphere at the end of the polar winter are thought to provide a surface on which the chemical reactions take place. Such clouds and ozone holes are not evident at lower latitudes.

The depletion of the ultraviolet-absorbing ozone layer will have some significant effects. Experts estimate that the cases of skin cancer will increase by 60 percent and that there will be many additional cases of cataracts. Crops and climate will also be affected. Sea levels will increase somewhere 1–4 m (2–12 ft) by the year 2100 as a result of global warming and melting of the polar ice caps. Also, CFCs are greenhouse gases and can contribute to warming in this manner.

The reduction of CFCs will not be done easily or quickly. In the United States alone products worth more than $125 billion rely on CFCs. Currently, over 35 nations have signed an agreement on a schedule for the cutback of CFC production. The 1989 production is to be held at 1986 levels, and by 1999 production should be about 50 percent of that level.

470°C (900°F), hot enough to melt lead. Of course, Venus is closer to the Sun than is Earth.

Possible effects of increasing temperature include the melting of the polar ice caps, as noted previously in the discussion on the effects of the depletion of the ozone layer. Also, there would be drier summers in the middle latitudes. In the United States there would be dry farmlands in the South and longer growing seasons in the North. Many other possibilities are not well understood, including the accelerated release of greenhouse gases, such as methane (CH_4), from swamps and bogs because of the warming climate.

Is it then possible that an increase in atmospheric CO_2 concentration led to the increase in global temperature (and subsequent changes in the climate) observed from 1880 to 1940? Scientists generally accept this possibility. However, note that a substantial portion of the change in the atmospheric CO_2 concentration has occurred in the past several decades, during which the contribution from the activities of human beings almost doubled. At the same time (since 1940), the average global temperature has shown a *decrease*. Evidently, air pollution also affects other mechanisms of the atmosphere's energy balance.

Thus air pollution may contribute to a variety of atmospheric effects—both local and global. The effects of pollution on local climatic conditions are somewhat clear, but there are too few data to understand or accurately predict global climatic effects. We are inclined to believe that increased atmospheric CO_2 concentrations and ozone-layer depletions would give rise to an increase in global temperature. This temperature increase would cause more water evaporation and an increase in the relative humidity. Particulate pollution could then give rise to increased cloud formation, which would increase the albedo and cause a decrease in global temperature.

Certainly, little comfort can be found in this speculative, counterbalancing pollution cycle. There are too many unanswered questions on the effects of pollution, which has occurred over a relatively short time, on the natural interacting cycles of the atmosphere and biosphere, which have taken millions of years to become established.

Learning Objectives

After reading and studying this chapter, you should be able to do the following without referring to the text:

1. Discuss and describe the major air pollutants.

2. Distinguish between regular smog and photochemical smog.

3. Discuss and describe the major land and water pollutants.

4. Define the term *biodegradable,* and explain how it is relevant to the pollution question.

5. State the major sources of air pollution.

6. State the sources and types of land and water pollution.

7. Explain some of the damaging effects of various air pollutants.

8. Explain eutrophication and the effects of pollution.

9. Distinguish between (a) combined and separated sewer systems and (b) primary and secondary sewage treatments.

10. Discuss the effects of air pollution on local weather conditions.

11. Discuss the effects of air pollution on global climate.

12. Explain the causes for the concern about the ozone layer.

13. Define and explain the other important words and terms listed in the next section.

Important Words and Terms

pollution	biodegradable	sanitary landfills
smog	carcinogens	air-fuel ratio
fossil fuels	ozone	pesticides
incomplete combustion	sulfur	fertilizers
impure fuels	sulfur dioxide	phosphates
carbon monoxide	acid rain	nitrates
nitrogen oxides	solid waste	animal wastes
photochemical smog	littering	

mercury pollution
thermal pollution
radiation pollution
detergent
detergent builders
eutrophication
combined sewer system

separated sewer systems
primary sewage treatment
secondary sewage treatment
trickle-bed filter process
activated sludge process
heat-island effect
dust dome

climate
El Niño
El Chichón
nuclear winter
particulate pollution
SST
CO_2 concentration

Questions

Pollutants

1. Define *air pollution.*

2. Is air pollution a relatively new problem?

3. What gives rise to the majority of air pollution?

4. What are the products of complete combustion? of incomplete combustion?

5. Are nitrogen oxides products of complete or incomplete combustion? Explain.

6. Distinguish between plain smog and photochemical smog.

7. What is one of the best indicators of photochemical smog?

8. What is the major fossil-fuel impurity?

9. What are the causes and effects of acid rain? In which areas is acid rain a major problem and why?

10. What is one of the chief sources of land pollution?

11. Approximately how much domestic solid waste is generated per capita in the United States?

12. What is the potential pollution danger of sanitary landfills?

13. How have human activities given rise to land and water pollution?

Sources of Pollution

14. The air-fuel mixture is important in an automobile engine. What is the proper air-fuel ratio, and what are the effects of an improper air-fuel ratio?

15. Name the major source of each of the following pollutants:
    (a) carbon monoxide   (b) sulfur dioxide
    (c) particulate matter   (d) nitrogen oxides

16. State the sources and types of water pollution.

17. What are the effects of pesticide pollution?

18. How do fertilizers contribute to water pollution?

19. Why are animal wastes a current potential pollution problem?

20. What are the effects of mercury poisoning?

21. What is thermal pollution, and what are its sources and effects?

22. What are the major sources of radiation pollution?

23. What are detergents, and how do they contribute to water pollution? What are detergent builders?

24. Describe the causes and effects of eutrophication resulting from pollution.

The Costs of Pollution

25. What is the estimated yearly cost of property damage resulting from the effects of air pollution in the United States?

26. Describe some of the costly effects of air pollution.

27. Why are plants particularly vulnerable to gaseous air pollutants?

28. What is the difference between combined and separated sewer systems?

29. What are the four main steps in primary sewage treatment?

30. How does secondary sewage treatment differ from primary sewage treatment?

31. Describe the operation of trickle-bed filters and the activated-sludge process.

Pollution, Weather, and Climate

32. What are the urban heat-island and dust-dome effects, and what are their effects on urban weather conditions?

33. How has the Earth's average temperature varied over the past 100 years?

34. What is (are) the possible effect(s) of increased atmospheric CO_2 concentrations?

35. What are possible explanations for changes in the Earth's average temperature?

36. What effects could CFCs have on the Earth's climate?

37. What is the concern about air pollution in the stratosphere?

38. How could CO_2 pollution be decreased? Consider solar and nuclear energy sources and economic effects.

Exercises

1. Taking the population of the United States to be about 250 million, estimate the amount of domestic solid waste generated each day in this country. (*Hint:* Consider the amount of waste per capita.)

   Answer: about 1.0 million tons

2. The average amounts of domestic water use are as follows:

	(avg. L)
Flushing a toilet (once)	15
Washing dishes	38
A washing machine load	94
A shower (5 min)	94
A tub bath	132
Watering lawn (1 h)	1135

   Using these values, estimate the (a) daily, (b) monthly, and (c) yearly amounts of water used by a single person and a family of four.

CHAPTER 25

Minerals and Rocks

> Touch the earth, love the earth, honour the earth, her plains, her valleys, her hills and her seas; rest your spirit in her solitary places.
>
> —Henry Beston

Historically, **geology** refers to the study of the planet Earth, and its composition, structure, and history. Today, this definition is expanded to include the study of the moon and other planets.

The planet Earth is unique in having abundant water, rich soil, and an atmosphere with the right amount of oxygen. These resources, plus the planet's position and orientation in respect to the Sun, provide energy and temperatures that support life. With the recent increasing emphasis on natural resources obtained from Earth, geology has taken on a new importance. Except for water and soil, mineral resources such as coal, gas, and petroleum are nonrenewable. The geologist must discover new mineral deposits to provide the energy-producing materials on which our civilization is based. An understanding of the various geologic processes is critical in locating and developing these mineral deposits.

To illustrate how our way of life depends on Earth's mineral deposits, consider what would happen if liquid fuels became unavailable. Our reserves of processed fuels would be consumed within a brief period of time, and every heat engine operating on gasoline or similar fuel would stop. Thus trucks, cars, airplanes, trains, farm tractors, and other machines, including industrial equipment, would cease to operate. The result would be chaos in our present way of life.

In this chapter and the following four chapters we introduce and discuss the concepts of geology necessary to comprehend the physical nature of the planet on which we live. The study begins with the outer layer of the planet. This outer layer, called the crust, is a thin shell only a few miles (it varies from 4.8 to 48 km) thick. The crust is composed of minerals and rocks, and this first chapter on geology begins with a discussion of minerals and is followed by a discussion of rocks and the activity that forms them. Next, we will concentrate on the structural geology of the planet, the interactions of the

planet's crust and internal processes, and the methods of geological dating and interpreting Earth's history. We conclude our study with the processes that wear away and level Earth's surface.

25.1 Minerals

A **mineral** is a naturally occurring, crystalline, inorganic substance (element or compound) that possesses a fairly definite chemical composition and a distinctive set of physical properties. Minerals are composed, for the most part, of eight elements. These elements, along with their relative abundance in Earth's crust, are listed in Table 25.1. Note that two elements, oxygen and silicon, make up about 75 percent of the crust. Over 2000 minerals have been found in Earth's crust; approximately 20 of them are common, and fewer than 10 account for over 90 percent of Earth's crust by weight.

The minerals are arranged naturally in groups to form a consolidated mixture called rock. A **rock** is defined as a natural aggregate of one or more minerals. Rocks are composed of 90 percent or more, by volume, oxygen atoms. Therefore, the oxygen atom (or ion) is the dominating influence controlling the number of possible element combinations in the formation of minerals.

Table 25.1 Relative Percentages of Elements in Earth's Crust

Element	Approx. Percentage (Weight)
Oxygen (O)	46.5
Silicon (Si)	27.5
Aluminum (Al)	8.1
Iron (Fe)	5.3
Calcium (Ca)	4.0
Magnesium (Mg)	2.7
Sodium (Na)	2.4
Potassium (K)	1.9
All Others	1.6
Total	100.0

◀ **Marble Gorge below Nankoweap Canyon, Grand Canyon National Park, Arizona.**

Most rock-forming minerals are composed mainly of oxygen and silicon. The fundamental silicon-oxygen compound is silicon dioxide, **silica**, which has the formula SiO_2. Quartz, a hard and brittle solid, is an example. Because carbon and silicon are in the same chemical group, one might think that SiO_2 would be a gas similar to carbon dioxide, CO_2. But the bonding is very different in the two compounds. The silicon-oxygen structure of quartz, SiO_2, is based on a network of SiO_4 tetrahedra (Fig. 25.1) with shared (covalent bond) oxygen atoms rather than SiO_2 molecules. See Fig. 25.2.

In silica the oxygen-to-silicon ratio is 2 to 1; however, the oxygen-to-silicon ratios in the **silicates** are greater than 2 to 1 and can vary greatly. The ratio varies greatly because the silicon-oxygen tetrahedra may exist as separate independent units or may share oxygen atoms at corners, edges, or sometimes faces in many dif-

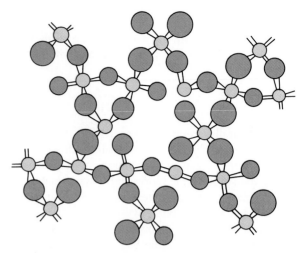

Figure 25.2a The structure of quartz, SiO_2.
The structure is based on interlocking SiO_4 tetrahedra, in which each oxygen atom (buff) is shared by two silicon atoms (blue).

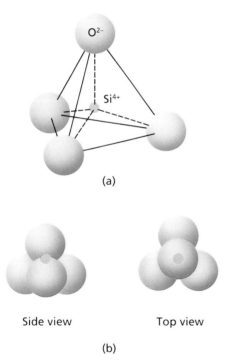

(a)

Side view   Top view

(b)

Figure 25.1 An illustration of the silicon-oxygen tetrahedron.
(a) An expanded view of the complex ion $(SiO_4)^{4-}$ showing the larger oxygen ions (O^{2-}) at the four corners of the tetrahedron (three-sided pyramid) equidistant from the smaller silicon ion (Si^{4+}) at the center. (b) Close-packed side and top views.

Figure 25.2b Quartz—one of the most common minerals in igneous, sedimentary, and metamorphic rocks.
Under favorable conditions, it occurs in hexagonal crystals. It is harder than glass.

ferent ways. Thus the structures of the silicate minerals are determined by the way the SiO_4 tetrahedra are arranged.

The rock-forming minerals are composed mainly of oxygen and silicon plus, in most cases, aluminum and at least one more element from Table 25.1. Feldspars are the most abundant minerals found in Earth's crust. There are two main types: (1) plagioclase feldspar, which contains oxygen, silicon, aluminum, and calcium or sodium (see Table 25.2 for the chemical formula of each);

(2) potassium feldspar, composed of oxygen, silicon, aluminum, and potassium. This feldspar is sometimes called orthoclase feldspar. The properties of each type of feldspar are given in Table 25.2.

Table 25.2 also lists several nonsilicate materials and their general or economic use. The nonsilicate materials include the carbonates (e.g., calcite), sulfides (galena), sulfates (gypsum), halides (halite), oxides (hematite), and native elements such as gold, silver, sulfur, and diamond.

The main rock-forming materials are all silicates. Some silicate structures are illustrated in Fig. 25.3a, which is a photograph of typical minerals. Olivine is a mineral with a single independent tetrahedron that is combined with either Mg^{2+} or Fe^{2+}. These metal ions can substitute for each other, depending on the conditions of mineral formation.

The silicon-oxygen tetrahedra can form single- and double-chain structures, as shown in Fig. 25.3b. Examples of minerals with these structures are pyroxene and hornblende. In the pyroxene chain each tetrahedron shares

Figure 25.3a Some common minerals.
Left to right: muscovite mica, pyroxene (acnite), plagioclase (albite), olivine, and amphibole (hornblende).

Silicate structure	Arrangement of tetrahedra (top view)	Typical mineral
Single independent tetrahedron		Olivine
Single chain		Pyroxene
Double chain	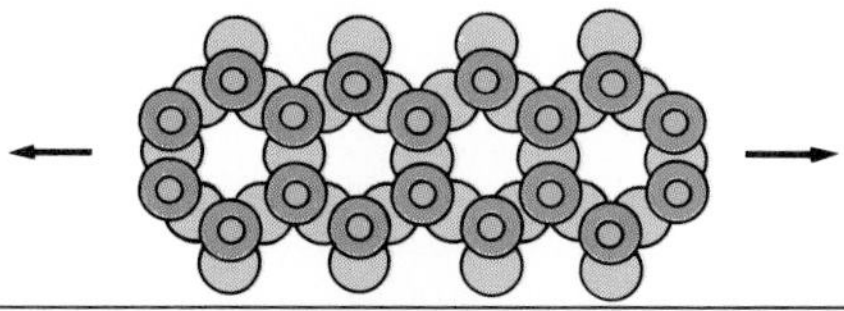	Hornblende
Continuous sheet	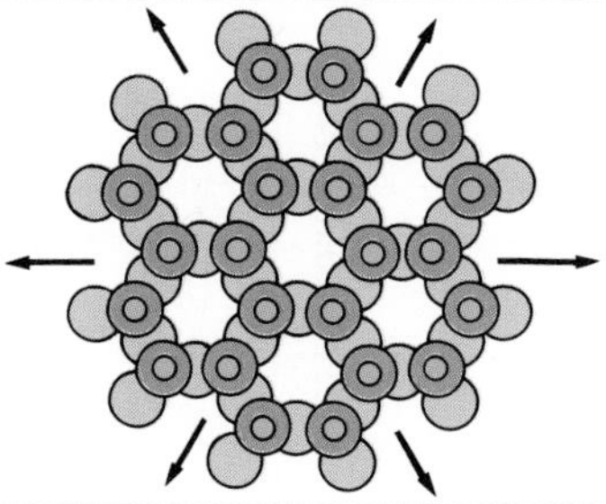	Mica
Three-dimensional work	Too complex to be shown by simple two-dimensional sketch	Quartz and feldspar

Figure 25.3b Molecular structures of several common minerals.

Table 25.2 Some Common Minerals and Their Properties

Minerals	Chemical Composition	Color	Luster	Hardness (Mohs' Scale)	Specific Gravity (approx.)	Cleavage	General Use or Occurrence
Asbestos (chrysotile)	Hydrous magnesium silicate, $H_4Mg_3Si_2O_9$	Green (different shades)	Silky	Low	2.5	Yes	Fireproofing; insulation
Calcite	Calcium carbonate, $CaCO_3$	White or colorless	Glassy to earthy	2.5–3	2.7	Yes	Cement manufacture; optical application; lime
Clay	Hydrous aluminum silicates containing Na, K, Fe, Mg, etc.	Light	Earthy	Very low	2.6	Fracture	Chinaware, pottery
Dolomite	Calcium (Mg, Fe, Mn, Ca) carbonate, e.g., $CaMg(CO_3)_2$	White-gray	Pearly	3.5–4.5	2.9	Yes	Cement manufacture; building materials; lime
Feldspar plagioclase	Calcium and sodium aluminum silicate, $CaAl_2Si_2O_8NaAlSi_3O_8$	White-gray	Pearly	6–6.5	2.7	Yes	Ceramic glazes
orthoclase	Potassium aluminum silicate, $KAlSi_3O_8$	White-pink	Pearly	6	2.5	Yes	Elongated crystals in igneous rock
Fluorite	Calcium fluoride, CaF_2	Clear and range of colors	Vitreous	4	3.2	Yes	Commonly found in metal ores
Galena	Lead sulfide, PbS	Gray	Metallic	2.5	7.4	Yes	Lead ore
Garnet	Orthosilicates	Commonly red	Vitreous	6.5–7.5	3.1–4.3	Fracture	Gemstones, abrasives
Gypsum	Hydrous calcium sulfate, $CaSO_4 \cdot 2H_2O$	Colorless or white	Silky-dull	2	2.3	Yes	Plaster of Paris; neutralizer of alkaline soils
Halite	Sodium chloride, NaCl	Colorless or white	Glassy	2.5	2.2	Yes	Table salt
Hematite	Iron oxide, Fe_2O_3	Gray reddish	Metallic, dull	5.5–6.5	5.1	Fracture	Iron ore
Hornblende	Hydrous Ca, Na, Mg, Fe, Al silicates	Green or blue-green to black	Vitreous	5–6	2.9–3.4	Yes	Rock-forming mineral
Magnetite	Iron oxide, Fe_3O_4	Black	Metallic, dull	6	5.2	Fracture	Magnetic iron ore

Mineral	Chemical composition	Color	Hardness	Luster	Specific gravity	Cleavage	Uses
Mica							
biotite	Group of hydrous silicates	Black-green	2.5–3	Pearly	3.0	Yes	"Isinglass" heat-proof windows; electrical insulator
muscovite		Reddish-brown	2.5–3	Pearly	2.9	Yes	
Olivine	Metallic silicate, $(Mg, Fe, Mn)_2SiO_4$	Green	6.5–7	Glassy	3.2–4.4	Fracture	Igneous rock mineral
Pyrite	Iron sulfide, FeS_2	Pale yellow	6–6.5	Metallic	5.0	Fracture	Source of sulfuric acid (mine acid); fool's gold
Pyroxene	Metallic aluminum silicate (Ca, Mg, Fe, Na)	Black-green	5–6	Glassy	3.4	Yes	Igneous rock mineral
Quartz	Silicon dioxide, SiO_2	Colorless when pure	7	Glassy	2.65	Fracture	Optical applications
Serpentine	$3\,MgO, 2\,SiO_2, 2\,H_2O$	Greenish; brownish	2–5	Greasy or waxy	2.5–7.65	None	Source of Mg; decorative stone
Sphalerite	Zinc sulfide, ZnS	Black, yellow-brown	3.5–4	Resinous	4.0	Yes	Zinc ore
Talc	Hydrous magnesium silicate, $Mg_3(Si_4O_{10})(OH)_2$	Green-white, silvery	1–2.5	Pearly, greasy	2.7	Yes	Cosmetics

two oxygens, whereas in the double hornblende chain half the tetrahedra share two oxygens and the other half share three oxygens. These structures have various metallic ion components, as in olivine. (See Table 25.2.) The two-dimensional sheet structure of tetrahedra shown in Fig. 25.3 is the structure of mica. The three-dimensional silicate structures are too complex to be shown in a simple illustration.

Everywhere around us we see minerals or the products of minerals. Some are quite valuable, while others are essentially worthless. For example, precious stones such as diamonds and rubies are valuable minerals. Also, we speak of a nation's mineral wealth when referring to natural raw materials such as ores or minerals containing iron, gold, silver, or copper. However, the minerals of common rock, like sandstone, have little monetary value.

The term *mineral* has also taken on popular meanings. For example, foods are said to contain vitamins and "minerals." In this case the term *mineral* refers to compounds in food that contain elements needed in small quantities by the human body, such as Fe, Na, I, Mn, Mg, and Cu. The chemical compositions of these compounds may be the same as those of naturally occurring minerals that make up Earth's crust, or they may actually be the naturally occurring minerals themselves. The names of minerals, like those of chemical elements, have historical connotations and may reflect the names of localities.

Mineral classification is advantageous because it is based on the physical and chemical properties of substances, which distinguish between different forms of minerals composed of the same element or compound. For example, graphite—a soft, black, slippery substance commonly used as a lubricant—and diamond are both composed of carbon. But because of different crystalline structures, their properties are quite different. (Graphite mixed with other substances to obtain various degrees of hardness is the "lead" in lead pencils.)

Minerals can be identified by chemical analysis, but most of these methods are detailed and costly and are not available to the average person. More commonly, the distinctive physical properties are used as the key to mineral identification. See Table 25.2 for a list of several minerals with their chemical composition and some of their more distinctive properties. These properties are well known to all serious rock and mineral collectors. Some of the physical properties used in mineral identification are described in the following paragraphs.

All crystalline substances crystallize in one of seven

major geometrical patterns. When a mineral grows in unrestricted space, the mineral develops the external shape of its crystal form. However, during the growth of most crystals the space is restricted, resulting in an intergrown mass of crystals that does not exhibit its crystal form. A branch of physical science called crystallography deals with the external shapes of crystals and with the geometrical relations between the atomic planes of crystals. The crystalline forms are studied in detail by means of X-ray analysis.

Hardness is a comparative property and so is indicated by a harder mineral being able to scratch a softer one. The varying degrees of hardness are represented on **Mohs' scale of hardness,** which runs from 1 to 10, soft to hard. This arbitrary scale is expressed by the 10 minerals listed in Table 25.3. Talc is the softest and diamond is the hardest. A particular mineral on the scale is harder than (can scratch) all those with lower numbers. Using these minerals as standards, one finds the following on the hardness scale: fingernail, 2–3; a penny, 3; window glass, 5–6.

Cleavage refers to the tendency of some minerals to break along definite smooth planes. The mineral may exhibit distinct cleavage along one or more planes, or it may exhibit indistinct cleavage or no cleavage. The degree of cleavage that a mineral exhibits is a clue to the identification of the mineral.

Color refers to the property of reflecting light of one or more wavelengths. Although the color of a mineral may be impressive, it is not a reliable property for identifying the mineral, because the presence of small amounts of impurities may cause drastic changes in the color of some minerals.

Streak refers to the color of the powder of the mineral. A mineral may exhibit an appearance of several colors, but it will always show the same streak. A mineral rubbed (streaked) across the surface of an unglazed porcelain tile will thereby be powdered and will show its true color.

Luster refers to the appearance of the mineral's surface in reflected light. Mineral surfaces appear to have a metallic or nonmetallic luster. A metallic luster has the appearance of polished metal; a nonmetallic appearance may be of varying lusters likened to the materials listed oppositely below.

Adamantine	appearance of a	diamond
Greasy	appearance of	oily glass
Pearly	appearance of a	pearl
Resinous	appearance of	yellow resins
Silky	appearance of	silk
Vitreous	appearance of	glass

Crystalline structure refers to the way the atoms or molecules that make up the mineral are arranged internally. This arrangement is a function of the size and shape of the molecules and the forces that bind them.

Fracture refers to the way a mineral breaks. The mineral may break into splinters, ragged or rough irregularly surfaced pieces, or shell-shaped forms known as conchoidal fractures.

Tenacity refers to the ability of the mineral to hold together. Some minerals are tough and durable; others are fragile and brittle.

Magnetism refers to the property of possessing a magnetic force field. A mineral possessing magnetism can be detected by a magnetic compass.

Fluorescence refers to the emission of light (to which the eyes are sensitive) by a mineral that is being stimulated by the absorption of ultraviolet or X-ray radiation.

Phosphorescence refers to the emission of light by a mineral after the stimulating source (rays or ultraviolet radiation) has been removed.

25.2 Rocks

A rock is defined as a natural aggregate of one or more minerals; rock is an essential and substantial part of Earth's crust. When we look at a mountain cliff, we see rock rather than individual minerals. The majestic mountains of our western states are made of rock. The Colorado River has carved the Grand Canyon through layers of rock, and the continents and ocean basins are composed of rock.

Rocks are classified into three major categories, based upon the way they originated. The first two types

Table 25.3 Mohs' Scale of Hardness

1. Talc	6. Feldspar (orthoclase)
2. Gypsum	7. Quartz
3. Calcite	8. Topaz
4. Fluorite	9. Corundum
5. Apatite	10. Diamond

originate deep within Earth, and the third is formed at Earth's surface. The two formed deep within Earth are:

1. **Igneous rocks,** which solidify from hot molten material called *magma*

2. **Metamorphic rocks,** which have been transformed from preexisting rocks by high temperature or pressure or both

3. **Sedimentary rocks,** which can form in three ways:
   a. The lithification of any preexisting sediment
   b. The precipitation of a mineral from a solution
   c. The consolidation of plant or animal remains

The rock cycle, shown in Fig. 25.4, is a graphic method used to illustrate the interrelationships among the processes that produce the three types of rock.

Magma is hot molten liquid located deep within Earth and composed of rock-forming materials. These materials are mostly silicates and steam held under great pressure. Because the magma is a hot liquid, the ions move freely in the melt. However, when the temperature decreases, the ions begin to crystallize and igneous rock is formed.

The general sequence of mineral formation (crystallization) with decreasing temperature is shown, in a simplified way, in Fig. 25.5. The diagram illustrates what is called the *Bowen reaction series.* A close look at the diagram shows that there are three separate series in which crystallization takes place. The Bowen reaction series can be applied only to certain basaltic magmas, but the series illustrates the formation of igneous geology in general.

Basaltic magma is a term applied to the hot melt that solidifies into a huge mass of rock called basalt. The word *basalt* is believed to have had an ancient Ethiopian origin, meaning "black stone." Although basalt rock is formed deep within Earth, the rock becomes exposed by uplift or erosion at the surface.

The diagram in Fig. 25.5 illustrates that minerals crystallize at different temperatures. The first material to crystallize from the hot melt is olivine, and in most cases, at approximately the same temperature calcic plagioclase begins to crystallize. The olivine-through-biotite series is a discontinuous series; that is, the formation of each mineral takes place in discrete steps. Plagioclase-through-feldspar formation is a continuous series that

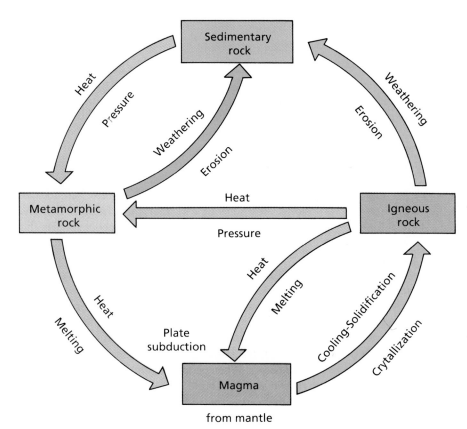

Figure 25.4 An illustration of the rock cycle.
Rock material in the lithosphere follows various routes in the cycle. Plate addition along mid-oceanic ridges and plate destruction in subduction zones correspond, respectively, to entrance to and exit from the rock cycle.

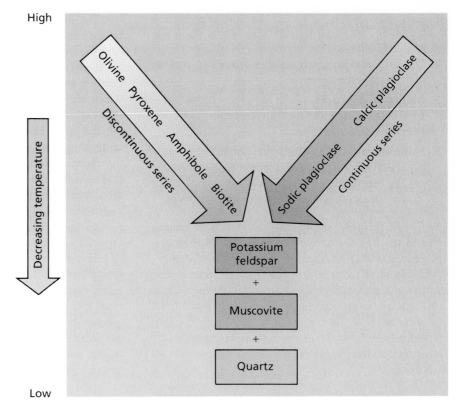

Figure 25.5 The Bowen Reaction Series.
The diagram illustrates the general sequence of crystallization of minerals with decreasing temperature in a basaltic melt. The minerals at the top crystallize first, the ones at the bottom last.

begins with calcium-rich plagioclase and, as the temperature falls, it gradually becomes more abundant in sodium ions. Several of these materials may exist simultaneously at any one time. The third series, which is not a real series, takes place after most of the magma has solidified. The melt that remains forms the minerals potassium feldspar, muscovite, and quartz. The whole process is more complex than the diagram indicates, however.

25.3 Igneous Rock

As mentioned above, igneous rock forms when molten material from somewhere deep beneath Earth's surface cools and solidifies. Accumulations of solid particles thrown from the throat of a volcano are also considered igneous rocks. The molten material is known as magma as long as it is beneath Earth's surface but becomes **lava** if it flows on the surface. The term *lava* may be a bit confusing, because it can be used to mean either the hot molten material or the resulting igneous rock.

Assuming that Earth was originally molten, the very first rocks of the continents and ocean basins must have been igneous. Geologic processes, however, long ago obscured any recognizable remnant of these most ancient materials of Earth. Nevertheless, igneous activity has continued so actively throughout the history of Earth that igneous rocks are by far the most abundant type; they are estimated to constitute as much as 80 percent of Earth's crust.

Most people, of course, know more about lava than magma. The eruption of a volcano is a spectacular geologic phenomenon and may be reported around the world. Any igneous rock that cools from molten lava is described as **extrusive rock.** Few people realize that the vast majority of magma (molten rock) never finds its way to the surface but cools to solid rock somewhere within Earth's interior as **intrusive rock.** We see intrusive rock only where erosion has stripped away the burden of overlying rock or where movement of Earth's crust has brought it to the surface.

Anyone who has seen a volcano erupt must wonder about Earth's interior from which such hot molten material comes. When people began to sink mines and deep wells in search of buried natural resources, they discovered that Earth grows 1°F hotter for each 150 ft of depth. If the increase in temperature continued at this same rate to the center of Earth, the temperature would be so high that rock more than a few tens of miles below Earth's surface would have to be molten regardless of pressure. This view was held by early geologists, who thought of the thin, solid outer layer as Earth's crust. We now know from several lines of evidence that outside the deep molten core Earth is solid and that its interior, though hot, is not nearly as hot as scientists once thought.

Why is Earth hot inside? Pressure, of course, produces heat, but even the immense pressure in Earth's interior cannot account for more than a small fraction of the heat found at great depths. Perhaps Earth conducts heat so poorly that it retains, in its interior, much of the heat from its molten origin. If we accept this view, as geologists did a century ago, then Earth must be cooling, and its inhabitants must face a dim, distant future when this heat will eventually be dissipated, Earth frozen and immobile, and all geologic processes extinct. That this view was incorrect became apparent with the discovery of radioactivity at the end of the nineteenth century. The decay of radioactive elements produces heat. We know from the distribution and abundance of these radioactive elements in Earth's interior that they supply at least enough heat to keep Earth at its present interior temperature.

The next question is more difficult and its answer very uncertain. If Earth as a whole is solid, why does it contain local pockets of molten material? We can only suggest certain possibilities. Earth produces molten magma only in those places where its interior processes are most active. An increase in pressure raises the melting temperature of rock, and a decrease lowers it. We believe that most rock deep below the surface, though solid, is very near its melting temperature. An increase in temperature or a decrease in pressure could melt the rock. Therefore, a local concentration of radioactive elements may provide the necessary additional temperature. A decrease in pressure when a rock fractures may also cause melting. Whatever the answer, we are still far from an understanding of how rocks melt.

It appears from the temperature (800°–1200°C) of lava that pours onto the surface that these magma pockets must be deep. What power drives the magma toward the surface? As rock melts, it expands and, therefore, has less weight per volume. Accordingly, it is squeezed toward the surface by the greater weight of the surrounding rock. This force can push fluid magma all the way to the surface but probably is much less effective with stiffer, less fluid magmas. Even stiff magma, how-

ever, can at times reach the surface, if it is driven by the enormous pressure of escaping gas. We will discuss later the role of gas in the eruption of a volcano.

Texture, which is grain size, is an important property of both intrusive and extrusive rocks. Grain size is determined primarily by the rate at which the molten rock cooled. An igneous body must cool slowly if its mineral grains are to grow large. Rapid cooling almost invariably yields small grains or perhaps none at all. The following factors control the rate of cooling:

1. *Location:* Lava exposed to the cool atmosphere loses heat quickly and, therefore, develops small grains. Magma deep within Earth loses heat very slowly because the cover of rock that overlies it is a very poor conductor of heat. Igneous rocks born at these great depths are almost invariably coarse grained.

2. *Size:* A small volume of magma can, of course, cool more quickly than a large one. The very largest of magma bodies may cool for millions of years before reaching the temperature of the surrounding rock.

3. *Shape:* Thin magma bodies lose heat more rapidly than those with the same volume but a more compact shape.

All objects must be classified and have names if they are to be discussed intelligently. A classification of igneous rocks must account for their widely differing physical and chemical characteristics and must be logical and useful. Geologists classify igneous rocks according to their mineral composition and their texture, as indicated in Table 25.4.

Igneous rocks can be roughly divided into those rich in silica (SiO_2) and those relatively low in silica. Rocks rich in silica contain minerals with abundant silicon, sodium, and potassium. These minerals are mostly light in color. Rocks low in silica are rich in minerals containing iron, magnesium, and calcium and are much darker. The color of an igneous rock is therefore a convenient guide to its chemical composition. Figure 25.6 shows the appearance of three igneous rocks.

25.4 Igneous Activity

Igneous rocks may be further named according to the size and shape of the intrusive bodies and according to their relation to the surrounding rock that they penetrate. An igneous rock is **discordant** if it cuts across the grain of the surrounding rock or **concordant** if it is parallel to the grain. The most important discordant igneous

Table 25.4 Classification of Common Igneous Rocks

Texture	High in Silica		Low in Silica	
	Orthoclase and Quartz Dominant; Some Hornblende, Biotite, and Muscovite	*Light (Na) Plagioclase Dominant; Abundant Hornblende*	*Dark (Ca) Plagioclase and Pyroxene Dominant*	*Augite and Olivine Dominant*
Coarse grained	Granite	Diorite	Gabbro	Peridotite
Fine grained	Felsite	Felsite	Basalt	
Glassy	Obsidian			
Glassy and porous	Pumice	Pumice		
Pyroclastic	{Volcanic tuff {Volcanic breccia	{Volcanic tuff {Volcanic breccia	Scoria	

Figure 25.6 Igneous Rocks.
Left to right: Granite, basalt, and obsidian.

rock by far is a **batholith,** a granitic rock whose most impressive characteristic is its enormous size. It must, by definition, be exposed in an area of at least 103 km² (40 mi²) but many are vastly larger than that. For example, the Coast Range Batholith in western Canada is more than 1600 km (1000 mi) long and in places more than 160 km (100 mi) wide. All surface exposures indicate that batholiths grow larger with depth, but the nature of their bottoms remains uncertain, because no canyons or mines have penetrated that deep. The invasion of a molten batholith into the rocks of Earth's crust occurs at great depth under high temperature and pressure and is an intimate part of the complex process of mountain building. Batholiths are exposed at the surface by uplift and where mountain chains have been deeply scarred by erosion. Figure 25.7 illustrates their shape and relation to the intruded rock.

Dikes are discordant rocks formed from magma that has filled fractures that are vertical or nearly so, and

their shape therefore is tabular—thin in one dimension and extensive in the other two, as illustrated in Fig. 25.7. The sizes of dikes are just as variable as the sizes of the fractures they fill. Dikes are quite common and have been recognized in many different kinds of geologic environments.

A **sill,** illustrated in Fig. 25.7, has the same shape as a dike but is concordant rather than discordant.

What comes out of a volcano? People who know that volcanoes produce lava may not be aware that lava is but one of three products of volcanic eruptions.

1. The expulsion of gas is the most widespread general characteristic of all volcanoes. A volcano may burp gas in its very earliest infancy, at the height of its activity, and during its final dying gasps, when all other signs of life are gone. Steam is by far the most abundant and may comprise as much as 90 percent of the gases.

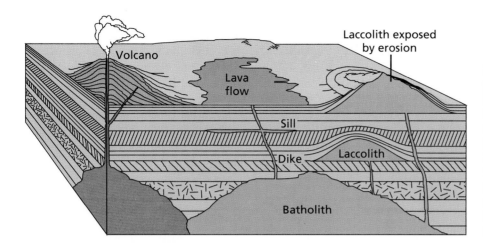

Volcano

Lava flow

Laccolith exposed by erosion

Sill

Dike

Laccolith

Batholith

Figure 25.7 An illustration of plutonic bodies.
Magma solidifying within the Earth forms intrusive igneous bodies. The batholith is the largest intrusive body. Sills and laccoliths are concordant bodies that lie parallel to existing rock formations. The discordant dike cuts across existing rock forma-tions. A volcanic stock is shown rising from the batholith.

2. Volcanoes may erupt lava in variable quantities and conditions. Some, but not all, produce vast amounts of very fluid lava, which flows easily and quietly with little explosive violence. If much gas escapes at the same time, there may be minor explosions, which hurl incandescent lava into the air. See Fig. 25.8. These lava fountains, impressive enough by day, form magnificent fireworks at night. Clots of lava sometimes harden in midair and strike the ground as spindle-shaped volcanic bombs of various sizes. Other volcanoes erupt relatively small volumes of lava so stiff and viscous that it can barely flow at all.

3. Many volcanoes spew enormous volumes of solids, which can range in size from the finest dust to huge boulders. Such particles are known collectively as **pyroclastic debris** and include not only fragments of rock but also gas-laden material that the volcano hurls out in molten form but that hits the ground as a solid. Such debris is blasted into the air by violently explosive volcanoes.

There have been few volcanoes in written history that can match the 1815 eruption of Tambora in violence and in the volume of solids hurled into the atmosphere. Tambora, on the island of Soembawa just east of Java, erupted with a thunderous roar from April 10 to 20. By the time the eruption was finished, there was a hole where before there had been a high mountain. The 145 km³ (35 mi³) of solid debris thrown into the air so darkened the sky that for several days total darkness reigned for hundreds of miles around. Enough fine volcanic dust was circulated around the Earth that the Sun's rays were partially blocked, and the year 1816 was uncommonly cold. The cold weather that accompanied

this and a few other similar eruptions have led some geologists to suggest that the recent ice age, whose great glaciers once blanketed much of Earth's surface, was caused by an unusually large number of violently erupt-

Figure 25.8
The force of gas escaping from the underground reservoir of magma throws molten lava high into the air over the crater of Hawaii's Kilauea volcano.

ing volcanoes. We must, however, learn much more about changes in climate before we can really know why the ice age began.

The behavior of a volcano depends to a considerable extent upon the viscosity of the lava that fills its throat. *Viscosity* is defined as the property of a fluid that resists the force tending to cause the fluid to flow. A liquid with a high viscosity is, therefore, one that is thick and stiff and does not flow readily—like "molasses in January."

Chemical composition and temperature are probably the most important of the several factors that determine the viscosity of a magma or a lava. As you learned in a previous section of this chapter, magmas (and lavas) can vary in their content of silica (SiO_2). Low-silica magmas, which cool to form basalt, are much more fluid and can flow more easily than high-silica magmas, which form felsite, as long as other factors, such as gas content and temperature, are the same. Temperatures are not always the same, however. Very hot lavas flow more easily than those just above their melting temperature. Because basaltic lavas are, in addition, usually much hotter than felsic lavas, the temperature differences reinforce the chemical differences to give basaltic lavas a much lower viscosity. Viscous silica-rich lavas tend to clog their vents and are removed only when enough pressure builds up that the vent is cleared by violent explosions.

Let us examine now the various types of eruptions. Some lava reaches the surface through long fractures in the surface rocks. More commonly, however, the lavas pour out of central vents or volcanic cones, which previous eruptions have constructed.

Fissure eruptions, which issue from long fractures, have been very rare in human history. Until very recently, the latest such volcanic event had affected Iceland in 1783. On January 23, 1973, however, a fissure eruption began that threatened to inundate the small island of Heimaey just off the south shore of Iceland. This eruption was of very great scientific interest, but it was also a matter of grave concern to residents of Iceland, whose economy depends heavily on the fishing industry centered in the town of Vestmannaejar on Heimaey.

Fissure eruptions have poured remarkably large volumes of basalt onto Earth's surface at various times in the geologic past. Virtually every continent has its own extensive area of flood basalts, as these thick accumulations are commonly called. The North American example, which is by no means the world's largest, is the great Columbia Plateau, a conspicuous landform over several of our northwestern states. The Columbia Plateau

basalt covers an area of 576,000 km² (225,000 mi²) to an average depth of 152 m (500 ft). Calculate the total volume of this lava, and you will realize its enormous size. The ocean basins are floored with even greater floods of basalt, estimated by some to be as much as 5 km (3 mi) thick.

The basaltic lavas that came from fissure eruptions were so extremely fluid that they flowed many miles over Earth's surface without constructing volcanoes. Individual lava flows were not thick, however, because they extended for many miles over the surface. As one lava flow followed another over millions of years, the entire landscape was eventually drowned in a sea of solid basalt, with perhaps here and there a high peak forming an island in the black and desolate ocean of rock. The Blue Mountains of Oregon are examples of such islands within the Columbia Plateau.

Volcanoes may erupt in a variety of ways. Some perform quietly, almost unobtrusively, and yet manage in their unspectacular way to pour great volumes of lava onto the surface. These eruptions may cause some destruction of property but very little loss of life. With only a few exceptions, basaltic volcanoes erupt quietly.

One of the most noted examples of such eruptions is in the chain of volcanoes that forms the foundation of the Hawaiian Islands (Fig. 25.9). Eruptions of huge quantities of lava from cracks in the floor of the Pacific Ocean have caused the formation of one of the most extensive mountain chains on Earth, most of which is beneath the ocean. The Hawaiian chain is nearly 2600 km (1600 mi) long, and the volcanic formations on the ocean's floor, nearly 5 km (3 mi) below sea level, rise to form islands that project far above sea level.

Other volcanoes erupt with explosive violence. The felsic lava that may accompany these eruptions is usually subordinate to the much greater volume of pyroclastic debris. An explosive volcano may lie dormant for years, centuries, or even thousands of years and then finally erupt with such violence that it blasts apart the volcanic cone that it had previously constructed and lays waste the surrounding countryside. Viscous felsite is largely responsible for this behavior because it clogs and seals the volcano's vents and causes pressure to build up, which can be relieved only by powerful blasts.

In 1902 geologists were reminded of the unusually destructive behavior of certain explosive volcanoes. After several months of violent, threatening activity, Mt. Pelée, a volcano at the northern end of Martinique in the Caribbean Sea, blasted out incandescent, cloudlike mixtures of superheated gas and pyroclastic debris,

Figure 25.9
Rivers of lava flow from Hawaii's Mauna Loa volcano.

Figure 25.10 The volcano Paricutin in Mexico.
This cinder-cone volcano erupted in a farmer's cornfield in 1943.

which swept down the side of the volcano and over the nearby city of St. Pierre. In a matter of seconds virtually all life and property were destroyed. Although nobody knows the exact loss of life because the city was crowded with refugees seeking shelter from Mt. Pelée, estimates are that between 28,000 and 40,000 people met almost instant death. Among the casualties was the governor of the island, who had come to St. Pierre to assure the inhabitants that they had nothing to fear from the nearby volcano. One of only two survivors of the volcanic holocaust was Auguste Ciparis, a convicted murderer, who was awaiting execution in a dungeon where he was shielded from the fiery blast.

In 1912 Mount Katmai, a dormant volcano on the Alaskan peninsula, erupted with a similar explosion that sent great quantities of dust and glassy volcanic ash into the air. This eruption occurred in an uninhabited area, so there was no loss of human life, as in the case of Mt. Pelée. The fine dust of the ejected material was carried around the world, but the larger particles dropped out within a few hundred miles. A hot blanket of material, almost 30 m (100 ft) thick, collected around the volcano. Areas of Kodiak Island, some 96 km (60 mi) away, were covered with several feet of dust. It took many years for this blanket to cool.

Steam and gaseous emissions from millions of fumaroles (gas vents) led to the name of "Valley of Ten Thousand Smokes" being applied to the area. Studies of this region have given much information about volcanic gaseous emissions and their contents. A great deal of

the gas from Katmai was hydrogen chloride, which gave rise to raindrops of hydrochloric acid.

From more recent times are the eruptions of Paricutin, Surtsey, Mount St. Helens, and El Chichón. In 1943 Paricutin erupted in a Mexican farmer's cornfield. In the first year the volcanic cone rose to a height of over 300 m (1000 ft) (Fig. 25.10). After about 10 years of subsiding activity, Paricutin is now a dormant volcano with a cone height of over 400 m (1300 ft).

In 1963 volcano Surtsey (after *Surtr*, the subterranean god of fire in Icelandic mythology) boiled up from the ocean floor off the coast of Iceland. Having sufficient lava to form a barrier against the sea, Surtsey is now a permanent volcanic island that may be found on the world map (Fig. 25.11). Geologists were able to study Surtsey from its "birth."

In almost everyone's recent memory is the eruption of Mount St. Helens. This volcano is located in southwestern Washington, about 64 km (40 mi) from Portland, Oregon. Prior to 1980 Mount St. Helens was a placid, snow-capped, dormant volcano, one of many such volcanoes in the Cascade Mountains in Oregon and Washington. Its last period of activity had been between 1800 and 1857.

In March 1980 a series of minor earthquakes occurred near Mount St. Helens, and the first eruptions took place. On May 18, 1980, a massive eruption occurred, devastating an area of more than 384 km² (150 mi²) and leaving more than 60 people dead or missing (Fig. 25.12). A column of ash rose to an altitude

Figure 25.11 The volcanic island Surtsey, off the southern coast of Iceland.

Figure 25.12 Mount St. Helens after the May 18, 1980, eruption.
More than 240 km² (150 mi²) were devastated.

of more than 19 km (12 mi). Nearby cities were blanketed with ash, and a light dusting of ash fell as far away as 1450 km (900 mi) to the east.

Huge mudflows from the ash caused flooding and silting in the rivers near the volcano, destroying and damaging many homes. Mount St. Helens became relatively quiet in 1981. However, it has shown minor activity in the years since, and geologists expect that intermittent activity may continue for years, and even decades, if the behavior of the volcano follows the pattern of its eruptions in the 1800s. Volcano-triggered mudflows also had a devastating effect in the 1985 eruption of a volcano in Colombia, South America. Over 20,000 people were killed.

Another significant volcanic eruption in this decade was that of El Chichón in Mexico. In late March 1982 El Chichón roared into life with a tremendous explosion

that sent a column of ash and gases 16 km (10 mi high) within an hour. Although the eruption of El Chichón was not very impressive compared with its earlier Mexican neighbor Paricutin, its significance arises from the projection of debris some 26 mi into the atmosphere. This debris in the stratosphere may have long-term climatic effects. See Section 24.4

The occurrence of volcanic activity is for the most part unpredictable. New volcanoes may be formed unexpectedly, while existing volcanoes lying dormant may suddenly erupt with practically no warning. However, the locations of eruptions and potential eruptions are known. From the theory of plate tectonics (Chapter 26), volcanic activity takes place predominantly at plate boundaries above subduction zones where one plate is deflected downward into Earth's interior (see Fig. 26.21). For example, the area of the Pacific Ocean is one of widespread volcanic activity. The outer rim of the Pacific Ocean is marked by a ring of volcanoes known as the **"Ring of Fire"** (Fig. 25.13). Comparing Figs. 25.13 and 26.17 clearly shows the marked correlation with plate boundaries.

The formation of the Hawaiian Islands in the central Pacific Ocean and on the central portion of the Pacific Plate is explained by the so-called **hot-spot theory.** This theory hypothesizes that as one of Earth's crustal plates moves over a fixed source of heat beneath it, a "hot spot," a succession of volcanoes, is generated. Each volcano slides away or moves with the plate, leaving room for the next.

According to the theory, volcanic island chains trace out the plates' movement over a hot spot. The Hawaiian Islands are cited as a classic example of this process, and studies (age dating of seafloor samples) have yielded data in support of the theory. However, not every island chain appears to fit the theory, so geologists have more work to do in explaining Earth's internal processes.

Volcanism in the United States has been confined mainly to the Hawaiian Islands, Alaska and the Aleutian Islands, and the Cascade Mountains in Oregon and Washington. Most of these volcanoes, particularly in the latter region, are now dormant. However, in addition to Mount St. Helens, an active volcano can be seen in Hawaii. The volcano Kilauea, on the island of Hawaii (the largest island in the Hawaiian chain), is one of the most active in the world (Figs. 25.8 and 25.9).

Because of its frequent, yet mild eruptions, the U. S. Geological Survey maintains a Volcano Observatory on the summit of Kilauea. It is of special interest to geologists because the erupted material comes from great depths and thus provides clues to the geochemical

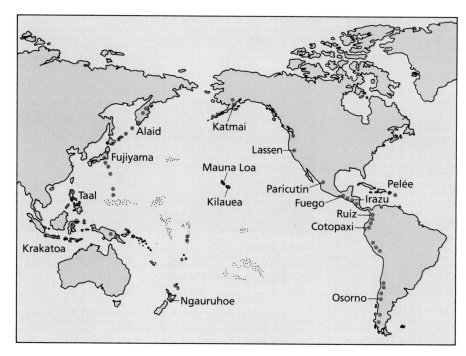

Figure 25.13 The Pacific "Ring of Fire."
The outer rim of the Pacific Ocean is circumscribed by a ring of volcanoes that tend to have violent eruptions, hence the name "Ring of Fire."

composition of Earth's interior. The observatory is located in the Hawaii Volcanoes National Park where more than 500,000 visitors come each year to view an active volcano.

As pointed out previously, many volcanoes do not erupt violently but relatively calmly eject liquid magma from fissures in Earth's surface. If the lava has a relatively low viscosity so that it flows easily, a gently sloping, low-profile **shield volcano** is formed by frequently repeated lava flows.

The classic example of a shield volcano is Mauna Loa in the Hawaiian island chain. In fact, Mauna Loa is a huge volcanic mountain—the largest single mountain on Earth in sheer bulk. Although not as tall as Mt. Everest (slightly over 8800 m—29,000 ft—above sea level), Mauna Loa rises 4600 m (15,000 ft) from the ocean floor to sea level and protrudes an additional 4150 m (13,600 ft) above see level for a total height of about 8700 m (28,600 ft). Its bulk comes from the fact that this partially submerged mountain has a base almost 160 km (100 mi) in diameter.

Volcanic eruptions of both lava and pyroclastic debris form a more steeply sloping, layered composite cone that is called a **stratovolcano** (also called a **composite volcano**). The lava of stratovolcanoes has a relatively high viscosity, and eruptions are more violent and generally less frequent than those of shield volcanoes. Many stratovolcanoes have an accumulation of material up to

Figure 25.14 Mount Fuji in Japan.
Mount Fuji is a dormant stratovolcano that was formed from eruptions of both lava and pyroclastic debris. Seen in the foreground of the photograph is Lake Kawaguchi, one of the five lakes around the base.

1800–2400 m (6000–8000 ft) above their base and have a characteristic symmetrical profile. Mount St. Helens (Fig. 25.12) is a stratovolcano. Dormant stratovolcanoes include Mt. Fuji in Japan (Fig. 25.14) and Mts. Shasta, Hood, and Rainier in the Cascade Mountains in Washington and Oregon.

Figure 25.15 Cinder-cone volcano Izalco in El Salvador.

Figure 25.16
A volcanic crater in Costa Rica is filled with water colored yellow by suspended sulphur.

Figure 25.17
Crater Lake on top of Mount Mazama in Oregon occupies a caldera about 10 km in diameter.

A volcanic eruption may also consist primarily of pyroclastic debris. In this case steeply sloped **cinder cones** are formed, which rarely exceed 300 m (1000 ft) in height. A cinder cone is shown in Fig. 25.15 rising above a previous lava flow. Volcano Izalco in El Salvador is another example of these small, steep, symmetric cones of pyroclastic debris.

Volcanic activity is usually not confined to the region of the central vent of a volcano. Fractures may split the cone, with volcanic material emitted along the flanks of the cone. Also, material and gases may emerge from small auxiliary vents forming small cones on the slope of the main central vent. A funnel-shaped depression called a **volcanic crater** exists near the summit of most volcanoes from which material and gases are ejected (Fig. 25.16).

Many volcanoes are marked by a much larger depression called a **caldera.** These roughly circular, steep-walled depressions may be up to several miles in diameter. Calderas result primarily from the collapse of the chamber at the volcano's summit from which lava and ash were emitted. The weight of the ejected material of the partially empty chamber causes its roof to collapse, much like the collapse of a snow-ladened roof of a building. Crater Lake on top of Mount Mazama in Oregon occupies the caldera formed by the collapse of the volcanic chamber of this once-active stratovolcano (Fig. 25.17). In some instances calderas are formed by a violent explosion rather than a collapse. Felsitic lava may clog the volcanic vent, and the resulting buildup of pressure causes the volcano to "blow its top," so to speak.

25.5 Sedimentary Rocks

Sedimentary rocks are the product of gradation, a process of such importance that it occupies a major portion of Chapter 29, which we must anticipate if we are to understand the origin of sedimentary rocks. The process begins when existing rock at Earth's surface is broken physically and decomposed chemically from its exposure to the atmosphere and then eroded by the system of streams that flow through the area. The streams, which transport the debris to the sea, drop the sediment as the river currents die within the standing waters of the ocean. Upon consolidation of the loose sediment, a sedimentary rock is formed.

Sediments and sedimentary rocks make up only 5 percent of Earth's crust, but their importance is entirely out of proportion to their limited abundance. One reason is that they cover about 75 percent of the continents and even more of the ocean basins. In a sense, they are a bit like clothing; they cover only the surface, but they are conspicuous. The North American continental interior is composed of a foundation of igneous and metamorphic rocks with a sedimentary veneer only a few thousand feet thick. Residents of many of the interior states of the United States may never have seen any rocks other than sedimentary ones. Our modern way of life owes much to sedimentary rocks, because they contain abundant petroleum, coal, and metal deposits and many of the materials essential to the construction industry.

Sedimentary rocks, with their many varieties, commonly form more interesting landscapes than do igneous and metamorphic rocks in the same setting. A comparison between the Grand Canyon of the Colorado River and the Snake River Canyon illustrates this point. Both are magnificent gorges in terms of sheer size, but the beauty and variety of the Grand Canyon with its staircase of brightly colored slopes and cliffs of sedimentary rock is famous around the world, while the Snake River Canyon, carved through a more monotonous sequence of drab gray to black volcanic rocks, is little known and seldom visited.

In order to better understand the classification of sedimentary rocks, let us consider very briefly how a river carries its load of debris to the ocean. Part of a stream's load is dissolved mineral matter. Another fraction is made of mineral and rock fragments so small that they are carried along in the main body of the stream. The brown, muddy waters of a stream in flood indicate a great load of silt and clay particles on their way to the ocean. Still a third part of a stream's load is the larger particles that can move only by rolling and bouncing along the river bottom.

The rock and mineral fragments eventually settle out as **clastic sediments.** The dissolved mineral matter may be extracted from seawater by plants or animals and accumulate on the seafloor as **organic sediments,** or it may be physically precipitated on the floor as **chemical sediments.** Sedimentary rocks, therefore, fit into three major categories—clastic, organic, and chemical. The most important types are listed in Table 25.5.

Geologists gain much information from sedimentary rocks because their characteristics tell so much about their origin and history. Color, rounding, sorting, bedding, fossil content, ripple marks, mud cracks, footprints, and even raindrop prints are common characteristics of sedimentary rocks that indicate the conditions under which the rock was formed. Figure 25.18 shows the appearance of three sedimentary rocks.

The color of a sedimentary rock is especially conspicuous in those drier parts of the world where there is little soil or vegetation to mantle the surface. The bright, variegated colors of the Painted Desert in Arizona draw crowds of tourists, but even rocks with dull colors are conspicuous in a dry landscape. Gray, the most common sedimentary color, usually reflects the rock's origin in shallow, well-aerated marine water. Rocks deposited above sea level in the presence of much oxygen usually are colored by iron oxides and are red-brown or yellow-brown—colors that are so striking in many of our western states. Those few sedimentary rocks that are dark gray to black contain carbon from the organic matter that accumulated in the stagnant water where the sediment was deposited.

The larger clastic sediments tend to be worn round and smooth as the currents and waves that move them grind the particles against the stream bottom and against each other. Angular fragments within the rocks called arkose and graywacke inform geologists that these sediments were not carried far and that they were dropped quickly. The rounder grains of quartzose sandstone tell of their long journey downstream and of the many hours they were shifted and rolled by the restless waves and currents of the sea.

A current or wave that can keep small grains in motion may be unable to move larger grains. Because wind and moving water can transport some particles more easily than others, the particles tend to become sorted or separated according to size. The greater the distance of transportation, the more effectively the grains are sorted. In a lake or ocean the largest particles come to rest near shore, but the finer grains are transported into the quieter water farther from shore, where they eventually settle to the bottom. Arkose and graywacke contain poorly sorted particles, which could not have been transported very far.

Table 25.5 Types of Sedimentary Rocks

Clastic Sedimentary Rocks

Sediment	Grain Size (mm)	Rock Name	Characteristics
Gravel	More than 2 mm	Conglomerate	Rounded pebbles
		Breccia	Angular pebbles
Sand	$\frac{1}{16}$ to 2 mm	Sandstone	
		Quartzone	Dominantly quartz
		Arkose	Quartz and angular feldspar
		Graywacke	Quartz, feldspar, chlorite, clay particles, volcanic debris, variable sizes
Mud	Less than $\frac{1}{16}$ mm	Shale	Silt and clay particles, rich in quartz
Shell fragments and calcite grains	Variable	Limestone	
		Coquina	Porous aggregate of shell fragments
		Oolitic limestone	Small, rounded grains

Organic Sedimentary Rocks

Rock Name	Characteristics
Bituminous coal	Compacted plant remains. Breaks into rectangular lumps.
Limestone	Compacted or cemented calcareous plant or animal remains. May be crystalline.
Fossiliferous limestone	Fossils present.

Chemical Sedimentary Rocks

Rock Name	Characteristics
Limestone	Compacted, cemented, calcareous material. May be crystalline.
Dolomite	Composed of dolomite instead of calcite.
Rock gypsum	Evaporite. Fine grained.
Anhydrite	Evaporite. Fine to coarsely granular.
Rock salt	Evaporite. Composed of halite.
Chert	Siliceous precipitate. Not crystalline.

Bedding, or stratification, as it is also called, is the layering that develops at the time the sediment is deposited. The bedding may be conpicuous, as shown in Fig 25.19, or vague. It may be in either thick or very thin layers. Because most sediment comes to rest on a level surface, most bedding is horizontal. There are many environments, however, where sediments accumulate in tilted layers known as **cross-bedding.** Much research and many pages have been devoted to the geometry and origin of the many kinds of cross-bedding. In every case the surface is sloping where the sediments come to rest. Cross-bedding is especially common where a river empties into a lake, where sediments fill in depressions scoured by floods along river channels, or where wind drapes sand down the flanks of a dune (Fig. 25.20). By recognizing the type of cross-bedding, the geologist makes an observation that helps in unraveling the origin and history of the rock.

The most distinctive and most interesting characteristic of a sedimentary rock is the fossils it sometimes contains. See Fig. 25.21. Although a mystery to those who lived several centuries ago, a **fossil** is now known to be the remains of an organism that lived in the past. The fossil organism need not be extinct, though many are, but it must have died in prehistoric times. A pet cat buried in a backyard, for example, could not be considered a fossil.

There are three principal reasons why the fossils in a sedimentary rock attract so much scientific attention:

1. Fossils are our only record of organic evolution. Biologists have found much convincing evidence of evolutionary change of plants and animals, but sedimentary rocks contain the only record of these slow organic changes.

Figure 25.18 Sedimentary Rocks.
Clockwise from top: Fossiliferous limestone, dolomite, and shale.

Figure 25.20 Cross-bedding in sandstone.

Figure 25.19 Bedding (stratification) in limestone.

Figure 25.21 Fossilized shellfish.

2. Because of slow changes wrought by organic evolution, ancient rocks contain fossils of ancient, primitive organisms, while younger sediments contain the remains of modern, more advanced forms of life. Geologists have collected so many fossils and have so refined their techniques that they can determine the relative age of a sedimentary rock from its fossil content.

3. Although there are a few exceptions, most organisms can thrive only in certain environments. If geologists can learn the environments of ancient plants and animals, they can use this information to tell more about the environment in which the sediments themselves were deposited. This field of science, known as paleoecology, has been making tremendous progress in the past twenty years.

Those who have looked along the bed of a briskly flowing creek or have seen the sandy bottom of a shallow pond may have noticed that movement or agitation of the water has developed **ripple marks** in the sediment, which look somewhat like waves. Current ripple marks form on stream bottoms in moving water, and oscillation ripple marks (Fig. 25.22a) form in shallow, standing water. Because these features are sometimes buried and preserved in sedimentary rock, they are still another clue in deciphering the history of ancient rocks.

Mud deposited in shallow water, or along a valley bottom, may be quickly exposed to the atmosphere when the water recedes. The mud may then be marked by the footprints of animals that walk across it or possibly pocked by the impact of raindrops; or as it dries, it may shrink to form **mud cracks** like those in Fig. 25.22b. The vast majority of these features do not even survive the season, but those rare ones that are buried and preserved become a part of the rock record and may later be exposed to the inquiring eye of a geologist.

The transformation of sediment into a clastic sedimentary rock is a process called **lithification.** During this process the loose, solid particles (sediment) are compacted by the weight of overlying material and eventually cemented together. Common cementing agents are silica (SiO_2), calcium carbonate ($CaCO_3$), and iron oxides, which are dissolved in groundwater that permeates the sediment. An example is the most common clastic sedimentary rock, shale. When fine-grained mud

Figure 25.22a Oscillation ripples in sandstone.

Figure 25.22b Mud cracks along the Rio Grande in Texas.

is subjected to pressure from overlying rock material, water is driven off and the clay minerals begin to compact (consolidate). As groundwater moves through the compacted sediment, materials dissolved in the water precipitate out around the individual mud-size particles, and cementation of the particles occurs. A clastic sedimentary rock results. Sandstones and conglomerates occur in the same manner but with differing sediment sizes.

Chemical sedimentary rocks are formed from the precipitation of a material from a solution, usually water. There are two subtypes of chemical rocks: indirect and direct chemical. Indirect chemical rocks are formed by biochemical reactions during the activities of plants and animals; that is, certain organisms, such as coral, extract calcium carbonate from seawater to build skeletal material. When coral dies, the collected skeletal deposits subsequently form biochemical limestone. Direct chemical rocks are formed when the evaporation of water with dissolved materials leaves behind a residue of chemical sediment, such as sodium chloride (rock salt) and calcium sulfate (gypsum). Another example of directly formed chemical sedimentary rock is cave dripstone, which is formed primarily by calcium carbonate precipitated from dripping water. Dripstone takes on a variety of forms, but most common are the icicle-shaped stalactites and cone-shaped stalagmites. See Fig. 25.23.

25.6 Metamorphic Rocks

About 15 percent of Earth's crust is metamorphic. Metamorphic rocks are those that have been changed under the influence of great temperature and/or pressure deep beneath Earth's surface. These changes that transform the parent material into metamorphic rocks result from a very fundamental property of virtually all minerals. Minerals (and therefore rocks) are stable only in the environment in which they form. The minerals remain stable and endure as long as the environment remains constant but become unstable and break down if the environment changes. Sedimentary rocks, which form under surface conditions, are especially susceptible to change by heat and pressure, but even igneous rocks may be affected. The conditions far beneath Earth's surface that produce metamorphic rock are not so fundamentally different from those that form magmas. Both intrusive igneous and metamorphic rocks are the products of processes that act at great depth and are, therefore, found in close association once they have been uncovered by erosion.

Two major kinds of change occur during metamorphism. The first is mechanical change and includes fracturing, crushing, and change in shape caused by intense pressure. The second type of change is chemical

Figure 25.23 Stalactites in a limestone cavern.

and consists of recrystallization and loss of water. Recrystallization, which is the growth of minerals, may involve reorganization and growth of minerals originally present; or it may, by more profound chemical changes, produce an entirely new set of minerals. The new minerals will be those that are stable in the changed environment.

Though it is always difficult to make arbitrary separations, we recognize several kinds of metamorphism: thermal, dynamic, regional, and hydrothermal.

Thermal metamorphism, commonly but not always accurately called *contact metamorphism,* is change brought about primarily by heat, with very little pressure being involved. Such changes commonly occur in shallow bedrock when it is subjected to the heat of a molten body of magma moving up from greater depths. Thermal metamorphism is most obvious at such a shallow depth, because bedrock near Earth's surface is normally cool, and the effects of great temperature changes are, therefore, quite pronounced. The rock immediately next to the molten magma experiences intense metamorphism and may be coarse grained, but it grades out into finer-grained rock that has been less severely transformed by the heat. This dark, fine-grained rock, containing recrystallized minerals with random orientation, is known as hornfels.

Rocks changed more by pressure than temperature are said to have undergone the effects of **dynamic metamorphism,** which is most common along fractures and shear planes where one rock unit slides past another. Mechanical deformation shatters the grains or changes their shapes plastically. Recrystallization accompanies the more intense forms of dynamic metamorphism, but the profound physical effects are more obvious.

Most metamorphic rocks have been affected by both high temperature and high pressure and have, therefore, experienced both mechanical deformation and chemical recrystallization. **Regional metamorphism,** as this type of change is known, receives its name from the extremely large areas it affects. The widely exposed metamorphic rocks in central Canada are of this sort. Although there is still much to learn about regional metamorphism, it appears mostly to affect rocks undergoing intense deformation by mountain building.

There are some conditions in which chemical changes during metamorphism are profound. Such conditions exist around the huge igneous bodies that are cooling deep below Earth's surface. These bodies release enormous quantities of fluids carrying sodium, potassium, and other similar metals, and the changes that these fluids help bring about in the overlying rock are known as **hydrothermal metamorphism.** More than in other forms of metamorphism, the hot vapors bring in new chemical constituents and drive out old ones, so that the surrounding rock changes not just by recrystallization of existing components but in its overall chemical composition.

A metamorphic rock itself is classified according to its texture, mineral composition, and foliation. **Foliation** is the ability of a metamorphic rock to split along a smooth plane. Foliated rocks have this property because they contain a large number of flat, platy, or sheetlike minerals, all oriented in the same direction. Minerals such as mica, chlorite, and hornblende are essential ingredients of foliated metamorphic rocks.

Geologists are fairly certain that foliation develops in response to pressure. It is tempting to give credit to the weight of the enormous overburden that bears down upon a rock undergoing metamorphism. This confining pressure, however, is exerted equally in all directions and could not possibly force the development of platy minerals with any one particular orientation. Foliation can develop under confining pressure only if there is an additional directed pressure, the sort that accompanies mountain building, when the rocks are squeezed under tremendous horizontal pressure.

The progressive metamorphism of shale is a good illustration of changes that occur as a sedimentary rock is subjected to more and more intense regional metamorphism. Shale, after relatively mild metamorphism, is transformed to slate (Fig. 25.24), a fine-grained metamorphic rock similar in many respects to shale but differing fundamentally in its excellent slaty cleavage, a variety of foliation. If the shale is subjected to more intense heat and pressure, it will change to schist, a metamorphic rock whose grains are visible to the un-

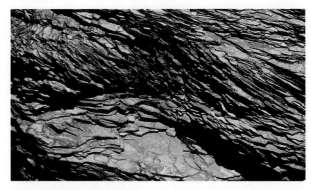

Figure 25.24 Slate bed.
Slates are fine-grained metamorphic rocks that possess a type of foliation known as slaty cleavage.

Figure 25.25 Gneiss.
A metamorphic rock, contorted under high temperature and pressure. Feldspar and quartz are the chief minerals.

Figure 25.26 Metamorphic rocks.
Clockwise from top: Marble, quartzite, and mica schist.

aided eye but whose foliation, though good, is less perfect than slate's. Very intense metamorphism produces gneiss (Fig. 25.25), an even coarser-grained rock, with a rough foliation characterized by distinct banding. The higher grades of metamorphism, therefore, produce larger grains but rougher foliation. Figure 25.26 shows the appearance of three metamorphic rocks.

Metamorphic rocks that lack foliation are said to be massive, and they include such well-known examples as marble, which is formed from limestone, and quartzite, which is metamorphosed sandstone. A simplified classification of metamorphic rocks is listed in Table 25.6.

Table 25.6 Metamorphic Rocks

	Foliated	
Rock Name	*Original Rock Name*	*Principal Minerals and Characteristics*
Slate	Shale, tuff	Mica, quartz—excellent foliation, very fine grained
Chlorite schist	Basalt, felsite, tuff	Chlorite, plagioclase—good foliation, coarse grained
Mica schist	Shale, felsite, tuff	Muscovite, quartz, biotite—very good foliation, coarse grained
Hornblende schist	Basalt, gabbro, felsite	Hornblende, plagioclase—good foliation, coarse grained
Gneiss	Granite, felsite, shale, mica schist	Feldspar, quartz, plagioclase, hornblende, etc.—rough foliation, coarse grained, banded
	Massive	
Rock Name	*Original Rock Name*	*Principal Minerals and Characteristics*
Hornfels	Any fine-grained rock, especially shale	Variable composition—fine grained, hard
Quartzite	Sandstone	Quartz—fine to coarse grained
Marble	Limestone, dolomite	Calcite—coarse grained
Anthracite	Bituminous coal	Carbon—conchoidal fracture

Learning Objectives

After reading and studying this chapter, you should be able to do the following without referring to the text:

1. Distinguish between minerals and rocks.

2. State the general composition and structure of silicates.

3. Name some nonsilicate materials.

4. Name some common minerals, and describe the methods of mineral identification.

5. State the three types of rocks, and describe how they originate.

6. Explain the rock cycle.

7. Distinguish between intrusive and extrusive rock.

8. Describe the characteristics and different types of volcanic activity.

9. Distinguish between clastic, organic, and chemical sediments.

10. Distinguish between thermal, dynamic, regional, and hydrothermal metamorphism.

11. Name some common igneous, sedimentary, and metamorphic rocks.

12. Define and explain the important words and terms in the next section.

Important Words and Terms

geology	discordant	organic sediments
mineral	concordant	chemical sediments
rock	batholith	bedding
silica	dikes	cross-bedding
silicates	sill	fossil
Mohs' scale of hardness	pyroclastic debris	ripple marks
igneous rocks	"Ring of Fire"	mud cracks
metamorphic rocks	hot-spot theory	lithification
sedimentary rocks	shield volcano	thermal metamorphism
magma	stratovolcano, composite volcano	dynamic metamorphism
lava	cinder cones	regional metamorphism
extrusive rock	volcanic crater	hydrothermal metamorphism
intrusive rock	caldera	foliation
texture	clastic sediments	

Questions

Minerals

1. Distinguish between silicon and silicates.

2. State the definition of mineral.

3. What two elements make up the majority of Earth's crust and in what percentages?

4. What is the silicon-oxygen tetrahedron? Sketch its structure.

5. What is the chemical composition of (a) asbestos, (b) galena, (c) gypsum, and (d) halite?

6. Name an ore of lead, of iron, and of zinc.

7. State some physical characteristics that are used to identify minerals.

Rocks

8. Define *rock.*

9. Name the three classifications of rock, and state how each is formed.

10. What is the Bowen reaction series?

11. Define *magma.*

12. What is basaltic magma?

13. What is the rock cycle?

Igneous Rock

14. Define *igneous rock.*

15. Distinguish between magma and lava.

16. Distinguish between intrusive and extrusive rock. State the physical characteristics of each.

17. What is the source of Earth's internal heat?

18. What two major characteristics are used to classify igneous rock?

19. How does the silica content of granite compare with that of basalt?

Igneous Activity

20. Distinguish between discordant and concordant igneous rock.

21. What is the most important discordant igneous rock?

22. Distinguish between a dike and a sill.

23. What two factors determine the viscosity of magma and lava?

24. What is the "Ring of Fire"?

25. What is the hot-spot theory?

26. Distinguish between (a) shield volcanoes, (b) stratovolcanoes, and (c) cinder cones.

27. What are caldera, and how are they formed?

Sedimentary Rocks

28. Distinguish between clastic, organic, and chemical sediments.

29. Name some of the characteristics geologists use to give information about the origin of sedimentary rocks.

30. Name three sedimentary rocks, and state the sediment from which they originate.

31. What is meant by *bedding?*

32. Describe lithification.

Metamorphic Rocks

33. Define *metamorphic rock.*

34. State the four kinds of metamorphism, and indicate the most essential differences among them.

35. Define *foliation.*

36. Name three metamorphic rocks, and state the original rock name.

CHAPTER 26

Structural Geology

> This earth, a spot, a grain, an atom.
>
> —Milton, *Paradise Lost*

ALTHOUGH EARTH is approximately 5 billion years old, the planet's internal driving forces remain active. These forces cause the crust of Earth to shake and tremble, producing what we call earthquakes. Of all natural disasters, earthquakes are the most destructive and terrifying, because they remain practically unpredictable. Also, the indirect effects of an earthquake, which include fire, tidal waves, and landslides, are in many instances more destructive than the shock waves of the quake.

Perhaps the most important advance in our understanding of the geology of Earth during the present century has been the development of the theory of continental drift and plate tectonics. This theory states that approximately 200 million years ago the present continents were one huge land mass, or supercontinent. This supercontinent broke up and drifted apart, leading to the present-day positions of the continents, which are still in motion.

Earthquakes, continental drift, and plate tectonics are presented and discussed in this chapter.

26.1 Earthquakes and Structural Geology

An earthquake, as everyone who has experienced a substantial one knows, is manifested by the vibrating and sometimes violent movement of Earth's surface. But waves are also propagated through Earth. It is these waves that provide one of the most useful ways of studying Earth's interior. **Seismology** is the branch of geophysics that studies these waves and uses them as probes to "see" below Earth's surface to determine its internal structure.

Earthquakes, which are shock waves traveling through Earth, rattle our globe perhaps a million times each year, but the vast majority are so mild that they can be detected only with delicate, sensitive instruments. A very powerful quake, however, can lay waste a large area.

───────────

◄ **Earthquake damage: Mexico 1985.**

Earthquakes may be caused by explosive volcanic eruptions or even explosions caused by humans, but most earthquakes appear to be associated with movements in Earth's crust. These movements in some instances will form large fractures in Earth called **faults.** According to the theory of plate tectonics (see Section 26.3), the optimum place for such movement is at the plate boundaries, and the major earthquake belts of the world are observed in these regions (see Fig. 26.1). Notice how the earthquake region around the Pacific Ocean is similar to that of the volcanic "Ring of Fire" (Fig. 25.13).

Movements of Earth's plates would certainly put strains on rock formations in Earth's crust near the plate boundaries. Rocks possess elastic properties, and energy is stored in the elastic deformations of the lithosphere. If the forces causing this deformation are great enough to overcome the force of friction along a plate boundary or some nearby fault, then the fault walls move suddenly and the stored energy in the rocks is released, causing an earthquake.

If the elastic limit of a stressed rock formation is not exceeded, the rock can rebound elastically and snap back to its original shape after a shift along a fault has relieved the stress—much like a spring. However, if the elastic limit of the rock is exceeded, the rock cannot fully recover when the strain is removed. If the elastic limit is greatly exceeded by the deformation force, the rock formation can even rupture.

Relative horizontal and parallel displacement of rock on each side of a fracture is called strike-slip faulting (see Section 27.2). Many large strike-slip faults are associated with plate boundaries and are called transform faults. An example of a transform fault is the famous San Andreas fault in California. It is the master fault of an intricate network of faults that runs along the coastal regions of California (Fig. 26.2). This huge fracture in Earth's crust is more than 960 km (600 mi) long and at least 32 km (20 mi) deep. Over much of its length a linear trough of narrow ridges reveals the fault's presence.

As you can see in Fig. 26.2, the San Andreas fault system lies on the boundary of the Pacific Plate and the

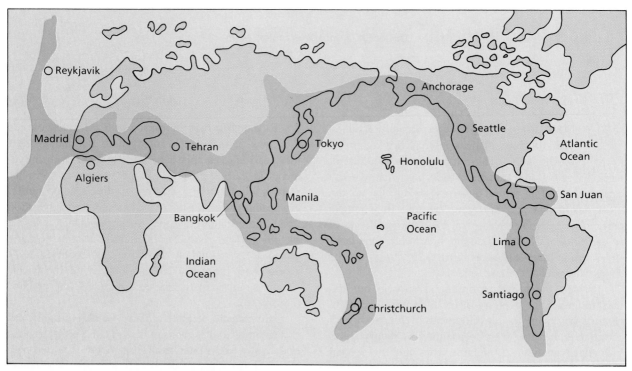

Figure 26.1 The earthquake belts of the world.
Notice the belt that circumscribes the Pacific Ocean in a pattern closely resembling that of the volcanic "Ring of Fire" in Fig. 25.13.

American Plate (see Fig. 26.17). Movement along the transform fault arises from the relative motion between these plates. The Pacific Plate is moving northward relative to the American Plate at a rate of several centimeters per year. At this current rate and direction, in about 10 million years Los Angeles will have moved northward to the same latitude as San Francisco. In another 50 million years the segment of continental crust on the Pacific Plate will have become completely separated from the continental land mass of North America.

The horizontal movement along the San Andreas fault is a cause for considerable concern for the populated San Francisco Bay area through which it runs. The famous San Francisco earthquake of April 18, 1906, which measured 8.3 on the Richter scale (earthquake scales will be discussed shortly), resulted in the loss of approximately 700 lives and in millions of dollars of property damage. There have been many other milder earthquakes since then along the San Andreas and other branch faults.

On December 7, 1988, at 11:41 A.M., a devastating earthquake struck the Soviet Republic of Armenia, killing over 25,000 people and leaving more than 500,000 homeless. Figure 26.3 shows some of the destruction and aftermath of this disastrous earthquake.

There is little that can be done about the San Andreas fault except to learn to live with it. Geologists measure the stresses along the fault, hoping to predict future earthquakes. Also, building codes in the area are strict and are concerned with construction that will lessen earthquake damage. However, increased population and an ever-increasing need for housing have caused land developers to use any available land, resulting in rows of houses that straddle the trace fault of 1906 and sit on hills where earthquake-induced landslides could occur. People live in these houses with the knowledge of the fracture in the crust below and the hope that any future movement along the fault will occur elsewhere.

When an earthquake does occur, the point or region of the initial energy release or slippage is called the **focus** of the quake. The vast majority of earthquakes originate in the crust or upper mantle, so an earthquake's focus generally lies at some depth—from a few to several hundred miles. Consequently, geologists designate the location on Earth's surface directly above the focus. This point is called the **epicenter** of the earthquake.

The energy released at the focus of an earthquake propagates outwardly as **seismic waves.** There are two general types of seismic waves produced by earthquake

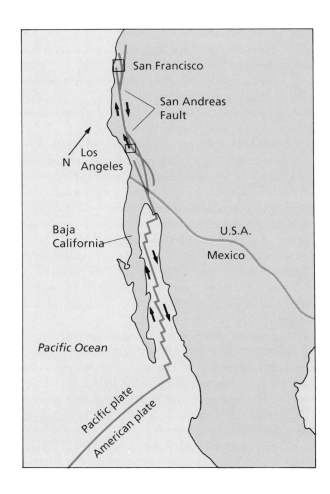

Figure 26.2 The San Andreas fault.
Above: An aerial view of the fault on the Carizzo Plain, California. *Left:* A map showing the boundary, marked by the fault, between the Pacific and American Plates, which are moving in opposite directions. This is a classic example of a transform fault.

Figure 26.3
Earthquake damage and aftermath caused by the December 1988 disaster that struck the Soviet Republic of Armenia, killing over 25,000 people and leaving more than 500,000 homeless.

vibrations: **surface waves**, which travel along Earth's surface or boundary within it; and **body waves**, which travel through Earth. Surface waves cause most earthquake damage because they move along Earth's surface. Body waves propagate through Earth's interior and so do less damage.

There are two types of body waves, P (for primary) or compressional waves and S (for secondary) or shear waves. The **P waves** are longitudinal compressional waves that are propagated by particles in the propagating material moving longitudinally back and forth in the same direction as the wave is traveling (see Chapter 6). Sound waves are an example of longitudinal compressional waves. In **S waves** the particles move at right angles to the direction of wave travel and hence are transverse waves. An example of transverse shear waves is the vibration in a plucked guitar string.

There are two important differences between P and S waves. S waves can travel only through solids. Liquids

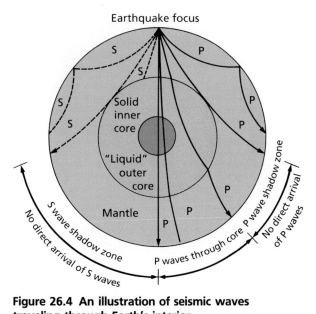

Figure 26.4 An illustration of seismic waves traveling through Earth's interior.
Because of the S and P wave shadow zones, Earth is believed to have a liquid outer core.

and gases cannot support a shear stress; that is, they have no elasticity in this direction and therefore their particles will not oscillate in the direction of a shearing force. For example, little or no resistance is felt when one shears a knife through a liquid or gas. The compressional P waves, on the other hand, can travel through any kind of material—solid, liquid, or gas. The other important difference is the velocity of the waves. P waves derive their name "primary" because they always travel faster than the "secondary" S waves in any particular solid material and hence arrive earlier at a seismic station. It is these differences between P and S waves that allow seismologists to locate the focus of an earthquake and learn about Earth's internal structure.

The speeds of the body waves depend on the density of the material, which generally increases with depth. As a result, the waves are curved or refracted. Also, the waves are refracted when they cross the boundary or discontinuity between different media—in the same manner light waves are refracted (Chapter 7). The refraction of seismic waves and the fact that S waves cannot travel through liquid media provide our present view of Earth's structure, as shown in Fig. 26.4. Because of the so-called *shadow zones* of the S and P waves, the outer core of Earth is believed to be a highly viscous liquid.

The seismic waves of an earthquake are monitored by an instrument called a **seismograph,** the principle of which is illustrated in Fig. 26.5. The recorded seismogram gives the time delay of the arrival of different types of waves, as well as an indication of their energy. (The greater the energy, the greater the amplitude of the traces on the seismograph.)

The severity of earthquakes is represented on different scales. The two most common are the **Richter scale,** which gives an absolute measure of the energy

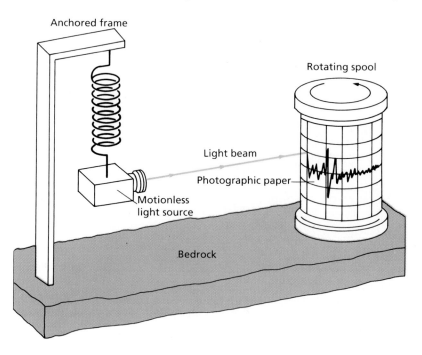

Figure 26.5 An illustration of the principle of the seismograph.
The rotating spool (anchored in bedrock) vibrates during the quake. A light beam from the relatively motionless source on a spring traces out a record or seismograph of the earthquake's energy on the light-sensitive photographic paper.

released by calculating the energy of seismic waves at a standard distance, and the modified **Mercalli scale,** which describes the severity of an earthquake by its observed effects. See Table 26.1. A comparison of these two scales is shown in Table 26.2. Notice that the 1985 Mexico earthquake was less severe on the Richter scale, but the death and destruction it caused gave it a high ranking on the Mercalli scale. The epicenter of the quake was off the Pacific coast of Mexico; however, it had devastating effects in Mexico City 400 km (250 mi)

Table 26.1 Modified Mercalli Scale

I	Not felt except by a very few under especially favorable circumstances.
II	Felt only by a few persons at rest, especially on upper floors of buildings.
III	Felt quite noticeably indoors, especially on upper floors of buildings, but many people do not recognize as an earthquake.
IV	During the day felt indoors by many, outdoors by few. Sensation like heavy truck striking building.
V	Felt by nearly everyone, many awakened. Disturbances of trees, poles, and other tall objects sometimes noticed.
VI	Felt by all; many frightened and run outdoors. Some heavy furniture moved; few instances of fallen plaster or damaged chimneys. Damage slight.
VII	Everybody runs outdoors. Damage negligible in buildings of good design and construction; slight to moderate in well-built ordinary structures; considerable in poorly built or badly designed structures.
VIII	Damage slight in specially designed structures; considerable in ordinary substantial buildings with partial collapse; great in poorly built structures. (Fall of chimneys, factory stacks, columns, monuments, and other vertically oriented features.)
IX	Damage considerable in specially designed structures. Buildings shifted off foundations. Ground cracked conspicuously.
X	Some well-built wooden structures destroyed. Most masonry and frame structures destroyed with foundations. Ground badly cracked.
XI	Few, if any, (masonry) structures remain standing. Bridges destroyed. Broad fissures in ground.
XII	Damage total. Waves seen on ground surfaces. Objects thrown upward into air.

Table 26.2 Scales of Earthquake Activity

Richter Scale, Magnitude	Modified Mercalli Scale, Maximum Expected Intensity (at Epicenter)	Description
1–2	I	Usually detected only by instruments
3–4	II–III	Can be slightly felt
4–5	IV–V	Generally felt; slight damage
6–7	VI–VIII	Moderately destructive
7–8	IX–X	Major earthquake
8 +	XI–XII	Great earthquake
Specific Earthquakes	*Richter*	*Mercalli*
1906 San Francisco	8.3	XI
1964 Alaska	8.4	XI
1985 Mexico	8.1	XI (estimated)

away. This city, the second most populous on Earth (Shanghai, China, is first), suffered thousands of deaths and injuries and billions of dollars of damage. The first earthquake was followed by an aftershock 36 h later that measured 7.5 on the Richter scale.

Of these two scales the Richter scale (developed in 1935 by Charles Richter of the California Institute of Technology) is used more often, with magnitudes expressed in whole numbers and decimals, usually between 3 and 9. However, the scale is a logarithmic function—that is, each whole-number step represents about 31 times more energy than the preceding whole-number step. For example, an earthquake that registers a magnitude of 5.5 on the Richter scale indicates the measuring seismograph received about 31 times as much energy as for an earthquake that registers a magnitude of 4.5.

Similarly, an earthquake with a magnitude of 8 does not represent twice as much energy as one with a magnitude of 4 but about one million times as much! (See Fig. 26.6.) As indicated in Table 26.2, an earthquake with a magnitude of 2 to 3 is the smallest tremor felt by human beings. The largest recorded earthquakes are in the magnitude range of 8.7 to 8.9.

The Richter scale gives no indication of the damage caused by an earthquake, only its potential for damage. Earthquake damage depends not only on the magnitude of the quake but also on the location of its focus and

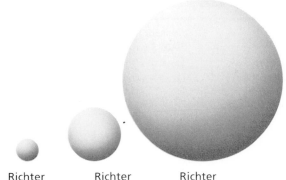

Richter magnitude 1   Richter magnitude 2   Richter magnitude 3

Figure 26.6 An illustration of the relationship between earthquake magnitude on the Richter scale and energy.
The volumes of the spheres are roughly proportional to the amount of energy released by earthquakes of the magnitudes given, and illustrate the exponential relationship between magnitude and energy.

epicenter and the environment of that region—specifically, the local geologic conditions, the density of population, and the construction designs of buildings. The modified Mercalli scale gives a better indication of earthquake effects, because the scale is based on actual observations. The Mercalli scale was developed by Giuseppe Mercalli in the 1890s before the advent of seismographs.

Earthquake damage may result directly from the vibrational tremors or indirectly from landslides and subsidence, as shown in Fig. 26.7. In populated areas a great deal of the property damage is caused by fires because of a lack of ability to fight them—a result of disrupted water mains and so on.

When the energy release of a quake occurs in the vicinity of or beneath the ocean floor, huge waves called **tsunamis** are sometimes generated. These waves travel across the oceans at speeds up to 960 km (600 mi)/h. In the open ocean the waves may be 160 km (100 mi) long and only 1 m (3 ft) high. However, as they travel in shallower water, the tsunamis grow in height and may be 15 m (50 ft) high or higher when they smash into shore with immense force (see Fig. 26.7). Tsunamis coming ashore are commonly and incorrectly referred to as *tidal waves,* although they have no relation to tides. The Hawaiian Islands have experienced many tsunamis, which have raced across the ocean from the Alaska-Bering Strait area or other locations of earthquake activity that circumscribe the Pacific Ocean.

Earthquakes, despite their potential for destruction,

Figure 26.7 *Above:* Subsidence damage from the 1964 Alaska earthquake. *Below: Tsunami* damage from the 1964 Alaska earthquake.

can be an aid to science. None of the remarkable scientific advances of the twentieth century has revealed with certainty the composition of the interior of the earth. Our ideas about the interior of our planet must rest upon indirect evidence provided by earthquake body waves, whose speed and direction reflect the type of materials they penetrate, by meteorites whose composition we believe is similar to earth's, and by laboratory experiments performed on rocks under very high temperature and pressure.

Most of our knowledge of Earth's interior structure comes from the monitoring of shock waves generated by earthquakes. The enormous releases of energy associated with movements along fractures in the solid exterior, which result in earthquakes, generate both transverse and longitudinal waves (Chapter 6) that propagate through Earth. By monitoring these waves at different locations on Earth and by applying their knowledge of wave properties in various types of Earth materials, such as wave velocity and refraction, scientists obtain information about Earth's interior structure.

From these indirect observations of Earth's interior, scientists believe that Earth is made up of a series of concentric shells, as illustrated in Fig. 26.8. There are three major shells: (1) the core, which is partly molten, (2) the mantle, and (3) the crust. The different shells are characterized by different composition and physical properties.

The two innermost regions are together called the **core** and have an average density of over 10 g/cm³. The density suggests a metallic composition, which is believed to be chiefly iron (80%) and nickel. This estimate is based on the behavior of seismic waves passing through the core, the known abundance of iron in meteorites, and the measured proportion of iron in the Sun and stars. The solid inner core has a radius of approximately 1200 km (750 mi) while the outer core, some 2250 km (1400 mi) thick, is believed to be composed of molten, highly viscous, "liquid" material. The magnetic field of Earth is thought to be related to the liquid nature of the outer core. As pointed out in Chapter 8, the magnetic field of Earth resembles that of a huge bar magnet within Earth. However, the interior temperatures of Earth are probably too high to have permanently magnetized ferromagnetic materials with the interior temperatures exceeding the Curie temperatures* of these materials. The slow change of the positions of Earth's magnetic poles suggests that the magnetic field is due to currents, which must arise somehow from motions of the molten metals in the outer core associated with Earth's rotation.

Around the core is the **mantle,** which is on the average about 2900 km (1800 mi) thick. The composition of the rocky mantle differs sharply from the metallic core, and their boundary is distinct. The average density of the mantle is of the order of 4.5 g/cm³, which indicates that its composition is an iron-magnesium, rock-type material.

Around the mantle is a thin, rocky, outer layer upon which we live, called the **crust.** It ranges in thickness from about 5–8 km (3–5 mi) beneath the ocean basin to about 24–48 km (15–30 mi) under the continents. About 65% of Earth's crust is oceanic crust; that is, about 65% of Earth's surface is made up of ocean basins.

An abrupt change in the behavior of body waves

---

* The temperature above which ferromagnetism disappears in a magnetic material.

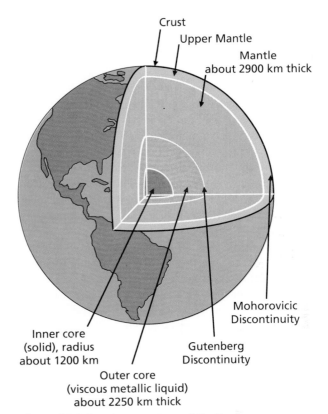

Figure 26.8 Interior structure of Earth.
The crustal thickness is greatly exaggerated.

as they travel in toward Earth's interior reveals the existence of the Mohorovicic discontinuity, a sharply defined surface that separates the crust from the mantle, the next layer down. Both scientists and students prefer the simplified term, Moho, for this important surface. Body waves make another important change in behavior at a depth of about 2900 km (1800 mi) as they cross the Gutenberg discontinuity, the surface that separates the mantle from the core.

There is much less certainty about the composition of these layers than about their thickness and depth. Direct observation and geophysical investigations have virtually confirmed that the ocean basins are made of basalt and that the continents are generally granitic in composition. Below the crust, however, we must depend upon indirect information. Earth scientists are in general agreement that the upper part of the mantle is probably composed of a rock near the composition of peridotite, a very-low-silica rock composed mostly of olivine and pyroxene. The lower mantle, with its much greater temperature and pressure, may contain minerals composed of magnesium silicate, magnesium oxide, and high-pressure forms of quartz. However, the conditions at this great depth are so different from what we experience at Earth's surface and are so difficult to duplicate in the laboratory that few Earth scientists are confident in their knowledge of the composition of the lower mantle. Equipment is now advanced enough that geologists can drill into and directly sample rock in the upper mantle. Such an undertaking, appropriately known as the Mohole Project, was conceived and then abandoned in the 1960s, a victim of competition from the urgency of military needs and the glamor of space exploration.

All evidence points to a core vastly different from the rest of Earth. Shear waves do not cross the Gutenberg discontinuity, and compression waves can do so only with sharply decreased velocity and a marked change in direction. The absorption of shear waves indicates a molten core, because these waves are unable to travel through fluids. But what sort of liquid is it? We do not know, but recent investigations have led many to believe in a core of molten iron. Still deeper within the interior of Earth, earthquake waves penetrate the inner core, which appears to be solid and which may be composed of iron or perhaps a mixture of iron and nickel. Figure 26.8 is a simplified summary of the current scientific theory concerning composition of Earth's interior.

If we view the interior of our Earth in terms of its behavior rather than its composition, we can divide it into somewhat different layers. The outer layer, which we call the **lithosphere,** extends to a depth of approximately 80 km (50 mi) and includes all of the crust and the uppermost part of the mantle. See Fig. 26.9. The lithosphere is rigid, brittle, and relatively resistant to deformation. Faults and earthquakes are mostly restricted to this layer. Below the lithosphere is the **asthenosphere,** which extends to a depth of roughly 700 km (135 mi) below Earth's surface. This rock plays essential roles in isostatic adjustments, continental drift, and seafloor spreading, which are discussed later in this chapter.

The difference in properties and behavior of the lithosphere and asthenosphere may well reflect the different ways in which they transmit heat from Earth's interior out to its surface. Heat moves through the lithosphere by conduction. Because this rock is a poor conductor of heat, there is a relatively high temperature gradient from its top to its bottom. The lithosphere, in other words, acts essentially as an insulating blanket, which traps much of the heat brought up to its bottom surface. Rock within the asthenosphere is therefore so hot that it is extremely near its melting temperature. Asthenospheric rock, because of its heat, is plastic and mobile enough to transmit heat mostly by convection. As you will learn in later sections, convection cells within

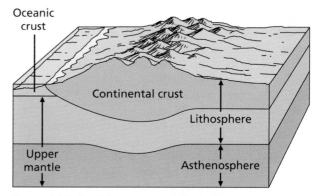

Figure 26.9 Earth's upper mantle and crust.
The upper mantle is about 700 km (435 mi) in thickness. The continental crust varies between 24 and 48 km (15 to 30 mi) in thickness, and the oceanic crust ranges in thickness from 5 to 8 km (3 to 5 mi).

the asthenosphere are considered by many to be the fundamental forces that have for virtually all of geologic time driven Earth in its restless internal activities. Below the asthenosphere Earth is of course even hotter, but the melting temperatures under these high pressures are still higher.

26.2 Continental Drift and Seafloor Spreading

When looking at a map of the world, we are tempted to speculate that the Atlantic coasts of Africa and North and South America could fit nicely together as though they were pieces of a jigsaw puzzle (Fig. 26.10). This observation has led scientists at various times in history to suggest that these continents, and perhaps the other continents, were once a single, giant supercontinent that broke and drifted apart. However, there was no evidence to support this theory other than the shapes of the continents.

In the early 1900s Alfred Wegener (1880–1930), a German meteorologist and geophysicist, revived the theory of **continental drift** and brought together various geological evidences for its support. Wegener's theory gave rise to considerable controversy, and only relatively recently has reasonably conclusive evidence been found that supports some aspects of the theory of continental drift.

Wegener's assumption was that the continents were once part of a single giant continent, which he called **Pangaea** (from the Greek, pronounced "pan-jee-ah" and meaning "all lands"). According to his theory, this hypothetical supercontinent somehow broke apart about

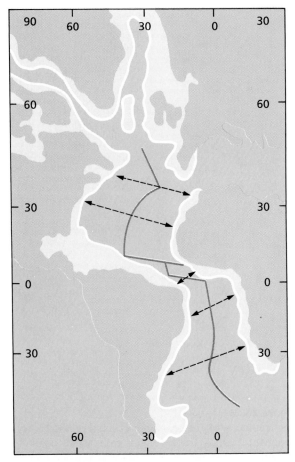

Figure 26.10 Continental jigsaw puzzle.
The coasts of Africa and North and South America appear as though these continents could fit together nicely, as if they were pieces of a jigsaw puzzle. The dashed lines show the paths the continents might have followed if they were once together and rifted and drifted apart. The heavy line indicating where the continents might once have been joined marks a present midoceanic ridge that runs the length of the Atlantic Ocean.

200 million years ago and its fragments drifted to their present positions and became today's continents (Fig. 26.11).

The geologic evidence supporting Wegener's theory takes on several different approaches. Some of these approaches are (1) similarities in biological species and fossils found on the various continents; (2) continuity of geologic structures such as mountain ranges and the distribution of rock types and ages; and (3) glaciation in the Southern Hemisphere. Let's consider each of these briefly.

1. Similarities in biological species and fossils suggest that there was formerly an exchange of these forms

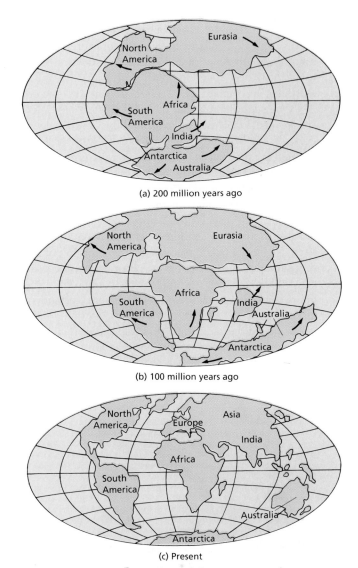

(a) 200 million years ago

(b) 100 million years ago

(c) Present

Figure 26.11 An illustration of the sequence of events in the theory of continental drift.
(a) 200 million years ago the giant supercontinent Pangaea rifted and the continental fragments began to drift apart. (b) Continental positions 100 million years ago. Notice that India was drifting northward on a collision course with the Asian continent. (c) Present position of the continents.

between continental regions when they were together as Pangaea. We would not expect that these biological species could traverse the present-day oceans. For example, a certain variety of garden snail is found only in the western part of Europe and the eastern part of North America. Also, a relatively young genus of earthworm is found in the same latitudes of Japan and the Asian and European continents, while it is found in similar latitudes on the east coast but not the west coast of North America. Similarly, fossils of identical reptiles are found in South America and Africa, and identical plant fossils have been found in South America, Africa, India, Australia, and Antarctica.

One suggestion is that the similarities in biological species and fossils can be explained by "land bridges" between the continents that eventually sank or were covered by the oceans. However, there is no evidence of such land bridges in the oceans.

2. As has been noted, it was the roughly interlocking shapes of the coastlines of the African and American continents that inspired the theory of continental drift. Imagine cutting the continental shapes from a printed page like jigsaw puzzle pieces and attemping to put the separated pieces together again. When we put the page pieces back together, we find that the continuity of the printed lines is common to the fitted pieces, as shown in Fig. 26.12.

If indeed the continents had rifted and drifted apart, we might expect some similar "printed lines" common to the pieces. Such evidence does occur in the form of geologic features. If the continents were put back to-

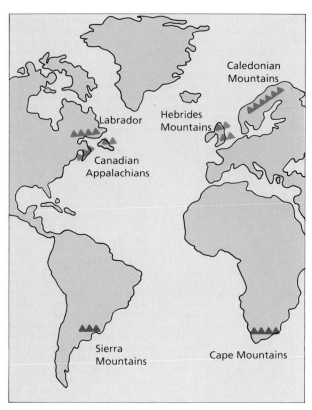

Figure 26.13
The continuity of geologic features supports the theory of continental drift. If the continents were fitted together, various mountain ranges of similar structure and rock composition on the different continents would line up analogous to the print in Fig. 26.12.

Figure 26.12 An illustration of the continuity of continental features.
Continental jigsaw pieces cut from a printed page would show the continuity of printed lines when put back together. If the continents were once together and drifted apart, we might expect some similar "printed lines" common to the separated continents in the form of continuities of geologic features such as mountains.

gether, the Cape mountain range in southern Africa would line up with the Sierra range near Buenos Aires, and these mountains are strikingly similar in geologic structure and rock composition (Fig. 26.13). In the Northern Hemisphere the Hebrides Mountains in Northern Scotland match up with similar formations in Labrador, and the Caledonian Mountains in Norway and Scotland have a logical extension in the Canadian Appalachians. Various other similarities in the plateau rock formations of Africa and South America have been found. More recently, with the development of radioactive-dating techniques, transatlantic areas of rocks of similar ages have been found.

3. There is solid geological evidence that a glacial ice sheet covered the southern parts of South America, Africa, India, and Australia about 300 million years ago, similar to the one that covers Antarctica today. Hence

a reasonable conclusion is that the southern portions of these continents must have been under the influence of a polar climate at this time. There are no traces of this ice age found in Europe and North America. In fact, fossil evidence indicates that a tropical climate prevailed in these regions during that period. Yet evidence of glaciation is found in India and in Africa near the equator. How could this be? It seems unreasonable that an ice sheet would cover the southern oceans and extend northward to the equator.

Wegener's theory suggests an answer and derives support from these observations. The direction of the glacier flow is easily determined by marks of erosion on rock floors and by moraines (deposits of rock and soil debris transported by glaciers). If the continents were once grouped together as Wegener's theory indicates, then the glaciation area was common to the various continents, as illustrated in Fig. 26.14. The glacial move-

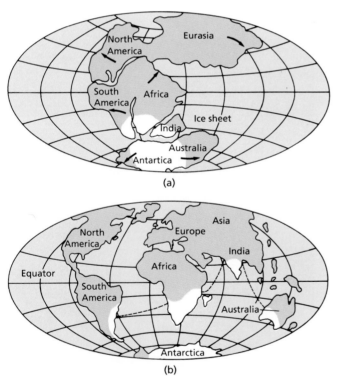

Figure 26.14 Glaciation evidence of continental drift. Geologic evidence shows that a glacial ice sheet covered parts of South America, Africa, India, and Australia some 300 million years ago, but there is no evidence of this glaciation in Europe and North America. Also, evidence of this glaciation is found near the equator in India and in Africa. This is explained if a single ice sheet covered the southern polar region of Pangaea and the continents subsequently drifted apart.

ments, as indicated by the arrows in the top drawing, support the idea of a single ice cap and subsequent continental drift.

Although there was evidence supporting Wegener's theory, it was not generally accepted, primarily because the proposed mechanism for continental drift was unsatisfactory. Wegener depicted the continents as giant rafts moving through the oceanic crust owing to Earth's rotation. This mechanism is unacceptable because the force associated with Earth's rotation is not strong enough to overcome the measured strength of the rocks, which would be disintegrated if the continental crust slid over or moved through the oceanic crust.

A more satisfactory theory for the mechanism behind continental drift was suggested in 1960 by H. H. Hess, an American geologist. Geologists know that a midoceanic ridge system stretches through the major oceans of the world. In particular, the Mid-Atlantic Ridge runs along the center of the Atlantic Ocean between the continents. This and other midoceanic ridges run along large fissures in Earth's crust, as evidenced by volcanic and earthquake activity along the ridges. Hess suggested a theory of **seafloor spreading,** where the seafloor spreads slowly and moves sideways away from the midoceanic ridges. This movement is believed to be accounted for by convection currents of subterranean molten materials that cause the formation of midoceanic ridges and surface motions in lateral directions from the fissure, as illustrated in Fig. 26.15a.

Support for this theory has come from the studies of remanent magnetism and from the determination of the ages of the rocks on each side of the midoceanic ridges. **Remanent magnetism** refers to the magnetism of rocks resulting from a special group of minerals called ferrites, one of which is ferrous (iron) ferrite, Fe_3O_4, commonly called magnetite or lodestone. When molten material containing ferrites is extruded upward from Earth's mantle (e.g., in a volcanic eruption) and solidifies in Earth's magnetic field, it becomes magnetized. The direction of the magnetization indicates the direction of Earth's magnetic field at the time.

This solidified rock, called igneous rock (Section 25.3), is worn down by erosion. The fragments are carried away by water, and they eventually settle in bodies of water, where they become layers in future sedimentary rock (Section 25.5). In the settling process the magnetized particles, which are in fact small magnets, become generally aligned with Earth's magnetic field. By studying the remanent magnetization of geologic rock formations and layers, scientists have learned about changes that have taken place in Earth's magnetic field.

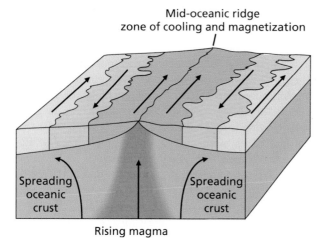

Mid-oceanic ridge
zone of cooling and magnetization

Spreading oceanic crust

Spreading oceanic crust

Rising magma

Figure 26.15 (a) Above: Diagram illustrating magnetic anomalies.
Magnetic anomalies are distributed in symmetrical, parallel bands on both sides of the Mid-Atlantic Ridge. Normal polarity is the direction of the present magnetic field at the ridge. Reverse polarity is in the opposite direction.

(b) Right: Mid-Atlantic Ridge.
This feature runs along the ocean floor between the continents. It is believed that seafloor spreading takes place along the ridge because of convection currents of molten material rising from the upper mantle and spreading outward from the ridge.

Measurements of the remanent magnetism of rock on the ocean floor revealed long, narrow, symmetric bands of **magnetic anomalies** (Fig. 26.15a) on both sides of the Mid-Atlantic Ridge (Fig. 26.15b). That is, the direction of the magnetization was reversed in adjacent parallel regions. Along with data on land rock, the magnetic anomalies indicate that Earth's magnetic field has abruptly reversed fairly frequently and regularly throughout recent geologic time. The most recent reversal occurred about 700,000 years ago. Why this and other reversals should occur is not known. We are evidently living in a period between pole reversals. However, the symmetry of the anomaly bands on either side of the ridge indicates a movement away from the ridge at the rate of a few centimeters per year and provides evidence for seafloor spreading.

Remanent magnetism also provides other support for continental drift. The remanent magnetism of a rock remains fixed throughout the history of the rock, even if forces within Earth move the rock. If the rock is moved, its remanent magnetism is no longer aligned with Earth's magnetic field. This situation is similar to the nonalign-

Figure 26.16 The JOIDES *Resolution*, an oceanographic research vessel.

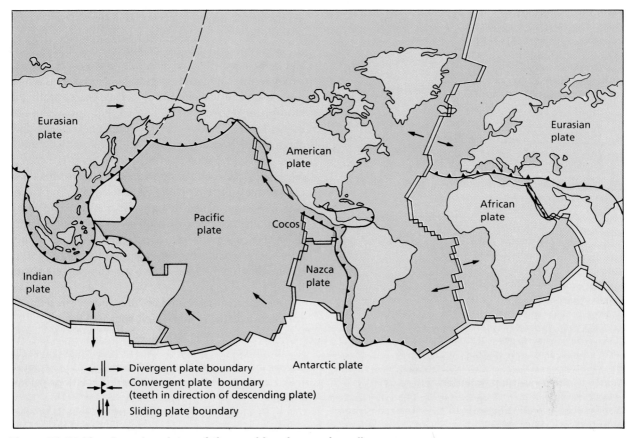

Figure 26.17 The six major plates of the world and several smaller ones.
The relative motions along the plate boundaries are indicated.

ment of printed lines on moved puzzle pieces, as described in the previous analogy on the continuity of geologic features. Scientists have studied the remanent magnetism of rock of varying ages and have found evidence of such changes of alignment. Moreover, they have found that rock of the same age on a continent has the same misalignment, but that the misalignment is different for different continents. This observation suggests that the entire continent must have moved as a unit and that each continent moved or "drifted" in its own separate direction.

In recent years investigative drilling into the ocean floor has been done from the oceanographic research vessel JOIDES (Joint Oceanographic Institutions for Deep Earth Sampling) *Resolution* (Fig. 26.16). These drillings have shown that the ocean floor between Africa and South America is covered by relatively young sediment strata. Also, the strata thicknesses increase away from the Mid-Atlantic Ridge, which implies that the older part of the ocean floor is farther away from the ridge. The older parts would be covered by a greater thickness of sediment because there would have been a longer time for the sediment to accumulate.

These observations support the idea of seafloor spreading as a mechanism for the theory of continental drift, which has culminated in the modern theory of plate tectonics.

26.3 Plate Tectonics

The view of ocean basins in a process of continual self-renewal has led to the acceptance of the concept of **plate tectonics.*** We now view the lithosphere not as one solid rock but as a series of solid sections or segments called plates, which are constantly in very slow motion and interacting with one another. There are about twenty plates over the surface of the globe. Some are very large, and some are small. The major plates are illustrated in Fig. 26.17.

---

* Tectonics (from the Greek *tekto*, "builder") is the study of Earth's general structural features and their changes.

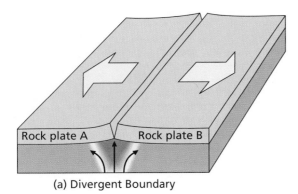

(a) Divergent Boundary

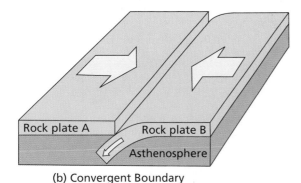

(b) Convergent Boundary

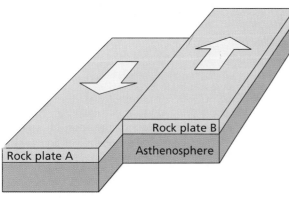

(c) Transform Boundary

Figure 26.18 Block diagrams illustrating the relationship between moving rock plates.
The interface between the two plates is called the plate boundary. Each boundary type has an identifying name.

The most active, restless parts of Earth's crust are located at the plate boundaries. Along the ridges where one plate is pulling away from another (a **divergent boundary**), new oceanic rock is formed. Where plates are driven together (a **convergent boundary**), rock is consumed. In still other parts of Earth's surface one plate slides along one side of another (a **transform bound-**

ary), and rock is neither produced nor destroyed. These relationships are illustrated in Fig. 26.18.

Figure 26.19 illustrates the structure of the lithosphere and asthenosphere. The interface between these two structures is significant in terms of internal geologic processes. The asthenosphere, which lies beneath the lithosphere, is essentially solid rock, but it is so close to its melting temperature that it contains pockets of molten magma and is relatively plastic. Therefore, it is much more easily deformed than the lithosphere. The movement of plastic rock during structural adjustments takes place essentially within the asthenosphere. The forces that move the plates one against another, one away from another, or one past another are also found within the asthenosphere. The plates, segments of the lithosphere, are actually passive bodies, driven into motion by the drag of the more active asthenosphere beneath them.

Most Earth scientists view this motion in terms of **convection cells,** with the basic source of heat provided by radioactive decay. See Fig. 26.19. The role of gravity in the process is to drag the cooler, heavier material deeper into the Earth's interior as the hotter, less dense rock rises toward the surface, where it can lose part of its heat.

Let us examine the role of convection cells in plate tectonics by focusing our attention first on the oceanic ridges where the plates are being drawn away from one another. (See Fig. 26.20.) Beneath these spreading oceanic ridges, convection currents lift material from the

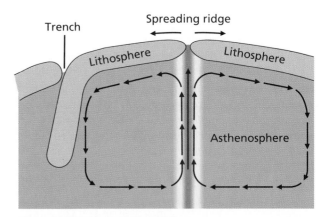

Figure 26.19 Convection Cells.
Unequal distribution of temperature within Earth causes the hot, lighter material to rise and the cooler, heavier material to sink. This generates convection currents in the asthenosphere. The lithosphere plates, resting upon the asthenosphere, are put in motion by the driving force of the convection cells.

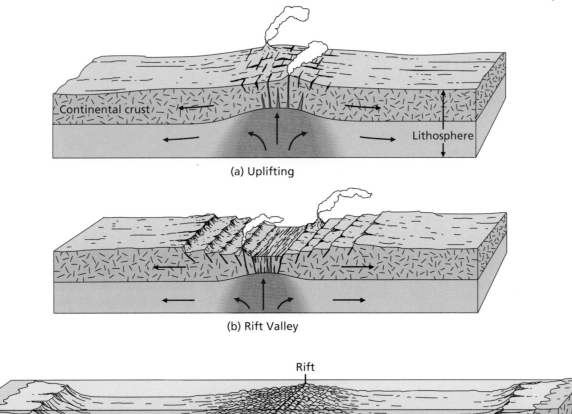

(a) Uplifting

(b) Rift Valley

(c) Mid-ocean ridge

Figure 26.20 Block diagrams illustrating the possible changes that occur when rock plates are forced apart.
The Mid-Atlantic ridge may have developed in this manner. See text for explanation.

hot mantle (asthenosphere) up into a region of lower pressure, where it begins to melt. The material only partially melts, however, forming a magma of basaltic composition.

As more mantle material wells up from below, the higher mantle material is shouldered to both sides and moves slowly in a horizontal direction beneath the lithospheric plates. It is the drag of the mobile asthenosphere against the bottom of the lithospheric plates that keeps the plates in motion and causes the fracturing at the spreading ridges. The fractures are filled in an area known as a rift valley, by the basaltic magma produced by the partial melting of the peridotite in the asthenosphere.

In its horizontal journey across the top of the convection cell, beneath the lithospheric plates, some of the asthenosphere cools to well below its melting temper-

ature and becomes part of the lithosphere. The lithospheric plate also cools as it moves slowly away from the hot spreading ridge, and therefore it loses volume. As a result, the top of the plate gradually subsides and causes the oceans to grow progressively deeper away from the spreading ridges. The slope of denser, cooler rock away from the midoceanic ridges also contributes to the motion of the plate toward the zones of compression.

The zones of compression occur at convergent plate boundaries, and there are three types: (1) oceanic-oceanic, (2) oceanic-continental, and (3) continental-continental.

Initially, when two oceanic plates collide, both begin to be subducted because each is very dense (3.0 g/cm³). Eventually, one plate is subducted more than the other. The place, or zone, where the plate descends into the

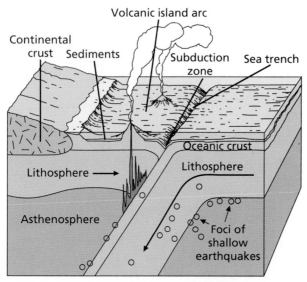

Figure 26.21 Diagram illustrating the effects produced by the convergence of two oceanic plates.
The subduction zone is the region in which one lithospheric plate plunges beneath another. Destructive earthquakes occur most often in the subduction zones.

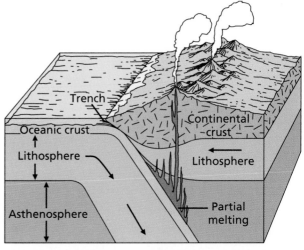

Figure 26.22 Block diagram illustrating the results produced by an oceanic-continental convergence.
An example is the Andes Mountains of South America.

asthenosphere is called the *subduction zone.* The descending oceanic plate, now in contact with the mantle, begins to melt. The molten material begins to rise, and a series of volcanoes develop in an arc shape on the overriding plate. (See Fig. 26.21.) Deep trenches lie in front of the arc system, marking the places where the plates are being subducted. The deepest trench known is the Marianas Trench in the western Pacific Ocean, which is 11 km (6.8 mi) below sea level. Examples of island arc systems are the Aleutian Islands and the Japanese Islands, both of which have deep trenches between the islands and the open ocean.

Whenever the oceanic crust collides with lighter-weight continental crust (2.7 g/cm³), the oceanic crust is always subducted beneath the continental crust. (See Fig. 26.22.) A trench will develop at the point at which the oceanic plate is being subducted. However, the trench is never as deep as the trench formed in the oceanic-oceanic convergence. The oceanic plate begins to melt as it descends into the asthenosphere. Molten-rock material then moves up into the overriding plate, causing large igneous intrusions and often volcanic mountains at the surface. An example is the Andes Mountains of South America.

When two continental plates, both being light-weight (2.7 g/cm³), collide, the boundary of rock be-

tween the continental edges is pushed and crumpled intensely to form folded-mountain belt systems. (See Fig. 26.23.) In this manner, continents grow in size by suturing themselves together along folded-mountain belt systems. Examples are the Alps, Appalachians, and Himalayas. Geologists believe that the Himalayas formed

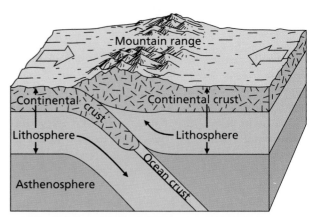

Figure 26.23 Diagram illustrating a continental-continental convergence.
The boundary of rock between the continental edges is pushed and crumpled to form folded mountain systems. The Appalachian mountain range is an example.

in this manner when the Indian Plate collided with the Eurasian Plate.

A zone of shear, or transform, fault is a point at which adjacent plates slide past each other without a gain or loss in surface area. This zone occurs along faults, which mark the plate boundaries. Movements and the release of energy along these boundaries give rise to earthquakes. Examples of such fault zones are along the San Andreas fault in California and the Anatolian fault in Turkey.

Learning Objectives

After reading and studying this chapter, you should be able to do the following without referring to the text:

1. Explain the causes of earthquakes.
2. State the difference between P and S waves, and explain how these waves provide information about the Earth's interior.
3. Define the Richter scale and the modified Mercalli scale.
4. Describe the early formulation of the theory of continental drift and the geologic evidence offered for its support.
5. Explain the mechanism and evidence for seafloor spreading and its relationship to continental drift.
6. Explain the theory of plate tectonics.
7. List the general relative motions of plates and the resulting geologic implications.
8. Define and explain the important words and terms listed in the next section.

Important Words and Terms

seismology	S waves	crust	magnetic anomalies
faults	seismograph	lithosphere	plate tectonics
epicenter	Richter scale	asthenosphere	divergent boundary
seismic waves	Mercalli scale	continental drift	convergent boundary
surface waves	tsunamis	Pangaea	transform boundary
body waves	core	seafloor spreading	convection cells
P waves	mantle	remanent magnetism	

Questions

Earthquakes and Structural Geology

1. What causes an earthquake?
2. What are the focus and epicenter of an earthquake?
3. Give the two general types of seismic waves and the subdivision of one of these types.
4. What is the difference between S and P waves? How do these waves allow seismologists to locate the focus of an earthquake and to investigate Earth's interior structure?
5. What is the basis of the (a) Richter scale and (b) modified Mercalli scale?
6. Why is *tidal wave* an incorrect and misleading term for a huge ocean wave generated by an earthquake? What is the correct term?

Continental Drift and Seafloor Spreading

7. Describe the geologic evidence that supported Wegener's theory of continental drift.

8. What is the theory of the mechanism for continental drift? Why was Wegener's explanation unacceptable?
9. What is meant by remanent magnetism?
10. What evidence supports the theory of seafloor spreading?
11. What is the average rate of seafloor spreading?

Plate Tectonics

12. What are the lithosphere and the asthenosphere?
13. What is a "plate" in the context of plate tectonics?
14. Describe the three general types of relative plate motions and the geologic results of each.
15. What is a subduction zone?
16. How are volcanic and earthquake activity explained in the theory of plate tectonics?
17. What are the names of the six major plates?

CHAPTER 27

Isostasy and Diastrophism

It is useful to be assured that the heavings of the earth are not the work of angry deities. These phenomena have causes of their own.

—Seneca

GEOLOGY BECAME A true science early in the nineteenth century when scientists first came to understand that the geologic processes that to-day form and change rocks on Earth's surface and within its interior are the same processes that have been at work throughout the very long history of Earth. Thus ancient rocks were formed in the same way as modern rocks and can be interpreted in the same manner. The present, therefore, is the key to the past. This basic doctrine, known as **uniformitarianism,** is the very foundation upon which the science of geology rests.

Geologic theory indicates that all rocks that compose Earth's lithosphere are in a condition of equilibrium and floating upon the upper mantle. Geologists label this concept *isostasy,* and we begin this chapter with a discussion of this concept and the changes it causes in the outer structure of Earth.

The movements of Earth's crust that result in relative changes of position and in the deformation of rocks is known as *diastrophism.* What are the shapes and structures of Earth's crust caused by the deformation of rocks? How is the contour of Earth's surface changed and formed by the deformation of rocks? Does diastrophism provide an explanation for the origin of mountains? This chapter provides answers to these questions.

27.1 Isostasy

The doctrine of **isostasy** is one of the most useful generalizations of geology because it provides an explanation for many geologic processes and phenomena. The doctrine states that there is a condition of equilibrium in Earth's crust so that all rock masses are in balance.

All rocks on Earth are drawn by gravity toward the center of Earth. If we consider entire Earth, we realize that the most stable possible arrangement of rock is a sphere. This shape is, of course, what space travellers have seen who look at Earth from the moon. Earth as a whole must also be in a condition of equilibrium with rocks as close to Earth's center as possible.

As we approach Earth and examine it more closely, we realize that it is not a perfect sphere. Its radius is somewhat greater at the equator than at the poles. Continents rise above sea level, while the ocean basins are below it; and even the continents and ocean basins contain rough, irregular topography.

If Earth's crust is in equilibrium, how do we explain its irregularities? The equatorial bulge is, of course, believed to be the result of the centrifugal force from Earth's rotation. The smaller irregularities within the continents and ocean basins persist because rock, unlike water, is a solid with strength and does not readily flow. A mountain peak or valley will endure, while a disturbance on a water surface will quickly diminish.

The existence of the vastly larger and long-enduring continents and ocean basins has an explanation that is central to the concept of isostasy. Continents and ocean basins stand at different elevations because they are composed of rock of different density. The continental granites have a specific gravity of about 2.7, compared with about 3.0 for the basalts of the ocean basins. Both are considered to "float" on still denser, deeper rock within the mantle, with the continents floating higher because of their lighter weight, as shown in Fig. 27.1. Isostatic balance can be maintained only because rock is plastic deep below the surface within the asthenosphere, where pressures and temperatures are very great. Rock within the asthenosphere, though still solid, has the capacity to flow and, therefore, to adjust itself to the distribution of mass above it.

A density difference accounts for the difference in elevation between ocean basins and continental masses but not for the differences in altitude within a continent. How can a plain barely above sea level and a towering mountain range both stand in isostatic balance, if they are both composed of basically the same rock type? This arrangement is possible because the continental masses vary markedly in thickness. Just as the top of an iceberg floats higher above water than an ice cube, so a thick mass of granite stands high as a mountain, while thin granite reaches only slightly above sea level. The continents, though irregular in elevation, are in isostatic balance.

◄ **Mount Fuji, Japan.**

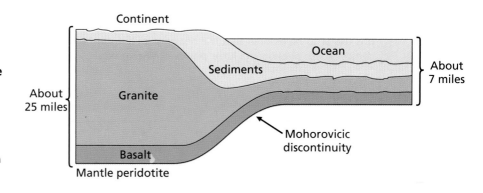

Figure 27.1 Isostatic balance of continents and ocean basins.
Continents stand higher than ocean basins because they are composed of granite, which is ligher than the basalt beneath the oceans.

Earth is such a vital, mobile body that its distribution of mass cannot remain constant. Rock removed from mountains by erosion is deposited elsewhere. Huge continental glaciers more than 3000 m (10,000 ft) thick advance and retreat across major portions of the continents. Diastrophic forces move rock masses both horizontally and vertically. Each of these changes represents a departure from isostatic balance. Within a short span of geologic time, isostasy would cease to exist were it not for continuing readjustments within the plastic rock of the upper mantle.

Nothing illustrates this principle more strikingly than the reaction to the arrival and departure of the recently departed, giant ice-age glaciers, whose 3000-m (10,000-ft) thickness imposed an immense additional burden upon the continents. Under this great weight of ice the rock surface subsided until isostasy was reestablished. The two areas where continental glaciers persist, Antarctica and Greenland, are still so deeply depressed that their rock surfaces are many feet below sea level. When the ice age glaciers melted away, the surfaces of Earth moved back up but were unable to keep pace with the retreat of the ice. Vast areas of North America and Europe are, therefore, still deeply depressed from the weight of ice sheets that melted thousands of years ago. The land surfaces continue to rise even today, and by the time they have reestablished equilibrium, they will have fashioned some major geographic changes. For example, the Baltic Sea is still rising about 1 m (3 ft) per century at its northern end and more slowly to the south, and will in time be lifted almost entirely above sea level and emptied of much of its water. Hudson Bay is moving upwards at an equal pace and is also doomed to extinction.

Isostatic adjustments to compensate for redistribution of mass by erosion and deposition are also important, but these adjustments have been slower and occurred earlier in time, and therefore the record is not so clear. As erosion strips rock away from mountains, the mountains should rise to compensate for their losses.

27.2 Rock Deformation

Diastrophism refers to the series of processes by which the major features of Earth's crust are formed and changed because of relative changes of positions and deformation of rock. The changes may be sudden and cataclysmic or so slow that they are not easily observed. Sudden changes produce earthquakes and cause great destruction, but the slow changes have more far-reaching effects on the history of Earth and its inhabitants.

Two major types of slow structural deformations are folding and faulting. Folding of Earth's crust takes place when extreme pressure is exerted horizontally. Rock can be compressed only a limited amount before it begins to buckle and fold. The folded rock layers can arch upward or downward, or they may fold and override an adjacent fold. Figures 27.2 and 27.3 illustrate the principal types of folded rock. The folding occurs during the early stages of mountain formation.

Faults are fractures in Earth's crust caused by extreme forces acting on the crust because of slow rock movement within the asthenosphere. The forces on the crust may be vertical, producing uplifts, or they may be horizontal and produce compression; or the forces may cause tension and bring about a lengthening of the crust. Figure 27.4 illustrates a few essential terms needed to describe fault geometry. The fault plane is the actual surface itself. The hanging wall describes the rock above the fault plane, and the footwall contains rock beneath it. The upthrown side has moved up relative to the downthrown side, as indicated by the arrows.

There are four types of faults: normal, reverse, strike-slip, and thrust (see Fig. 27.5). Faults may occur in any type of rock, and they may be vertical, horizontal, or inclined at an angle. Most faults occur at an angle.

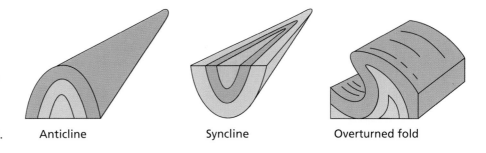

Figure 27.2 Crustal Folding.
Folded rock layers can occur arching downward (anticline) or upward (syncline), or they can override (overturned fold).

Anticline   Syncline   Overturned fold

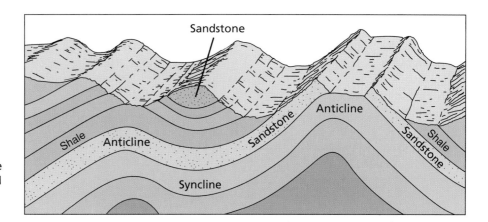

Figure 27.3
Some of the more common re lations between anticlines and synclines and the land forms above them.

Sandstone

Shale   Anticline   Sandstone   Anticline   Sandstone   Shale

Syncline

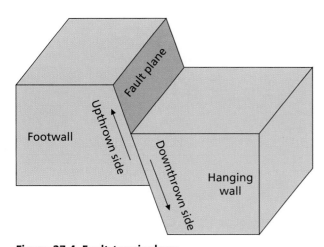

Fault plane

Footwall

Upthrown side

Downthrown side

Downthrown side

Hanging wall

Figure 27.4 Fault terminology.
The surface along which the rocks move is the fault plane, and the footwall is the rock beneath it. The upthrown side moves up with respect to the downthrown side.

A **normal fault** occurs as the result of expansive forces that cause the overlying side of the fault to move downward relative to the side beneath it. In this case the stress forces are in opposite directions, and the fault-ing tends to expand or pull the crust apart. Figure 27.6 illustrates the formation and development of a normal fault in Earth's crust.

A **reverse fault** occurs as the result of compressional stress forces that cause the overlying side of the fault to move upward relative to the side beneath it. A special case of reverse faulting, called **thrust faulting,** describes the faulting when the fault plane is at a small angle to the horizontal. As might be expected, this type of fault-ing occurs in subduction zones, in which one plate slides over a descending plate, or along convergent plate boundaries. Figure 27.7 illustrates the formation and development of a reverse fault.

Finally, **strike-slip faulting** occurs when the stresses are parallel to the fault boundary, such that the fault slip is horizontal. This type of faulting takes place along the boundary of two plates that *strike* and *slip* by each other without an appreciable gain or loss of surface area. Strike-slip faulting occurs along what are commonly called **transform faults.** Transform faults occur near plate boundaries that do not move directly away from or toward each other.

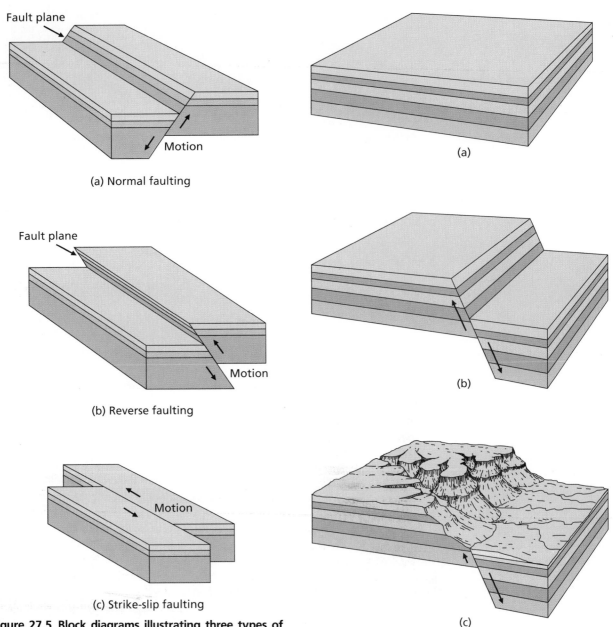

Fault plane

Motion

(a) Normal faulting

Fault plane

Motion

(b) Reverse faulting

Motion

(c) Strike-slip faulting

Figure 27.5 Block diagrams illustrating three types of faulting.

(a)

(b)

(c)

Figure 27.6 Block diagrams illustrating the development of a normal fault.
Block c shows the effects of erosion.

27.3 **Mountain Building**

The sheer massiveness and heights of mountains have always fascinated and awed observers. To ancient cultures their lofty summits were the dwelling places of gods. Throughout history, and even today, superstitions surround mountains. To some, mountains present a challenge that must be met by scaling the highest peak.

However, to the geologist mountains are a key to Earth's geologic history. In an effort to know this history, geologists study and theorize how mountains were built or formed.

Mountains differ greatly. Some are singular masses rising majestically above plains, such as the volcanic

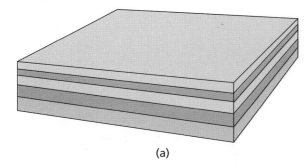

(a)

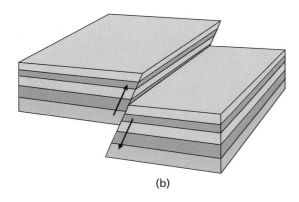

(b)

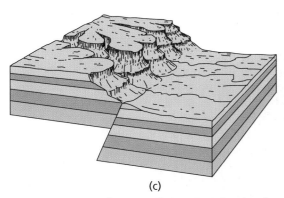

(c)

Figure 27.7 Block diagrams illustrating the development of a reverse fault.
Block c shows the effects of erosion.

cone of Mt. Kilimanjaro in Africa. Other mountains belong to a geologic unit or series of mountains that make up a **mountain range,** such as the Sierra Nevada range in the western United States. A group of similar ranges form a **mountain system.** The Sierra Nevada range, along with several other ranges, form the Rocky Mountain system. An elongated unit of ranges is referred to as a **mountain chain.**

Mountains are generally classified into three principal kinds, based on characteristic features: (a) volcanic mountains, (b) fault-block mountains, and (c) folded mountains.

Volcanic mountains have been built by volcanic eruptions. As discussed in Chapter 26, most volcanoes, and hence volcanic mountains, are located above the subduction zones of plate boundaries. If the colliding plates are both oceanic, volcanic mountain chains are formed on the ocean floor of the overlying plate. New mountains are being formed beneath the sea today by this process. Chains of oceanic volcanic mountains are evidenced above sea level by island arcs. Japan and the islands of the West Indies are good examples of island arcs. Island arcs lie adjacent to ocean trenches that mark the boundaries of oceanic plates.

Continental volcanic mountain chains occur when the overlying plate of the subduction zone is continental crust. The Andes Mountains along the western coast of South America are an example of such a continental mountain range. The Cascade Mountains in Washington and Oregon are the only volcanic mountain range in the United States (Fig. 27.8).

Fault-block mountains are believed to have been built by normal faulting in which giant pieces of Earth's crust were tilted and uplifted (Fig. 27.9). These mountains evidence great stresses within Earth's crustal plates.

Figure 27.8 The Cascade mountains in Washington and Oregon.

Figure 27.9 Fault-block mountains.
Mount Whitney, in California.

Figure 27.10 Folded mountains.
Alberta, Canada.

Fault-block mountains rise sharply above the surrounding plains. The Sierra Nevada range in California, the Grand Teton Mountains in Wyoming, and the Wasatch range in Utah are examples of fault-block mountains in the United States.

Folded mountains, as the name implies, are characterized by folded rock strata. Examples of folded mountains include the Alps, the Himalayas, and the Appalachian Mountains (Fig. 27.10). Although folding is the main feature of folded mountains, these complex structures also contain external evidence of faulting and central evidence of igneous metamorphic activity. Folded mountains are also characterized by exceptionally thick sedimentary strata, which indicates that the material of these mountains was once at the bottom of an ocean basin. Indeed, marine fossils have been found at high elevations in the Himalayas and other folded mountain systems.

Two eminent American geologists, James D. Dana (1813–1895) and his contemporary James Hall (1811–1898), advanced many of the original concepts of how such mountains were formed. These concepts considered the mountains on Earth's surface to be a result of the cooling and contracting of Earth during its formation—much like the wrinkling of the skin of an apple as it dries out. A good statement of this theory is presented in the conclusion of an article by Dana that appeared in the *American Journal of Science and Arts* in 1873 (Series 3, Vol. 106):

> The views on mountain building now sustained suppose the existence, through a large part of geologic time, of a thin crust, and of liquid rock beneath that crust so as to make its oscillations possible and refers the chief oscillations, whether of elevation or subsidence (sinking), to lateral pressure from the contraction of that crust; and this accords with my former view, and with that earlier presented by the clear-sighted French geologist, Prevost. I hold also, as before, that the prevailing position of mountains on the borders of continents, with the like location of volcanoes and of the greater earthquakes, is due to the fact that the oceanic areas were much the largest, and were the areas of greatest subsidence under the continued general contraction of the globe.

However, Dana and Hall disagreed on how the subsidence of ocean basins occurred. From a study of fossils once contained in **geosynclines,** which are long, narrow, subsiding ocean troughs containing large accumulations of sediment, Hall found evidence that over geologic time periods a basin sank at approximately the same rate as the sediment accumulated. He believed that the basin sank because of the accumulating burden of sediment.

Dana, on the other hand, thought that the sinking was the result of diastrophism—that is, the movement of Earth's crust. Most of the present-day geological data support Dana's view. Certainly, the weight of the sediment contributes to subsidence, but the fundamental mechanism of mountain building appears to be diastrophism.

The theory of plate tectonics offers an explanation of the mechanism of diastrophism and the building of folded mountains. Oceanic geosynclines containing large accumulations of sediment occur along the margins of continents, which supply the eroded sediment. If these geosynclines occur in regions of descending ocean plates, the downward movement of a plate pushes the stratified layers in the geosyncline against the continental crust (Fig. 27.11). The deep-water strata are dragged downward by the moving plate, where they are heated and subjected to metamorphism. Some of the deepest strata may melt and rise to form igneous intrusions in the strata above, becoming the core of a new chain of folded mountains. The shallow-water strata are pushed aside and are folded and faulted along the marginal regions of the folded chain.

Such mountain building occurs today along the western coasts of North and South America. Large geosynclines lie along the Atlantic coasts of North and South America, but these regions are not along plate boundaries. If, perhaps, the growing accumulations of sediment in the geosynclines cause a downward plunge into the lithosphere at some future geologic time, then a folded mountain chain may form along the Atlantic Coast.

Let's consider the formation of some of the folded mountain ranges of Earth within the framework of plate tectonics. The Himalayas are believed to have formed after the supercontinent Pangaea broke up some 200 million years ago. In terms of surface features, India broke away from Africa and ran into Asia, with the collision resulting in the Himalayas.

In more detail, the northward movement of the Indian Plate after its separation from Africa resulted in the loss of oceanic lithosphere formerly separating India and Asia, with the eventual collision of these two continental masses. The Indian Plate descended under the Eurasian Plate, and the geosynclines along the Eurasian Plate were folded into mountains, as illustrated in Fig. 27.12. When the two continental crusts met, the edge of the Eurasian Plate was lifted, with the spectacular result of the highest mountain range on Earth. Present-day analyses of plate movements indicate that the Indian Plate is still moving northward relative to the Eurasian Plate, causing the slow, continuing uplift of the Himalayas.

The Alps are another example of what is believed to be the result of plate collision. In this case the colliding plates were the African Plate and the Eurasian Plate.

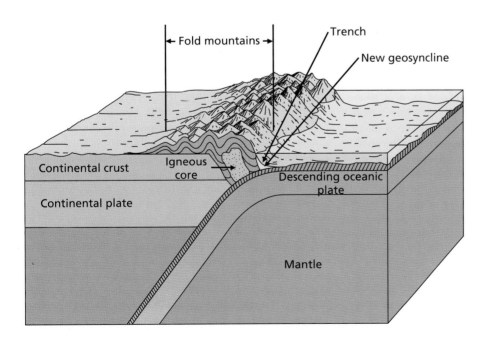

Figure 27.11 An illustration of the formation of folded mountains.
The downward movement of the descending oceanic plate pushes the geosyncline strata against the continental plate. The deep-water strata are dragged downward by the moving plate, where they are heated. The deepest strata melt and rise to form the core of the folded-mountain chain, while the shallow-water strata are folded and faulted along the marginal regions of the chain.

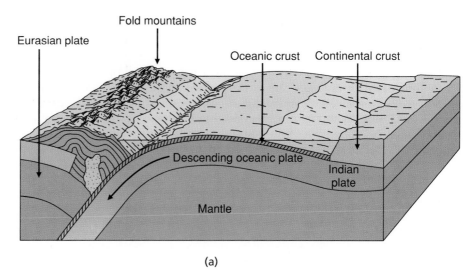

(a)

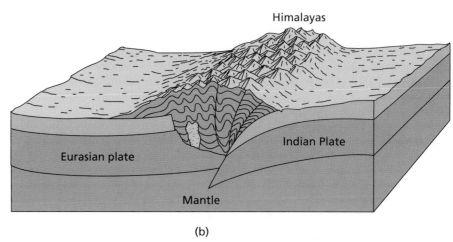

(b)

Figure 27.12 An illustration of the formation of the Himalayas.
(a) After its separation from Africa, the northward-moving Indian Plate descended under the Eurasian Plate and the geosynclines along the Eurasian Plate were folded into mountains. (b) When the two continental crusts met, the edge of the Eurasian Plate was lifted, giving rise to the lofty Himalayas.

The Appalachian Mountains are quite old compared with the Himalayas and the Alps and were apparently once joined to mountains in northern Europe (see Fig. 26.13). How did these folded mountains form? That is, where was the geosyncline? Geologists speculate that before Pangaea an ancient ocean existed between North America and Europe. The ocean disappeared as the plates of these landmasses converged to form Pangaea. In the collision the Appalachians were formed. Thus the Appalachians were once an interior mountain range of Pangaea. When Pangaea broke up, a plate boundary cut across the mountain range and part went with the new American Plate and part with the new Eurasian Plate.

A similar explanation of the collision of pre-Pangaea plates is given for the Ural mountain range, which runs north-south between Europe and Asia. In this case the European and Asian plates remained intact as the Eurasian Plate when Pangaea broke up.

Thus we see that the theory of plate tectonics provides a beautiful and simplified explanation of mountain building. However, keep in mind that this theory is relatively new: and although there is a great deal of supporting evidence, it has not really been examined and tested fully. There are other theories of mountain building that also give acceptable explanations.

One such theory suggests that mountains can arise independently of horizontal plate movements. According to this theory, gravity causes heavier rock to sink in the lithosphere while lighter rock rises to form mountains such as the Alps. Any theory must stand the test of time and evidence; and because of the complex nature of Earth, geologic evidence is slow in coming. There is a great deal of work to be done before it is fully established whether the sequence of events described by the theory of plate tectonics happened in the manner described by the theory or in some other way.

Learning Objectives

After reading and studying this chapter, you should be able to do the following without referring to the text:

1. Define *isostasy*, and describe how the concept is used to explain certain geologic processes.

2. State and explain the different types of rock deformations.

3. Explain how mountains are formed.

4. Describe the principal types of mountains, and give an example of each.

5. Define and explain the important words and terms listed in the next section without referring to the text.

Important Words and Terms

uniformitarianism	reverse fault	mountain range	folded mountains
isostasy	thrust faulting	mountain system	geosynclines
diastrophism	strike-slip faulting	volcanic mountains	anticlines
normal fault	transform faults	fault-block mountains	synclines

Questions

Isostasy

1. Explain the concept of isostasy.

2. How does the continental crust differ fundamentally from the ocean crust?

3. Why does a continent stand higher than an ocean basin?

4. What evidence do geologists have that supports the concept of a floating lithosphere?

5. What happens to the elevation of a floating rock mass when additional weight is placed on it because of the deposition of sediments? Explain your answer.

6. What happens to the elevation of a rock mass when the weight is reduced by erosion? Explain your answer.

Rock Deformation

7. Explain each of the following terms: (a) normal faulting, (b) reverse faulting, (c) thrust faulting, (d) strike-slip faulting.

8. What types of relative plate motions correspond to the different types of faulting?

9. What type of faulting is associated with a transform fault?

10. How do normal faults and reverse faults reflect fundamentally different stresses on Earth's crust?

11. State the difference between anticlines and synclines.

12. What is the fault plane?

Mountain Building

13. What is diastrophism?

14. What are the three principal types of mountains, and how are they distinguished? Give an example of each.

15. How is the formation of each type of mountain explained by the theory of plate tectonics?

16. What is a geosyncline?

17. Describe how the formation of each of the following mountain ranges is explained by the theory of plate tectonics: (a) the Himalayas, (b) the Alps, (c) the Appalachians.

18. Name the type of plate boundary that is most likely to produce mountain building.

CHAPTER 28

Geologic Time

Oh earth, what changes thou hast seen!

—Tennyson

MOST GEOLOGIC PROCESSES are so slow that there is little apparent progress in a human lifetime. It is only when we recognize the enormous length of time these processes have been operating that we can appreciate the vast amount of geologic change that has taken place since the origin of Earth. Earth as we know it today has developed by gradual evolution through an incredibly long span of geologic time.

We learned in Chapter 16 that planet Earth was formed about 5 billion years ago. During the planet's long history many changes have taken place. The history of many of these changes are recorded in the rock that makes up the crust of Earth. In this chapter a chronological history of the planet is given in both relative and absolute time.

The history of Earth is recorded in geologic time and events. Geologists deal with processes that have taken place over millions of years, a time interval that is very difficult for us to comprehend. However, if natural processes are constant and the same processes operate today, on and within Earth, the present may be considered the key to the past. This is an important geologic concept called the **principle of uniformity,** or uniformitarianism. Changes occurring on and within Earth today are clues to the long-term total picture of the history of our planet.

28.1 Relative Geologic Time

From the very birth of geology as a science, geologists have seen the need to measure geologic time. The first and simplest task was to establish the sequence of geologic events within a local area. Figure 28.1 is a cross-sectional diagram illustrating several geologic events that occurred in sequence. The sedimentary rock and igneous rock are identified in the legend. The relative

age of the sedimentary rock is indicated by numbers, with the number 1 indicating the oldest. The numbers are not really necessary, however. You can determine the relative ages yourself by using the **law of superposition.** This so-called law is the simple observation that in a succession of stratified deposits the younger layers lie on top of the older layers. Geologists established a **relative geologic time** scale based on the characteristic fossils of each stratum. When you realize, in addition, that sedimentary rocks were originally deposited in horizontal layers and that younger geologic structures cut across older ones, you are ready to unravel the history of this local area.

Let us, however, first identify the structures in Fig. 28.1. The structure B-B′ is a dike, and D-D′ is a combination dike and sill. We know that C-C′ is a normal fault, because the hanging wall has moved down. The structure A-A′ is an unconformity, and E-E′ is a modern erosion surface.

The structure B-B′ cuts across C-C′ and is therefore the younger of the two. Structure A-A′ is younger than B-B′ for the same reason. Structure D-D′ cuts across A-A′ and therefore is younger. Structure E-E′ must be the most recent because it cuts off all older rock and structures. One final question before we outline the geologic history of this area: Which occurred first, the folding of the rock beneath the angular unconformity (A-A′), or the normal fault (C-C′)? The folding must have come before the faulting; otherwise, the fault plane would have been folded.

In outline form the geologic history of Fig. 28.1 is as follows:

1. Deposition of sedimentary rock 1 through 8
2. Folding of sedimentary rock to form anticlines and synclines
3. Normal faulting (C-C′)
4. Intrusion of dike (B-B′)
5. Uplift and erosion to surface A-A′
6. Deposition of sedimentary rock 12 to 15 on surface A-A′

◄ Bryce Canyon, Utah.

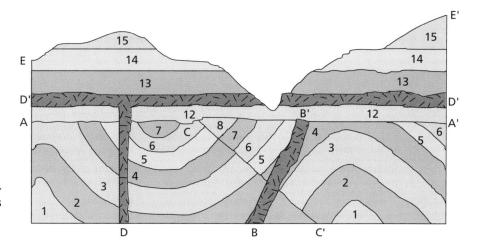

Figure 28.1 Diagram illustrating several geologic events that occurred in sequence. See text for an explanation.

7. Intrusion of dike and sill (D-D')

8. Uplift and erosion to surface E-E'

The individual histories of the many local areas must be related to one another, or geologic history can be little more than a random collection of isolated interpretations. Using a procedure called **correlation,** geologists have long been busy tying the loose ends together and constructing an organized picture of the geology of large areas and even entire continents. Geologists correlate rock from one area to another by determining whether the separate rock sequences were formed at the same time or different times. There are many ways to correlate rock, but only a few ways are reliable for rock separated by great distances. One way that has been successful is the use of fossils. A **fossil** is a remnant or trace of an organism preserved from a past geological age, such as a skeleton, footprint, or leaf imprint embedded in Earth's crust.

Fossils are useful in determining the age of rock because life has changed through geologic time. Rocks of different ages contain different fossil assemblages, even if deposited in the same environment. The age of a sedimentary rock can therefore be determined from the fossils contained within it, provided, of course, the age of the fossils is known. The history of life is far from simple, however. The various trends and patterns of organic evolution make a fascinating story of their own. A very thick book would be required to trace the long and involved history of ancient life. The evolutionary changes have differed from one family to another. Certain animals and plants have changed little through time. These creatures in fossil form reveal little of the age of the rock in which they lie buried. Other

forms of life, however, have changed in response to their shifting environments and evolved rapidly through geologic history. If these latter creatures become fossilized and are also numerous and widespread, they are very useful in rock-age determinations and are known as **index fossils.**

Using index fossils and all other tools of correlation at their command, geologists through long, patient labor have been able to construct a relative geologic time scale similar to our human time scale. Just as human history is divided into units and subdivided into smaller units, so is geologic time. The largest unit of geologic time is the **era,** which is subdivided into smaller time units known as **periods.** Periods may be subdivided into **epochs** and epochs, in turn, into **ages.** Each era has a name, as has each period, epoch, and age. The geologic time units are shown in Fig. 28.2.

Unconformities are breaks or local unrecorded intervals in the geological record. They represent a period of time when deposition stopped, erosion removed surface rock, and deposition of sediments resumed. Unconformities may take several different forms; three are illustrated in Figs. 28.3, 28.4, and 28.5. As illustrated in Fig. 28.3, the unconformity represents an interval of erosion, followed by a period of deposition. The difference in age between the rock above and below the unconformity may be relatively small or hundreds of millions of years.

One of the simplest types of unconformity has beds above and below the erosion surface parallel to each other. Such a structure can reflect uplift of the surface above sea level with erosions and then subsidence, but the fall and rise of sea level can have the same effect. Some unconformities (Fig. 28.4) have the beds below

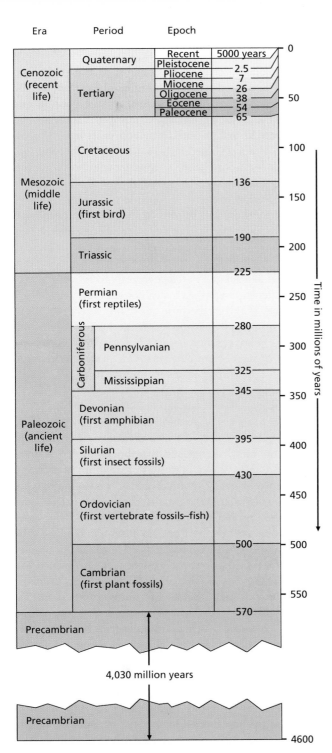

Era	Period	Epoch		
Cenozoic (recent life)	Quaternary	Recent	5000 years	0
		Pleistocene	2.5	
	Tertiary	Pliocene	7	50
		Miocene	26	
		Oligocene	38	
		Eocene	54	
		Paleocene	65	
Mesozoic (middle life)	Cretaceous		100	
			136	150
	Jurassic (first bird)		190	200
	Triassic		225	
Paleozoic (ancient life)	Permian (first reptiles)		250	
	Carboniferous	Pennsylvanian	280	300
		Mississippian	325	
	Devonian (first amphibian)		345	350
	Silurian (first insect fossils)		395	400
	Ordovician (first vertebrate fossils–fish)		430	450
	Cambrian (first plant fossils)		500	500
			550	
			570	
Precambrian				

Time in millions of years

4,030 million years

Precambrian

4600

Figure 28.2 The Geologic Time Scale.
The largest time units are the eras. The Precambrian era represents more than four thousand million years, or over 87 percent of the total time scale. The other eras—Paleozoic, Mesozoic, and Cenozoic—refer to the ancient, middle, and recent periods, respectively, in the history of life on Earth.

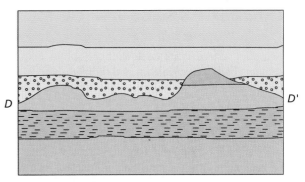

D   D'

Figure 28.3 One of three types of unconformity.
The sedimentary rock above and below the erosion surface (D-D') is parallel.

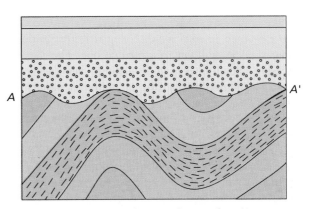

A   A'

Figure 28.4 A second type of unconformity.
Rock below the erosion surface (A-A') is at an angle to that above it.

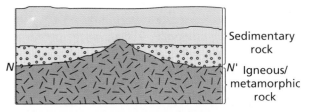

N   N' Igneous/metamorphic rock

Sedimentary rock

Figure 28.5 A third type of unconformity.
The surface (N-N') separates the older, eroded igneous or metamorphic rock from the younger sedimentary rock above it.

the erosion surface at an angle to those above, because the older beds were folded and eroded before the younger ones were deposited. A third type of unconformity contains sedimentary rock resting on older, eroded igneous or metamorphic rock (Fig. 28.5). Unconformities are important because they portray noteworthy geologic events in the local history of Earth's crust and play an important part in relative geologic dating.

28.2 Absolute (Atomic) Geologic Time

The establishment and refinement of the relative geologic time scale was a most significant scientific achievement but one that left geologists with a gnawing sense of frustration. They were not content to know only the order in which geologic events occurred; they needed to know how long ago *in years* these events occurred. They yearned, in other words, for an **absolute geologic time** scale in addition to a relative one. The need to measure the absolute age of geologic events became apparent when the doctrine of uniformitarianism revealed the great age of Earth, and it became truly urgent when Charles Darwin's theory of organic evolution redoubled interest in the long early history of our planet.

Space in this book does not permit a review of all the early attempts to measure the absolute ages of rocks or the dates of events in the geologic past. For the most part they all involved the same principle. The age can be computed if the investigator divides the total amount of geologic work accomplished by the rate at which the geologic process is currently operating. For example, we can hope to measure the age of a delta by dividing the total volume of the delta by the volume of sediment deposited on the delta each year. In a similar manner, we could attempt to determine the age of the oceans by dividing their total salt content by the volume of salt the many rivers of the world dump into the seas each year.

For a variety of reasons, most of these early efforts were completely inaccurate. The age of the oceans was incorrectly calculated to be 90 million years. From the total thickness of sedimentary rocks, minimum ages from 17 million to more than 1 billion years were estimated for Earth. The most impressive and influential calculations were those of Lord Kelvin, the distinguished nineteenth-century physicist, who attempted to determine the age of Earth from the rate of heat loss from Earth's interior. Kelvin assumed that Earth had its beginning as a hot molten body, which became solid as it cooled and which continued to lose its residual heat from its still hot interior. From a measurement of the existing rate of heat loss, Kelvin calculated that Earth became solid between 20 million and 40 million years ago. This estimate, although based on actual measurements and supported by Lord Kelvin's considerable prestige, was mostly unacceptable to geologists and biologists alike. Both sciences preferred a much larger figure and were forced by these calculations to the reluctant conclusion that the pace of geologic processes and organic evolution must have been much quicker in the past than at present. Such was the weight of Lord Kelvin's authority that Darwin felt obliged in his later years to retract some of his earlier arguments. Actually, Lord Kelvin's calculations were badly in error, because one of his basic assumptions was wrong. As we know now, the heat radiated from Earth's interior is not residual but is generated by the decay of radioactive elements. Our planet has its own built-in furnace that keeps its interior hot.

Radioactivity, the phenomenon whose existence made Lord Kelvin's calculations incorrect, has proved to be the very key to the ages of many rocks. You will recall from Chapter 10 that a radioactive element contains an unstable nucleus that breaks down spontaneously to a lower energy state and to a different chemical element. These radioactive transformations are of several sorts. During alpha decay the nucleus loses two protons and two neutrons and therefore decreases the atomic number by 2 and the mass number by 4. The nuclear disturbance is accompanied by an additional loss of energy in the form of gamma radiation, which like light, X-rays, radio, and other similar waves, travels at an enormous speed. Beta decay, another form of nuclear decay, is the emission of an electron from one of the atom's neutrons. Although an electron has very little mass and there is, accordingly, no change in mass number, beta decay results from the transformation of a neutron to a proton and electron and therefore increases the atomic number by 1. Gamma radiation accompanies this nuclear change as well. During the third type of radioactive transformation, the nucleus captures an electron from the atom's inner electron shell and thereby transforms a proton to a neutron. Electron capture, as this process is called, causes no change in mass number, but it decreases the atomic number by 1. This change is also accompanied by gamma radiation. Physicists use the term *branching decay* to describe those situations where a radioactive element has two modes of decay, as we shall see in the decay of potassium-40.

The decay product—or daughter element, as it is commonly known—may have a stable nucleus and therefore be subject to no further radioactive decay, the transformation being completed in a single step. Elements whose atomic number is less than 82 are most likely to behave this way. The daughter elements of some other radioactive elements, however, are themselves unstable and may in turn break down into another daughter element, which may also be radioactive. This radioactive series can be long and complex, with several branching steps before finally achieving a stable end product. The radioactive transformation of uranium 238, for example, involves eight alpha steps and six beta steps to become lead 206, the stable end product.

Physicists usually express the rate of decay of an element in terms of what is known as its half-life. The **half-life** is the number of years required for half of the parent element to decay. Thus if we start with 8 g of a radioactive element whose half-life is 12 years, there will be 4 g left at the end of 12 years, 2 g at the end of 24 years, and 1 g in 36 years. The number of nuclei that break down is a function of the number of atoms still present, and therefore, the radioactive parent element is totally depleted only after a very long time.

Because the rate of decay remains constant—unaffected by temperature, pressure, and chemical environment—we can use radioactivity as a clock to measure the march of geologic time. The older the rock, the greater is the ratio of the daughter to the radioactive parent. Ideally, a radioactive element can tell the age of the rock that contains it under the following conditions:

1. There has been no addition or subtraction of the parent or daughter other than that caused by radioactive decay for the entire life of the rock.

2. There was no trace of the daughter element in the rock when it was formed.

3. The age of the rock differs from the half-life of the element by no more than a factor of 10.

Usually, satisfying condition 1 is not difficult. Condition 2 is more of a problem. If, for example, there happened to be some lead present at the birth of a uranium-bearing mineral, the rock would appear older than its actual age. The amount of contamination must be estimated and corrected for.

The earliest radioactive age measurements made use of uranium and lead and were based on the ratio of the two. Later, more precise analyses revealed that what had been once considered as simply uranium was actually two isotopes, uranium-238 and uranium-235, in addition to thorium-232. The lead in turn was discovered to consist of three isotopes. Because the half-lives of the parents are fundamentally different, an accurate age determination requires that all six constituents be separated and their quantities be measured accurately.

Uranium-238, whose half-life is 4510 million years, decays to lead-206 through eight alpha steps and six beta steps. Uranium-235 has a half-life of 713 million years and decays to lead-207 through seven alpha steps and five beta steps. Thorium-232, with a half-life of 14,100 million years, decays to lead-208, also through a series of steps. Common lead, which may have contaminated the original uranium-bearing minerals, consists of all three isotopes in addition to lead-204, which does not form by radioactive decay. It is the presence of this last isotope that enables the analyst to correct for any common lead. Uranium-238 and uranium-235 are clocks that have timed the birth of many rocks on Earth, but thorium is not so widely employed. Most of the early measurements used uranium-bearing minerals such as uraninite, but this mineral is so rare and forms under such unusual geologic conditions that a sufficient number of significant dates seemed unlikely. Fortunately, the very delicate procedures now possible enable the analyst to measure the small traces of uranium in zircon ($ZrSiO_4$), a much more abundant and representative mineral.

Potassium, one of the most abundant and widespread chemical elements, has a rare radioactive isotope, potassium-40. This isotope is found in many rocks that do not contain measurable quantities of uranium, and so potassium-40 has come to be one of the most useful tools in determining where a rock falls on the calendar of geologic time. Potassium-40 illustrates well the principle of branching decay. A potassium-40 atom decays by a single step, either by beta radiation to calcium-40 or by electron capture to argon-40. Although 88 percent of the atoms become calcium-40, it is not practical to use for age determinations because the end product is lost in the vast sea of ordinary calcium, an extremely common, widespread element, most of which happens also to be calcium-40. All potassium age determinations, therefore, use argon-40, whose abundance can be measured with great precision.

The half-life of potassium-40 is 1.3 billion years. It is therefore, like uranium, valuable in dating the most ancient rocks on Earth. Still, the geologist must apply these dates cautiously, because argon is an inert gas and may escape from the mineral in which it accumulates. So much leaks from orthoclase, for example, that geologists have been forced to abandon the use of this very abundant, widespread mineral, which once seemed so promising in age determinations. Other potassium-bearing minerals such as biotite, muscovite, and hornblende seem better able to retain their argon and are widely analyzed for their age. Biotite from normal igneous rocks and from volcanic ash deposits has proven to be especially useful. Glauconite, a most interesting granular green mineral, seemed promising because it forms in sedimentary rock at the time the rock is deposited. Here is one mineral that might provide a direct measure of the age of sedimentary rock. All of these minerals, however, are susceptible to argon leakage if they have been heated. This heating effect, unfortunately, appears to apply especially to glauconite. A potassium-argon date may merely reveal the last time a rock was heated, rather than its true age. Geologists, therefore, tend to regard potassium-40 ages as minimum rather than true dates and to require that they be consistent with other geologic evidence before being accepted.

Rubidium-87 deteriorates in a single step to strontium-87 through beta decay, with a half-life of 47 billion years. Rubidium-87, which commonly occurs with potassium, is more abundant than potassium-40 and has the further advantage of decaying to a daughter element that is not a gas. A disadvantage is the great abundance of strontium-87, which has not formed from radioactive decay and which may have to be corrected for. Geologists often use rubidium-strontium ages to compare with potassium-argon determinations from the same rock.

Carbon-14 is a radioactive element whose relatively short half-life of 5730 years puts a second hand on the radioactive time clock. Obviously, an element with such a short half-life exists today only because it is in continuous production. The fuel for this radiocarbon factory is cosmic radiation from outer space, which bombards the nuclei of oxygen and nitrogen atoms in the upper atmosphere. The shattered oxygen and nitrogen nuclei release neutrons, which, traveling at high speed, collide with other nitrogen atoms and transform them into radioactive carbon-14. The carbon-14 reverts by beta radiation to nitrogen-14.

Before its death, however, the radiocarbon can be described as "leading a full life." Carbon-14 combines with oxygen to form carbon dioxide, which is carried by wind and water currents until it is evenly distributed around the world. Some finds its way into the tissues

of plants and from there into animals. Assuming that the rate of cosmic radiation remains constant, the ratio of carbon-14 to normal carbon-12 also remains constant, even in living organisms. A balance is struck between the rate of production and the rate of decay. Once an organism dies, however, its intake of radiocarbon ceases, while decay of carbon-14 continues; and so its radioactive supply dwindles. This decrease in the proportion of carbon-14 to carbon-12 provides the clock by which we time the death of an organism and, of course, determine the age of the sediment in which it lies buried.

The short half-life of carbon-14 allows us to place dates on comparatively young rock. Precise analyses have permitted us to obtain the radiocarbon age of some sediments deposited as recently as 40,000 to 50,000 years. Working from the other end of the geologic time scale, potassium-40 analyses have, under ideal circumstances, determined the age of rock as young as 100,000 years. But this leaves a great, irritating chasm of at least 50,000 years, a chasm geologists are most anxious to bridge. Thorium-230 and protactinium-231 are already promising to fill in the gap for ocean basin sediments, but they cannot be used for continental sediments. Perhaps scientists on land must await further refinements in existing techniques.

Humans, who devised the radiocarbon-dating techniques, are also responsible for complicating the picture with their technology. Radiocarbon dates assume a constant ratio of carbon-14 to carbon-12 in the atmosphere. As we burn coal and petroleum products, we add more carbon-12 to the atmosphere and thus reduce the ratio. As we explode nuclear bombs, we release additional carbon-14 with exactly the opposite effect. Fortunately, the carbon imbalance in our present atmosphere has little or no effect on the radiocarbon content of organic remains dead thousands of years.

28.3 Geologic Time Scale

Relative geologic time and atomic geologic time both are combined to give the **geologic time scale,** which is shown in Fig. 28.2. The time, in million of years as determined by radioactive methods, has been added to the right side of the relative time units.

Although radioactive dating has provided geologists with many essential dates for the ages of rocks, there are many time intervals where data are questionable. The questionable data arise because geologists have principally used sedimentary rocks to establish the rel-

Figure 28.6 *Top:* **A discordant dike formation.**
This exposed dike shows how the forming magma intruded across existing rock formations. *Bottom:* **A concordant sill formation.**
The intrusion of magma that formed this sill in Glacier National Park, Montana, was parallel to existing rock formations.

ative geologic time scale, while most of their radioactive determinations have been made on igneous rocks. Even though sedimentary rocks contain radioactive isotopes, their ages cannot be determined accurately because the ingredients that compose them are not the same age as the rock layer in which they are found. The sediments

that form the rock have been weathered from rocks of different ages.

Finding the proper place for igneous rocks in the relative time scale has not always been easy. Figures 28.6, 28.7, and 28.8 illustrate two such problems. In Figs. 28.6 (top) and 28.7, igneous rock has intruded

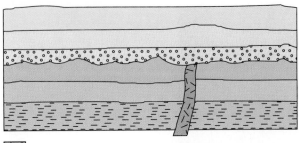

 Igneous rock

Figure 28.7 Buried intrusive igneous rock.
The intrusive rock is younger than the rock it intrudes and older than the rock that buries it.

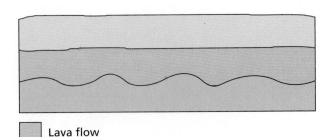

Lava flow

Figure 28.8 Buried lava flow.
The lava is younger than the rock beneath it and older than the rock above it.

sedimentary rock and has been, in turn, cut off by erosion, with the erosion surface buried by younger sedimentary rock. If there is very little difference in age between the rock above and below the unconformity, the ages of the sedimentary rock must be very nearly the absolute age of the igneous rock. If the two layers of sedimentary rock differ widely in age, we can calculate their absolute ages with much less precision. Figures 28.6 (bottom) and 28.8 illustrate a lava flow interbedded with sedimentary rock. The chance of great age difference is much smaller in this case, and therefore, the age of the igneous rock establishes the absolute age of the sedimentary rock.

Learning Objectives

After reading and studying this chapter, you should be able to do the following without referring to the text:

1. Explain the concept of uniformitarianism.
2. State the law of superposition.
3. Describe the methods used to determine relative geologic time.
4. Describe the methods used to determine absolute (atomic) geologic time.
5. Compare relative and absolute geologic time by using the geologic time scale.
6. Define or explain the important words and terms listed in the next section.

Important Words and Terms

principle of uniformity	index fossils	unconformities
law of superposition	era	absolute geologic time
relative geologic time	periods	half-life
correlation	epochs	geologic time scale
fossil	ages	

Questions

Relative Geologic Time

1. Define *relative geologic time.*
2. State the law of superposition.

3. On what is the relative geologic time scale based?
4. What is correlation?

5. What are fossils? How are fossils used to date rocks?

6. What are unconformities? Name three types of unconformities.

Atomic Geologic Time

7. Distinguish between relative and absolute (atomic) geologic dating of rocks.

8. Define *half-life.*

9. What are the limitations on the use of carbon-14 in determining the age of sediment?

10. What are the limitations in the use of potassium-40 for determining the ages of rocks?

Geologic Time Scale

11. State the meaning of Cenozoic, Mesozoic, and Paleozoic eras. See Fig. 28.2.

12. What era, on the geologic time scale, represents the longest length of time on the relative time scale? What percentage of total Earth history does this era represent?

13. Name the period of the first plant fossils.

14. Name the period of the first insect fossils.

15. Why is it difficult to assign absolute times to layers of sedimentary rock?

CHAPTER 29

Surface Processes

> The same regions do not remain always seas or always land, but all change their condition in the course of time.
>
> —Aristotle

NOTHING IS PERMANENT on Earth's surface—in the context of geologic time. Since early times people have erected edifices and monuments as memorials to persons and peoples and as symbols of various cultures. In doing so, they used the most durable rock materials; however, even recent erections show the inevitable signs of deterioration. Buildings, statues, and tombstones will eventually crumble when exposed to the elements; and on a larger scale, nature's mountains are continually being leveled.

An integral part of the rock cycle, as discussed in Chapter 25, is the decomposition of rock by weathering and other natural processes. As soon as rock is exposed at Earth's surface, its destruction begins. Many of the resulting particles of rock are washed away by surface water and are transported in streams and rivers as sediment toward the oceans. Over geologic time this process constitutes a leveling of Earth by the wearing away of high places and transporting of sediment to lower elevations.

We say that the wearing away and leveling of Earth's surface is due to **erosion,** a group of interrelated processes by which rock is broken down and the products are removed. While volcanism and diastrophism account for the uplifting of landmasses, erosion accounts for their leveling. This chapter will consider some of the processes and agents of erosion, including "unnatural" agents— human beings.

Also, the oceans, ocean currents, and seafloor topography will be considered. Important geologic processes take place on the ocean floor, which accounts for about 65% of Earth's surface.

29.1 Weathering and Mass Wasting

Weathering

The erosion process begins with weathering, which is the physical disintegration and chemical decomposition of rock. Weathering depends on a number of factors,

such as the type of rock, moisture, temperature, and overall climate.

Physical weathering involves the physical disintegration or fracture of rock, primarily as a result of pressure. For example, a common type of physical weathering in some regions is **frost wedging.** When rock is formed by solidification, lithification, or metamorphism, internal stresses are produced in it. One common result of these stresses is cracks or crevices in the rock, called joints. Joints provide an access route for water to penetrate into the rock. If the water freezes, the less dense ice requires more space than the water and exerts a strong pressure on the surrounding rock. As a result, the rock may break apart, as illustrated in Fig. 29.1. The expansive force of freezing water is readily observed by freezing water in a glass container, which usually cracks. Frost wedging is most effective in regions where freezing and thawing occur daily.

The action of freezing water is also effective in loosening the soil, which promotes its erosion. In the winter, water in the soil freezes as the ground cools. An accumulation of freezing water from the upward migration of water by capillary action or from rainwater or melting snow can give rise to an upward expansion that heaves up the material above and creates a bulge at the surface. This so-called **frost heaving** can result in fractures and bulges in street pavement. Highway joints and cracks are usually "sealed" with waterproofing materials to prevent accumulations of water and frost heaving. Also, building foundations are laid below the surface frost zone to prevent damage from frost heaving.

In cold upper latitudes the subsurface soil may remain frozen permanently, giving rise to a **permafrost** layer. During a few weeks in the summer, the top soil may thaw to a depth of a few inches to a few feet. The subsurface permafrost provides a stable base and prohibits the melted water from draining. As a result, the ground surface becomes wet and spongy (Fig. 29.2). The permafrost in Alaska caused many problems in the construction of the Alaskan oil pipeline. On the North Slope where the oil fields were discovered, the thawed surface is especially unstable due to a relatively high moisture

---

◄ **Precambrian rocks in the Grand Canyon.**

Figure 29.1 Frost Wedging.
The expansive forces of freezing water in rock joints cause the rock to fracture.

Figure 29.2 A "roller-coaster" railroad near Strelna, Alaska.
This condition was caused by differential subsidence resulting from the thawing of top soil over subsurface permafrost.

content. Thus the thawed surface does not recover easily if disturbed and if removed or compressed, the subsurface permafrost is susceptible to progressive thawing and erosion.

Frost wedging and frost heaving account for little if any physical weathering in hot, arid climates, but a similar process can. This weathering arises from the growth of salt crystals rather than water crystals. Groundwater with high salt concentrations seeps into porous rock—sandstone is a prime example. When the water evaporates and the salt crystallizes, the salt crystals exert pressure on the surrounding rock. This process of **salt wedging** can loosen and fracture the cementing material holding the sand grains together. The loosened sand grains can then be carried away by wind and water from rainstorms.

Plants and animals play a relatively small role in physical weathering. Most notable is the fracture of rock by plant root systems when they invade and grow in rock crevices. Burrowing animals—earthworms, in particular—loosen and bring soil to the surface. This action promotes aeration and access to moisture, which are important factors in chemical weathering. Thus one type of weathering promotes another. The activities of human beings also give rise to weathering and erosion that are often unwanted, as we will learn later in the chapter.

In all cases of physical weathering the disintegrated rock still has the same chemical composition. **Chemical weathering,** however, involves a chemical change in the rock's composition. Because heat and moisture are two important factors in chemical reactions, this type of weathering is most prevalent in hot, moist climates. One of the most common types of chemical weathering involves limestone, which is made up of the mineral calcite ($CaCO_3$). Rain can absorb and combine with carbon dioxide (CO_2) in the atmosphere to form a weak solution of carbonic acid:

$$H_2O + CO_2 \longrightarrow H_2CO_3$$
Carbonic acid

Also, as water moves downward through soil, it can take up even more carbon dioxide that is released by soil bacteria involved in plant decay. Recall that carbonic acid (carbonated water) is the weak acid in carbonated drinks.

When carbonic acid comes in contact with limestone, it reacts with the limestone to produce calcium hydrogen carbonate or calcium bicarbonate:

$$H_2CO_3 \quad + \quad CaCO_3 \longrightarrow \quad Ca(HCO_3)_2$$

Carbonic acid   Limestone   Calcium bicarbonate

Calcium bicarbonate dissolves readily in water and is carried away in solution. Because limestone is generally impermeable to water (and dilute carbonic acid), this type of chemical weathering acts primarily on the surfaces of limestone rock along which water flows.

Water flowing through underground limestone formations can carve out large caverns over millions of years. The cavern ceilings may collapse, causing **sinkholes** to appear on the land surface (Fig. 29.3). In many stable caverns, as water seeps through the cavern ceiling and drips to its floor, it loses its carbon dioxide, and minute amounts of calcium carbonate are precipitated. These precipitations build up to form icicle-shaped stalactite, stalagmite, and column dripstone. Do you recall which formation extends down from the ceiling and which protrudes up from the floor? See Fig. 25.23.

The rate of chemical weathering depends on many factors. The principal factors are climate and the mineral content of rock, with humidity and temperature being their chief climate controls. Chemical weathering is relatively rapid in hot humid climates, as compared with chemical decomposition in polar regions. On the other hand, chemical weathering is quite slow in hot dry climates. Egyptian pyramids and statues have stood for millenia with relatively minimal weathering. Some rock minerals are more susceptible to chemical weathering than others, as evidenced in the weathering of tombstones (Fig. 29.4). Marble tombstones, which consist of soluble calcite, may show a great deal of chemical weathering; sandstone tombstones, which consist of relatively insoluble quartz, will show little chemical weathering.

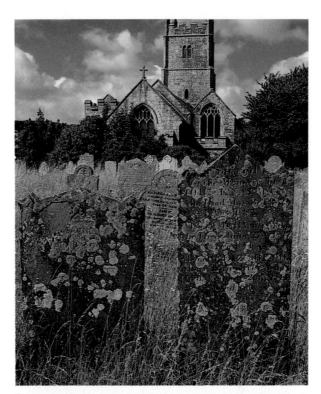

Figure 29.4 Weathered tombstones in Devon, England.

However, the weathering of sandstone tombstones can cause a loosening of the cementing materials holding the sand grains together and give rise to the erosion of large blocks of rock.

As we mentioned previously, burrowing animals, both large and small, bring soil or rock particles to the surface, where the exposed particles are more susceptible to chemical weathering. The mounds of soil particles built up by ants and earthworms are common sights.

Mass Wasting

The weathered material that accumulates on base rock is called **overburden.** This general term refers to rock and rock fragments or soil that is free of the base rock and so may start its descent toward lower elevations. The downslope movement of overburden under the influence of gravity is called **mass wasting.** If you drop a rock from a cliff, it is, in the geological sense, mass wasted. However, mass movement usually occurs with the aid of a transport medium such as running water. Mass wasting is divided into two categories, fast and slow, which are based on the comparative times involved in the mass-movement process. As we might expect, the

Figure 29.3 A sinkhole collapse in Florida.
Water flowing through limestone formations can carve out a large underground cavern. The cavern ceiling can then collapse, causing a sinkhole to appear on the land surface.

angle of the land slope is a critical factor in the distinction between fast and slow mass wasting.

Two important types of *fast mass wasting* are landslides and mudflows. **Landslides** involve the downslope movement of large blocks of weathered materials. Spectacular landslides occur in mountainous areas when large quantities of rock break off and move rapidly down the steep slopes. This type of landslide is termed a **rockslide.**

A comparatively slower form of landslide is a **slump,** the downslope movement of an unbroken block of overburden, which leaves a curved depression on the slope (Fig. 29.5). Slumps are commonly accompanied by debris flows that consist of a mixture of rock fragments, mud, and water that flows downslope as a viscous liquid. Small slumps are commonly observed on the bare slopes of new road constructions.

Mudflows are the movements of large masses of soil that have accumulated on steep slopes and become unstable with the absorption of large quantities of water from melting snows and heavy rains. Mudflows are likely to occur on slopes lacking vegetation, which hinders mass movement. Consequently, mudflows are common in hilly and mountainous regions where land is cleared for development without regard to soil conservation.

Mudflows are also associated with volcanoes. Volcanic mudflows result from loose ash on the slopes of the volcano that absorbs water from thunderstorms commonly associated with volcanic activity. The mudflows of ash can flow long distances and build up thick deposits of sediment. In some instances volcanic mudflows have buried nearby towns.

Fast mass wasting is quite dramatic, but *slow mass wasting* is a more effective geological transport process. In contrast to the rates of fast mass wasting, the rates of slow mass wasting processes are generally imperceptible. An important type of slow mass wasting is called creep. **Creep** is the slow particle-by-particle movement of weathered debris down a slope, taking place year after year. It cannot be seen happening, but the manifestations of creep are evident (Fig. 29.6). Although spectacular landslides and slumps involve large quantities of mass movement, this is but a fraction of the cumulative total of mass movement by creep over a period of time.

Another type of slow mass wasting is **solifluction,** or soil flow, which is the "flow" of weathered material over a solid, impermeable base. This type of mass wasting is common in cold climates where permanently frozen ground (permafrost) forms a solid base for the solifluction of a surface layer that thaws during the short summers. Moisture in the thawing surface and from summer rains is trapped in the surface layer because the ground below is frozen. As a result, the surface layer becomes saturated, and the water-soaked soil "flows" slowly downhill.

Figure 29.5 Slump—a common form of mass wasting.

Figure 29.6 A result of creep.
Tilted fence posts indicate the imperceptively slow mass wasting due to creep.

29.2 Agents of Erosion

The interacting factors in erosion processes are quite complex. To simplify matters, geologists talk about the **agents of erosion**—that is, the major physical phenomena and mechanisms that supply the energy for erosion. These agents are running water, ice, wind, and waves.

Running Water

Running water refers primarily to the waters of streams and rivers that erode the land surface and transport and deposit eroded materials. Rainfall that is not returned to the atmosphere by evaporation either sinks into Earth as groundwater or flows over the surface as runoff. Runoff or overland flow occurs whenever rainfall exceeds the amount of water that can be immediately absorbed in the ground. Overland flow usually occurs only for short distances before the water ends up in a stream. The erosion performed by overland flow is referred to as **sheet erosion.** The flowing "sheets" of water can move a surprisingly large amount of sediment, particularly from unirrigated slopes and cultivated fields (see Figure 29.7).

Streamflow is the flow of water occurring between well-defined banks. The eroded material in a stream is referred to as **stream load.** The load of a stream varies from dissolved minerals and fine particles to large rocks, and the transportation process, as well as erosion, depends on the volume and swiftness (discharge) of the stream's current. A stream's load is divided into three components: a dissolved load, a suspended load, and a bed load.

The *dissolved load* consists of dissolved water-soluble minerals that are carried along by a stream in solution. As much as 20 percent of the material reaching the oceans is transported in solution. Fine particles not heavy enough to sink to the bottom are carried along in suspension. The *suspended load* is quite evident after a heavy rain when the stream appears muddy. Coarse particles and rocks along or close to the bed of the stream constitute its *bed load.* These rocks and particles are rolled and bounced along by the current. This transport mechanism of bed load is referred to as **traction.**

As it is moved along by traction, the coarser load material near the bottom of a stream is further broken into smaller rocks and particles by **abrasion.** Stream abrasion is evidenced by smooth, rounded rocks and pebbles (Fig. 29.8). During the transport process the bed load is worn finer and finer and eventually may be

Figure 29.7 Sheet erosion.
The overland flow of water removes a large amount of sediment.

Figure 29.8 Stream abrasion.
The rocks and pebbles are smooth and rounded during the transport process.

transported by suspension. The final products of a long journey are sand (rock particles) or a mixture of finer particles and organic material picked up along the way, which is collectively referred to as **silt** and clay.

The action of the flowing water in a river causes the erosion of its bed. By studying the development of this process, geologists can determine the age of a river. Over geologic time a river goes through three general stages: youth, maturity, and old age. In its youth it is characterized by a V-shaped valley or canyon, which reflects downcutting erosion. The depth of an eroded valley is limited by the level of the body of water into which the river flows. The limiting level below which a stream cannot erode the land is called its **base level.** In general, the ultimate base level for rivers is sea level.

As a stream or river flows overland, it twists and turns, following the path of least resistance. A looplike bend in a river channel is called a **meander** (Fig. 29.9). Meanders shift or migrate because of greater erosion on the side of the stream bed on the outside of the curved loops. The speed of the streamflow is greater in this region than near the inside bank of the meander where sediment can deposit. When the erosion along two sharp meanders causes the stream to meet itself, it may abandon the water-filled meander, which is then known as an oxbow lake.

As the river matures, it becomes **graded,** which is the condition when its erosion and transport capabilities are in balance. A mature, graded river no longer has much tendency to deepen its channel. Instead, it begins to widen its valley, forming a **flood plain.** This widening continues until in old age the river has a very wide flood plain.

The amount of material eroded and transported by streams and rivers is enormous. The oceans receive billions of tons of sediment each year as a result of the action of running water. The Mississippi River alone discharges approximately 500 million tons of sediment yearly into the Gulf of Mexico. A river's suspended and bed loads may accumulate at its mouth and form a **delta,** such as the Nile River Delta (Fig. 29.10). The rich sediment makes delta areas important for agriculture.

However, some sediment deposits are not always to our benefit, as shown in Fig. 29.11. Many harbors must be dredged to remove sediment deposits to keep them navigable, and the sediment buildup behind dams poses operational problems. Our activities that remove erosion-preventing vegetation from the land may give rise to sediment pollution in rivers and streams and result in environmental problems.

Figure 29.9 Meandering river.
Goosenecks of the San Juan River, Utah.

Figure 29.10 A high-altitude infrared photograph of the Nile River Delta.

Ice

One of the eroding actions of ice, frost wedging, has been mentioned previously. Many of us are familiar with the ice of winter, but parts of Earth are covered with large masses of ice the year round. These large masses of ice are called **glaciers.** To most of us, the term glacier usually brings to mind the thought of the ice age. Indeed,

Figure 29.11 The deposit of sediments.
The unwanted sediments must be removed to keep the
river navigable.

large areas of Greenland and Antarctica are presently
covered with glacial ice sheets or **continental glaciers**
similar in size to those that covered Europe and North
America during the last ice age over 10,000 years ago.

But are there any glaciers in the United States today? The answer is yes. In fact, you may be surprised
to learn that there are about 1100 glaciers in the western
part of the conterminous (48 states) United States and
that 3% of the land area of Alaska (43,520 km², or
17,000 mi²) is covered by glaciers. However, these glaciers differ greatly in size and form from the ice sheets
of Greenland and Antarctica.

Glaciers are formed when, over a number of years,
more snow falls than melts. As the snow accumulates
and becomes deeper, it is compressed by its own weight
into solid ice. When enough ice accumulates, the glacier
"flows" downhill or out from its center, if it is on a flat
region. The icebergs commonly found in the North Atlantic and Antarctic oceans are huge chunks of ice that

have broken off from the edges of the glacial ice sheets
of Greenland and Antarctica, respectively.

Small glaciers, called **cirque glaciers,** form in hollow
depressions along mountains that are protected from the
Sun. A majority of the glaciers in the United States are
of this variety. The ice movement further erodes the
land and forms an amphitheaterlike depression called a
cirque. When the ice melts, these glacier-eroded cirques
often become lakes.

If snow accumulates in a valley, the valley floor
may be covered with compressed glacial ice, and a **valley
glacier** or mountain glacier is formed, which flows down
the valley (Fig. 29.12). At lower and warmer elevations
the ice melts, and where the melting rate equals the
glacier's flow rate, the glacier becomes stationary. The
flow rate of a glacier may be a few inches to over 30 m
(100 ft) per day, depending on the glacier's size and
other conditions. The end of a glacier is called its *snout,*
which may advance and recede with the seasons.

The erosion action of a valley glacier is not unlike
that of a stream. As the ice flows, it loosens and carries
away materials, or bed load, that will be ground fine by
abrasion. Although much slower than a stream, a glacier
can pick up huge boulders and gouge deep holes in the
valley. The paths of vigorous, preexisting mountain glaciers are well marked by the deep U-shaped valleys they
leave.

Glaciers, like streams, also deposit the material they
carry. The general term **drift** is applied to any type of
glacial sediment deposit. Material that is transported and
deposited by ice, in contrast to meltwater, is called **till.**
Till deposits are not layered or sorted as would be
sediment carried away by the melted water of a glacier.
Near the end and sides of a glacier the till may form
ridges known as **moraines** (Fig. 29.13). The terminal
moraine marks the farthest advance of the glacier. Terminal moraines give us an indication of the extent and
advance of the glacial ice sheet in North America, which
retreated about 10,000 years ago. These moraines lie as
far south as Indiana, Ohio, and Long Island, New York.

Not only do glaciers erode the land surface, but
they also supply an estimated 2.12×10^9 m³ (560 billion
gal) of water to streamflow during the summer months
in the coterminous United States alone. These huge
freshwater reservoirs are being eyed as water sources
for the heavily populated western areas. The glacier cycle
fits well into our needs. Glaciers accumulate and store
water in the winter when most areas have sufficient
water supplies and release it in the summer when many
city reservoirs are low.

Figure 29.12 Valley glacier near Juneau, Alaska.

Figure 29.13 Alpine glacier showing lateral, medial, and contorted moraines in Alaska.

Wind

The action of wind as an agent of erosion is a slow process, but nevertheless wind contributes significantly to the leveling of Earth. Dust particles that are small enough are transported great distances by the wind, while larger particles are moved short distances by rolling or bouncing along the surface. The transport action of the wind is not unlike that of traction in a stream, only on a broader and less confined scale. This action is quite evident in areas with large quantities of loose, weathered debris. Dust storms may darken the sky and be of such intensity that visibility is reduced to almost zero (Fig. 29.14). During the 1930s drought conditions created areas known as *dust bowls*, in which layers of fertile topsoil were blown away by the wind.

Because sand grains are relatively heavy, the transport action of sand by the wind is usually within a few feet of the ground, similar to the coarse-sediment transportation near the bed of a stream. On the ground the sand forms drifts. The resulting abrasion from such sandblast action is evident in nature.

Waves

The waves of large bodies of water also erode the land along their shorelines. This erosion is most evident where the ocean surf pounds the shoreline. Some coasts are rocky and jagged, which evidences that only the hardest materials can withstand the unrelenting wave action over

long periods of time. Along other coastlines cliffs are formed on the shoreline, terraced from the eroding action of waves (Fig. 29.15).

29.3 Earth's Water Supply

All rivers run into the sea, yet the sea is not full; into the place whence the rivers come, thither they return again.

Eccles. 1:7

Water is often referred to as the basis of life. The human body is composed of 65–70% water by weight, and water is necessary to maintain our body functions. This common chemical compound is an essential part of our physical environment, not only in life processes but also in other areas, such as agriculture, industry, sanitation, firefighting, and even religious ceremonies. Early civilizations developed in valleys where water was in abundance, and even today the distribution of water is a critical issue. Consider how your life would be affected without an adequate water supply.

Earth's water supply, some 1.25×10^{18} m^3 (300 million mi^3), may seem inexhaustible, because it is one of our most abundant natural resources. Approximately 70% of Earth's surface is covered with water. However, about 98% of the water on Earth is salt water, and only about 2% is fresh water (Fig. 29.16). Most of the fresh water is frozen in the glacial ice sheets of Greenland and Antarctica. Even so, there are about 3.8×10^{12} m^3 (10^{15}

Figure 29.14 Sandstorm in Tibet.

Figure 29.15 Eroded cliffs, Point Reyes National Seashore, California.

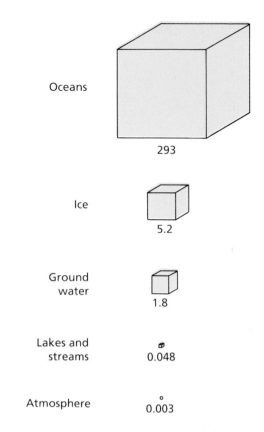

Millions of cubic miles

Figure 29.16 Distribution of Earth's water supply.

gal) of fresh water available each year. The oceans contain another 3.8×10^{17} m^3 (10^{20} gal) of water that is easily accessible, but this water has 3%–4% salt content—principally sodium chloride—which makes it unfit for human consumption without a costly desalination process.

Although the daily withdrawal of fresh water in the United States is over 1.7×10^9 m^3 (450 billion gal), the supply of fresh water is generally adequate in most areas. Regional problems arise because of population concentrations, but these are human problems that involve water management.

Earth's water supply is a reusable resource that is constantly being redistributed over Earth. There are many factors that enter into this redistribution, but in general it is a movement of moisture from large reservoirs of water, such as oceans and seas, to the higher-elevated inland regions. As we discussed in the preceding section, the water flows back to the sea, eroding as it goes. This gigantic cyclic process is known as the **hydrologic cycle** (Fig. 29.17).

Moisture evaporated from the oceans moves over the continents through atmospheric processes and falls as precipitation. Some of this water evaporates and returns to the oceans via atmospheric processes, but a large part of it soaks into Earth to become groundwater.

Some of the precipitation falling on the land may be lost in direct runoff. How much water goes into the ground depends on the permeability of the soil and rocks near Earth's surface. **Permeability** is a measure of a material's capacity to transmit fluids. Naturally, loosely packed soil components, such as sand and gravel, permit

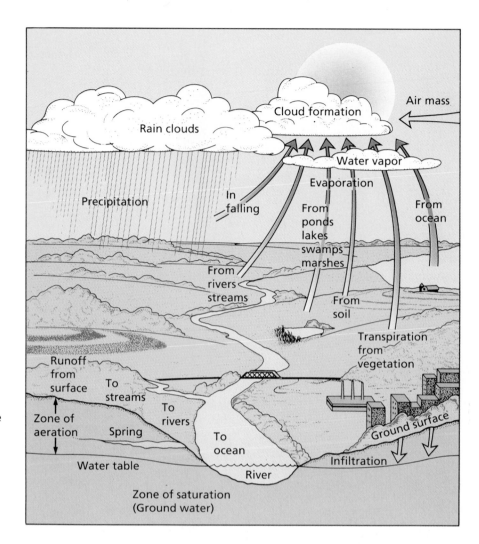

Figure 29.17 An illustration of the hydrologic cycle.
Evaporated moisture from oceans, rivers, lakes, and soil is distributed by atmospheric processes. It eventually falls as precipitation and returns to the soil and bodies of water. The groundwater percolates downward through the zone of aeration until it reaches the level (water table) at which the ground is saturated with water.

a greater movement of water. Clay, on the other hand, has fine openings and relatively low permeability.

Porosity is closely related to permeability. **Porosity** is the percentage volume of unoccupied space in the total volume of a substance. The porosity of rocks and soil near Earth's surface determines the capacity of the ground for water storage. In total, ground porosity is so great that, except for glaciers, the ground is our largest reservoir of fresh water.

Under the influence of gravity, water percolates downward through the soil until at some level the ground becomes saturated with water. The upper boundary of this **zone of saturation** is called the **water table.** The unsaturated zone above the water table is called the **zone of aeration** (Fig. 29.17).

In the zone of aeration the pores of the soil and rocks are partially or completely filled with air. However,

in the zone of saturation all the voids are saturated with groundwater, forming a reservoir from which we obtain part of our water supply by drilling wells to depths below the surface of the water table. Lakes, rivers, and springs occur where the water table is at the same level as Earth's surface. Springs may also form above the water table as a result of impervious rock layers, but these tend to dry up in seasons of light precipitation. Also, the level of the water table shows seasonal variations, and shallow wells may go dry in late summer.

A body of permeable rock through which groundwater moves is called an **aquifer** (from the Latin for "water carrier"). Sand, gravel, and loose sedimentary rock are good aquifer materials. Aquifers are found under more than half of the area of the coterminous United States, and it is in these aquifers that we find wells and springs. Ordinary wells and springs fill with water be-

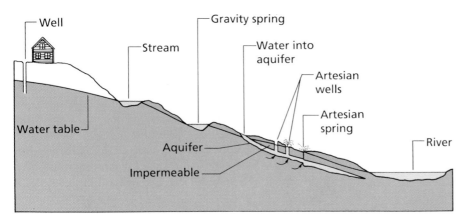

Figure 29.18 An illustration of wells, streams, and springs relative to the water table.
A special geometry of impermeable rock layers gives rise to artesian wells and springs. If the groundwater is not recharged by precipitation, the level of the water table will fall; then wells, springs, and streams can go dry. This often occurs during a hot, dry summer.

cause they intersect the water table. A *spring* is simply a flow of groundwater that emerges naturally on Earth's surface. Ordinary springs result from the gravitational flow of water and are hence sometimes called gravity springs (Fig. 29.18).

The special geometry of impermeable rock layers gives rise to what are called **artesian wells** and **artesian springs**, which are wells and springs in which water rises above the aquifer. As illustrated in Fig. 29.18, this geometry can occur when an aquifer is sandwiched between sloping impermeable rock strata. The pressure of gravity can cause water to spurt or bubble onto the surface above the aquifer. The name *artesian* comes from the French province of Artois, where such wells and springs are common.

Thermal or hot springs are found in many areas. They transmit heat energy from Earth's interior. Groundwater seeping several thousand feet underground is heated as a result of Earth's geothermal gradient 1°C per 25 m (1°F per 150 ft of depth). Also, at these depths the water can come in contact with hot igneous rocks and may be heated to 204°C (400°F) or higher (superheated) without boiling because of the great underground pressures.

The superheated water expands and rises to Earth's surface through fissures, where it forms thermal springs. Heat energy is still lost during the water's journey to the surface, but some springs are boiling hot. The reduction of pressure near the surface can cause the water to boil. The resulting stream causes some hot springs to erupt in the form of a **geyser** (from the Icelandic *geysir*, "gush or rage").

Geysers from time to time spurt water into the air to heights of 46 m (150 ft) or more. Most geysers have very irregular eruptions, but a few, such as Old Faithful in Yellowstone National Park, are regular enough to

Figure 29.19 Old Faithful.
This geyser erupts about every 65 min.

satisfy impatient tourists (Fig. 29.19). Old Faithful's eruptions vary between 30 and 90 min apart, with an average time interval of 65 min.

Geothermal areas are found around the world, principally in volcanic regions associated with plate boundaries (Fig. 29.20). Hot springs have been used as baths or spas down through history. In Iceland, hot water from geyser fields is used for domestic heating. A significant commercial application occurred in 1905 with the building of the first geothermal electric generation plant in Larderello, Italy. The first geothermal generation facility in the United States was built by the Pacific Gas and Electric Company in 1960 at The Geysers, California, some 90 mi north of San Francisco (Fig. 29.21). Other geothermal areas in the western United States are being eyed as possible energy sources.

Geothermal fields are of two general types—hot springs at the surface and deep, insulated, superheated reservoirs with little surface leakage. It is the latter type that is important in electrical generation application. When such reservoirs are tapped by drilling, the water boils and the expanding steam causes the hot water to be propelled to the surface. The drilled well acts as a continuously erupting geyser, so there is very little pumping cost. In some wells "flash" boiling occurs and only steam erupts, with very little water discharge. This type of well is suitable for electrical generation, as is the case with the wells at The Geysers and Larderello.

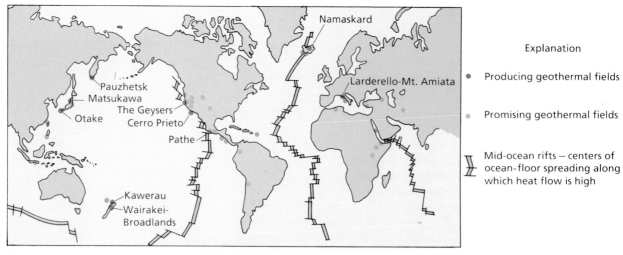

Figure 29.20 The geothermal areas of the world.
Notice how the areas are associated with the regions of divergent plate boundaries.

Figure 29.21 The Geysers, California.
A commercial geothermal electricity-generation plant is located at the Geysers in Sonoma County, 90 mi north of San Francisco.

Although these geothermal sources eliminate the air pollution associated with combustive electrical generation, they are not without problems. Gases in the steam and dissolved minerals in the hot water cause equipment to erode rapidly. The cooling and disposal of the salty, mineral-rich water from the spring may also present a problem. This water could, and possibly should, be returned to the water table of the area rather than allowing it to run off. There is evidence that the groundwater in areas where geothermal energy is used must be replaced to prevent subsidence.

Thus we see that our water supply, within the context of the hydrologic cycle, is an intricate system that depends on many factors. The quality of the local water supply depends to a great extent on the mineral composition of the soil and rocks through which it percolates in the zone of aeration. Water-soluble minerals are dissolved and carried along with the water in solution.

In some cases water containing bicarbonates (hydrogen carbonates) and chlorides is sold commercially as "mineral water." However, dissolved minerals in "hard" water have undesirable effects. **Hard water** is due to dissolved calcium and magnesium salts (bicarbonates, chlorides, and sulfates). Iron salts also contribute to hard water. Such dissolved minerals not only affect the taste of the water but also cause use problems. The salts combine with the organic acids in soaps to form insoluble compounds, thereby reducing the lathering and cleansing qualities of the soap. It is because of these insoluble compounds from soap that clothing does not come out "whiter white" on wash day and also that there is a ring around the bathtub.

At high temperatures the metal ions present may form insoluble compounds (e.g., carbonates and sulfates) and precipitate. This precipitation tends to clog pipes and forms scale in teakettles and boilers. Boiler scale is a poor conductor of heat and reduces the boiler efficiency and wastes fuel. To avoid these conditions, people "soften" the hard water by removing the mineral salts prior to use. Water softening is an active business in many parts of the country.

29.4 The Oceans and Seafloor Topography

The Oceans

The vastness of the restless oceans imparts an awesome and humbling feeling to most observers. And rightly so, because the oceans cover about 65% of Earth's surface. There are five major oceans. In order of decreasing size, they are the Pacific, the Atlantic, the Indian, the Antarctic, and the Arctic oceans. The average depth of the oceans is about 4 km (2.5 mi); the greatest measured depth is about 11 km (7 mi), in the Marianas Trench in the western Pacific (see Section 26.3).

To the novice beachgoer, one of the first things noticed on an initial swim is the saltiness of the seawater. About 3.5% of the average seawater by weight consists of dissolved mineral salts. Through the hydrologic cycle, the dissolved salts of over 2 billion years of erosion have found their way into the ocean waters. Some ions of the dissolved salts are effectively removed from the oceans by biological processes and some by precipitation or absorption by other minerals. (Recall from Chapter 14 that dissolved chemical salts dissociate into metallic and nonmetallic ions).

These salts become part of the ocean sediment and reenter the rock cycle from which they originated. Scientists believe that during the course of geologic time these processes have reached a steady state, so that the present salt composition of seawater has been maintained for millions of years. The quantities of salts in the seas may vary at some locations because of variations in the size of the mineral deposits of the drainage regions whose rivers feed the oceans. However, the same salts are found everywhere, generally in the same proportions.

The saltiness of the sea is measured in terms of its **salinity,** which is commonly expressed in parts per thousand (ppt) rather than percentage or parts per hundred. Parts per thousand is the same as the grams of dissolved salt per kilogram of seawater (g/kg). The salinity of average seawater is therefore 35 ppt or 35 g/kg (equivalent to 3.5%). The range of seawater salinity in the open sea is from 33 to 38 ppt. As indicated in Table 29.1, the principal substances that contribute to the salinity of seawater are sodium and chloride ions.

The ocean is sometimes referred to as a mineral storehouse from which we might extract minerals. This is true in part, but one must not overlook the concentrations and the difficulties of extracting these minerals from solution. It is a fact that NaCl has been taken from the ocean since ancient times, and that needs prompted the extraction of magnesium and bromine from seawater during World War II. Even today, much of our output of these substances still comes from seawater. The extraction of metals like gold, however, is a different matter. There are 10^{10} tons of gold in the oceans, but the concentration is only about 6×10^{-11} oz/gal. Thus with 100% recovery about 6.4×10^7 m^3(1.7×10^{10} g) of seawater would have to be processed to obtain one ounce of gold. This amount would hardly pay for the effort.

Table 29.1 Some Chemical Constituents of Seawater (Average Concentrations)

Ion or Element	Concentration (ppt)	Ion or Element	Concentration (ppt)
Chloride	19.3	Nitrate	3.5×10^{-4}
Sodium	10.7	Iodine	5×10^{-5}
Sulfate	2.7	Iron	1×10^{-5}
Magnesium	1.3	Copper	5×10^{-6}
Calcium	0.4	Zinc	5×10^{-6}
Potassium	0.39	Lead	4×10^{-6}
Bicarbonate	0.14	Silver	3×10^{-7}
Bromine	0.070	Nickel	1×10^{-7}
Strontium	0.013	Mercury	3×10^{-8}
Fluorine	0.001	Gold	6×10^{-9}

Note: Over 60 chemical elements have been found in seawater.

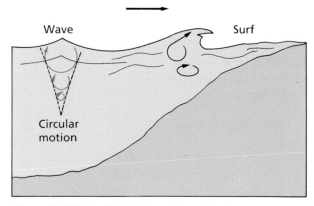

Figure 29.22 An illustration of a surface wave approaching the shore.
The water particles move in more-or-less circular paths, the diameters of which decrease with increasing depth. In the shallow water near the shore, the water particles are forced into more elliptical paths. This causes the wave to grow higher and steeper and to eventually break, forming a surf.

Another feature noted when one is at the beach is the restless motion of the ocean waters. There are three types of seawater movements that are quite noticeable at the beach or along the coast. They are surface waves, long shore currents, and tidal currents. **Surface waves** continually lap the shore. In general, the waves near the surface are a combination of longitudinal and transverse wave motions. It is tempting to think of water waves as being transverse because the sinusoidal profile is clearly evident. However, the water "particles" move in more or less circular paths, as illustrated in Fig. 29.22. Circular particle motion is a combination of longitudinal and transverse motions.

The diameter of the circular paths of the water particles decreases rapidly with depth, and the wave motion is hardly observed when one dives several feet below the surface. As a wave approaches shallower water near the shore, the water particles experience difficulty in completing their circular paths and are forced into more elliptical paths. The surface wave then grows higher and steeper. Finally, when the depth becomes too shallow, the water particles can no longer move through the bottom part of their paths and the wave breaks, with the crest of the wave falling forward to form a **surf.**

When at the beach, you may have noticed debris moving along the shore as it bobs up and down. This movement is an indication of a **long shore current** flowing along the shore. These currents arise from incoming ocean waves that break at an angle to the shore.

The component of water motion along the shore causes a current in that direction. Waves and their resulting long shore currents are important agents of erosion along coastlines.

The periodic rise and fall of the tides are also quite evident at the beach. These **tidal currents** result from the two tidal bulges that "move" around Earth daily as a result of the gravitational attractions of the moon and Sun and the rotation of Earth (see Chapter 18). The water level may rise as much as 12 m (40 ft) in some regions at high tide.

In the open ocean tidal currents are of little significance, but then as they approach the shore, the rise of the water level is evident, particularly when they are confined within the boundaries of a bay or a river outlet. The water level can rise rapidly, and seawater can back up into rivers that open to the sea. In France, tidal currents are used to generate electrical power on the Rance River. Power is generated not only from the incoming tidal current as the seawater flows rapidly into the Rance River estuary but also from the outgoing tidal current as the seawater flows back into the Gulf of St. Malo.

In the open ocean the major movement of seawater results from two general types of currents: surface currents and density currents. **Surface currents** in the ocean are broad drifts of surface water that are set in motion by the prevailing surface winds. These currents rarely extend more than 100 m in depth (about 100 yards, or the length of one football field).

As might be expected, surface currents are influenced primarily by the prevailing winds of Earth's atmospheric semipermanent circulation structure, which we discussed in Section 21.2. A comparison of the global surface currents shown in Fig. 29.23 with the prevailing wind zones in Fig. 21.8 should make this evident. Surface currents, like air movements, are affected by the Coriolis force (see Section 21.1). As a result, the general circulations of surface currents are clockwise in the Northern Hemisphere (deflection to the right) and counterclockwise in the Southern Hemisphere (deflection to the left). The surface current circulations are larger in the Southern Hemisphere because there is more open ocean—two-thirds of Earth's land surface is north of the equator. The lack of landforms in high southern latitudes makes possible a global circulating surface current—the West-Wind Drift. Water can move from one ocean to another in this drift.

The surface-current circulation patterns are impor-tant climatic factors. As shown in Fig. 29.23, the circulations form seawater convection cycles and the currents carry warm water from equatorial regions to higher latitudes. For example, in the Atlantic Ocean the North Equatorial Current carries warm water westward into the Gulf of Mexico. It emerges as the Florida Current and flows northward as the Gulf Stream. The circulation continues as the North Atlantic Current flowing toward Europe, where it is diverted to the north and south.

The warm water that the northward-deflected current brings to the British Isles and to the northern European coast is an important factor in moderating the climate of this northern region. The current deflected to the south cools and returns to the equatorial regions via the Canary Current, where it is warmed and starts on another cycle. In a similar manner, the surface-current circulation in the northern Pacific brings a warming influence to Japan via the Kuroshio Current, and the returning cool California Current moderates the climate along the coast of southern California.

Deep ocean currents are called density currents. A **density current** is the motion of seawater caused by

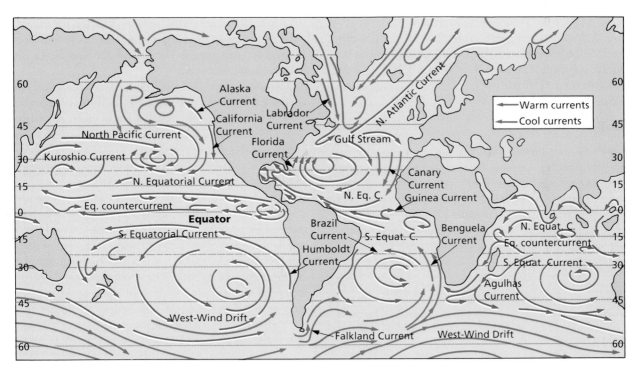

Figure 29.23 Major ocean surface currents.
Notice that the circulation pattern of the surface currents is clockwise in the northern hemisphere and counter-clockwise in the southern hemisphere. Like atmospheric currents, these patterns result from the Coriolis force.

dense water sinking through less dense water. The density of seawater is dependent on its temperature, salinity, and sediment load. The cooler the water and the greater its salinity, the greater is its density.* In the polar regions the surface water in contact with the atmosphere is cooled and its density increases. Also, when seawater freezes, the dissolved salts are not incorporated in the ice, which is fresh water. The residual salts from the frozen water increase the salinity of the surrounding seawater.

By these methods the polar water becomes denser and sinks, giving rise to deep-density currents that may extend to the tropics. Such a deep-water current starts in the North Atlantic with the surface water brought northward by the surface North Atlantic Current. This water already has a high salinity because of evaporation in the equatorial region where it began its journey. This north Atlantic Deep-Water Current flows southward along the ocean bottom far past the equator before mixing decreases its density and halts its flow.

Deep-water currents flow northward from Antarctica, where cooling and freezing give rise to some of the densest water in the oceans. These currents flow northward into the Atlantic, Pacific, and the Indian oceans. However, there is no counterpart of the south-bound North Atlantic Deep-Water Current in the Pacific, because there is no large, deep, cold-water source in the northern Pacific. A shallow barrier in the Bering Strait prevents the cold Arctic water from entering this region.

Seafloor Topography

Scientists once thought that the surface features or topography of the ocean basins consisted of an occasional mountainous island arc on a relatively smooth sediment-covered floor. This incorrect view resulted from the lack of direct observation. Surprisingly enough, the surface of a major portion of Earth—the oceanic crust—was not explored in great detail until after World War II.

With the advent of modern technology, sounding and drilling operations revealed that the ocean floor is about as irregular as the surfaces of the continents, if not more so (Fig. 29.24).

We now know that the seafloor has a system of midoceanic ridges, which mark divergent plate boundaries. These rocky submarine mountain chains are along fracture zones through which magma rises from below to form new oceanic crust. Oceanic volcanism is widespread. Large volcanic mountains also rise from the ocean floor in the midst of plates, such as those in the Hawaiian island chain. As pointed out in Chapter 25, geologists believe that these mountains arise from plates moving over internal "hot spots."

Many isolated, submarine, volcanic mountains have also been discovered. They are known as **seamounts** and are individual mountains that may extend to heights of over 1.6 km (1 mi) above the seafloor. Some seamounts have flat tops and are given the special name of **guyots***
(Fig. 29.24). Their shapes suggest that the tops were once islands that were eroded away by wave action. However, many of the guyot tops are several thousand feet below sea level. Evidently, the eroded seamounts must have subsided and sunk below sea level. This subsidence in the oceanic crust perhaps occurs as the oceanic crust moves away from a spreading ridge or results from an isostatic adjustment because of the weight of the large volcanic seamount.

Another marked feature of seafloor topography is **seafloor trenches,** which mark the locations of plunging plate boundaries. These trenches are as much as 240 km (150 mi) in width and 24,000 km (15,000 mi) or more in length. The deepest trenches show little evidence of sediment accumulation, whereas other trenches near land areas are partially filled with sediment.

The huge volumes of sediment flowing into the oceans from continental regions do have an effect on seafloor topography. Distributed by ocean currents, sediment accumulates in some regions such that a sedimentary layer covers and masks the irregular features of the rocky ocean floor. The resulting large flat areas are called **abyssal plains.** Abyssal plains are most common near the continents, which supply the sediment.

Although 70% of Earth's surface is covered with water, the oceanic crust basins account for only about

---

* The density of water increases with decreasing temperatures to a maximum of 1.0 g/cm^3 at 4°C. Above and below 4°C, the water is slightly less dense. For this reason, lakes freeze at the top rather than at the bottom. As the water near the surface loses heat to the atmosphere, it becomes denser and sinks (causing a density current), displacing the warm water below. This process continues until the air and surface water temperature is below 4°C. The cool surface water is then less dense than the water below and hence remains at the surface, where freezing begins when the temperature goes below 0°C.

---

* Named in honor of Arnold Guyot, the first geologist at Princeton University, by Professor Harry Hess, a geologist at Princeton University who discovered the first flat-topped seamounts in the 1950s.

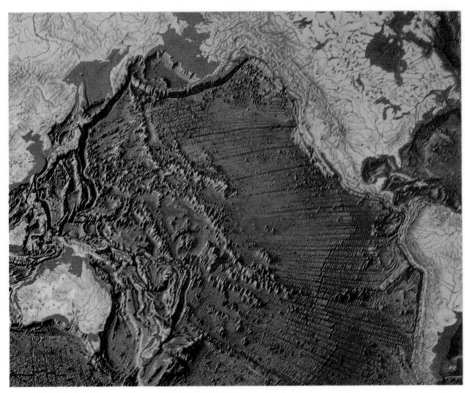

Figure 29.24 The Pacific Ocean floor.
The seafloor topography was once thought to be smooth because of the accumulation of sediment deposits. However, relatively recent explorations have shown the seafloor surfaces to be as irregular as the surfaces of the continents. See also Fig. 26.15b for the topography of the Atlantic Ocean floor.

65% of the surface area. Thus the continental crust comprises 35% of Earth's surface area; but since only about 30% of Earth's surface is land area, 5% of the continental crust must be submerged. This submerged area occurs along the continental margins, and the shallowly submerged borders of the continental masses are called **continental shelves** (Fig. 29.25).

The widths of these shelves vary greatly. Along the Pacific coast of South America, there is almost no continental shelf—only a relatively sharp, abrupt continental slope. However, off the north coast of Siberia, the continental shelf extends outward into the ocean for about 1280 km (800 mi). The average width of the continental shelves is on the order of 64–80 km (40–50 mi).

The continental shelves have recently become a point of international interest and dispute. The majority of commercial fishing is done in the waters above the continental shelves. Also, the continental shelves are the locations of oil deposits that are now being tapped by offshore drilling. As a result, many countries including the United States have extended their territorial claims to an offshore 320-km (200-mi) limit. In fact, one of the reasons for the Argentine-British conflict over the Falkland Islands was potential offshore oil deposits.

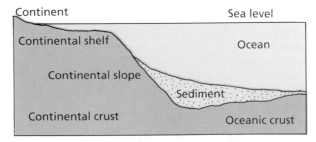

Figure 29.25 A cross-sectional illustration of a continental shelf.
Continental shelves are the shallowly submerged borders of the continents. Continental slopes define the true edges of the continents. Sediment from continental erosion collects at the bases of the continental slopes.

Beyond a continental shelf, the surface of the continental landmass slopes downward to the floor of the ocean basin. The **continental slopes** define the true edges of the continental landmasses. Erosion along these slopes gives rise to deep submarine canyons that extend downward toward the ocean basins. And near the edges of the ocean basins, the sediment collects in geosyncline troughs. Thus we come back to one of the factors in continental mountain building.

Undersea exploration is a relatively new phase of scientific investigation. Indeed, there is a great deal more to be learned about this vast region that makes up almost 70% of Earth's surface. As advances in technology provide more data on this previously inaccessible part of Earth's surface, we may expect our knowledge of Earth and its geologic processes to grow. Perhaps one day we will truly understand the dynamics of our planet.

Learning Objectives

After reading and studying this chapter, you should be able to do the following without referring to the text:

1. Define *erosion.*
2. Distinguish between physical and chemical weathering, and give examples of each.
3. Define *mass wasting,* and give two important types of mass wasting, with examples of each.
4. Describe the agents of erosion.
5. Distinguish between the three components of a stream's load.
6. Explain how glaciers are formed, and distinguish between the different types of glaciers.
7. Discuss the distribution of Earth's water supply.
8. Describe the hydrologic cycle.
9. Explain why the sea is salty.
10. State and describe the types of ocean currents.
11. Explain how ocean currents affect climate.
12. Describe the major features of seafloor topography.
13. Define and explain the important words and terms listed in the next section.

Important Words and Terms

erosion	solifluction	valley glacier	salinity
physical weathering	agents of erosion	drift	surface waves
frost wedging	sheet erosion	till	surf
frost heaving	stream load	moraines	long shore current
permafrost	traction	hydrologic cycle	tidal currents
salt wedging	abrasion	permeability	surface currents
chemical weathering	silt	porosity	density current
sinkholes	base level	zone of saturation	seamounts
overburden	meander	water table	guyots
mass wasting	graded river	zone of aeration	seafloor trenches
landslides	flood plain	aquifer	abyssal plains
rockslide	delta	artesian wells	continental shelves
slump	glaciers	artesian springs	continental slopes
mudflows	continental glaciers	geyser	
creep	cirque glaciers	hard water	

Questions

Weathering and Mass Wasting

1. What is erosion?

2. Distinguish between physical and chemical weathering.

3. What are frost wedging and frost heaving?

4. What type of weathering takes place in hot, arid climates that is similar to frost wedging?

5. How do plants and animals contribute to weathering?

6. On what factors does chemical weathering depend?

7. Describe in detail the chemical weathering process of limestone and the formation of caverns and dripstone.

8. What is a sinkhole?

9. What is mass wasting?

10. Explain each of the following and state whether it is a fast or slow type of mass wasting: (a) rockslide, (b) creep, (c) slump, (d) solifluction, (e) mudflow.

Agents of Erosion

11. State the four major agents of erosion.

12. What is sheet erosion?

13. State and explain the three components of a stream's load.

14. Explain stream abrasion and traction.

15. What is silt?

16. State and describe the three general stages of river development.

17. Why does a river meander, and what is a graded river?

18. Distinguish between continental and valley glaciers.

19. Are there any glaciers in the United States? Explain.

20. Describe each of the following: (a) drift, (b) till, and (c) a moraine.

Earth's Water Supply

21. How much of Earth's water supply is fresh water, and where is most of it located?

22. Explain the hydrologic cycle.

23. Explain each of the following: (a) permeability, (b) porosity, (c) the zone of aeration, (d) the zone of saturation, and (e) the water table.

24. What is an aquifer?

25. Distinguish between gravity springs and artesian wells and springs.

26. What causes hot springs and geysers?

27. What are the two general types of geothermal fields? Which of the types is more important in the generation of electricity?

28. What is the cause of "hard" water, and what are some of its effects?

The Oceans and Seafloor Topography

29. What is salinity, and in what units is it measured? What is the average salinity of the oceans?

30. Describe the three types of seawater movements along a coast.

31. Why do waves form a surf?

32. What causes surface currents in the open ocean, and what are some of the major surface currents?

33. Why is the surface current circulation larger in the Southern Hemisphere? Are there any global circulating surface currents?

34. Explain how the climate of the British Isles and northern Europe is moderated by surface currents.

35. What are density currents? Describe their effect on ocean circulation.

36. Why is there no counterpart of the North Atlantic Deep-Water Current in the Pacific Ocean?

37. Define and explain the formation of each of the following: (a) seamounts, (b) guyots, (c) seafloor trenches, (d) abyssal plains.

38. What are continental shelves and continental slopes? What defines the true edges of the continental landmasses?

39. Why are the continental shelves the focus of current international interest?

Appendixes

Appendix I

The Seven Base Units of the International System of Units (SI)

Meter, m (length)	The meter is defined in reference to the standard unit of time. One meter is the length of the path traveled by light in a vacuum during a time interval of 1/299,792,458 of a second. That is, the speed of light is a universal constant of nature whose value is defined to be 299,792,458 meters per second.
Kilogram, kg (mass)	The kilogram is a cylinder of platinum-iridium alloy kept by the International Bureau of Weights and Measures in Paris. A duplicate in the custody of the National Institute of Standards and Technology serves as the mass standard for the United States. This is the only base unit still defined by an artifact.
Second, s (time)	The second is defined as the duration of 9,192,631,770 cycles of the radiation associated with a specified transition of the cesium-133 atom.
Ampere, A (electric current)	The ampere is defined as that current that, if maintained in each of two long parallel wires separated by one meter in free space, would produce a force between the two wires (due to their magnetic fields) of 2×10^{-7} newton for each meter of length.
Kelvin, K (temperature)	The kelvin is defined as the fraction 1/273.16 of the thermodynamic temperature of the triple point of water. The temperature 0 K is called *absolute zero*.
Mole, mol (amount of substance)	The mole is the amount of substance of a system that contains as many elementary entities as there are atoms in 0.012 kilogram of carbon-12.
Candela, cd (luminous intensity)	The candela is defined as the luminous intensity of 1/600,000 of a square meter of a black body at the temperature of freezing platinum (2045 K).

Appendix II

Measurement and Significant Figures

A **measurement** is a comparison of an unknown quantity with a precisely specified quantity called a standard unit. Measurements should be made and recorded as accurately as possible. The term **accuracy** refers to how close the measurement comes to the exact value. Accuracy depends upon the precision of the measuring instrument and the ability of the individual taking the measurement. Errors will be made in every measurement, but the magnitude of the errors can be kept small when the observer is careful and uses an instrument with high precision.

Precision refers to the degree of reproducibility of a measurement—that is, the maximum possible error of the measurement. Precision may be expressed as a plus or minus correction. The accuracy of a measurement depends on the precision of the measuring instrument. All measurements are approximate. It is impossible to know the "exact" length, mass, or amount of anything. The observer and the measuring instrument place a limit on the accuracy of all measurements.

Instruments for taking measurements are constructed with a calibrated scale for obtaining numerical values concerning the property being measured. The smallest division on the calibrated scale that can be read by the observer, without guessing, is known as the **least count** of the instrument. For example, a meterstick is divided into 100 equal divisions, marked on the stick as centimeters. Each centimeter is further divided into 10 equal divisions. Thus the least count of the meterstick is one-tenth of one centimeter, or one-thousandth of one meter.

When the meterstick is used to take a measurement of an unknown quantity (e.g., length), the observer can always obtain a value within one-thousandth (0.001) of a meter of the exact value of the unknown length, without guessing. But an additional step taken can increase the accuracy by one doubtful digit. That is, the observer can estimate a fractional part of the smallest division on the meterstick. For example, if one end of the meterstick is placed at one end of the object to be measured, and the other end of the object falls between two of the smallest divisions, the observer estimates this additional value and adds it to the known scale reading. This estimated digit is significant and is the last digit recorded when one is taking a measurement.

When the digits are recorded, the number recorded contains all known digits of the measurement plus the doubtful digit. This recorded number is a significant figure. By definition, a **significant figure** is a number that contains all known digits plus one doubtful digit.

Recorded numbers may contain the digit zero (0), which may or may not be significant. When a zero digit is used to locate the decimal point, it is not significant. For example, the numbers 0.048, 0.0032, and 0.00057 each have two significant digits. When a zero appears between two nonzero digits in a number, it is significant. For example, 2.04 has three significant digits; 8.002 has four significant digits. Zeroes appearing at the end of a number may or may not be significant. For example, in the number 5480, if the digit 8 is a doubtful digit, then the zero is not significant. If the 8 is a known digit and the zero is a doubtful digit, then the zero is significant.

The powers-of-10 notation can be used to remove the ambiguity concerning a number like 5480. When using powers-of-10 notation, we first write all the significant digits of the number. Then we write 10 to the correct power to locate the decimal point. For example, for three significant digits we write 5.48×10^3; for four significant digits we write 5.480×10^3.

The following procedure is usually used to round off significant figures to fewer digits. If the last significant digit on the right is less than 5, drop it and insert zero instead. If the last significant digit is 5 or greater, drop it and increase the preceding digit by one.

EXAMPLE 1

Round off the following numbers to two significant digits: 247, 243.

Solution In 247, because the last digit on the right is greater than 5, drop it and increase the preceding digit by one, for the result 250.

In 243, because the last digit on the right is less than 5, drop it and insert zero instead, for the result 240.

As a general rule, the number of significant digits of the product or the division of two or more measurements should be no greater than that of the measurement with the least number of significant digits. For example, suppose we want to determine the area of a table. We measure the length and width of the table, and then multiply one value by the other. In this procedure it is inaccurate and meaningless to calculate and give an answer indicating greater accuracy than justified by the original data. For example, the length of a table is measured with a meterstick as 1.8245 m and the width as 0.3720 m. The area $A = 1.8245 \text{ m} \times 0.3720 \text{ m} = 0.678714 \text{ m}^2$, as shown on a calculator. This six-figure number is not justified as the correct area of the table as given by the two measurements. The correct value for the area is 0.6787 m². This value has four significant digits, corresponding to the least number of significant digits in the original data.

When adding or subtracting significant figures, use the following two rules:

Rule 1. A known digit plus or minus a doubtful digit will give a doubtful digit.

Rule 2. Only one doubtful digit is allowed in a significant figure.

EXAMPLE 2

Add 2.34 and 16.5.

Solution
$$
\begin{array}{ll}
2.34 & \text{the 4 is doubtful} \\
+16.5 & \text{the 5 is doubtful} \\
\hline
28.84 & \text{the 8 and the 4 are doubtful}
\end{array}
$$

From Rule 1 the known digit 3 plus the doubtful digit 5 equals a doubtful digit 8. Therefore, the correct answer is 28.8, which is a significant figure with only one doubtful number, satisfying Rule 2.

Appendix III

Rules and Examples of the Four Basic Mathematical Operations Using Powers-of-10 Notation

1. In addition or subtraction, only like terms can be used; that is, the exponent for the number 10 must be of the same value. Thus:

$$
\begin{array}{ll}
4.6 \times 10^{-8} & 4.8 \times 10^{-8} \\
+1.2 \times 10^{-8} \quad\text{and} & -2.5 \times 10^{-8} \\
\hline
5.8 \times 10^{-8} & 2.3 \times 10^{-8}
\end{array}
$$

2. In multiplication, the exponents are added. Thus:

$$(2 \times 10^4)(4 \times 10^3) = 8 \times 10^7$$

$$(1.2 \times 10^{-2})(3 \times 10^6) = 3.6 \times 10^4$$

3. In division, the exponents are subtracted.

Thus:
$$\frac{4.8 \times 10^8}{2.4 \times 10^2} = 2.0 \times 10^6$$

and
$$\frac{3.4 \times 10^{-8}}{1.7 \times 10^{-2}} = 2.0 \times 10^{-6}$$

4. A power of 10 may be transferred from the numerator to the denominator, or vice versa, by changing the sign of the exponent. Thus:

$$\frac{3.4 \times 10^{-8}}{1.7 \times 10^{-2}} = \frac{3.4 \times 10^{-8} \times 10^2}{1.7} = 2.0 \times 10^{-6}$$

Appendix IV

Problem Solving: Five Major Steps

1. Read the problem carefully to find out what is given and what is inferred.
2. Determine the unknown or unknowns; that is, what are you looking for?
3. Apply the mathematical relationship that connects the unknown quantity or quantities with the given data.
4. Check data for correct units.
5. Solve mathematically for the unknown value or values.

EXAMPLE 1

A ball is thrown upward from the ground with an original velocity of 128 ft/s. How long does the ball remain in the air? What was the maximum height attained by the ball?

Solution
Step 1. Write the given data,

$$v_o = 128 \text{ ft/s}$$

and the inferred data,

(acceleration) $a = -32$ ft/s^2
(the minus sign is used because v_o and
the acceleration are in opposite directions)
(final velocity) $v_f = 0$ at maximum height

Step 2. Write down the unknown term or terms:

$t = ?$ seconds (total time the ball is in the air)

$s = ?$ feet (maximum height or displacement
of the ball)

Because the original velocity is in ft/s, find the unknown time in seconds and the unknown height in feet.

Step 3. The terms v_o, v_f, and a are known; therefore, use the relation

$$a = \frac{v_f - v_o}{t}$$

which contains these terms, plus the time, which is unknown.

Step 4. All the quantities used in solving a problem should be in the same system of units. The units of the result will then also be in that system. In this example the units are in the British system; hence the height will be in feet (Step 5). Units are often mixed in a problem. For example, velocity may be given in feet per second, while an acceleration may be given in meters per second^2. One of these units must be converted before being used in an equation.

Step 5. Solve the equation for t, and obtain

$$t = \frac{v_f - v_o}{a}$$

Substitute the given data and solve to obtain

$$t = \frac{0 - 128 \text{ ft/s}}{-32 \text{ ft/s}^2}$$

$$= 4 \text{ s (time for ball to reach maximum height)}$$

The total time in the air is twice this value, because the ball will require the same amount of time to return to the ground.

After we know the value of t, v_o, v_f, and a, the unknown height or distance can be determined by

$$d = v_o t + \frac{at^2}{2}$$

Because v_o at the top of the ball's flight is 0,

$$d = \frac{at^2}{2}$$

Because a in this case is equal to g, which is 32 ft/s^2, we can substitute the data and find

$$d = \frac{32 \text{ ft/s}^2 \times 4^2 \text{ s}^2}{2}$$

$$= \frac{512}{2} \text{ ft}$$

or $\qquad d = 256 \text{ft}$

In solving problems in grade school, you learned not to add apples to oranges, because doing so gives an answer that has no meaning. In physical science you must remember that only like terms can be added or subtracted. When you have a mathematical relationship (i.e., an equation), terms on the left side of the equal sign must be equal to the terms on the right side, not only in numerical value but also in units. For example, in the equation

$$a = \frac{v_f - v_o}{t}$$

both sides of the equation must have like dimensions. A **dimensional analysis** of an equation is done by substituting the fundamental units of length, mass, and time in the equation for the given terms.

EXAMPLE 2

We can use dimensional analysis to show that the above relation is an equality.

Solution Although it has no mass term, the expression has terms that can be traced back to the fundamental units of length and time. Therefore, by substituting, we obtain

$$\frac{L}{t^2} = \frac{\dfrac{L}{t} - \dfrac{L}{t}}{t}$$

Solving just for the numerator on the right, and remembering that we are subtracting units rather than numerical or algebraic values, we obtain

$$\frac{L}{t} - \frac{L}{t} = \frac{L}{t}$$

The top expression now can be written

$$\frac{L}{t^2} = \frac{\dfrac{L}{t}}{t}$$

Using negative exponents to bring our denominators above the line, we find

$$Lt^{-2} = Lt^{-1} \times t^{-1}$$

or

$$Lt^{-2} = Lt^{-2}$$

The left side of the expression has the same dimensions as the right side, showing the original to be an equality. If the two sides were not the same dimensionally, it would not be an equation that would give a correct solution.

Data found in physics problems often must be converted to different units in order to give the answer desired. For instance, measurements for rates of travel may be given in miles per hour and the answer requested in feet per second. How can we obtain the proper result?

EXAMPLE 3

What is the distance in feet traveled by an automobile in 10 s if the average speed of the car is 60 mi/h?

Solution Because we want to know the distance traveled in feet, we must convert mi/h to ft/s. This conversion can be done as follows:

1. Write the term you want to convert and set it equal to itself. At the same time, make a mental or written note of the terms you wish to convert.

$$\frac{60 \text{ mi}}{h} = \frac{60 \text{ mi}}{h}$$

2. Multiply the right side of the equation by 1 or 1/1, which is the same thing.

$$\frac{60 \text{ mi}}{h} = \frac{60 \text{ mi}}{h} \times \frac{1}{1}$$

3. Place a unit in the numerator or denominator of the 1/1 that will cancel an unwanted unit in the original term. In this case we place the unit "mile" in the denominator. At the same time, we place an equal term above our newly introduced denominator unit, in order to keep the value equal to 1/1. Because we are after feet in our answer for this problem, we place 5280 ft in the numerator of the 1/1, since that is equal to 1 mi. Then we cancel like units.

$$\frac{60 \text{ mi}}{h} = \frac{60 \text{ mi}}{h} \times \frac{5280 \text{ ft}}{1 \text{ mi}}$$

4. To eliminate the unit "hour" in the denominator, multiply again by 1/1 and place the unit "hour" in the numerator of the 1/1 term. At the same time, place an equal unit in the denominator of the 1/1 term in order to keep the value equal to 1/1. Because 60 minutes equal 1 hour, use it; then cancel both "hour" terms.

$$\frac{60 \text{ mi}}{h} = \frac{60 \text{ mi}}{h} \times \frac{5280 \text{ ft}}{1 \text{ mi}} \times \frac{1 \text{ h}}{60 \text{ min}}$$

5. At this point we have feet per minute, but we want feet per second; therefore, we must convert the minutes to seconds. Again, we multiply by 1 or 1/1 and add the appropriate units for conversion. Because 60 seconds equal 1 minute, we obtain

$$\frac{60 \text{ mi}}{h} = \frac{60 \text{ mi}}{h} \times \frac{5280 \text{ ft}}{1 \text{ mi}} \times \frac{1 \text{ h}}{60 \text{ min}} \times \frac{1 \text{ min}}{60 \text{ s}}$$

6. After all like units are canceled, the numbers are canceled as much as is possible. We arrive at

$$\frac{60 \text{ mi}}{h} = \frac{528 \text{ ft}}{6 \text{ s}}$$

or, after dividing,

$$\frac{60 \text{ mi}}{h} = \frac{88 \text{ ft}}{s}$$

The problem can now be solved, using 88 ft/s for 60 mi/h. Knowing that

$$d = vt$$

and substituting

$$d = \frac{88 \text{ ft}}{s} \times 10 \text{ s}$$

we obtain

$$d = 880 \text{ ft}$$

Derivation of Displacement Equation

The displacement of an object moving with constant acceleration can be determined by

$$d = v_o t + \frac{at^2}{2}$$

This relationship is derived as follows:

$$\bar{v} = \frac{d}{t}$$

Therefore,

$$d = \bar{v}t$$

We know that
$$\bar{v} = \frac{v_f + v_o}{2}$$

when the acceleration is constant. Substituting, we obtain

$$d = \left(\frac{v_f + v_o}{2}\right)t$$

Knowing that $v_f = v_o + at$, we may substitute for v_f and obtain

$$d = \left(\frac{v_o + at + v_o}{2}\right)t$$

or
$$d = \left(\frac{2v_o + at}{2}\right)t$$

or
$$d = v_o t + \frac{at^2}{2} \qquad \text{(A.1)}$$

The simplest way to obtain an equation of motion that does not include time is to take two equations that include time and eliminate the time factor by cancellation. By definition,

$$a = \frac{v_f - v_o}{t}$$

From the derivation of Eq. A.1 we know that

$$d = \left(\frac{v_f + v_o}{2}\right)t$$

Multiplying these equations,

$$\left(a = \frac{v_f - v_o}{t}\right) \times \left[d = \left(\frac{v_f + v_o}{2}\right)t\right]$$

and canceling the time, we obtain

$$ad = (v_f - v_o)\left(\frac{v_f + v_o}{2}\right)$$

or
$$2ad = v_f^2 - v_o^2 \qquad \text{(A.2)}$$

Derivation of Work–Kinetic Energy Relationship

In Chapter 4 we use the fact that a change in kinetic energy is related to work done against inertia. This equation can be derived as follows: First, we write the equation for the work done by an unbalanced force in accelerating a mass through a horizontal distance:

$$W = Fd \qquad \text{(A.3)}$$

The unbalanced force F, obtained from Newton's second law, is

$$F = ma$$

Substituting in Eq. A.3 for F, we have

$$W = mad \qquad \text{(A.4)}$$

Acceleration has been defined as

$$a = \frac{v_f - v_o}{t}$$

Substituting in Eq. A.4 for a, we have

$$W = m\left(\frac{v_f - v_o}{t}\right)d \qquad \text{(A.5)}$$

The distance traveled can be obtained from

$$d = \bar{v}t$$

or
$$d = \left(\frac{v_f + v_o}{2}\right)t$$

Substituting in Eq. A.5 for d, we obtain

$$W = m\left(\frac{v_f - v_o}{t}\right) \times \left(\frac{v_f + v_o}{2}\right)t$$

Canceling the t, we have

$$W = m\left(\frac{v_f - v_o}{1}\right) \times \left(\frac{v_f + v_o}{2}\right)$$

Multiplying $(v_f - v_o)$ by $(v_f + v_o)$, we have

$$W = \left(\frac{m}{2}\right)(v_f^2 - v_f v_o + v_f v_o - v_o^2)$$

or
$$W = m\left(\frac{v_f^2 - v_o^2}{2}\right)$$

or
$$W = \left(\frac{mv_f^2}{2}\right) - \left(\frac{mv_o^2}{2}\right)$$

which is the equation used in Chapter 4.

Appendix V

Derivation of Bohr's Equations

The force between the nuclear proton and the electron in circular orbit in the Bohr model is given by Coulomb's equation:

$$F = \frac{kq^2}{r^2}$$

This force supplies the necessary centripetal force ($F_c = mv^2/r$) for uniform circular motion. Then

$$F = F_c$$

$$\frac{kq^2}{r^2} = \frac{mv^2}{r} \qquad (A.6)$$

The total energy E of the electron is the sum of the kinetic energy E_k and the potential energy E_p—i.e.,

$$E = E_k + E_p$$

The kinetic energy of the electron is just

$$E_k = \tfrac{1}{2}mv^2$$

Then rearranging the force equation, we obtain

$$\frac{kq^2}{r^2} = \frac{mv^2}{r}$$

or

$$\frac{kq^2}{2r} = \frac{mv^2}{2} = E_k$$

The potential energy can be shown, using advanced mathematics, to be given by

$$E_p = \frac{-kq^2}{r}$$

where the potential energy is negative—or as we say, the electron is in a potential "well." Hence we have, for the total energy of the electron,

$$E = E_k + E_p$$

$$= \frac{kq^2}{2r} - \frac{kq^2}{r} = \frac{-kq^2}{r} \qquad (A.7)$$

Bohr's quantum assumption in his theory was that the angular momentum of the electron was quantized. Thus only certain discrete values of the angular momentum (mvr) are allowed. Specifically,

$$mvr = \frac{nh}{2\pi} \qquad (A.8)$$

That is, only integer values of $h/2\pi$ are allowed—$h/2\pi$, $2(h/2\pi)$, $3(h/2\pi)$, and so on.

Equation (A.8) can be rewritten as

$$v = \frac{nh}{2\pi mr}$$

Then substituting this expression into Eq. A.6 yields

$$\frac{kq^2}{r^2} = \frac{m(v^2)}{r}$$

$$k\frac{q^2}{r^2} = \frac{m}{r}\left(\frac{n^2h^2}{4\pi^2m^2r^2}\right)$$

or

$$r = \left[\frac{h^2}{kq^24\pi^2m}\right]n^2 \qquad (A.9)$$

We can now solve for E:

$$E = -\frac{kq^2}{2}\left(\frac{1}{r}\right)$$

$$= -\frac{kq^2}{2}\left(\frac{kq^24\pi^2m}{h^2n^2}\right)$$

$$= -\left[\frac{k^2q^44\pi^2m}{2h^2}\right]\frac{1}{n^2} \qquad (A.10)$$

The quantities in brackets in Eqs. A.9 and A.10 are all constants and can be evaluated. The results are Eqs. 9.3 and 9.4:

$$r = 0.53n^2 \text{ Å}$$

$$E = \frac{-13.6}{n^2} \text{ eV}$$

Appendix VI

Length Contraction, Time Dilation, and Relativistic Mass Increase

The special theory of relativity is based on two principles:

1. The relativity principle states that the laws of physics are the same in all inertial reference frames. An inertial reference frame is one in which Newton's law of inertia is valid. Inertial reference frames can move at constant velocity relative to one another. Noninertial reference frames exhibit acceleration.

2. The principle of constancy of the speed of light states that the speed of light in empty space has the same value in all inertial reference frames.

The phenomena called *length contraction, time dilation,* and *relativistic mass increase* are effects that become significant when the relative velocity between the object being measured and the observer is an appreciable fraction of the velocity of light.

In length contraction the length of a moving object is measured to be less in a direction parallel to the direction of motion than that of a similar object in the rest frame. The equation that relates the lengths of objects in motion to their lengths when at rest is

$$L = L_o \sqrt{1 - \frac{v^2}{c^2}}$$

where L_o = length of object at rest,
L = length of object in motion,
v = velocity of object with respect to observer,
c = velocity of light = 3×10^8 m/s.

EXAMPLE 1

What is the length contraction of a meterstick when it is moving in the direction of its length at half the velocity of light?

Solution

$$L = (1 \text{ m}) \sqrt{1 - \frac{(1.5 \times 10^8 \text{ m/s})^2}{(3.0 \times 10^8 \text{ m/s})^2}}$$

$$= (1 \text{ m}) \sqrt{1 - \frac{2.25}{9}} = (1 \text{ m})(0.866) = 0.866 \text{ m}$$

In time dilation (an "increase" of time), moving clocks run slower than clocks at rest with respect to an observer. The equation for relativistic time dilation is

$$t = \frac{t_o}{\sqrt{1 - v^2/c^2}}$$

where t_o = time interval of clock at rest,
t = time interval of clock in relative motion as determined by outside observer,
v = velocity of relative motion,
c = velocity of light.

In the effect of relativistic mass increase, the measured mass of an object at rest is not the same as the measured mass of the object in relative motion. The mass of the moving object is larger, or increased, and is given by the following equation:

$$m = \frac{m_o}{\sqrt{1 - v^2/c^2}}$$

where m_o = rest mass as measured by stationary observer,
m = mass of object moving relative to observer,
v = velocity of relative motion,
c = velocity of light.

EXAMPLE 2

What is the relativistic mass increase of a 1-kg mass moving at half the velocity of light?

Solution

$$m = \frac{1 \text{ kg}}{\sqrt{1 - \frac{(1.5 \times 10^8 \text{ m/s})^2}{(3.0 \times 10^8 \text{ m/s})^2}}} = \frac{1 \text{ kg}}{\sqrt{1 - \frac{2.25 \times 10^{16} \text{ m}^2/\text{s}^2}{9 \times 10^{16} \text{ m}^2/\text{s}^2}}}$$

$$= \frac{1 \text{ kg}}{\sqrt{1 - 0.25}} = \frac{1 \text{ kg}}{0.866} = 1.154 \text{ kg}$$

Appendix VII

Psychrometric Tables (Pressure: 30 in Hg)

Table 1 Relative Humidity (%) and Maximum Moisture Capacity

Air Temp. (°F) (Dry Bulb)	Max. Moisture Capacity (gr/ft³)	Degrees Depression of Wet-Bulb Thermometer (°F)													
		1	2	3	4	5	6	7	8	9	10	15	20	25	30
25	1.6	87	74	62	49	37	25	13	1						
30	1.9	89	78	67	56	46	36	26	16	6					
35	2.4	91	81	72	63	54	45	36	27	19	10				
40	2.8	92	83	75	68	60	52	45	37	29	22				
45	3.4	93	86	78	71	64	57	51	44	38	31				
50	4.1	93	87	80	74	67	61	55	49	43	38	10			
55	4.8	94	88	82	76	70	65	59	54	49	43	19			
60	5.7	94	89	83	78	73	68	63	58	53	48	26	5		
65	6.8	95	90	85	80	75	70	66	61	56	52	31	12		
70	7.8	95	90	86	81	77	72	68	64	59	55	36	19	3	
75	9.4	96	91	86	82	78	74	70	66	62	58	40	24	9	
80	10.9	96	91	87	83	79	75	72	68	64	61	44	29	15	3
85	12.7	96	92	88	84	80	76	73	69	66	62	46	32	20	8
90	14.8	96	92	89	85	81	78	74	71	68	65	49	36	24	13
95	17.1	96	93	89	85	82	79	75	72	69	66	51	38	27	17
100	19.8	96	93	89	86	83	80	77	73	70	68	54	41	30	21
105	23.4	97	93	90	87	83	80	77	74	71	69	55	43	33	23
110	26.0	97	93	90	87	84	81	78	75	73	70	57	46	36	26

Note: To use the table, determine the air temperature with a dry-bulb thermometer and degrees depressed on the wet-bulb thermometer. Read the maximum capacity directly. Read the relative humidity (in percent) opposite and below these values.

Table 2 Dew Point (°F)

Air Temp. (°F) (Dry Bulb)	Degrees Depression of Wet-Bulb Thermometer (°F)													
	1	2	3	4	5	6	7	8	9	10	15	20	25	30
25	22	19	15	10	5	−3	−15	−51						
30	27	25	21	18	14	8	2	−7	−25					
35	33	30	28	25	21	17	13	7	0	−11				
40	38	35	33	30	28	25	21	18	13	7				
45	43	41	38	36	34	31	28	25	22	18				
50	48	46	44	42	40	37	34	32	29	26	0			
55	53	51	50	48	45	43	41	38	36	33	15			
60	58	57	55	53	51	49	47	45	43	40	25	−8		
65	63	62	60	59	57	55	53	51	49	47	34	14		
70	69	67	65	64	62	61	59	57	55	53	42	26	−11	
75	74	72	71	69	68	66	64	63	61	59	49	36	15	
80	79	77	76	74	73	72	70	68	67	65	56	44	28	−7
85	84	82	81	80	78	77	75	74	72	71	62	52	39	19
90	89	87	86	85	83	82	81	79	78	76	69	59	48	32
95	94	93	91	90	89	87	86	85	83	82	74	66	56	43
100	99	98	96	95	94	93	91	90	89	87	80	72	63	52
105	104	103	101	100	99	98	96	95	94	93	86	78	70	61
110	109	108	106	105	104	103	102	100	99	98	91	84	77	68

Note: To use the table, determine the air temperature with a dry-bulb thermometer and degrees depressed on the wet-bulb thermometer. Find the dew point opposite and below these values.

Photo Credits

Glossary

Absolute geologic time   the time of past geologic events based on the radioactive decay of certain atomic nuclei.

Absolute humidity   the amount of water vapor in a specific volume of air. The amount is measured in grain/ft^3 or grams/m^3.

Abyssal plain   a large flat area on the ocean floor where layers of sediment have covered the original seafloor topography.

Acceleration   the change in velocity divided by the change in time: $a = \Delta v/\Delta t$.

Acceleration of gravity   usually given as the symbol g; equal to 32 ft/s^2, 980 cm/s^2, or 9.8 m/s^2.

Accuracy   refers to how close the measurement comes to the exact or true value.

Acid   a substance that acts as a proton donor.

Acid mine drainage   water drainage from mining areas that contain sulfuric acid because of the reaction in the air (oxygen) and water with sulfur-bearing minerals.

Acid rain   rain that has a relatively low pH (acid) due to air pollution.

Actinide series   a group of 14 elements (thorium through lawrencium) following actinium in the periodic table in which the $5f$ orbitals are being filled—the second inner transition series.

Activated-sludge process   a sewage-treatment process in which aerated sewage effluent is mixed with bacteria-laden sludge for the removal of organic matter.

Activation energy   the energy necessary to get a chemical reaction started.

Adiabatic   a process in which no heat is added or removed from the system.

Advection fog   a fog formed as a result of air movement over a cold surface.

Air current   vertical air movement.

Air-fuel ratio   the ratio of air-to-fuel mixture that is important in the operation of the gasoline internal-combustion engine.

Air mass   a mass of air with physical characteristics that distinguish it from other air.

Air pollution   any atypical contributions to the atmosphere resulting from human activities.

Air-pollution potential   weather conditions that are favorable for potential pollution conditions—e.g., stagnation conditions as a result of a temperature inversion.

Albedo   the reflectivity or average fraction of light a body reflects.

Alcohol   any compound that contains an —OH group and a hydrocarbon group.

Aldehydes   any compound that contains a —C=O group.
|
H

Alkali metal   a Group 1A metal.

Alkaline earth   a Group 2A metal.

Alkanes   hydrocarbons with a composition that satisifes the general formula C_nH_{2n+2}.

Alkenes   a hydrocarbon containing a double bond; the general formula for the alkenes is C_nH_{2n}.

Alkynes   a hydrocarbon containing a triple bond; the general formula for the alkynes is C_nH_{2n-2}.

Alpha decay   the disintegration of a nucleus into an alpha particle ($^4_2$He nucleus) and the nucleus of another element.

Alpha particle   the nucleus of a helium-4 atom ($^4_2$He).

Altitude   the angle measured from the horizon to a celestial object.

Alto   the prefix associated with the middle cloud family.

Alveoli   tiny air spaces, or sacs, in the lungs.

Amino acids   compounds that contain both —NH$_2$ and —COOH groups.

Ampere   the unit of electric current defined as that current which, if maintained in each of two long parallel wires separated by one meter in free space, would produce a magnetic force between the two wires of 2×10^{-7} newton for each meter of length.

Amplitude   the maximum displacement of a wave from its equilibrium position.

amu   one atomic mass unit— $\frac{1}{12}$ the mass of the ^{12}C isotope.

Analgesic   a drug that relieves pain without dulling consciousness.

Anemometer   an instrument used to measure wind speed.

Angstrom   a unit of length equal to 10^{-8} cm.

Angular momentum mvr for a mass m going at a speed v in a circle of radius r.

Anion   a negatively charged ion.

Annular eclipse   an eclipse of the Sun in which the moon blocks out all of the Sun except for a ring around the outer edge of the Sun.

Anode   the electrode in an electrochemical cell at which oxidation takes place.

Anticyclone   a high-pressure area characterized by clockwise air circulation (in the northern hemisphere).

Apogee   the point in its orbit at which a satellite is farthest from Earth's center.

Aquifer   a body of permeable rock through which ground water moves.

Arrhenius acid-base concept   the concept stating that acids produce hydrogen ions in water solutions and bases produce hydroxide ions.

Artesian wells and springs   wells and springs in which water rises above an aquifer; the pressure due to gravity can cause the water to bubble or spurt onto the surface.

Asteroids   large chunks of matter that orbit the Sun (usually between Mars and Jupiter) and that are too small to be labeled as planets.

Asthenosphere   the rocky substratum below the lithosphere that is hot enough to be deformed and capable of internal flow.

Astronomical unit   the mean distance between Earth and the Sun—93,000,000 mi.

Atmosphere   the gases that surround a planet or celestial object.

Atmospheric science   the investigation of every aspect of the atmosphere, from the ground to the edge of outer space.

Atom   the smallest particle of an element that can enter into a chemical combination.

Atomic number   the number of protons in the atom.

Atomic time scale   a geologic time scale based on radioactive dating.

Atomic weight   the weight of an average atom of an element with reference to the ^{12}C isotope. ^{12}C is 12 exactly.

Autumnal equinox   the time (near September 21) when the Sun's declination crosses the equator moving south (for the norther hemisphere).

Avogadro's law   equal volumes of gases at the same temperature and pressure contain an equal number of molecules.

Avogadro's number   the number of molecules in a mole of gas, liquid or solid—6.02×10^{23} molecules per mole.

Barometer   a device used to measure atmospheric pressure.

Basaltic lava   lava with a low silica content and low viscosity.

Base   a substance that acts as a proton acceptor.

Batholith   a large intrusive igneous rock formation that has an area of at least 40 mi^2.

Bedding   the stratification of sedimentary rock formations.

Beta decay   the disintegration of a nucleus into a beta particle (electron) and the nucleus of another element.

Beta particle   an electron.

Big Bang Theory   theory of the origin of the universe that states that the known universe was concentrated in a massive glob of extremely dense material which exploded approximately 15 billion years ago.

Binding energy   total binding energy—the amount of energy necessary to completely separate the protons and neutrons of the nucleus of an atom.

Biodegradable   capable of being broken down or degraded by bacterial action.

Black hole   a very dense collapsed star from which no light can escape.

Blizzard   a snow accompanied by high winds that whip the snow into blinding swirls and drifts.

Boiling point   the temperature at which a substance changes from the liquid to the gas phase.

Boyle's gas law   the volume of a perfect gas varies inversely as the absolute pressure, if the temperature remains constant.

British system   the system of measurement used in the United States, which uses the foot, pound, second, and coulomb as the standards of length, weight, time, and electric charge, respectively.

Btu   the amount of heat required to raise one pound of water one degree Fahrenheit at normal atmospheric pressure.

Caldera   a roughly circular, steep-walled depression formed as a result of the collapse of a volcanic chamber.

Calorie   the amount of heat necessary to raise one gram of pure liquid water one degree Celsius at normal atmospheric pressure.

Cancer   a condition of unregulated cell growth.

Carbohydrate   compounds composed of carbon, hydrogen, and oxygen with the hydrogen-to-oxygen ratio usually 2 to 1; sugars and starches are typical.

Carboxyl group   the —COOH functional group in an organic acid.

Carcinogen   a cancer-producing agent.

Catalyst   a substance that changes the rate of a chemical reaction without undergoing a permanent change itself.

Cathode   the electrode in an electrochemical cell at which reduction takes place.

Cation   a positively charged ion.

Ceilometer   an instrument used to measure the heights of clouds by light reflection.

Celestial sphere   the imaginary sphere on which all the stars seem to hang.

Centripetal force   a "center-seeking" force that causes an object to travel in a circle.

Cepheid variables   stars that vary in magnitude with a fixed period of between 1 and 100 days.

cgs system   a metric system, used throughout most of the world, which has the centimeter, gram, second, and coulomb as the standard units of length, mass, time, and electric charge, respectively.

Chain reaction   a self-propagating process in which product neutrons from one fission are used to induce fission in other nuclei.

Charles' gas law   the volume of a perfect gas varies in direct proportion to the absolute temperature, if the pressure remains constant.

Chemical properties   the properties involved in the transformation of one substance into another.

Chemical sedimentary rocks   rocks formed by the precipitation of minerals dissolved in water.

Chinook the name (literally, snow eater) applied to the warm leeward winds on the eastern slopes of the Rocky Mountains that give rise to the rapid melting of snow; in Europe the term *föhn* is used.

Chromosphere an outer layer of the Sun, which lies just outside the photosphere.

Cilia tiny, hairlike projections in the respiratory tract.

Cinder cone a volcano with a steeply sloped cinder cone formed by eruptions of pyroclastic debris.

Cirque glacier a small glacier formed in a hollow depression along a mountain.

Cirrus a root name used to describe wispy, fibrous cloud forms.

Clastic sedimentary rocks rocks formed from sediment composed of fragments of preexisting rocks.

Cleavage the splitting of a mineral along an internal molecular plane.

Climate the long-term average weather conditions of a region or the world.

Cloud buoyant masses of visible droplets of water vapor and ice crystals in the lower troposphere.

Coalescence the combining of small droplets of water vapor to make larger drops.

Cold front the boundary of an advancing cold mass over a warmer surface.

Column the cavern dripstone formation consisting of a joined stalactite and stalagmite.

Combined sewer system a system that uses the same network as both a sanitary and storm sewer.

Comet a chunk of matter that displays a long tail as it passes near the Sun and has a highly elliptical orbit.

Composite volcano a volcano with a steeply sloping symmetrical cone formed by eruptions of high-viscosity lava and pyroclastic debris; also called a stratovolcano.

Compound a substance composed of two or more elements chemically combined in a definite proportion.

Concave mirror a mirror shaped like the inside of a small section of a sphere.

Concept a meaningful idea that can be used to describe and explain phenomena.

Conduction the transfer of heat energy by molecular transfer.

Conductor a material that easily conducts an electric current because some electrons in the material are free to move.

Conjugate acid the substance formed when a base gains a proton; the substance is considered an acid, because it can lose a proton to reform the base.

Conjugate base the substance formed when an acid loses a proton; the substance is considered a base, because it can gain a proton to reform the acid.

Conjunction the time at which a planet and sun occur on the same meridian.

Conservation of angular momentum, law of the angular momentum of an object remains constant unless acted upon by an external torque.

Conservation of energy, law of the total energy of an isolated system remains constant.

Conservation of linear momentum, law of the total linear momentum of a system remains constant if there are no external unbalanced forces acting on the system.

Conservation of mass, law of there is no detectable change in the total mass during a chemical process.

Consolidation the process of forming sedimentary rock from sediment. Also called lithification.

Continental drift the theory that continents move, drifting apart or together.

Continental glacier (ice sheet) a large mass of ice that covers a large surface region and flows outward toward the sea.

Continental shelf the shallowly submerged margin of a continental land mass.

Continental slope the seaward slope beyond the continental shelf that extends downward to the ocean basin.

Convection the transfer of heat through the movement of a substance.

Convection cycle the cyclic movement of matter—e.g., air—due to localized heating and convectional heat transfer.

Convex mirror a mirror shaped like the outside of a small section of a sphere.

Core the innermost region of Earth, which is composed of two parts—a solid inner core and a molten, highly viscous, "liquid" outer core.

Coriolis force a pseudoforce arising in an accelerated reference frame on the rotating (accelerating) Earth. The apparent deflection of objects is attributed to the Coriolis force.

Correlation establishing the equivalence of rocks in separate regions: correlation by fossils is an example.

Cosmic background radiation the microwave radiation that fills all space and is believed to be the redshifted glow from the Big Bang.

Cosmological principle on the large scale, the universe is both homogeneous and isotropic.

Cosmological red shift the Doppler shift toward longer wavelengths caused by the expansion of the universe.

Cosmology the study of the structure and evolution of the universe.

Coulomb the unit of electric charge equal to one ampere-second ($A \cdot s$).

Coulomb's law the force of attraction or repulsion between two charged bodies is directly proportional to the product of the two charges and inversely proportional to the square of the distance between them.

Covalent bond a chemical bond in which electron pairs are shared by atoms.

Covalent compounds compounds formed by the electron-sharing process.

Crater the funnel-shaped depression of the summit of a volcano.

Creep a type of slow mass wasting involving the slow,

particle-by-particle movement of weathered debris down a slope, which takes place year after year.

CRT a cathode ray tube, or picture tube. It consists of an electron beam that is deflected by electric or magnetic fields and strikes a phosphorescent screen to form a picture.

Crust the thin outer layer of Earth.

Crystal an orderly arrangement of atoms.

Cumulus a root name used to describe billowy, round cloud forms.

Curie a unit of radioactivity from a radioactive source. One curie is arbitrarily defined as 3.7×10^{10} disintegrations per second.

Curie temperature a high temperature, above which ferromagnetic materials cease to be magnetic.

Current rate of flow of electric charge.

Cyclone a low-pressure area characterized by counterclockwise air circulation (in the northern hemisphere).

Daylight saving time time advanced one hour from standard time, adopted during the spring and summer months to take advantage of longer evening daylight hours.

Declination the overhead position of a celestial object, such as the Sun, measured in degrees latitude.

Definite proportions, law of different samples of a pure compound always contain the same elements in the same proportions by weight.

Degree a unit of angle; there are 360 degrees in a circle.

Density current a deep ocean current that is due to dense water sinking through less dense water.

Detergent a general term for a cleansing agent.

Detergent builders substances added to a detergent to make its cleansing action more efficient; phosphates are common detergent builders.

Deuteron the nucleus of a deuterium atom (^2_1H).

Dew point the temperature at which a sample of air becomes saturated—i.e., has a relative humidity of 100 percent.

Diastrophism a geologic term meaning the movement of the Earth's crust.

Diffraction the bending of waves when an opening or obstacle has a size smaller than or equal to the wave length.

Dike a discordant pluton formation that is formed when magma fills a nearly vertical fracture in rock layers.

Diode an electronic device that allows current to flow in only one direction.

Dispersion the fact that different frequencies are refracted at slightly different angles.

DNA deoxyribonucleic acid, a nucleic acid located primarily in the nucleus of a cell. DNA has the molecular structure of a double-stranded helix.

Doldrums the low-pressure region near the equator.

Doppler effect an apparent change in frequency resulting from the relative motion of the source or the observer.

Drift glacial deposits of eroded material that are deposited by either ice or meltwater.

Drug a compound that may produce a physiological change in humans or other animals and may become a poison when used in excessive amounts.

Dust dome a concentration of dust over a city due to the self-contained thermal circulation cell set up as a result of the heat-island effect.

Dyne a unit of force, $1 \text{ g} \cdot \text{cm/s}^2$.

Earthquake the sudden release or transfer of energy because of sudden movement resulting from stresses in the Earth's lithosphere.

Eclipse an occurence in which one celestial object is partially or totally blocked from view by another.

Ecliptic plane the plane of Earth's orbit around the Sun.

Electric charge a fundamental property of matter that can be either positive or negative and gives rise to electrical forces.

Electric field a set of imaginary lines that indicate the direction that a small positive charge would move if it were placed at a particular spot.

Electricity the effects produced by moving charges.

Electrolysis the production of an oxidation-reduction reaction by means of an electric current.

Electrolyte a compound in the molten or dissolved form that conducts an electric current.

Electromagnet a current-carrying coil of insulated wire wrapped around a piece of soft iron that creates a magnetic field inside the iron only when the wire conducts a current.

Electromagnetic wave a wave caused by oscillations of electric and magnetic fields.

Electromagnetism the interaction of electric and magnetic effects.

Electromotive series a list of standard oxidation potentials at a given temperature.

Electron an elementary subatomic particle, with a very small mass of 9.01×10^{-31} kilograms and a negative charge of 1.602×10^{-19} coulombs, that orbits the atomic nucleus.

Electron affinity a measure of the energy released when an electron is added to a gaseous atom to form a negative ion.

Electronegativity the measure of the ability of an atom to attract electrons in the presence of another atom.

Electron period a set of energy levels, all of which have approximately the same energy.

Electron shell consists of all the orbits of electrons with the same principal quantum number (n)

Electron subshell an energy level; all electrons in the same subshell have the same energy.

Electron volt the amount of kinetic energy an electron (or proton) acquires when it is accelerated through an electric potential of one volt.

Element a substance that has the same number of protons in all of its atoms.

Emphysema a lung condition characterized by large air sacs due to the breakdown of the walls of the alveoli.

Endothermic reaction   a reaction in which energy is absorbed, as in the melting of ice.

Energy   the capacity to do work.

Entropy   a measure of the disorder of a system.

Enzymes   organic substances of high molecular weight that catalyze reactions in living organisms.

Epicenter   the point on the surface of Earth directly above the focus of an earthquake.

Epoch   an interval of geologic time that is a subdivision of a period.

Era   an interval of geologic time made up of periods and epochs.

Erg   a unit of energy, 1 dyne·cm or $1 \text{ g·cm}^2/\text{s}^2$.

Erosion   the group of interrelated processes by which rock is broken down and the products removed.

Ester   any compound that conforms to the general formula R—C—O—R′.
$$\overset{\|}{\underset{O}{}}$$

Eutrophication   the natural aging process of lakes that is accelerated by phosphate pollution.

Event horizon   the position in space at which the escape velocity from a black hole equals the speed of light.

Excited state   a state of the atom with energies above the ground state; *see* Ground state.

Exothermic reaction   a reaction in which energy is released, as in the burning of a candle.

Fault   a fracture along which a relative displacement of the sides has occurred.

Fault-block mountains   mountains that were built by normal faulting, in which giant pieces of the Earth's crust were uplifted.

Faulting   the relative motion along a fracture or fault that results in the displacement of rock masses on one side of the fracture relative to those of the other side.

Felsitic lava   lava with a high silica content and high viscosity.

First-quarter moon   the phase of the moon between the new and full moon, in which an observer in the United States sees the right half of the moon bright and overhead at 6 P.M., local solar time.

Fission   the splitting of the nucleus of an atom into two nuclei of approximately equal size, with an accompanying release of energy.

Focal length   the point at which light rays from a distant source will converge after being reflected from a mirror or refracted by a lens.

Focus (earthquake)   the point within Earth at which the initial energy release of an earthquake occurs.

Fog   a maze of visible droplets of water vapor near Earth's surface.

Föhn   the name applied to the warm leeward winds on mountain slopes that give rise to the rapid melting of snow; in the Rocky Mountain region, the term *chinook* is used.

Folded mountains   mountains characterized by folded rock strata, with external evidence of faulting and central evidence of igneous metamorphic activity; folded mountains are believed to be formed at convergent plate boundaries.

Foliation   the orientation characteristic of some metamorphic rocks due to formational directional pressures.

Force   any quantity capable of producing motion.

Formula weight   the sum of the atomic weights given in the formula of the compound; if the formula is the molecular formula, the formula weight is also its molecular weight.

Fossil fuels   fuels of organic origin—e.g., coal, gas, and oil.

Foucault pendulum   a pendulum with a very long length that swings for a long period of time and demonstrates the rotation of Earth.

Frequency   the number of oscillations of a wave in a given period of time.

Friction   the opposing force that exists when contact surfaces tend to slide past one another.

Front   the boundary between two air masses.

Frost heaving   the uplifting of rock and soil resulting from accumulations of freezing water.

Frost wedging   a form of physical weathering by which rocks are pushed apart and fractured because of the pressure of freezing water within the rocks.

Full moon   the phase of the moon that occurs when the moon is on the opposite side of Earth from the Sun.

Fundamental quantities   physical quantities that serve as the basis for other physical concepts; length, mass, and time are examples.

Fusion   the transmutation of lightweight atomic nuclei into heavier nuclei with the release of energy.

G   the universal gravitational constant—
$$G = 6.67 \times 10^{-11} \text{ N·m}^2/\text{kg}^2.$$

Galaxy   A large-scale aggregate of stars plus some gas and dust, held together gravitationally. They have a spiral, elliptical, or irregular structure, and contain, on the average, one hundred billion solar masses.

Gamma decay   the emission of electromagnetic energy from the nucleus of an atom. The atomic number, mass number, and neutron number of the atom remain the same, but the nucleus decreases in energy.

Gamma ray   high-energy electromagnetic radiation with wavelengths ranging from 3×10^{-14} to 3×10^{-12} m.

Gas   matter that has no definite volume or shape.

Gay-Lussac's Law (of Combining Volumes)   the volumes of gases taking part in a chemical reaction, at constant temperature and pressure, can be expressed as a ratio of small whole numbers.

General theory of relativity   a theory of relativity true for systems that are accelerating with respect to one another.

Generator   a device that converts mechanical work or energy into electrical energy.

Geocentric theory   the old false theory of the solar system, which placed Earth at the center of the universe.

Geologic time scale   a relative time scale based on the fossil index of rock strata.

Geology   the study of Earth, its processes, and its history.

Geosyncline   long, narrow ocean troughs containing large accumulations of sediment.

Geothermal gradient   the increase of temperature with depth that occurs in the outer portion of Earth's crust. The rate of increase is about 1°F per 150 feet of depth.

Geyser   a surface opening through which steam and boiling water erupt intermittently.

Glacier   a large ice mass, consisting of recrystallized snow, that flows on a land surface under the influence of gravity.

Graded river   the condition of a river when its erosion and transport capabilities are in balance.

Gram   a unit of mass in the metric cgs system of units. One gram is the mass of one cubic centimeter of pure water at its maximum density (4°C).

Gram atomic weight   the weight of an atom of an element relative to ^{12}C, expressed in grams.

Gram molecular weight   a mass in grams equal to the molecular weight of the molecule.

Gravitational collapse   the collapse of a very massive body because of its attraction for itself.

Greenhouse effect   the heat-retaining process of atmospheric gases—viz., water vapor and CO_2—due to the selective absorption of long-wavelength terrestial radiation.

Greenwich meridian   the reference meridian of longitude, which passes through the old Royal Greenwich Observatory near London.

Ground state   the lowest energy level of an atom.

Group   a number of elements appearing in any one column of the periodic table.

Guyot   a seamount with a flat top.

Half-life   the time required for half of any sample of a radioactive element to disintegrate.

Hard water   water containing dissolved calcium and/or magnesium salts.

Heat   a form of energy; energy in transit.

Heat capacity   the amount of heat energy in calories required to raise the temperature of a substance one degree Celsius.

Heat engine   a device that uses heat energy to perform useful work.

Heat-island effect   the condition of the temperature and heat content of a city being higher than the surrounding rural areas because of greater radiation absorption, etc.

Heat lightning   lightning that occurs below the horizon or behind a cloud, which illuminates the cloud with flickering flashes of light.

Heat pump   a device used to transfer heat from a low-temperature reservoir to a high-temperature reservoir.

Heisenberg uncertainty principle   it is impossible to know simultaneously the exact velocity and position of a particle.

Heliocentric theory   the current theory of the solar system, which places the Sun in the center.

Herbicides   chemicals used to kill unwanted plant species.

Hertz   one cycle per second.

Heterogeneous   matter that is of nonuniform composition.

Heterosphere   a region of the atmosphere based on the heterogeneity of the atmospheric gases between approximately 60 and several hundred miles in altitude.

Homogeneous   matter that is of uniform composition throughout.

Homosphere   a region of the atmosphere based on the homogeneity of the atmospheric gases between approximately 0 and 60 miles in altitude.

Horse latitudes   the high-pressure region near 30° latitude.

Horsepower   a unit of power, 550 ft·lb/s.

Hot-spot theory   the theory that explains central-plate volcanic chains as being formed as a result of plate movement over a hot spot beneath it.

H-R diagram   a plot of the absolute magnitude versus the temperature of stars.

Hubble's law   the recessional speed of a distant galaxy is proportional to its distance away.

Humidity   a measure of the water vapor in the air.

Hurricane   a tropical storm with winds of 74 mi/h or greater.

Hurricane warning   an alert that hurricane conditions are expected within 24 hours.

Hurricane watch   an advisory alert that hurricane conditions are a definite possibility.

Hydrocarbons   compounds composed of carbon and hydrogen—e.g., methane, CH_4.

Hydrologic cycle   the cyclic movement of Earth's water supply from the oceans to the mountains and back again to the oceans.

Hydronium ion H_3O^+

Hygroscopic nuclei   particulate matter that acts as nuclei in the condensation process.

Ice point   the temperature of a mixture of ice and air-saturated water at normal atmospheric pressure.

Igneous rock   rock formed by the cooling and solidification of hot, molten material.

Incomplete combustion   combustion in which the carbon element of the fuel is not completely reacted to form CO_2, but rather CO.

Index fossil   a fossil that is related to a specific span of geologic time.

Index of refraction   the ratio of the speed of light in a vacuum and the speed of light in a medium.

Indian summer   a period of warm weather in the late fall.

Inertia   the property of matter to resist any change of motion.

Inner planets   the four planets closest to the Sun (Mercury, Venus, Earth, and Mars).

Insolation   the solar radiation received by Earth and its atmosphere—*incoming solar radiation*.

Insulator a material that does not conduct an electric current because the electrons in the material are not free to move.

International date line the meridian that is 180° E or W of the prime meridian.

Inner transition elements elements in which the third shell from the outer shell is increasing from 18 to 32 electrons.

Intrusive rock igneous rock that formed below Earth's surface.

Ion an atom or group of atoms with a net electric charge.

Ionic bond a chemical bond in which electrons have been transferred from atoms of low ionization potential to atoms of high electron affinity.

Ionic compounds compounds that are formed by the process of electron transfer.

Ionization potential the electrical voltage required to remove an electron from an atom.

Ionosphere a region of the atmosphere, between 50 and several hundred miles in altitude, characterized by ion concentrations.

Isobar a line drawn through points of equal pressure.

Isobaric process a constant-pressure process.

Isomers molecules with the same molecular formula but a different structure or arrangement of atoms, hence slightly different physical properties.

Isostasy the concept that Earth's crustal material "floats" in gravitational equilibrium on a "fluid" substratum.

Isothermal a constant-temperature process.

Isotope atoms whose nuclei have the same number of protons but a different number of neutrons.

Jet streams fast-moving "rivers" of air in the upper troposphere.

Joule a unit of energy. 1 N·m or kg·m^2/s^2.

Kepler's harmonic law the ratio of the square of the period to the cube of the semimajor axis (one-half the larger axis of an ellipse), is the same for all the planets.

Kepler's law of elliptical paths all planets (asteroids, comets, etc.) revolve around the Sun in elliptical orbits.

Kepler's law of equal areas as a planet (or asteroid or comet) revolves around the Sun, an imaginary line joining the planet to the Sun sweeps out equal areas in equal periods of time.

Kilo prefix that means 10^3 or one thousand.

Kilogram the unit of mass in the mks system; one kilogram has an equivalent weight of 2.2 pounds.

Kinetic energy energy of motion equal to $\frac{1}{2}mv^2$.

Laccolith a concordant pluton formation formed from a blisterlike intrusion that has pushed up the overlying rock layers.

Landslide a type of fast mass wasting that involves the downslope movement of large blocks of weathered material.

Laser an acronym for *l*ight *a*mplification by *s*timulated *e*mission of *r*adiation; it is coherent, monochromatic light.

Lanthanides the 14 elements following lanthanum in the periodic table—the first inner transition series.

Lapse rate the rate of temperature change with altitude; in the troposphere the normal lapse rate is $-3\frac{1}{2}°F$ per 1000 ft.

Last-quarter moon (or third-quarter moon) the phase of the moon between the full and new moon in which an observer in the United States sees the left half of the moon bright and overhead at 6 A.M. local solar time.

Latent heat of fusion the amount of heat required to change one gram of a substance from the solid to the liquid phase at the same temperature.

Latent heat of vaporization the amount of heat required to change one gram of a substance from the liquid to the gas phase at the same temperature.

Latitude for a point on the surface of Earth, the angular measurement, in degrees, north or south of the equator.

Lava magma that reaches Earth's surface through a volcanic vent.

Length the measurement of space in any direction.

Leukemia a cancerous condition characterized by an abnormal increase in the white cells (leucocytes) of the blood.

Lightning an electric discharge in the atmosphere.

Light-year the distance light travels in one year.

Linear momentum mass × velocity.

Linearly polarized transverse waves that vibrate in only one direction.

Line squalls a series of storms along a front.

Lipid a general term that includes such substances as fats, oils, and waxes.

Liquid matter that has a definite volume but no definite shape.

Lithification the process of forming sedimentary rock from sediment; also called consolidation.

Lithosphere the outermost solid portion of Earth, which includes the crust and part of the upper mantle.

Local Group the cluster of galaxies that includes our own Milky Way galaxy.

Longitude for a point on the surface of Earth, the angular measurement, in degrees, east or west of the prime meridian.

Longitudinal wave a wave in which the vibrations are in the same direction as the wave velocity.

Long-shore current a current along a shore due to the waves that break at an angle to the shore line.

Lorentz-Fitzgerald contraction the length of a moving object appears to be shorter to an observer who is at rest than to an observer moving with the object.

Lunar eclipse an eclipse of the moon caused by Earth's blocking of the Sun's rays to the moon.

Magma hot, molten rock material.

Magnetic anomalies adjacent regions of rocks with remanent magnetism of opposite polarities. That is, the directions of the magnetism are reserved.

Magnetic declination the angular variation of a compass from geographic north.

Magnetic field a set of imaginary lines that indicate the direction a small compass needle would point if it were placed at a particular spot.

Magnetic monopole a single magnetic north or south pole without the other—as yet undiscovered.

Main sequence a narrow band on the H-R diagram on which most stars fall.

Mantle the interior region of Earth between the core and the crust.

Maria the large dark areas on the moon.

Mass a quantity of matter and a measurement of the amount of inertia that a body possesses.

Mass number the number of protons plus the number of neutrons in an atom.

Mass wasting the downslope movement of overburden under the influence of gravity.

Matter anything that exists in time, occupies space, and has mass.

Matter (de Broglie) waves the waves produced by moving particles.

Meandering the looping, ribbonlike path of a river channel that results from accumulated deposits of eroded material having diverted the stream flow.

Mean solar day the average length of a solar day. One solar day is the elapsed time between two successive crossings of the same meridian by the Sun.

Measurement the comparison of an unknown quantity to a standard.

Mega prefix that means 10^6 or one million.

Mercalli scale a scale of earthquake severity based on the physical effects produced by an earthquake.

Mesosphere a region of the atmosphere, based on temperature, that lies between approximately 35 and 60 miles in altitude.

Metal an element that tends to lose its valence electrons.

Metalloid an element that exhibits the properties of both metals and nonmetals.

Melting point the temperature of a substance at which the substance changes from a solid to a liquid.

Metamorphic rock rock that results from a change or metamorphism in pre-existing rock because of heat and pressure.

Meteor a small chunk of matter that burns up as it flies through Earth's atmosphere and appears to be a shooting star.

Meteorites chunks of matter from the solar system that fall through the atmosphere and strike Earth's surface.

Meteoroids small, interplanetary objects in space before they encounter Earth.

Meteorology the study of atmospheric phenomena.

Meter the standard of length in the mks system. It is equal to 39.37 inches, or 3.28 feet.

MeV million electron volts, a unit of energy.

Micro prefix that means 10^{-6} or one one-millionth.

Milky Way the name of our galaxy.

Milli prefix that means 10^{-3} or one one-thousandth.

Millibar a unit of pressure; 1 mb = 10^3 dyne/cm^2.

Mineral any naturally occurring, inorganic, crystalline substance.

Mixture a nonchemical combination of two or more substances of varying proportions.

mks system the metric system that has the meter, kilogram, second, and coulomb as the standard units of length, mass, time, and electric charge.

Mohorovicic discontinuity the boundary between the Earth's crust and mantle.

Mohs' scale a 10-point scale of hardness based on mineral standards, diamond being the hardest and talc being the softest.

Molarity the number of moles of solute in one liter of solution; molarity is designated by the letter M.

Mole a gram formula weight.

Molecule an uncharged particle of an element or compound.

Molecular weight the weight of one molecule of a substance relative to ^{12}C, expressed in grams.

Monomer the fundamental repeating unit of a long chain molecule.

Monsoons winds associated with seasonal convectional cycles set up between continents and oceans.

Motion the changing of position.

Moraine a ridge of glacial till.

Motor a device that converts electrical energy into mechanical energy.

Mountain range a geologic unit or series of mountains.

Mountain system a group of similar mountain ranges.

Mudflow the movement of large masses of soil that have accumulated on steep slopes and become unstable with the absorption of large quantities of water from melting snows and heavy rains.

Multiple proportions, law of Whenever two elements combine to form a series of compounds, the various amounts of one element and a constant amount of another element that it can be combined with will form a ratio of small whole numbers.

Muons mu mesons, which are particles with properties like the electron but are more massive than the electron; 1 mu meson mass = 207 electron masses.

National Meteorological Center the nerve center of the National Weather Service where weather data are received and processed and forecasts are made.

National Weather Service a federal organization under NOAA that provides national weather information.

Neap tide moderate tides with the least variation between high and low.

Nephanalysis a satellite cloud-cover photograph with a map grid.

Neutralization the mutual disappearance of the H^+ and the OH^- ions. They combine to form water (HOH).

Neutrino a subatomic particle that has no rest mass or electric charge but does possess energy and momentum.

Neutron number the number of neutrons in the atom.

Neutrons elementary particles that have approximately the same mass (1.6×10^{-27} kg) as protons but have no charge. They are one constituent of the atomic nucleus.

New moon the phase of the moon that occurs when the moon is between Earth and the Sun.

Newton 1 kg $\times$ m/s^2.

Newton's first law of motion a body will move at a constant velocity unless acted upon by an external unbalanced force.

Newton's law of universal gravitation $F = Gm_1m_2/r^2$.

Newton's second law of motion the acceleration of an object is equal to the unbalanced force on the object divided by the mass of the object ($a = F/m$).

Newton's third law of motion whenever one mass exerts a force upon a second mass, the second mass exerts an equal and opposite force upon the first mass.

NOAA the National Oceanic and Atmospheric Administration of the U.S. Department of Commerce, which is responsible for the study and monitoring of oceanic and atmospheric phenomena.

Node a point where the moon's path crosses the ecliptic.

Nonmetal an element that tends to gain electrons to complete its outer shell.

Normal faulting movement along a nonvertical fault in which the overlying side moves downward relative to the side beneath it.

Novae stars that suddenly increase dramatically in brightness for a brief period of time.

Nucleic acid very-high-molecular-weight polymers that are present in living cells.

Nucleus the central core of the atom.

Occluded front a front that is occluded, or closed off, from Earth's surface because of forced ascension.

Octet rule the tendency of atoms to have eight electrons in their outer shell when forming molecules or ions. There are many exceptions to the rule.

Ohm the unit of resistance; equal to one volt per ampere.

Ohm's law the voltage across two points is equal to the current flowing between the points times the resistance between the points.

Oort comet cloud the cloud of cometary objects believed to be orbiting the sun at 50,000 AU and from which comets originate.

Opposition the time at which a planet is on the opposite side of Earth from the Sun.

Organic acid any compound that contains the molecular arrangement $-C\overset{\displaystyle O}{\underset{\displaystyle OH}{}}$ called the carboxyl group.

Organic compounds compounds of carbon.

Orographic precipitation precipitation resulting from moisture-laden air forced to ascend because of a mountain; the ascending air cools adiabatically and precipitation may occur.

Outer planets the five planets farthest from the Sun (Jupiter, Saturn, Uranus, Neptune, and Pluto).

Overburden the weathered material that accumulates on base rock.

Oxidation the process by which electrons are lost.

Oxidation number a positive or negative number assigned to an atom as a measure of the electric charge that an atom in a molecule would have if it were ionically bound.

Oxidation state a term used to indicate the oxidation number.

Oxide a compound containing oxygen and one or more other elements.

Ozone the compound O_3. It is found naturally in the atmosphere in the ozonosphere and is also a constituent of photochemical smog.

Ozonosphere a region of the atmosphere characterized by an ozone concentration between approximately 0 and 50 miles in altitude.

Pangaea the single giant supercontinent that is believed to have existed over 200 million years ago.

Parhelia two bright-colored patches that appear on each side of the Sun due to scattering by the ice crystals of cirrostratus clouds. (Sometimes called side suns.)

Parallax the apparent motion, or shift, that occurs between two fixed objects when the observer changes position.

Parsec the distance to a star when the star exhibits a parallax of one second. This distance is equal to 3.26 light years or 206,265 astronomical units.

Pauli exclusion principle no two electrons can have the same set of quantum numbers.

Penumbra a region of partial shadow. During an eclipse, an observer in the penumbra sees only a partial eclipse.

Perfect gas law the relationship that exists between the volume, absolute pressure, and temperature of a gas. $PV/T = K$.

Perigee the point in its orbit at which a satellite is closest to Earth's center.

Period an interval of geologic time that is a subdivision of an era and made up of epochs; the time for a complete cycle of motion; a horizontal row of the periodic table—therefore elements that have approximately the same energy.

Periodic law the properties of elements are periodic functions of the atomic number.

Permafrost ground that is permanently frozen.

Permeability a measure of a material's capacity to transmit fluids.

Pesticides chemicals used to kill unwanted insect species.

*p*H   the exponent of the negative power to which 10 is raised when used to express H^+ ion concentration.

Photochemical smog   air-pollution conditions resulting from the photochemical reactions of hydrocarbons with oxygen in the air and other pollutants in the presence of sunlight.

Photon   a "particle" of electromagnetic radiation.

Photosphere   the Sun's outer surface, visible to the eye.

Photosynthesis   the process by which plants convert CO_2 and H_2O to sugars with the release of oxygen.

Physical change   a change in a substance that does not involve a change in chemical composition.

Physical properties   those that do not involve a change in the chemical composition of the substance.

Pi(π)   the ratio of the circumference of a circle to its diameter. It is equal to 3.14159. . . .

Pitch   the highness or lowness of a sound. It is a consequence of the frequency of the sound waves received by the ear.

Planck's constant   a constant of proportionality relating the energy and frequency of a photon. The constant has the value of 6.225×10^{-27} erg·s.

Plasma   matter that is a disordered population of ionized atoms, in which the ion cores and the free electrons are not in thermal equilibrium—a state in which the temperature is so high that electrons and atomic nuclei are separated and move very rapidly, constantly colliding with each other.

Plate   a huge slab of rock that makes up a portion of the outer layer of Earth and is in relative motion with respect to other plates.

Plate tectonics   the theory that the outer layer of Earth is made up of rigid plates that are in relative motion with respect to each other.

Pluton   a large body of intrusive igneous rock.

Polar covalent bond   a covalent bond in which electrons are unequally shared.

Polar easterlies   the prevailing winds in the latitudes from 60° to 90°.

Polarization   the restriction of the electric vector of a wave to one direction.

Polymer   the resulting chainlike structure produced by many monomers.

Porosity   the percentage of unoccupied space in the total volume of a substance.

Prophyritic   a rock texture characterized by a structure of coarse mineral grains scattered through a mixture of fine mineral grains.

Position   the location of an object with respect to another object.

Potential energy   the energy a body possesses because of its position in a force field.

Potential well   a term used to indicate a negative potential energy.

Power   work per unit time.

Precession   the slow rotation of the axis of spin of Earth around an axis perpendicular to the ecliptic plane.

Precision   the degree of reproducibility of a measurement—that is, the maximum possible error of the measurement.

Pressure   force per unit area.

Pressure gradient   a variation of pressure with position.

Pressure trough   a region with a pressure minimum.

Primary sewage treatment   the basic method of sewage treatment consisting of four main steps—screening, grit removal, sedimentation, and chlorination.

Principal quantum number   the numbers (1, 2, 3, . . .) used to designate the various principal energy levels that an electron may occupy in an atom.

Protein   the complex combination of amino acids. Proteins are composed mainly of carbon, oxygen, nitrogen, and hydrogen.

Protons   elementary particles that are 1836 times as massive (1.67×10^{-27} kg) as electrons and have a positive charge (1.6×10^{-19} coulomb). They are one constituent of the atomic nucleus.

Psychrometer   an instrument used to measure relative humidity.

Pulsar   a star that has a radio signal pulses regularly and very rapidly.

P waves   primary (P) waves, so called because they reach a seismic station before the S waves. P waves are longitudinal or compressional waves—i.e., their particle oscillations are in the direction of propagation and are transmitted by solids, liquids, and gases.

Pyroclastic debris   solid material emitted by volcanoes that range in size from fine dust to large boulders.

Quantum   a discrete amount.

Quantum mechanics (or wave mechanics)   a field of physics used to solve problems in which geometrical sizes are comparable to the wavelengths of particles. Schrödinger's equations forms the basis of wave mechanics.

Quantum number   a number assigned to one of the various values of a quantized quantity.

Quarks   particles with fractional charges such as $+\frac{2}{3}$ or $-\frac{1}{3}$ that combine together (three at a time) to form protons and neutrons. So far, a single free quark has not been found.

Quasar   a shortened term for *quasi*-stel*lar* radio source.

Radar   an instrument that sends out electromagnetic (radio) waves, monitors the returning wave that is reflected by some object, and thereby locates the object. Radar stands for *Radio Detecting And Ranging*. Radar is used to detect and monitor precipitation and severe storms.

Radian   a unit of angle equal to 57.4°. There are 2π radians in a circle.

Radiation   the transfer of energy by means of electromagnetic waves.

Radiation fog   a fog formed as a result of radiative heat loss (sometimes called a valley fog).

Radioactive the spontaneous disintegration of certain atomic nuclei, generating one or more of three types of radiation: alpha, beta, and/or gamma.

Radiosonde a small package of meteorological instruments with a radio transmitter that is carried aloft by a balloon.

Rays streaks of light-colored material extending outward from craters.

Rayleigh scattering the preferential scattering of light by air molecules and particles that accounts for the blueness of the sky. The scattering is proportional to $1/\lambda^4$.

Real image an image from a mirror or lens that can be brought to focus on a screen.

Red giant a red star that has a diameter much larger than average.

Red shift a Doppler effect caused when a light source, such as a galaxy, moves away from the observer and shifts the light frequencies lower or toward the red end of the electromagnetic spectrum.

Reduction the process by which electrons are gained.

Reflection the change in the direction of a wave because of a boundary.

Refraction the bending of light waves caused by a velocity change as light goes from one medium to another.

Relative humidity the ratio of the actual moisture content of a volume of air to its maximum moisture capacity at a given temperature.

Remanent magnetism the magnetism retained in rocks containing ferrite minerals after solidifying in Earth's magnetic field.

Representative elements elements in which the added electron enters the outermost shell in which the outermost shell is incomplete—the A groups in the periodic table.

Resistance the opposition to the flow of electric current.

Resonance a wave effect that occurs when an object has a natural frequency that corresponds to an external frequency.

Rest mass the mass of an object at zero velocity.

Retina the part of the human eye onto which light is focused by the lens of the eye.

Reverse faulting movement along a nonvertical fault in which the overlying side moves downward relative to the side beneath it.

Revolution the movement of one mass around another.

Richter scale a scale of earthquake severity based on the amplitude or intensity of seismic waves.

Right ascension a coordinate for measuring the east-west positions of celestial objects; the angle is measured eastward from the vernal equinox in hours, minutes, and seconds.

Rill a narrow trench or valley on the moon.

Ring of Fire the area generally circumscribing the Pacific Ocean that is characterized by volcanic activity.

RNA ribonucleic acid, a nucleic acid primarily in the cytoplasm or outer structure of the cell. RNA has the structure of a double-stranded helix.

Rock any naturally occurring, solid, mineral mass that makes up part of Earth's lithosphere.

Rock cycle the cyclic movement of rock, during which the rock is created, destroyed, and metamorphosed by Earth's internal and external geologic processes.

Rockslide a type of landslide occurring in mountain areas when large quantities of rock break off and move rapidly down steep slopes.

Rotation a spinning motion.

Salinity a measure of saltiness.

Salt an ionic compound containing the cation of a base and the anion of an acid.

Sanitary landfill a method of solid waste disposal in which refuse is covered with a layer of soil within a day after being deposited at the landfill site.

Saturated hydrocarbons contain only single bonds and their hydrogen content is at a maximum.

Saturated solution when the maximum amount of solute possible is dissolved in the solvent.

Scalar quantity that has a magnitude and units but no direction is associated with it.

Seafloor spreading the theory that the seafloor slowly spreads and moves sideways away from mid-ocean ridges. The spreading is believed to be due to convection cycles of subterranean molten material that causes the formation of the ridges and a surface motion in a lateral direction from the ridges.

Seamount an isolated submarine volcanic structure.

Second the standard unit of time. It is now defined in terms of the frequency of a certain transition in the cesium atom.

Secondary sewage treatment the treatment that adds an additional step of either a trickling filter or an activated sludge process to the primary treatment for the removal of suspended organic matter.

Sedimentary rock rock formed from the consolidation of layers of sediment.

Seismic waves the waves generated by the energy release of an earthquake.

Seismograph an instrument that records the intensity of seismic waves.

Seismology the geophysical science of earthquakes.

Separated sewer system a system with separate networks for sanitation and storm-water transportation.

Sheet erosion erosion resulting from runoff or the overland flow of water.

Shield volcano a volcano with a low, gently sloping profile formed by a fissure eruption of low-viscosity lava.

Sidereal day the rotation period of Earth with respect to the vernal equinox; one sidereal day is 23h 56min 4.091s.

Sidereal period the orbital or rotation period of any object with respect to the stars.

Silicon tetrahedron the pyramid-shaped structure of the complex silicate ion $(SiO_4)^{4-}$.

Sill a pluton formation that lies between and parallel to existing rock layers.

Silt   a mixture of fine particles and organic material delivered by a stream.

Singularity   the center of a black hole. The point to which the entire mass of a star has contracted.

Slug   a unit of mass, 1 lb/ft/s^2.

Slump   a type of landslide involving the downslope movement of an unbroken block of overburden that leaves a curved depression on the slope.

Smog   a contraction of *smoke-fog* used to describe the combination of these conditions.

Solar constant   the average amount of solar energy received per area at the top of the atmosphere per time, 2.0 cal/cm^2·min.

Solar eclipse   an eclipse of the Sun caused by the moon's blocking of the Sun's rays to an observer on Earth.

Solid   matter that has a definite volume and a definite shape.

Solid waste   any normally solid material resulting from human or animal activities that is useless or unwanted.

Solifluction (soil flow)   a type of slow mass wasting involving the "flow" of weathered material over a solid, impermeable base; common in permafrost areas.

Solubility   the amount of solute that will dissolve in a specified volume of solvent (at a given temperature) to produce a saturated solution.

Solute   the dissolved substance in a solution.

Solution   a homogeneous mixture.

Solvent   the substance in excess in a solution.

Sound   a wave phenomenon caused by variations in pressure in a medium such as air.

Source region   the region or surface from which an air mass derives its physical characteristics.

Special theory of relativity   a theory of relativity true for systems that are moving at a constant velocity with respect to one another.

Specific heat   the amount of heat energy in calories necessary to raise the temperature of one gram of the substance one degree Celsius.

Spectroscope   an instrument used to separate electromagnetic waves into their component wavelengths.

Spectrum   an ordered arrangement of various frequencies or wavelengths of electromagnetic radiation.

Speed of light   186,000 mi/s or 3 × 10^{10} cm/s or 3 × 10^8 m/s.

Spring tide   the tides of greatest variation between high and low.

Squall line   a region along a front characterized by storms and turbulent weather.

SST   an acronym for "supersonic transport" aircraft.

Stable layer   a layer of air with uniform temperature and density.

Stalactite   an icicle-shaped dripstone formation extending downward from a cavern roof.

Stalagmite   a blunt icicle-shaped dripstone formation extending upward from a cavern floor.

Standard conditions   0°C and sea-level atmospheric pressure.

Standard unit   a fixed and reproducible reference value used for the purpose of taking accurate measurements.

Steam point   the temperature at which pure water, at normal atmospheric pressure, boils.

STP   standard temperature 273.16 K (0°C) and pressure 1 atmosphere (760 mm of Hg).

Stratosphere   a region of the atmosphere based on temperature, approximately 10–35 miles in altitude.

Stratovolcano   a volcano with a steeply sloping symmetric cone formed by eruption of high-viscosity lava and pyroclastic debris. A stratovolcano is also called a composite volcano.

Stratus   a root name used to describe stratified or layered cloud forms.

Streak   the color of a powdered mineral on a streak plate (unglazed porcelain).

Stream load   the eroded material transported by a stream.

Strike-slip faulting   movement along a horizontal transform fault in which the two sides of the fault strike and slip by each other.

Subduction   the process in which one plate is deflected downward beneath another plate into the asthenosphere.

Substance   a homogeneous sample of matter, all specimens of which have identical properties and identical composition.

Summary solstice   the farthest point of the Sun's declination north of the equator (for the northern hemisphere).

Sunspots   patches of cooler, darker material on the surface of the Sun.

Supercooled   same as supersaturated—see below.

Supernova   an exploding star.

Supersaturated   the condition of air when its temperature is lowered below the dew point without condensation.

Supersaturated solution   a solution that contains more than the normal maximum amount of dissolved solute.

Surf   the breaking of surface waves along a shore.

Surface current   a broad drift of surface water in the ocean that is set in motion by prevailing winds.

Surface wave (oceanic)   a wave on the surface of the ocean near a shore line.

Surface wave (seismic)   a seismic wave that travels along Earth's surface or a boundary within it.

Superposition, law of   the principle that in a succession of stratified deposits the younger layers lie over the older layers.

S waves   secondary (S) seismic waves, so called because they reach a seismic station after the P waves. S waves are transverse or shear waves—i.e., their particle oscillations are at right angles to the direction of propagation, and are transmitted only by solids.

Syncline   a rock fold that curves downward—a basin-like fold; the opposite of anticline.

Synergism   the combined effect of two or more influences, neither of which would produce the same result by itself.

Synodic period the orbital or rotational period of an object as seen by an observer on Earth.

Synoptic weather charts charts or maps that present a synopsis of weather data.

Tectonics the study of Earth's general structural features and their changes.

Temperature a measure of the average kinetic energy of the molecules.

Temperature inversion a condition characterized by an inverted lapse rate.

Ten-degree rule a 10-degree-Celsius increase in temperature leads to a doubling of the chemical reaction rate.

Thermal circulation the cyclic movement of matter (e.g., air) due to localized heating and convectional heat transfer.

Thermal conductivity a measure of the ability of a substance to conduct heat energy.

Thermal pollution pollution resulting from warm water, chiefly from industrial cooling processes, that is discharged into natural waterways.

Thermodynamics, first law of the heat energy added to a system must go into increasing the internal energy of the system, or any work done by the system, or both. The law also states that heat energy removed from a system must produce a decrease in the internal energy of the system, or any work done on the system, or both. The law is based upon the law of conservation of energy.

Thermodynamics, second law of it is impossible for heat to flow spontaneously from an object having a lower temperature to an object having a higher temperature.

Thermodynamics, third law of a temperature of absolute zero can never be attained.

...mosphere a region of the atmosphere, based on temperature, between approximately 60 and several hundred ...s in altitude.

...faulting a special case of reverse faulting in which ...ult plane is at a small angle to the horizontal.

... the sound associated with lightning that arises ...e explosive release of electrical energy.

...ent a current due to the tidal movement of sea

... material that is transported and deposited by ...rather than by meltwater.

Time ...ntinuous forward-flowing of events.

...time (the stretching-out of time) the passage of ...oving object appears to be longer to an observer... object... at rest than to an observer moving with the at rest. ...ions appear to be slower than to an observer

Titius-Bode ...approximat... a numerical sequence that gives a good ... Sun in astro... ance of the first seven planets from the ...al units.

Tornado a vio... ...storm characterized by a funnel-shaped cloud and high ...ds.

Tornado warning the alert issued when a tornado has actually been sighted or indicated on radar.

Tornado watch the alert issued when conditions are favorable for tornado formation.

Total internal reflection a phenomenon in which light is totally reflected because refraction is impossible.

Torque a force about an axis; the product of the magnitude of the force and the perpendicular distance from the line of action of the force to the axis.

Traction the transport of a stream's bed load as a result of rocks and particles being rolled and bounced along by the stream's current.

Trade winds the prevailing winds in the latitudes from 0° to 30°.

Transformer a device that increases or decreases the voltage of alternating current.

Transform fault a fault with a horizontal fault plane along which strike-slip faulting occurs.

Transistor an electronic device whose primary purpose is to amplify an input signal.

Transition elements elements in which the second shell from the outer shell is increasing from 8 to 18 electrons— the B groups in the periodic table.

Transverse wave a wave in which the vibrations are perpendicular to the wave velocity.

Trickling filter a bed of bacteria-laden stones through which sewage effluent trickles for the removal of organic matter.

Tropical year the time interval from one vernal equinox to the next.

Troposphere a region of the atmosphere, based on temperature, between Earth's surface and 10 miles in altitude.

Umbra a region of total darkness in a shadow. During an eclipse, an observer in the umbra sees a total eclipse.

Unconformity a break in the geologic rock record.

Uniformitarianism *see* Uniformity.

Uniformity, principle of the principle that the same processes operate today on and within Earth as in the past. Hence, the present is considered the key to the past.

Unsaturated hydrocarbons compounds that can take on hydrogen atoms to form saturated hydrocarbons or atoms of substances other than hydrogen to form derivatives.

Valence the net electric charge of an atom or the number of electrons an atom can give up (or acquire) to achieve a filled outer shell.

Valence shell the outermost shell (or shells in some higher-atomic-number atoms) that have valence electrons.

Valence electrons the electrons that are involved in bond formation, usually the outermost electrons.

Valley glacier a glacier that flows downward, creating a valley.

Vector a quantity that has not only a magnitude and units but also a direction associated with it.

Velocity, average   the change in displacement divided by the change in time, $v = \Delta s / \Delta t$.

Velocity, instantaneous   the velocity at a particular instant of time.

Vernal equinox   the time (near March 21) when the Sun's declination crosses the equator moving north (for the northern hemisphere).

Virtual image   an image from a lens or mirror that cannot be brought to focus on a screen.

Viscosity   the internal property of a substance that offers resistance to flow.

Vitamins   organic substances that are needed in minute amounts to perform specific functions for normal growth and for the nutritional needs of the human body.

Volcanic mountains   mountains that have been built by volcanic eruptions.

Volt   the unit of voltage equal to one joule per coulomb.

Voltage   the amount of work it would take to move an electric charge between two points, divided by the value of the charge—i.e., work per unit charge.

Warm front   the boundary of an advancing warm air mass over a colder surface.

Waterspout   a tornado over water.

Water table   the boundary between the zone of aeration and the zone of saturation.

Watt   a unit of power, $1 \text{ kg} \cdot \text{m}^2 / \text{s}^3$ or 1 J/s.

Wave cyclone   cyclonic or rotational disturbances along a front that are formed by air motion in opposite directions.

Wavelength   the distance from any point on a wave to an identical point on the adjacent wave.

Wave motion   the emanation of energy from the disturbance of matter.

Weather   the atmospheric conditions of the lower troposphere.

Weathering   the physical disintegration and chemical decomposition of rock.

Weight   the force of gravity on Earth's surface.

Westerlies   the prevailing winds in the latitudes from 30° to 60°.

White dwarf   a white star that has a much smaller diameter than average. It is believed to be the end stage of small- and average-mass stars.

Wind   horizontal air motion.

Winter solstice   the farthest point of the Sun's declination south of the equator (for the northern hemisphere).

Work   the product of a force and the parallel distance through which it acts.

Zenith   the position directly overhead of an observer on Earth.

Zenith angle   the angle between the zenith and the Sun at noon.

Zodiac   a section of the sky extending around the ecliptic 8° above and below the ecliptic plane.

Zone of aeration   the zone of rock and soil above the water table whose open spaces are filled mainly with air.

Zone of saturation   the subsurface zone of rock and soil in which all the openings are completely filled or saturated with water.

Index

Conversion Factors

Mass
$1 \text{ gram} = 10^{-3} \text{ kg} = 6.85 \times 10^{-5} \text{ slug}$
$1 \text{ kg} = 10^3 \text{ g} = 6.85 \times 10^{-2} \text{ slug}$
$1 \text{ slug} = 1.46 \times 10^4 \text{ g} = 14.6 \text{ kg}$
$1 \text{ amu} = 1.66 \times 10^{-24} \text{ g} = 1.66 \times 10^{-27} \text{ kg}$

Length
$1 \text{ cm} = 10^{-2} \text{ m} = 0.394 \text{ in}$
$1 \text{ m} = 10^{-3} \text{ km} = 3.28 \text{ ft} = 39.4 \text{ in}$
$1 \text{ km} = 10^3 \text{ m} = 0.62 \text{ mi}$
$1 \text{ in} = 2.54 \text{ cm} = 2.54 \times 10^{-2} \text{ m}$
$1 \text{ ft} = 12 \text{ in} = 30.48 \text{ cm} = 0.3048 \text{ m}$
$1 \text{ mi} = 5280 \text{ ft} = 1609 \text{ m} = 1.609 \text{ km}$
$1 \text{ Å} = 10^{-10} \text{ m} = 10^{-8} \text{ cm}$
$1 \text{ parsec} = 3.26 \text{ light years} = 205{,}265 \text{ a.u.}$

Time
$1 \text{ h} = 60 \text{ min} = 3600 \text{ s}$
$1 \text{ day} = 24 \text{ h} = 1440 \text{ min} = 8.64 \times 10^4 \text{ s}$
$1 \text{ year} = 365 \text{ days} = 8.76 \times 10^3 \text{ h}$
$= 5.26 \times 10^5 \text{ min} = 3.16 \times 10^7 \text{ s}$

Energy
$1 \text{ joule} = 10^7 \text{ erg} = 0.738 \text{ ft} \cdot \text{lb}$
$= 0.239 \text{ cal} = 9.48 \times 10^{-4} \text{ Btu}$
$= 6.24 \times 10^{18} \text{ eV}$
$1 \text{ kcal} = 4186 \text{ J} = 4.186 \times 10^{10} \text{ erg}$
$= 3.968 \text{ Btu}$
$1 \text{ Btu} = 1055 \text{ J} = 1.055 \times 10^{10} \text{ erg}$
$= 778 \text{ ft} \cdot \text{lb} = 0.252 \text{ kcal}$
$1 \text{ cal} = 4.186 \text{ J} = 3.97 \times 10^{-3} \text{ Btu}$
$= 3.09 \text{ ft} \cdot \text{lb}$
$1 \text{ ft} \cdot \text{lb} = 1.36 \text{ J} = 1.36 \times 10^7 \text{ erg}$
$= 1.29 \times 10^{-3} \text{ Btu}$
$1 \text{ eV} = 1.60 \times 10^{-19} \text{ J} = 1.60 \times 10^{-12} \text{ erg}$
$1 \text{ kWh} = 3.60 \times 10^6 \text{ J} = 3.413 \times 10^3 \text{ Btu}$

Speed
$1 \text{ m/s} = 3.6 \text{ km/h} = 3.28 \text{ ft/s}$
$= 2.24 \text{ mi/h}$
$1 \text{ km/h} = 0.278 \text{ m/s} = 0.621 \text{ mi/h}$
$= 0.911 \text{ ft/s}$
$1 \text{ ft/s} = 0.682 \text{ mi/h} = 0.305 \text{ m/s}$
$= 1.10 \text{ km/h}$
$1 \text{ mi/h} = 1.467 \text{ ft/s} = 1.609 \text{ km/h}$
$= 0.447 \text{ m/s}$
$60 \text{ mi/h} = 88 \text{ ft/s}$

Force
$1 \text{ newton} = 10^5 \text{ dynes} = 0.225 \text{ lb}$
$1 \text{ dyne} = 10^{-5} \text{ N} = 2.25 \times 10^{-6} \text{ lb}$
$1 \text{ lb} = 4.45 \times 10^5 \text{ dynes} = 4.45 \text{ N}$
Equivalent weight of 1 kg mass $= 2.2 \text{ lb}$
$= 9.8 \text{ N}$

Pressure
$1 \text{ atm} = 14.7 \text{ lb/in}^2 = 1.013 \times 10^5 \text{ N/m}^2$
$= 1.013 \times 10^6 \text{ dyne/cm}^2$
$= 30 \text{ in Hg} = 76 \text{ cm Hg}$
$1 \text{ bar} = 10^6 \text{ dyne/cm}^2 = 10^5 \text{ Pa}$
$1 \text{ millibar} = 10^3 \text{ dyne/cm}^2 = 10^2 \text{ Pa}$
$1 \text{ Pa} = 1 \text{ N/m}^2 = 10^{-2} \text{ millibar}$

Power
$1 \text{ watt} = 0.738 \text{ ft} \cdot \text{lb/s} = 1.34 \times 10^{-3} \text{ hp}$
$= 3.41 \text{ Btu/h}$
$1 \text{ ft} \cdot \text{lb/s} = 1.36 \text{ W} = 1.82 \times 10^{-3} \text{ hp}$
$1 \text{ hp} = 550 \text{ ft} \cdot \text{lb/s} = 745.7 \text{ watt}$
$= 2545 \text{ Btu/h}$